TRANSITIONS

AF772711

17

TRANSITIONS

Pathways Towards
Sustainable Urban Development in Australia

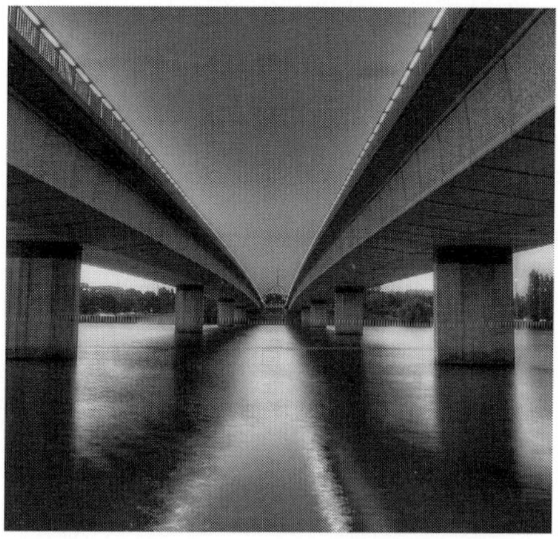

EDITOR: PETER W NEWTON

CSIRO
PUBLISHING

Springer

© CSIRO 2008

No part of this work may be reproduced, stored in a retrieval system, or transmitted in any form or by any means, electronic, mechanical, photocopying, microfilming, recording or otherwise, without written permission from the Publisher, with the exception of any material supplied specifically for the purpose of being entered and executed on a computer system, for exclusive use by the purchaser of the work.

Co-published by Springer Science + Business Media B.V., Dordrecht, The Netherlands and **CSIRO** PUBLISHING, Collingwood, Australia

Sold and distributed:
In the Americas, Europe and Rest of the World (excluding Australia and New Zealand)
by Springer Science + Business Media B.V., with ISBN 978-1-4020-8812-4
springer.com

In Australia and New Zealand by **CSIRO** PUBLISHING, with ISBN 978-0-6430-9419-2
www.publish.csiro.au

Front cover
Photo by Kristof Meder, istockphoto

Set in 10.5/12.5 Adobe Minion
Cover and text design by James Kelly
Typeset by Desktop Concepts Pty Ltd, Melbourne
Index by Russell Brooks
Printed in Australia by Ligare

Every effort has been made to contact copyright holders for their permission to reprint/reproduce material in this book. The publishers would be grateful to hear from any copyright holder who is not acknowledged here and will undertake to rectify any errors or omissions in future editions of this book.

Contents

Foreword

Tim O'Riordan

The urban setting already houses over half of the 6.5 billion people of the world's population and will swell in the coming generation. Up to 60% of global humanity will exist beyond any decent contact with the natural resource base from which all of their economic gains flow, and their social comfort is protected. Yet we see from many of the chapters presented in this book that the ecosystem services supporting this half of the world's population are overstretched. We also recognise that we are using natural resources, energy and water in a manner that is unsustainable. Yet the urban dweller is detached from all of this and, for the most part, oblivious to it.

The Spring 2007 issue of the *Journal of Industrial Ecology* (2007) entitled 'Industrial Ecology and the Global Impact of Cities' reinforces this broad conclusion. In essence, cities are becoming the 'battleground for sustainability'. The essays in this collection show that the amount of resources used to service a modern and expanding city is growing to the point where future waste streams arising from this huge concentration of material flows will create serious economic, social and environmental difficulties. There is, as yet, no coherent plan for managing future waste flows from resource use across cities. Even a wealthy city, such as Singapore, shows no measurable sign of using less material resources per capita, despite its reputation as a leading-edge city in the emerging knowledge economy. Changing the income-generating basis of a metropolitan economy, if serious, does not make its residents more resource efficient or environmentally friendly. To make matters worse, rapidly industrialising and urbanising China may well face acute natural resource shortages long before the economic revolution of new urbanisation is over.

We live in a human economy that relies more on production than on the renewable webs of life that the planet can offer. The major issue here is that all cities are increasingly vulnerable to some sort of failure—flood, drought, salt intrusion from sea-level rise, failed infrastructure (water, sewerage, electricity, communications), transport congestion, food shortages, oil shortages, material bottlenecks, and the like. No city could function without all of these services if deprived for more than three days, and many would show signs of strain within 24 hours.

Vulnerability of this sort is particularly troublesome if we also throw in the prospect of terror attack, social conflict over groups of residents and incomes that cannot share each others' needs because of competition over jobs, housing, social services or lack of common identity. Solecki and Leichenko (2006, p.13) comment that some 6% of Americans live in barricaded properties, and that many of the affluent in emerging economy cities, such as Caracas and Manilla, are moving to protected zones of housing. All of this adds to the resource density of the new metropolis, expanding and consuming and polarising.

The growing multiculturalism of cities—spurred by economic migration and social linkages causing clusters of nationals in confined spaces—may contribute to increased competition and conflict over jobs, cultural assimilation and a sense of social harmony. If cities stagnate because of material shortages or ecosystem stress, then any underlying social tensions will inevitably become more pronounced.

In short, the two fundamentals of a sustainable city—managing for nature's limits and maintaining social cohesion and resilience—may not be met by a large number of cities within the coming generation. This possible crisis may well affect the affluent cities much sooner and more dramatically than the poor urban complexes. Generally speaking, poverty creates resilience: patterns of sharing and stoic response that are not so evident in the wealthier cities.

What we may be witnessing is the emergence of two 'types' of urban dynamic: one that is broadly resilient to various forms of environmental and social stress; and one that could apply its wealth and many economic skills, plus its capacity to learn from the already resilient societies, to create a more robust form of urban governance for the future, sustainable city.

This is the challenge that is currently being faced by London. Its feisty Mayor, Ken Livingstone, is determined to establish a new form of governing for carbon reduction, user-friendly transport and local distributed renewable energy schemes. He has established a London Climate Change Agency with a remit to reduce London's carbon dioxide emissions by 30% by 2020. This will be addressed by radical pricing measures for transport (making the car expensive and the public transport/cycling/walking much more accessible). It will also involve major incentives for carbon neutral properties, including incentives for getting carbon neutrality into social housing. And the Agency is also planning to provide community-based renewable energy supplies feeding directly into buildings.

Lying behind the London purpose is a way of looking at resource flows so that the city is managed on a basis of lower consumption. To achieve this, one aim is to improve again involvement and participation in local management of resources and materials flows. London is beginning to experiment with possible new forms of governing for sustainability. Charting the progress of its new institutions will be most revealing. If London can do it, then any city can.

So, one particular need is to design resilience mechanisms for cities. This would require better accounting for resource flows, more community-based economies, housing and services, and the capacity to protect and manage the hazard of climate vulnerability, flood and drought, and much more localised food supply and waste reconstitution (compost, re-use and re-assembly, and energy from combustibles).

Another major issue is the physical infrastructure of the future city—which must be based on sustainability principles—especially in the critical areas of energy, water, communications and food provision. As yet we have no clear blueprint as to what the infrastructure of a truly sustainable city might look like in various parts of the globe. For example, what would we need for a hydrogen economy, for 50% renewable energy, or for distributed energy generation? To make such choices requires a lead time of 20 years, yet we have little idea of what any city might look like in 20 years if sustainability is the genuine outcome.

This challenge offers the scope for major urban areas to bring together key policy makers and movers and shakers to map out what a future sustainable society and economy might look like. It will doubtless be a process with no particular blueprint. It will be a matter of building from the existing urban economy, governing bodies and economy, yet also promoting the kinds of radical ideas and institutional frameworks being tried out in London.

This leads to a third proposition: building on urban resource audits and infrastructure scenarios—namely how to bring in an urban-based democracy that cares for the sustainable urban livelihoods of two generations hence.

Democracies are beginning to take longer views, and to be sensitive to uncertainty and improbability.

What we need are scenarios for various possible urban social, economic, environmental and institutional futures. These might range from step change, crisis-change, options with huge catastrophic short-term repercussions, to better ways of visioning and encouraging community dialogue linking the local to the metropolitan and mega-metropolitan.

Above all, we may need to see the emergence of the 'virtuous' urban citizen: individuals who act and think collectively, who can cope with huge amounts of ambiguity by preparing their behaviour for a collective sustainable 'good' and who feel strong about preparing for the wellbeing of their grandchildren.

The notion of a virtuous citizen begins with the ideas of Aristotle, who argued that the act of being a citizen means to be responsible for the collective wellbeing of all neighbours. There is the essence of a functioning democracy. What we now need to establish is a sense of genuine shared responsibility among citizens who can see themselves as neighbours, not only to those nearby, but to the offspring of future generations. As yet, it is not at all easy to imagine how a modern city—which is stressed and possibly insecure—can create a binding and enduring sense of virtue and collective responsibility for a sustainable future.

Possibly a combination of scenarios of opportunity and crisis, of forecasts of resources and material shortage, may enable the emerging virtuous citizen to see the overwhelming advantage of living for a common good. If every school can become a laboratory of sustainable living, with appropriate measures of wellbeing to go with it, then there is scope for such a virtuous shift in the urban culture.

All of this is possible, and indeed is beginning to happen. The pages of this text provide a basis for hope and the prospects for real opportunity. The key is to enable a degree of power to be genuinely directed to the 'urban-village' level and to distributed, rather than centralised, systems and economies with real scope for generating and distributing resources locally for community-based and community-owned sustainable futures.

Already, political responses are being prepared for a socially resilient city for a decade hence. These chapters provide hope, as well as an image of what might befall us if we do not respond effectively and in time.

Reference

Solecki WD and Leichenko RM (2006). Urbanization and the metropolitan environment: lessons from New York and Shanghai. *Environment* **48(4)**, 8–23.

Preface

The *raison d'être* for this book is to highlight the formidable challenges and vulnerabilities that confront Australia and its human settlements, while at the same time identifying sustainable solutions to these critical problems. The solutions will require fundamental *transitions* in many instances: to our urban infrastructures, our urban planning, to our institutions and, perhaps most of all, to ourselves in terms of our lifestyles and consumption patterns.

Australia's urban development will be an explicit focus for the book. 2007 marked the year in which most of the world's population would live in urban, rather than rural, environments—one of the key global transitions of this century. Australia was among the first countries to make this transition—in 1882. Urban settlements are now where the vast majority of Australians live, where 95% work, and where over 80% of the nation's GDP is generated. Their design, planning, construction and operation are fundamental to the productivity and competitiveness of the economy, the quality of life of all citizens and the ecological sustainability of the continent. Their current trajectory is, however, unsustainable.

For many Australians, there is no connection between the natural environment and the built environment. The nation has experienced 200 years of urban settlement and unfettered growth. The impacts of growth have begun to accelerate since the mid 20th century, and feedback loops related to environmental impact are only beginning to be felt–in domains such as water, energy, biodiversity and material flows, among many others. We are only beginning to develop the ability to understand and adapt to the limits of our environment. To date there has been a measure of profligacy in this respect, accentuated by the fact that Australian urban dwellers have long been disconnected from their natural environment. There is almost a complete disconnection between the city-dweller's consumption of manufactured products and an understanding of the environmental impact associated with the life cycles of those products. The metabolism of our cities is also on a collision course with the metabolism of their supporting ecosystems. The evidence for much of this can be found in consecutive Australia State of Environment Reports (www.environment.gov.au/soe) and *Australia's Scorecard* (Newton 2007: which found materialism to be increasing, concern for environment declining and personal happiness flatlined). As well as in contributions to this book.

We also need to start thinking and planning outside of the paradigms that dominated 20th century urban development. The 21st century has ushered in both a carbon- and a resource-constrained world. Climate change, and especially the challenge of extremes in temperature, rainfall (or lack of) and sea-level rise, will demand new mitigation and adaptation responses. The good news is that there are solutions to these challenges. These solutions are outlined in Sections 2 through 10 of this book and cover technological, urban planning and behaviour change elements.Governments and the key urban utility and service agencies have, however, lagged in initiating the much needed transitions. It is argued in Section 1 that, to be sustainable, 21st century cities need to be able to appropriate from a pipeline of innovative technologies, products, designs and processes that can be substituted for those currently in operation that are beginning to show signs of failure. These innovations can be designated as Horizon 1, 2 or 3—depending upon the period of time required for implementation, the significance of the barriers to be overcome, and the magnitude of impact on sustainability performance that is expected to follow from widespread diffusion. To date, most H2 and H3 innovation in key areas such as water, energy, transport and waste management have been ignored or shelved. Short electoral cycles appear to be a major inhibitor to long term thinking.

There is an urgency for change in the manner in which planning for 21st century Australian cities needs to occur that has not been previously reflected in the actions of government and industry over the past decade and more. Several key inquiries have been held by the Howard federal government (House of Representatives Standing Committee on Environment and Heritage 2005, 2007) and submissions made to government by peak industry associations (CIE 2006) with no tangible result. In part, this may be due to the fact that Australia's major capital cities are regularly rated by international agencies such as The Economist Intelligence Unit as among the world's ten most liveable cities, and that there is a measure of complacency, as well as pride, that our city managers and governments take from that. Liveability, however, does not equate to sustainability. The ecological footprints of Australia's capitals are all of the order of 7 hectares per capita—over three times the global average.

Despite the significance that the built environment in general, and cities in particular, has as a national asset (the built environment represents approximately 40% of Australia's total asset base; natural capital represents 33%), and despite their roles as economic engines and demographic absorbers, and the impact they have on the nation's environment, there remains no federal agency with the mandate and responsibility for providing strategic long-term direction for sustainable urban development across the nation. The vision and leadership displayed by US President John F. Kennedy in establishing the Department of Urban Affairs and Housing as a cabinet-level agency in January 1962—and which continues to this day—and when he stated '...we will neglect our cities to our peril, for in neglecting them we neglect the nation' (http://home.att.net/~jrhsc/jfk.html)—is currently absent in Australia.

In such a vacuum, we saw an absence of a demand for new ideas, leadership and investment by the Howard federal government and state counter-parts in major new transformational nation-building projects that are currently sitting as Horizon 2 and Horizon 3 opportunities in the pipeline of urban innovation. Cities are highly complex systems, yet there is comparatively little investment by government or industry in research designed to understand how they function and change. Given this knowledge vacuum, there has been a reticence on the part of governments to set targets for the performance of our buildings, urban infrastructures and cities, including how they are operated. All combine to inhibit transitioning to a more sustainable built environment.

The paradox is, however, that there is now a broad and diverse demand, from all levels of government, industry and the community for the knowledge and guidance necessary to begin accelerating the transition to more sustainable urban development. Australians are *aware* that change is needed—but the link to *action* is missing. The convergence of the threats of peak oil, climate change, security of water supply, biosecurity, loss of biodiversity, decline in food supply relative to demand, to name just half a dozen—all of which will significantly impact cities, their industries and their residents—has intensified the demand for sustainable solutions. Australia has the expertise to chart pathways for this sustainability transition—witness the contributions by the 92 researchers, who are all leaders in their field, assembled for this book. But the expertise is fragmented. If ever the timing was right for a Co-operative Research Centre for Sustainable Cities (Australia invented this highly innovative CRC model for applied research), involving the three tiers of government, industry and research providers, it is now.

Professor Peter W. Newton
Swinburne University of Technology
Melbourne
January 2008

References

CIE (Centre for International Economics) (2006), 'Sustainable Communities. A National Action Plan for Urban Australia'. Report prepared for the Planning Institute of Australia, the Property Council of Australia, the Royal Australian Institute of Architects and the Intergovernmental Planning Officials Group for the Sustainable Communities Roundtable, Canberra.

House of Representatives Standing Committee on Environment and Heritage (2007) *Sustainability for Survival: Creating a Climate for Change.* Commonwealth Government, Canberra <www.aph.gov.au>

House of Representatives Standing Committee on Environment and Heritage (2005) *Sustainable Cities.* Commonwealth Government, Canberra <www.aph.gov.au>

Newton PW (2007). Horizon 3 Planning: Meshing Liveability with Sustainability, *Environment and Planning B: Planning and Design* **34**, 571–575.

Acknowledgements

Transitions is one of the products to emerge from the time the Editor of this book spent directing CSIRO's research program in Sustainable Built Environments over a period of 15 years from 1992 to 2006. I am indebted to CSIRO and its excellent cadre of scientists, such as those who feature in this book, and most of whom I have personally worked with at some stage of my career—in particular to Dr John Brotchie (1929–2007) who was a mentor for much of my time at CSIRO. As well, I am greatly appreciative of the commitment made by colleagues and fellow researchers—some 92 in total and all experts in their field—to contribute chapters to this collection. I also wish to thank Professor Ian Young, Vice-Chancellor at Swinburne University of Technology, for his invitation to join the Institute for Social Research in 2007 and help lead the University's Sustainable Futures initiative. At an operational level, I have greatly appreciated the contribution of my editorial team at CSIRO Publishing (John Manger, Briana Elwood and Peter Storer) and my local support at Swinburne (Sue Lester). What follows are acknowledgements specific to individual chapters in the book.

Chapter 2: Dr Barrie Pittock (CSIRO) provided helpful comments on this chapter.

Chapter 11: Data on which the first section of this paper is based was assembled by the author in the process of preparing the 2006 Australia Statement of Environment Report: Human Settlements for the Department of Environment and Heritage, Canberra (www.environment.gov.au/soe).

Chapter 12: This chapter is a contribution to the CSIRO Climate Change Research Program and the CSIRO Wealth from Oceans Flagship and was supported by the Australian Government's Cooperative Research Centres Programme through the Antarctic Climate and Ecosystems Cooperative Research Centre. The authors were partly funded by the Australian Climate Change Science Program. Hourly sea-level data were provided by the National Tidal Centre, Bureau of Meteorology, Australia, Copyright reserved.

Chapter 13: This chapter is a contribution to the CSIRO Climate Adaptation Flagship. The author was partly funded by the Australian Climate Change Science Program.

Chapter 14: We would like to thank CSIRO Sustainable Ecosystems Internal Venture Capital Fund and the CSIRO Wealth from Oceans Flagship for supporting this work. ACTEW generously provided the water demand data and Dr Stephen Charles provided the climate change data. We would like to thank Dr Greg McKeon for providing constructive ideas early in the project and Drs Xuemei Bai and Matthew Inman and Professor Peter Newton for helpful comments on the manuscript and Dr Guy Barnett for his support early in the project.

Chapter 15: The authors are indebted to the Bushfire Cooperative Research Centre, Australia, for sponsoring much of the research reported herein.

Chapter 16: Some material in this Chapter was previously published in McDonnell (2005). The authors wish to thank Amy Hahs and Nick Williams for their assistance in compiling the vegetation map. Amy Hahs also provided significant input into the creation of Figure 1. The Baker Foundation provided generous support for this research.

Chapter 17: The authors would like to thank Dr Steven Hatfield-Dodds and Dr Rohan Nelson of CSIRO Sustainable Ecosystems for their helpful comments, thoughts and ideas on adaptive biosecurity.

Chapter 18: Risk Frontiers at Macquarie University is a research centre devoted to helping insurers better understand and price natural catastrophe risks in the Asia Pacific Region. By performing research into the likelihood and costs of natural disasters, Risk Frontiers aims to assist the insurance industry in setting realistic premiums; build safer communities; and encourage the responsible management of natural hazard risks. It is sponsored by Aon Re, the Australian Reinsurance Pool Corporation, Benfield Australia, Guy Carpenter, IAG Insurance, QBE, Suncorp Group and Swiss Re.

Chapter 19: This project was funded by the Australian Institute of Health and Welfare (AIHW) through its Asthma, Arthritis and Environmental Health work program. The authors would like to thank Ms Susan Killion, Senior Executive, Health and Functioning Group, AIHW and Dr Penny Allbon, Director of AIHW for their support in undertaking this project, Mr Andrew Phillips for providing helpful comments, and Ms Faith Margrie for assisting with the preparation of the manuscript.

Chapter 20: Part of this chapter was based on an earlier paper: Western J, Stimson R, Baum S and van Gellecum Y (2005). Measuring community strength and human capital. *Regional Studies* **39(8)**, 1095–1109.

Chapter 22: This work was performed for The Energy Transformed National Research Flagship of CSIRO. We acknowledge the assistance of Michael O'Connor, Peter K Campbell and Tim Grant, and financial support from the Department of Environment and Water Resources (as it then was).

Chapter 35: The authors would like to acknowledge VicUrban, who commissioned the work on the Dandenong LOGIS project, and in particular Malcolm Druce (Senior Project Manager) and Ayman Allaf (Development Manager). Special thanks go to Sandra George (South East Network) and Paul Dowling (South East Melbourne Manufacturing Alliance) for their role in identifying suitable Dandenong businesses and facilitating contact with those courteous people who responded to the survey or agreed to an interview. The Centre for Sustainable Resource Processing (www.csrp.com.au) is established and supported under the Australian Commonwealth Cooperative Research Centres Program, and funds the regional synergies research in Kwinana and Gladstone and the development of the Regional Synergies Toolkit. The CSRP acknowledges the many and diverse contributions of the Kwinana Industries Council (KIC), Gladstone Area Industry Network (GAIN), and Kwinana and Gladstone industries.

Chapter 37: Figures 2 and 3 are slightly-modified versions of figures published in the *NSW Public Health Bulletin*, 2007, **18** (3–4).

Chapter 38: This chapter was written in conjunction with research funded by an Australian Research Council Grant (#LP0455453) and the RESIDE study team is gratefully acknowledged. The first author is supported by an Australian Research Council Postgraduate Award (Industry), with the support of the Heart Foundation as Industry Partner. The second author is supported by a NHMRC Research Fellowship (#503712).

Chapter 41: The authors would like to thank ICLEI Local Governments for Sustainability Oceania for incubating the Centre for Public Agency Sustainability Reporting and the Centre's collaborators (Global Reporting Initiative (GRI), the State Government of Victoria and the City of Melbourne) for supporting the Centre's work. The work of the Centre for Public Agency Sustainability Reporting continues in the Integrated Sustainability Services unit of ICLEI Oceania.

Chapter 43: The authors are indebted to the Outer Eastern Melbourne Indicators Project team which includes staff members for Knox, Maroondah and Yarra Ranges Councils and to Local Government Victoria for funding the most recent stage of the indicators project.

Contributors

The editor

Professor Peter W Newton is a Research Professor at Swinburne University of Technology. Prior to joining Swinburne in December 2006, Dr Newton held senior research and management positions in CSIRO as a Chief Research Scientist, where he directed the program of research related to sustainable built environments. Between 2001 and 2007 he also directed the sustainability research of the CRC for Construction Innovation.

The authors

Dr Deborah Abbs, CSIRO Marine and Atmospheric Research, Melbourne

Michael Ambrose, CSIRO Sustainable Ecosystems, Melbourne

Dr Xuemei Bai, CSIRO Sustainable Ecosystems, Canberra

Dr David Batten, CSIRO Energy Transformed Flagship and The Temaplan Group, Melbourne

Assoc Professor Scott Baum, Urban Research Program, Griffith University, Brisbane

Dr Tom Beer, CSIRO Energy Tranformed Flagship, CSIRO Marine and Atmospheric Research, Melbourne

Dr Kuldeep Bhatia, Australian Institute of Health and Welfare, Canberra

Jane Blackmore, CSIRO Sustainable Ecosystems, Sydney

Charlotte Brunt, Centre for Regional Development, Swinburne University of Technology, Melbourne

Dr Keping Chen, Risk Frontiers – Natural Hazards Research Centre, Macquarie University, Sydney

Josephine Cheng, CSIRO Energy Technology, Newcastle

Dr John Church, Centre for Australian Weather and Climate Research, Wealth from Oceans Flagship and Antarctic Climate and Ecosystem CRC

Dr David Cook, CSIRO Entomology, Canberra, and Visiting Fellow, Fenner School of Environment and Society, The Australian National University, Canberra

Steve Cook, CSIRO Land and Water, Melbourne

Dr Peter Coppin, Wind Energy Research Unit, CSIRO Marine and Atmospheric Research, Canberra

Dr Glen Corder, Centre for Social Responsibility in Mining, The University of Queensland, Brisbane

Dr Peter Cowell, School of Geosciences, The University of Sydney, Sydney

Steve Crimp, CSIRO Sustainable Ecosystems, Canberra

Ryan Crompton, Risk Frontiers – Natural Hazards Research Centre, Macquarie University, Sydney

Dr Clare Diaper, CSIRO Land and Water, Melbourne

Dr Andrew Dicks, School of Engineering, The University of Queensland, Brisbane

Dr Jago Dodson, Urban Research Program, Griffith University, Brisbane

Dr Michael Dunlop, CSIRO Sustainable Ecosystems, Canberra

Professor Larry Dwyer, School of Marketing, Australian School of Business, The University of New South Wales, Sydney

Dr Deborah Edwards, School of Leisure, Sport and Tourism, University of Technology, Sydney

Dr Peter Ellyard, Preferred Futures Group, Sustainable Prosperity Foundation, Melbourne

Ryan Falconer, Institute for Sustainability and Technology Policy, Murdoch University, Perth

Dr Chris Fell, CSIRO Energy Technology, Newcastle

Barney Foran, Institute for Land, Water and Society, Charles Sturt University, Albury

Professor Billie Giles-Corti, Centre for the Built Environment and Health, School of Population Health, University of Western Australia, Perth

Justin Graf, Australian Institute of Health and Welfare, Canberra

Paul Graham, Energy Transformed Flagship, CSIRO Energy Technology, Newcastle

Francis Grey, Sustainability Asset Management Group, Melbourne

Professor John Handmer, Bushfire CRC and Centre for Risk and Community Safety, RMIT University, Melbourne

Professor Ann Harding, The National Centre for Social and Economic Modelling, University of Canberra, Canberra

Leanne Hardwicke, Engineers Australia, Canberra

Kevin Hennessy, CSIRO Marine and Atmospheric Research, Melbourne

Dr Dave Henry, CSIRO Livestock Industries, Geelong

Dr Jim Hinkley, CSIRO Energy Technology, Newcastle

Kelly Holland, Australian Research Centre for Urban Ecology, The University of Melbourne, Melbourne

Dr Mark Howden, CSIRO Sustainable Ecosystems, Canberra

Phil Hughes, The Centre for Public Agency Sustainability Reporting, Melbourne

Professor Graeme Hugo, Department of Geographical and Environmental Studies and National Centre for Social Applications in GIS, The University of Adelaide, Adelaide

Dr John Hunter, Antarctic Climate and Ecosystems CRC, Hobart

Laraine Hunter, Risk Frontiers – Natural Hazards Research Centre, Macquarie University, Sydney

Scott Ibbotson, School of Mechanical and Manufacturing Engineering, The University of New South Wales, Sydney

Dr Anne Imenes, CSIRO Energy Technology, Newcastle

Dr Matthew Inman, CSIRO Sustainable Ecosystems, Sydney

Terry Jones, Centre for Distributed Energy and Power, CSIRO Energy Transformed Flagship, Melbourne

Kirrily Jordan, Faculty of Arts, The University of Sydney, Sydney

Professor Hartmut Kaebernick, School of Mechanical and Manufacturing Engineering, The University of New South Wales, Sydney

Dr Sami Kara, School of Mechanical and Manufacturing Engineering, The University of New South Wales, Sydney

David Lamb, CSIRO Energy Transformed Flagship, Melbourne

Anne Langworthy, Centre for Regional Development, Swinburne University of Technology, Melbourne

Dr Bob Leicester, Bushfire CRC and CSIRO Sustainable Ecosystems, Melbourne

Roy Leigh, Risk Frontiers – Natural Hazards Research Centre, Macquarie University, Sydney

Dr Mark Lonsdale, CSIRO Entomology, Canberra

Dr Shiroma Maheepala, CSIRO Land and Water, Melbourne

Professor John McAneney, Risk Frontiers – Natural Hazards Research Centre, Macquarie University, Sydney

Professor Peter McDonald, Australian Demographic and Social Research Institute, The Australian National University, Canberra

Dr Mark McDowell, Australian Research Centre for Urban Ecology, The University of Melbourne, Melbourne

Dr Kathleen McInnes, Centre for Australian Weather and Climate Research, Wealth from Oceans Flagship and Antarctic Climate and Ecosystem CRC

Professor Anthony McMichael, National Centre for Epidemiology and Population Health, Australian National University, Canberra

Dr Nina Mistilis, School of Marketing, Australian School of Business, The University of New South Wales, Sydney

Blair Nancarrow, CSIRO Land and Water, Perth

Professor Peter Newman, Professor of Sustainability, Curtin University, Perth

Professor Peter Newton, Institute for Social Research, Swinburne University of Technology, Melbourne

Marcia Nugent, The Centre for Public Agency Sustainability Reporting, Melbourne

Dr Deborah O'Connell, CSIRO Sustainable Ecosystems, Canberra

Dr Siobhan O'Farrell, Centre for Australian Weather and Climate Research, Wealth from Oceans Flagship and Antarctic Climate and Ecosystem CRC

Professor Tim O'Riordan, University of East Anglia, Norwich, UK

Alicia Payne, The National Centre for Social and Economic Modelling, University of Canberra, Canberra

Dr David Rand, CSIRO Energy Technology, Melbourne

Dr Luke Reedman, CSIRO Energy Technology, Newcastle

Assoc. Professor Michael Regan, School of Sustainable Development, Bond University, Gold Coast

Carolina Roman, School of Geography and Environmental Sciences, Monash University, Melbourne

Dr Jan Scheurer, Australian Housing and Urban Research Institute, RMIT University, Melbourne

Dr Ashok Sharma, CSIRO Land and Water, Melbourne

Dr Andy Sheppard, CSIRO Entomology, Canberra

Dr Neil Sipe, Urban Research Program, Griffith University, Brisbane

Andrew Speers, Member, Management Committee, International Water Association's Specialist Group on Sustainability in the Water Sector, Sydney

Wesley Stein, CSIRO Energy Technology, Newcastle

Professor Frank Stilwell, Faculty of Arts, The University of Sydney, Sydney

Dr Geoffrey Syme, CSIRO Land and Water, Perth

Dr Perri Timmins, Australian Institute of Health and Welfare, Canberra

Grace Tjandraatmadja, CSIRO Land and Water, Melbourne

Dr Ted Trainer, School of Social Work, The University of New South Wales, Sydney

Dr Graham Turner, CSIRO Sustainable Ecosystems, Canberra

Dick van Beers, Centre of Excellence in Cleaner Production, Curtin University of Technology, Perth

Dr Quoc Ngu Vu, The National Centre for Social and Economic Modelling, University of Canberra, Canberra

Dr Neil White, Centre for Australian Weather and Climate Research, Wealth from Oceans Flagship and Antarctic Climate and Ecosystem CRC

Assoc Professor Geoff Woolcock, Urban Research Program, Griffith University, Brisbane

SECTION 1

Transitions: An Introduction

A transition to more sustainable urban systems, in Australia and internationally, needs to occur in the first half of the 21st century given the problems and vulnerabilities that surround current trajectories of urban development. The key challenges relate primarily to the impacts of *unsustainable consumption and production.*

The next stage of urban environmental evolution—*the eco-city*—is introduced in Chapter 1. A transition to eco-city stage from the present consumption stage (developed cities) or production stage (developing cities), characteristic of a majority of cities worldwide, will require the adoption of sustainable development goals for cities akin to those associated with the *urban metabolism model.* The goals involve:

- using fewer resources more efficiently
- using wastes as resources
- restoring and maintaining urban environmental ecosystem quality
- enhancing human well-being
- implementing highly efficient and effective urban and industrial planning, design and management systems, and
- not confusing liveability with sustainability.

Current approaches for delivering urban water and energy systems, waste management, transport, planning and governance are neither appropriate nor sufficiently resilient in the face of 21st century challenges, which include resource constraints (shortages of water, oil, agricultural land and food, selected materials), climate change impacts, and globalisation- and population-led development pressures (Sections 2 and 3).

Twenty-first century cities need to be able to appropriate a pipeline of innovative technologies, products, designs and processes that can be substituted when existing ones begin to show signs of failure. A *Three Horizons process of urban innovation* is introduced as being necessary for transitioning to a sustainable built environment. Horizon Two and Horizon Three (H2, H3) innovations in particular are those with the capacity for more fundamental urban transformation. Horizon 1 innovations, in contrast, are linked primarily to driving efficiencies in existing systems and technologies, providing societies with the temporal windows necessary for transitioning to the more sustainable H2 and H3 solutions. The latter feature in the set of transitions listed below:

- *Energy*: moving from highly inefficient centralised fossil-fuel-based systems to decentralised, distributed energy systems based on renewables (Section 5)
- *Water*: moving from wasteful and environmentally damaging centralised systems—that divert environmental flows, dam, distribute, use once and then dispose them to receiving waters—to integrated urban water systems that utilise wastewater and stormwater as substi-

tutes for potable water in gardens, laundries and toilets (Section 6)

- *Waste*: moving from the current extract-process-assemble-use-dispose paradigm to one where waste, once minimised, is seen as a resource stream for cradle-to-cradle and eco-industrial manufacturing processes (Section 7)
- *Urban systems*: must become more resilient and adaptive in response to the threats of climate change (sea level rise, flooding, heat and drought, bushfires), peak oil challenges and the need to recycle the city via smart redevelopment strategies that minimise sprawl, reduce automobile dependence and create healthier, more liveable and walkable mixed-use neighbourhoods
- *Urban governance* also needs to be appropriate for the planning and management of large, complex, culturally and economically diverse mega-metropolitan regions and their bioregions (Sections 4 and 8).
- *Human behaviour*: moving from lifestyles where materialism, indebtedness and consumption are a dominant feature of urban society to lifestyles associated with smaller ecological footprints and a concern for community (Section 9).

The *barriers to transition* within existing urban systems and institutions are not to be under-estimated. They are associated with path dependency, organisational inertia and a bunch of NIMBY attitudes. Barriers are identified and pathways to implementation of socio-technical innovation are explored in all Sections, but most explicitly in Sections 9 and 10 where focus is on a need for fundamental change in values, attitudes and behaviour among those in industry, government and community around issues of production and consumption.

Transformation of Australia's urban systems in the 21st century is possible. There **are** solutions for a sustainable future. But the window of opportunity for transitioning without significant dislocation and hardship is diminishing. There is an urgency to this whole process.

PWN

1

Transitioning to Sustainable Urban Development

Peter Newton and Xuemei Bai

Transition
A change from one form or type to another, or the process by which this happens
(http://dictionary.cambridge.org)

Formidable challenges confront Australia in relation to the type of urban development required in the 21st century. Most of these challenges are consumption-related and are linked to our world-leading demands on key resources, such as water and energy, to our dependency on cars and oil, to our pursuit of ever larger homes for smaller families, to our increasing levels of material acquisition, and to our growing levels of waste generation. (Hamilton and Denniss 2005; Beeton 2006; Newton 2006, 2007). Paradoxically, Australians also live in some of the most liveable cities in the world that have resulted from this pattern of resource consumption (The Economist Intelligence Unit 2004). Liveability is, however, only part of the equation that defines the sustainability of urban settlements. Based on the ecological footprint metric created by Wackernagel and Rees (1996), the average Australian urban resident requires in the order of 7 to 8 hectares of land to support their contemporary lifestyle compared with approximately 2 hectares per person globally. If the entire world's population adopted a similar level of consumption to that of Australia, then three to four Earths would be required to support it. Inherent in the footprint concept is the idea that for the economic development of a nation or region to be sustainable, it must respect the limits and capacities of the natural and social systems that support it. This transition in thinking about sustainable development over the past half century is outlined in Figure 1. Based on the writings of Daly (1996) and Lowe (1996) among others, it identifies the need for decision making about development to integrate ecological thinking into the overall social and economic planning process.

What makes this transition in thinking critical is that 21st century urban development in Australia will be undertaken in a setting radically different to that of the 20th century. Climate change—linked to escalating concentrations of greenhouse gas (GHG) in the Earth's atmosphere—has ushered in a *carbon-constrained* world. There is an increasing public and political recognition that radical changes are required in relation to energy use, including greater efficiencies, more conservation, a major switch to renewables and ultimately transitioning to a solar–hydrogen economy. This will require urban planning and design that creates more resilient cities in the face of new climate-induced vulnerabilities. Twenty-first century development will also be *resource-constrained* in key areas such as water, oil and gas, food and selected building and manufacturing materials. Key transitions will be needed in relation to recycling and re-use of water and materials as well as to the introduction of renewable energy that supports stationary and transport energy needs. The technology platform capable of supporting many of these transitions already exists. The challenge becomes a matter of how the knowledge that underpins the implementation of these sustainable technologies can be effectively adopted by cities and the rate that incumbent technologies can be replaced by sustainable technologies in order to reduce excessive resource use.

The 21st century is also one in which the globe's population will become predominantly urban

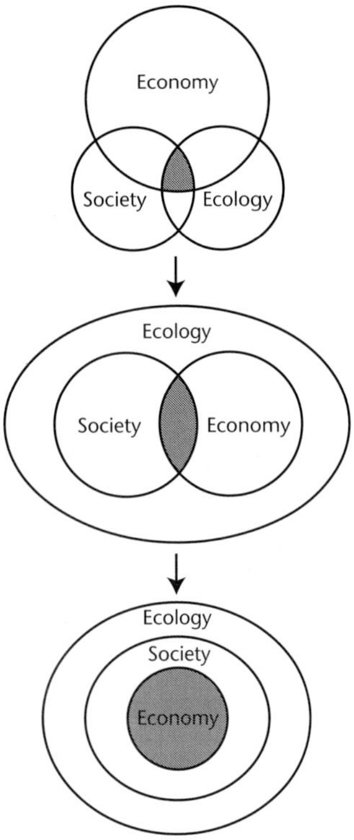

The predominant model of decision making in Australia until the 1980s. It gives primacy to economic decisions and assumes that environmental problems can always be solved if the economy is sound.

This model of decision making is a variant of the previous one. It characterises the piecemeal approach towards ESD.

This is the decision making model needed for an ecologically sustainable future for Australia. It recognises that the economy is a sub-set of society, since many important aspects of society do not involve economic activity. Similarly, human society is totally constrained by the natural ecology of our planet. It requires integration of ecological thinking into all social and economic planning.

Figure 1: Decision-making framework for ecological sustainability (Source: Lowe 1996).

(Figure 2). This transition reflects the fact that cities have become the key economic engines as well as the principal demographic absorbers for planet Earth—via international, as well as internal, migration. The planning and design of most post-1950s urban growth, however, was undertaken in an era when practitioners and populations alike foresaw little or no resource or environmental constraints on urban development (Rees and Roseland 1991). Consequently, significant levels of resource consumption have been designed into western cities in the form of its housing stock, land-use mix, transport systems and density of development (Newton 1997). Cities are not only a major source of global environmental problems and the main bearer of their consequences, they are also the locus of solutions for these environmental and sustainability challenges (Grimm *et al.* 2008). Consequently, the challenge of achieving sustainable development in the 21st century will be won or lost in the worlds' urban areas.

Transitioning to more sustainable cities requires the conceptualisation of an appropriate systems framework for their representation and analysis, such as that provided by the extended metabolism model of urban settlement (Figure 3). This model enables a representation and assessment of complex built environments in terms of:

1. *resource inputs*: whether renewable or non-renewable; involves linking the impact of urban development to the draw-down of stocks and endowments of natural capital, as well as on human, financial and built capital).
2. *urban systems and processes*: the engine or central nervous system of human settlements, including all urban infrastructure—water, energy, communications, transport; their demand–supply interactions; as well as the impacts of governance, technological sophistication, and quality of design, planning and management on system performance. It is in this domain that sustainability has emerged since Brundtland (1987) as a new and powerful driving force that is engaging government, industry and the community in a new way of thinking about urban development. It is the domain in which significant *transitions* are

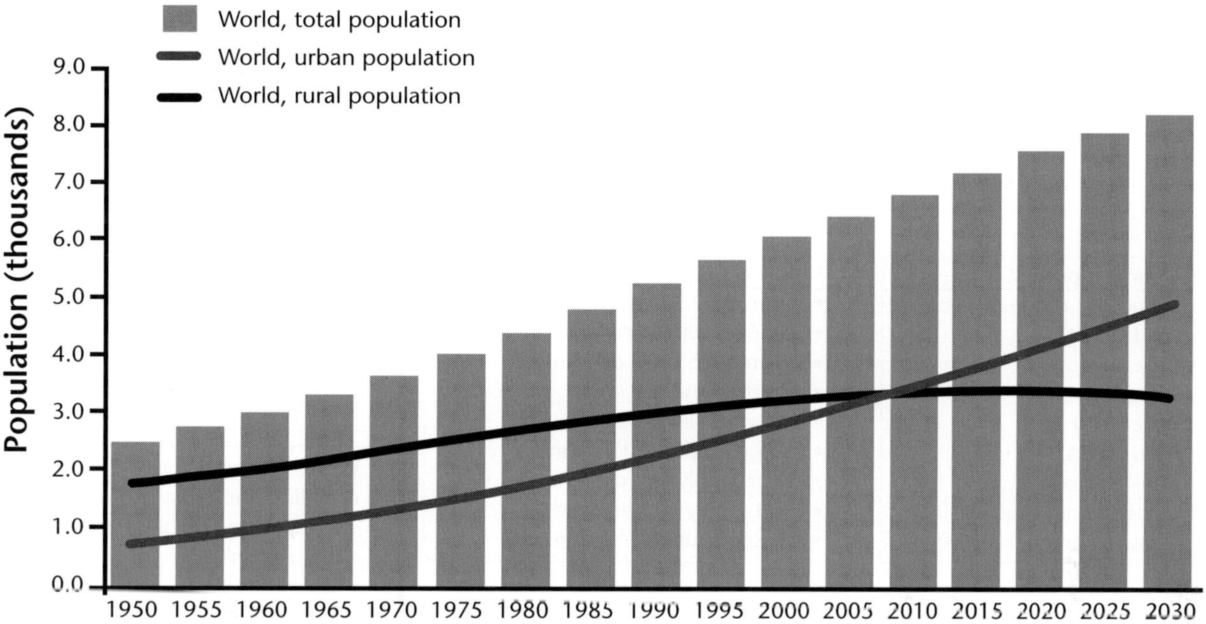

Figure 2: The urban and rural population of the world 1950–2030 (Source: United Nations 2006).

required in relation to how future urban systems are planned, designed and operated.

3. *urban environmental quality*: addresses the condition of the bioregion in which the urban settlement is physically located: the water catchments, the airshed, the waterways, the biodiversity; as well as urban amenity, which includes environmental noise, odours, green space, streetscape quality, and so on.
4. *liveability and human wellbeing*: are broad concepts that include both human and social capital, environmental health and access to key urban services, such as housing, transport, health, education and employment, as well as more qualitative factors such as 'happiness'.
5. *waste and emissions*: a concept originally linked to linear models of industrial production that involved 'extraction – processing – manufacture – distribution – use – disposal'; but which now needs to redefine waste streams as 'resource streams' (for example, stormwater and cities as catchments; wastewater re-use; recycling construction, demolition and household wastes; and the emergence of industrial ecology principles that are linking manufacturers into symbiotic waste streams in order to create new products) while continuing to minimise generation of waste.
6. *exogenous pressures*: include those linked to globalisation of production, international migration, international tourism, flows of capital, resources (such as oil), goods and services, information and knowledge, invasive species, infectious diseases and perhaps most significantly, climate change—all of which have a fundamental impact on human settlements.

The urban metabolism model can also be used to communicate key sustainable development *goals* for cities, such as:

- using resources more efficiently
- using wastes as resources
- restoring and maintaining urban environmental quality
- enhancing human wellbeing
- promoting highly efficient and effective urban and industrial planning, design and management systems.

These goals should be used to drive the strategic planning of all urban developments.

The ultimate objective of pursuing these goals through to fruition would be the development of

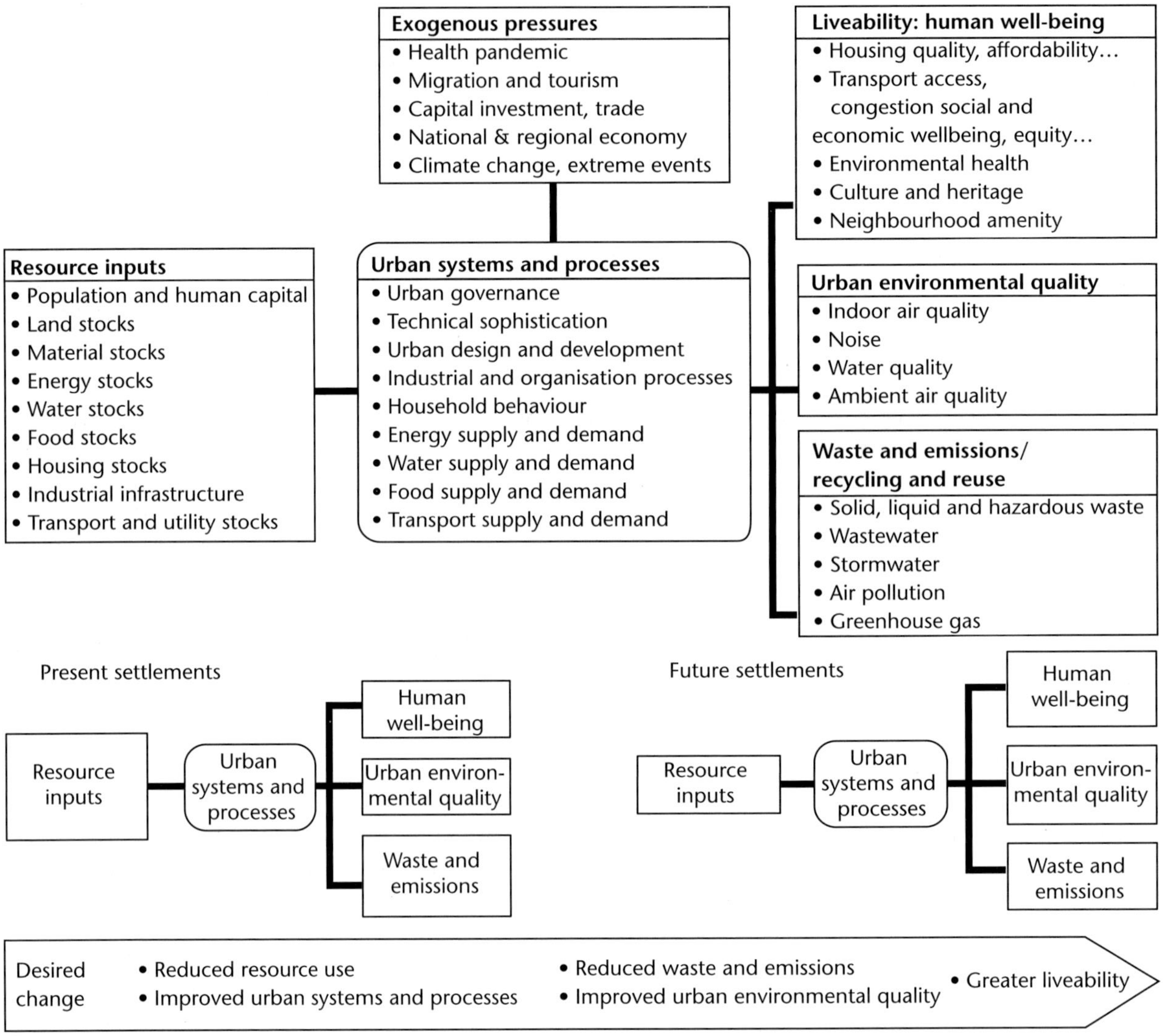

Figure 3: Extended urban metabolism model of human settlements (Source: modified from State of the Environment Advisory Council 1996).

healthy and sustainable cities, or what has been referred to the achievement of eco-city status—the next stage of urban systems evolution from an environmental perspective (Bai and Imura 2000; see Figures 4 and 5). Figure 4 illustrates the typical generalised evolution trajectory that has been observed in many cities globally, especially in developed countries, in relation to the urban development–environment quality nexus: from poverty-associated environmental problems to those associated with production and then to the more recent consumption-related challenges. The transition model is speculative on the eco-city stage and assumes:

> *…that as eco-efficiency improves, citizens will change their lifestyles and develop greater environmental consciousness, and that currently growing environmental impacts of consumption-related issues will not surpass the limitations of the earth's natural ecosystem. (Bai and Imura 2000, p.153)*

How likely it is that this key *transition*—to Stage 4 eco-city status—is achievable for Australia's cities in the 21st century is the principal question for examination in this book. Before introducing the substantive chapters that explore possible *pathways* to a more sustainable future, we will briefly review

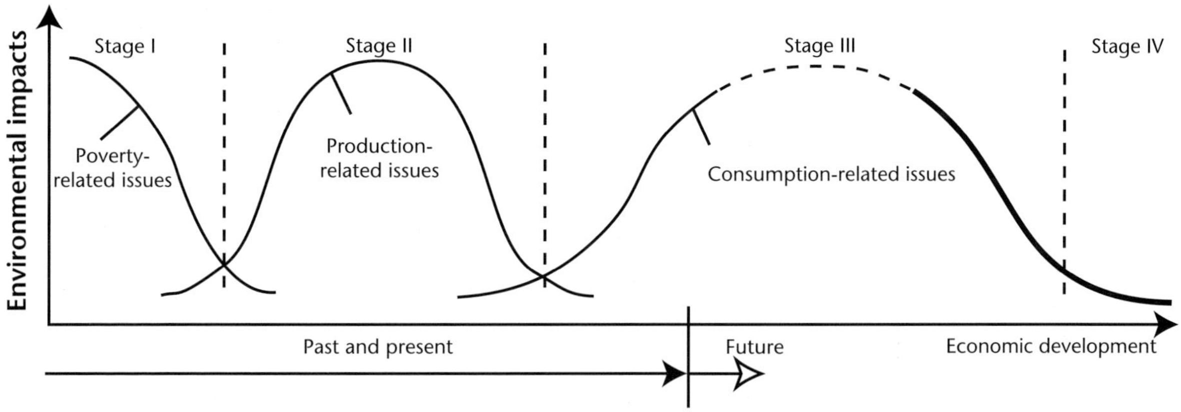

Figure 4: Conceptual illustration of stage model in relation to major urban environmental problems (Source: Bai 2003).

a range of concepts and models that are relevant to understanding transition issues and questions, especially those that are related to transition in an urban context.

Transition concepts

Limits to growth and collapse

Why do we need thought and action on transitions?

Limits to Growth (Meadows *et al.* 1972) was a ground-breaking book that provoked a wide range of discussion on environmental and resource issues that human society could be expected to face in the 21st century. It was the first attempt to systematically analyse and report that environmental and resource constraints would have a significant influence on economic development in the foreseeable future. The book was vastly successful in bringing the attention of the public to the consequences of economic growth on the environment and the potential for over-use of limited resources. The book was one of the first that called for a transition away from current economic development trajectories.

Employing system dynamics modelling, *Limits to Growth* presented a series of scenarios on future resource requirements and environmental impacts, and warned that if our society kept growing in population while pursuing narrow national, corporate or individual economic interests, there was prospect for depleting key resources leading to a collapse of economic and social systems. In the 1992 update, *Beyond the Limits* (Meadows *et al.* 1992), the authors found that in many aspects humanity had already overshot the limit of earth's supporting capacity and that humanity is facing a crucial choice of collapse or creating a sustainable future. In the most recent update *Limits to Growth: The 30-year Update* (Meadows *et al.* 2004), the authors emphasise that the trend is still on an unsustainable track, that the global ecological footprint is continuing to grow and that the outlook is even more pessimistic than 30 years ago.

The work was criticised as not taking into account the capability of societies to respond to such a situation, through individual choices, market mechanisms and technological development; and the authors were dismissed as doomsayers. However, as the books repeatedly emphasised, the purpose was to avoid a collapse rather to

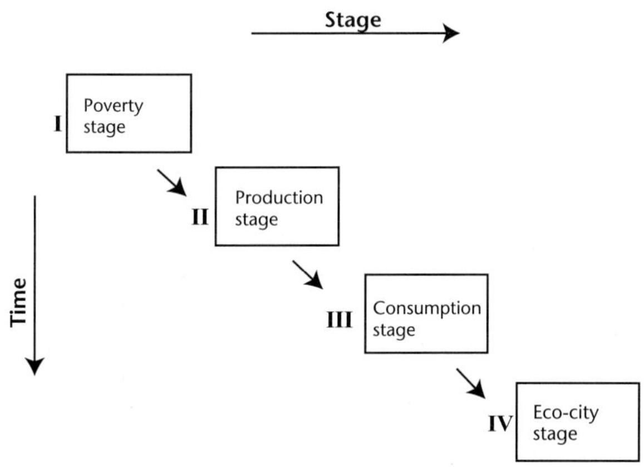

Figure 5: Stages of urban environmental evolution (Source: Bai and Imura 2000).

successfully predict one. It was to warn about the danger of current business-as-usual practices and to call for a transition or revolution equivalent to that of the agricultural and industrial revolutions.

A recent book by Diamond (2005) entitled *Collapse: How Societies Choose to Fail or Succeed* suggests that there are many examples in human history where societies have collapsed. Diamond has listed the following factors as contributing to historical collapse in human societies:

- deforestation and habitat destruction
- soil erosion, salination and fertility loss
- water resource management problems
- over-hunting
- over-fishing
- effects of non-native species on native species
- population growth
- increased per-capita impact of people.

Almost all these factors are related to the growth of human population or wealth—or over-extraction of natural resources—all of which were highlighted in the *Limits to Growth*. In addition to the examples of failure, Diamond also provided success stories where societies have established a more harmonious relationship with their environment.

These books highlight the need for more thought and action on urban transitions and what might constitute the elements of an evolution towards more sustainable cities. One of the key questions requiring examination is whether there are limits to growth in terms of the size of cities. To date, cities have continued to grow physically, demographically, functionally and in spatial extent and impact, without many questions being posed as to the limits to city size itself. From a management point of view, megacities were once cited as a culprit of urban problems, but increasingly it is considered to be more important to examine how well cities are managed rather than their size (Lee 1994; Devas and Rakodi 1993). The failure appears to lie in the management, and not in the size, of cities. However, natural resource constraints, such as water, might become a limiting factor in urban development for some regions in Australia and globally. Behaviour change (for example, demand management) can push the boundary of sustainable water yield further, however, as can technology via high level treatment of recycled water injected back into the water supply system.

In *Beyond Growth*, Daly (1996) examines the philosophies underlying the policies and processes of major global financial institutions involved in urban development (such as the World Bank) and argues for a transition in their theoretical and ideological perspective to one which views the economy as a sub-system of the total ecosystem (namely, the environment of planet Earth, now increasingly affected by the growth of human activities); that growth is ultimately physically limited; and that focus needs to shift from growth to sustainable development.

Environmental Kuznets curve

Does economic development bring about transition?

The *Limits to Growth* advocates argue that the world is on a collision course between economic growth and resource depletion plus environmental pollution. With regard to pollution, it has been proposed that countries progress through an 'inverted-U curve', or Kuznets curve, of environmental development (see Figure 6):

They start poor and clean, then they industrialise and get rich and dirty, and then they are rich enough so that they can afford pollution control so that they ultimately end up rich and clean. (von Weizsacker 2005, p. 1)

The existence of this inverted U-shaped pattern implies that even though a city might be facing serious environmental problems now, it may eventually outgrow and overcome these problems if its economy can keep growing. This has profound impact on the policy commitment for pollution abatement, especially in the developing world. In a sense, some policy makers in developing countries 'have interpreted such results as conveying a message about priorities: "Grow first, then clean up." ' (Dasgupta *et al.* 2002) However, this seems to be blind optimism due to two factors:

1. many issues don't follow the inverted U-shape—most notably greenhouse gas emissions and biodiversity loss
2. even for issues that seem to follow the pattern, it doesn't guarantee the same pattern might replicate in all places automatically.

Consumption-related issues don't seem to be following the inverted U-shape pattern as statistics on per capita resource consumption are revealing regarding water and energy use, housing, travel, waste generation and so on (World Bank 1992; Rothman 1998; de Bruyn and Opschoor 1997; Suri and Chapman 1998; Newton 2006). Bai and Imura (2000) point to several possible reasons for this worsening pattern in the urban context, such as:

- these problems usually cause no visible local harm to urban populations who are geographically removed from the location of resource extraction and sometimes product manufacture, or their negative impacts are limited to a small group of people
- the impacts can easily be externalised (*Tragedy of the Commons*)
- municipal government and industries have few incentives to take countermeasures and regulations remain relatively permissive from an environmental perspective.

The future trend for this type of issue is uncertain; it is unknown if consumption will continue increasing, stabilize at a certain level, or decrease after reaching a certain level. However, studies show that ceilings on consumption seem far beyond the current income level of today's richest cities, considering the fact that there is no evidence showing a voluntary slowing of consumption and associated lifestyles in any cities to date, despite various government initiatives to change consumption patterns (OECD 1998). Indeed, increasing indebtedness appears to have been fuelling growth in consumption over the past decade in countries such as Australia and the United States (Newton 2007). Suri and Chapman (1998) suggest that the improvement in environmental quality in OECD countries might have been achieved simply through globalisation and the associated 'export' of polluting industries.

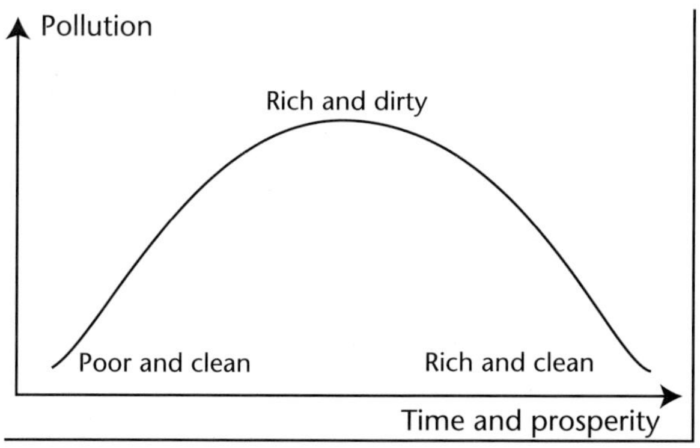

Figure 6: The inverted-U paradigm (Source: von Weizsacker 2005).

Urban environmental evolution

How do urban environments change over time? Are there commonalities in such change? And, if so, what are the drivers and mechanisms of such change?

These are some of the questions behind the development of the urban environmental evolution concept. Urban environmental evolution (Bai and Imura 2000; Bai 2003) is a concept emerging from empirical analyses; for example, in-depth case studies as well as comparative analyses relating to the processes, driving forces and mechanisms of urban environmental change in Asian cities. There are four major components in the urban environmental evolution concept:

1. cities can be viewed as complex systems that are subject to constant change, which constitutes a dynamic evolutionary process
2. urban environmental profiles of cities are diverse, but there are certain commonalities in the longitudinal dynamics among different cities
3. there is a strong non-linearity in the trajectories of the environmental evolution of cities, rather than there being a fixed common pattern
4. the trajectory is shaped by a unique combination of endogenous and exogenous forces, reflecting both the external pressures and the responses within the city (Bai 2003).

Evidence of shifting environmental problems in cities was highlighted by Kirk Smith (1993) and was followed by other studies on urban environmental performance (Lee and Smith 1993; Pugh 1996; Bai and Imura 2000; McGranahan *et al.* 2001; McGranahan *et al.* 2005; Smith and Ezzati 2005). Bai and Imura (2001) argue that urban environmental problems can be categorised into three groups: poverty-related issues, industrial production-related issues, and consumption-related issues (refer again to Figure 4). Empirical evidence shows that in many developed cities, the urban environment proceeds through stages where dominant environmental issues change from poverty-related to industrial pollution-related issues, and then consumption-related issues. In many of the rapidly developing cities in Asia, this pattern tends to be followed (Yedla 2001), but it is much more compressed and 'telescoped'. Consequently, these cities are facing all three categories of problems at the same time, but often with much lower human and financial capital levels to draw upon (Bai and Imura 2000).

It is important to note that the evolutionary pattern shown in Figure 4—especially the first three generations of curves—don't automatically translate into a sustainability transition. Recent work by Ooi (2007) in South-East Asian cities re-emphasises this point. The challenge is to understand how it may be possible to bend the ever increasing curve of consumption-related issues and enter an eco-city stage of urban development. As stated above, this represents a core challenge for Australian cities.

Tipping point

Why and how do some systems achieve sudden and significant change over a short period of time?

The concept of 'tipping point' helps to explain part of the question. It has become popularised following the publication of Gladwell's *The Tipping Point* (2000). The concept describes a well-known phenomenon in physics that refers to the point of sudden and chaotic change from one state to another, and in epidemiology as the moment in an epidemic when a virus reaches critical mass. It has also been used in urban sociology to describe the point at which a neighbourhood will significantly shift in character (most notably in relation to changing ethnic or socio-economic residential patterns associated with 'invasion-succession'; gentrification) Tipping point in a social systems context can be linked to the concept of social epidemics, such as ideas, behaviours, messages and products, and hopefully environmental improvement, where ideas, policies and new technologies can sometimes behave like outbreaks of infectious disease.

The concept is illustrated in Figure 7, which features three hypothetical alternative processes or pathways designed to achieve some desired outcomes (indeed these are illustrative of literally hundreds, if not thousands, of attempts to achieve more sustainable outcomes across all sectors of the economy). Path 2 represents a process that could be considered to have reached a successful outcome (desired state)—but somewhat later than that represented by Path 1. The difference could be attributed to the emergence of some event, person, program or regulation that is capable of propagating change in behaviour at a rapid rate—that can be recognised as a tipping point. Tipping point theory sets out to identify the conditions under which this process operates (Foliente *et al.* 2007).

If the behaviour of a system is predictable, radical changes can be identified and navigated. However, it seems to be extremely difficult to predict when a system will reach tipping point. Indeed it has been stated that: 'Anyone claiming to know for sure when a particular tipping point will be reached would be treated with suspicion' (Nature 2006).

Nonetheless, in the field of Earth systems science, some argue that the Earth is approaching a tipping point in terms of concentrations of CO_2 in the atmosphere and its link to irreversible climate change (Kluger 2006; Hennessy, this volume). These types of concept are difficult to convey to non-scientists as a trigger for behaviour change. Partly in response to this, the Potsdam Research Institute has produced the '*Tipping Point Map*', nominating regions 'where the balance of a particular system has reached the critical point at which potentially irreversible change is imminent, or actually occurring'—such as the melting of the Greenland ice sheet, the Indian monsoon transformation and the collapse of the Amazon forest (Kemp 2005).

There are several implications of tipping-point transitions. On the one hand, it is about little things making big differences if they occur at the right time and in the right place. In terms of inducing a positive transition, it has significant meaning in terms of using minimum efforts or resources to reach the maximum outcome. On the other hand, the fact that radical changes can occur by even the smallest additional disturbance means significant caution is needed in managing such a system. While it is very appealing to think about inducing a system to reach tipping point, or

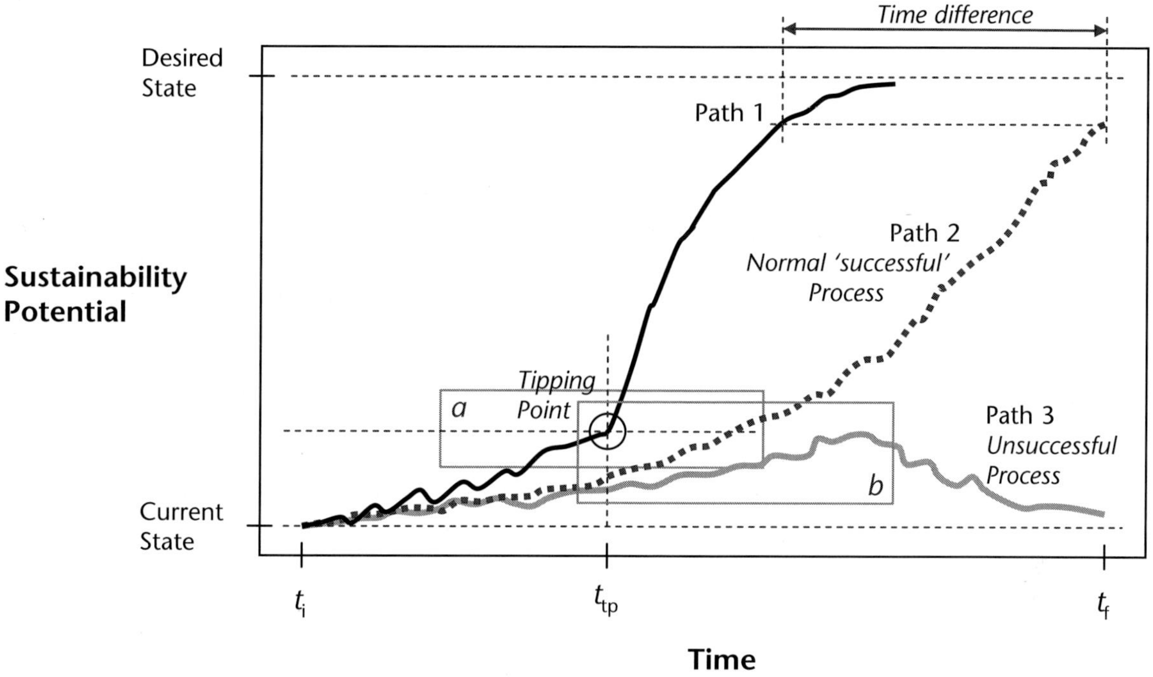

Figure 7: The tipping point process (Source: Foliente *et al.* 2007).

preventing it from approaching one, there are concerns as well: primarily that the pathway to capitalising on the tipping point and inducing widespread social change remains unclear and could divert our focus away from pursuing multiple initiatives now to achieve change, albeit incrementally (Rosa 2006).

System innovation

System innovation theory deals with one of the fundamental questions in transition: how changes occur to a particular social-technological or social-ecological system, what induces the change, and what are the barriers to change?

The innovation and change referred to here includes both incremental and radical forms (Berkhout 2004; Geels 2005), but it is the more radical forms of change that are of particular interest in considering transition, because it is these kinds of innovation that transform the entire structure of an existing system (Elzen *et al.* 2004) and thus bring with it the opportunity of transition to a different, and significantly improved level of performance.

The logistic curve (Figure 8) is commonly used to depict the path of socio-technical innovation in a process sense (that is, from invention to full adoption). It can also be used to describe the five main types of actors or participants and their role in the diffusion process (namely, innovators, early adopters, early majority, late majority and laggards). Given the inertia that is usual across most sectors of the economy, the initial levels of resistance to new products, processes and policies can be quite high. There can be good reasons for this that relate to the time and cost required to translate an idea into an invention and then into a prototype that can be 'road tested'. Many of the technologies discussed in Sections 5 through 7 of this volume offer considerable potential for supply-side innovation in urban infrastructure transition, and can be mapped onto this logistic curve. An early understanding of the barriers to innovation adoption—at the community level and within public and private institutions—will be important if the time to reach tipping point and widespread implementation and adoption is to be minimised. Recent studies examining sustainability transition from a systems innovation point of view emphasise that in addition to the importance of technology availability, the existence of effective mechanisms that

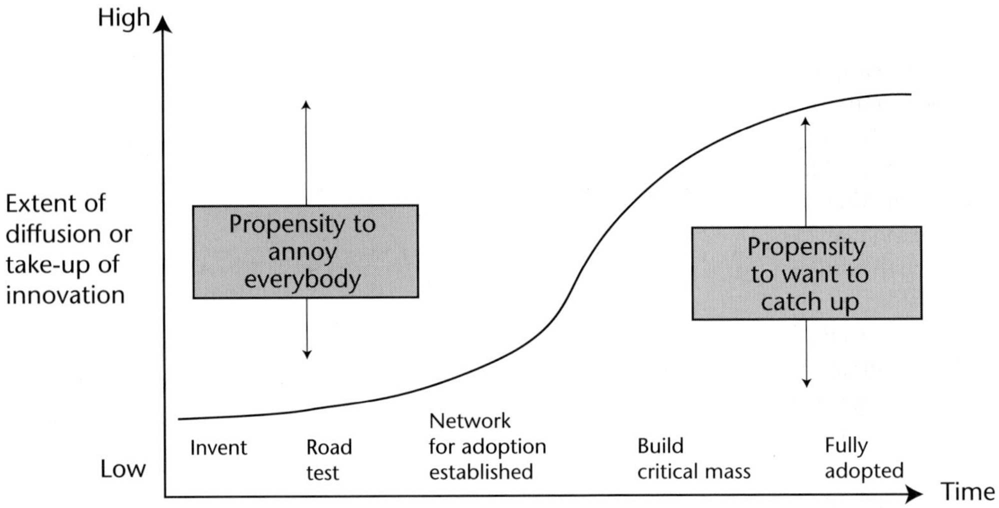

Figure 8: Diffusion of innovation—typical trajectory (Source: CSIRO Environment Sector Advisory Committee 2004).

provide functioning linkages among different levels of the institutional settings of the system are also critical (Bai *et al.* forthcoming). Many of the urban challenges outlined in the body of this book have been evident for at least the past decade or longer, but due to some combination of government, industry and community inertia, now have diminishing windows of opportunity within which to achieve a sustainable response—and as the Stern Report (2006) foreshadowed, they are now likely to come at a much higher cost.

Resilience

Why do some systems change more easily than others under the same level of disturbance? How do we measure this system attribute?

Resilience is a measure of the ability for systems to absorb changes and persist; that is, to adapt and adjust to changing internal or external processes (Walker *et al.* 2004). It is a concept that has grown to include a broad range of system attributes and features such as:

1. the amount of disturbance a system can absorb and still remain within the same state or general condition
2. the degree to which the system is capable of self-organisation
3. the degree to which the system can build a capacity for learning and adaptation (Carpenter *et al.* 2001; Klein *et al.* 2003).

It is about staying 'in the game' even if there are shocks and disturbances from outside (Pickett *et al.* 2004), and therefore is closely related to concepts of sustainability and sustainability transition (Folke *et al.* 2002).

Resilience has been an important theoretical underpinning for understanding transformation (Gunderson and Holling 2002). Walker *et al.* (2004) argue that resilience, together with adaptability and transformability, are system attributes that govern the dynamics of a system, and together determine the future trajectories of that system. While resilience represents the capacity of a system to absorb disturbance and reorganise, adaptability is the capacity of actors in the system to influence resilience, and transformability is the capacity to create a fundamentally new system when existing systems become untenable.

Despite significant diffusion of the resilience concept, there is as yet no single established way of measuring resilience. Carpenter *et al.* (2001), for example, argue that it is important to ask the question of resilience 'of what to what'?—what system state is being considered; what perturbations are of interest, over what time period and at what scale?

The resilience concept has been widely discussed and adopted in studying natural ecosystems. Recently, there has been interest in applying the resilience concept to cities as complex adaptive systems. Pickett *et al.* (2004) argue that the metaphor of 'cities of resilience' can be a new approach for integrating the ecological, socioeconomic and planning realms together. Based on an empirical study of the relationship between urban vegetation pattern and biodiversity in the Puget Sound region, Alberti and Marzluff (2004) found that resilience in urban ecosystems is closely related to patterns of human activity and their impact on natural habitats. Although being cautious about using the concept of resilience without specific definition, Klein *et al.* (2003) consider resilience to be a useful concept to apply to mega-city hazard management, in terms of analysing present and future hazard scenarios, and enhancing capacity for disaster prevention and preparedness with disaster recovery.

Aiming at a more systematic exploration of resilience within urban ecosystems, the Urban Resilience node within the Resilience Alliance (www.resiliencealliance.org/) has proposed a research framework of resilience in cities that includes a focus on metabolic flows, social dynamics, governance networks and the built environment. It also proposed a set of core research questions to be explored under this framework:

- What are the key drivers of change and threats to cities at global, regional and local scales?
- Can we develop typologies of cities and their developmental trajectories based on social, economic, and ecological resilience to help decision-makers avoid common traps/pitfalls?
- How does the emergence of a new type of social–ecological system—the extended urban region or 'megalapolitan' region, comprising one or more core cities, nearby interdependent towns, and interspersed rural landscapes—influence system resilience at multiple scales?
- How could further urbanisation and urban transformation be re-directed so that cities can be harnessed as generators of innovation and solutions to issues of global sustainability and thereby contribute to delivery of the UN's Millennium Development Goals for 2015? (www.un.org/millenniumgoals)

Issues of urban resilience in relation to natural, built, social and human capital are central to Sections 2 through 4 of this book, especially concerning the impacts of globalisation, population growth and urbanisation, peak oil and the impacts of and adaptations to climate change.

Sustainability transition

The above-mentioned transition theories wouldn't make much sense in practice if they are not linked to the transition towards sustainability. The science of sustainable development has been recognised as an independent area of science by the US Association for the Advancement of Science (Raven 2002). It has been suggested that the field is 'defined by the problems it addresses rather than by the disciplines it employs' (Clark 2007)—which are complex and trans-disciplinary in nature. Sustainability science asks the core question of 'how are long-term trends in environment and development, including consumption and population, reshaping nature–society interactions in ways relevant to sustainability?' (Kates *et al.* 2001) Sustainability science is the study of complex systems interactions, and its central aim is to facilitate a 'transition toward sustainability' (National Research Council 1999).

There are different pathways for achieving sustainability transition: changing the course of the trend (for example, bending the curve), accelerating positive transition and slowing down negative trends, and understanding complex trends (Raskin *et al.* 1998). Kates and Parris (2003) argue that examining major long-term trends in nature and society often reveal that good news and bad news co-exist, and that a central task is to accelerate the positive trends that favour a transition and slow the negative trends that impede them.

As a highly interdisciplinary and policy relevant field of study, sustainability science has been enriched by experiences and lessons learnt through case studies from all over the world. These learnings come from real world successes and failures in attempting sustainability transitions (such as the 10 year effort by the Chinese government in pollution control in Huai River Basin; Bai and Shi 2006). Lessons from rapidly developing countries and regions are of special significance. For example, China has been undergoing rapid economic growth and rapid urbanisation, and 'is just entering the difficult straits of a journey toward sustainability' (O'Riordan 2006). The sustainability transition in China is an exercise that both the country and the world cannot afford to fail.

Urbanisation represents one of the major trends in human society, and has been recognised as 'a battleground for global sustainability (Clark 2003). Whether the focus is on 'sustainable cities' or 'how cities contribute to sustainability' (Satterthwaite, 2001), can be a point for academic debate, but there is no doubt about the importance of cities in achieving sustainability transition in the 21st century. Urbanisation seems to be an inevitable and, in most cases, irreversible trend, and is still progressing in Australia (one of the most urbanised societies on Earth—see Section 3 of this book) and therefore it is important to talk about how to influence the trend towards achieving more sustainable outcomes, rather than trying to assess whether the current trend is sustainable or not (Bai 2006).

A high-level overview of a range of transitions that shape cities (see Newton's discussion of settlement transitions in Section 3) provides some measure of optimism for more sustainable urban development pathways into the future. Throughout history, cities have been strongly shaped for ill or for good (but usually the latter) by the application of key technologies that have permitted particular trajectories of development to occur as a means of accommodating growth of population, industry and housing. The auto-city configuration was a 20th century development that now dominates most early 21st century cities—but is now almost universally recognised as unsustainable in a carbon- and resource-constrained word. Moreover, the range of key infrastructures that have supported urban development up to the present are now being re-examined in terms of their capacity to deliver their services in a sustainable manner into the future. These include:

- urban water systems, and the prospect of transitioning from a linear system (divert–dam–distribute–use once–dispose) to an integrated closed-loop system that uses a portfolio of treated stormwater and wastewater, and possibly desalination, in the context of water-sensitive urban design principles of urban development and re-development
- urban energy systems that are based on distributed renewable sources, compared with highly inefficient and polluting centralised, fossil fuel-based networks
- eco-industrial development that is based on cradle-to-cradle processes for individual organisations and multi-firm clusters based on accessing synergistic waste, energy and material streams appropriate for the regional development of new green industries.

Sections 5 to 7 of this book explore each of these domains in more detail.

Pathways towards more sustainable cities: 3-Horizon thinking

We have argued elsewhere (Newton 2007) that for 21st century cities to be sustainable they need to be able to appropriate from a pipeline of innovative technologies, products, designs and processes that can be substituted for those currently in operation that are beginning to show signs of failure. In Section 8 of this book there are clear signals that:

1. key urban infrastructures are underperforming as a result of decades of under-investment
2. new models for financing and managing urban infrastructure need to be adopted
3. a major re-think of metropolitan planning is required in order to accommodate mega-metropolitan development that is ecologically and socially sustainable.

How to deliver transformational change that is capable of lifting the sustainability performance of Australia's cities is the raison d'être for the ideas and material assembled in this book.

Three horizon thinking has been advanced as a framework for identifying innovations that have capacity for delivering sustainability transitions (see Figure 9; Newton 2007). Horizon 1 initiatives include those that are currently available, have demonstrable benefit, yet have not been widely implemented. Horizon 2 innovations exist as a model, prototype product or system, scoped process or similar, but require some challenging testing, extension, real-world case study application, in order to develop a convincing 'business case' for wider application. Horizon 3 innovations tend to be more radical, based on concepts and technologies that are vastly different from those currently operating and where there are currently major barriers (of a technical, social, corporate or governance nature) to be overcome, and where widespread implementation is unlikely inside 20 years or so. Their impact on system performance is, however, likely to be more substantial.

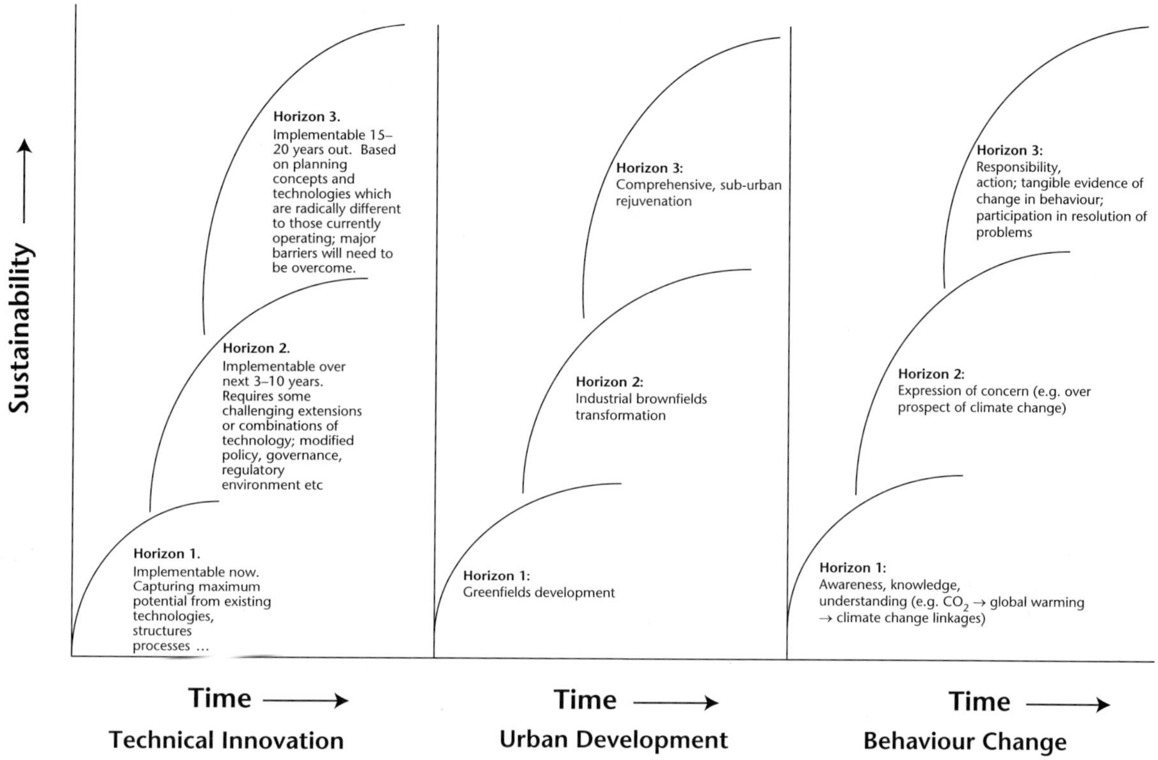

Figure 9: 3-Horizon thinking (Source: Newton 2008).

There are three principal domains where 3-Horizon thinking is required for more sustainable urban development: technological innovation, innovation in planning and design of urban development and behaviour change.

Technological innovation and adoption

Technological change can be seen as a powerful force that has changed the form and function of cities through history. (Urban Institute 1971; Jones 1982; Brotchie *et al.* 1985, 1987, 1991, 1997; ter Heid 1990; OECD 1992; Troy 1995; US Congress Office of Technology Assessment 1995). Technical innovation continues to offer the basis for re-shaping cities, and the chapters assembled in this book for the key infrastructures of energy and water (Sections 5 and 6) outline the potential transformations that are possible, as do the chapters in Section 7 related to manufacturing and waste. Most of the initiatives outlined are Horizon 2 (such as water sensitive urban design, distributed energy and hybrid vehicles) or Horizon 3 in nature (such as integrated urban water systems, the hydrogen economy and intelligent transport systems), although there remains much mileage to be gained from Horizon 1 efficiency-based initiatives.

Planning, design and management of urban development

The science, engineering and technology capacity to deliver more sustainable urban development in greenfield sites is increasing significantly, and is classed here as Horizon 1 innovation—activity which should be occurring now. There are emerging opportunities through new technologies, knowledge access and innovation to integrate sustainable urban developments from the dwelling scale (www.yourbuilding.org; www.greenhouse.gov.au/yourhome) to the neighbourhood scale (www.yourdevelopment.org). Some examples from Newton (2006) include:

- water wise house—water sensitive urban design of neighbourhood
- energy efficient house—distributed generation at precinct level using renewables and smart

grid (including the ability to sell surplus electricity to the grid)
- smart house—neighbourhood broadband access
- zero-waste house—comprehensive neighbourhood waste collection
- secure house—safe neighbourhood
- accessible house—resident access to public and active transport modes
- healthy house—good outdoor air quality
- quiet house—level of environmental noise abatement
- durable house—environmental hazards.

The revitalisation of industrial brownfield sites is also being advanced (www.brownfields.org) and is classed here as Horizon 2 innovation, given the added challenges that site remediation often brings to a development.

The rejuvenation and re-development of suburbs built during the 19th and 20th centuries, however, represents the greatest (Horizon 3) challenge to planners and designers attempting to create more sustainable 21st century cities. The deficiencies of contemporary cities have been documented and discussed throughout the volume. Creating more visually distinctive neighbourhoods with mixed-use development and a greater spectrum of affordable housing that is accessible via public and active transport modes needs to be the norm—that is, *smart redevelopment* (see Section 8).

Changing human behaviour—individual and collective

The ultimate Horizon 3 transition is not on the technology supply side. Nor does it rest entirely with urban planners and designers. Rather, it involves individuals and their attitudes towards consumption and the natural and built environment—attitudes that are failing to link personal consumption habits directly with environmental depletion and pollution and a concern for the environment (Newton 2007). In models of behaviour change (Voronoff 2005), three stages are typically identified:

1. an *awareness* stage where individuals grow in the amount of knowledge and understanding about a particular product or situation (Horizon 1)
2. a *concern* stage where knowledge is translated into an articulation of some individual expression of disquiet or apprehension about the issue in question (Horizon 2)
3. a final stage characterised by the taking of some *action* that reflects a tangible change in behaviour (Horizon 3).

If the Australian Bureau of Statistics' surveys of resident concern for the environment over the last 15 years are any guide (Newton 2006, 2007; also Chapter 11); then Horizon 3 behaviour change in relation to resource consumption represents a formidable challenge for contemporary society. Section 9 of this book explores some of the progress to date in changing attitudes to sustainability among government, the corporate sector and the community. The volume concludes (Section 10) with two key challenges to change current human value systems: transitioning to a *simpler way of living* (with particular reference to consumption) and transitioning to a *planetist perspective on living* in the 21st century—coming to terms with the reality that there is only one Earth.

References

Alberti M and Marzluff JM (2004). Ecological resilience in urban ecosystems: linking urban patterns to human and ecological functions. *Urban Ecosystems* **7**, 241–265.

Bai XM, Wieczorek A, Kaneko S, Lisson S and Contreras A (forthcoming). Enabling sustainability transition in Asia: the importance of vertical and horizontal linkages. *Technology Forecasting and Societal Change*, in press.

Bai XM (2003). The process and mechanism of urban environmental change: an evolutionary view. *International Journal of Environment and Pollution* **19(5)**, 528–541.

Bai XM (2006). 'Urbanization: is the Chinese model sustainable?' Speech at the Final Special Session on Future Cities, World Urban Forum III, Vancouver, 23 June.

Bai XM and Imura H (2000). A comparative study of urban environment in East Asia: stage model of urban environmental evolution. *International Review for Global Environmental Strategies* **1(1)**, 135–158.

Bai XM and Imura H (2001). Towards sustainable urban water resource management: a case study in Tianjin, China. *Sustainable Development* **9**, 24–35.

Bai XM and Shi PJ (2006). Pollution control in China's Huai River basin: what lessons for sustainability? *Environment* **48(7)**, 22–38.

Beeton RJ, Buckley KI, Jones GJ, Morgan D, Reichelt RE and Trewin D (2006). *Australian State of the Environment 2006.* Department of the Environment and Heritage, Canberra.

Berkhout F, Smith A and Stirling A (2004). Socio-technological regimes and transition contexts. In *System Innovation and the Transition to Sustainability: Theory, Evidence and Policy.* (Eds B Elzen, FW Geels and K Green) pp. 48–75. Edward Elgar, Cheltenham, UK.

Brotchie J, Batty M, Hall P and Newton P (Eds) (1991). *Cities of the 21st Century: New Technologies and Spatial Systems.* Longman Cheshire, Melbourne.

Brotchie J, Hall P and Newton P (Eds) (1987). *The spatial impact of technological change.* Croom Helm, London.

Brotchie J, Newton P, Hall P and Dickey J (Eds) (1997). *East-West Perspectives on 21st Century Urban Development.* Ashgate, Aldershot, UK.

Brotchie J, Newton P, Hall P and Nijkamp P (Eds) (1985). *The Future of Urban Form: the Impact of New Technology.* Croom Helm, London.

Brundtland GH (1987). *Our Common Future.* World Commission on Economic Development, New York

Carpenter S, Walker B, Anderies JM and Abel N (2001). From metaphor to measurement: resilience of what to what? *Ecosystems* **4**, 765–781.

Clark WC (2003). Urban environments: battlegrounds for global sustainability. *Environment* **45(7)**, 1.

Clark WC (2007). Sustainability science: a room of its own. *Proceedings of the National Academy of Sciences of the United States of America* **104(6)**, 1737–1738.

CSIRO Environment Sector Advisory Committee Melbourne (2004). Powerpoint presentation by Greg Bourne, CEO BP Australia.

Daly HE (1996). *Beyond Growth: the Economics of Sustainable Development.* Beaco, Boston, Massachuseets.

Dasgupta S, Laplante B, Wang H and Wheeler D (2002). Confronting the environmental Kuznets curve. *Journal of Economic Perspectives* **16(1)**, 147–168.

de Bruyn SM and Opschoor JB (1997). Developments in the throughput-income relationship: theoretical and empirical observations. *Ecological Economics* **20(3)**, 255–268.

Devas N and Rakodi C (1993). *Managing Fast Growing Cities.* Longman: Harlow, UK.

Diamond JM (2005). *Collapse: How Societies Choose to Fail or Succeed.* Penguin: New York.

Elzen B, Geels FW and Green K (Eds) (2004). *System Innovation and the Transition to Sustainability: Theory, Evidence and Policy.* Edgar Elgar, Cheltenham, UK.

Foliente G, Rogers A, Blutstein H and Wang X (2007). 'Urban sustainability transition – a tipping point approach'. Paper presented to State of Australian Cities national conference, Adelaide, 28–30 November.

Folke C, Carpenter S, Elmqvist T, Gunderson L, Holling CS and Walker B (2002). *Resilience and Sustainable Development: Building Adaptive Capacity in a World of Transformations.* Environmental Advisory Council to the Swedish Government, Stockholm.

Geels FW (2005). *Technological Transitions and System Innovations: a Co-evolutionary and Socio-Technical Analysis.* Edward Elgar, Cheltenham, UK.

Gladwell M (2000). *The Tipping Point: How Little Things Can Make a Big Difference.* Little, Brown, New York.

Grimm NB, Faeth SH, Golubiewski NE, Redman CL, Wu JG, Bai XM and Briggs JM (2008). Global change and the ecology of cities. *Science* **319**, 756–760.

Gunderson LH and Holling CS (Eds) (2002). *Panarchy: understanding transformations in human and natural systems.* Island Press, Washington DC.

Hamilton C and Denniss R (2005). *Affluenza: When Too Much is Never Enough.* Allen and Unwin, Sydney.

Johnston RJ (1971). *Urban Residential Patterns.* Bell, London.

Jones B (1982). *Sleepers, Wake! Technology and the Future of Work.* Oxford University Press, Melbourne

Kates R and Parris T (2003). Long-term trends and a sustainability transition. *Proceedings of the National Academy of Sciences of the United States of America* **100(14)**, 8062–8067.

Kates R, Clark W, Corell R *et al.* (2001). Sustainability science. *Science* **292**, 641–642.

Kemp M (2005). Inventing an icon. *Nature* **437**, 1238.

Klein RJT, Nicholls RJ and Thomalla F (2003). The resilience of coastal megacities to weather-related hazards. In *Building safer cities: the future of disaster risk.* (Eds A Kreimer, M Arnold and A Carlin) pp. 101–120. World Bank Disaster Management Facility, Washington, DC.

Kluger J (2006). Earth is at the tipping point. *Time*, 3 April, <http://reference.aol.com/global-warming/timemagazine/_a/earth-is-at-the-tipping-point/20060327120109990001>.

Lee Y (1994). Myths of environment management and the urban poor. In *Mega-City Growth and the Future.* (Eds RJ Fuchs, E Brennan, J Chamie, F-C Lo and JI Uitto) pp. 390–411. United Nations University Press, Tokyo.

Lowe I (1996). Towards ecological sustainability. In *Australia State of the Environment*, State of the Environment Advisory Council. pp. 10–30. CSIRO Publishing, Melbourne.

McGranahan G, Jacobi P, Songsore J, Surjadi C and Kjellen M (2001). *The Citizens at Risk.* Earthscan, London.

McGranahan G, Marcotullio P, Bai XM, Balk D, Braga T, Douglas I, Elmqvist T, Rees W, Satterthwaite D, Songsore J, Zlotnick H, Eades J, Ezcurra EI and White A (2005). Urban systems. In *Ecosystems and Human Well-Being: Current State and Trends* (Millennium Ecosystem Assessment Series). (Eds R Hassan, R Scholes and N Ash). Island Press, Washington DC.

Meadows D, Randers J and Meadows DL (2004). *Limits to Growth: the 30-Year Update.* Chelsea Green, White River Junction, Vermont.

Meadows DH, Meadows DL and Randers J (1992). *Beyond the Limits: Global Collapse or a Sustainable Future.* Earthscan, London.

Meadows DH, Meadows DL, Randers J and Behrens W (1972). *The Limits to Growth.* Universe, New York.

National Research Council (1999). *Our CommonJ-journey.* National Academies Press, Washington, DC.

Nature (2006). Editorial: reaching a tipping point. **441(7095)**, 785.

Newton PW (Ed.) (1997). *Re-shaping Cities for a more Sustainable Future*, Research Monograph No. 6. Australian Housing and Urban Research Institute, Melbourne.

Newton PW (2006). 'Australia state of the environment: human settlements: theme report'. Department of Environment and Heritage: Canberra. <http://www.environment.gov.au/soe/2006/publications/commentaries/settlements/index.html>.

Newton PW (2007). Horizon 3 planning: meshing liveability with sustainability. *Environment and Planning B: Planning and Design*, **34**, 571–575.

O'Riordan T (2006). China and the agony of the sustainable transition. *Environment* **48**(7), 1.

OECD (1992). *Cities and New Technologies.* OECD, Paris

OECD (1998). *Towards Sustainable Consumption Patterns: a Progress Report on Member Country Initiatives.* OECD, Paris.

Ooi GL (2007). Urbanization in Southeast Asia: assessing policy process and progress toward sustainability. *Journal of Industrial Ecology* **11**(2), 31–42.

Pickett STA, Cadenasso ML and Grove JM (2004). Resilient cities: meaning, models, and metaphor for integrating the ecological, socio-economic, and planning realms. *Landscape and Urban Planning* **69**, 369–384.

Pugh C (1996). *Sustainability, the Environment and Urbanization.* Earthscan, London.

Raskin P, Gallopin G, Gutman P, Hammond A and Stewart R (1998). *Bending the Curve.* Stockholm Environmental Institute, Boston, Massachusetts.

Raven PH (2002). Science, sustainability, and the human prospect. *Science* **297(5583)**, 954–958.

Rees WE and Roseland M (1991). Sustainable communities: planning the 21st century. *Plan Canada* **31**(3), 15–26.

Rosa EA (2006). Sticking points in the push for change. *Nature* **442**, 627.

Rothman DS (1998). Environmental Kuznets curves – real progress or passing the buck? A case for consumption-based approaches. *Ecological Economics* **25**(2), 177–194.

Satterthwaite D (2001). Sustainable cities or cities that contribute to sustainable development? In *The Earthscan reader in sustainable cities*. (Ed. D Satterthwaite) pp. 80–106. Earthscan, London.

Smith KR and Ezzati M (2005). How environmental health risks change with development: the epidemiologic and environmental risk transition revisited. *Annual Review of Environment and Resources* **30**, 291–333.

Smith KR and Lee Y-S (1993). Urbanization and the environmental risk transition. In *Third World Cities: Problems, Policies, and Prospects*. (Eds JD Kasarda and AM Parnell) pp. 161–179. Sage, London.

State of the Environment Advisory Council (1996). *Australia State of the Environment 1996*. pp. 2–5. CSIRO Publishing, Melbourne.

Stern N (2006). *Stern Review on the Economics of Climate Change*. HM-Treasury, London

Suri V and Chapman D (1998). Economic growth, trade and energy: implications for the environmental Kuznets curve. *Ecological Economics* **25(2)**, 195–208.

ter Heide H (Ed.) (1990). *Technological Change and Spatial Policy, Netherlands Geographical Studies 112*. Royal Netherlands Geographical Society, Amsterdam.

The Economist Intelligence Unit (2004). Melbourne World's Top City, *The Age*, February 6.

The Urban Institute (1971) *The Struggle to Bring Technology to Cities*, The Urban Institute, Washington DC.

Troy P (Ed.) (1995). *Technological Change and the City*. The Federation Press, Sydney.

United Nations (2006). 'World urbanization prospects: the 2005 revision'. Working Paper No. ESA/P/WP/200. Department of Economic and Social Affairs, Population Division, New York.

Urban Resilience Working Group (2007). *Urban resilience research prospectus*. <http://www.resalliance.org/1610.php>.

US Congress Office of Technology Assessment (1995). *The Technological Reshaping of Metropolitan America*, OTA,Washington DC.

von Weizsacker EU (2005). 'Buildings technology in the vanguard of eco-efficiency'. Keynote speech, World Sustainable Buildings conference, Tokyo, 27 September, <http://www.sb05.com/plenary/WeizsaeckerKeynote.pdf>.

Voronoff D (2005). *Community Sustainability: a Review of What Works and How it is Practiced in Victoria*. Environment Victoria, Melbourne.

Wackernagel M and Rees W (1996). *Our Ecological Footprint: Reducing Human Impacts on the Earth*. New Society Publishers, Gabriola Island, British Columbia.

Walker B, Holling CS, Carpenter SR and Kinzig A (2004). Resilience, adaptability and transformability in social–ecological systems. *Ecology and Society* **9**(2), <http://www.ecologyandsociety.org/vol9/iss2/art5/>.

World Bank (1992). *World Development Report 1992: Development and the environment*. Oxford University Press: New York.

Yedla (2002). 'Urban environmental evolution: the case of Mumbai'. Commissioned paper for Institute for Global Environmental Strategies, Hayama, Japan.

SECTION 2

Global and Local Forces Re-Shaping Cities

The 21st century brings a different set of factors into play that will shape our cities and regions, compared with those of the 20th century. Two of the principal driving forces of *globalisation* outlined by Stilwell and Jordan—the international *movement of capital* and the *role of technology* in decentralising production and spatially concentrating control functions—have implications for Australia's evolving urban regions as they compete for capital, production and producer services—and tourism (Dwyer *et al.*) nationally, as well as globally. There is evidence of *growing spatial inequalities* between Australia's cities and regions, as well as within cities, in relation to human capital, jobs and services—especially the gentrified core versus the industrial periphery.

The 21st century will be a *resource-constrained world* (Turner and Foran) as well as a *carbon-constrained world* (Hennessy and Section 4). The impact of *peak oil* will be considerable, as it re-shapes mobility and accessibility for residents and businesses alike within cities (Newman; Dodson and Sipe)—until the transition to a society based on renewables and solar–hydrogen becomes more firmly established. The intervening bridging period of several decades calls for increased energy efficiencies, diversification of energy sources and a re-shaping of the structure of our cities in relation to compactness, density and transport, (especially active and public modes). Residents in outer metropolitan areas are particularly vulnerable to the challenges that will follow peak oil.

The 21st century will also witness a transition from baby boomer-led patterns of living, working and consumption to those of generation X, Y and Z (Payne *et al.*)—where key signals and trends are yet to clearly emerge.

PWN

2

Climate Change

Kevin Hennessy

Climate change is one of many issues being considered within the context of sustainable development. Climate change can slow the pace of progress toward sustainable development, either directly through increased exposure to adverse impact or indirectly through erosion of the capacity to adapt (IPCC 2007b). Conversely, sustainable development can reduce vulnerability to climate change by enhancing adaptive capacity and increasing resilience (IPCC 2007b). This chapter will review the evidence for observed climate change and the likely causes, the projections for future climate change and impacts, and the options for risk management through reducing greenhouse gas emissions and enhancing adaptation.

Most of the information is derived from the Intergovernmental Panel on Climate Change (IPCC), which was established in 1988 by the World Meteorological Organisation and the United Nations Environment Program. Its role is to assess the best available peer-reviewed scientific, technical and socioeconomic information on climate change from around the world. The assessments help governments and communities understand the scientific basis of the risk of human-induced climate change, its potential impacts and options for adaptation and mitigation. IPCC assessments were published in 1990, 1996, 2001 and 2007.

The 2007 IPCC assessment

In early 2007, the IPCC published its Fourth Assessment in three Working Group reports. The first dealt with *The physical science basis*; the second covered *Impacts, adaptation and vulnerability*; and the third looked at *Mitigation* (reducing net greenhouse gas emissions). These reports were based on literature published between 2001 and 2006. They represent a comprehensive, balanced and transparent assessment of our current understanding about climate change, including uncertainties and research priorities.

A few aspects of the IPCC process are worth highlighting, because they demonstrate the degree of rigour involved. IPCC reports are written by teams of authors, nominated by governments and international organisations. They come from universities, research centres, business and environmental associations across the world. More than 450 lead authors and more than 800 contributing authors were involved. Four drafts of each report were produced over four years. Review is an essential part of the IPCC process—to ensure an objective and complete assessment. Comments were sought from over 2500 scientific expert reviewers for the first three drafts, and from government reviewers for the last three drafts. Thousands of comments were received on each chapter, and the relevant authors were required to deal with every comment in a standard spreadsheet. Independent review editors were appointed to ensure that comments were handled properly; that is, if comments were accepted, changes had to be implemented in the chapter; if comments were simply observations, they had to be noted; and if comments were considered invalid, reasons had to be given. The IPCC reports do not attempt to provide a consensus view—they simply report the balance of evidence and uncertainty in the available literature, based on expert judgement.

A strong emphasis was placed on assessment of peer-reviewed literature in scientific journals, technical reports and books. Where necessary, non-peer-reviewed literature was also assessed—these included government reports and consultancy reports. The IPCC report on *The physical science basis* drew heavily on journal papers because they provided most of the relevant information. This was also the case for the reports on *Impacts, adaptation and vulnerability* and *Mitigation*, but they included a good deal of non-peer-reviewed literature because this also provided much of the required information.

Some of the IPCC conclusions have been criticised by a handful of scientists, statisticians, economists and media commentators. Few, if any of these criticisms, withstand the scrutiny of the peer-review process. Indeed, many of the alternative theories have not been submitted to scientific journals—they tend to appear on websites and in the media where peer review is absent. A few alternative theories have been published in scientific journals, but have been subsequently challenged and put to rest in follow-up journal papers that expose flaws in those theories (such as Mann *et al.* 2003). A media example is the BBC documentary called the *Great Global Warming Swindle*. On the basis of a number of incorrect/incomplete graphs and theories, the program claimed that the late 20th century was cooler than the Medieval Warm Period around 1100–1300 AD, and the late 20th century warming was primarily due to solar variability rather than increases in greenhouse gases. Neither of these contentions is supported by papers in reputable scientific journals (BAMOS 2007; Lockwood and Frohlich 2007). Some sceptics also claim that climate models are unreliable for climate projections. The IPCC (2007a) dedicates a whole chapter to model evaluation and concludes that progress in the simulation of important modes of climate variability has increased the overall confidence in the models' representation of important climate processes, and the ability of models to simulate extreme events has improved.

Answers to commonly asked questions, many of which deal with issues raised by sceptics, can be found in reviews by Le Page (2007) and AGO (2006a, 2006b). The IPCC reports remain the best source of information about climate change up to the end of 2006. The next section summarises the key findings of the IPCC reports [with other relevant material published in 2007 shown in square brackets].

Key findings of the IPCC reports

The physical science basis (IPCC 2007a)

Global atmospheric concentrations of carbon dioxide, methane and nitrous oxide have increased markedly as a result of human activities since 1750 and now far exceed pre-industrial values determined from ice cores spanning many thousands of years. Carbon dioxide concentrations have risen 35%, methane 150% and nitrous oxide 18%. The global increases in carbon dioxide concentration are due primarily to fossil fuel use and land-use change, including deforestation, while those of methane and nitrous oxide are primarily due to agriculture.

There is very high confidence that the globally averaged net effect of human activities since 1750 (around the Industrial Revolution) has been one of warming, with a net radiative forcing of +1.6 Wm^{-2} (Figure 1). This is mainly due to increases in carbon dioxide, methane and nitrous oxide, which overwhelm the cooling effect of increases in aerosols (mainly sulphate, organic carbon, black carbon, nitrate and dust). The warming contribution from solar variations is 1/20th of the contribution from greenhouse gases. It is very likely that greenhouse gases caused more global warming over the last 50 years than changes in solar irradiance. [All trends in solar activity that could have influenced the Earth's climate have been in the opposite direction to that required to explain the observed global warming of the past 20 years (Lockwood and Frohlich 2007).]

Warming of the climate system is unequivocal and is now evident from a variety of observations:

- increases in global average air and ocean temperatures (0.7°C from 1906–2005)
- the past 11 years were among the 12 warmest years since 1850 (and 2007 makes it 12 of the past 13)
- warming of the upper 3000 m of ocean
- warming in the lowest 8 km of atmosphere similar to surface warming
- an increase in extremely high temperatures; a decrease in extremely low temperatures

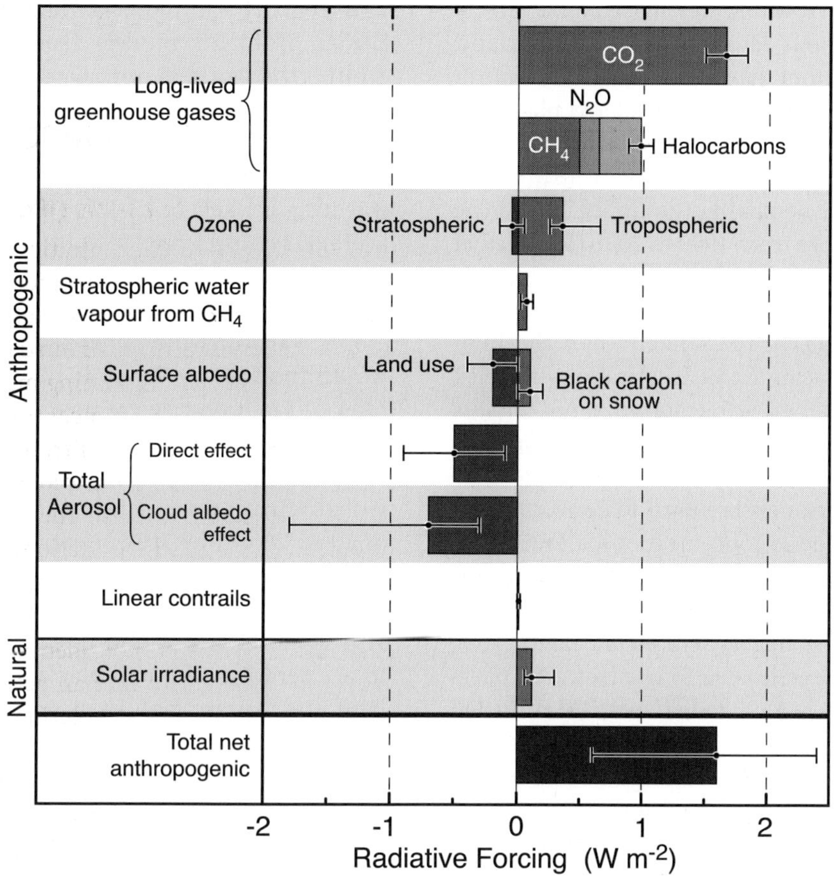

Figure 1: Global average radiative forcing estimates and ranges in 2005 for anthropogenic carbon dioxide (CO_2), methane (CH_4), nitrous oxide (N_2O), halocarbons, ozone, water vapour, surface albedo (reflectivity), aerosols, aircraft contrails and solar irradiance. The net anthropogenic radiative forcing requires summing asymmetric uncertainty estimates from the component terms, and cannot be obtained by simple addition (Source: simplified from IPCC 2007a).

- widespread melting of snow and ice
- rising global average sea level (17 cm since 1900).

Other changes include:

- increase in water vapour (a greenhouse gas) since at least 1980, consistent with theory that warmer air can hold more moisture
- more intense rainfall
- more intense cyclones in North Atlantic since 1970, but no clear trend in cyclone numbers
- oceans have become more acidic due to higher dissolved CO_2
- glaciers and Arctic sea-ice extent have decreased, with little change in Antarctic sea-ice
- shifts in plant and animal locations and behaviour consistent with warming.

Most of the observed increase in globally averaged temperatures since the mid-20th century is very likely (90% likelihood) to have been caused by the observed increase in anthropogenic greenhouse gas concentrations. Discernible human influences now extend to other aspects of climate, including ocean warming, continental-average temperatures, temperature extremes and wind patterns. The warmth of the last half century is unusual in at least the previous 1300 years. The last time the polar regions were significantly warmer than present for an extended period (about 125,000 years ago), reductions in the volume of polar ice led to a sea level rise 4 to 6 metres.

To provide a basis for estimating future climate change, the IPCC (SRES 2000) prepared 40 greenhouse gas and sulfate aerosol emission scenarios for the 21st century, which were based on a variety of assumptions about demographic, economic and technological factors likely to influence future emissions. Each scenario was considered plausible. [However, observed carbon dioxide emissions and concentrations, global mean temperatures and sea-level rise have been tracking the upper end of the IPCC scenario range from 1990 to 2006 (Rahmstorf *et al.* 2007; Raupach *et al.* 2007). Although this 17-year period is very short, it suggests that the mid and low projections are less likely than the high projections, with significant implications for risk management. The growth rate in emissions is strongest in rapidly developing economies, such as China (Raupach *et al.* 2007).]

Carbon cycle models are used to convert emissions into atmospheric concentrations, allowing for uptake of emissions by the land and ocean, land and ocean climate feedback, and transport and chemical reactions in the atmosphere. Atmospheric CO_2 concentrations are estimated to rise from 352 parts per million (ppm) in 1990 to between 500 and 1200 ppm by the year 2100. Methane concentrations are projected to change from 1700 parts per billion (ppb) in 1990 to between 1400 and 3500 ppb by 2100, and nitrous oxide concentrations are projected to increase from 308 ppb in 1990 to between 360 and 460 ppb by 2100. Concentrations of other greenhouse gases, such as tropospheric ozone, hydrofluorocarbons and perfluorocarbons are also projected to increase. Sulfur dioxide aerosols are projected to generally fall during the 21st century.

Increasing concentrations of greenhouse gases and changes in aerosol emissions affect the radiative balance of the Earth. The balance between incoming solar radiation and outgoing heat radiation defines the Earth's radiation budget (Bryant 1997) and determines the Earth's average temperature. Projected changes in greenhouse gases and aerosols are converted to a radiative forcing. Positive forcing warms the Earth, while negative forcing cools the Earth. The projected increases in CO_2 dominate the radiative forcing and have a warming effect.

The climatic effects of projected changes in radiative forcing can be simulated using a climate model. This is a mathematical representation of the Earth's climate system based on well-established laws of physics, such as conservation of mass, energy and momentum. A new set of simulations from 23 models from research groups around the world is now available. These simulations indicate that a global-average warming of about 0.2°C per decade is likely until 2025 for the range of emission scenarios. By the year 2100, the warming is likely to be 1.1 to 6.4°C, with a global-average sea-level rise of 0.18–0.59 m, relative to 1990. If the 1993–2003 ice flow rates from Greenland and Antarctica were to grow linearly with global warming, the upper ranges of sea level rise would increase by 0.1 to 0.2 m. [However, larger rates of sea-level rise are possible. The correlation between observed global average temperature and sea-level rise has been applied to the IPCC (2007) global warming predictions, giving a sea-level rise of 13 to 20 cm by 2030, relative to 1990, 32 to 56 cm by 2070, and 49 to 89 cm by 2095 (Harton *et al.* 2008). If the relationship between global warming and sea-level rise is non-linear (for example, due to acceleration of outlet glacier melt from Greenland and Antarctica), then sea-level rise by 2100 could be much greater (Hansen 2005, 2007). These findings heighten the risk of more rapid and damaging climate change impacts, and make both adaptation and mitigation more urgent.]

Regional patterns of climate change for the late 21st century include:

- Greatest warming will occur over land and at high northern latitudes, and less warming will occur over the southern oceans and North Atlantic, consistent with observations during the latter part of the 20th century.
- Precipitation will generally increase in the wet tropics (such as the monsoon regions) and over the tropical Pacific in particular, with general decreases expected in the sub-tropics, and increases at high latitudes due to the intensification of the hydrological cycle.
- It is very likely that the duration, intensity and frequency of heat waves will increase, and cold periods and frosts will decline.
- Intensity of precipitation events is projected to increase, particularly in tropical and high latitude areas that experience increases in mean precipitation.
- Even in areas where mean precipitation decreases (most subtropical and mid-latitude regions),

precipitation intensity is projected to increase but there would be longer periods between rainfall events.

- More intense tropical cyclones, and associated extreme wind and rain are projected.
- A tendency for drying of the mid-continental areas during summer is expected to increase the risk of droughts in those regions.
- Run-off is projected to decline in many mid-latitude regions (southern Europe, the middle east, southern USA, southern South America, South Africa and southern Australia) and some tropical regions (central America, northern Africa), with increases in South-East Asia, northern Australia and at high latitudes.
- Snow cover and sea ice extent are expected to decrease, glaciers and ice caps are expected to lose mass, and this will contribute to further sea-level rise.

[The main goal of the United Nations Framework Convention on Climate Change is to stabilise greenhouse gas concentrations at a level that would prevent dangerous impacts. The term 'dangerous' is not well defined and will involve a mixture of scientific, economic, political, ethical and cultural considerations. The IPCC (2007a) projections of global warming and sea-level rise focus on the range of change by the year 2100. These long-term projections do not provide sufficient guidance for shorter-term regional adaptation, nor the probability of 'dangerous' climate change to properly inform mitigation policy (Cox and Stephenson 2007). CSIRO and the Bureau of Meteorology have tried to tackle this issue by fitting probability distributions to the IPCC (2007a) global warming projections for 2030, 2050 and 2070, based on the global warming simulations from 23 climate models. This has enabled the development of probabilistic projections of Australian climate change (CSIRO and BoM 2007).]

Impacts, adaptation and vulnerability (IPCC 2007b)

Many natural systems are being affected by regional climate changes, particularly temperature increases. These include:

- enlargement and increased numbers of glacial lakes
- increasing ground instability in permafrost regions, and rock avalanches in mountain regions
- changes in some Arctic and Antarctic ecosystems
- increased run-off and earlier spring peak discharge in many glacier-fed and snow-fed rivers
- warming of lakes and rivers in many regions, with effects on thermal structure and water quality
- earlier timing of spring events, such as leaf-unfolding, bird migration and egg-laying
- poleward and upward shifts in ranges in plant and animal species
- earlier fish migrations in rivers.

Since 1970 it is likely that human-induced warming has had a discernible influence on many physical and biological systems. Most of the observed warming since the mid-20th century is very likely due to the increase in greenhouse gases from human activities. More than 89% of observed changes in physical and biological systems since 1970 are consistent with the expected response to warming, based on more than 29,000 data sets and 75 studies.

Other effects of regional climate changes on natural and human systems are emerging, although these are more difficult to discern due to adaptation and non-climatic factors. These effects include earlier spring planting of crops in Northern Hemisphere, greater heat-related deaths in Europe, increases in infectious disease vectors in some areas, loss of coastal wetlands and mangroves, and increased coastal flood damage in many areas.

Regarding future impacts, more specific information is now available across a wide range of systems and sectors, such as water resources, ecosystems, food, forests, coasts, industries, cities and health. For example:

- By mid-century, annual-average river run-off and water availability are projected to increase by 10–40% at high latitudes and in some wet tropical areas, and decrease by 10–30% over some dry regions at mid-latitudes and in the dry tropics.
- Drought-affected areas and flood-risk areas will expand.

- 20–30% of plant and animal species assessed so far are likely to be at increased risk of extinction if global warming exceeds 1.5–2.5°C.
- Crop productivity is projected to increase slightly at mid to high latitudes for local warming of 1–3°C depending on the crop, then decrease for greater warming.
- Crop productivity is projected to decrease for local warmings of 1–2°C at lower latitudes (such as the tropics), which would increase the risk of hunger.
- Global commercial timber productivity is projected to rise modestly in the short to medium term.
- By the 2080s, many millions of people are projected to be flooded every year due to sea-level rise. The numbers affected will be largest in the mega-deltas of Asia and Africa. Small islands are highly vulnerable.
- Most vulnerable industries and cities are generally those in coastal and river flood plains, those with climate-sensitive resources and those in areas that are prone to extreme weather events.
- Malnutrition will increase, with implications for child growth and development.
- Deaths, disease and injury due to heat waves, floods, storms, fires and droughts will increase, but there will be fewer deaths from cold exposure.
- The frequency of cardio-respiratory disease will increase due to higher concentrations of ground-level ozone.
- The spatial distribution of some infectious diseases will be altered.

As mentioned in the previous section, the concept of 'dangerous' climate change (or tipping points) is not well-defined and involves value judgements. However, science can facilitate informed debate by assessing global and regional risks. Potential key vulnerabilities were identified using seven criteria:

- magnitude of impacts
- timing of impacts
- persistence and reversibility of impacts
- likelihood of impacts and vulnerabilities and confidence in those estimates
- potential for adaptation
- distributional aspects of impacts and vulnerabilities
- importance of the system(s) at risk.

General conclusions are:

- Some observed key impacts have been at least partly attributed to anthropogenic climate change. Among these are increases in human mortality, loss of glaciers and increases in the frequency and/or intensity of extreme events.
- Global mean temperature changes of up to 2°C above 1990–2000 levels would exacerbate current key impacts, such as those listed above, and trigger others, such as reduced food security in many low latitude nations. At the same time, some systems, such as global agricultural productivity, could benefit.
- Global mean temperature changes of 2 to 4°C above 1990–2000 levels would result in an increasing number of key impacts at all scales, such as widespread loss of biodiversity, decreasing global agricultural productivity and widespread deglaciation of Greenland and West Antarctic ice sheets. Sea-level rises of up to 7 m from melting of the Greenland ice sheet, and 5 m from melting of the West Antarctic ice sheet, are possible over subsequent centuries to millennia.
- Global mean temperature changes greater than 4°C above 1990–2000 levels would lead to major increases in vulnerability, exceeding the adaptive capacity of many systems.
- Regions that are already at high risk from observed climate variability and climate change are more likely to be adversely affected in the near future by projected changes in climate and increases in the magnitude and/or frequency of already damaging extreme events.

Australia is already experiencing impacts from recent climate change. Since 1950, there has been 0.8°C warming, with more heatwaves, fewer frosts, more rain in north-west Australia, less rain in southern and eastern Australia, an increase in the intensity of droughts, and a rise in sea level of about 70 mm. Impacts are now evident in increasing stresses on water supply and agriculture, changed natural ecosystems and reduced seasonal snow cover. Australia's future climate is almost certain to be warmer, with changes in extreme events. Heatwaves, fires, floods, landslides, droughts and storm surges are expected to increase in intensity and frequency. Snow and frost are very likely to become less frequent. In the absence of significantly enhanced adaptation, the following impacts are likely (Table 1):

- As a result of reduced precipitation and increased evaporation, water security problems are projected to intensify by 2030 in southern and eastern Australia.
- Ongoing coastal development and population growth are projected to exacerbate risks from sea-level rise and increases in the severity and frequency of storms and coastal flooding by 2050.
- Significant loss of biodiversity is projected to occur by 2020 in some ecologically rich sites, including the Great Barrier Reef and Queensland Wet Tropics. Other sites at risk include the Kakadu wetlands, south-west Australia, sub-Antarctic islands and alpine areas.
- Risks to major infrastructure are likely to increase. By 2030, design criteria for extreme events are very likely to be exceeded more frequently. Risks include failure of floodplain protection and urban drainage/sewerage, increased storm and fire damage, and more heatwaves, causing more deaths and more blackouts.
- Production from agriculture and forestry is projected to decline by 2030 over much of southern and eastern Australia due to increased drought and fire.

Adaptation will be necessary to deal with impacts resulting from the warming, which is already unavoidable due to past and likely future emissions. A wide array of adaptation options is available (Table 2), but more extensive adaptation than is currently occurring is required to reduce vulnerability. There are formidable environmental, economic, social, informational, attitudinal and behavioural barriers to implementation of adaptation. For developing countries, the availability of resources and building adaptive capacity are very

Table 1: Six key risks in Australia (assuming no new adaptation) and associated regional hotspots. Underlying climate projections include higher temperatures, sea-level rise, heavier rainfall, greater fire risk, less snow cover, reduced run-off over southern and eastern Australia, more intense tropical cyclones and larger storm surges.

System	Impacts	Hotspots
Natural systems	Damage to coral reefs, coasts, rainforests, wetlands and alpine areas. Increased disturbance, loss of biodiversity including possible extinctions, changed species ranges and interactions, loss of ecosystem services (e.g., for tourism and water). Potentially catastrophic for some systems (e.g., reefs may be dominated by macroalgae by 2050, extinctions of endemic vertebrates in Queensland Wet Tropics).	Great Barrier Reef, Kakadu, east Queensland, alpine zones, Murray-Darling Basin and south-western Australia
Water security	Reduction in water supply for irrigation, cities, industry and riverine environment in those areas where stream flow declines, (e.g., in the Murray-Darling Basin, annual mean flow may drop 10 to 25% by 2050 and 16 to 48% by 2100).	Southern and eastern Australia
Coastal communities	Greater coastal inundation and erosion, especially in regions exposed to cyclones and storm surges. Coastal development is exacerbating the climate risks.	Tropical and South-East Queensland
Agriculture and forestry	Reduced crop, pastoral and rangeland production over much of southern and eastern Australia. Reduced grain and grape quality. A southward shift of pests and disease vectors. Increased fire risk for forests.	Southern and eastern Australia
Major infrastructure	Design criteria for extreme climatic events, floods and storm surges very likely to be exceeded more frequently. Increased damage likely for buildings, transport structures, telecommunications, energy services and water services.	Large cities, floodplains of major rivers, coastal communities
Health	By 2050, 3200 to 5200 more heat-related deaths/year, and 0.6 to 1.4 million more people exposed to dengue fever.	Large cities

Table 2: Examples of current and potential options for climate change adaptation.

	Food, fibre and forestry	Water	Human health	Industry and settlement
Drying/ drought	Development of drought-resistant crop varieties, intercropping, crop residue retention, weed management and better irrigation efficiency Supplementary livestock feeding, change in stocking rate and rotation of pasture Better extension services to farmers, debt relief and diversification of income	Rainwater harvesting and storage, piping irrigation water to reduce leakage and evaporation, water demand management through metering and pricing, soil moisture and groundwater conservation, desalination, education for sustainable water use	Grain storage and provision of emergency feeding stations Provision of safe drinking water and sanitation Strengthening of public institutions and health systems Access to international food markets	Improve adaptation capacities, especially for livelihoods Incorporate climate change in development programmes Improved water supply systems and coordination between jurisdictions
Increased rainfall/ flooding	Improved soil drainage, development and promotion of alternative crops, adjustment of planting and harvesting schedule, improved extension services for farmers	Better protection measures, including flood forecasting and warning, regulation through planning legislation and zoning, promotion of insurance, relocation of vulnerable assets	Early warning systems, disaster preparedness planning, effective post-event emergency relief	Improved flood protection infrastructure, 'flood proof' buildings, change land use in high-risk areas, flood hazard mapping and flood warning systems, empower community institutions
Warming/ hot spells	Development of new heat-resistant crop varieties, altered timing of cropping activities, pest control and surveillance Shade for livestock, change to heat-tolerant breeds Forest fire management through altered stand layout, landscape planning, dead timber salvaging, prescribed burning, and the introduction of improved fire suppression capacity. Insect control through prescribed burning, non-chemical pest control	Water demand management through metering and pricing Education for sustainable water use	International surveillance systems for disease emergence Strengthening of public institutions and health systems Heat warning systems Measures to reduce urban heat island effects through creating green spaces Adjusting clothing and activity levels, and increasing fluid intake	Assistance programmes for especially vulnerable groups Improve adaptive capacities Technological change
Storms	Development of wind-resistant crops	Coastal defence design and implementation to protect water supply against contamination	Early warning systems; disaster preparedness; effective post-event emergency relief	Emergency preparedness (e.g. early warning systems, more resilient infrastructure, financial risk management)

Source: modified from IPCC (2007b) Table TS 5.1.

Table 3: Stabilisation scenarios for global warming, CO_2 concentration and CO_2-equivalent concentration (including the effect of non-CO_2 greenhouse gases).

Global warming relative to 1750	Global warming relative to 2000	CO_2 concentration by 2100–2150	CO_2-equivalent concentration by 2100-2150
2.0–2.4°C	1.3–1.7°C	350–400 ppm	445–490 ppm
2.4–2.8°C	1.7–2.1°C	400–440 ppm	490–535 ppm
2.8–3.2°C	2.1–2.5°C	440–485 ppm	535–590 ppm
3.2–4.0°C	2.5–3.3°C	485–570 ppm	590–710 ppm

Source: IPCC (2007c) Table SPM-5.

important. Adaptation alone is not expected to cope with all the projected impacts of climate change, especially over the long term. Many impacts can be avoided, reduced or delayed by mitigation (net emission reductions). A portfolio of adaptation and mitigation measures can diminish the risks associated with climate change.

Mitigation (IPCC 2007c)

As noted in the previous section, negative impacts become more widespread and more intense with greater global warming, especially above 2°C, and many impacts can be avoided, reduced or delayed by mitigation. Stabilising global warming at 1.3 to 1.7°C relative to the year 2000 would require stabilising CO_2-equivalent concentrations (including the effect of non-CO_2 greenhouse gases) at 445 to 490 ppm between 2100 and 2150 (Table 3). This would require a 50 to 85% reduction in emissions by 2050.

In order to stabilise CO_2-equivalent concentrations below 490 ppm, emissions would need to peak between 2000 and 2010 and decline thereafter. The lower the stabilisation level, the more quickly this peak and decline would need to occur. Mitigation efforts over the next two to three decades will have a large impact on opportunities to achieve lower stabilisation levels.

Emission reductions can be achieved by deployment of a portfolio of technologies that are currently available and those that are expected to be commercialised in coming decades. This assumes that appropriate and effective incentives are in place for development, acquisition, deployment and diffusion of technologies and for overcoming related barriers.

In 2030, macro-economic costs for multi-gas mitigation, consistent with CO_2-equivalent stabilisation between 445 and 710 ppm are estimated at between a 3% decrease of global GDP and a small increase, compared with the baseline. By 2050, the costs range between a 5.5% loss and a 1% gain in GDP, which is only 0.05–0.12% of the projected growth in GDP by 2050.

Choosing an appropriate level of global mitigation is a risk management process. It includes mitigation and adaptation—taking into account actual and avoided climate change damages, co-benefits, sustainability, equity and attitudes to risk. Delayed emission reductions lead to investments that lock in more emission-intensive infrastructure and development pathways. This significantly constrains the opportunities to achieve lower stabilisation levels and increases the risk of more severe climate change impacts and associated costs of adaptation and damage.

There is substantial economic potential for the mitigation of global greenhouse emissions over the coming decades. Key technologies for reducing emissions are available in various sectors (Table 4). These technologies can have co-benefits, such as reduced air pollution, increased energy security, balance of trade improvement, provision of modern energy services to rural areas, employment, increased agricultural production and reduced pressure on natural ecosystems.

Changes in lifestyle and behaviour patterns can contribute to climate change mitigation across all sectors, especially those that emphasise resource conservation and can contribute to developing a low-carbon economy that is both equitable and sustainable. Education and training programs can help to overcome barriers to the market acceptance

Table 4: Key mitigation technologies and practices by sector, listed in no particular order.

Sector	Key mitigation technologies and practices currently commercially available	Key mitigation technologies and practices projected to be commercialised before 2030
Energy supply	Improved supply and distribution efficiency; fuel switching from coal to gas; nuclear power; renewable heat and power (hydropower, solar, wind, geothermal and bioenergy); combined heat and power; early applications of CCS (e.g. storage of removed CO2 from natural gas).	Carbon capture and storage (CCS) for gas, biomass and coal-fired electricity generating facilities; advanced nuclear power; advanced renewable energy, including tidal and waves energy, concentrating solar, and solar photovoltaics
Transport	More fuel-efficient vehicles; hybrid vehicles; cleaner diesel vehicles; biofuels; modal shifts from road transport to rail and public transport systems; non-motorised transport (cycling, walking); land-use and transport planning	Second generation biofuels; higher efficiency aircraft; advanced electric and hybrid vehicles with more powerful and reliable batteries
Buildings	Efficient lighting and better use of daylight; more efficient electrical appliances and heating and cooling devices; improved cooking appliances, improved insulation; passive and active solar design for heating and cooling; alternative refrigeration fluids, recovery and recycle of fluorinated gases	Integrated design of commercial buildings including technologies, such as intelligent meters that provide feedback and control; solar PV integrated in buildings
Industry	More efficient end-use electrical equipment; heat and power recovery; material recycling and substitution; control of non-CO2 gas emissions; and a wide array of process-specific technologies	Advanced energy efficiency; CCS for cement, ammonia, and iron manufacture; inert electrodes for aluminium manufacture
Agriculture	Improved crop and grazing land management to increase soil carbon storage; restoration of cultivated peaty soils and degraded lands; improved rice cultivation techniques and livestock and manure management to reduce methane emissions; improved nitrogen fertiliser application techniques to reduce nitrous oxide emissions; dedicated energy crops to replace fossil fuel use; improved energy efficiency	Improvement of crops yields
Forestry and forests	Afforestation; reforestation; forest management; reduced deforestation; harvested wood product management; use of forestry products for bioenergy to replace fossil fuel use	Tree species improvement to increase biomass productivity and carbon sequestration. Improved remote sensing technologies for analysis of vegetation/ soil carbon sequestration potential and mapping land use change
Waste	Landfill methane recovery; waste incineration with energy recovery; composting of organic waste; controlled waste water treatment; recycling and waste minimisation	Biocovers and biofilters to optimise methane oxidation

Source: IPCC (2007c) Table SPM 3. Each has pros and cons.

of energy efficiency. In industry, management tools that include staff training, reward systems, regular feedback and documentation of existing practices can help overcome industrial organisation barriers, reduce energy use and lower greenhouse gas emissions.

There are many national policies and instruments available to governments to create the incentives for mitigation action. Examples include integrating climate policies into development policies, regulations and standards, taxes and charges, tradeable permits and other incentives. Policies that provide a price on carbon could create incentives for producers and consumers to significantly invest in low-emission products, technologies and processes. Carbon prices of 20 to 50 US$/tonne of CO_2 equivalent, sustained or increased over decades, could lead to a power generation sector with low emissions by 2050 and make many mitigation options in the end-use sectors economically attractive. The United Nations Framework Convention on Climate Change and its Kyoto protocol have stimulated an array of national policies, created an international carbon market and established new institutional mechanisms that may provide the foundation for future mitigation efforts.

References

AGO (Australian Greenhouse Office) (2006a). *Climate Change Science: Questions Answered.* Australian Greenhouse Office, <http://www.greenhouse.gov.au/science/qa/pubs/science-qa.pdf>

AGO (2006b). *Hot Topics in Climate Change Science.* Australian Greenhouse Office, <http://www.greenhouse.gov.au/science/hottopics/index.html>

Bryant E (1997). *Climate Processes and Change.* Cambridge University Press, Cambridge, UK.

Cox P and Stephenson D (2007). A changing climate for prediction. *Science* **317**, 207–208.

CSIRO and BoM (2007). *Climate Change in Australia.* CSIRO and Bureau of Meteorology Technical Report. <www.climatechangeinaustralia.gov.au>

Hansen JE (2005). A slippery slope: how much global warming constitutes 'dangerous anthropogenic interference'? *Climatic Change* **68**, 269–279.

Horton, R, Herweijer C, Rosenzweig C, Liu J, Gornitz V and Ruane AC (2008). Sea level rise projections for current generation CGCMs based on the semi-empirical method. *Geophy. Res. Lett.* **35**, L02715. doi:10.1029/2007GL032486.

Hansen J (2007). Scientific reticence and sea level rise. *Environmental Research Letters* **2**, 1–6.

IPCC (2007a). Summary for policymakers. In *Climate Change 2007: The Physical Science Basis. Contribution of Working Group I to the Fourth Assessment Report of the Intergovernmental Panel on Climate Change* (Eds S Solomon, D Qin, M Manning, Z Chen, M Marquis, KB Averyt, M.Tignor and HL Miller). Cambridge University Press, Cambridge, UK <www.ipcc.ch>

IPCC (2007b). *Climate Change 2007: Impacts, Adaptation and Vulnerability. Contribution of Working Group II to the Fourth Assessment Report of the Intergovernmental Panel on Climate Change* (Eds ML Parry, OF Canziani, JP Palutikof, PJ van der Linden and CE Hanson), Cambridge University Press, Cambridge, UK. <www.ipcc.ch>

IPCC (2007c). *Climate change 2007: Mitigation. Contribution of Working group III to the Fourth Assessment Report of the Intergovernmental Panel on Climate Change* (Eds B Metz, OR Davidson, PR Bosch, R Dave, LA Meyer), Cambridge University Press, Cambridge, UK. <www.ipcc.ch>

Jones D, Watkins A, Braganza K and Coughlan M (2007). *Bulletin of the Australian Meteorological and Oceanographic Society* **20(3)**, 63–72. <http://www.amos.org.au/BAMOS_GGWS_SUBMISSION_final.htm>

Le Page M. (2007). Climate change: a guide for the perplexed. *New Scientist* **16 May**, 34–42. <http://environment.newscientist.com/channel/earth/dn11462>

Lockwood M and Frohlich C (2007). Recent oppositely directed trends in solar climate forcings and the global mean surface air temperature. Proceedings of the Royal Society A, doi:10.1098/rspa.2007.1880, 1–14.

Mann ME, Ammann CM, Bradley RS, Briffa KR, Crowley TJ, Jones PD, Oppenheimer M, Osborn TJ, Overpeck JT, Rutherford S, Trenberth KE and Wigley TML (2003). On past temperatures and anomalous late-20th century warmth. *EOS* **84**, 256.

Rahmstorf S, Cazenave A, Church JA, Hansen JE, Keeling RF, Parker DE and Somerville RCJ (2007). Recent climate observations compared to projections. *Science* **316**, 709–710.

Raupach M, Marland G, Ciais P, Le Quere C, Canadell JG, Kleppe G and Field CB (2007). Global and regional drivers of accelerating CO_2 emissions. *Proceedings of the National Academy of Sciences* **104(24)**, 10288–10293.

3

Resource Consumption and Resource Depletion

Graham Turner and Barney Foran

This chapter describes physical aspects of future environmental challenges in Australia and explores the physical feasibility of potential pathways for change. It opens with a global perspective on environmental sustainability issues as an introduction to a more detailed examination of national issues relating to resource challenges, based largely on the Australian Stocks and Flows Framework (ASFF) (Poldy *et al.* 2000; Turner and Poldy 2001; Foran and Poldy 2002). The use of ASFF provides a whole-economy context even though the challenges are presented separately. The second section describes how the sustainability challenges are made less tractable due to system interactions, both in time and space. Finer scale analysis—at the city level—shows how urban settlements draw resources from, and emit wastes, beyond their boundaries. We look in particular at land, water and energy for Melbourne.

In the third and final section, some pathways for change that focus on sustainable development are explored. It is beyond the scope of this chapter to consider the economic and institutional mechanisms or impacts of such change. Key messages in this chapter are:

- cities cannot be considered individually or in a sectoral or regional boundary—there are no sustainable parts of an unsustainable whole
- likewise, resource and environmental challenges also cannot be considered in isolation—tackling these separately tends to compound problems elsewhere
- continually growing consumption is related to technological progress on efficiency, and is a key cause of current sustainability challenges
- sustainability can be achieved physically through a combination of curbing consumption, creating urban settlements that are more closely linked with their surroundings and substantially employing green technology.

The challenges

Global perspectives

The global sustainability implication of Australian lifestyles and economy is summarised in this section using the 'ecological footprint' concept (Wackernagel *et al.* 1999). The footprint estimates the land area that would be required to sustain consumption activity indefinitely using sustainable land-based processes, such as forests for C-sequestration. Current estimates of the footprint for the household consumption activities of an average Australian citizen are 6.5 hectares per capita (see Table 1) (ACF 2007). If that citizen is in full time work then there is an additional footprint quota that varies from 2 to 6 hectares or more, depending on whether the work is totally office based or also involves substantial car and plane travel. Thus it is not difficult for an average upwardly mobile worker to have a combined home and work footprint of more than 12 hectares.

The average per capita footprints of Australian cities range from 5.7 hectares for Hobart to 7.4 hectares for Perth. These footprint differences are driven mainly by the volume of consumption (Sydney is more affluent than Hobart, for example) but there are also state differences in the carbon intensity of electricity and meat production. There is a strong income effect for footprint

Table 1: A comparison of footprints in hectares per capita for capital cities, states and the Australian average.

City	City footprint (ha p.c.)	State footprint (ha p.c.)	National average (ha p.c.)
Sydney	6.6	6.3	6.5
Melbourne	6.4	6.2	
Brisbane	6.7	6.5	
Adelaide	6.8	6.7	
Perth	7.4	7.3	
Hobart	5.7	5.5	
Darwin	7.1	6.7	
Canberra	7.0	7.0	

Source: anon, 2007a

size—with less affluent suburbs having footprints of five hectares and most affluent suburbs twelve hectares per capita. However there is also a consumption choice and behaviour effect shown by a difference of 2 hectares per capita for the same income size.

Estimates of the entire globe's footprint in 2005 was 0.8 hectares per capita compared with its bio-capacity (the capacity of the globe to provide food and fibre) of 0.9 hectares per capita (Lenzen *et al.* 2007). Thus, for the current human population and its average consumption levels, there is a productivity buffer of one tenth of a hectare per capita. This buffer will shrink by about 20% over the next 50 years under the competing forces of growth in human population and consumption versus crop yield increases and some positive effects of global climate change.

Given the global situation, there is a stark comparisons between the average Australian citizen, with a footprint of 6.5 hectares per capita, and the average world citizen, with 0.8 hectares per capita—a factor of eight times difference. Aspirations for global equity will require that poorer, developing countries may have to double their footprint in the search for reasonable levels of human development. In order to stay within the globe's bio-capacity, this will require that Australian cities reduce their footprints by more than 80% over the next two human generations.

National resource context

The past and possible future development of pressures on our global footprint is examined in this section from a domestic perspective, and aggregated to the national level. This is based on the CSIRO report *Future Dilemmas* (Foran and Poldy 2002) and other research using the Australian Stocks and Flows Framework (ASFF) (Poldy *et al.* 2000). This section places subsequent urban analysis within a national context of resource sustainability dilemmas. Not only are the dilemmas pressing on each urban area, but they are also a direct result of the aggregation of all urban and rural area activity.

The national context can be presented in terms of a series of dilemmas, or 'tensions' that typically arise between requirements of our society and economy and how they are met. Given the ubiquitous dependence of modern economies on oil and other fuels, we start with the tension between production and requirements for transport fuel. Obviously, water, food and shelter are also essential for urban or rural life, and increasing pressure on water, productive land and fisheries will diminish our capacity for providing more of these inputs. Other issues are associated with factors involved in using inputs or transforming them—such as labour and infrastructure (particularly the ageing of both of these stocks). Emissions resulting from economic activity also present society with challenges, such as climate change impacts.

Importantly, attempts to deal with the dilemmas need to be examined in terms of their interactions—solving one can make other dilemmas worse. This aspect is taken up in the second section of this chapter.

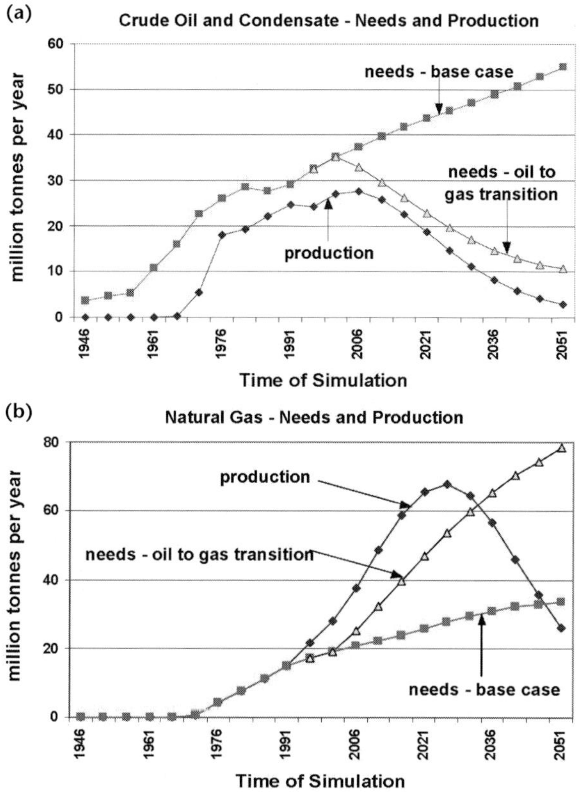

Figure 1: (a) Domestic production and consumption of crude oil. Historical data and two alternative scenarios are depicted. The upper curve is the oil required assuming 'business as usual'; the middle curve (yellow triangles) shows oil required if there is a transition to gas-based transport fuel. (b) Domestic production and consumption of gas for the corresponding scenarios (Source: Foran and Poldy 2002).

Transport fuels

Like most cities, Australian cities depend heavily on motor vehicles. The majority (more than 90%) of commuting is undertaken by private car (Foran and Poldy 2002) and inter-city freight transport is dominated by trucks. This creates a particular dependence on access to oil as a transport fuel, and a substantial challenge in maintaining current transport systems into the future.

In the past few decades, Australia has produced sufficient oil to satisfy about three-quarters of domestic requirements (Figure 1a). However, domestic production is understood to be peaking and predictions suggest declining production over coming decades (lower curve in Figure 1a). A scenario of continued dependence on oil-based transport (upper curve in Figure 1a) implies a large and rapidly increasing requirement for imported oil—reaching 50% of requirements within just 15 years.

At the very least this will have a negative impact on the Australian trade balance, even if the global oil supply is plentiful. However, it appears that this supply may also be peaking, or nearing this critical stage (Deffeyes 2001; NPC 2007), and with demand for oil rising around the world the cost of importing oil into Australia could be substantial. There is probably no correct way to estimate the potential cost because modern economies have not previously experienced such a severe shortage of such a crucial resource.

An alternative to attempting to increase oil imports—assuming a continuing dependence on cars and trucks—is to make the transition to another fuel. Natural gas is an attractive option for Australia because there are large domestic resources (of about the same size in energy units as domestic oil), predominantly in the North West Shelf. Using gas is also attractive because changes to infrastructure and engine technology would be reasonably minor compared with other alternatives.

Estimates of production (upper curve in Figure 1b) indicate that surplus gas could be produced well above business-as-usual (BAU) projections of domestic requirements (lower curve in Figure 1b), providing Australia with substantial export opportunities until about 2045. Switching domestic transport fuel to gas would approximately double total requirements (middle curve in Figure 1b) and lead to reduced export potential—by about 2030 Australia would become a net importer of gas.

Consequently, a continuing reliance on fossil-fuel based car and truck transport leaves Australia exposed to potential international price increases and supply disruptions. Additionally, there are substantial implications in terms of the greenhouse gas (GHG) emissions from this transport system (see later section on emissions).

Water

Total water use in Australian has increased rapidly to over 20,000 GL per annum, with most of the consumption associated with crop and animal production (Figure 2). Some proportion of irrigation water is used for food consumed directly by the domestic population—the remainder is for

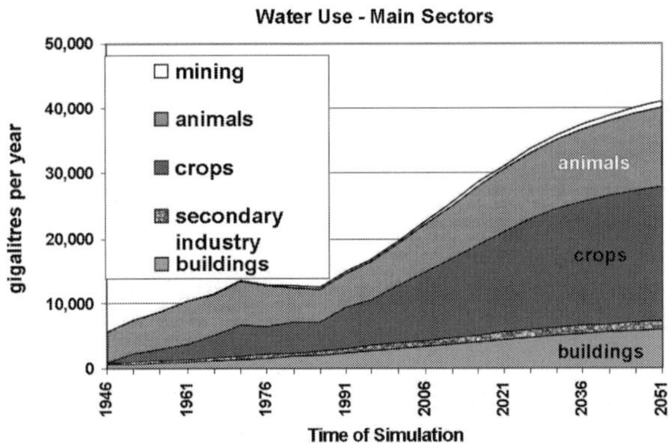

Figure 2: Total water use in Australia, by broad sector use (Source: Foran and Poldy 2002).

exported commodities, which earn foreign currency (for example, over $1 billion in cotton exports (ABARE 2006a)), providing indirect benefits to all Australians.

Projections of continuing growth in population and the economy will lead to further demands on water (Figure 2). To accommodate growth in water-use sectors (agriculture, electricity and urban) it has been suggested that the large volumes of northern Australia, could either be exploited by developing new industries in the north or by moving the water south (see, for example, Government of South Australia 2005).

Irrespective of future developments, it is unlikely that domestic food production is totally threatened by water constraints, unless more extreme climate change scenarios eventuate (which is not out of the question, given recent observations of GHG emissions tracking on worst case IPCC scenarios (Raupach *et al.* 2007) and growing evidence of climate change impacts accelerating faster than some scientific expectations (Spratt 2007)). However, it would appear that further expansion of irrigated agriculture in southern Australia is possibly constrained to short-term gains made through efficiencies in water transfer and application (Dunlop and Turner 2003).

While water used directly in urban areas is a relatively small component of the total, recent drought has brought to the attention of water authorities the high level of dependence of future urban development on current local water resources. Possible impacts of climate change significantly heighten the pressure on urban water (see below).

Food

Production of food for domestic consumption or export markets may be constrained by issues other than water availability.

Various types of land degradation (dryland and irrigation salinity, acidification and soil structural decline) have been recognised as factors likely to reduce agricultural productivity (NLWRA 2001; Dunlop and Turner 2003; Dunlop *et al.* 2002). Salinity in particular was a high profile issue before droughts in the early years of this century lowered water tables and eased the immediate impacts of land clearing and irrigation.

Nevertheless, land degradation remains a persistent issue because the fundamental causes have not been tackled. Assembling a history of agricultural practice over the past one-and-a-half centuries reveals exponential growth in total land area of about 2% pa on average (the upper curve of Figure 3 shows the area of the Midlands statistical division in WA that represents about one-tenth of the Australian total of cropping and grazing land) (Dunlop and Turner 2003). This growth in land area contributes about equally with yield improvements to the growth in total production volume.

Parcels of land show signs of loss of condition (by at least 20%) after having been in use for about 40–60 years (resulting in the lower curve of Figure 3). In the past, loss of productivity has effectively been masked by the addition of new un-degraded land with higher productivity. The cessation of landing clearing, combined with attempts to maintain all present agricultural land in production, is likely

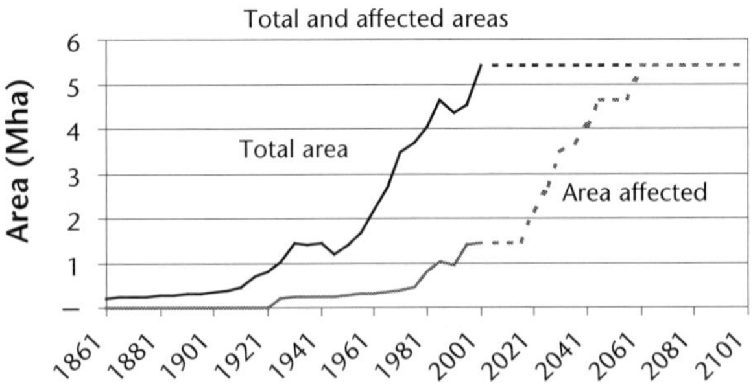

Figure 3: The total area of crops and pasture in Midlands statistical division region of WA, and the area with at least 20% reduction in yield due to loss of landscape condition. The dashed lines indicate a future scenario with no new agricultural land. A 60-year delay is evident (Source: Dunlop and Turner 2003).

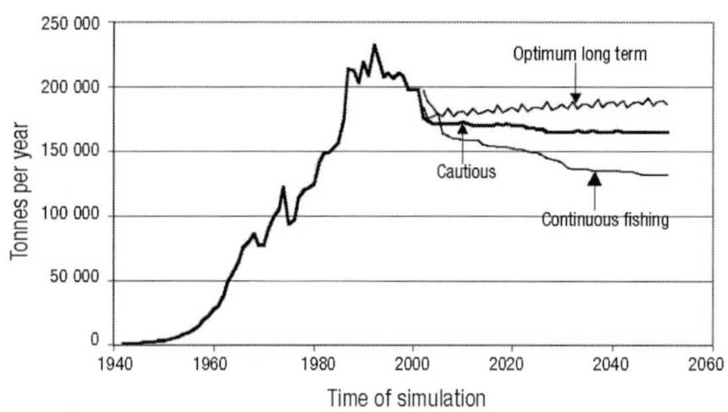

Figure 4: Historical and simulated future total catch for 1942 to 2050 from the commercial fishery for 'cautious', 'continuous fishing' and 'optimum long-term' management scenarios (Source: Kearney *et al.* 2003).

to result in a majority of the land area returning 20% decreased productivity within several decades.

Such degradation should not threaten domestic food production in the near to medium term. More likely are indirect implications such as significant displacement of rural population and businesses, loss of export volume, or pressures to ameliorate lost production through increasing inputs such as fertilizers and conditions.

It is not likely that domestic fisheries would be able to fill any gap in nutrition caused by lower crop and livestock production. Australian wild-catch rates peaked in the 1990s (Figure 4), and stocks of several long-lived species are now significantly depleted (Lowe *et al.* 2003a). Attempts to maintain fishing rates would result in continuously decreasing catches, while a more cautious approach, such as the recent imposition of quotas on selected fisheries could essentially maintain catch rates. It is possible over the long term to somewhat increase catch rates if monitoring of stocks and adaptive management was used to limit fishing when stocks reach critical levels.

Like land-based domestic food production, any shortfall in fisheries catch does not appear to be a serious direct threat to food supply for the Australian population. Instead, indirect implications are more likely, such as:

- pressure to increase imports (in competition with demand from other nations)
- expansion of aquaculture production (with consequences for local water quality and increased inputs of fish feed and energy)
- halting past increases in per capita consumption of fish (with potential implications on health) (Lowe *et al.* 2003b).

Labour

Another key factor of economic output—and of less formalised activities such as personal care—is the provision of sufficient able and skilled people. Australia's population, like many other developed countries, will continue to age. Even with recent increases in the immigration rate, the number of people aged over 65 years is projected to increase from about 2.5 million currently to about 6.5 million in 2050 (Foran and Poldy 2002). Due to growth in the total population, those over 65 years will form between 20% and 25% of the total population in 2050, compared with about 12% today—the exact figure depends on a range of future population parameters (fertility, mortality and immigration).

The working population must support both this increasing proportion of older people as well as those too young to be part of the workforce. In recent times, the ratio of people dependent on those of working age (the dependency ratio) has been at historic lows of about 47% (Figure 5). By mid-century this dependency ratio may increase to 57–67%, depending on immigration rates (the lower and middle curves of Figure 5).

This means that to maintain economic output per capita at current levels, the productivity of the working population must increase by 20–40%, and

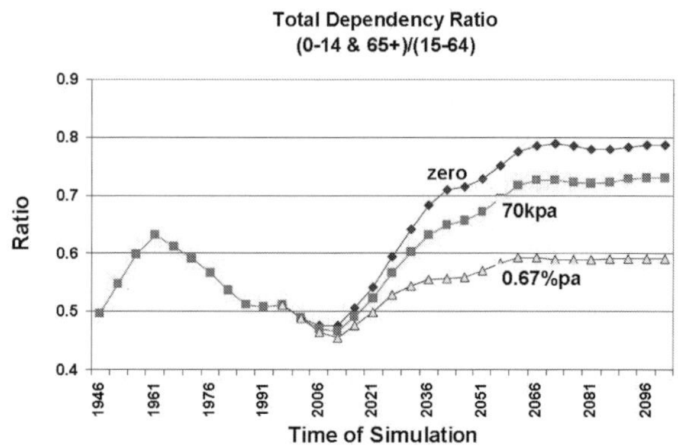

Figure 5: Ratio of young (0–14 years) and old (65+ years) to the remainder of the population who must support the young and old (Source: Foran and Poldy 2002).

even higher to support further economic growth per capita (past productivity increases of 1.8% per annum have been achieved over several decades (Commonwealth of Australia 2007b), although this was at a time when the dependency ratio also decreased). Future overall labour output could be achieved through a combination of factors, such as advances in information technology, or simply by working beyond 65 years, or by working harder or longer days—in the extreme it could mean working the equivalent of 13 hour days.

The pressure on the workforce could also be alleviated somewhat by encouraging older people to remain in employment beyond 65 years. Subsequent implications on education sector requirements for re-training, or the health sector, are not explored here.

In addition to these basic demographic facts, there may be compounding issues related to the health of the population. Obesity is recognised as a major problem, which is due in part to the high level of obese young people that are likely to present with significant health problems later in life. This creates two pressures: fewer active participants in the labour force and greater demand for health services (Lattimore 2005).

Infrastructure

Like labour, infrastructure is an important factor for economic output—and it also ages. Overall, Australia's infrastructure has been rated as 'adequate' in 2005, generally requiring 'major changes to enable infrastructure … to be fit for its current and anticipated purpose' (Hardwicke 2006)—receiving a rating of C+ in 2005, up from C in 2001 and D+ in 1999 (Engineers Australia 2006). Despite this improvement, significant challenges loom when turnover rates associated with age profiles are considered (Figure 6). Possibilities for improving the performance of infrastructure are linked with the age of the plant, because retro-fitting new technology into old plant can be technically difficult or considered uneconomic. Similarly, the maintenance and replacement of infrastructure may remain negligible for some decades, suddenly re-emerging as an issue as the end of the lifespan of the stock is reached.

Electricity plant is an important case. The ageing of Victoria's brown-coal power stations is

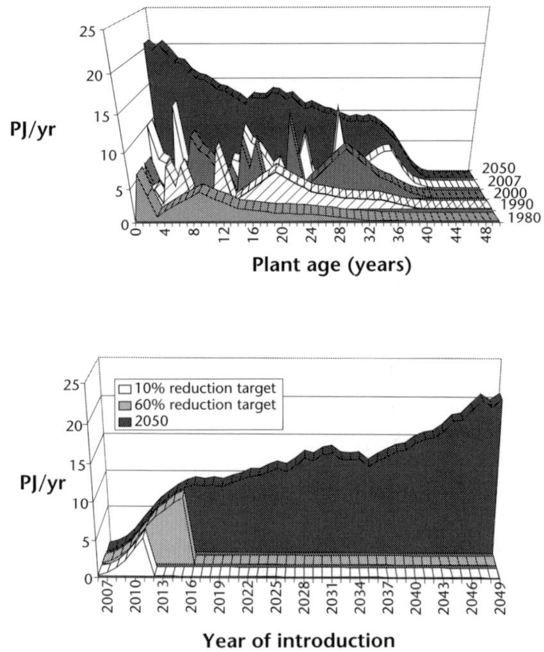

Figure 6: Age profiles for electricity plants (expressed as electricity generated in PJ/yr graphed against plant age)—(top) brown-coal power stations in Victoria for several historical years and a projection at 2050 based on a 'business-as-usual' scenario; (bottom) total projected electricity produced in 2050 graphed against vintage of the brown-coal power station (that is, the year of introduction); the portions on the left show the time window for building power stations that can freely emit CO_2 (assuming all later plants have zero emissions) to achieve proposed 60% and 90% GHG reduction targets (relative to 1990 levels) to avoid dangerous climate change. (1990 was the original reference year, but subsequently separate announcements and analysis has used more current years, such as 2000, as the reference, perhaps without changing the percentage reduction).

evident in Figure 6 (left), where past age profiles (reconstructed in the stocks and flows from historical data of new and retired plant) become progressively extended toward the older end of the scale. Power stations built decades ago are likely to continue operating for some decades, and so will contribute to increasing GHG emissions because they are generally deemed unsuitable for CO_2 capture (because the CO_2 is diluted with other exhaust gases).

Implications for meeting possible GHG reduction targets in 2050 also require knowledge of plant age and turnover. Using an average lifetime of 40 years for electricity plant—and assuming a 'business-as-usual' growth in electricity demand compounding at 2% pa per capita (or about 3% growth in final demand)—results in the age profile of brown-coal stations shown for 2050 (Turner *et al.* 2007c). This is mirrored in Figure 6 (bottom), to show the year that new capacity of coal plant is introduced. Even though none of the current suite of power stations would be operating in 2050, the potential for reduced CO_2 emissions in 2050 relies on making rapid infrastructure changes in the near future. This is evident from calculating what proportion of the 2050 power stations could operate without any CO_2 capture and storage, assuming that all other stations are completely emission-free.

A target of 60% reduction of 1990 emissions—which has increasing currency as a target for avoiding dangerous climate change (Allen Consulting 2006c)—requires that only power stations constructed over about the next 10 years can emit CO_2, as shown by the segment of Figure 6 (bottom) ending at 2015. This is only several years after the proposed introduction of an Australian carbon trading scheme (Department of the Prime Minister 2007). Clearly this represents a very rapid response to the scheme—assuming it is even technologically feasible, which remains uncertain. Other more demanding reduction targets for developed nations have been proposed—such as 90% relative to 1990—to allow developing countries to catch up in technology while growing economically. This target would mean only those stations constructed in the next five years could emit CO_2.

Such pressing demands on infrastructure turnover will be heightened by some factors and ameliorated by others. Ongoing efficiency improvements (evident in Figure 6 as the vertical increase in the broad peak that moves from age 10 year in 1980 to 30 year in 2000) will marginally extend the date when electricity plant must become totally zero emitting—though gains are limited, as shown below Figure 7. Conversely, lifetimes of plant greater than the 40 years assumed above will bring the date closer to the present because more of these polluting plants will operate in 2050. In the later section on pathways for change, we examine other alternatives for achieving GHG reduction targets.

Emissions

Given the ubiquitous use of energy in all sectors of society and the economy, it is not surprising that emissions from these activities, such as CO_2, have risen strongly over past decades (Figure 7). This compounding growth has largely been in line with economic growth when viewed over the long term (Turner and Poldy 2001), even though energy intensity has decreased somewhat in the last few decades. Unlike most of the other sustainability issues presented in this chapter, these emissions are global pollutants that cause global warming and climate change, though they are also likely to have local effects (see below regarding water supply for Melbourne).

Future management of GHG emissions may be particularly challenging, as indicated in Figure 7. The upper curves show emissions for three population scenarios using conventional technology for power generation and the transport fleet. In the three scenarios, the population reaches 32 million, 25 million and 20 million at 2050—associated with different immigration rates. Currently the growth of the national population is tracking between the 32 and 25 million scenarios. The emissions seriously exceed the 1990 levels associated with the Kyoto Protocol targets, despite other settings in the scenarios reflecting conservative growth of the economy. Such targets are considerably above levels recommended recently to avoid dangerous climate

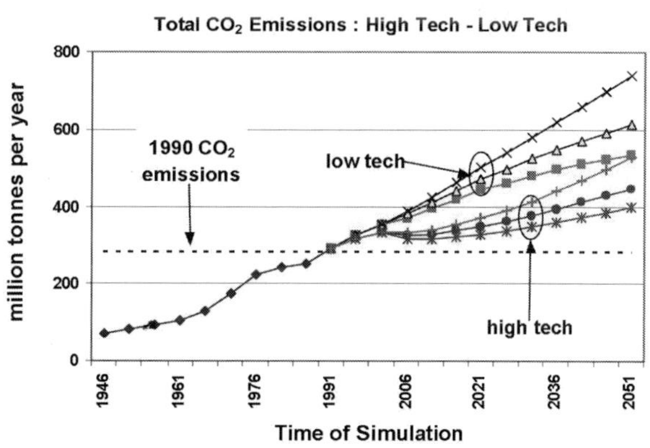

Figure 7: Annual CO_2 emissions from fuel burnt for electricity generation and transport across Australia. Several scenarios show the implications of three population scenarios and the effects of continuing with low-technology generation and transport (upper curves), compared with higher efficiency alternatives (Source: Foran and Poldy 2002).

change—these levels would be equivalent to at least 1970 emissions or substantially lower.

Even implementing significant levels of new technology into electricity generation and transport does little to avert the situation: the lower curves combine the population scenarios with fuel-efficient cars (3L per 100km compared with about 10L), power stations that are 50% more efficient (approaching thermodynamic limits), and cutting energy use in buildings by 50% by 2020. This has the effect of stalling growth in emissions for several decades, but ultimately economic growth overrides the efficiency gains.

Further alternatives pathways for change in the city context are explored below.

Convergence and connectivity

Concurrent and linked issues

In addition to each sustainability issue identified above, another major challenge is understanding, and dealing with, the interaction among these issues—trying to solve one issue in isolation can frequently make others worse. This follows from the fact that the economic system is highly integrated both geographically and in terms of dependencies between various sectors. It is also a result of many of the issues becoming critical within similar timeframes—typically by mid-21st century or earlier, as indicated previously.

Exploring the various potential interactions might be best achieved on a detailed case-by-case basis, using the stocks-and-flows framework to quantify the implications for more accurate evaluation. The number of interactions, and the number of possible alternatives for each issue, is substantial, especially when all combinations are represented.

Table 2 summarises some of the potential interactions when trying to solve sectoral issues. The issues described above have been listed, along with a selection of potential options that are commonly raised in analyses of each sector. Each of these options will have further implications for other sectors.

In many cases, because the solutions proffered often involve large technological change, there are generally requirements for substantial infrastructure development. For example, making the transition to another transport fuel (NPC 2007) requires new processing plants, distribution lines, refuelling stations and new or modified engines in the transport fleet. Accompanying this is the expanded need for labour to develop, construct, operate and maintain this infrastructure. While the labour force will grow in future, the above analyses show that there are likely to be constraints and pressures due to population ageing, and workforce health. There are also other linked resource inputs required for the expanded infrastructure, such as land and water for biofuels.

Consequently, the proposed solutions often lead to increased pressure on other sectors. Multiple system feedbacks ensue, and their exact nature requires more in-depth research than the summary of Table 2. A key message is that resource and environmental challenges cannot be considered in isolation.

It must be recognised that increasing efficiency, or other technological progress, can act to counter or relieve the positive feedbacks. However, empirical evidence shows that such progress leads to further growth in economic throughput, causing the environment pressures to grow, rather than subside, as discussed in the following sections.

Connectivity of urban areas

Mass Flow Analysis (MFA) of several settlements across Australia has identified the considerable extent to which urban areas are connected to other areas, both rural and urban (Lennox and Turner 2005). The MFA was based on detailed collection of data—including transport freight statistics to help identify movement of mass in and out of the urban centres—for settlements as diverse as the growing coastal conurbation in South-East Queensland, centred on Brisbane, to the dormant inland town of Shepparton in Victoria.

Figure 8 (and similar MFA flows for other settlements) shows settlements appear strongly linked in material flows to other settlements, as illustrated by the relatively large material imports and exports to and from the settlement region. A very large proportion of settlement mass flows are associated with internal consumption and production for export beyond the hinterland.

A small proportion (about 10%) annually is incorporated in the stocks of the settlement (this

Table 2: Summary of sectoral issues, with some options proposed for how these might be remedied, and subsequent implications for other sectors.

'Sector'	Potential issue	Proposed option (to remedy the issue)	Implications or requirements (possibly resulting from the options)
Fuel (transport)	oil constraints	shift to alternatives (gas, biofuels, electricity (via fuel-cells))	more infrastructure; more materials; more water consumption; more skilled labour; competition (with food) for land for biofuels
		gas-to-liquids and coal-to-liquids	more infrastructure; more materials; more skilled labour; more emissions
		import more oil	trade deficit
Water	demand-induced scarcity (and quality) climate induced scarcity	desalinate recycle wastewater harvest stormwater; rural-urban water trading; piping water	more energy; new infrastructure; compete for use
		new dams	compete for land; more GHG emissions
Food	healthy land availability	increase land area increase productivity	more GHG emissions; more water use—lower river health more fertilisers and other chemicals; more fuel use; greater emissions; further land degradation
		decrease exports	increase trade deficit; rural decline
Labour	insufficient numbers to support aged population and economic growth long-term health	continue to increase immigration;	increase demand for energy, water, materials, food, education, health service...
		raise productivity	increased workforce stress; increased demand for education, health services; more infrastructure
Emissions	emissions (GHG) climate change	diversify away from fossil fuels	other wastes (e.g. radioactive material); more skilled labour; new infrastructure; more water
		C-sequestration	new technology and more infrastructure; uncertain environmental impacts
Infrastructure	ageing and slow turnover	build new plants extend the transport network regenerate infrastructure	more skilled labour; more materials, energy, water, land

mirrors a currently held view that national infrastructure assets, such as power supply and port facilities, lack sufficient investment). The relatively low level of flows into stocks indicates a possible focus on the shorter term processes of production and consumption rather than longer term investment in structures and functions that support settlements. However, further analysis is required to establish whether the small proportion of mass flows into stocks identified is sufficient or not for the sustainability of the settlement. Suitable analysis requires a longer timeframe and consideration of turnover rates and other factors associated with infrastructure.

Recognising the extent of connectivity is important because drawing a boundary around a particular region, such as a city, and limiting sustainability analyses to activities within that region means that

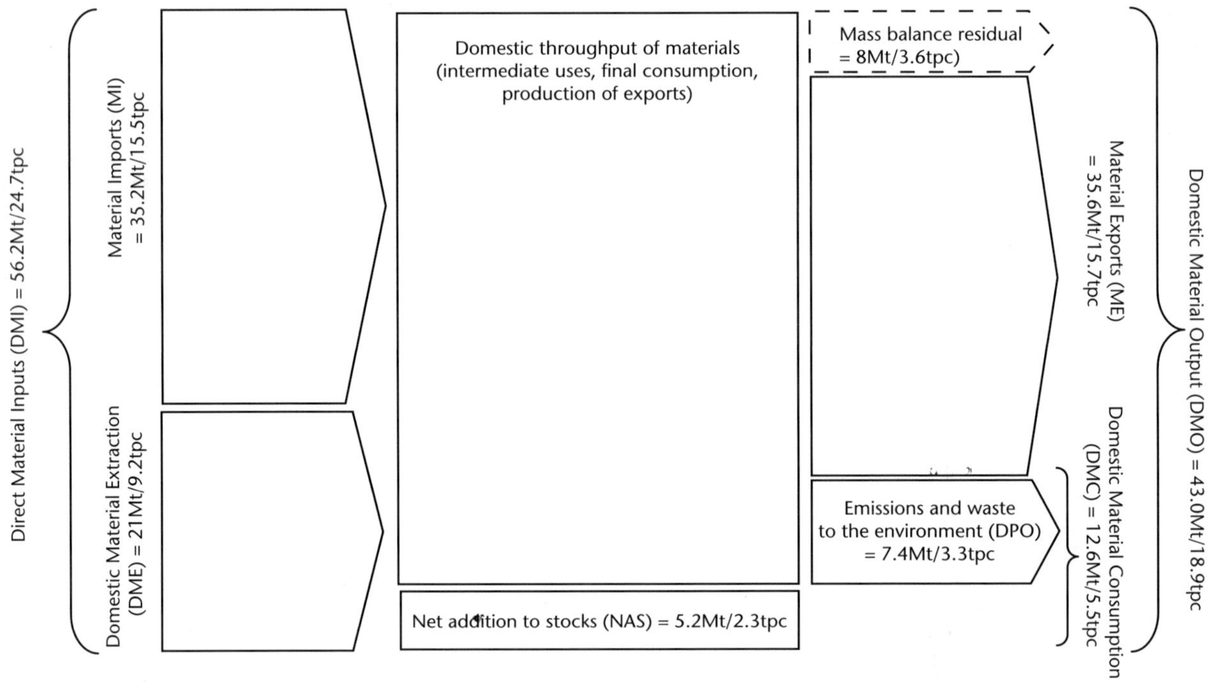

Figure 8: Mass Flow Analysis schematic for South-East Queensland, circa 2001 (Source: Lennox and Turner 2005).

any issues beyond the boundary (no matter how significant) are largely ignored. Nevertheless, this approach is often applied—using the argument that it is more pragmatic in terms of data collection and analysis, and that the analyses correspond to the same scale as the overseeing local management authorities so that policy development and implementation will be enhanced. These reasons may seem attractive, but are inappropriate if the key sustainability issues are geographically dispersed, with impacts on the city, rest of state, rest of nation and globally. A key message is that cities cannot be considered in isolation.

Additionally the MFA results presented here—like many other analyses—are static and contemporary. Applying such a 'temporal boundary' is directly analogous to the problems associated with drawing geographical boundaries.

The remainder of the analysis in this chapter therefore examines key sustainability issues using a dynamic simulation of the bio-physical operation of cities (using Melbourne as an example). This is based on the Victorian Regional Stocks and Flows Framework (VRSFF) (West *et al.* 2005; Turner *et al.* 2007c). The VRSFF is a more detailed account than the national framework (ASFF) of the demography, land-use, electricity and water sectors of Victoria.

Dynamics of cities

Growth in urban areas occurs incrementally, but accumulates to cover large areas over time (Figure 9). In any one year, announcements of individual land releases for new house construction are of the order of 10,000 dwellings, which does not seem significant compared with the existing stock and does not of itself immediately suggest sustainability issues. However, the impacts of continual growth become evident over longer timeframes, including:

- transport congestion associated with commuters travelling greater distances as the city expands at low density at the fringe
- loss of productive agricultural land
- capacity constraints of water supply, stormwater and sewage distribution networks based around centralised treatment plants.

Of themselves, these issues do not fundamentally undermine sustainability, but they do contribute to the pressures that are felt in other areas

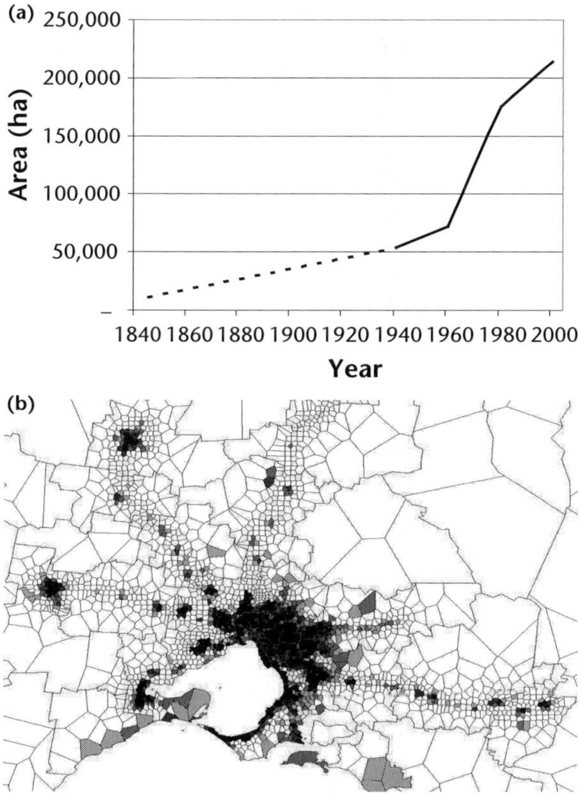

Figure 9: Development of Melbourne, showing (a) area developed derived from census data and historical maps (dashed line shows large interpolation period) (b) map of low-density dwellings in Melbourne (darker shades indicate higher proportion of area used for dwellings).

or sectors of society and the economy. For instance, while commuting congestion may be both detrimental in economic terms and frustrating for commuters, the additional emissions of GHGs (Figure 7) and consumption of oil (see Figure 1) are more serious issues that are directly associated with the growth in distance travelled and dependence on cars (and congestion may add a further burden).

Electricity consumption in Melbourne has grown exponentially—by about 2.0% per annum per capita in recent decades (Akmal and Riwoe 2005; Turner *et al.* 2007c), in comparison with a national average of about 3% per annum per capita. This electricity is predominately supplied from brown-coal-fired thermal power stations located to the east of Melbourne in the Latrobe Valley. Maintaining past Victorian growth rates into the future implies an eight fold increase in production by 2100 (Figure 10). As a comparison, a per capita consumption growth of rate of just 0.4% pa results in a doubling of production. The growth of electricity production also factors in current policy aspirations for population growth (*Victoria in Future*—DSE 2004) and electricity consumed in intermediate (that is, processes that indirectly support final demand) and material production.

Clearly, unless substantial changes are made—either using technology or lowering end-use consumption, or some combination—future CO_2 emissions will also grow exponentially. The conflict with recent scientific calls for reductions in CO_2 emissions by 2050 of 60–90% relative to 1990 levels is glaring. The section on 'Pathways for Change' below examines some of the possibilities for change and their implications.

In terms of resource depletion, the production of electricity for Melbourne is not limited by reserves of brown coal—there are many hundreds of years of brown coal reserves in Victoria. However, supply of fresh water in the Latrobe Valley for steam generation is not so certain, particularly under the combined pressures of projected growth in power generation and a decline in availability due to climate change.

Similarly, a large portion of water used directly in Melbourne is also sourced beyond the urban boundary. The Thomson Reservoir lies to the north-east of Melbourne and supplies the majority of the cities' potable water. Historically low reservoir levels have been recorded recently, which are associated with drought conditions—levels less than 30% of capacity imply most of Melbourne has

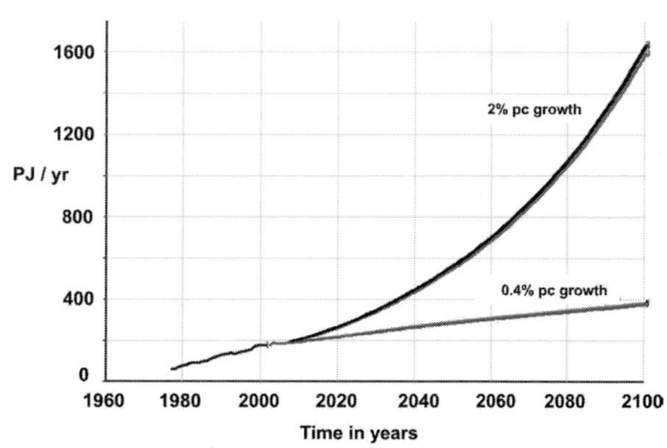

Figure 10: Comparison of electricity production for 2.0% p.a. growth in material and energy intensity *per capita* against 0.4% p.a. per capita growth scenarios.

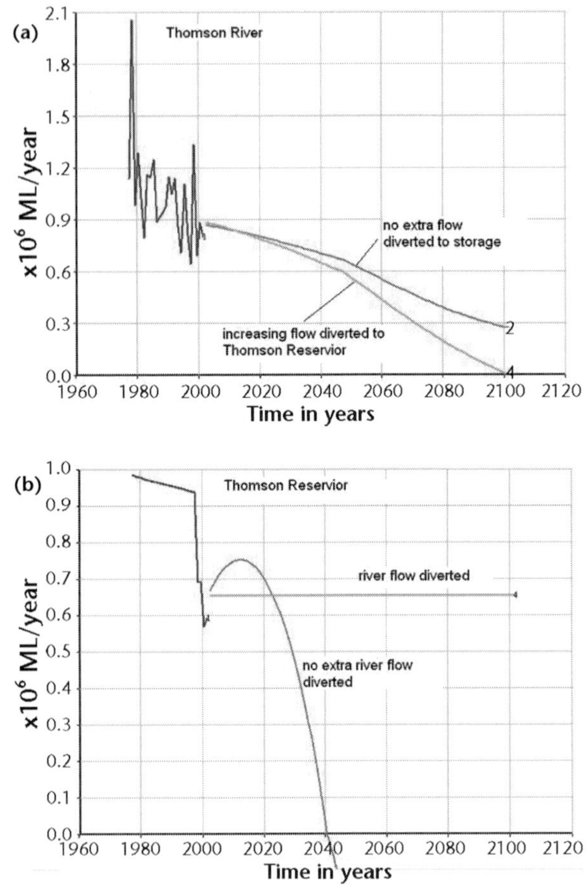

Figure 11: Thomson River flow (a) and Thomson Reservoir levels (b) under a climate change scenario (2.2°C at 2050). Two alternative management scenarios are shown—the first where the proportion of river flow diverted to the Reservoir is kept constant at recent values, which sees storage levels subsequently plummet while river flows are higher. The second shows the Thomson Reservoir maintained at relatively secure levels by drawing on increasing diversions from the Thomson River (Source: Turner *et al.* 2007c).

about 1 year of water security unless significant rainfall occurs.

Future potential trajectories of urban water supply are also cause for serious concern (Figure 11). Using assumptions about population growth, water conservation policies and climate change impacts (Jones and Durack 2005) on rainfall, evaporation and watering behaviour (Moran 2005), calculations reveal how river health and water storage security may not be achievable simultaneously (Turner *et al.* 2007b; Turner *et al.* 2007c) if global average temperatures rises by about 2.2°C by 2050. Such climate scenarios were previously categorised as the upper end of the envelope, but are increasingly being considered as likely, given recent updates to global CO_2 emission trends and climate impacts.

Assuming diversions to the Thomson Reservoir are maintained at recent proportions of river flow, then average flows of the Thomson River may fall to below 30% of the natural flow by about 2070, and natural variations in rainfall may cause the river to cease flowing in some years. At the same time, the average annual levels of the Thomson Reservoir drop precipitously (though in keeping with some historical rates of change) after 2020, with the Reservoir forecast to be empty by 2040.

Alternatively, if the Reservoir is maintained at about 50% of capacity with increasing fractions of river flow diverted to storage, then flow is increasingly affected beyond mid-century—the years when average flow drops to thresholds of 30% and zero advance by about one to three decades.

This is just one illustration of how a city draws on resources and emits wastes beyond its political and functional boundary. Other examples include: the production of food from across Australia and imported from overseas; and the consumption of oil as a transport fuel, which is sourced largely from the North West Shelf and increasingly imported from overseas.

Pathways for change—local (urban) solutions

Causes of sustainability challenges

The previous sections indicate that serious challenges to the viability of Australia's current way of life are likely to reach critical levels within the first half of the 21st century. In many cases, the physical evidence of the unsustainable levels of urban resource consumption is remote geographically from the majority of the Australian population living in urban areas. In other words, signs of problems are distant from the system, in both space and time. Without recognisable signals of a need for change, the economic system is more likely to pursue continued growth similar to that of the past.

Clearly this growth has brought substantial benefits to the population—longer and healthier lives, better and diverse food, universal education, broader cultural experiences, stimulating entertainment, financial well-being and remarkable material wealth, to name a few. It is not surprising therefore that there is a general desire among the population for more economic growth, despite studies that indicate a potential plateau or decrease in happiness once a population has reached a certain level of material well-being. Indeed, if happiness has decreased or reached a plateau, but the perception remains that further consumption and material possessions would lead to increased happiness, then clearly this creates a vicious cycle or positive feedback encouraging yet more material consumption (Newton 2007).

Various indicators of per capita consumption in the Australian population show little sign of abating (Newton 2007). Indeed, most are increasing and are documented in the most recent Australian State of Environment Report on Human Settlement (Newton 2006). Our desire for more food, such as all types of seafood (Figure 12) has increased steadily (Lowe *et al.* 2003a); the average size of our houses is still growing (Figure 12), even though family size has been on the decline and car ownership has increased dramatically over the previous century (Figure 12) (Foran and Poldy 2002).

Continual economic growth is also desired by business and, perhaps more importantly, government. This has been attributed to the lesson learnt from the Great Depression of the 20th century: avoid social upheaval associated with mass unemployment when economic growth falters significantly (Homer-Dixon 2006). Fundamentally, our technological innovation is a key underlying cause for growth (Grossman and Helpman 1991). As Homer-Dixon argues, when an industry becomes more efficient through better technology it tends to displace labour with capital; that is, people are replaced by machinery and become unemployed. If this situation were to continue without adjustment, there would be mass unemployment with subsequent social unrest. Alternatively, if consumers can be encouraged (by means not explored here) to purchase more or different goods and services, then the increased demand creates an opportunity to employ the displaced labour.

Consequently there are bottom-up and top-down pressures to maintain economic growth—individual desire for material well-being and avoidance of social unrest, respectively. The positive feedback of continual compounding growth leads to exponential escalation of environmental effects and, ironically, technological efficiency tends to enhance this rather than alleviate this 'rebound effect' (Polimeni and Polimeni 2006). This can be seen for instance in the continuous decrease of carbon intensity (or CO_2 emissions per

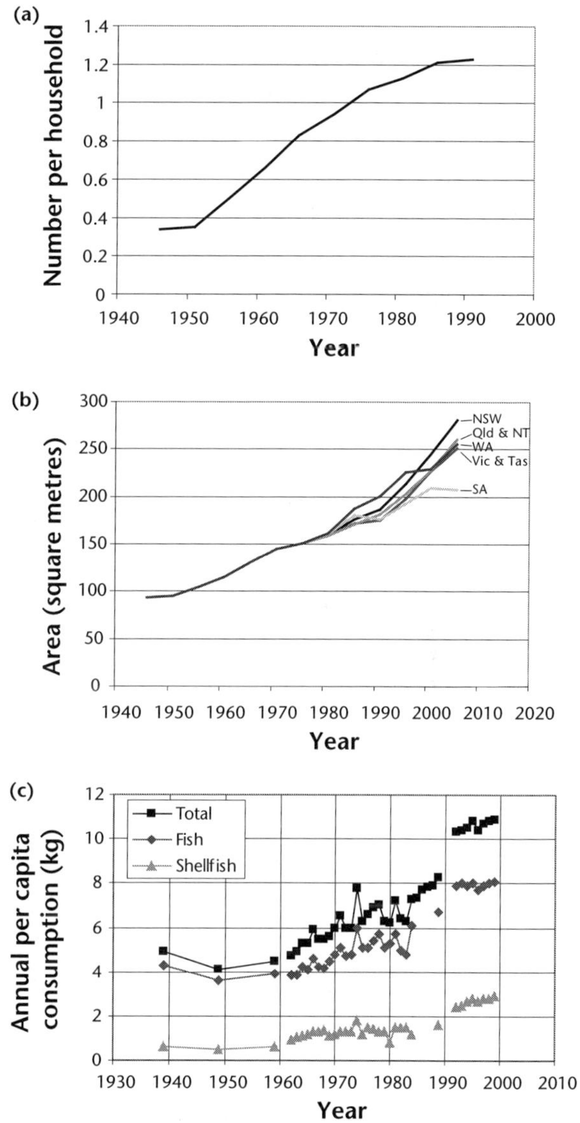

Figure 12: Trends of increasing consumption: (a) cars per household (Source: ASFF); (b) floor area of dwellings by State (Sources: based on ASFF and Centre for Design 2006); (c) per capita consumption of seafood by major groups (live weight) (Source: Lowe *et al.* 2003a).

unit primary energy) globally over 150 years, while absolute emissions have increased exponentially over the same period (Grubler 1998).

A critical issue for many people is the emission of CO_2 from the burning of fossil fuels. Even in Victoria, where end-use consumption of energy is growing at only 2% per capita per annum (somewhat less than the national average), a 'business-as-usual' (BAU) scenario would see CO_2 emissions from electricity generation increase about 850% by the end of the century (Figure 13). If growth in end-use consumption is not curbed then CO_2 emissions will continue to increase in the long term (three upper curves at 2100 in Figure 13). This is true even when aggressive technology is introduced, such as a 25% increase in brown-coal-power efficiency and a steady switch to 90% electricity supplied from a mix of gas combined cycle and renewable energy as coal stations are retired. While this scenario produces an initial marginal decrease in CO_2 emissions for a couple of decades, growth in demand for electricity inevitably overtakes any of the technological savings and so emissions rise steadily.

In comparison, if growth in end-use consumption of materials and energy could be held at 0.4% per capita per annum, then CO_2 emissions would be substantially reduced (three lower curves at 2100 in Figure 13). Even when no change is made to the way electricity is produced in Victoria (that is, predominantly by brown coal), the eventual emission rate at 2100 is limited to less than double current rates. When low growth in consumption is combined with the aggressive technology scenario, CO_2 emissions reduce most rapidly and remain at relatively low levels—by 2050 the emissions are about 50% of the 1990 levels (lowest curve in Figure 13).

Evidently the only scenario that realistically approaches suggested targets of reductions by 60–90% of 1990 emissions is one that involves substantial lowering of end-use consumption combined with technological progress.

Proposed carbon geo-sequestration may provide a means to substantially reduce CO_2 emissions—assuming that economic and environmental viability are established. Physical and economic analysis of this option should include consideration of pipeline and pumping infrastructure and energy required, as well as proximity of electricity production to suitable reservoirs for storing CO_2

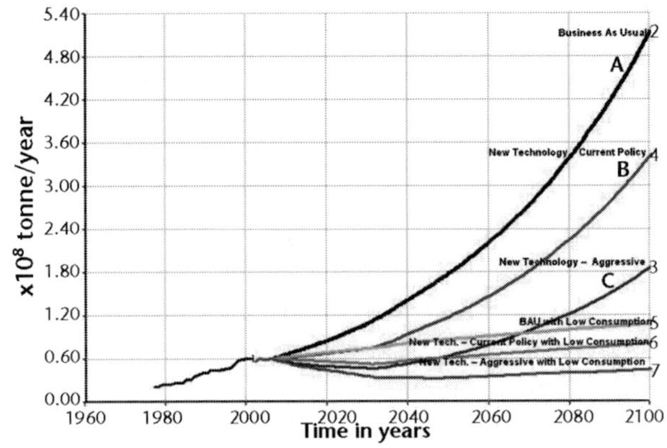

Figure 13: CO_2 emissions associated with electricity generation for the Business As Usual (A), New Technology–Current Policy (B) and New Technology–Aggressive (C) scenarios on 2.0% per capita growth in consumption (upper three curves at 2100). The same three technology scenarios are then played out over low (0.4% pa pc) growth in consumption.

and the stability and size of these reservoirs. The BAU curve provides some idea of the amount of storage required over time with continued growth in energy consumption. The storage volume required would have to be multiples of the current CO_2 emissions, rather than large fractions of it, to make any significant difference. It is not clear whether such reservoir volumes are available and would remain stable indefinately (national estimates based on cost and risk factors are of the order of 100 Mt/y of CO_2 (Georgiou 2007; Bradshaw *et al.* 2003), which is just a fraction of Victoria's total emissions under BAU).

Achieving low growth rates in end-use consumption—as in the 0.4% per annum scenarios (see Figure 13)—under our current economic and social system is not likely. It is often proposed that total end-use of electricity could be reduced by increasing efficiencies in end-uses, such as lighting, white goods, computers, machinery and better designed and insulated building, or by using solar heating and cooling systems. Such improvements typically lead to lower total prices paid by consumers for the (electrical) services delivered, and consequently the savings are available for expenditure elsewhere in the economy. This stimulates growth in the overall economy, which therefore generally requires increased energy production—an example of the 'rebound effect' (Wackernagel and Rees

1997; Polimeni and Polimeni 2006). Any such rebound resulting in additional energy use and associated CO_2 emissions have not been incorporated in the scenario calculations.

Likewise, increasing efficiency in the production of electricity, such as that assumed in the scenarios above, typically lowers the cost of production. Consequently, the industry has higher profits and can invest the increase in expanding its own capital or make investments in other sectors. A key message from this section is that continually growing consumption is related to technological progress on efficiency, and is a key cause of current sustainability challenges.

Rebound would not occur if the financial savings are used in ways that do not contribute to economic activities involving energy use, or if the lower end-use consumption was achieved simply by less per capita demand for goods and services, rather than efficiencies (Wackernagel and Rees 1997). Clearly, such approaches would involve substantial change to the fundamentals of the current economic structure.

Change

In the previous section we have argued that both increased material consumption *and* progress in technological efficiency are key drivers of economic growth and associated environmental impacts, including depleting natural resource stocks and ecosystem degradation from pollution or over-use. The analysis above suggests that substantive efforts to achieve sustainability imply *constraining consumption or efficiency* (the latter because it is linked to increasing consumption as described above).

One potential means is to evolve the social and economic system to one where the working population is not employed for some substantial fraction, say 50%, of the time. For this to be feasible without creating social unrest, it would be necessary for people to be engaged in meaningful activities that have little or no material production outputs, such as some artistic pursuits or caring for the elderly and the young. The latter is in keeping with the demographic 'dependency' pressure described above (see Figure 5)—as long as material growth in the economy was constrained so that other pressures are avoided. This would be a form of the 'service' economy.

The possibility of creating such a system obviously begs the question of how such a shift in values away from materialistic goods might come about. Imposing the change in a top-down manner may require strong leadership—that is, implementing policies that are not otherwise perceived as required by the general public (Diamond suggests that 17–19th century Japan achieved a top-down change through a non-democratic process (Diamond 2005)). A bottom-up change requires a majority of people to recognise a need for change, which is more likely in a small population (Diamond presents research suggesting this may have occurred, for example, in the small Pacific island of Tikopia).

Providing environmental signals sufficient to cause attitude and behavioural change among the Australian population remains a major challenge. Unfortunately, it appears that appropriate programs have not yet been established for educating the population about energy use and climate change—despite the IPCC and prominent people (such as Al Gore) raising awareness of climate change, GHG emissions rates are at the upper end of the IPCC scenarios, and climate change impacts have accelerated faster than expected. Nor is it clear that any changes would be maintained indefinitely.

Another approach explored here is for environmental signals to be provided more directly to the social and economic system, rather than through the media or other institutions. This could be achieved, for example, by decentralising both the supply of resources (such as food, water and energy), and the sites for waste assimilation. In other words, organise local communities to be more self-reliant, rather than relying on the import of resources from outside their locality and the export of wastes beyond their boundaries. If externalised costs of GHG from transportation were imposed for instance, then local self-reliance may be more economic. Biophysical limits within each local area provide more immediate checks on over-exploitation of the environment. By using modern technology, it may be possible to achieve such a locally based society without regressing to pre-industrial lifestyles.

Several aspects of such an approach are presented here as examples of what might be possible. In the first, a scenario of decentralised energy

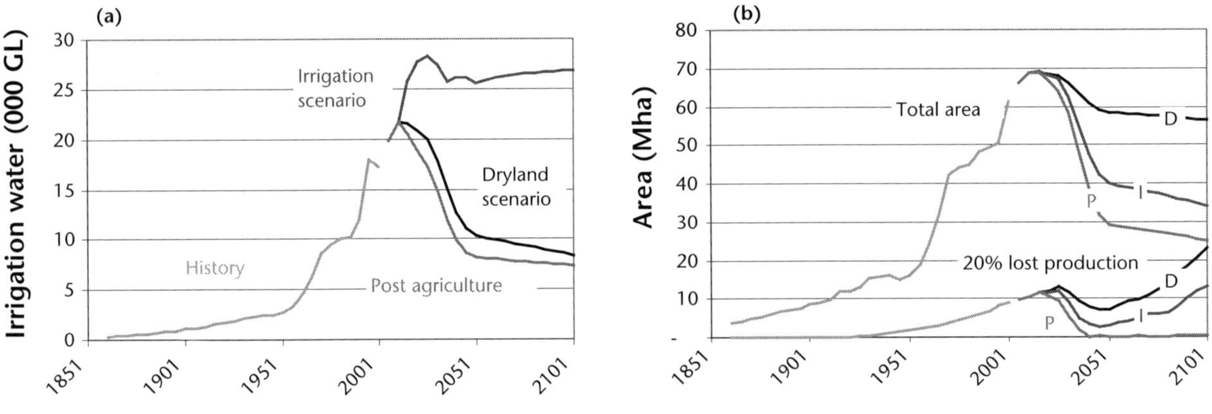

Figure 15: (a) Total irrigation water use over history and for three future scenarios and (b) total land area and that area with at least 20% yield loss due to degrading landscape condition (Source: Dunlop and Turner 2003).

there are major additional savings to energy use (illustrating the interactions between sectors).

Other benefits of more compact urban form would include potential for reducing transport energy consumption and associated emissions through the use of mass transit systems that are practical when servicing higher concentrations of population (Newton 1997). Additionally, surrounding land can be maintained for agricultural production for local consumption. Not only are the transport energy requirements reduced, but the productivity, and consequently the quality, of that land becomes of direct interest to the urban population—providing a significant and more visible bridge between the built and natural environments. A key message from this section is that sustainability can be achieved physically through a combination of curbing consumption, urban settlements that are more closely linked with their surroundings, and substantially employing green technology.

Conclusion

The analyses presented here support policy development by exploring the physical implications of alternative scenarios of the national economy and urban development. This analysis could be

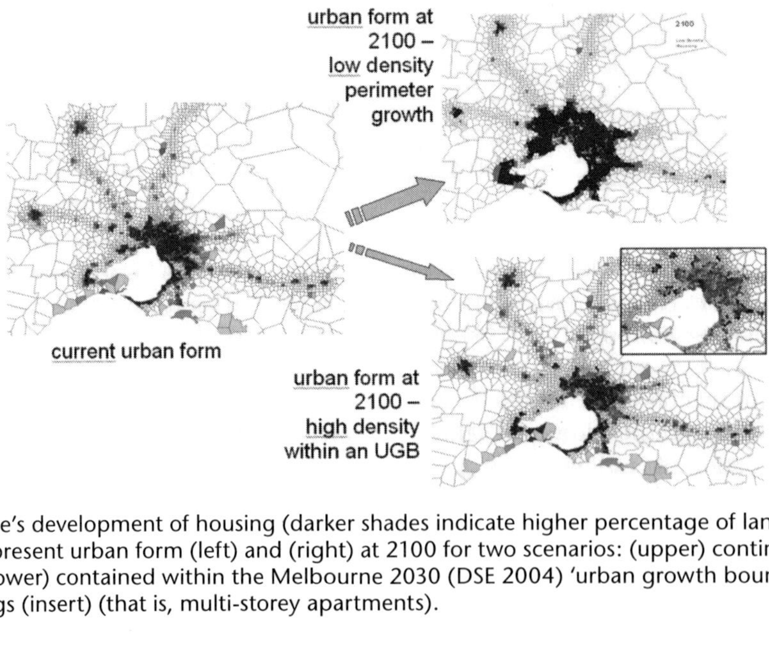

Figure 16: Melbourne's development of housing (darker shades indicate higher percentage of land area for residential dwellings) from the present urban form (left) and (right) at 2100 for two scenarios: (upper) continued low density perimeter growth; (lower) contained within the Melbourne 2030 (DSE 2004) 'urban growth boundary' using high-density dwellings (insert) (that is, multi-storey apartments).

complemented with similarly strategic economic assessments.

This chapter opened with how Australia's ecological footprint is several times higher than the sustainable global average. More detailed analysis based on the Australian Stocks and Flows Framework showed how the resource and environmental situation may worsen in future. A number of specific challenges were highlighted:

- supplying enough oil or alternative fuel for transport
- growing demand for water resources
- averting further degradation of agricultural land and fish stocks
- pressures on the labour force from an ageing population
- constraints on introducing new technology due to infrastructure inertia
- the difficulty of achieving safe levels of GHG emissions.

The challenges will be made all the more difficult because they are linked. More detailed analysis of scenarios at the city level illustrated how such resource and environmental issues arise at finer scales.

The scenarios explored the positive environmental impacts that are possible by decentralising society—local communities living within their local means—such as renewable electricity production, reduced GHG emissions and preserving land condition. What is also clear is the substantial change that would be involved, in terms of urban form, economic structure and consumption behaviour. Technology may help make this feasible, while maintaining some selected positive benefits of a modern society, such as health services and communications.

However, a simple reliance on technological progress is not likely to be enough, because higher growth in consumption offsets the gains made against resource depletion and wastes/emissions. The rebound effect means that improvements in efficiency may even make things worse. This is based on the understanding that standard economic growth is underpinned by both efficiency gains and increasing material consumption, which results in escalating environmental pressures. Our physical analysis of urban and national challenges shows that unless ingrained social habits move away from material consumption, it is possible that the sustainability challenges will worsen and converge over the first half of this century.

References

ABARE (Australian Bureau of Agricultural and Resource Economics) (2006). *Australian Commodity Statistics.* ABARE, Canberra.

ACF (Australian Conservation Foundation) (2007) *Consumption Atlas: What is the Environmental Cost of Our Spending?* ACF. <http://www.acfonline.org.au/custom_atlas/index.html>.

Akmal M and Riwoe D (2005). *Australian Energy: National and State Projections to 2029–30.* ABARE (Australian Bureau of Agricultural and Resource Economics).

Allen Consulting (2006). *Deep Cuts in Greenhouse Gas Emissions.* The Allen Consulting Group, Sydney.

Bradshaw J, Allinson G, Bradshaw BE, Nguyen V, Rigg AJ, Spencer L and Wilson P (2003). *Australia's CO2 Geological Storage Potential and Matching of Emission Sources to Potential Sinks.* GEODISC program of the Australian Petroleum Cooperative Research Centre,Perth

Centre for Design (2006). *Measures for Improving the Environmental Sustainability of Building Materials - Scoping study to Investigate Measures for Improving the Environmental Sustainability of Building Materials.* Centre for Design, RMIT University, Melbourne.

Commonwealth of Australia (2007). 'Intergenerational Report 2007'. Treasurer of the Commonwealth of Australia, Canberra.

Deffeyes KS (2001). *Hubbert's Peak: The Impending World Oil Shortage,* Princeton University Press, Princeton, New Jersey.

Department of the Prime Minister (2007) 'Report of the Task Group on Emissions Trading'. The Department of the Prime Minister and Cabinet, Canberra.

Diamond J (2005). *Collapse: How Societies Choose to Fail or Survive.* Penguin Group,New York.

DOI (Department of Infrastructure) (2002). *Melbourne 2030: Planning for Sustainable Growth.* Victorian State Government Department of Infrastructure, Melbourne.

DSE (Department of Sustainability and the Environment) (2004). *Victoria in Future 2004.* Victorian State Government Department of Sustainability and the Environment, Melbourne.

Dunlop M and Turner G (2003). 'Future sustainability of the Australian grains industry'. Report for the Grains Council of Australia'. CSIRO Sustainable Ecosystems, Resource Futures Program, National Futures, Canberra.

Dunlop M, Turner G, Foran B and Poldy F (2002). 'Decision points for land and water futures'. CSIRO Sustainable Ecosystems, Resource Futures Program, National Futures, Canberra.

Engineers Australia (2006) *Australian Infrastructure Report Card: comparison of all Infrastructure Report Card results form 1999-2005.* Engineers Australia, <http://www.engineersaustralia.org.au/shadomx/apps/fms/fmsdownload.cfm?file_uuid=1D0E6F28-AAE3-10C5-7519-DC22AE24FB75&siteName=ieaust>.

Foran B and Poldy F (2002). 'Future dilemmas: options to 2050 for Australia's population, technology, resources and environment'. Report to the Department of Immigration and Multicultural and Indigenous Affairs. CSIRO Sustainable Ecosystems, Canberra.

Georgiou P (2007). 'Between a rock and a hard place: the science of geosequestration'. House of Representatives Standing Committee on Science and Innovation, Canberra.

Government of South Australia (2005). 'Water Proofing Adelaide – Large Scale Water Supply Schemes Information Sheet'. Government of South Australia.

Grossman GM and Helpman E (1991). *Innovation and Growth in the Global Economy,* MIT Press, Cambridge Massachussetts.

Grubler A (1998). *Technology and Global Change,* Cambridge University Press, Cambridge, UK.

Hardwicke L (2006). '2005 Australian Infrastructure Report Card'. Engineers Australia, Canberra.

Homer-Dixon T (2006). *The Upside of Down-Catastrophe, Creativity and the Renewal of Civilization,* Island Press, Washington DC.

Jones RN and Durack PJ (2005). *Estimating the Impacts of Climate Change on Victoria's Runoff using a Hydrological Sensitivity Model.* CSIRO Atmospheric Research, Melbourne.

Kearney B, Foran B, Poldy F and Lowe DB (2003). *Modelling Australia's Fisheries to 2050: Policy and Management Implications.* Fisheries Research and Development Corporation, Hobart.

Kenway S, Turner G, Cook S and Baynes TM (2007). 'Water-Energy Futures – Policy levers and their impacts on future water and energy use for Melbourne '. CSIRO Water for a Healthy Country. Canberra.

Lattimore R (2005). *Economic Implications of an Ageing Australia.* Productivity Commission, Canberra.

Lennox JA and Turner G (2005). 'State of the Environment Report on Human Settlements: Stocks and Flows Indicators'. CSIRO Sustainable Ecosystems, Resource Futures, Canberra.

Lenzen, M, Wiedmann T, Foran B, Dey C, Widmer-Cooper A, Williams M and Ohlemüller R (2007). *Forecasting the Ecological Footprint of Nations: A blueprint for a dynamic approach.* Centre for Integrated Sustainability Analysis, University of Sydney. <http://www.isa.org.usyd.edu.au/publications/DEF.pdf>.

Lowe D, Poldy F, Kearney B, Foran B and Turner G (2003a). 'Australian Fish Futures: 2020 and Beyond'. Report to the Fisheries Research and Development Corporation. CSIRO Sustainable Ecosystems, Resource Futures Program, National Futures.

Lowe DB, Poldy F, Foran B, Kearney RE and Turner G (2003b). 'Australian Fisheries Futures: 2020 and Beyond'. CSIRO Sustainable Ecosystem, Resource Futures Program, National Futures, Canberra.

Mackenroth T (2005). *South East Queensland Regional Plan 2005–2026.* Office of Urban Management, Queensland Government.

Moran R (2005). *Scenarios of Urban Demand Change as a Result of Climate Change for Use in Water Supply-Demand Strategies.* Department of Sustainability and Environment, Melbourne.

Newton P (Ed.) (1997). *Reshaping Cities for a More Sustainable Future. Exploring the Link between Urban Form, Air Quality and Greenhouse Gas*

Emissions, Australian Housing and Urban Research Institute, Melbourne.

Newton P (2006). *State of Environment 2006: Human Settlements.* Department of Environment and Heritage, Canberra.

Newton P (2007). Horizon 3 planning: Meshing liveability with sustainability. *Environment and Planning B* **34,** 517–575.

NLWRA (National Land and Water Resources Audit) (2001). *Australian Dryland Salinity Assessment 2000. Extent, Impacts, Processes, Monitoring and Management.* National Land and Water Resources Audit, Canberra.

NPC (National Petroleum Council) (2007). *Facing the Hard Truths about Energy - A Comprehensive View to 2030 of Global Oil and Natural Gas.* National Petroleum Council, Canberra.

Poldy F, Foran B and Conroy J (2000). 'Future Options to 2050: Australian Stocks and Flows Framework'. Report to Department of Immigration and Multicultural Affairs; 00/04. CSIRO Sustainable Ecosystems (previously Wildlife and Ecology), Resource Futures Program, National Futures, Canberra.

Polimeni JM and Polimeni RI (2006). Jevons' Paradox and the myth of technological innovation. *Ecological Complexity* **3**, 344–353.

Raupach MR, Marland G, Ciais P, Le Quere C, Canadell JG, Klepper G and Field CB (2007). Global and regional drivers of accelerating CO_2 emissions. *Proceedings of the National Academy of Sciences of the United States of America* **104**, 10288–10293.

Saddler, H., Diesendorf, M. and Denniss, R. (2004). *A Clean Energy Future for Australia.* World Wildlife Fund, Sydney.

Spratt D (2007). 'The big melt: lessons from the Arctic summer of 2007'. Carbon Equity. Available online at <www.carbonequity.info/PDFs/arctic.pdf.>

Turner G and Poldy F (2001). Let's get physical: creating a stocks and flows view of the Australian economy. In *MODSIM 2001 Congress.* (Eds F Ghassemi, M McAleer, L Oxley and M Scoccimarro) pp. 1637–1642, Australian National University, Canberra.

Turner GM, Baynes TM, McInnis BC, West J and Hoffman M (2007a). Water accounting system for strategic resource management. Paper presented to *MODSIM 2007 Conference.* Christchurch, New Zealand.

Turner GM., Baynes TM and West J (2007b). Combined pressures and climate change impacts on the Victorian water system and possible responses. Paper presented to *MODSIM 2007 Conference.* Christchurch, New Zealand.

Turner GM, Baynes TM and West J (2007c). 'Victorian Regional Scenarios – Interactions of demographics, land-use, electricity, water and climate systems'. CSIRO Sustainable Ecosystems, Canberra.

Wackernagel M, Lewan L and Hansson (1999). Evaluating the use of natural capital with the ecological footprint. *Ambio* **28**, 604–612.

Wackernagel M and Rees,WE (1997). Perceptual and structural barriers to investing in natural capital: Economics from an ecological footprint perspective. *Ecological Economics* **20**, 3–24.

West J, Turner G and Baynes TM (2005). 'Technical Manual – Melbourne Region Stocks and Flows Framework'. CSIRO Sustainable Ecosystems, Canberra.

4

Energy Security and Oil Vulnerability

Jago Dodson and Neil Sipe

This chapter examines the problems, challenges and opportunities that changes in global petroleum energy security will pose for Australian cities, their residents, communities, planners and politicians. Australian cities are among the most petroleum-dependent in the world, as a result of over 50 years of support for private motor vehicles as the preferred mode of transport. High levels of reliance on cars inevitably imply high levels of dependence on petroleum. With rising concerns about the future security of global petroleum supplies, it is essential for Australia's urban managers to begin contemplating the risks and vulnerabilities that have been created by rising dependence on cars.

This chapter examines Australia's dependence on petroleum in three ways. Firstly, the chapter examines the strategic global petroleum energy context, the outlook for petroleum supplies over the medium and longer term and the threats and risks that global petroleum production faces. Secondly, the chapter considers the current status of Australian cities in terms of their petroleum dependence, the risks and vulnerabilities that this dependence poses for cities generally, as well as for specific areas and subgroups. This section draws on existing empirical research to identify these challenges. Thirdly, the chapter considers the current state of Australian metropolitan planning policies in terms of the degree to which they incorporate an understanding of energy security and oil vulnerability and the extent to which they seek to reduce petroleum risk. The chapter concludes by detailing the policy steps that must be taken to ensure the social and economic sustainability of Australian cities in light of changing global petroleum energy security.

Energy security

The past four years have seen a marked increase in the global price of oil. Global petroleum prices have risen substantially from the level they have generally held during the past two decades. The price of around US$65 per barrel reached in mid-2006 contrasts sharply with the price earlier in the decade, both in real and nominal terms (Figure 1). At the time of final revision of this book the price of oil had risen to near US$110, which is an inflation adjusted record high.

The sharp rise in the price of oil since 2004 has sparked a great deal of commentary about the causes and dynamics of the price increases. Explanations have included a mix of geopolitical concerns, production constraints and threats, and disequilibrium between growing global demand and production capacity. Further concerns relate to the notion of 'peak oil' and the possibility of the depletion of global oil reserves.

Growing demand for petroleum has been among the major factors underpinning the recent growth in fuel prices. The global economy has demonstrated a trend toward higher growth in the past two decades (Figure 2). This expansion has been particularly strong since the turn of this century with global growth reaching historical highs of 5.3% in 2004 and 5.4% in 2006. International Monetary Fund projections suggest this strong growth is likely to continue for some time, with global GDP anticipated to expand by 4.9% during each of 2007 and 2008. High growth rates in the large developing economies, such as China and India, have drawn particular attention because this growth is creating substantial additional demand

Towards a 'peak' in global oil production?

The price rises of the past four years can be largely explained by a combination of rising demand in the face of supply constraints and by concerns about geopolitical risks to supply. However, the marked increase in oil prices and concerns about production has drawn attention to the longer term issues of the global supply of petroleum, the estimated volume of global oil reserves and the pace of global oil consumption. Some petroleum industry commentators suggest that we may experience a point at which total oil production reaches a maximum 'peak' before declining as global oil reserves become depleted.

The theory of 'peak oil' is drawn from observations of the production behaviour of individual oil reservoirs aggregated to the global scale. When drilling commences in a new oil reservoir the rate of extraction typically accelerates quickly as high pressures drive the extrusion of crude. The addition of new wells further speeds this production pace. However the reservoir eventually reaches a point of diminishing marginal gains in extraction rates relative to increases in extractive capacity. Typically a point of maximum extraction occurs after which production rates begin to slow and then decline irrespective of the addition of further extractive capacity (Deffeyes 2001). When production is plotted over time, the resultant graph is broadly 'bell-shaped'. This is known as a 'Hubbert' curve, which was named after the petroleum geologist who first posited the theory (Hubbert 1956). A number of oil regions have already experienced production peaks. The United States saw its production peak in 1972 while the North Sea (UK and Norway) peaked in 1999. Australian oil production peaked at 214 million barrels in 2000 and has declined each year since then to 124 million barrels in 2006—the lowest output since 1972 (APPEA 2007). In 2004 the Australian Government estimated that there is a 90% chance that Australian crude oil production will drop to just 72,000 barrels per day by 2020 (Geoscience Australia 2004).

The notion of a global 'peak' in oil production is drawn from the aggregation of individual oil fields into larger oil regions and, in turn, into a single global curve. The notion of peak oil is extremely important for energy security because it implies an eventual decline in global oil production. Given that most economic projections anticipate growing world consumption of petroleum over the coming decades, the possibility that global oil production might peak and then decline implies an eventual disequilibrium between global demand and global supply of this commodity. In relatively open international markets, this will result in rising oil prices.

The theory of a peak in global oil production followed by an inexorable decline is controversial for a number of reasons. Firstly, there is limited petroleum industry agreement on the volume of remaining oil reserves, largely due to lack of transparent and reliable data sources. Much of the data concerning oil reserves and the condition of oil fields is commercially confidential and thus highly guarded by oil companies, is retained as a state secret by oil-producing nations, or is manipulated for political and economic reasons (Bentley 2002; Hirsch 2007). Together these unknowns make accurate predictions of future reserves very difficult.

Secondly, and in part due to the lack of transparent reserves data, there is a great deal of industry debate over the timing of a global oil peak, if at all. If global reserves are large relative to present production rates, then a global peak is likely to occur much later than if remaining reserves are modest, but an earlier peak is more likely if reserves are much lower than is assumed. The peak oil literature contains a large number of analyses that have sought to assess the point of global peak oil, which range in timescale from November 2005 (Deffeyes 2001) or 2006/2007 (Bakhtiari 2004) to after 2030 (Yergin 2006) or not at all (Lynch 2003). Of the 27 significant and credible peak oil predictions reviewed by Hirsch (2007), 21 anticipated that peak oil would occur by 2020. Jackson (2007) has argued that global reserves are much higher than most figures suggest implying a much later and less-pronounced peak in global oil production than other analyses. It is unlikely that any consensus will be reached on the timing of peak oil—if it occurs at all—in part because the peak moment is heavily dependent upon unpredictable variables such as global economic growth and oil consumption rates, petroleum production investment levels and geopolitical events. Campbell (2005) and

4

Energy Security and Oil Vulnerability

Jago Dodson and Neil Sipe

This chapter examines the problems, challenges and opportunities that changes in global petroleum energy security will pose for Australian cities, their residents, communities, planners and politicians. Australian cities are among the most petroleum-dependent in the world, as a result of over 50 years of support for private motor vehicles as the preferred mode of transport. High levels of reliance on cars inevitably imply high levels of dependence on petroleum. With rising concerns about the future security of global petroleum supplies, it is essential for Australia's urban managers to begin contemplating the risks and vulnerabilities that have been created by rising dependence on cars.

This chapter examines Australia's dependence on petroleum in three ways. Firstly, the chapter examines the strategic global petroleum energy context, the outlook for petroleum supplies over the medium and longer term and the threats and risks that global petroleum production faces. Secondly, the chapter considers the current status of Australian cities in terms of their petroleum dependence, the risks and vulnerabilities that this dependence poses for cities generally, as well as for specific areas and subgroups. This section draws on existing empirical research to identify these challenges. Thirdly, the chapter considers the current state of Australian metropolitan planning policies in terms of the degree to which they incorporate an understanding of energy security and oil vulnerability and the extent to which they seek to reduce petroleum risk. The chapter concludes by detailing the policy steps that must be taken to ensure the social and economic sustainability of Australian cities in light of changing global petroleum energy security.

Energy security

The past four years have seen a marked increase in the global price of oil. Global petroleum prices have risen substantially from the level they have generally held during the past two decades. The price of around US$65 per barrel reached in mid-2006 contrasts sharply with the price earlier in the decade, both in real and nominal terms (Figure 1). At the timc of final revision of this book the price of oil had risen to near US$110, which is an inflation adjusted record high.

The sharp rise in the price of oil since 2004 has sparked a great deal of commentary about the causes and dynamics of the price increases. Explanations have included a mix of geopolitical concerns, production constraints and threats, and disequilibrium between growing global demand and production capacity. Further concerns relate to the notion of 'peak oil' and the possibility of the depletion of global oil reserves.

Growing demand for petroleum has been among the major factors underpinning the recent growth in fuel prices. The global economy has demonstrated a trend toward higher growth in the past two decades (Figure 2). This expansion has been particularly strong since the turn of this century with global growth reaching historical highs of 5.3% in 2004 and 5.4% in 2006. International Monetary Fund projections suggest this strong growth is likely to continue for some time, with global GDP anticipated to expand by 4.9% during each of 2007 and 2008. High growth rates in the large developing economies, such as China and India, have drawn particular attention because this growth is creating substantial additional demand

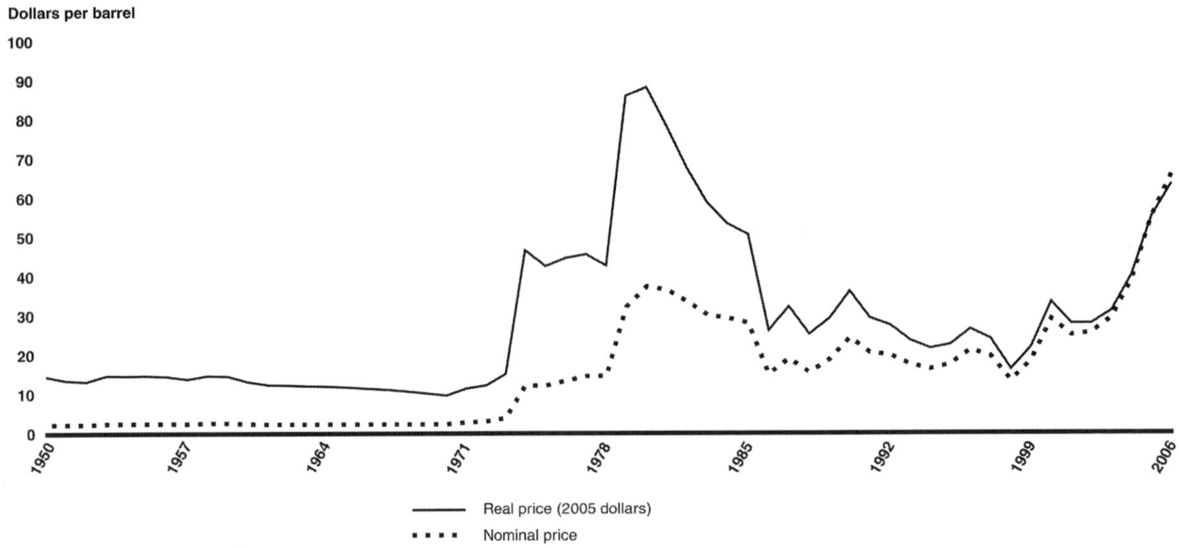

Figure 1: World real and nominal prices for petroleum, 1950–2006 (in 2006 US dollars); (Source: US Government Accountability Office 2007, p.11).

for petroleum to that expected from developed economies. Annual GDP growth in China has been more than 7.5% since 1991, and has not fallen below 9% since 2002 (International Monetary Fund 2007). India's GDP growth has not been as great, but has consistently been above 4% since 1991 and has exceeded 7% since 2002 (International Monetary Fund 2007).

World petroleum demand has grown steadily over the past decade—from 69.5 million barrels per day in 1995 to over 82.5 million barrels per day in 2005, which equates to an average 1.2% annual growth (Yi-chong 2007). Both India and China have large populations and the energy needs of their growing economies have added new demand to global petroleum markets. Chinese petroleum demand almost doubled between 1995 and 2005—from 3.4 million barrels per day to 7.4 million barrels per day (Yi-chong 2007). China is also increasingly dependent on oil imports, which rose from 6.4 percent of consumption in 1993 to 46% in 2004 (Yi-chong 2007). India already sources 67% of its crude oil internationally. The addition of China and India to global petroleum consumption places new demands on already constrained oil production systems.

Production capacity constraints—wells, pipelines and refineries—are a further dimension of petroleum insecurity. Global petroleum production capacity has not expanded at a sufficient rate to meet anticipated future demand growth – a deficit that is expected to increase in future. There are few detailed analyses of this problem although some reports have suggested that the cost of global investment necessary to meet petroleum demand growth by 2030 is approximately US$3 trillion, or US$100 billion per year (LeBlond 2005). The International Energy Agency (IEA 2006, p. 40) has estimated the production investment required to meet 2030 demand at approximately US$4.3 trillion. The task of assembling this scale of investment capital is viewed by many industry analysts as a major challenge, even before the question of the security of such investment is considered (IEA 2006). The distribution of current petroleum production suggests global oil reserves are becoming increasingly concentrated within a small number of Middle East nations (IEA 2006). Investment in new production facilities at the magnitude necessary to supply future global petroleum demand is unlikely to be readily forthcoming in a region that remains beset by political and economic instability and thus investment risk (Klare 2005). Further oil price rises are highly likely if global petroleum production does not expand to supply projected demand over the period to 2030.

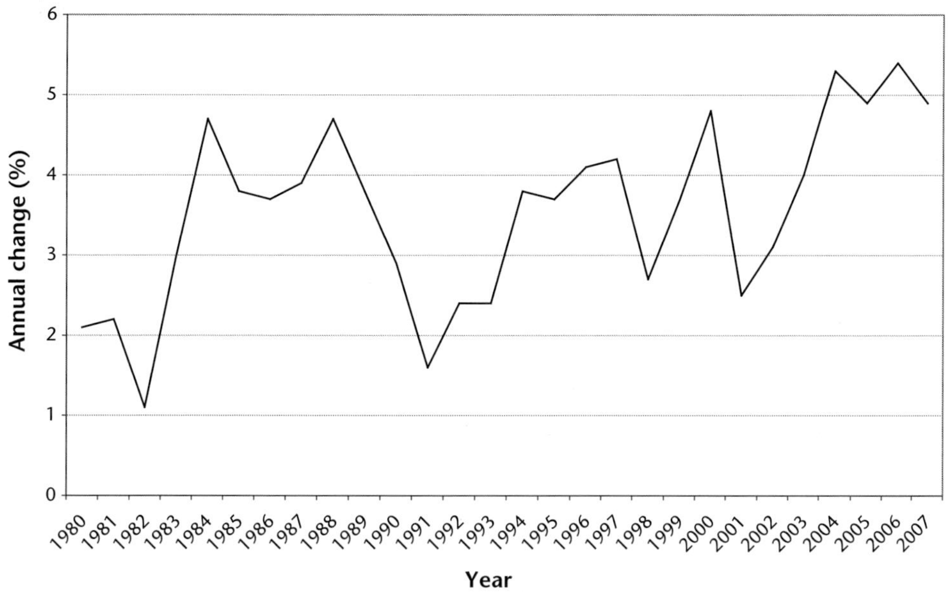

Figure 2: Global GDP growth, 1980–2006 (%); (Source: International Monetary Fund 2007).

Geopolitical instability is a further factor behind recent fuel price rises, in part, because it increases the investment risks for exploration and production. Many of the major oil-producing regions of the world are geopolitically volatile. The Middle East is the most significant area of geopolitical uncertainty, combining a tense mix of brittle political regimes intersecting with a strong military presence and high levels of intervention by some major oil-consuming nations. Political tensions also persist in major oil-producing nations beyond the Middle East, such as Nigeria, Venezuela and the former Soviet republics of Central Asia. Political instability raises the 'risk premium' in oil prices and can amplify price increases if combined with production constraints. The most recent example of this effect was the petroleum market anxiety caused by the month-long war between Israel and Lebanon in August 2006, which pushed price of crude oil up to almost US$77 per barrel. Sea-lane 'choke points' are another vulnerable link in the petroleum production chain. Approximately 35% of global oil deliveries pass through either the narrow Straits of Hormuz in the Persian Gulf or the Straits of Malacca between Malaysia and Indonesia (EIA 2005). Any restrictions on tanker passage through these and other important sea-lanes would inevitably place strong pressure on global oil prices.

The likelihood of geopolitical insecurity diminishing over the longer term in some oil-producing regions seems unlikely. The Middle East already contains 62% of global petroleum reserves, with this proportion expected to increase over time (Klare 2005). This concentration of reserves may increase competition for oil resources between major oil-consuming nations. The United States, Russia and increasingly China and Japan have major oil interests in the Middle East leading Kleveman (2003) to describe the emerging strategic situation in this region as a new 'great game'. While China's petroleum security strategy appears partly intended to avoid frictions in the procurement of petroleum resources, greater tension is possible over the longer term (Yi-chong 2007). Klare (2005) has argued that petroleum considerations already lie behind much of the United States' recent interventions in Iraq. With political tensions continuing to beset the Middle East, the prospects for increased petroleum security globally remain uncertain. A number of other assessments suggest that considerable ambiguity surrounds the longer term availability of petroleum beyond the immediate uncertainties generated by geopolitical risks, demand growth and supply constraint.

Towards a 'peak' in global oil production?

The price rises of the past four years can be largely explained by a combination of rising demand in the face of supply constraints and by concerns about geopolitical risks to supply. However, the marked increase in oil prices and concerns about production has drawn attention to the longer term issues of the global supply of petroleum, the estimated volume of global oil reserves and the pace of global oil consumption. Some petroleum industry commentators suggest that we may experience a point at which total oil production reaches a maximum 'peak' before declining as global oil reserves become depleted.

The theory of 'peak oil' is drawn from observations of the production behaviour of individual oil reservoirs aggregated to the global scale. When drilling commences in a new oil reservoir the rate of extraction typically accelerates quickly as high pressures drive the extrusion of crude. The addition of new wells further speeds this production pace. However the reservoir eventually reaches a point of diminishing marginal gains in extraction rates relative to increases in extractive capacity. Typically a point of maximum extraction occurs after which production rates begin to slow and then decline irrespective of the addition of further extractive capacity (Deffeyes 2001). When production is plotted over time, the resultant graph is broadly 'bell-shaped'. This is known as a 'Hubbert' curve, which was named after the petroleum geologist who first posited the theory (Hubbert 1956). A number of oil regions have already experienced production peaks. The United States saw its production peak in 1972 while the North Sea (UK and Norway) peaked in 1999. Australian oil production peaked at 214 million barrels in 2000 and has declined each year since then to 124 million barrels in 2006—the lowest output since 1972 (APPEA 2007). In 2004 the Australian Government estimated that there is a 90% chance that Australian crude oil production will drop to just 72,000 barrels per day by 2020 (Geoscience Australia 2004).

The notion of a global 'peak' in oil production is drawn from the aggregation of individual oil fields into larger oil regions and, in turn, into a single global curve. The notion of peak oil is extremely important for energy security because it implies an eventual decline in global oil production. Given that most economic projections anticipate growing world consumption of petroleum over the coming decades, the possibility that global oil production might peak and then decline implies an eventual disequilibrium between global demand and global supply of this commodity. In relatively open international markets, this will result in rising oil prices.

The theory of a peak in global oil production followed by an inexorable decline is controversial for a number of reasons. Firstly, there is limited petroleum industry agreement on the volume of remaining oil reserves, largely due to lack of transparent and reliable data sources. Much of the data concerning oil reserves and the condition of oil fields is commercially confidential and thus highly guarded by oil companies, is retained as a state secret by oil-producing nations, or is manipulated for political and economic reasons (Bentley 2002; Hirsch 2007). Together these unknowns make accurate predictions of future reserves very difficult.

Secondly, and in part due to the lack of transparent reserves data, there is a great deal of industry debate over the timing of a global oil peak, if at all. If global reserves are large relative to present production rates, then a global peak is likely to occur much later than if remaining reserves are modest, but an earlier peak is more likely if reserves are much lower than is assumed. The peak oil literature contains a large number of analyses that have sought to assess the point of global peak oil, which range in timescale from November 2005 (Deffeyes 2001) or 2006/2007 (Bakhtiari 2004) to after 2030 (Yergin 2006) or not at all (Lynch 2003). Of the 27 significant and credible peak oil predictions reviewed by Hirsch (2007), 21 anticipated that peak oil would occur by 2020. Jackson (2007) has argued that global reserves are much higher than most figures suggest implying a much later and less-pronounced peak in global oil production than other analyses. It is unlikely that any consensus will be reached on the timing of peak oil—if it occurs at all—in part because the peak moment is heavily dependent upon unpredictable variables such as global economic growth and oil consumption rates, petroleum production investment levels and geopolitical events. Campbell (2005) and

Heinberg (2003) have argued the actual timing of peak oil will probably only be recognised long after the event has occurred as marginal growth investment in exploration and production fails to produce marginal gains in global oil output.

Finally, there is considerable debate over the potential impacts of peak oil. Some observers argue that a petroleum production peak will not be a problem because the world contains many undiscovered petroleum and other fossil-fuel energy sources that can be exploited by better exploration and production techniques (Huber and Mills 2005). Others suggest that production scarcity will increase prices and bring forward new investment in production or stimulate new investment in efficient or alternative fuel technology that will moderate or minimise any effects of rising petroleum scarcity (Clarke 2007).

More pessimistic peak oil commentators suggest that petroleum is essential to economic growth and that declining supplies will reduce economic capacity (Heinberg 2004; Kunstler 2005; Simmons 2005). Such writers argue that the success of future exploration ventures cannot be assumed, no matter how high the price of oil rises, nor that technological advances will necessarily occur. Some speculative assessments posit a highly adverse longer term outlook. Kunstler (2005) is perhaps the most pessimistic amongst this group, arguing that economies that are dependent on cheap and abundant petroleum will become dysfunctional under a peak oil scenario. Kunstler suggests that the economies and spatial structures of developed nations, such as the USA, will transform into low-growth agrarian societies based on small networks of regional communities.

Despite the growth of literature on the potential for, and implications of, peak oil, there is only a modest body of scholarly literature on this topic. Deffeyes (2001, 2005) is one of the few scholars who has written about peak oil. Bentley (2002) has provided a useful overview of the issue while predicting a peak in oil production within 5 to 10 years. Greene *et al* (2006) suggested that

Peaking of conventional oil production is almost certain to occur soon enough to deserve immediate and serious attention. If peaking is already underway and oil supplies are as limited as the pessimists believe, the world is facing a drastic transition for which it is unprepared. If peaking is one to three decades away, it is not too soon to begin efforts to understand and prepare for the transition to other energy sources. (p. 15)

Much of the literature on peak oil is from a range of authors, particularly members of the oil industry (Campbell 2005; Simmons 2005). There is also a flourishing independent internet-based literature on peak oil (for example www.theoildrum.com; www.lifeaftertheoilcrash; www.peakoil.org). Some of the peak oil material has been analytical, in that it examines the patterns of global petroleum production and future prospects (Campbell 2005; Greene *et al.* 2006). A wider, more speculative body of literature considers the implications of declining petroleum supply for societies and economies (Heinberg 2003; Kunstler 2005).

The lack of a broad scholarly engagement with the question of whether peak oil will occur and when it might eventuate makes assessing the importance of the problem difficult. It is clear that the future of petroleum security and peak oil pose enormous questions for cities, but the likely future trajectory of petroleum supply is radically uncertain. This uncertainty contrasts with the strong scientific consensus on climate change manifested in the reports issued by the Intergovernmental Panel on Climate Change. Unlike biophysical systems, petroleum supply and demand are the outcome of complex economic, political and social processes that are not directly amenable to scientific assessment. In the absence of a clear scholarly assessment of peak oil, the task of a public evaluation of the potential for an oil production peak has fallen on government and political authorities.

A number of observers have noted that discussion of peak oil questions poses enormous problems for both the petroleum industry and for governments because the implications for economies and oil-dependent societies are likely to be considerable (Bentley 2002). Conveying these implications to markets and the public is highly problematic because of the impact that an oil-constrained future may hold for business investment and for government economic and social policies. Many of the governmental and oil industry assessments of the likelihood and risk of peak oil, such as Jackson (2007), have been either sceptical of the theory or conservative in assessing its importance. The International Energy Agency (IEA) held a cautious view on energy security and oil depletion until recently. Thus, for example, the IEA's 2004 *World Energy Outlook* argued that:

The world is not running out of oil just yet. Most estimates of proven oil reserves are high enough to meet the cumulative world demand we project over the next three decades. Our analysis suggests that global production of conventional oil will not peak before 2030 if the necessary investments are made. (p. 29)

In 2004, the IEA (2004, p. 119) estimated the level of investment necessary to meet this level of demand at US$3 trillion. Only two years later this estimate had grown to US $4.3 trillion (2006, p. 102). The IEA's long-term projection for petroleum prices also grew during this period from US$35 per barrel to US$50. While it continued to downplay the possibility of a peak in global oil production before 2030 the IEA's 2006 *World Energy Outlook* warned that growing demand and geopolitical uncertainties would nonetheless have a major impact on petroleum security over the next two decades.

Other government reports have also attempted to illustrate the range of issues around peak oil. A significant contribution was provided in a report by Hirsch *et al* (2005) to the US Department of Energy, which evaluated twelve significant or credible peak projections and the impacts these might have on the US economy. These projections suggested a global oil production peak would occur between 2005 and 2025, with a majority suggesting a peak before 2020. The authors argued that planning for mitigation of an oil production peak should begin immediately through pursuit of fuel and energy efficiencies and diversification of petroleum sources. The US military—one of the world's largest single institutional consumers of petroleum—has also examined the possibility and challenges of petroleum depletion. Fournier and Westervelt (2005), writing for the US Army Corps of Engineers, advised:

The doubling of oil prices from 2003-2005 is not an anomaly, but a picture of the future. Oil production is approaching its peak; low growth in availability can be expected for the next 5 to 10 years. As world-wide petroleum production peaks, geopolitics and market economics will cause even more significant price increases and security risks. (p. vii)

The US Department of Defence has since moved to achieve a 30% reduction in the energy intensity of its operations by 2015 and is exploring further efficiencies in fuel use (Crowley *et al.* 2007).

In early February 2007 the US Government Accountability Office (equivalent to an Auditor General in Australia) released a report that called for the US government to begin developing a national strategy to deal with oil depletion, given the immense challenge that peak oil would pose for the world's greatest oil-consuming nation. The Government Accountability Office report (GAO 2007) expressed particular concern that a peak oil scenario involving successive annual declines in global oil production would pose particular problems because any adaptation would need to keep pace with the rate of depletion. Some local governments in the USA have also begun to address peak oil questions. The most notable of these is the City of Portland (2007), which has prepared a peak oil strategy to deal with the challenge and consequences of a peak in petroleum production for that city. Among the Portland objectives are a reduction in petroleum consumption within the city of 50% by 2032.

Major oil companies have also begun to acknowledge concerns about longer term petroleum security. Chevron Oil (2005) suggested that 'the age of easy oil is over', while US Energy Secretary, Samuel Bodman, has been reported as saying that 'the era of cheap and abundant petroleum may now be over' (cited in Hirsch 2007). Former Australian Prime Minister John Howard also seemed to appreciate the uncertain outlook for petroleum supply when he observed during the recent oil price increases that 'I'm afraid we're not going to go back to the days of having petrol below a dollar a litre for quite some time, if at all' (Howard 2005). While many long-term oil production assessments do not necessarily point to the acceptance of peak oil theory, they are in general agreement that the issue of global petroleum security will have to be dealt with over the longer term. Even if peak oil does not eventuate within the next few decades, it appears almost certain that the cost of petroleum will continue to increase.

Australian assessments of petroleum security

Official petroleum security assessments in Australia have painted an inconsistent picture of future prospects. The Australian Government's Energy White Paper *Securing Australia's Energy Future* (DPMC 2004) suggested that Australia had a high level of overall energy security due to its large endowments of coal and natural gas. The strategy did not directly address peak oil. However, the outlook for Australia's petroleum security was conservatively ambiguous:

Despite increasing demand for oil, there are sufficient reserves to supply world demand for around 40 years...The dominance of the Middle East as the primary oil-producing region, the ability of these countries to act as a cartel through OPEC, and the political instability of some countries in the region have been major factors behind concerns about transport fuel security. In the longer term, concerns also exist about the longevity of oil supplies. (DPMC 2004, p. 119)

The most significant evaluation of energy security and peak oil in Australia has come from the Australian Senate Rural and Regional Affairs and Transport Committee inquiry into Australia's oil supply and alternative fuels. The committee produced a consensus final report which agreed that global oil production would likely peak by 2030 (Australian Senate 2007). The committee recommended few specific actions because of the differing political views of senators about the necessary responses from markets and governments. The main recommendation from the inquiry was for the Australian government to engage the Australian Bureau of Regional and Agricultural Economics to conduct a detailed analysis of the potential impacts of peak oil on Australia's regions and cities. The Senate Committee (2007) nonetheless argued that:

[T]he possibility of a peak of conventional oil production before 2030 should be a matter of concern. Exactly when it occurs (which is very uncertain) is not the important point. In view of the enormous changes that will be needed to move to a less oil dependent future, Australia should be planning for it now. (p. 55)

Since the Senate reported on energy security, the Brisbane City Council—Australia's largest local government—has also produced its own independent assessment of potential petroleum futures, along with climate change (BCC 2007). The BCC's report did not attempt to predict the timing of a global oil production peak. However, the report presented a 2004 projection of the Association for the Study of Peak Oil (ASPO), which suggested an oil production peak would occur in the later part of this decade, before 2010 (BCC 2007, p. 6). The BCC report advocated for immediate planning for peak oil and climate change mitigation and made a large set of recommendations including major reductions in energy use for transport. At the time of writing, the BCC was still considering its response to the taskforce recommendations.

Australian governments have not yet begun to recognise or address petroleum energy security questions in any systematic or deliberate way through urban policy or metropolitan planning. This failing can be attributed to the relatively recent renewal of interest in petroleum security, and the preparation of most recent Australian metropolitan plans either before, or shortly after, the period in 2004 when global oil prices began to increase to recent high levels. This lack of attention to petroleum security must change, however, given the implications of potential petroleum supply disruptions or constraints for cities that are highly petroleum-dependent. While the literature is ambivalent on many dimensions of energy security, there is general expectation that petroleum will become much more expensive over the medium and long term. The remainder of this chapter examines the risks that declining energy security and higher fuel prices pose for Australian cities, and the shifts in urban policy that may be necessary to mitigate or avoid these risks.

Energy security and Australian cities

Australia's cities and transport systems face major challenges from declining petroleum security due to their overwhelming and entrenched dependence on petroleum and continued growth in private motor vehicle use. In Sydney, private motor vehicles are used for 70% of all trips, while vehicle kilometres travelled grew 11.6% between 1991 and

2001 (DIPNR 2003). In South-East Queensland 80% of trips in 2003 were made by private vehicle compared with 78% in 1992 (Queensland Transport 2005). Vehicle kilometres travelled within the region will grow approximately 50% between 2001 and 2026 (Office of Urban Management 2005b, p. 107), which implies an equivalent increase in the demand for petroleum. While private vehicle travel is increasing, there has been no recent improvement in the fuel efficiency of private motor vehicles. The average petrol consumption of Australian private passenger vehicles increased from 11.3 L/100km in 2000 (ABS 2000, p. 15) to 11.4 L/100km in 2005 (ABS 2005, p. 13), which was equivalent to the 1963 level (Mees 2000b, p.61).

Aggregate urban transport figures also hide important sub-metropolitan differences in automobile dependence, as Newman *et al.* (1985; 1990) demonstrated in their ground-breaking work on motor vehicle dependence. These divergences are persistent and growing. In Sydney, for example, households in outer suburban areas have much higher levels of car dependence than those in middle or inner urban areas, despite making fewer trips. Residents of Sydney's outer north-west made, on average, 3.3 trips per day, compared with 3.8 trips for inner eastern residents but travelled an average of 23.2 vehicle kilometres daily compared with inner eastern residents who recorded an average of 10.1 daily vehicle kilometres travelled (DIPNR 2003). Similar patterns are found in Melbourne (Morris *et al.* 2002).

The high level of dependence on cars in Australian cities has combined with other factors to amplify differences in socioeconomic opportunity between households. Widely dispersed suburbs necessitate the use of cars to access employment and services, due to a shortage in public transport. Housing markets have increasingly allocated lower-income groups, especially those seeking home ownership, to outer suburban locations where land prices are cheaper. These patterns have produced transport and spatial disadvantage for many urban households. Morris (1981) demonstrated that the lack of adequate public transport in the outer suburbs of Australian cities meant residents of these areas had poorer access to employment and community services than more centrally located households. Maher *et al.* (1992, 1994) argued that many households in Australian cities experience 'locational disadvantage' due to dependence on cars and the dispersed nature of employment and services in outer suburbs. Burnley *et al.* (1997) demonstrated that groups on modest incomes often experienced socioeconomic disadvantage in relocating to outer suburban areas to achieve home ownership in Australian cities, largely because of the additional travel time and cost burdens associated with these moves.

Many of the differences in the relative rates of usage of public transport and private motor vehicles in Australian cities can be attributed to the extent and quality of public transport networks. Areas where public transport services are poor or inadequate to meet travel demands are overwhelmingly those in middle and outer suburban zones where the provision and integration of public transport services has not kept pace with development (Dodson *et al.* 2007). Where services exist in such areas they tend also to be dominated by private operators with poor inter-modal integration (Mees 2000a; Mees 2000b).

There have been few empirical assessments of the impacts of rising petroleum prices on the car dependence of urban households in Australia. Some preliminary work was undertaken in response to the 1970s oil price shocks (Figure 1), but is now largely out of date (Ogborn and Bannister 1978; Bannister 1979; Davies and Glazebrook 1980). The most recent attempt to assess potential impacts of rising fuel prices on urban households is the work of Dodson and Sipe (2006, 2007) who assessed the spatial distribution of household exposure to adverse impacts from rising fuel costs, in terms of socioeconomic status and dependence on cars. Their assessment for Sydney is presented in Figure 3. The study indicates that a higher degree of household socioeconomic exposure to rising transport fuel costs is found in middle and outer suburban areas where car dependence is high and public transport services are of poor quality. This contrasts with the relative lower exposure of less car-reliant inner urban areas that are serviced by higher quality public transport. While Dodson and Sipe's study provides a useful insight into the question of petroleum exposure of Australia's cities, more research is needed on these questions to better comprehend

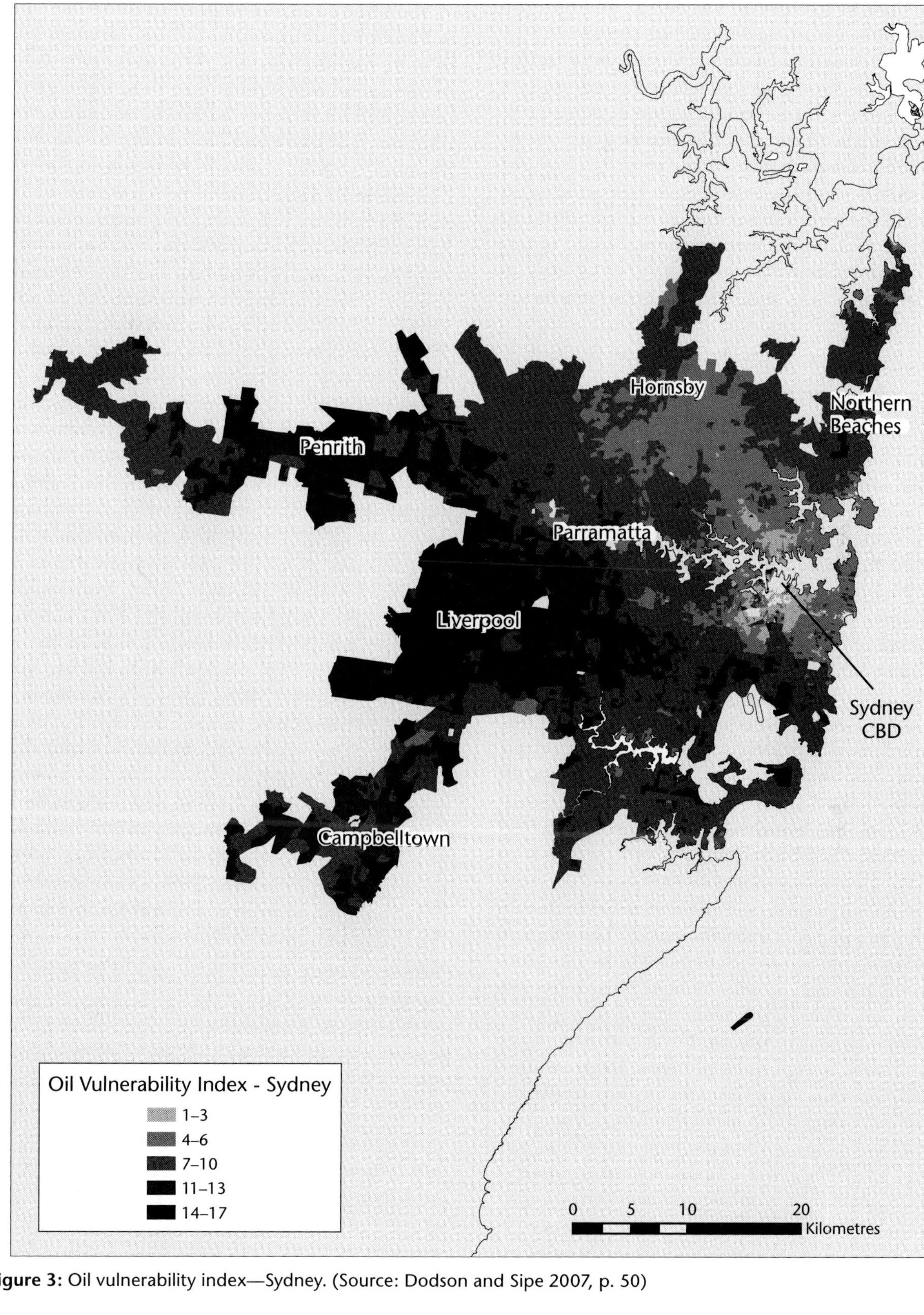

Figure 3: Oil vulnerability index—Sydney. (Source: Dodson and Sipe 2007, p. 50)

the impacts that rising fuel costs will have on households and, in turn, on urban economies.

A pressing task is to determine what policy shifts might be required to reduce the exposure of urban households to declining global energy security and higher fuel prices. Undertaking this assessment requires knowledge of the urban development and planning processes that made Australian cities highly dependent on cars and petroleum. The next section briefly discusses these problems, before outlining the steps that will be needed to begin to secure Australian cities from rising petroleum energy risks.

Current planning for energy security in Australian cities

Australian cities were not always as dependent on motor vehicles as they are today. Until the 1950s, walking, cycling and public transport were the modes used for most trips in Australian cities (Manning 1984). Rail and tram lines provided the spatial structure for Australian cities and most suburban development occurred within walking distance of these services. From the 1950s onwards, however, the form and structure of urban development shifted with accelerating suburbanisation stimulated by the need to house growing populations. Planning models shifted to support private motor vehicle transport. Investment in public transport either failed to keep pace with the spatial growth of increasingly dispersed suburbs and, in the 1990s, was delegated to the private sector, which lacked sufficient coordination or investment capacity to provide a quality of service needed to sustain high levels of patronage (Mees 2000). Government funding priorities shifted to roads with the major cities developing expensive and expansive freeway plans. The favouring of roads and cars in government plans drew investment and patronage away from public transport. Institutional fragmentation between public and private sectors has continued to limit the integration of public transport services. The result is the car-dependent travel patterns now found in contemporary Australian cities, particularly in areas developed after the late-1940s.

Australian metropolitan planning continues to promote car dependence. An example is South-East Queensland Infrastructure Plan (SEQIP) which will allocate 60% of anticipated $8.6 billion transport funding for greater Brisbane to roads, while public transport is to receive only 28% (OUM 2005a, p. 15). Walking and cycling, which are not dependent on petroleum, will receive only 2.3% of the greater Brisbane SEQIP funding (OUM 2005a, p. 15). Any intent to reduce the region's reliance on oil-dependent private motor vehicles seems, at best, rhetorical and is certainly not being treated seriously by the region's planners. Similar problems are apparent in the Victorian Government's *Melbourne 2030* metropolitan strategy (DOI 2002) in which the main project has been the Mitcham–Frankston tollway (East-Link). Sydney has pursued a tollways-based urban transport strategy for over a decade, while its rail system has languished. Perth's *Network City* metropolitan strategy has perhaps come closest to seriously addressing reliance on motor vehicles by decisively describing car dependence as the primary transport challenge facing the city and instituting policies that seek to 'improve the efficiency and management of the existing transport network rather than building more roads' (WAPC 2004, p. 69). This approach seems to be supported by funding allocations—for example, the new 70 km Mandurah rail line comprises the largest transport project currently being undertaken in Perth.

Few recent Australian urban strategies have identified petroleum energy security as a planning challenge for cities. Neither the Melbourne or Sydney metropolitan strategies nor the South-East Queensland Regional Plan mention energy security in their discussion of key planning issues. Again, Perth's *Network City* is the exception to this pattern even if its engagement is cursory:

Vulnerability to changes in oil supply (potential for significantly more expensive oil and supply shocks within 5 or 10 years perhaps)... are potent grounds for reducing the need to travel and shifting a greater share of trips from car to other modes. (WAPC 2004, p. 69)

The failure of most recent Australian metropolitan plans to tackle energy security could be explained on the grounds that most were prepared or were in preparation prior to the growth in oil prices that has occurred since 2004. But this would be a weak excuse. Sufficient scholarship (Fleay

1995; Newman and Kenworthy 1999), petroleum industry commentary (Campbell and Laherrere 1998) and government agency literature has been available for at least a decade that should have at least raised critical questions about future petroleum security for Australia's planners—especially in the context of their developing long-term strategies for Australia's four largest metropolitan regions. Given that a sustained or even episodic decline in petroleum security would have significant implications for Australian cities through disruption to transport systems and to economic and social activity, the lack of engagement on this issue by planning strategies seems a dangerous oversight. Australian metropolitan planning must engage with energy security questions to ensure our cities are adequately protected from the risks of rising fuel costs. The remainder of this chapter examines what the options for energy-secure urban planning might be and the potential steps that are necessary to achieve this.

Planning for petroleum energy security in Australian cities

Two tasks confront Australian and state governments in dealing with uncertainty over petroleum security. The first is to recognise and understand the problem, while the second is to commence the process of re-shaping urban structure and transport systems to reduce their vulnerability to future petroleum supply constraints.

Assessing energy security risks

As yet, only Perth's metropolitan plan acknowledges the possibility of less secure future petroleum scenarios, and even this is cursory. The first step needed to secure Australian cities against the adverse impacts of global oil supply shocks is for state governments to undertake formal and open analyses of the current state of global petroleum security and supply scenarios. Detailed assessments of how alternative petroleum supply constraint scenarios would impact on the economies and social fabric of Australian cities are then required. The final step would involve assessments of the most efficient means of adapting or restructuring current urban spatial patterns and infrastructures to mitigate potential impacts.

The risk to Australia's cities of a rise in the price of oil has not been comprehensively assessed by Australian state and local governments. There is a need to begin assessing this risk and the costs that may be borne by individuals, households, businesses and governments under various adverse energy security scenarios, including a peak in global oil production. This assessment should also evaluate what impacts rising costs of private motor vehicle travel would have for alternative modes such as public transport. It is unlikely that Australia's urban public transport systems would cope with a sustained medium or long-term growth in patronage. Brisbane and Melbourne are both currently struggling to source new fleet to satisfy current demand growth and Australian bus and train production is at capacity. There is a further need to assess the current petroleum dependence of Australian public transport networks and to re-design them to operate at maximum petroleum energy efficiency. This would probably involve reconfiguring public transport systems to reduce wasteful competition between bus and rail services wherever this occurs, especially on trunk radial routes. Given the higher passenger capacity and lower petroleum dependence of heavy rail compared with buses, it is likely that many line-haul buses would be re-deployed to feed upgraded rail networks. Any such risk assessment must also account for likely changes to travel patterns under alternative energy security scenarios.

There are presently few methods for assessing petroleum security risk for cities. Part of the risk assessment task for governments must be to develop appropriate methods for assessing petroleum risk, including the use of urban models to establish the economic and socio-spatial distribution of impacts. There is further need for broader community input into the assessment of petroleum security risk. The political, economic and social ramifications of rising energy insecurity and the potential for peak oil pose an enormous leadership challenge for elected officials who face resistance from various stakeholders who have an interest in the perpetuation of the status quo. Members of urban communities must be involved in the development of an energy security response strategy.

Urban energy security interventions

The second task to be undertaken in responding to petroleum risk is to take steps to reduce the oil vulnerability of Australian cities. Inevitably this involves reducing reliance on petroleum for urban travel which implies limiting support for, or minimising the use of, private motor vehicles in cities. Transport scholars have consistently shown that road capacity expansion is a key factor in encouraging private motor vehicle use in cities (Mogridge 1997; Downs 2004). Most contemporary Australian urban plans seek to facilitate growth in the use of private automobiles by expanding road networks, particularly freeways and arterial roads. Given the uncertainties about future petroleum security, the road capacity expansion plans embedded in current metropolitan strategies should be postponed until a comprehensive petroleum security risk assessment has been undertaken. Such assessments should take account of the aggregate impacts of rising fuel costs on household travel behaviour and should examine the distribution of sub-metropolitan impacts and the spatial cost and socioeconomic impact burdens on sub-regions. There is a need for new discounting rates to be established for petroleum-dependent transport projects to build oil risk into long-term economic value assessments.

Significantly higher global petroleum prices would increase the cost of motor vehicle travel. Higher costs for private motor vehicle travel could reduce the number of discretionary trips made by many households, especially those in outer suburban zones with high average daily travel distances. The most likely result of higher fuel prices would be reduced demand for private motor vehicle travel and growth in demand for bus and rail services. Some assessments of travel mode demand elasticities relative to fuel prices have suggested values of between 0.17 (Taplin *et al.* 1999) and 0.26 (de Jong *et al.* 1998) over the long term. Currie and Phung (2006) examined transport demand since fuel prices began to rise in 2004 and found an elasticity for heavy rail of 0.475. This suggests that a 10% increase in fuel prices would generate a 4.75% growth in rail patronage.

Public transport is in a parlous state in many Australian cities. It is unlikely that our urban rail systems could cope with a growth in patronage of 20 or 40% over the medium term if global oil prices were to rise quickly. There is a critical task to reduce the oil vulnerability of Australian cities by improving the service quality of public transport networks so that they might adequately respond to higher demand This task is most critical in middle and outer suburban areas where services are presently poor. The responsibility of improving public transport must be shared by the Commonwealth through transport funding programs such as Auslink.

Support for walking and cycling are further areas where there is a clear need for planning interventions to reduce the oil vulnerability of Australian cities. Most current transport models operate at the metropolitan scale and do not adequately account for local-scale walking and cycling. Walking and cycling do not use oil—and form a part of almost all public transport trips—yet receive little recognition or funding in current urban plans and budgets. New models of transport assessment must be developed that include walking and cycling, so that their contribution to reducing energy vulnerability can be better quantified and appreciated. Investment in walking and cycling infrastructure needs to be increased by an order of magnitude throughout Australian cities.

A further challenge for policy makers is to discontinue subsidy regimes that encourage the use of private motor vehicles. Queensland provides a petrol subsidy of 8.54 cents per litre and thus artificially reduces the direct cost of fuel to motorists. The Federal Government's Fringe Benefit Tax (FBT) scheme provides increased subsidies with increased use of motor vehicles (Riedy and Diesendorf 2003). These schemes encourage car use and thus work against petroleum security goals. Abolishing both the Queensland Fuel Subsidy and the FBT are necessary steps in reducing Australian oil vulnerability. Revisiting fuel excise regimes is also desirable, given that Australian motor vehicle fuels are among the least taxed in the developed world (DPMC 2004, p. 96). Increasing fuel excise would accelerate price increases for petroleum fuels and increase the rate of market adaptation to higher fuel prices. A proportion of the increased fuel costs could be allocated to supporting public transport improvements in presently poorly served suburbs of Australian cities.

The challenge of reducing petroleum use has underpinned, in part, the pursuit of urban consolidation in Australian planning schemes. A number of scholars have identified higher urban densities as necessary to achieve reduced petroleum reliance in Australian cities (Newman and Kenworthy 1999). Planning models such as higher density 'transit-oriented development' have been claimed by some to contribute to reduced private motor vehicle use. There are a number of reasons to be cautious about the use of urban density as a means of responding to declining energy security. Firstly, the evidence on the relationship between urban density and transport energy use remains underdeveloped, particularly at the sub-regional scale. There is scant research evidence to support the assumption that transit-oriented development reduces either private vehicle trip rates or mode share, at the regional level. Secondly, urban consolidation in its present form is dependent on private investment in housing and is thus subject to market cycles and long development timeframes. Recent experience has demonstrated that energy security can be highly volatile and subject to short- and medium-term fluctuations. Market-led urban consolidation would be unlikely to provide much relief in the event of a medium-term peak in global oil production or other sudden decline in energy security. A housing market phase of modest increases in higher density housing supply could similarly slow planning for reduced oil reliance if densification was the primary policy instrument. Furthermore, higher density development is typically more economically viable in higher value land markets, especially those in city centres. These zones are already served by high-quality public transport in Australian cities. The areas of Australian cities most dependent on cars are the outer suburbs where public transport is poor, yet current plans anticipate little allocation of higher density housing to such zones—in part because of relatively lower land prices, which make market-led densification uneconomic. Planning based on market-led densification begs questions about how petroleum constraint would affect existing dispersed low-density outer suburban areas in the absence of other adaptation or mitigation measures. Finally, while new housing stock could be developed at higher densities, the majority of Australia's existing urban housing stock remains in dispersed low-density suburbs. More directive planning is needed in these areas to reduce oil vulnerability than simply the relaxation of density limits.

How can we respond to the challenge of petroleum security if we do not pursue dramatic changes in urban densities? Australian research has demonstrated that the quality of public transport networks is as significant as urban density in supporting public transport use (Mees 2000b; Newman 2006). Further research has shown that comprehensive planning of public transport networks, based on rail networks operating at high frequencies with close integration with other modes, is crucial to higher modal shares for public transport (Mees 2000b). This has been recognised by supporters of compact city and transit-oriented development planning as demonstrated by the dense grids of public transport routes underpinning the suburban areas within 'network city' schemes (Newman and Kenworthy 1996, p. 12; Newman and Kenworthy 1999, p. 185). Recent best-practice guides (Nielsen *et al.* 2005) developed by the European Union based on Mees (2000b) research, identify the 'network effect' generated by a grid of high-frequency integrated public transport lines as the key basis for successful public transport systems.

A combination of the recognition of the importance of public transport services in supporting reduced reliance on motor vehicles and Dodson and Sipe's (2006, 2007) observations of the distribution of oil vulnerability in Australian cities demonstrates the need for a major expansion of public transport networks within middle and outer suburban areas to support 'go anywhere anytime' travel. Much of this expansion could be achieved through reorganisation and addition of bus routes to serve upgraded heavy rail networks and employment centres. Basically, this would only require investment in fleet, bus stops and priority traffic measures, rather major expenditures on heavy infrastructure.

Finally, to reduce oil vulnerability, Australia's urban planners must implement a decisive and sustained shift in planning practices. The failure of recent plans to arrest and reverse accelerating car dependence within cities suggests that our planning agencies and elected representatives do not have the ability or imagination to adequately grasp the problem of car dependence and the steps required to reduce this problem. The capacity of privatised models of public transport to support patronage growth is also in serious doubt (Mees 2005) and such arrangements may obstruct attempts to overcome oil vulnerability. Agencies and institutions that continue to impede the shift towards energy-secure urbanisation in Australia should face the possibility of radical restructuring to ensure they develop a planned response to oil vulnerability.

Conclusions

Australian cities are among the most petroleum-dependent in the world, yet the future for petroleum appears increasingly uncertain. Even if the theory of peak oil is unproven, the strategic outlook for global petroleum supplies suggests a declining level of security and increasing risk. The challenge of achieving petroleum security now stands among the biggest tasks needed to protect Australian cities from the uncertainty of future global petroleum supply. The Australian Senate has already stated that planning to reduce the oil vulnerability of Australian cities must begin immediately. The role of policy makers and planners is to begin actively pursuing this task—now.

Reducing Australia's vulnerability to uncertain future petroleum security is a conceptually simple task, and some of the measures needed to achieve greater petroleum security in our cities already exist in our urban plans. However, policies and practices that increase oil vulnerability—such as freeway construction, taxation exemptions and fuel subsidies that support the use private motor vehicles—must be abandoned.

Perhaps the most critical task, however, is for government, industry and researchers to begin debating petroleum vulnerability and assessing the future trajectory of petroleum supplies. The emerging literature on this issue must be debated publicly so that it can be democratically assessed and evaluated. This public discussion must begin so that we can undertake a measured, timely and democratic process of considering the planning options for oil-vulnerable cities in a petroleum-constrained world. We must all examine the structure, form and function of Australian cities under conditions of costlier and increasingly scarce petroleum.

References

Australian Bureau of Statistics (ABS) (2000). *Survey of Motor Vehicle Use*. (Catalog No. 9208.0). Australian Bureau of Statistics, Canberra.

Australian Bureau of Statistics (ABS) (2005). *Survey of Motor Vehicle Use*. (Catalog No. 9208.0). Australian Bureau of Statistics, Canberra.

Australian Petroleum Production and Exploration Association (APPEA). (2007). Australian Petroleum Production Statistics (Microsoft Excel Spreadsheet). Retrieved 31 March, 2007, from <http://www.appea.com.au/Statistics/documents/AnnualProductionStatistics.xls.>

Australian Senate (2007). 'Inquiry into Australia's future oil supply and alternative transport fuels'. Final Report. Australian Senate, Canberra.

Bakhtiari AMS (2004). World Oil Production Capacity Model Suggests Output Peak by 2006-07. *Oil and Gas Journal* **102(16)**, 18.

Bannister H (1979). The Effect of Rising Petrol Prices on Residential and Employment Markets. In *Proceedings of 5th Australian Transportation Research Forum*. 18–20 April, ATRF, Sydney.

Bentley RW (2002). Global oil and gas depletion: an overview. *Energy Policy* **30**, 189–205.

Brisbane City Council (BCC) (2007). *Climate Change and Energy Taskforce Final Report*. Brisbane City Council and Maunsell Australia, Brisbane.

Burnley I, Murphy P and Jenner A (1997). Selecting Suburbia: Residential Relocation to Outer Sydney. *Urban Studies* **34(7)**, 1109–1127.

Campbell C (2005). *Oil Crisis*. Multi-Science Publishing, London.

Campbell C and Laherrere JH (1998). The End of Cheap Oil. *Scientific American* **278(3)**, 78–83.

Chevron Oil Ltd (2005). Advertisement in *The Economist* magazine, 16 July, pp. 6–7.

City of Portland (2007). *Descending the Oil Peak: Navigating the Transition from Oil and Natural Gas*. City of Portland.

Clarke D (2007). *The Battle for Barrels: Peak Oil Myths and World Oil Futures*. Profile Books Ltd, London.

Crowley TD, Corrie TD, Diamond, DB, Funk, SD, Hansen, WA, Stenhoff, AD and Swift, DC (2007). *Transforming the Way DOD Looks at Energy: An*

Approach to Establishing an Energy Strategy. Department of Defence, Washington DC.

Currie G and Phung J (2006). Exploring the Impacts of Fuel Price Increases on Public Transport Use in Melbourne. In *Proceedings of the 29th Australasian Transportation Research Forum*, 27–29 September, Crowne Plaza, Gold Coast.

Davies A and Glazebrook G (1980). Transport Energy and Equity: Winners and Losers. In Proceedings of 6th Australian Transportation Research Forum, 22–24 October. ATRF, Sydney.

de Jong GC, Biggiero L, Coppola P, Gunn HF, Tegge O, De Palma A, deVillemeur E, Duchateau H. and Gayda S (1998). Elasticity Handbook for Prototypical Contexts. Costs of private road travel and their effects on demand, including short and long term elasticities (TRACE Project). European Commission Directorate-General for Transport, Brussels.

Deffeyes K (2001). *Hubbert's Peak: The Impending World Oil Shortage*. Princeton University Press, New Jersey.

Deffeyes K (2005). *Beyond Oil: The view from Hubbert's peak.* Farrar, Strauss and Giroux, New York.

Department of Infrastructure (2002). *Melbourne 2030: Planning for Sustainable Growth*. Department of Infrastructure, Melbourne.

Department of Infrastructure Planning and Natural Resources (DIPNR) (2003). *Regional Transport Indicators for Sydney.* Transport and Population Data Centre, NSW Government, Sydney.

Department of Prime Minister and Cabinet (DPMC) (2004). *Securing Australia's Energy Future*. Australian Government, Canberra.

Dodson J, Gleeson B, Evans R and Sipe N (2007). Investigating the Social Dimensions of Transport Disadvantage II: From Concepts to Methods Through an Empirical Case Study. *Urban Policy and Research* **25(1),** 63–89.

Dodson J and Sipe N (2006). Suburban shocks: Assessing locational vulnerability to rising household fuel and mortgage interest costs. Proceedings of 29th Australasian Transportation Research Forum, 27–29 September, Crowne Plaza, Gold Coast City, ATRF.

Dodson J and Sipe N (2007). Oil vulnerability in the Australian city: Assessing socio-economic risks from higher urban fuel prices. *Urban Studies* **44**, 37-62.

Downs A (2004). *Still Stuck in Traffic*. The Brookings Institution, Washington DC.

Energy Information Administration (EIA) (2005). *World Oil Transit Chokepoints*. Department of Energy, Washington D.C.

Fleay BJ (1995). *The Decline of the Age of Oil: Petrol Politics and Australia's Road Ahead*. Pluto Press Australia, Sydney.

Fournier DF and Westervelt ET (2005). *Energy Trends and Their Implications for U.S. Army Installations*. Washington, US Army Corps of Engineers, Washington DC.

Government Accountability Office (2007). *Crude Oil: Uncertainty about Future Oil Supply Makes It Important to Develop a Strategy for Addressing a Peak and Decline in Oil Production*. United States Government, Washington DC.

Greene DL, Hopson JL and Li J (2006). Have we run out of oil yet? Oil peaking analysis from an optimist's perspective. *Energy Policy* **34**, 515–531.

Heinberg R (2003). *The Party's Over: Oil, War and The Fate of Industrial Societies*. New Society Publishers, Gabriola Island BC, Canada.

Heinberg R (2004). *Powerdown: Options and actions for a post-carbon world*. New Society Books, Gabriola Island BC, Canada.

Hirsch R (2007). *Peaking of World Oil Production: Recent Forecasts*. Department of Energy and National Energy Technology Laboratory, Washington DC.

Hirsch R, Bezdek R and Wendling R (2005). *The Peaking of World Oil Production: Impacts, Mitigation and Risk Management*. Department of Energy, Washington DC.

Howard J (2005). Radio Interview with Tanya Nolan on ABC Radio '*World Today*' Show, 26 September, 12:18 pm. Australia.

Hubbert MK (1956). Nuclear Energy and Fossil Fuels. Spring Meeting of the American Petroleum Institute, 7–9 March. American Petroleum Institute, San Antonio, Texas. (Also published as Shell Production Division Publication No.95; Houston, Texas.)

Huber P and Mills MP (2005). *The Bottomless Well: The Twilight Of Fuel, The Virtue Of Waste, And Why We Will Never Run Out Of Energy.* Basic Books, New York.

International Energy Agency (IEA) (2004). *World Energy Outlook 2004*. International Energy Agency and Organisation for Economic Cooperation and Development, Paris.

International Energy Agency (IEA) (2006). *World Energy Outlook 2006*. International Energy Agency and Organisation for Economic Cooperation and Development, Paris.

International Monetary Fund. (2007). World Economic Outlook Database, April 2007. from <http://www.imf.org/external/pubs/ft/weo/2007/01/data/download.aspx>.

Jackson P (2007). Peak oil theory could distort energy policy and debate. *Journal of Petroleum Technology* Online (February): n.p.

Klare M (2005). *Blood and Oil: The Dangers and Consequences of America's Growing Petroleum Dependency*. Penguin, London and New York.

Kleveman L. (2003). *The New Great Game: Blood and Oil in Central Asia*. Atlantic Monthly Press, New York.

Kunstler JH (2005). *The Long Emergency: Surviving the Converging Catastrophes of the Twenty-first Century*. Grove/Atlantic, New york.

LeBlond D (2005). Investment cited as most likely limit to oil supply. *Oil and Gas Journal*, **103(27)**, 25.

Lynch MC (2003). Petroleum resources pessimism debunked in Hubbert model and Hubbert modelers' assessment. *Oil and Gas Journal* **101(27)**, 38–47.

Maher C (1994). Residential mobility, locational disadvantage and spatial inequality in Australian cities. *Urban Policy and Research* **12(3)**, 185–191.

Maher C, Whitelaw J, McAllister A, Francis R, Palmer J, Chee E and Taylor P (1992). *Mobility and Locational Disadvantage within Australian Cities*. Department of Prime Minister and Cabinet Social Justice Research Program into Locational Disadvantage, Canberra.

Manning I (1984). *Beyond Walking Distance: The gains from speed in Australian urban travel*. Australian National University Press, Canberra.

Mees P (2000a). 'Rethinking Public Transport in Sydney'. UFP Issues Paper 5. Urban Frontiers Program Issues Papers. Urban Frontiers Program, University of Western Sydney.

Mees P (2000b). *A Very Public Solution: Transport in the Dispersed City*. Melbourne University Press, Melbourne.

Mees P (2005). Privatization of rail and tram services in Melbourne: What went wrong? *Transport Reviews* **24(4)**, 433–449.

Mogridge M (1997). The self-defeating nature of urban road capacity policy. *Transport Policy* **4(1)**, 5–23.

Morris J (1981). Urban Public Transport. In *Equity in the City*. (Ed. P Troy). George Allen and Unwin, Sydney.

Morris J, Wang F and Berry M (2002). Planning for public transport in the future: challenges of a changing metropolitan Melbourne. Australasian Transport Research Forum. 2-4 October. Australian Bureau of Transport and Regional Economics, Canberra.

Newman P (2006). Transport greenhouse gas and Australian suburbs: what planners can do. Australian Planner **43(2)**, 6–7.

Newman P and Kenworthy J (1996). The land use-transport connection. *Land Use Policy* **13(1)**, 1–22.

Newman P and Kenworthy J (1999). *Sustainability and Cities: Overcoming Automobile Dependence*. Island Press, Washington DC.

Newman P, Kenworthy J and Lyons TJ (1985). Transport energy use in the Perth metropolitan region: Some urban policy implications. *Urban Policy and Research* **3(2)**, 4–15.

Newman P, Kenworthy J and Lyons TJ (1990). *Transport Energy Conservation Strategies for Australian Cities: Strategies for Reducing Automobile Dependence*. Institute for Science and Technology Policy, Murdoch University, Perth.

Nielsen G, Nelson JD, Mulley C, Tegner G, Lind G, Lange T (2005) *Public transport: planning the networks*, Hi-Trans Best Practice Guide 2. Hi-Trans and EU Interreg IIIB, Norway.

Office of Urban Management (2005a). *South East Queensland Infrastructure Plan and Program 2005-2006*. Office of Urban Management, Queensland Government, Brisbane.

Office of Urban Management (2005b). *South East Queensland Regional Plan 2005–2006*. Office of Urban Management, Queensland Government, Brisbane.

Ogborn K and Bannister H (1978). *Urban Areas and Energy Use: Rising petrol prices and employment markets in urban areas.* Department of Environment, Housing and Community Development, Canberra.

Queensland Transport (2005). *Smart Travel Choices for South East Queensland: A Transport Green Paper.* Queensland Government, Brisbane.

Riedy C and Diesendorf M (2003). Financial subsidies to the Australian fossil fuel industry. *Energy Policy* **31**, 125–137.

Simmons M (2005). *Twilight in the Desert: The Coming Saudi Oil Shock and the World Economy.* John Wiley and Sons, Hoboken, New Jersey.

Taplin JHE, Hensher D and Smith B (1999). Preserving the symmetry of estimated commuter travel elasticities. *Transportation Research B*, **33(B)**, 215–232.

Yergin D (2006). *Why the 'Peak Oil' Theory Falls Down—Myths, Legends, and the Future of Oil Resources.* Cambridge Energy Research Associates, Cambridge, MA.

Yi-chong X (2007). China's Energy Security. In *Energy Security in Asia*. (Ed. M Wesley) Routledge, London and New York.

5

The Oil Transition

Peter Newman

The global oil peak

In 1956, M. King Hubbert first suggested that the US would peak in oil production in 1970 and, despite being greeted with derision by many economists, it happened. The US began to import oil, which led to the strengthening of OPEC and the geopolitics of oil in the last part of the 20th century. The US now imports half its oil and by March 2008 the cost was $3706 billion (the website zfacts.com shows a continuous scrolling total of what oil imports are costing the US). This chapter is premised on an even deeper crisis than that faced by the US in the past few decades—the global oil production peak and what this may mean for our cities.

Global discoveries of oil peaked in the 1960s as shown in Figure 1. The graph also shows how global oil production has followed the general shape of the Hubbert curve, apart from reductions due to the three oil crises of 1973, 1979 and 1991, before following a general upward trend until 2005. Oil production fell in 2006, despite demand for oil growing at around 2 to 3% (mainly due to China and India, although together they still only take 12% of world oil production, while the US takes 26%). Some people believe that the peak in conventional oil production has already happened.

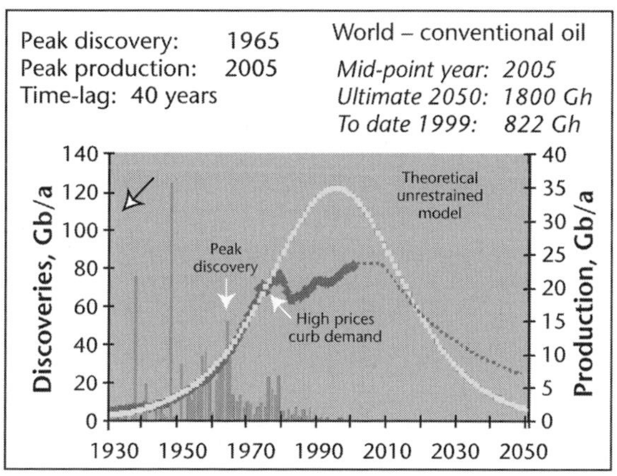

Figure 1: Peak oil in discoveries and production, as well as the Hubbert Model (Source: Association for the Study of Peak Oil (Australia) www.aspo-australia.org.au).

Enormous effort has gone into trying to discover more oil and to extract more from the reserves that remain. Oil reserves are now growing at a much slower rate than oil is being consumed. A peak in global oil production is therefore inevitable. There are various estimates of when the peak will occur, but they cluster around the early part of this century, with a few (now mostly discredited) way out into the future. C J Campbell—an oil geophysicist who founded the Association for the Study of Peak Oil—says that conventional oil peaked in 2004 and 'all oil liquids' will peak in 2010. Deffyes (2006) calculated an exact date for the peak as 16 December 2005 and later found that he was just a few months out, as May 2006 appears to have been the peak in conventional oil (see Figure 2).

There is much debate about whether the peak is only short term, whether we bounce along on the top of the curve for a decade or so, or whether we begin a precipitous decline. Campbell (2004) says:

The real point is not so much the exact date of peak but the statement that the First Half of the Oil Age, which was characterised by growing production, is about to be followed by the Second Half when oil production is set to decline along with all that depends upon it. On that at least we can stand firm.

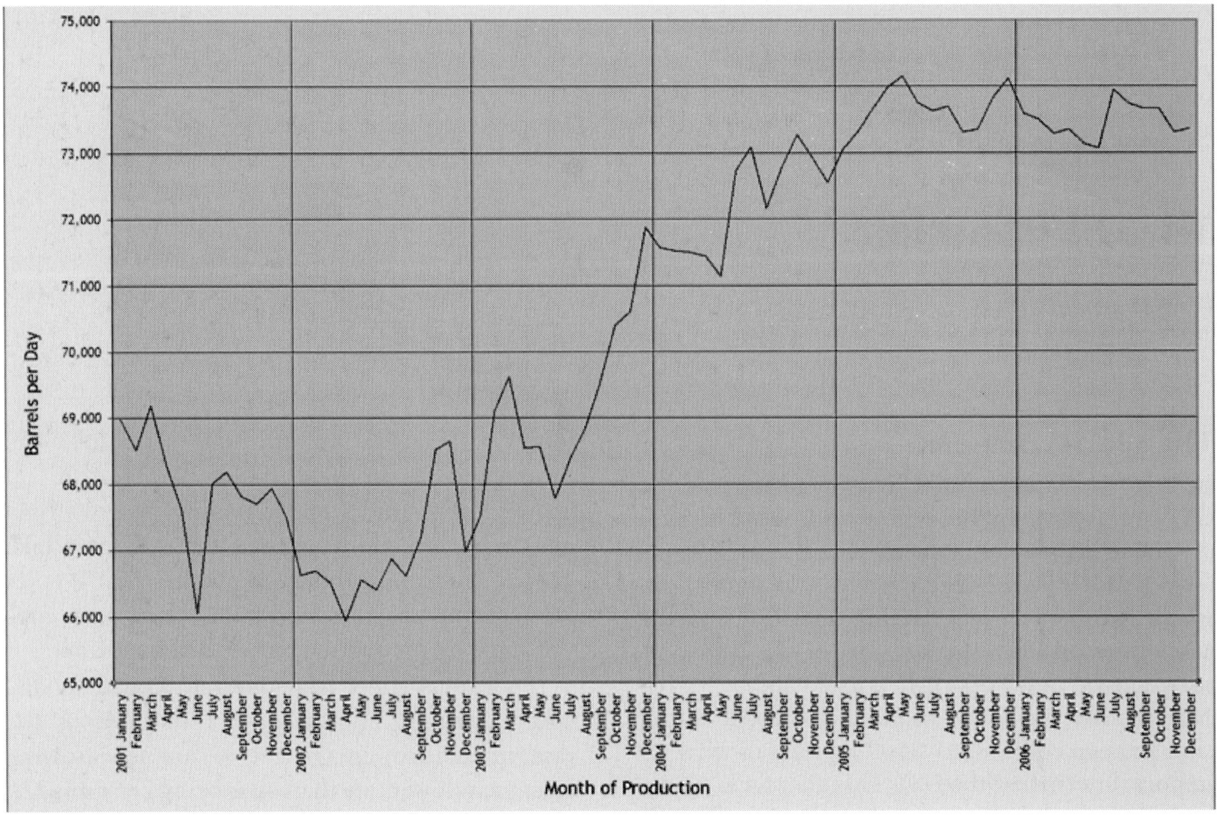

Figure 2: World oil production. Daily average from January 2001 to December 2006 (Source: www.princeton.edu/hubbert/).

An Australian television show on peak oil (ABC Catalyst, November 2005) reported an Australian oil company CEO who had asked the attendees at the previous Australian Petroleum Production and Exploration Association conference whether they thought the global oil production peak had already occurred—half raised their hands. Global oil production declined by about 1% globally in 2006/7.

More conservative estimates from government sources in the UK and US are saying that the peak (including non-conventional deep-sea oil) will be sometime around 2010 to 2020 (Morita *et al.* 2001; IEA 2006). Michael Pacheco from NREL National Biotechnology Center says that 'we need to start working toward replacement fuels 20 years before that peak' (Consumer Reports 2006). So, whatever the date of the peak (and it is probably earlier rather than later), we are already well behind schedule in responding adequately to this challenge.

Oil companies tend to give statistics on oil availability based on how long it will last at present rates of consumption. This tends to be around 30 to 40 years, but this is misleading because oil will not keep being produced at the same rate then suddenly stop. The second half of the production cycle is more difficult, so there is physically less and less oil available each year once the peak is passed. In an economy used to increases in oil each year, this is a significant difference. The peak is the critical change point.

Despite OPEC's recent attempts to increase production, global oil seems to be at, or close to, its full capacity. This level may hold for a few years then it will inevitably fall. This is happening just as China and India have entered the global oil market in a big way—China is now the world's second biggest consumer of oil. Supply to consumers in the West will be constrained by physical capacity and by the new players in the global market. The reality is the Golden

Age of oil is over. A large number of academic and popular articles has supported this perspective, including some published in respected trade journals such as *Oil and Gas International* (a number of the key books are listed in the references).

Why has there been a growing awareness and concern around this issue? A number of geophysicists and industry insiders have 'blown the whistle' with their misgivings, especially on how 'real' the oil reserves actually are. Most oil reserves are in the hands of OPEC, but analysts are concerned about the realism of some Middle Eastern reserves (especially Saudi Arabia)—they were able to demonstrate that 'phantom reserves' had been created in the past decades in order to allow higher OPEC quotas on production. Quotas on production were a license to make money in a world willing to pay for more oil. Re-evaluating the known geophysical data by people such as Campbell shows we are close to the peak—if we are not already there—and that oil will soon be subject to major physical shortages.

There are many conferences on this topic now, especially those held by the Association for Peak Oil. A key presenter at their International Oil Depletion Workshop in 2002 was Matthew Simmons—a prominent energy-sector investment banker from Houston—who is author of *Twilight in the Desert*, which shows that Saudi Arabia's reserves are not as large as the world had been relying on. Simmons said:

I have studied the depletion issue intensely for too long now to have any remaining doubts as to the severity of the issue. But I am still amazed at the limited knowledge that exists, even in the US or within our major oil and gas company's senior management about this topic and its dire consequences. (Simmons 2002)

Most serious scientists worry that the world oil supplies will peak [and then decline]. Peaking of oil cannot be predicted accurately, but the event will occur. Peaking turns out to only be clear through a 'rear-view mirror'. By then, an alternative or solution is too late. My analysis leads me to worry that peaking is at hand, not years away. If I am right, the unforeseen consequences are devastating. The facts are too serious to ignore. (Simmons 2003)

Bentley (2002), from the Oil Depletion Analysis Centre in London, published an overview article on oil in *Energy Policy* and concluded:

The date of this resource-limited global peak depends on the size of Middle-East reserves, which are poorly known, and unreliably reported. Best estimates put the physical peak of global conventional oil production between 5 and 10 years from now. The world contains large quantities of non-conventional oil, and various oil substitutes. But the rapidity of the decline in the production of conventional oil makes it probable that these non-conventional sources cannot come on-stream fast enough to fully compensate. The result will be a sustained global oil shortage.

A report prepared for the US Department of Energy by Hirsch *et al.* (2005) confirmed such warnings:

... the world has never faced a problem like this. Without massive mitigation more than a decade before the fact, the problem will be pervasive and will not be temporary. Previous energy transitions were gradual and evolutionary. Oil peaking will be abrupt and revolutionary.

Dr Samsam Bakhtiari, of the National Iranian Oil Company, provided a pessimistic view of future oil supply decline and of its effects:

Seen from a Middle Eastern perspective, the present global oil situation can be summarised within five major and inescapable trends:

- *The world's super giant and giant oil fields are dying off*
- *There are no more major frontier regions left to explore besides the Earth's poles*
- *Production of non-conventional crude oil has been initiated at great costs—in Venezuela's Orinoco belt, Canada's Athabasca tar sands and ultra-deep waters;*
- *Even OPEC's oil production has its limits;*
- *No major primary energy rival can possibly take over from oil and gas in the medium term.*

Adding up these five trends, one can envision a global oil crunch at the horizon—most probably within the present decade ... It would take a number of miracles to thwart such a rational scenario. Now, a single miracle is always a possibility, but a series of simultaneous miracles is not—for there are limits even to God Almighty's mercifulness. (Samsam Bakhtiari 2002)

Dr Bakhtiari has since published simulations of the World Oil Production Capacity model that suggest that global oil production will peak before the end of the decade, most likely by 2006–07 (Samsam Bakhtiari 2004). Dr Bakhtiari presented seminars in four Australian cities in 2004 and also briefed the Western Australia Cabinet about oil depletion risks—one of the few times a 'peaker' has been able to speak directly to decision makers.

Oil companies are now beginning to break with their traditional approach of only giving 'years of oil remaining'. A paper in December 2002 by Exxon Mobil Vice President, Harry J. Longwell (Longwell 2002) contains the world oil discovery decline curve, which agrees closely with those published in Aleklett and Campbell (2002) and shown earlier. Acknowledgment of this by a major oil company is very significant, because it is clear that oil production can only follow oil discoveries. Longwell also showed a peak of global gas discovery in about 1970, with a sharp decline in natural gas discovery rates since then.

Denial?

Although there is growing awareness of the problem, there is also widespread ignorance and denial in most sectors of the community. Why? The reality of the dominance of economics, with its inherent optimism about resources, is the core reason why most serious people don't take 'peakers' seriously. Obviously oil companies don't want to talk up looming shortages due to the impact on their share price. Twenty-three per cent of Shell's reserves stated from 1997 to 2004 were found to be 'phantom reserves', which led to the resignation of their CEO in 2005 and a plunging share price after they were fined $122 million for accounting fraud by the Securities Exchange Commission.

Governments can also take their share of responsibility for not taking up this issue —it is rarely in their interest to disturb people about something as fundamental as their oil dependence, despite the odd statement like President Bush's 2006 State of the Union comment about the US 'addiction to oil'. But, most of all, the community at large seems unwilling to think about the implications of increasing oil scarcity and just wants the price to be kept down. An article in *The Guardian* (2 December 2003) called 'Bottom of the barrel—the world is running out of oil – so why do politicians refuse to talk about it?' concluded:

Every generation has its taboo, and ours is this: that the resource upon which our lives have been built is running out. We don't talk about it because we cannot imagine it. This is a civilisation in denial.

This is an 'elephant in the bedroom' issue. We prefer to go on with life as though it weren't there. It challenges every aspect of life. We have spent the past 50 or more years building our cities and rural regions around the free availability of cheap oil.

Are there alternatives to oil?

The alternatives to oil are set out in Box 1.

The response from most economists to peak oil is that supply and demand will create the necessary change and these alternatives will seamlessly fill the gap— technological modernism and the idea of substitutability of resources due to price, is still the most powerful belief of our age. 'The Stone Age didn't end with a shortage of stones' is a statement frequently heard when oil depletion is debated. However, there seems to be something quite different and challenging about oil as it is so fundamental to everything we do and is likely to involve a considerably more expensive set of alternatives that have not yet fully emerged despite 30 years of 'oil crisis' in the global economy.

BP Exploration Manager Richard Miller (2004, p. 10), in response to the statement about 'stone shortages', said that:

This is the classical economist's view: something will turn up, when the price of oil is high enough, because something always does. But there isn't anything conceivable that could replace conventional oil, in the same quantities or energy densities, at any meaningful price. We can't mine the oil sands in sufficient quantity because there isn't enough water to process them. We can't grow bio-fuels because there would be no land left to grow food.

Box 1. Alternatives to oil.

1. **Non-conventional oil,** which is mostly from deeper wells and some heavier oils (this is usually included in most scenarios, even though it still requires considerable technological development to extract economically)
2. **'Dirty oil',** which is from tar sands and oil shale and which have serious environmental problems as well as requiring huge amounts of water and natural gas to extract them from sand and rock, making them large greenhouse contributors (these are also usually included, but are only small contributors)
3. **Biofuels,** such as ethanol from corn or sugar cane, and biodiesel from soybean and waste oil (promising, but limited as the source soon creates conflicts with food (see below); there needs to be a breakthrough in cellulose to fuel technology)
4. **GTL and CTL** (gas-to-liquids and coal-to-liquids), which are known technologies. These are small operations that could be increased (though each takes a lot of energy to create the diesel, is capital intensive and creates greenhouse gases);
5. **Hydrogen,** which requires an energy source to split it from water, can then be used in hydrogen-fuel-cell vehicles. However this is not yet past the experimental phase in most vehicles and would require a complete overhaul in infrastructure to provide a reliable and safe supply. In the long term, this technology is likely to be favored as long as it comes from renewable sources.

Solar, hydro, wind, and geothermal don't yield enough energy, hydrogen (from water) takes more energy to make than it can yield, and nuclear fission and fusion are presently off most political agendas... When oil gets too expensive, surviving Americans will still obtain energy from alternative sources, but in much smaller amounts and at much higher prices.

The true size of the gap that is opening up between demand and supply gives some idea of the task that is required if we are to continue by switching to alternatives. The amount of energy required to make up the shortfall by 2030 is equivalent to 6000 500MW power plants. None of the current alternatives get anywhere near being able to provide even a reasonable proportion of this.

In the US there has been a sudden growth in ethanol as the possible replacement for gasoline, and biodiesel for diesel. This has meant that production of ethanol (mostly from corn) rose to 4.8 billion gallons in 2006 and biodiesel (mostly from soybean oil and waste grease) to 75 million gallons. However it is worth noting that this is less than 2% of the gasoline consumption and a mere 0.2% of the diesel consumed. Neither ethanol nor biodiesel are significant players yet.

Biofuels have been widely viewed as a possible alternative, mainly because Brazil has achieved a much higher proportion of their fuel from distilling sugarcane (Worldwatch 2006). Lester Brown (2006) has suggested that there is a growing conflict developing between food for people and food for cars. He shows that grain reserves worldwide have been depleted by this sudden alternative use of US corn and that a doubling of ethanol production (which seems almost certain) will induce a major crisis in the price of grain—and this is occurring when nearly a billion people in the world remain malnourished. In 2007, 30% of the US grain crop went into biofuel—leading to a tripling of world grain prices. To convert the entire grain crop of the US to biofuel would produce just 12% of the gasoline and diesel supply in the US, but, in any case this will not happen as the amount of food used to make fuel will be tightly regulated.

The UK Minister for the Environment from 1997 to 2003 Michael Meacher says:

Alternatives like biofuels, ethanol or biomass can play a marginal role but nowhere near on the scale required. When the oil runs out the economic and social dislocation will be unprecedented. (Meacher 2006)

He goes on to quote Exxon Mobil's John Thompson who foresees that by 2015 we will need to find and develop 8 out of the 10 barrels of oil (and gas equivalent) that we are now producing from. This is just 'not available', he concludes.

What all this suggests is that conventional oil and natural gas are in a league of their own. As they become less available, the alternatives are going to take a lot more effort to provide and hence are bound to be more expensive. The era of cheap oil seems to be incontrovertibly over.

Alternatives, such as the 'dirty oils' and coal to liquids, are likely to be highly marginal because it takes so much energy to produce them. The Canadian tar sands, for example, require substantial quantities of natural gas to extract the oil from the sand, and hence require subsidies. And this does not include the external costs of their environmental impacts or their greenhouse gas emission impacts, which are substantial.

The biofuel alternatives are also quite marginal because substantial energy is needed to distill them from their carbohydrate base. Some potential exists for reductions in this ratio if the conversion of cellulosic material (agricultural and forestry waste, for example) to ethanol can be commercially developed (WorldWatch 2006).

Critiques of peak oil

For the past few decades of debate on peak oil, the International Energy Agency (IEA) has been a peak oil sceptic. The IEA has always said that there is much more oil to be found that will enable us to proceed for 50 years as we have been doing. In the IEA (2005) report, 'Resources to Reserves: Oil & Gas Technologies for the Energy Markets of the Future', Claude Mandil, the executive director, says:

Soaring oil prices have again spotlighted the old question. Are we running out of oil? The doomsayers are again conveying grim messages through the front pages of major newspapers. 'Peak oil' is now part of the general public's vocabulary, along with the notion that oil production may have peaked already, heralding a period of inevitable decline. (page 6)

Mandil, however, says 'the IEA has long maintained that none of this is a cause for concern' Then the report looks at the data and says most countries outside of OPEC:

...have passed their peaks in conventional oil production, or will do so shortly...Their world is one of maturing oil fields. Their exploration and production costs are typically higher but they limit OPEC's monopoly effect, thus operating with smaller margins. Cost reduction is therefore a constant concern. Proven reserves to production ratios are small, averaging around 15 years and production in the older fields is declining. (page 46)

This is the exact problem that 'peakers' have been trying to highlight for some time. The report goes on to confirm that the peak oil theory did describe accurately what happened in the US in the 1970s:

The striking success of Hubbert in predicting the peak of US production suggests that such conditions were more or less met in the US during that time period. (page 42)

This has never been admitted before by the IEA.

The report then questions the current relevance of Hubbert to global oil production suggesting that:

The controversies surrounding peak oil in the literature revolve around four main points. Does the Hubbert model apply to oil production worldwide? If the Hubbert model does apply, when will the peak in worldwide oil production be? What happens after the peak? How fast will the decrease of production be? What role does technology play in such models? (page 42)

After admitting that a peak could be approaching (maybe even as early as 2010) the report opts for new technologies and increased investment as the way to counter any production inflection. But the level of investment suggested as being necessary is almost beyond belief. By 2030, it will need around $5 trillion—or about half a billion dollars a day between now and then. It is hard to imagine such numbers, but $5 trillion is half of the entire Gross National Product of the United States. The report concludes that:

Neither private enterprise nor national companies necessarily have the incentive to assume the risk ... Furthermore, private industry cannot be relied upon to invest in research on technologies that are too far from being economical. (page 119)

If this is the good news about peak oil then it is understandable why many people are getting worried. The IEA have only confirmed the unease that we have about the oil situation. The peak oil crisis to them is a capital crisis to create more oil from expensive, hard-to-get sources and various alternatives. They, too, can see that the era of cheap oil is ending. The IEA have finally recognised that *conventional oil is peaking* and may decline by as much as 5%. This is a remarkable admission. But the IEA then suggest that, unlike all the other oil fields that have been declining, the next batch will be pushed much harder so that they will not decline in the traditional manner. They believe much more can be squeezed out of conventional wells to ease us through the next few years and then the alternatives will flow. The potential for pushing conventional oil was questioned by critics of this report (such as Brown 2006), but the amount of alternative fuels that can be produced to fill the looming gap received the biggest criticism (Cohen 2006; Kunstler 2006).

This is the motivation for casting this chapter around peak oil. The only real solutions being proffered by groups such as the IEA are part of the problem. I would suggest that there are many more cost-effective solutions facing us in our cities and rural regions. A capital crisis that is beyond our capability is a real crisis for the future of our cities—we have pushed our cities into a new realm of potential vulnerability, if not possible collapse.

Natural gas ... saviour or same problem?

Natural gas is the obvious transition fuel to help ease the problem of oil depletion. It can be converted to diesel, as well as being used in vehicles itself. Natural gas has already been used to replace heating oil in many homes and to replace oil used in commerce and industry. In Australia, oil went from 57% of the fuel mix in 1980/81 to 48% in 1997/98, while natural gas went from 13% to 20% as boilers, kilns, stoves and heaters all shifted to natural gas (Productivity Commission 2005).

The next stage would appear to be to switch to greater use of natural gas in transport because trucks and trains can use CNG or LNG in their diesel engines and cars can be switched over as well (particularly if the manufacturer makes them as standard, as occurred in Sweden when the government committed to gas cars for their vehicle fleet). The attraction is that natural gas distribution infrastructure is already in place, although suitable filling stations are not in place.

The conversion to natural gas is an obvious step in places such as Australia where there is a good supply of natural gas available. However, in Europe and in the US this is not the case. Europe is sourcing natural gas from far away places in the east and there are already some signs of an OPEC-like protection of the resource developing. In the US, natural gas has already peaked and it is now looking to import gas using LNG tankers—starting an overseas dependence similar to when local oil production peaked in 1970.

Global natural gas production has had similar estimates on its peak to oil. They range from 2010 to 2030 (for example, Fleahy 2002; Caruso 2003). The peak in discoveries was in the late 1960s to early 1970s, so it seems to be following the same pattern as oil. It is not surprising that oil and natural gas have similar patterns as they have similar geological origins in marine sediment (unlike coal, which comes from ancient forests). Also, oil prices and natural gas prices are closely linked, so as oil goes up in price, natural gas is likely to follow. Natural gas can only be a small part of the transitional arrangements for oil; it cannot be seen as the long-term replacement.

With natural gas in short supply, the world must come to grips with another energy source for use in commercial and residential heating, cooking, water heating and even electric power ... as that has replaced oil in many parts of the world. Natural gas is much more critical in industry and is likely to attract a premium for its use there. But the priority now must be to build houses and offices that can use solar and renewable fuels, as well as designing to use less fuel overall. This will be one of the key focus points for the chapter, along with how petrol and diesel can be reduced in transport.

Industrial and petrochemical uses of oil and natural gas—freight, diesel and aviation—would appear to be premium applications of oil. As oil peaks, these premium uses will mean that the decline in availability for private transport and buildings will be even more obvious, and the need to do something about replacing them will become even more critical.

It is sometimes stated that nuclear power and coal can take up the slack. However nuclear and coal are used to produce electricity and bear little relationship to the issues being considered here.

How quickly will the decline set in?

As suggested in Figure 4, the conventional oil decline could be as much as 5% per year. Campbell (2001) suggests that our goal should be to cut demand by 5% per year through to 2050. Even if a more conservative 3% per year decline in availability of conventional oil and natural gas became the normal setting for cities, what would this mean? To begin with, there will be a clear case for some users to continue to use oil and gas as they are so critical to the functioning of cities. As outlined above, it is likely that cities will have to use dramatically less oil and natural gas in buildings and transport because other users capable of paying premium prices will take up the supply. Thus, instead of growth in using these fuels for transport and buildings, we are facing perhaps more than 5% per year declines. This will mean that not only will the use of alternative fuels be essential, but also we will also need to better manage the design of cities for these functions. We will rapidly need different buildings that don't need as much heating or cooling, and cities will need to be configured differently and rapidly to reduce the need for cars.

Responding to this challenge would be necessary even if the oil crisis was not happening (see other chapters about health and environmental issues associated with car dependence), but it makes the agenda of reducing dependence upon the car and making more use of the sun even more vital, if not critical, to the future of cities. If we don't respond then what is likely to happen?

City futures

Cities have grown strongly in the age of cheap oil. Cities are presently growing globally at 2.3% per year and rural areas at 0.1%, which means people are still leaving the country for the city. Today around half the world's population lives in cities. Nearly all cities participate in a global economy that has been built around the availability of cheap oil.

There are no models we can easily use to understand how cities will manage in the age of reduced oil availability. All Australian cities have recently had strategic planning studies done for the next 30 years of development. Although the studies have recognised that there is a need to reduce motor vehicle dependence and save on oil, they have not intervened in any radical way to stop oil-consuming behaviours, and they have assumed continuing growth in population and jobs—growth that peak oil could throw into doubt. What will happen to such cities as oil peaks? I am suggesting four possible scenarios.

Scenario 1: Collapse

Many forecasts of the consequences of peak oil are taking on an increasingly apocalyptic tinge. The work of Jared Diamond (2005) on how civilisations in the past have collapsed is often related to this issue. A Republican Senator, Roscoe Bartlett, after discovering peak oil said 'Civilisation as we know it will end soon.' Websites that discuss this possibility talk about the 'die-off' scenario where billions will die as the world's cities are unable to adapt to the decline phase of the oil cycle.

I would not want to discount this as a possibility—history is littered with cities that did not adapt. But I would point out that there are many cities that use hardly any fuel (such as Chinese and Indian cities at around 2 GJ per person per year), many that are very wealthy use only a modicum of oil and could easily adapt to almost nothing (such as Tokyo and Barcelona that use 8 GJ per person) while most US, Canadian and Australian cities use frightening amounts of fuel (Australian and Canadian cities average around 30 GJ, US cities average 56 GJ with Atlanta the highest at 103 GJ per person; see also Newman and Kenworthy 2007).

Thus, some cities could easily collapse, but most are likely to be able to adapt. The emerging subprime mortgage meltdown in the US has started the collapse of some outer and fringe suburbs, partly because people cannot afford to re-pay their mortgages due to increased loan repayments in an environment of increasing private transport costs.

Scenario 2: Ruralised City

Some believe that the people in our cities will respond to this possibility of 'collapse' by dispersing to then create a more sustainable semi-rural lifestyle where communities will be responsible for a large proportion of their own food. This would be reflected in suburban agriculture, permaculture villages or hobby farms (Trainer 1995; Gunther 2001; Holmgren 2002, 2005) where most of the economy we now see would be devolved down to small groups or even individual households. It is a ruralisation of cities. Holmgren (2005) sees the suburbs as a farming opportunity waiting to happen:

Suburban sprawl in fact gives us an advantage. Detached houses are easy to retrofit, and the space around them allows for solar access and space for food production. A water supply is already in place, our pampered, unproductive ornamental gardens have fertile soils and ready access to nutrients, and we live in ideal areas with mild climates, access to the sea, the city and inland country.

Fundamental to this scenario is population decline—as a quick calculation shows that the available land to enable small-scale production would quickly disappear if all of our cities were pushed out onto agricultural land. The transition is never described but it would seem that the 'Killing Fields' would be a playground by comparison. It is sometimes euphemistically called 'the die-off scenario'. It is, however, often discussed with some relish—the anti-population and anti-urban movements often link together to imagine a ruralised city as the preferred future. Peak oil is thus sometimes perceived as a good thing. Such ideas are rarely challenged and their currency grows daily under the frightening scenarios of peak oil.

I do not believe that scenario 2 is likely. The broad sweep of history shows that although cities have collapsed as they depleted their soils, or were unable to manage their settlements or were destroyed by invaders, they have not gone back to a 'rural Eden': cities tend to be rebuilt, and have endured and grown.

The reason we rebuild and adapt cities is that our choices for returning to nature are very limited and, to most people, are not acceptable. Not only do we not want to become totally dependent on foraging or hunting for our food, we mostly do not want the responsibility for food production at all. The attraction of doing other things, which are only possible if we are freed from food production, drives us to cities.

The example of Havana in Cuba is often presented as a city that had to develop its own agricultural response to being cut off from oil when the Soviet regime collapsed. This was an important phenomenon, but the reality is that Havana kept going as a city it did not just evolve into a 'permaculture paradise'. We are going to have come to terms with a new kind of agriculture and a new kind of city, but we are not likely to reverse 13,000 years of urban history. Nor would we want to—the consequences for rural production as occurred under the forced ruralisation programs of Mao and Pol Pot were disastrous.

Cities are our likely habitat for the foreseeable future. We need to constantly adapt our urban technology and lifestyles to ensure cities are more sustainable in their resources, in their ecological base and in their human livability. They could collapse if we don't adapt. In the same way, rural production will still need to be the source of food, minerals and materials. Such activity, however, will need to adjust to being done without oil, or it also will collapse. The core argument of Jared Diamond is not that physical limits exist, but that our ability to adapt to these is the real test of our civilisation.

Scenario 3: Divided City

The response in this scenario is that some people (mostly those with the economic means) will move to centres where electric transit is available and destinations are easily accessed by walking. Highly efficient, healthy buildings full of solar technologies ensure that those who can afford to can live even better than before. Outside these dense eco-centres are dispersed suburbs, which become poorer and poorer—descending into a kind of collapse that is depicted in the *Mad Max* films with petrol being the resource that warring gangs fight over (George Miller's movies were apparently inspired by reading about peak oil).

There are many signs of this possibility occurring. The wealthy are moving to central locations, especially to transit-oriented-developments (TODs). The poor increasingly have to pay a higher and higher proportion of their income on transport (STPP 2005). The market will move quickly to enhance this option as fuel prices rise and we are beginning to see the first signs in the US of abandoned outer suburban developments and rapidly accelerating inner urban enclaves of wealth. This Divided City is not hard to imagine and could become significantly worse quite quickly.

Scenario 4: Resilient, Sustainable Solar City

This is obviously how we must try to move. It is a city where the things that the wealthy are seeking in the Divided City are provided for across the whole city, which has walkable centres being built in all of the suburbs that are linked by a quality mass transit system. It may mean abandoning some of the scattered rural hobby farms around cities, where no real urban services can be provided without excessive driving. These areas are seen to be ideal land uses by some in the permaculture movement, but invariably they are heavily car-dependent suburbs, where neither rural nor urban functions survive well. It is a conflict that must be resolved before the price of oil causes real pain for such residents.

So what should we do to achieve such cities?

Take it seriously—create peak oil strategies

Awareness of this issue needs to go through a similar process to the one that the world has adopted over climate change. Global conferences of nations need to be established through the UN, in order to set goals and begin sharing information on how the world economy can be weaned off oil. States and cities need to create peak oil strategies. In Western Australia, in the State Sustainability Strategy, there is a section on oil vulnerability, the gas transition and the hydrogen economy (Government of WA 2003). This described the peak oil issue and began a process of examining what it could mean. It was a major reason why the state backed a hydrogen fuel cell bus trial, as well as a revamped electric rail system, as it helped to begin the transition away from oil. Two cities have now developed peak oil strategies—Portland and Brisbane (City of Portland Peak Oil Taskforce 2007; Maunsell 2007). Both were developed by a peak oil taskforce, which was set up to contain activists, industry and government. The Brisbane Taskforce also included climate change, because the two issues overlap—although peak oil may be more imminent. Such reports need to be followed up with new government structures, such as an Office of Oil Vulnerability.

Plan and build compact cities based around transit.

The world's cities existed and thrived before the age of oil—it is possible to imagine them transitioning to a future where oil is no longer needed. Car dependence is the problem that drives oil vulnerability, so cities must plan and build to overcome this (Newman and Kenworthy 1999).

The solution to car dependence is a combination of transport infrastructure and land-use policy, as well as household education programs (such as TravelSmart) that have already successfully reduced car use. This set of policy approaches has been recognised by all Australian cities in their metropolitan strategies, and all suggest a policy of centres and transit-oriented corridors that can provide travel alternatives in every suburb.

The major problems are not in the central and inner-city areas—as these have similar fuel consumption per person to Asian and European cities (6 to 17 GJ per person per year), they are comparatively well off in terms of public transport infrastructure, and they are dense and mixed in their land use. They will easily adapt to peak oil. However, the newer suburbs built in the past 4 or 5 decades in the outer areas of cities across the world are heavily car dependent, with fuel consumption similar to US cities (average 56 GJ per person per year). There are significant equity issues here, because the wealthy live mostly in inner areas and the poor are increasingly trapped on the urban fringe. There are signs that this is now happening in US cities, with some outer area households using 40% of their income just to travel around to jobs and services. This will become intolerable after peak oil.

The need for more compact, mixed-use development in centres can also assist with the need for buildings to use less energy. Newton *et al.* (2000) have shown that, in Australian conditions, denser development will also save on building energy—mainly through a shared insulating effect. Direct attempts to increase these savings, and to use green materials, green appliances and renewable power in the buildings, will also be required.

This strategy also requires ensuring social housing initiatives are incorporated in any new centres and a priority program in public transport infrastructure for the middle and outer suburbs. Extensions of electric rail lines are the obvious way to go along, with integrated local buses that can provide a service at least as quick as that provided by cars. Time savings will remain a bigger factor in determining mode of transport, even under very high fuel costs.

The Australian House of Representatives Report on *Sustainable Cities* (2005) recognises all of these things and recommends that infrastructure funds (especially for rail) be provided for cities, particularly in the middle and outer suburbs. Most national governments do this—even US cities—and all countries need such a mechanism to cope with peak oil. Partnership funding of the required rail systems and integrated transport programs would follow if an Urban Infrastructure Program was begun.

Perth's new rail system is unusual because its $2 billion cost (which has given the city a 180 km modern electric rail system with 72 stations) was built without any federal funds, although the Freeway it passes down was funded almost entirely from federal coffers. This railway (which is much cheaper than in many other cities because government owned the land and only two bridges had to be built) has been justified over many elections as a way of 'oil-proofing' the city. There are many new developments now being planned around its stations to take advantage of this insurance and obvious amenity, which will become even more critical in the future. The new rail system has removed 100,000 cars from the road system. This figure is set to double again in the next decade. But there is still much of Perth that remains highly vulnerable to peak oil—just as in most Australian and North American cities.

We cannot afford to build further and further out from cities because this inevitably increases car dependence. Most city plans now recognise that the main task is to redevelop in the present urban area, especially in Transit Oriented Developments around rail lines. Cities need to be more urban and the countryside more rural. However, there are many still who believe cities should be dispersed into rural areas as outlined above. Getting serious about oil depletion means that rural lifestyles on the edge of the city cannot be facilitated and subsidised as they have been. There is a real clash between those who want to ruralise cities and those who realise the problems this creates in car dependence.

This does not mean that we do not try to create local eco-villages where people in cities can be more reliant on each other, with reduced oil consumption as a result. Such places—like Christie Place in Adelaide, Somerville in Perth or Vaubun in Frieburg—are essential to help us through this transition. But, if they are used as an excuse to extend the city into areas that are better left rural, then they are not going to help.

Rebuild peri-urban agriculture

A lot of the literature on peak oil suggests that the biggest impact is going to be on agriculture. Certainly agriculture has become very dependent on diesel. In the analysis above I have tried to show that I do not expect our cities to begin taking over from rural producers in the production of food and fibre. In fact, the loss of good agricultural soils to fringe urban suburbs and hobby farms has to stop because they are generally very unproductive as well as being highly car dependent. The fringe areas of Australian cities, where agricultural soils have been sacrificed to scattered suburbs and rural hobby farms, such as the Central Coast north of Sydney and Mornington Peninsular in Melbourne, exhibit three to four times the average fuel consumption per person and eight times that of the inner city. If we are serious about oil depletion and sustainability in general, we need to ensure that the countryside is more rural as well as the city being more urban. This would also sustain the current attractiveness of Australia's cities and rural hinterlands from a tourist point of view.

One of the ways we can do this in Australian cities—and I suspect in any city across the world—is to establish horticultural precincts immediately adjacent to cities. These areas of good soil need to be set aside to ensure they are retained in perpetuity for horticulture, rather than always being seen as growth corridors for the next suburb. We can then get serious about recycling wastewater in these areas. Water Corporations cannot invest in the pipes and technology for this important new approach to urban technology unless they have certainty about the future for an area.

Agriculture will need to adapt by growing its own biodiesel, using gas and switching to more efficient rail transport, rather than trucks. This transition will need assistance, but subsidising diesel is not helping. A rapid phase-out of diesel could change peri-urban agricultural oil vulnerability in a 5-year period.

One of the obvious ways that agriculture will need to change is to become more localised. There may be a reduction in choice—as regional produce will be favoured over imported produce—but this can be part of regional identity and the slow food movement, rather than deprivation. Cities should not need to import vegetables from other regions, especially by plane. Wheat and other grains, however, are bulk commodities that can be transported by train and ship at low oil, or no, oil cost. Agriculture needs to be more localised for the particular cities in its bioregion, as well as creating surpluses that are traded for the benefit of regional and urban opportunity.

Facilitate localism

James H. Kunstler (2005) in the Long Emergency says that in response to peak oil 'Our lives will become profoundly and intensely local.' Localism is the required modus operandi for the post peak oil world—just as globalisation was for the cheap oil era. Globalisation of the economy began with the first cities that began to trade beyond their immediate region—probably 4000 years ago—and it will continue under a post peak oil world. But its character will alter as the extent of trade and movement cannot be expected to continue as though its transport costs were unimportant. As mentioned above, there are social movements that are beginning to push us more towards localism anyway:

- the need for local identity and sense of place
- the slow food movement and its base in local foods
- the eco-city movement with its desire to enable local community to be the basis for managing local resources and local infrastructure
- the local enterprise movement that has shown business development is best done with local social capital as its base (Sirolli 2001).

The value of the internet and video conference facilities on our phones will become even more obvious to maintain the global interaction that we have tasted and will not easily give up. But in the same way that governments have facilitated businesses to export globally and have pushed international tourism, we now need to facilitate localism. I would suggest we need an Office of Localism with a program to fund innovations in localism. This can help to fund demonstrations of new urban technologies where there is a need to create:

- distributed energy and water infrastructure (involving less energy for distribution and pumping)
- localised industrial ecology of businesses that can share their wastes as resources or work together to ensure local resources are used and re-used
- local food linkages that link between peri-urban growers and urban communities to take direct supply of whatever is fresh
- local enterprises that can be facilitated based on local resources and talents
- local tourism that can be marketed to local people.

Regulate for the post-oil transition

A systematic review of regulations will show that at present we subsidise oil consumption, whether by fuel rebates to producers, tax rules on cars and fuel, the subsidy on land development at the urban fringe (around $40,000 per block in most Australian cities without considering health and education services, see Newman and Kenworthy 1999), as well as the subsidy given to road users in the form of state-based road-building grants. Proactive regulation is also required to phase out the excessive use of four-wheel drives and other gas guzzling cars with a clear phase-in program for gas-based electric hybrid cars.

The past 20 years saw a move away from regulation on vehicles and the result was declining overall fuel efficiency in the fleet. This cannot be allowed again—we must regulate for motor vehicles to transition away from oil—the first signs of this are now apparent in California. Governments can begin by regulating their own fleets.

Aviation is a special case. There seems to be no alternative to oil on the horizon. The only solution it seems will be to allow gradual price increases to reduce unnecessary travel, to switch to fast trains for medium distance journeys, and to do more and more interaction by internet conferencing rather than face to face (including family events).

However, there seems to be a case for ensuring that aviation has some priority on remaining fuel—this would also require some kind of regulation.

Conclusions

Australia—and probably most advanced countries—has developed highly complex scenarios for dealing with terrorism. There are no such scenarios for dealing with oil vulnerability and its implications for our cities. Strategic analysts and policy leaders need to take oil depletion seriously and see what must be done in short-, medium- and long-term scenarios for reductions in oil supplies. A process was started with the development of thinking about the hydrogen economy, but this is only in the long term. We need to see how we can reach the future without oil in a series of steps in urban development. We cannot afford to depend on the market to handle this because it will always seem that cheap supplies of oil from the Middle East are the only thing we need to worry about and, while they were expanding, we were not in trouble. Global supplies of oil are not expanding any longer and we cannot just hope that they will continue to grow again. A new urban world is emerging where we must adapt to using less non-renewable resources and using different technology and different urban forms or our vulnerability to oil will be seriously exposed.

References

Bentley RW (2002). Global oil and gas depletion: an overview. *Energy Policy* **30(3)**, 189–205.

Barlow R, McNelis B and Derrick A (1993). *Solar Pumping.* IT Publications, Intermediate Technology Development Group, London.

Brown L (2006). *Plan B 2.0 Rescuing a Planet Under Stress and a Civilisation in Trouble.* W W Norton, New York.

Campbell CJ (1991). *The Golden Age of Oil, 1950–2050: the Depletion of a Resource.* Kluwer, Dordrecht.

City of Portland Peak Oil Taskforce (2007). *Descending the Oil Peak: Navigating the Transition from Oil and Natural Gas.* Office of Sustainable Development, City of Portland, Oregon.

Cohen D (2006). Inside the IEA's medium term oil market. *Energy Bulletin*, **25th July**. <http://www.energybulletin.net/32514.html>.

Consumer Reports (2006). *The Ethanol Myth.* October, pp. 15–19. <http://www.consumerreports.org/cro/cars/new-cars/ethanol-10-06/overview/1006_ethanol_ov1_1.htm>

Diamond J (2005). *Collapse: How Societies Choose to Fail or Succeed.* Viking Books, New York.

Deffeyes KS (2005). *Beyond Oil : The View from Hubbert's Peak.* Hill and Wang, New York.

Deffeyes KS (2007). *Beyond Oil: The View from Hubbert's Peak*, January 19th, <http://www.princeton.edu/hubbert/current-events-07-01.html>

Government of Western Australia (2003). *Focus on the Future: The State Sustainability Strategy.* Department of the Premier and Cabinet, Perth.

Gunther F (2001). Fossil Energy and Food Security. *Energy and Environment* **12(4)**, 253–275.

Hirsch RL, Bezdek RH, Wendling RM (2005). *Peaking of World Oil Production: Impacts, Mitigation and Risk Management.* Department of Energy/National Energy Technology Laboratory, February.

Holmgren D (2002). *Permaculture: Principles & Pathways Beyond Sustainability.* Holmgren Design Services, Hepburn, Victoria.

Holmgren D (2005). *Retrofitting the Suburbs for Sustainability.* Published on 30 Mar 2005 by CSIRO Sustainability Network. Archived on 5 Apr 2005 by Energy Bulletin (energybulletin.net). <http://www.bml.csiro.au/susnetnl/netwl49E.pdf>

House of Representatives (2005). *Sustainable Cities.* Parliament House, Canberra.

Hubbert MK (1956) 'Nuclear energy and the fossil fuels'. Report presented before the Spring Meeting of the Southern District, American Petroleum Institute, Plaza Hotel, San Antonio, Texas, March 7–9, 1956.

IEA (International Energy Agency) (2005). *Resources to Reserves: Oil and Gas Technologies for the Energy Markets of the Future.* IEA, Vienna.

Kunstler JH (2005). *The Long Emergency: Surviving the End of the Oil Age, Climate Change, and Other Converging Catastrophes of the*

economic relations can be typified as one of 'financialisation' or 'financial capitalism' (Frankel 2001, Vertova 2006). In many respects, the trade in financial instruments, such as options and futures that are based on commodities, has become as important as the trade in commodities themselves. Much of this trade is speculative in nature. Combined with the greater speed and volume of transactions, this increases the exposure of national economies to the vagaries of international financial investors. The collapse of financial markets, particularly those based on foreign exchange in Thailand, was a key factor in the East Asian financial crisis in 1997.

A second feature of globalisation is the restructuring of trade, foreign investment and production through the growth of transnational corporations (TNCs). TNCs are companies with subsidiaries or affiliates in more than one country. Their growth is a key feature distinguishing current forms of international commerce from those of earlier eras. In the 1970s there were about 7000 TNCs (Buckman 2004). By 2004 there were around 70,000, with over 690,000 foreign affiliates (UNCTAD 2005). In terms of market capitalisation, many TNCs are now bigger than nations (Sheehan 2005).

The growth of TNCs has dramatically altered the organisation of international trade and investment. Not only have TNCs proliferated, but many have grown in size and market share, through mergers and takeovers, so that international trade and investment are increasingly controlled by the largest TNCs. Moreover, much of the global volume of international trade and investment occurs within different branches of the same company. The largest 500 TNCs control 70% of global trade. They are also responsible for 80% of the world's foreign investment (Buckman 2004). This includes foreign direct investment (FDI), which occurs when a parent company within the TNC invests in a new or existing foreign affiliate.

In terms of production, TNCs may have firms producing similar products or services in several countries, or have facilities in one country to produce inputs for manufacturing operations overseas. They may be conglomerates with productive activities in a range of unrelated industries. They may even have financial sources, production facilities, administrative offices and marketing operations in different locations around the world. While the international organisation of production within large corporations is not new, it is the rapid proliferation of TNCs over the last few decades and their increasing global dominance that has dramatically altered the way in which production is organised internationally.

Having the ability to spread their operations around the world allows TNCs to invest in places where the costs of production will be lowest. For example, they can strategically locate their activities where factors of production are most accessible, wages are lowest, tax burdens are lightest and environmental regulations least stringent. So the decisions of TNCs regarding location become key factors influencing the global distribution of industry and employment. For example, many manufacturing industries have relocated from the wealthier nations to less economically developed countries, including to export-processing zones that offer incentives such as reduced tariffs and quotas.

Of course, TNCs are not the only actors that are internationally mobile. Globalisation has also seen the greater mobility of labour, particularly professionals, managers and highly skilled workers. These workers are increasingly sought out by national governments, such as Australia's, with the skilled migrant stream forming a growing proportion of Australia's migrant intake. Unskilled and semi-skilled workers have also remained internationally mobile, albeit restricted more heavily by national immigration laws and border controls, such as the US–Mexico border. Guest-worker programs and international contracting and subcontracting arrangements also facilitate the movement of workers around the globe.

The changing patterns of labour mobility and the increased internationalisation of finance, trade, investment and production give the current era of globalisation its distinctive features. In particular, each of these developments has reinforced the growth and dominance of 'global cities' in both the more-developed countries and emerging markets. These are the cities that increasingly control the international infrastructure for trade and financial flows, as well as being the major destinations for professional and highly skilled migrants (Hedrick-Wong 2007).

Drivers of change

What has been driving these changes? The intensification of international economic relations has been facilitated by rapid innovations in technology, particularly in transport, information and telecommunications. These innovations have dramatically reduced the time and costs involved in international travel and commerce. They have also allowed more decentralised forms of production, such as outsourcing and subcontracting, that can be spread out across the world. The reduction of the temporal and spatial barriers to international commerce creates what David Harvey (1989) calls 'time–space compression', allowing profit-maximising firms to take advantage of new markets and cheaper production costs offshore. Time–space compression has been particularly significant for Australia because of its relative geographic isolation from the major industrialised nations that has, historically, acted to insulate it from full exposure to the international economy.

While technological advancement has allowed the international decentralisation of production, it has also encouraged the spatial concentration of control functions. That is, as production activities have become more dispersed, the need for central management and control has grown. This has encouraged the growth of global cities, which are characterised by the spatial concentration of business services and the centralisation of financial instruments that facilitate the international commodities trade (Hedrick-Wong 2007). The growth of producer services in particular is widely regarded as one of the hallmarks of globalising cities (Sassen 2000).

As well as changes in technologies and production processes, globalisation has also been facilitated by the interaction of the international economy and the political decisions of nation-states. For example, the collapse of the Bretton Woods system in the early 1970s, and the resultant removal of restrictions on capital mobility, has allowed the massive volumes of financial flows that characterise today's international economy. The volume and speed of these flows, as well as their predominantly speculative nature, have limited the capacities of governments in implementing some forms of national economic management, such as independent monetary policy (Vertova 2006). The relative power of capital in bargaining with both labour and the state has grown considerably.

However, to claim that national governments are beholden to international capital is to overlook their crucial and active role in facilitating greater international economic integration. For example, policy decisions made by successive Australian governments in the last three decades of the twentieth century actively dismantled the existing systems of trade protectionism and financial regulation. These changes were inspired as much by ideology as by economic imperative. They reflect the widespread adoption among Australia's politicians and bureaucrats of the ideology of 'economic rationalism', which espouses the benefits of 'free trade' and competitive international markets. This is not to deny the constraints on government policy generated by concerns about capital mobility. Rather, it highlights the active role of governments in facilitating and strategically responding to changes in the international economy.

This distinctive combination of increased capital mobility and the changing role of the state as it is shaped and constrained by those forces has diverse spatial effects, contributing to a changing economic geography that affects both cities and regions. We will now look at these spatial consequences of globalisation.

What are the spatial impacts?

As economic relations become more densely internationalised there are major consequences for the spatial distribution of economic power and resources. Individual nations, and cities and regions within them, compete to attract mobile capital and skilled labour from across the globe. The decisions made by TNCs about where to locate their headquarters and production facilities can have dramatic impacts on local economies, with the geographic patterns of industry and investment generating spatial disparities in employment, income and wealth. The extent to which these changes can be directly attributable to globalisation is impossible to determine, because we cannot isolate its effects from other possible causal factors such as historical patterns of growth, national economic policies or demographic change. However, in general terms, the spatial competition necessitated by globalisation can be expected to intensify regional disparities.

It is possible to identify several key dimensions of this competition. These include spatial competition between regions, between cities and within cities. Examining these dimensions in turn illustrates the ways in which globalisation creates a distinctive set of pressures for social-spatial change.

The ability of TNCs to shift the location of their firms to take advantage of changes in national policy, albeit with a significant investment in new facilities abroad, means that there has been increasing competition to attract and retain firms producing goods and services for global markets. This competition tends to heighten regional disparities, as some regions attract significant foreign investment while others are left behind. Indeed, some regions of the world, including most of sub-Saharan Africa, seem largely untouched by globalisation. The dramatically uneven patterns of regional development have led some to conclude that globalisation is simply a continuation of Western imperialism, with the global redistribution of wealth favouring the already wealthy Western nations (see Vertova 2006).

Within nations, increased capital mobility has often meant that regional economies have become disarticulated. The linkages with other regional economies may be strengthened, but at the expense of severing the internal linkages that provide for stability and diversification of employment opportunities. Local industries may be bought out or destroyed by TNCs. Local economies may become increasingly dependent on decisions made by international businesses with no roots in the area, rather than being integrated into local production chains. As mentioned earlier, the decentralisation of production has been accompanied by the greater concentration of control functions in the major cities. This further entrenches the economic disparities between cities and regions.

Spatial disparities are also heightened between cities. For example, there is intense competition between urban areas to attract the regional headquarters of TNCs. These institutions typically establish themselves in downtown office sites and boost local property markets. Because they don't need to be located near the productive activities of the TNC, they are potentially transient—requiring the state to offer infrastructure and other incentives to attract and retain them. Moreover, the centralisation of control functions is accompanied by the surrounding concentration of associated business services, such as accounting and legal services, that are needed for the effective administration of transnational commerce. Command functions also tend to be concentrated in the strongest financial centres—usually the cities that house major stock exchanges (Vertova 2006). The typical outcome is the increasing dominance of one city in each nation over both regional areas and other cities as it becomes disproportionately integrated into the transnational economy.

While some cities have an advantage as major financial centres, others compete on the basis of lifestyle. For example, strategies for attracting company headquarters and other command functions may involve significant investment in entertainment, leisure and tourism services to appeal to highly mobile professional workers and the multinational business elite. The promotion of cultural services and the increasing commoditisation of culture and leisure also reinforce the competition between cities for consumption expenditures. As manufacturing industries have moved out of the major cities, these cultural industries have become crucial elements in local economic growth.

The shift from manufacturing to service industries has also encouraged urban restructuring within cities. For example, while manufacturing enterprises in the inner areas of major Australian cities have closed down or moved offshore, the outer suburbs are now home to the remaining industrial operations. As a result, the cities have experienced a growing divide between a 'gentrified' inner core and a more industrial periphery. Marcuse and van Kempen (2000) suggest that globalisation has been a key factor in strengthening such divisions along industrial and residential lines.

Not surprisingly, these changes have also affected the spatial distribution of incomes and wealth. As manufacturing operations have moved out of formerly industrial areas, employment in service industries has increased. But employment in these industries tends to be concentrated at the upper and lower ends of the pay scale, with jobs created in high-technology and knowledge-intensive industries, such as financial and legal services, as well as in lower paid jobs such as cleaning and hospitality (Marcuse and van Kempen 2000). The less-skilled workers are most vulnerable to wages pressure created by government policies seeking to attract mobile capital through wage restraint and the removal of collective bargaining and centralised wage determination. At the other end of the scale, the massive incomes of top corporate executives are justified on the basis of having to attract professionals in the highly competitive international market.

These changes in employment and incomes are reflected in the spatial socioeconomic divide. For example, within some cities, the now gentrified inner city areas are increasingly occupied by the high-income workers in knowledge-intensive service industries, while poorer residents have been displaced to the remaining industrial suburbs and the urban fringe (Stilwell 1998). In other cases, wealthier households have left the inner city for the suburbs or for 'gated communities' in search of better housing and a greater sense of security (Marcuse and van Kempen 2000).

Clearly, while there are some common trends, globalisation does not necessarily lead to predictable spatial patterns in cities and regions around the world. That is, 'reasoning from megatrends through social interdependence to spatial patterns should be done with great care' (Marcuse and van Kempen 2000). The remainder of this chapter investigates these broad trends in the Australian context.

Australia's cities and regions

It is important first to consider Australia within the international context. Not surprisingly, foreign investment in Australia has increased markedly over the last three decades as the processes of corporate globalisation have taken hold. Figure 1 shows the rapid growth of foreign investment in Australia since 1980. The growth since the late 1990s has been particularly pronounced, with the total investment doubling from $717 billion in June 2000 to reach $1433 billion by 2006. Foreign direct investment has grown rather more slowly to reach $312 billion in 2006 (ABS 2007).

Clearly, Australia is strongly integrated into the international economy. But where is this investment going and what have been the differential effects on cities and regions? It is not easy to disentangle the effects of the current period of globalisation from more long-standing changes. As discussed in Stilwell (1998), the spatial development of Australian cities and regions has always been shaped by processes of international capital accumulation. The colonies of New South Wales, Victoria, Tasmania, Queensland, South Australia and Western Australia served Britain's quest for additional territory and

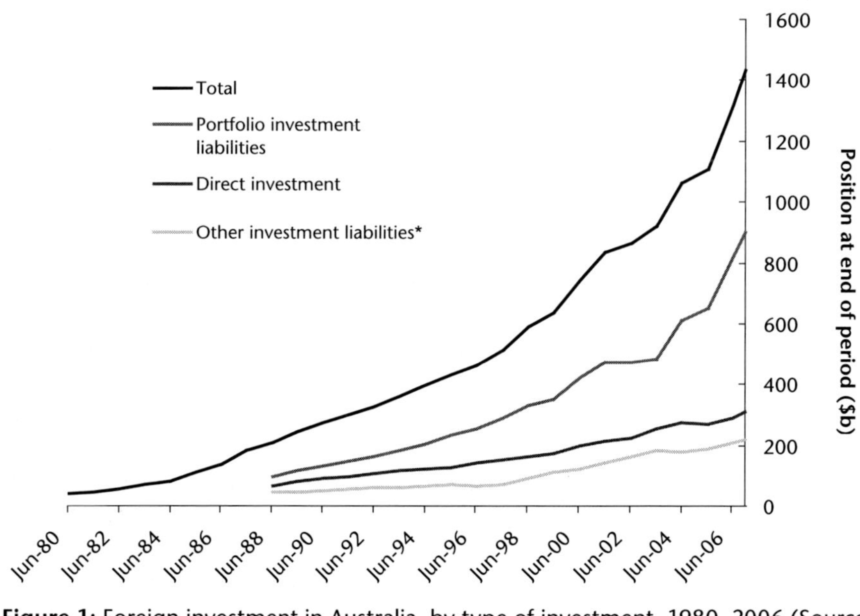

Figure 1: Foreign investment in Australia, by type of investment, 1980–2006 (Source: ABS 2007).

natural resources. The early role of the capital cities within each state involved the management and disposition of the economic surplus generated by resource exploitation in the surrounding regions.

Since that time, the cities have maintained their political and economic primacy, not least because such a high proportion of the Australian population has always lived within them. With the development of manufacturing and service industries in the 20th century, the cities began to generate their own surpluses and were well integrated into the international economy through investment, trade and finance. Of course, local economic and political factors, as well as topographical features, have also been instrumental in shaping Australia's cities and regions. But, over time, the changing interaction between intra-national and international forces has been equally important in re-shaping them. The patterns of this change are evident at various spatial levels: in competition between cities and regional areas, between cities, and within them.

Globalisation has shifted the balance of power between Australian cities and regions. For example, they have been affected very differently by processes of industrial restructuring. The Bureau of Transport and Regional Economics (BTRE) classifies regions into five categories of 'remoteness'—major cities, inner regional, outer regional, remote and very remote—by taking into account their population size and the distance to the nearest urban centre. Examples of places in each remoteness class include: Major cities (Melbourne CBD, Newcastle CBD, Fremantle, Surfers Paradise); Inner regional (Hobart, Noosa, Tamworth, Bunbury); Outer regional (Darwin, Cairns, Gunnedah); Remote (Alice Springs, Mount Isa, Esperance); Very remote (Longreach, Coober Pedy, Kununurra) (BTRE 2006). The share of employment in technology- and knowledge-intensive industries, including business services, declines sharply by a region's remoteness.

Technology and knowledge-intensive industries, as defined by the BTRE, include:

- Education
- Health and community services
- Finance and insurance
- Communication services
- Business services
- Services to mining.

They also include various medium and high-technology manufacturing industries, although it is noted that the technology or knowledge-intensity of these industries is not necessarily consistent across regions (BTRE 2003).

In 2001, 40% of workers in major cities were employed in these industries, compared with 32% in inner regional areas, 26% in outer regional and 22% in remote and very remote areas combined (BTRE 2003: 19). The more remote a region, the higher the share of employment in agriculture, mining and hospitality industries (BTRE 2003).

Table 1 illustrates the change in the industrial structure of employment in regions during the 1990s. The biggest change was the pronounced growth in employment in business services in major cities, indicating an increased spatial concentration of this industry in the cities over time.

These trends have important labour market implications because they are associated with growth in employment opportunities for professionals and skilled workers, and also, at the other extreme, for low-wage workers, such as cleaners and routine service workers. These have been partly met through migration. Over the last few decades, the Australian government has continually changed immigration laws and processes to seek out skilled migrants. In 2005–06, there were 137,745 permanent additions under the Migration Program, of which 91,501 (or 66%) were under the Skill Stream (DIAC 2006a: 2). Permanent additions include those already living in Australia with temporary residency who are granted permanent residence, and new arrivals entitled to permanent residency. Figure 2 illustrates that the proportion of permanent additions coming under the Skill Stream has been increasing over time.

Migration patterns have tended to exacerbate inequalities between regions, with the spatial patterns of settlement being quite marked. In the year 2005–06, almost 36% of permanent additions intended to settle in New South Wales, followed by Victoria (25%) and Queensland (17%) (DIAC 2006a; settler arrivals indicate their intended state or territory of residence but may not settle there). Figure 3 shows the intended state of residence of settler arrivals in the skilled migrant category for the financial year 2005–06. The clear majority of

Table 1: Change in industry's share of employment by remoteness class, selected industries, 1991–2001.

Industry	Major cities	Inner regional	Outer regional	Remote and Very remote
Agriculture	0.0%	–1.5%	**–2.9%**	**–2.6%**
Metal ore mining	0.0%	0.0%	–0.2%	**–3.4%**
TCF manufacturing	–0.6%	–0.4%	0.0%	0.1%
Petroleum, coal, chemical and associated product manufacturing	0.6%	0.3%	0.2%	0.1%
Machinery and equipment manufacturing	–0.7%	0.1%	0.2%	–0.1%
Food retailing	**2.4%**	**2.9%**	**2.3%**	1.5%
Personal and household good retailing	–1.2%	–1.1%	–0.8%	–0.8%
Accommodation, cafes and restaurants	0.8%	0.7%	0.6%	0.3%
Finance	–1.5%	–1.0%	–0.8%	–0.6%
Services to finance and insurance	–0.8%	–0.3%	–0.2%	–0.3%
Property services	0.4%	0.4%	0.4%	0.5%
Business services	**5.8%**	**2.7%**	**2.2%**	1.7%
Government administration	–1.3%	–1.0%	–0.7%	**4.5%**
Defence	–0.6%	–0.6%	0.2%	–0.2%
Education	0.1%	0.1%	0.2%	0.5%

Note: Only industries that experienced more than a 0.5 percentage point change in at least one column are included in the table. Changes of greater than 2 percentage points are in bold.

Source: BTRE 2003, p. 16

migrants intended to settle in either New South Wales or Victoria.

Not only are new arrivals unevenly distributed across the states and territories, but most prefer to settle in the major cities. In an effort to reduce this bias, the federal government has introduced several schemes to attract intending migrants to regional areas or metropolitan areas with low population growth. The Regional Sponsored Migration Scheme, introduced in 1995/96, allows employers in regional areas to sponsor overseas workers to fill positions that they can not fill with local labour. The Skilled–Independent Regional visa was introduced in 2004, giving intending migrants who are ineligible for permanent residency a 3-year visa provided they agree to live and work outside of the major metropolitan areas. After the 3 years, the holder of a Skilled–Independent Regional visa may apply for permanent residency. There are also a number of visa categories that make it easier for intending family-sponsored migrants and business migrants to obtain residency if they agree to settle in regional areas (DIAC 2006b).

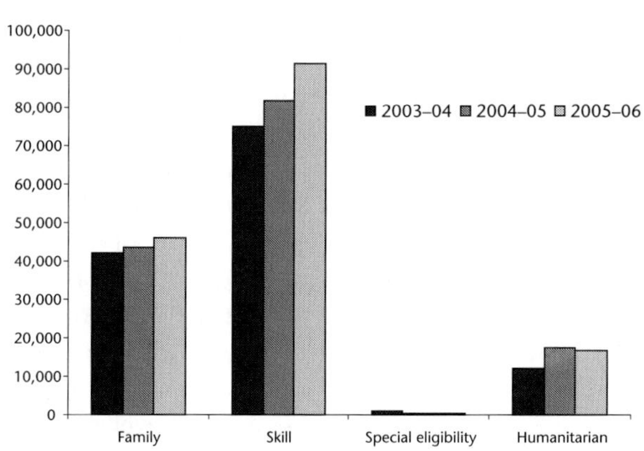

Figure 2: Permanent immigration by eligibility category, financial years 2003–04 to 2005–06 (Source: DIAC 2006a).

As well as contributing to a changing balance of industry, employment and settlement between Australia's cities and regions, globalisation has also reinvigorated competition between the major cities. As in other Western nations, the relative decline of traditional manufacturing industries in Australia's cities has been accompanied by a rapid growth in

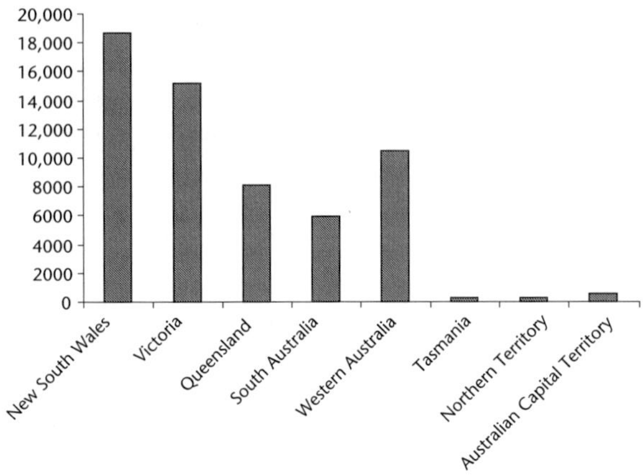

Figure 3: Settler arrivals in Skill Stream by State or Territory of intended residence, financial year 2005–06 (Source: DIAC 2006a).

services such as communications, finance, insurance, property services and business services. But some locations have been much more successful at securing the rewards of that growth. Sydney is Australia's largest city, so it has been well placed to capitalise on the increasingly important service industries, particularly business and financial services. But the competition for international capital has tended to increase its dominance over other Australian cities. For example, Sydney has become Australia's clear financial centre. It has housed the Australian Stock Exchange since it replaced the separate state-based exchanges in 1987. This has attracted further financial capital and elevated Sydney as the city of choice for most of the regional headquarters of TNCs that have established themselves in Australia. For this reason, Searle (1996) argues that Sydney, more than any other Australian city, has become a 'global city'.

Australian cities also compete vigorously to maximise consumption expenditures. For example, as Sydney has de-industrialised, inner-city warehouses and industrial areas have been converted to expensive apartments, retailing, boutique hotels and entertainment venues. International capital has featured prominently in these redevelopments. Similarly, Melbourne's formerly industrial Docklands has been redeveloped into a waterfront entertainment precinct, with the city marketing itself heavily as a destination for leisure and consumption.

Not surprisingly, these changes have also affected the economic geographies within Australian cities. Large regional cities, such as Newcastle and Wollongong, have borne the brunt of industrial decline, with some smaller regional towns also being acutely affected by the closure of particular manufacturing firms on which their economies previously depended. The metropolitan areas have experienced deindustrialisation, with what remains of the manufacturing industries now being concentrated in outlying suburbs, particularly those with residents of predominantly lower socioeconomic status, such as in western and south-western Sydney. Other suburbs, such as French's Forest and North Ryde in Sydney's north and north-west, have become focal points for 'business parks' of high-technology firms, while inner city suburbs have rapidly gentrified.

The desire to attract and retain productive capital has further affected processes of urban restructuring. For example, the push for urban consolidation that encourages higher residential densities, usually justified in terms of concerns with environmental sustainability and infrastructure costs, also frees some capital investment from housing so that it can flow into the internationally traded goods and services sectors.

In Australia, spatial competition over international capital has had clear implications for the distribution of income and wealth. Of course, the spatial inequalities in income and wealth long predate the onset of the current era of globalisation. But, the changing spatial patterns of industry, occupation and employment interact with the housing market to produce a changing geography of socioeconomic disparities. Over time, these patterns become more entrenched, as rising housing costs mean that the wealthier areas become increasingly inaccessible to all but the most affluent elite.

These spatial patterns are compounded by the policies and activities of various levels of government. For example, at the local level, wealthier urban areas provide the greatest revenue-raising potential for local governments. Hence, the affluence of these areas is rewarded with the greater provision of public services. At the level of state governments, urban and regional policies have become geared to beggar-thy-neighbour tendencies that are dominated by the desire to attract and

Table 2: Income differences in Australian regions, 2000–01

Remoteness class	Aggregate Real Taxable Income (ARTI) ($m)	ARTI (percentage of total)	Real Income per Taxpayer (RIPT) ($)	RIPT average annual growth, 1990–91 to 2000–01 (%)
Major cities	248 723	72.9	43 104	2.0
Inner regional	56 371	16.5	35 790	1.6
Outer regional	28 255	8.3	35 187	1.6
Remote	5 026	1.5	39 262	1.5
Very remote	2 112	0.6	39 522	1.2
Australia	341 272	100	40 865	1.9

Source: BTRE 2004, p. 34

retain mobile capital but undermine the state's capacity to tackle local social needs (Stilwell 1998). For example, public funds may be targeted towards new communications technologies and transport infrastructure in order to improve a region's competitive advantage in attracting investment. This may boost the region's standing within the 'national settlement hierarchy' (Newton 1995). But, at the same time, state governments may withdraw funding for social services in an effort to improve their credit ratings—again with the goal of improving business confidence and attracting national and international investment.

There are consequential changes in the spatial inequalities in income. One means of assessing these is by reference to the Bureau of Transport and Regional Economics (BTRE) estimates of the Real Income Per Taxpayer (RIPT) of residents across the country (RIPT excludes people with taxable incomes below the tax–free threshold and people who did not submit a tax return, so is likely to be an overestimate; BTRE 2004). According to these estimates, throughout the 1980s and 1990s the income inequality between Australia's regions grew substantially. This is particularly evident in the growing gap between capital cities and the rest of Australia, and between suburbs within capital cities. Outside of the capital cities, the RIPT of regions has tended to converge (BTRE 2004).

RIPT is above the national average only in Major cities. It is lowest in Inner regional and Outer regional areas. Major cities have the highest average annual growth in RIPT, with RIPT in Inner regional and Outer regional areas falling as a proportion of the Australian average. Hence, there is a 'widening income differential between the major cities and all other classes' (BTRE 2004). RIPT is relatively high in remote and very remote areas, but in the few years prior to 2000–01 this levelled or even fell, perhaps reflecting the trend towards 'fly-in fly-out' employment in mining operations (BTRE 2004). Table 2 illustrates RIPT by remoteness class in 2000–01. It also indicates Aggregate Real Taxable Income (ARTI), showing that almost 73% of the total income goes to residents of the major cities (BTRE 2004). This is disproportionately high, with just over 65% of the total Australian population living in major urban

Table 3: Mean household net worth ($'000) by state and region, 2003–04

	NSW	Vic.	QLD	SA	WA	Tas.	NT(a)	ACT(b)	Aust. (c)
Capital cities	640.6	505.7	402.3	362.1	394.1	374.5	377.0	504.9	503.7
Balance of State	439.1	409.7	371.4	392.7	452.6	289.2			405.1

Notes: (a) Data not available for NT outside of capital city
(b) Capital city estimate for ACT relates to whole of ACT
(c) Includes ACT and NT households

Source: ABS 2006, pp. 37–39

Table 4: Mean value of household property assets ($'000) by state and region, 2003–04

	NSW	Vic.	QLD	SA	WA	Tas.	NT(a)	ACT(b)	Aust. (c)
Capital cities	499.2	357.2	284.4	232.9	275.7	224.5	245.1	349.4	367.0
Balance of state	276.2	218.4	229.2	198.0	226.6	154.4			237.9

Notes: (a) Data not available for NT outside of capital city
(b) Capital city estimate for ACT relates to whole of ACT
(c) Includes ACT and NT households

Source: ABS 2006, pp. 36–38

centres (ABS 2001c; that is, those with at least 100,000 people).

The spatial distribution of household wealth follows a similar pattern to the distribution of income. As Table 3 demonstrates, mean household net worth is much higher in capital cities than in the rest of each state. It is also highest in the largest cities of Sydney and Melbourne.

Not surprisingly, as property forms the major component of most households' net worth, the spatial distribution of property values follows roughly the same pattern, as illustrated in Table 4. Globalisation is not the direct driver of these spatial wealth inequalities, although the influx of foreign capital into urban real estate has been a factor.

Conclusion

The evidence and analysis developed in this chapter illustrates how Australian cities and regions are being influenced by globalisation. They are being restructured in the interests of transnational capital, combined with local interests and with the policies of federal, state and local governments. The consequences are profound, as they result in changes in patterns of investment and industry, employment opportunities, land uses and inequality in the distribution of income and wealth. Issues of efficiency, equity and sustainability are raised. Developing policies that are appropriate for dealing with the economic, social and ecological consequences of this restructuring are necessarily a high priority.

References

ABS (Australian Bureau of Statistics) (2001a). *Australian Census of Population and Housing 2001.* <http://www.abs.gov.au/websitedbs/d3310114.nsf/home/Census%20data>.

ABS (2001b). *Balance of Payments and International Investment Position 1999–2000 (5363.0).* <http://www.ausstats.abs.gov.au/ausstats/subscriber.nsf/0/6C011D5B37246D3DCA256A5300045875/$File/53630_1999-2000.pdf>.

ABS (2001c). *Census of Population and Housing, Australia in Profile, A Regional Analysis (2032.0)*, Table 1.1. <http://www.ausstats.abs.gov.au/ausstats/subscriber.nsf/0/6E673B244F83579CCA257156007B9D31/$File/20320_2001.pdf>.

ABS (2006). *Household Wealth and Wealth Distribution, 2003-04 (6554.0).* <http://www.ausstats.abs.gov.au/ausstats/subscriber.nsf/0/EF361428DC5A4C76CA25715C001374B3/$File/65540_2003-04.pdf>.

ABS (2007). *Balance of Payments and International Investment Position, December Quarter 2006 (5302.0)*, Time series spreadsheets. Table 29. <http://www.abs.gov.au/AUSSTATS/abs@.nsf/DetailsPage/5302.0Dec%202006?OpenDocument>.

BTRE (Bureau of Transport and Regional Economics) (2003). 'Focus on Regions No.1: Industry Structure', Information paper 49, Commonwealth of Australia, Canberra.

BTRE (2004). *Focus on Regions No.3: Taxable Income*, Information paper 54, Commonwealth of Australia, Canberra.

BTRE (2006). *About Australia's Regions.* <http://www.btre.gov.au/statistics/regional/regstats06/Australias_Regions_06.pdf>.

Buckman G (2004). *Globalization: Tame It or Scrap It?* Zed Books, London and New York.

Department of Immigration and Citizenship (DIAC) (formerly Department of Immigration and Multicultural Affairs) (2006a). *Immigration*

Update 2005–2006. <http://www.immi.gov.au/media/publications/pdf/Update_June06.pdf>.

Department of Immigration and Citizenship (DIAC) (formerly Department of Immigration and Multicultural Affairs) (2006b). *Visas and Immigration; Regional Initiatives; Visa Options*. <http://www.immi.gov.au/skilled/regional-employment/visa-options.htm>.

Frankel B (2001). *When the Boat Comes In: Transforming Australia in the Age of Globalisation*. Pluto Press, Sydney.

Hedrick-Wong Y (2007). *The Dynamics of Global Cities and Global Commerce*, MasterCard Worldwide Centers of Commerce™ Insights. <http://www.mastercard.com/us/company/en/wcoc/pdf/insights_Q2_2007.pdf>.

Marcuse P and van Kempen R (2000). Introduction. In *Globalizing Cities: A new spatial order?* (Eds P Marcuse and R van Kempen). Blackwell Publishers, Oxford.

Newton P (1995). Changing places? Households, firms and urban hierarchies in the Information Age. In *Cities in Competition: Productive and sustainable cities for the 21st century* (Eds J Brotchie, M Batty, E Blakely, P Hall and P Newton) pp. 161–190. Longman Australia, Melbourne.

Sassen S (2000). *Cities in a World Economy*. Pine Forge Press, Thousand Oaks California.

Sheehan P (2005). All aboard the big red juggernaut. *The Sydney Morning Herald*, **January 22**, 41

Stilwell F (1998) *Going Global? Australian Cities and the Challenge of Globalisation*. Transnational Corporations Research Project, Department of Geography, University of Sydney.

United Nations Conference on Trade and Development (UNCTAD) (2005). *World Investment Report: Transnational Corporations and the Internationalization of RandD*. <www.unctad.org/en/docs/wir2s005_en.pdf>.

Vertova G (2006) Reinventing space. In *The Changing Economic Geography of Globalization: Reinventing Space*. (Ed. G Vertova) pp. 1–19. Routledge, Oxford.

7

Tourism

Larry Dwyer, Deborah Edwards, Nina Mistilis and Carolina Roman

Cities are always shaped by mobility—or by flows—of people, money and goods. Changes are taking place globally that will influence the future shape of urban destinations. The coming decade and a half should see major shifts in the leisure and tourism environment: reflecting changing consumer values, political forces, environmental changes and the explosive growth of information and communication technology. How urban destination managers and tourism operators respond to these changes will influence the way in which these destinations are developed. Because the future cannot be known with certainty, public and private sector tourism organisations must use today's information to allocate resources, to maintain or achieve competitive advantage and to inform strategic planning for their destinations.

A key element of a sustainable urban tourism industry is the ability to recognise and deal with change across a wide range of key factors. The challenge for urban tourism destination managers is to deal with these changes to achieve and maintain competitive advantage for their destinations. To realise these goals, it is important to know how world events influence consumers and suppliers of goods and services and, consequently, how this shapes urban tourism environments. There is increasing competition in the tourism and hospitality industries—between urban destinations worldwide (between established markets and from new markets), between urban destinations domestically, and between firms within an urban destination. Achieving competitive advantage in times of rapid change requires tourism stakeholders to have a clear understanding of the direction of change and its implications for destination management.

Because tourism is essentially integrated with other sectors in the economy, tourism trends cannot be considered in isolation from key drivers that will shape the world of the future. This chapter argues that in a world in which a number of broad global trends are evident, their influences on urban destinations can be evaluated. In these circumstances, research can play a major role in assisting enterprises that operate at the destination and helping urban destination managers with strategic planning. The first section of this chapter briefly considers five key trends that will have an influence on urban tourism destinations over the next 15 years:

- globalisation and long-term economic development
- social trends
- political trends
- environmental trends
- science and technology trends.

The implications of these forces on urban tourism environments are the subject of the second part of this chapter.

Key trends affecting tourism

Globalisation and long-term economic trends

Globalisation, which is the growing interconnectedness of the world, as reflected in the expanded flows of information, technology, capital, goods, services and people, is seen by many as an overarching 'mega-trend'—a force so ubiquitous that it will substantially shape all the other major trends over the next 15 years. Globalisation is a product of the revolutions in information communication technology (ICT), computing, the growth of free trade, the decline of communism and the democratisation of financial markets. Increasing international connectivity will continue to shape modern life, further diminishing the constraints of physical boundaries and extending the geographical scope of social networks. The extent to which local events and conditions can be attributed to globalisation is not as important as the fact that globalisation is used to explain patterns of change and many features of urban life (United Nations Human Settlements Programme 2005). It is a powerful force shaping national and regional economies, which are interdependently linked as never before.

There are six main factors that will combine to promote widespread economic dynamism and growth and continue to drive globalisation (National Intelligence Council 1999):

- There is political pressure for higher living standards in developing societies. A growing global middle class—now 2 billion strong—is creating a cycle of rising aspirations. Democracy has spread to countries previously under the control of communist, military led or otherwise authoritarian regimes giving political power to formerly disenfranchised citizens. If successful, many countries will enjoy unprecedented prosperity.
- There will be improved macroeconomic policies. In recent years, widespread improvements in economic policy and management are setting the stage for future growth. Inflation rates have been dramatically lowered across a wide range of economies and are expected to remain at low levels. The abandonment of unsustainable fixed exchange rate regimes in Asia and the creation of the European Monetary Union (EMU) are both contributing to economic growth.
- The continued deregulation and liberalisation of air transport and 'open skies' policy is increasingly recognised as enhancing the growth of trade and tourism. Efforts to remove barriers to international travel by means of the ongoing liberalisation of transport, and other forms of deregulation, are expected to continue to remove artificial barriers. Meanwhile, political realignments in the European Union and North American free-trade zone are encouraging greater travel within each region.
- International trade and investment flows will continue to grow, spurring increases in world Gross Domestic Product (GDP). International capital flows, which have risen dramatically in the past decade, will remain bouyant, especially for emerging market countries that demonstrate good governance and economic management.
- The pervasive incorporation of information technology in economies will continue to produce significant efficiency gains in developed and developing societies. The growing two-way flow of skilled workers between the developing world and the West, the increasing size of the computer-literate, information-oriented workforce in some developing countries, and efforts by global corporations to diversify their high-tech operations will fuel the spread of new technologies. In the next 15 years, the greatest benefits of globalisation will accrue to those countries and groups that can access and adopt new technologies. Indeed, a nation's level of technological achievement generally will be defined in terms of its investment in integrating and applying the new, globally available technologies.
- A rapid expansion of the private sector in many emerging market economies—along with deregulation and privatisation in Europe and Japan—will spur economic growth by generating competitive pressures to use resources more efficiently. The impact of improved efficiencies will be multiplied as the information revolution enhances the ability of firms around the world to adopt 'best practices' from the most successful enterprises. This dynamism is expected to be broadly based worldwide, but should be strongest among so-called 'emerging markets'—especially in China and India. However, all countries are integrally locked into the global economy; and no aspiring market leader in any industry can succeed without operating (or establishing networks in) most major industrialised and emerging markets.

Without doubt, more firms will become global, and those operating in the global arena will become more diverse, both in size and origin, more Asian and less Western in orientation. The gradual integration of China, India and other emerging countries into the global economy will result in hundreds of millions of working-age adults becoming available for employment in what is evolving into a more integrated and interrelated world labour market. Labour mobility will increase as the Internet provides greater opportunities to gain jobs internationally. The implications of this will include an increased demand for people with language skills, for language training and for training that provides people with the skills to work in different cultural environments (UNWTO 2002). Those countries with endemic internal and/or regional conflicts and that fail to diversify their economies will be at risk of falling behind economically.

Social trends

Unprecedented demographic shifts are having profound effects on most social institutions. North America, Australia and New Zealand—traditional magnets for migrants—will continue to have the highest rates of population growth among developed countries, with annual population growth rates between 0.7 and 1.0% (National Intelligence Council 2004). With increasing longevity, the growing world population is ageing, especially in rapidly expanding urban areas (UNWTO 2002; National Intelligence Council 2004). However, while there will be an increase in world population overall, declining birth rates will cause a marked fall in the population of developed countries where, within 15 years, one out of every eight people will be 60 years old or over.

Ageing 'baby boomers' will demand products and services that cater to their needs (Weston *et al.* 2001). High standards of public health in developed countries have contributed to increased longevity. This will result in increased demand for a combination of health and travel products and there will be a blurring of working life and retirement in developed countries (Weston *et al.* 2001). Household types are diversifying away from the traditional 'nuclear' family in contemporary western society. In developed countries, family households are getting smaller: sole-parent and single-person households are becoming more common, partly due to lower fertility rates and higher divorce rates. A new definition of 'family' is evolving to include any combination of two or more people living in a domestic household comprising, a minimum of two adults, or one adult and one child.

The growing complexity of 21st century life is driving the need for greater clarity and a quest for a deeper personal sense of meaning, not only in private, spiritual arenas but in all aspects of people's lives, including those of work and consumption. We now talk about the 'experience economy', mirroring these changes in society. Some argue that we have entered a new era where experiences are the economic offerings that are in highest demand and generate the highest value returns (Pine and Gilmore 1999). Consumers in a search of a wider meaning will seek newer, richer or deeper experiences. In terms of values and lifestyles, ethical consumption will continue increasing. The distinction between work and leisure will blur. Society and men will become more feminised, with women having increasing influence on all the key decisions, including those involving travel.

Increasingly, education will be a determinant of success for destinations and businesses. Innovative businesses that are well attuned to their customers needs and staffed with highly educated workers valued as 'human capital', and organisations with external knowledge focusing on organisational culture that enshrines life-long learning, will be the most successful. Also relevant are social infrastructures that maximises opportunities for individuals and businesses to be innovative, learn and develop knowledge, skills and access knowledge and creative age services.

There is a worldwide trend towards urbanisation. More than 95% of the additional world population will be located in developing countries—nearly all in rapidly expanding urban areas. By 2020, more than 60% of global population will live in cities. Over the next 15 years the number of very large cities—megalopolises of more than 10 million people—will double to about 30 and managing them will be a significant new problem (National Intelligence Council 2004). As a consequence, travellers will tend to favour holidays away from crowds. Subsequently, tourism firms in cities will need to work hard to demonstrate that they are destinations that are worth visiting for more than a short break, and for more than one visit (Ulbrick 2006).

Political trends

Political trends can be grouped under four main themes: international power, security, conflict and health. In terms of international power, the USA is likely to remain an important influence, but with its relative position eroded. An enlarged Europe will wield increased power internationally. Japan faces an ageing crisis, which could challenge its regional status. China and India will emerge as new major global players and will transform the geopolitical landscape. Russia still has international potential due to its gas and oil, but will be limited by social and political challenges (National Intelligence Council 2004).

In terms of security, the key factors that spawned international terrorism show no signs of abating over the next 15 years and the threat of terrorism is likely to become more decentralised, due to Internet use and international migration (National Intelligence Council 2004). Today, the world faces the challenge of confronting networked terrorism. The issues and fallout from ongoing heightened security and terror will continue, and cyber terrorism is expected increase. Security issues will herald stricter border controls, thus creating barriers or deterrents to tourism. In this environment, destination regions and individual tourism operators must maintain safety and security, including cyber security and biosecurity. For the tourist, the destination image based on safety is increasingly important.

Whether real or perceived, risks associated with international tourism place serious constraints on tourist behaviour...it is likely that behaviour is as constrained by perceived risk as it is by actual risk. The tourism industry's vulnerability to terrorism is proportionate to its reliance on peace and stability. (Sonmez and Graefe 1998)

In terms of urban health risks, the steady increase in international travel is a driving force in the global emergence and spread of infectious diseases (Perz *et al.* 2001). Growth in international travel and tourism will bring with it the unprecedented risk of infectious disease and other health-related crises, especially in cities. Economic pressures to promote tourism 'mean that even the most dangerous, infected, and poverty-stricken countries are still vying for tourists. Such nations are often reluctant to report, or are incapable of monitoring, infectious outbreaks' as they don't want to discourage tourism (Ritcher 2003, p. 341).

Environmental trends

Contemporary environmental problems will persist, and in many instances grow, over the next 15 years and beyond. All countries are likely to face intensified environmental problems as a result of population growth, economic development and rapid urbanisation (Dwyer 2004). Tourism is closely linked to the environment. The natural environment and climate are very important in determining the viability and attractiveness of a region as a tourist destination. Tourist developments tend to be based near attractive or unique natural features, enabling visitors to gain easy access. Typically the concern of tourism stakeholders, including researchers, has focussed on how tourism development may provide enhanced opportunities for the effective management of environmentally sensitive areas and the preservation of unique environments, particularly when these are major sources of attractions for visitors.

In recent years, however, the focus has shifted toward the part that tourism can play in reducing environmental pollution and demands on resource use. Climate change will influence which destinations will be preferred by tourists and which ones will cease to be as attractive. As environmental changes take place, the development of 'artificial' (indoor) environments for tourism may be expected to increase. Diminishing supplies of oil will affect fuel costs, which will in turn affect transport costs and hence tourism flows. In addition to its direct effect on natural tourist attractions, such as increasing temperatures, climate change could reduce the profitability of the industry through its effects on water and energy use. Roughly 50 per cent of all energy consumption in hotels is used for air-conditioning, so rising temperatures may affect the 'bottom line' of operators in many areas of tourism operations, including in Australia. Globally, policies to tackle climate change, such as carbon pricing, will inevitably affect the aviation sector. While the consequences for particular destinations remain uncertain, it is likely that long-haul destinations such as Australia will suffer the worst effects.

Science and technology trends

The world will encounter more quantum leaps in information and communication technology (ICT) and in other areas of science and technology. Technology is by far the most important management tool in achieving results and competitiveness in the business environment. Business to business e-commerce (B2B) will grow much faster than business to consumer e-commerce (B2C). Networking is the most important element of the ICT revolution because synergies between departments and organisations reduce labour cost, increase efficiency and enable flexible and differentiated product development.

ICT is fundamental to organisational competitive strategy. Technology must complement traditional ways of competing—the Internet rarely nullifies the most important sources of competitive advantage in any industry (Mistilis and Daniele 2004). The key question is *how* to deploy information technology for competitive advantage. While the Internet provides new opportunities to develop distinctive a competitive positioning, it must be used to build on 'proven principles of effective strategy'.

Knowledge is increasingly recognised as essential to the competitive advantage of any destination and organisation. Tourism's vulnerability raises the need for disaster-management strategies, particularly at the destination level (Faulkner 2001; Mistilis and Sheldon 2006). Knowledge can be collected, stored, processed and disseminated for preventative planning and action at the time of a disaster, and beyond to recovery planning. Such an approach would lead to the development of a *learning* or *knowledge* destination—one with a dynamically changing knowledge base to minimise the impact of disastrous extreme events on tourism and facilitate the best response (Mistilis and Sheldon 2006).

Framework

Each of the global trends has relevance to tourism both conceptually and empirically. Each trend influences the others to a greater or lesser degree and will have varying impacts on tourism destinations. Taken together the trends set the context in which the global tourism industry may be expected to develop to 2020. For example, the incidence of terrorism in the future is dependent on political trends (Tarlow 2002). Social trends are also affecting tourist values and needs (Cetron 2001). These trends provide a useful framework for considering the future demand and supply for tourism and the type of urban destination that will be successful in the future. For all established and emerging markets, the values and attitudes of 'tomorrow's tourist' have implications for both the tourism destination and for suppliers of tourism goods and services. Figure 1 depicts the framework in which key drivers affect the tourism industry. The five megatrends (economic, social, political, environmental and technological) influence three key tourism facets – the destination (destination management), enterprise (enterprise management) and tourist (values, needs, flows). In addition, the influences of the three facets on new product development are all interrelated. The spheres of influences are graphically depicted as circles, indicating the fluidity of the situation. An understanding of the key trends and their potential impacts should guide decision processes and resource allocation in the tourism sector. Each of these trends has sub-components, as well as occasional counter-trends. Only by understanding and acting upon reliable trend forecasts will the tourism industry be able to avoid the most common cause of bad decisions—failure to 'read' the future external demographic, economic, political, technological and natural environment contexts. The effects of the key trends outlined above will vary in different regions and countries. The effects of some trends may be limited to particular industries, segments or even firms.

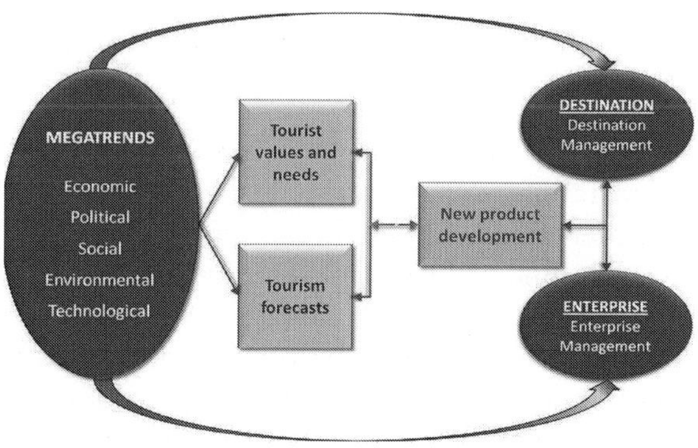

Figure 1: Influence of mega trends on tourism

Implications of key drivers on urban tourism environments

Globalisation and long-term economic trends will influence tourism in a number of ways:

- Rising income is the most powerful *generator of tourism flows*(Crouch 1994; Lim 1999; Tse 2001). The deregulation associated with globalisation will also positively influence inbound and outbound tourism flows.
- *The diffusion of ICT* drives globalisation and can enhance the competitive advantage of destinations and tourism businesses.
- Greater market share will be obtained by the relatively small number of large global *travel and tourism networks.*
- *Destination choice* will be determined by economic factors as well as the availability of time and the perceived desirability of the destinations.
- Globalisation *and the increased internationalisation of business* will lead to high rates of growth in business travel.

Urban-based economic activities account for 55% of GNP in all countries, and up to 85% in high-income countries such as Europe (United Nations Human Settlements Programme 2007) and Australia. This economic contribution of cities—indeed, their importance to the economy as a whole—is normally ignored or overlooked. There is also a bias in tourism research towards understanding and promoting rural tourism over urban tourism and cities (Edwards *et al.* 2007). This neglect is unfortunate because cities and towns are not only the loci of production, but they are also the loci of the most important impacts of globalisation and, hence, are places of change and expectations for the future. A better understanding of the dynamics of globalisation and cities can therefore help to develop new strategies for protecting and sustaining urban tourism destinations over the longer term.

Contemporary trends will influence the emergence of secondary cities, both as origins of, and destinations for, tourism. To date, most of the world's tourists have originated in the major cities and headed for other major cities, usually as an initial stopping off point en route to somewhere else. However, worldwide, there are hundreds of state and provincial capitals with sizeable populations, which are all seeking to access the tourist dollar. Increasingly, low-cost airline and transportation links will improve access to secondary cities that are keen to increase their participation in tourism (PATA 1999). As awareness of these cities increases, stronger airline and transportation links will help them to become markets in their own right.

Urban destinations will increasingly undertake regeneration strategies so that leisure, enjoyment, spectacle and pleasure can be effectively produced, packaged, marketed and consumed. Ideally these features should come together in a way that acknowledges the uniqueness of the destination. All too frequently, however, cities exhibit replication and sameness, so that it is often difficult for the visitor to differentiate one urban space from another. The same retail, and even cultural, products can be purchased in cities as geographically separated as Sydney, Paris and Hong Kong. The expanded flows of people, products and images arising from globalisation can reduce absolute differences of culture, space and attractions. For permanent migrants, cities present challenges for integration into the indigenous or existing urban culture:

Current trends suggest that if one of the great strengths of cities is their openness to the economic functions that strangers fulfil, their great weakness is a slowness to absorb them in the micropolitics of everyday life, in both public spaces and private institutions. (United Nations Human Settlements Programme 2007, p. 2)

Very often, a lack of money and knowledge of the local language both pulls and pushes immigrants into neighbourhoods with similar ethnic backgrounds. Ironically, these districts can develop into attractive places of difference within the city—such as Chinatown in Sydney and Lygon Street in Melbourne—becoming magnets for attracting visitors.

The political climate of a country can have significant relevance for tourism:

- Destinations perceived as less safe and secure will be avoided. Cities are particularly vulnerable sites and need to possess and be able to effectively implement disaster-management strategies (Mistilis and Sheldon 2006).

- Destinations that do not advance economically because of political constraints will generate fewer outbound and domestic tourist numbers.
- Conflicts between countries will constrain tourism flows overall, but can also benefit those destinations perceived to be 'safe'.
- Conflicts in and around the Middle East have the capacity to substantially increase fuel costs. This affects the demand for travel and in turn places increased pressures on airlines, many of which operate already on low margins.
- The destination region as a whole, as well as the individual operators who deliver the tourism product, must play a role in sustaining the safety and security of their visitors and, collectively, the tourism destination. Increased crowding and urbanisation makes cities more vulnerable to health threats transmitted by visitors. Urban destinations will need to take a number of steps to protect their populations, such as reducing the risk of health crises through adequate health checks at borders, publication of up-to-date information, directing funds at health-related tourist issues, development of coordinated emergency management protocols by health officials. Health units at all levels need to be carefully coordinated, with other basic policies focused on sustainable tourism development (Ritcher 2003).
- Marketers of destinations affected by real or imagined risk factors need to target tourists who have a higher tolerance for risk and seek markets that don't have poor security standards.

Tourism is closely linked to the environment, which underlines the latter's extreme importance. Climate change will influence which destinations will be preferred by tourists and which ones will cease to be as attractive. Climate change will have an impact on the profitability of tourism through increasing energy costs that will seriously erode the 'bottom line' of operators. Tourism is transport-intensive, and so it is heavily exposed to oil prices. In the face of future declining oil stocks, basic transport needs, such as for economic development and commuting to work, may take precedence over discretionary transport needs, such as recreational tourism. Australian cities may become much less price competitive if carbon taxes are imposed on international aviation. Without such action, costs to economies from growth in emissions will increase much faster than loss of benefits to individual tourism industry stakeholders (representing something akin to the tragedy of the commons). The rapid growth in aviation emissions contrasts with the success of many other sectors of the economy in reducing emissions. Today, the question is not whether we should change our business operations and lifestyles to accommodate the environment, but how radically we must change current environmental conservation practices; how we adapt to new environmental conditions; how quickly we must act; and how much it will cost us.

The growing urban congestion in both industrialised and developing worlds leads people to increasingly feel the need to engage in tourism to escape and/or to indulge. Cities must seek innovative ways to capitalise on their role as tourism gateways by decongesting and de-stressing the urban environment. Making the urban environment more welcoming—by creating physical forms that inspire, as well as personal, safe and hospitable spaces—will create environments that will suit the type of tourism that will be demanded in the future. One of the world's fastest growing tourism markets is that of meetings, incentives, conventions and exhibitions (MICE). It is also typically one of the highest yielding inbound tourism segments because of the high per-delegate spend. Worldwide, there is strong competition among destinations for such business and, with substantial government support for marketing activity and infrastructure development, this trend is expected to continue. Business tourism comprising corporate meetings, incentives, conventions and exhibitions, is expected to provide increased opportunities for the Australian tourism industry over the next 10 years, with strong appeal across all key markets. Products that focus on ease of accessibility to the destination, sound infrastructure that supports the product, quality facilities and services, food and beverages that meet visitors' dietary considerations, safety and creative experiences will be attractive to the business market.

The most pervasive technological changes that will be applied to the tourism and recreation industry are both predictable and manageable. The success of urban destinations will continue to hinge on the efforts of enterprises to add value to products and services through the application of technology to produce competitive advantage. Using ICT, the most innovative tourism organisations and enterprises can re-define both their own structure and their relationships with their partner organisations. Interactive access to product offering via the Internet gives tourists unprecedented control over how they spend their time and money. Destinations and tourism organisations that do not embrace the Internet, and make the associated investments in information technology (hardware and software) and employee training, will lose competitiveness.

The gradual development of ambient technologies and 3G mobile devices providing 'the mobile Internet' will empower consumers by providing location- and time-dependent information, interpretation and service offerings. Cities therefore can upgrade their ICT capability to meet the needs of tourists and citizens; for example by providing free wireless access and m-tour guides in public spaces. The smartest e-destinations include photos, blogs, hotels, attractions, transport and other information, maps, videos, travel guides, user reviews and trip planners. For example, trip planners may create themed drives for special interest groups, which are integrated with accommodation and attractions.

Conclusion

This chapter has sought to establish the implications for cities of global economic, social, political, environmental and technological change. For managers of urban destinations, achieving competitive advantage in times of rapid change requires a clear understanding of the key trends and their implications for urban environments.

Not so many years ago, cities in Australia attracted few tourists, as there was a limited range of recreational and leisure activities. Today, at the weekend, most Australian city centres are alive with activity: shops are open, restaurants and cafes have spilled out onto pavements, similar to their European counterparts, while recreational and leisure opportunities have expanded. Every city in Australia is incorporating what are now considered universal urban redevelopment features, such as waterfront precincts, promenades, casinos, museums, urban tourism precincts and convention centres. Australian cities have re-invented themselves to attract global travellers. Today, key trends are creating another set of challenges that cities must face if they are going to become urban environments that recognise the need to preserve their diversity, both within and between cities. But cities are also the locations of wealth generation, political decision making and new technologies, all of which can combine to create opportunities to attract and retain future visitors.

References

Buhalis D (2000a). Trends in information technology and tourism. In *Trends in Outdoor Recreation, Leisure and Tourism.* (Eds WC Gartner and DW Lime) pp. 47–61. CABI Publishing, Cambridge, MA.

Buhalis D (2000b). Tourism in an era of information technology. In *Tourism in the 21 Century: Lessons from Experience, Continuum.* (Eds B Faulkner, G Moscardo and E Laws), pp. 163–180. London, New York.

Butler R and Jones P (2001). Conclusions-problems, challenges and solutions. In *Tourism and Hospitality in the 21st Century* (Eds A Lockwood and S Medlik). pp. 296–310, Butterworth-Heinemann, Oxford.

Cetron M (2001) The world of today and tomorrow: The global view. In *Tourism and Hospitality in the 21st Century.* (Eds A Lockwood and S Medlik) pp. 18–28. Butterworth-Heinemann, Oxford.

Crouch GI (1994). The study of international tourism demand: a review of findings, *Journal of Travel Research* **33**, 12–23.

Dwyer L (2004). Trends underpinning global tourism in the coming decade. In *Global Tourism* (Ed. W Theobald). pp. 529–545. Butterworth-Heineman, New York.

Edwards D, Griffin T and Hayllar B (2007).'Development of an Australian urban tourism research agenda'. Technical Report. CRC for Sustainable Tourism Pty Ltd, Gold Coast, Queensland.

Faulkner B (2001). Towards a framework for tourism disaster management. *Tourism Management* **22(2)**, 135–147.

Johnson AL (2004). Why not expect a 'World State'? *Cross-Cultural Research* **38**, 119–132.

Johnson G and Scholes K (1997). *Exploring Corporate Strategy.* 4th edn. Prentice Hall, London.

Lim C (1999). A meta-analytic review of international tourism demand. *Journal of Travel Research* **37(3)**, 273–84.

Mistilis N and Sheldon P (2006). Knowledge management for tourism crises and disasters. *Tourism Review International* **10(1/2)**, 39–46.

Mistilis N and Daniele R (2004). Challenges for competitive strategy in public and private sector partnerships in electronic national tourist destination marketing systems. *Journal of Travel & Tourism Marketing* **17(4)**, 63–73.

National Intelligence Council (2000). *Global Trends 2015: A Dialogue about the Future with Non-government Experts.* US National Foreign Intelligence Board, Washington DC.

National Intelligence Council (2004). *Mapping the Global Future.* US: National Intelligence Council, Washington DC.

Olsen M (1999). Macroforces driving change into the new millennium – major challenges for the hospitality professional, *International Journal of Hospitality Management* **18(4)**, 371–385.

Perz JE, Allen CS and Schaffner W (2001). Mixed outbreak of parainfluenza type 1 and influenza B associated with tourism and air travel. *International Journal of Infectious Diseases* **5**, 189–191.

Profound R and Duerbeck K (2005). *Natural Ingredients For Cosmetics.* Centre for Promotion of Imports from Developing Countries (CBI). London, UK.

Raynolds P (2005). *The Arctic's New GoldRush.* BBC News, London.

Ritcher LK (2003). International tourism and its global public health consequences. *Journal of Travel Research* **41**, 340–347.

Sonmez, SF and Graefe AR (1998) Influence of terrorism risk on foreign tourism decisions. *Annals of Tourism Research* **25**, 112–144.

Stephens G (2003). Global trends in crime. *The Futurist* **37**, 41–46.

Stern N (2006). *Stern Review on the Economics of Climate Change. An Independent Review for The Treasury, United Kingdom Government.* <http://www.hm-treasury.gov.uk> Accessed 18 November 2006.

Sutton A and Wilson D (2004). Open-street CCTV in Australia: the politics of resistance and expansion. *Surveillance & Society,* **2**, 310–322.

Tarlow PE (2002). Tourism in the twenty-first century. *The Futurist* **36**, 48–51.

Tse R (2001).Estimating the impact of economic factors on tourism: evidence from Hong Kong'. *Tourism Economics* **7(3)**, 277–293.

Ulbrick D (2006). *Growing Yield in the City Holiday Market: Finding a Competitive Edge.* Fifth National Conference on Tourism Futures Melbourne, December 4–6. Roy Morgan Research, Melbourne.

UNEP (2002). *Global Environment Outlook 3 - Past, Present, and Future Perspectives,* United Nations Environment Program. Geneva, Switzerland.

United Nations Human Settlements Programme (2005). UN-HABITAT Annual Report 2004: Responding to the challenges of an urbanizing world. Kenya: United Nations Human Settlements Programme (UN-HABITAT). Geneva, Switzerland

United Nations Human Settlements Programme (2007). UN-HABITAT Globalization and Urban Culture. Kenya: United Nations Human Settlements Programme (UN-HABITAT). Geneva, Switzerland.

Wilder-Smith A (2005). The Severe Acute Respiratory Syndrome: impact on travel and tourism. *Travel Medicine and Infectious Disease* **20,** 1–8.

Viner D and Nicholls S (2005). Climate change and its implications for international tourism. In *Tourism Management Dynamics: Trends, Management and Tools.* (Eds D Buhalis and C Costa) pp. 156–169. Elsevier, Amsterdam, Netherlands.

Weston R, Qu L and Soriano G (2001). *The Changing Shape of Australia's Population.* Australian Institute of Family Studies, Commonwealth of Australia, Melbourne.

Willmott M and Graham S (2001). The world of today and tomorrow: the European picture. In *Tourism and Hospitality in the 21st century* (Eds A Lockwood and S Medlik) pp. 29–38, Butterworth-Heinemann, Oxford,

World Tourism Organisation (2002). *Tourism 2020 Vision*, WTO, Madrid.

8

Consumers in 2020

Alicia Payne, Ann Harding and Quoc Ngu Vu

Due to changing patterns of fertility and increasing life expectancy, the profile of Australia's population is changing significantly, with profound implications for the economy in coming decades. The ageing of the population is a key issue for government planning for the costs of a slowly growing workforce and increasing proportion of older people compared with younger people (Treasury 2007). This change will also affect consumer habits—as the number of older consumers increases much faster than that of younger consumers—with implications for which goods and services will experience high and low rates of expenditure growth over coming years.

Using the most recent data on household expenditure in Australia, this study projects what Australians will be spending their money on in 2020, taking into account the current habits of consumers and the demographic, earnings and labour force changes projected to take place by the year 2020. The lifestyles and consumer behaviour of different 'generational' age groups will also have implications for the mix of goods and services set to experience high increases in demand, as periods of life are 're-defined' with generational change.

There is a growing body of study of the 'generations' and their characteristics in terms of attitudes, labour market and social behaviour and consumer habits. The generations have been given various names and age definitions by commentators, but are generally similar to the following:

- **Generation Y** are those aged about 15–29 years old who are just beginning to emerge as a group of influence in the labour market.
- **Generation X** gained their name from Douglas Coupland's book (1991) and—aged about 30–45—the group is often perceived to have missed out on a lot of the achievable dreams of their parents, such as home ownership and the bright futures of their Generation Y cousins.
- Before Generation X came the numerous and influential **Baby Boomers**–those born following the end of World War II and now aged 45-59, the eldest of whom are now entering their retirement.
- The parents and older brothers and sisters of the Boomers—sometimes referred to as the '**Builders**'—now aged over 60. These Australians grew up during two world wars and the Depression and worked hard to set the economic and political foundations for the prosperity of their children to come.

The major historical, social, political and economic climates in which these groups of people have grown up have shaped them to have different attitudes, lifestyles and expenditure profiles as consumers. People demand different mixes of goods and services throughout their lives, and the changing lifestyles of progressive generations have 'rewritten' what sorts of expenditure Australians will be making at different ages. For example, Generations X and Y have shown a trend towards delaying home ownership (Harding *et al.* 2003; Cassells and Harding 2007), while the Baby Boomers have redefined retirement as a time to devote to recreation and leisure. Increasing lifespan and awareness of health mean that sport-related expenditure, for example, extends to later ages, and that different sorts of foods and entertainment are demanded. Also, decreasing rates of fertility and the delay of child rearing to later in life for many people means that related types of expenditure, such as toys and school fees, are now dominated by different age groups than they once were. As Australians spend more of their young adult life studying— and thus have lower incomes for longer than their parents might have—this similarly generates changes in their demand for substantial expenditure items, such as new cars, home furnishings and white goods. If these 'rewritten' life phases remain the same, these emerging life cycle expenditure patterns, along with population ageing and the distribution of expenditure across age groups, will result in a vastly different economy in 2020.

Following a brief explanation of the methodology and data used in this study we present predicted changes in population structure between 2003–04 and 2020 and predicted broad changes in expenditure behaviour. This is followed by analysis of the likely expenditure habits of each generational group, from Generation Y to those over 75.

Methodology and data

The methodology used in this study is based on that used in Harding *et al.* 2006—a report commissioned by AMP. The base data for this project was the most recent 2003–04 *Household Income and Expenditure Survey* (HIES) confidentialised unit record file, released by the Australian Bureau of Statistics (ABS) in 2006. This is a representative national survey of 6,957 households that contains extensive socio-demographic, labour force, income and expenditure details about individuals, families and households.

In order to estimate expenditure in 2020, it is assumed that households in 2020 will spend the same proportion of their income on each of the 625 detailed level expenditure items recorded in the survey as similar households in the 2003–04 survey. By 2020 the number of households in Australia will have grown, and the composition of household types will have changed, with some types of households projected to grow as a proportion of all households, while other types will lose share of the population. The change in expenditure projected in this study is driven by this growth in number and changing profile of Australian households, as well as income growth over the period. Thus, our projections estimate what expenditure will be like in 2020 if households maintain the same spending habits, taking into account of estimated population change and income growth.

The major task carried out within this modelling is to 'age' this dataset to a future world, which is achieved through two key steps—'reweighting' and 'uprating'. The reweighting stage involves using population projections to adjust the survey 'weights' (numbers given to each record in the sample that enable estimates to be made for the whole population), so that estimates can be made for the population in 2020 using the information in 2003–04 survey. Various benchmarks were used including age, sex, labour force status and state and territory. Uprating involves estimating income growth over the period. For a detailed description of the methodology used in this analysis see Harding *et al.* 2006 at <http://www.amp.com.au/group/3column/0,2449,CH37252%255FSI3,00.html>.

Projected changes in population structure and overall expenditure

Australia is currently experiencing population ageing. This has been caused by a combination of the sharp fall in fertility combined with lengthening life spans. While in 1960 every person aged 65 years and over could potentially be supported by 7.3 working age Australians, that ratio is expected to fall to one retiree for every 2.4 working age Australians by 2040. Population ageing is a major concern for governments in terms of the fiscal burden, as a growing proportion of older Australians will require the support of a relatively slowly growing workforce (Productivity Commission 2005; Treasury 2007).

Government research has examined the fall in labour force participation rates that population ageing is likely to produce. This is one of the key factors driving the forecast fall in economic growth in Australia, with Treasury expecting the average rate of growth in real per capita GDP to decline from 2.1% a year over the past 40 years to only 1.6% over the coming 40 years (Treasury 2007). The projected decline in the rate of economic growth is not only of concern to government, but also for industry—as it means less growth in consumption and profound changes in the types of goods and services demanded by consumers. This section looks at the projected changes in Australia's

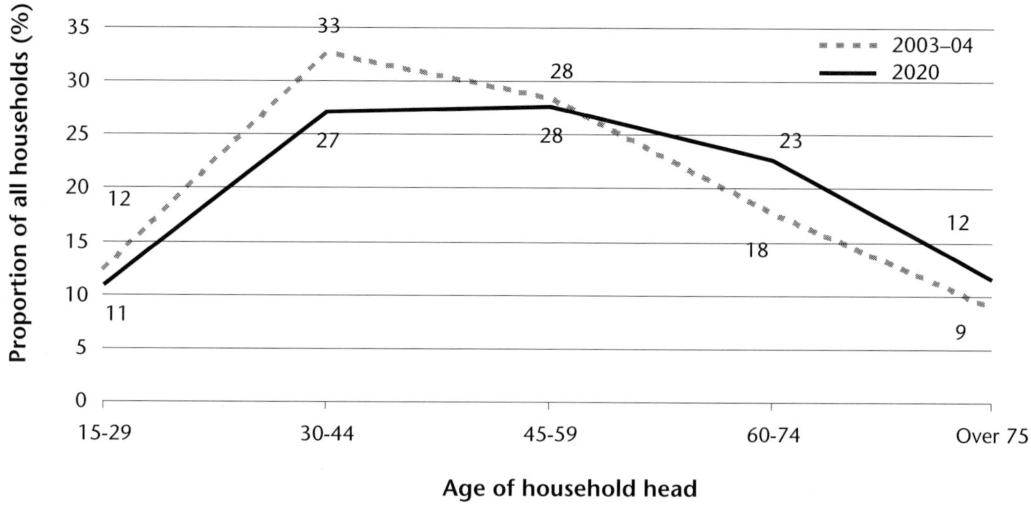

Figure 1: Forecast composition of households by age of head, 2003–04 and 2020

population by 2020 and the broad level effects this is likely to have on consumer expenditure.

Between 2003–04 and 2020 the total number of households in Australia is expected to grow from 7.7 million to about 9.5 million. There will also be major changes in the *composition* of households. There will be more households headed by older people. As Figure 1 shows, while in 2003–04 thirty-three per cent of households had a household head aged 30 to 44 years, by 2020 this is expected to fall to about 27%. In 2020 it is projected that almost a quarter of households (23%) will be headed by someone aged between 60 and 74, and a further 12% headed by someone aged over 75. Overall, there will be fewer households headed by people of general working age, with about 57% of all households in 2020 being headed by a person aged 15 to 54 years, down eight percentage points from 2004. In contrast, the proportion of all households headed by a person aged 55 years and over will grow from about 35% in 2004 to 43% in 2020. Because people in this age group tend to have lower incomes and poorer health, this has major implications for industry and our economy.

The strong growth in the number of households headed by older Australians is shown clearly in Figure 2. Of the almost 1.8 million new households expected to be created between 2004 and 2020,

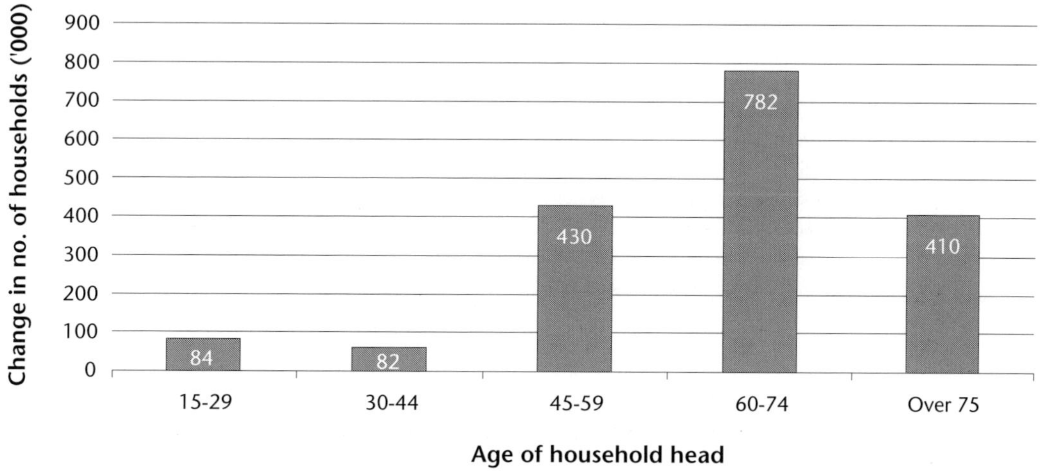

Figure 2: Forecast change in total number of households by age of head, 2003–04 to 2020

only 145,000 of them will be headed by a person aged less than 45 years. In contrast, over 780,000 of the new households will be headed by someone aged 60–74. About 410,000 will be headed by a person aged 75 plus.

Figure 3 provides additional detail about the forecast changes in household type. While a notable feature of the past 10 years has been the growth in two-income working age couples with children (Harding and Robinson 1999), population ageing is bringing this trend to its end. While the proportion of households headed by older people is projected to grow, younger households and families with children are expected to represent a smaller proportion of all households in 2020 than they did in 2003–04.

As clearly shown in Figure 3, in 2020 the greatest proportion of households (nearly 12%) will be couples on fewer than two incomes where the head is aged 60–74. This group will probably be composed primarily of couples where one or both partners receive the age pension. While the second greatest proportion of households (just below 10%) will be couples with children on two incomes where the head is aged between 30 and 44, this group's share of the total number of households is projected to decrease by 2020. The proportion of households in this group with fewer than two incomes is also projected to fall by 2020, to less than 4% of all households. The next greatest proportion of households in 2020 is projected to be 'other families', which include group households and multiple family households with and without dependent children.

Other types of households expected to grow as a proportion of all households by 2020 include single people aged over 60, couples aged over 75 and couples without children on two incomes aged 45–74. Types of households for which the share of the total households is expected to fall include singles aged under 60, couples without children on two incomes aged 15–34, and couples with children of all ages and income types (except couples with children where the head is aged 15–29 on two

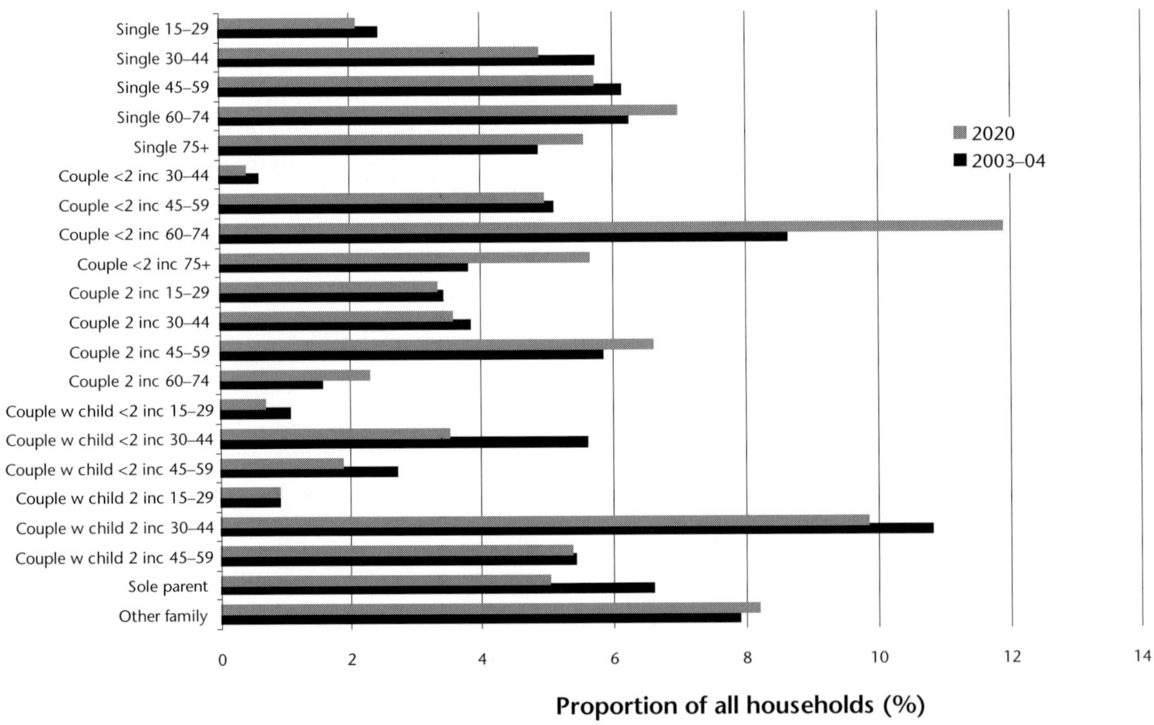

Figure 3: Forecast change in composition of households by household type and age of head, 2003–04 to 2020. Note: Results are not shown where the age-family group contained a sample of less than 30 households (for example couple with two incomes where the head is aged over 75 years). The categorisation of families as being on two or fewer than two incomes is based on the labour force status of the reference person and spouse in each household.

Table 1: Forecast average and aggregate income and expenditure of households in 2003–04 and 2020

	Weekly average per household			Annual total for all households			
	2003–04 $	2020 $	Change %	2003–04 $billion	2020 $billion	Difference $billion	Change %
Gross income[a]	1,124	1,388	24	452.1	685.8	233.8	52
Current housing costs	143	169	18	57.7	83.4	25.7	45
Domestic fuel and power	24	29	23	9.5	14.4	4.9	51
Food and non-alcoholic beverages	153	189	24	61.5	93.6	32.1	52
Alcoholic beverages	23	29	25	9.4	14.4	5.0	54
Tobacco products	12	14	18	4.6	6.7	2.1	45
Clothing and footwear	35	43	21	14.2	21.0	6.8	48
Household furnishings and equipment	52	64	24	20.9	31.7	10.8	52
Household services and operation	54	67	23	21.8	32.9	11.1	51
Medical care and health expenses	46	60	31	18.4	29.5	11.1	60
Transport	139	172	23	56.0	84.7	28.7	51
Recreation	115	143	24	46.1	70.5	24.4	53
Personal care	17	22	25	6.9	10.6	3.7	54
Miscellaneous goods and services	79	95	19	32.0	46.9	14.9	47
Life insurance	3	4	21	1.2	1.8	0.6	49
Superannuation	20	26	29	8.0	12.6	4.7	59
Total expenditure	1,010	1,233	22	406.2	609.4	203.3	50

Note: All dollars are in 2003–04 dollars (that is, the table shows trends in real income and expenditure, after adjustment for inflation). Note that there are concerns about the accuracy of the superannuation figures when compared with other benchmark sources (as explained).

[a] Gross income is total weekly household income from all sources, such as wages and salary, government cash benefits and investments, before tax and the Medicare levy have been deducted.

incomes, who are projected to experience a slight increase in number), and sole parent families. Such changes in the structure of the population are likely to have a major impact on the economy, as older households, renowned for having smaller incomes and spending less, make up a greater share of the economy. As the proportion of families with children falls, there will be relatively slow growth for industries reliant on this group - for example manufacturers of children's products and providers of goods and services aimed at the family market (Harding *et al.* 2006).

This change in household composition is one reason why the growth in total spending is slightly lower than the forecast growth in total real income. Based on projections of earnings growth, the real total income of the average household is expected to increase by 24% between 2003–04 and 2020. Table 1 compares the 2003–04 data with projections of income and expenditure on selected broad items for all Australian households in 2020. The first three columns in Table 1 show household averages, while the next four columns show the aggregates for the entire economy. The percentage increases in demand for the whole economy are much higher than those for the average household. The reason is that there are expected to be substantially more households in 2020, with growth in demand generated by the greater number of households as well as the rise in projected expenditure within each household.

Because of the expected increase in the number of households, the 24% increase in the income of the average household translates into a 52% growth in total income within Australia (Table 1). That is, total income for all households living in private dwellings is projected to increase from about $452 billion in 2003–04 to about $686 billion by 2020. Total household spending is projected to increase from $406 billion to almost $610 billion a year.

The differences in spending growth between the various types of goods and services are more dramatic. Spending on housing is forecast to increase by only 45%, reflecting the lower housing costs of older Australians (who are more likely to have paid off all or the majority of their mortgage). Similarly, spending on tobacco products is set to grow by only 45%, due to the increasing dominance of older households, who are less likely to smoke than younger households (Harding and Percival 1997). In contrast, household spending on medical care and health is expected to grow by 60% by 2020. Other key areas where spending is expected to increase by more than the average include alcohol, household furnishings and equipment (such as carpets, cutlery, and linen), recreation and personal care services (such as hairdressing).

While expenditure on superannuation appears to grow substantially by 2020, it is likely that these figures underestimate both the current and projected future amount that will be invested in superannuation by Australians. Compared with the $8 billion of voluntary contributions estimated to be put into superannuation in 2003–04 in the HIES (as shown in Table 1), the Australian Prudential Regulation Authority (APRA) reported in their 2005 Bulletin that Australians put $22.9 billion into their super accounts between July 2004 and June 2005 (APRA 2005). This refers to the amount contributed by employees (and thus excludes that contributed by their employers). As the ABS survey requires people to state how much they 'voluntarily' put into super over the survey period, it is likely that this low figure could be the result of people not really knowing how much they contribute to superannuation each pay.

Figure 4 compares the projected change in the share of expenditure with the share in the total number of households for each generational age group. The chart shows that expected changes in the share of expenditure and of total households do not change at the same rate for each generational age group. While the share of older households will increase, this does not translate into equal growth in the share of expenditure taken up by this group. While the proportion of households headed by someone under 59 will decrease, the share of expenditure for these groups does not decrease by as much as the share of households. This indicates that population ageing is not a boon for the economy, because the lower incomes and spending of older households mean that their greater numbers are not automatically translated into an equivalent growth in market share. Thus, while the proportion of all households headed by a person aged 75 and over is expected to rise by 2.6 percentage points, their share of total household spending is forecast to grow by only 1.5 percentage points.

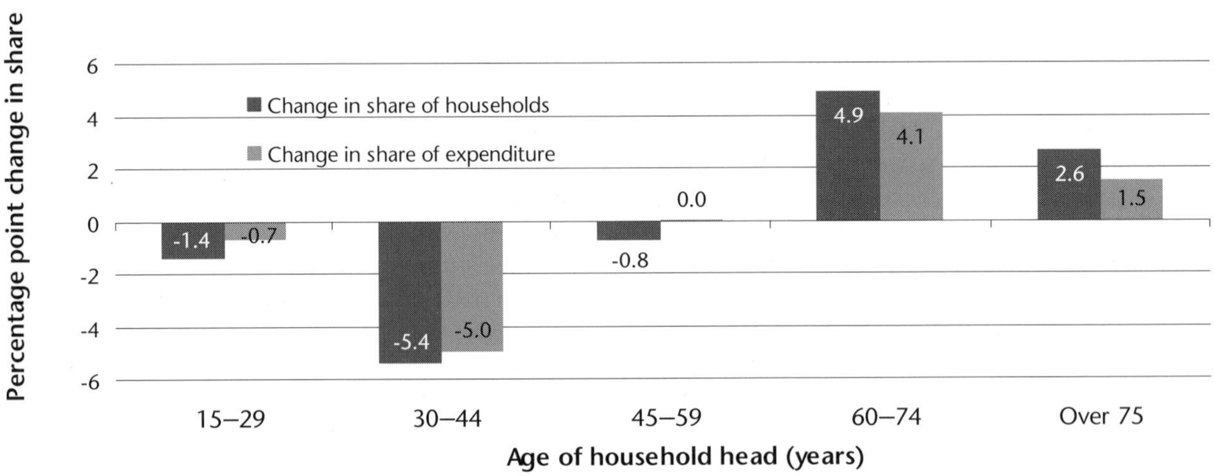

Figure 4: Forecast change in share of households and total expenditure by age of reference person, 2003–04 to 2020

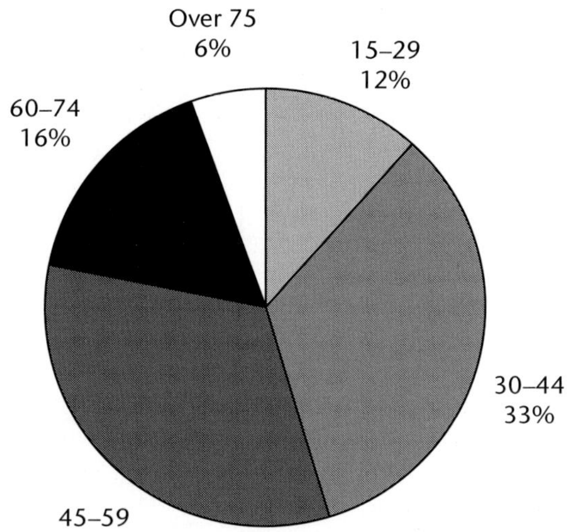

Figure 5: Forecast share of total expenditure by age of household head, 2020

This point is emphasised in Figure 5, which shows that working age households will continue to represent the largest markets for industry. About one-third of all spending in 2020 will be from the pockets of households headed by 30 to 44 year olds, and a further third will be from those headed by a 45 to 59 year old. Thus, despite the relatively rapid forecast growth in households headed by a person aged 75 years or over, this group will still only make up 6% of estimated total household spending in 2020. The key point for businesses planning their futures will be to take into account which market segments are growing most rapidly.

Expenditure and age group

Table 2 shows the projected percentage real (inflation-adjusted) increase in expenditure for all households between 2003–04 and 2020, and the projected aggregate expenditure of all households in 2020 (in real 2003–04 dollars). As discussed, the ageing population will have dramatic impacts on the mix of goods and services that will experience high growth over the next 13 years. Table 2 breaks down expenditure on broad expenditure categories by generational age group. The top row shows the predicted percentage increase in *total* expenditure for households headed by people in each age group.

Clearly there is substantial difference in these rates across age groups, attributable to the relatively much higher growth in numbers of households headed by older consumers, and slower growth in the number of younger households. While the whole population's total expenditure is expected to increase by 50%, the aggregate expenditure of households headed by someone aged over 75 is expected to more than double to 106%. The expenditure of households headed by someone aged 60–74 is also expected to double—increasing by 100%. The total expenditure of younger households is set to increase by less than the average for the whole population. The expenditure of households headed by a 15–29 year old is expected to grow by a relatively slow 42% and that of households headed by someone 30–44 by 31%.

The shaded cells in Table 2 indicate goods and services for which it is projected that the age group's expenditure on the category will exceed the growth in total expenditure for the age group. Thus, this indicates relative 'winners' within each age group—the broad expenditure categories that are in particular demand for the age group. Notably for those in the 60–74 and 75 and over age groups, expenditure on all broad categories (including those set to increase below the age group averages), is projected to increase more than average for the total population. For these age groups, the broad categories projected to increase by less than the age groups' average are still projected to increase substantially, some by near to and even more than 100%. Australians over 60 will clearly be the greatest market for expenditure growth over the next 13 years, and are thus a key target group for industry.

Alcoholic beverages are the only broad category projected to experience above average growth for all age groups—a 'winner' with all generations. In contrast, expenditure on tobacco products is predicted to grow at a below average rate for all age groups. Expenditure on housing is also expected to experience below-average growth across all age groups, although, as noted above, these projections do not incorporate projected price changes.

Recreation is expected to generally be a winner across age groups. Expenditure on recreation is projected to increase more than the age groups' average expenditure for age groups between 30 and 74. For households headed by 15–29 year olds,

Table 2: Percentage increase in broad expenditure categories, 2003–04 to 2020, by age of household head

	15–29	30–44	45–59	60–74	Over 75	Whole population
Total expenditure						
Growth in aggregate expenditure 2003–04 to 2020 (%)	42	31	50	100	106	50
Aggregate expenditure in 2020 ($ billion)	71.02	204.72	200.38	99.37	33.94	609.44
Current housing costs						
Growth in aggregate expenditure 2003–04 to 2020 (%)	39	30	46	94	101	45
Aggregate expenditure in 2020 ($ billion)	13.69	32.78	21.73	10.53	4.71	83.44
Domestic fuel and power						
Growth in aggregate expenditure 2003–04 to 2020 (%)	38	28	48	97	104	51
Aggregate expenditure in 2020 ($ billion)	1.31	4.28	4.64	2.88	1.25	14.36
Food and non-alcoholic beverages						
Growth in aggregate expenditure 2003–04 to 2020 (%)	40	30	50	101	108	52
Aggregate expenditure in 2020 ($ billion)	9.42	29.02	30.64	17.95	6.54	93.56
Alcoholic beverages						
Growth in aggregate expenditure 2003–04 to 2020 (%)	44	35	52	102	109	54
Aggregate expenditure in 2020 ($ billion)	2.13	4.08	4.94	2.67	0.61	14.43
Tobacco products						
Growth in aggregate expenditure 2003–04 to 2020 (%)	36	30	46	84	95	45
Aggregate expenditure in 2020 ($ billion)	0.91	2.28	2.21	1.13	0.19	6.72
Clothing and footwear						
Growth in aggregate expenditure 2003–04 to 2020 (%)	39	31	47	97	112	48
Aggregate expenditure in 2020 ($ billion)	2.41	7.11	7.07	3.48	0.93	21.00
Household furnishings and equipment						
Growth in aggregate expenditure 2003–04 to 2020 (%)	41	32	53	100	109	52
Aggregate expenditure in 2020 ($ billion)	3.44	10.78	10.11	5.76	1.65	31.74
Household services and operation						
Growth in aggregate expenditure 2003–04 to 2020 (%)	42	31	50	99	103	51
Aggregate expenditure in 2020 ($ billion)	3.56	11.35	9.94	5.72	2.32	32.89
Medical care and health expenses						
Growth in aggregate expenditure 2003–04 to 2020 (%)	48	31	53	108	101	60
Aggregate expenditure in 2020 ($ billion)	1.89	7.17	10.14	7.24	3.08	29.52
Transport						
Growth in aggregate expenditure 2003–04 to 2020 (%)	47	29	50	100	111	51
Aggregate expenditure in 2020 ($ billion)	11.06	25.15	29.35	15.31	3.87	84.75
Recreation						
Growth in aggregate expenditure 2003–04 to 2020 (%)	42	31	52	102	105	53
Aggregate expenditure in 2020 ($ billion)	7.89	21.37	23.80	13.90	3.53	70.48
Personal care						
Growth in aggregate expenditure 2003–04 to 2020 (%)	47	31	52	98	104	54
Aggregate expenditure in 2020 ($ billion)	1.23	3.04	3.68	1.98	0.70	10.63

	15–29	30–44	45–59	60–74	Over 75	Whole population
Miscellaneous goods and services						
Growth in aggregate expenditure 2003–04 to 2020 (%)	34	30	46	101	112	47
Aggregate expenditure in 2020 ($ billion)	5.68	15.61	16.75	6.97	1.87	46.87
Life insurance						
Growth in aggregate expenditure 2003–04 to 2020 (%)	29	35	42	122	117	49
Aggregate expenditure in 2020 ($ billion)	0.05	0.60	0.78	0.29	0.03	1.75
Superannuation						
Growth in aggregate expenditure 2003–04 to 2020 (%)	45	32	60	112	225	59
Aggregate expenditure in 2020 ($ billion)	0.58	3.15	6.49	2.29	0.11	12.63

Note: All aggregates given in real 2003–04 dollars.

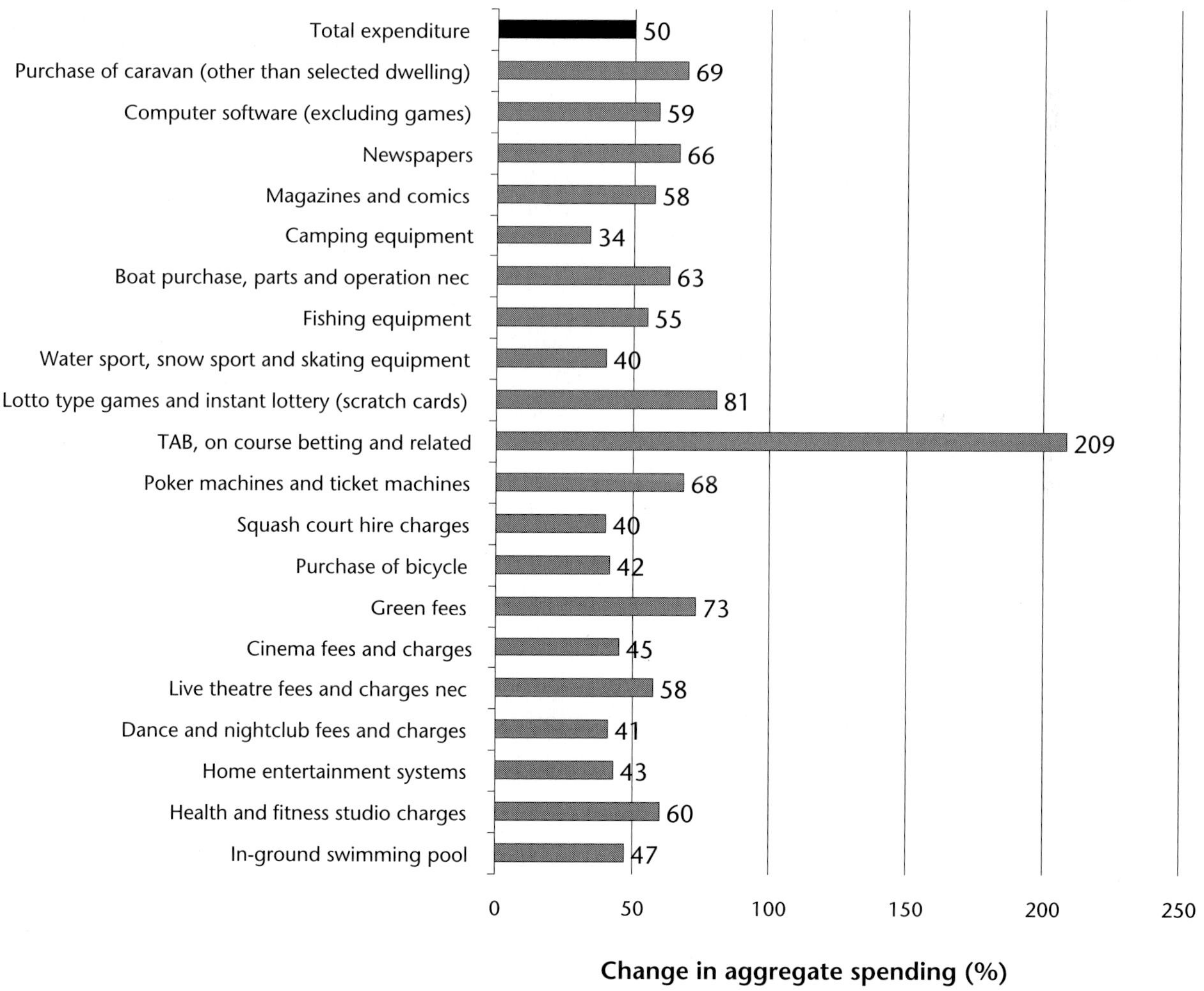

Figure 6: Projected growth in leisure expenditure items from 2003–04 to 2020 (Source: Harding *et al.* 2006).

leisure expenditure is projected to increase at the average rate for that age group (42%) and, for over 75 year olds total leisure spending will more than double. The significant growth in leisure demand projected to come from older consumers will affect the types of leisure goods and services likely to benefit most from this growth (Harding *et al.* 2006). As shown in Figure 6, the recreation goods and services projected to experience high growth are those perhaps associated with older consumers, such as gambling, purchase of caravans and boats, green fees for golf bowls and croquet, and live theatre, while those associated with younger leisure-makers—such as squash court hire, purchase of bicycles, dance and nightclub fees and charges and camping equipment—will experience below average growth.

Over the next 13 years, medical expenditure will also experience substantial growth—associated with the growth in older households and the generally higher demand for health care of this group. While this growth is associated with the ageing of the population, within age groups healthcare expenses are expected to be above average for all groups between 15 and 74 years, indicating there will be growth in this area across age groups.

Expenditure on household furnishings and equipment is projected to increase more than average expenditure for the 30–44 and 45–59 age groups, although is slightly below average for households headed by those aged 15–29. This evidences the trend for Australians to settle into their 'own homes' later, as they struggle against high housing costs, start families later and devote time to study in their younger years (Cassells and Harding 2007). Household goods also experience high growth for households headed by someone over 60.

These projections for 2020 are based on the expenditure of each generational group in 2003–04, and thus help to paint the picture of the spending habits of each generation now, and what the expenditure of Australians will look like in 2020 should their life phases be characterised by similar expenditure as their predecessors in 2003–04. The remainder of this section looks at the spending profile of each generation in 2003–04 and how Australians in the same age groups in 2020 will spend their money if habits remain the same.

Generation Y and households headed by 15–29 year olds in 2020

Generation Y are those Australians aged 15–29, who are entering their late teens and early adulthood and just beginning to be noticed in the work and marketplace. Generation Y's lifestyle, and thus consumer habits, are significantly different to those of their elders. Huntley (2006) explains that they were the first generation to be viewed as consumers in their own right and targeted by advertising since birth, which has led them to be savvy in making purchase choices. The living arrangements and career paths of Generation Y Australians affect on their consumer profile also. The group has been stereotyped as spending frivolously in lieu of saving for investments that are out of reach, such as a home. However, recent research shows that, like Generation X before them, Generation Y spend the greatest proportion of their income on housing and transport (Cassells and Harding 2007). A major difference between Generations X and Y is that a much lower proportion of Generation Y live in couples with children. Only 22% of Australians aged 25–29 were in this family type in 2004 compared with 42% of Generation X when they were in the same age group in 1989. Also a greater proportion of Generation Y, especially those aged 25–29 still live in the family home as they delay investment, and increased labour force participation, in exchange for study. Around one-third of Generation Y were full-time students in 2004 and only 2.6% were home owners (Cassells and Harding 2007). Entering full-time labour force participation, settling into their own 'home' and starting a family are among the most influential factors on a consumer's profile.

In 2003–04 over 950,000 households were headed by Generation Y Australians. By 2020, over one million households will be headed by Australians in the age group that were Generation Y in 2003–04. The following section shows the projected expenditure profile of young Australians (not yet born) heading their own households in 2020 (assuming the trends for this age group as represented by Generation Y continue).

In Tables 3 to 7, the 'broad' expenditure categories shown in Table 1 have been disaggregated into the 625 'detailed' expenditure items available in the ABS *Household Expenditure Survey 2003–04*. Thus, while the aggregate expenditure items that appear in Tables 3–7 are the *detailed* items of greatest expenditure, the age group might be spending a greater aggregate amount on goods and services that have been disaggregated to a fine level. For example, while 'purchase of motor vehicle' is the detailed item of greatest expenditure for households headed by a 45–59 year old as shown in Table 5, the group spends a greater amount on 'food and non-alcoholic beverages' as shown in Table 2. Food does not make the top ten detailed items of greatest expenditure because the category is disaggregated to such a fine level; for example, 'bananas'.

Table 3 shows the ten detailed items that will attract the greatest aggregate expenditure of households headed by 15–29 year olds in 2020. The table gives a picture of expenditure indicative of the lifestyle of the young Australians described above—a mixture of the burdens of the high cost of living, the value placed on education, and the fun and social aspects of this group's lifestyle. The greatest aggregate expenditure for households headed by a 15–29 year old in 2020 is predicted to be rent payments, followed by the interest component of mortgage repayments on a home. Repayment of the principal component of a mortgage is the fourth greatest expenditure item for the group, after the purchase of a vehicle. Young households will be spending $2.2 billion on petrol, although the placement of this item is probably more meaningful than this figure, as the study does not take account of price changes, that are like to apply to fuel. Servicing of the vehicle is the eighth highest total expenditure for the group.

Indicative of Generation Y's busy and social lifestyle and often relatively high incomes (Cassells and Harding 2007), the sixth and seventh items of greatest expenditure are takeaway and fast food and restaurant meals. Should the trend towards high levels of educational qualification continue for Australians in this age group, households headed by someone aged 15–29 will spend an estimated $1.15 billion on 'higher education institution fees not elsewhere classified' which includes fees for university and other tertiary institutions (excluding TAFE). The item of tenth greatest expenditure is the mobile telephone account—a Generation Y 'must' that is likely to continue into the future.

Generation X and households headed by 30–44 year olds in 2020

Generation X—who are defined in this study as being aged 30–44 in 2003–04—have also been noted for delaying home ownership and having children and giving priority to education. They have also been described as being more financially prudent than their Baby Boomer predecessors, and to have missed out on some of the benefits they enjoyed, such as more affordable housing, free education and job security (Harding *et al.* 2003). Table 4 shows the ten items of greatest aggregate expenditure for households headed by

Table 3: The ten items of highest aggregate expenditure for 15–29 year olds in 2020

	Expenditure items	Expenditure ($ billion—2003–04 real dollars)
1	Rent payments	7.10
2	Mortgage repayments—interest component (selected dwelling)	4.31
3	Purchase of motor vehicle (other than motor cycle)	3.74
4	Mortgage repayments—principal component (selected dwelling)	2.72
5	Petrol	2.20
6	Fast food and takeaway (not frozen)	1.95
7	Meals in restaurants, hotels, clubs and related	1.40
8	Vehicle servicing (including parts and labour)	1.23
9	Higher education institution fees not elsewhere classified	1.15
10	Mobile telephone account	1.08

Table 4: The ten items of highest aggregate expenditure for 30–44 year olds in 2020

	Expenditure items	Expenditure ($ billion—2003-04 real dollars)
1	Mortgage repayments—interest component (selected dwelling)	13.81
2	Mortgage repayments—principal component (selected dwelling)	10.27
3	Rent payments	10.00
4	Purchase of motor vehicle (other than motor cycle)	8.77
5	Petrol	5.48
6	Fast food and takeaway (not frozen)	4.79
7	Additions and extensions	4.38
8	Purchase of selected dwelling or other property (excluding mortgage repayments but including outright purchase, deposit, net of sales)	3.98
9	Internal renovations	3.54
10	Meals in restaurants, hotels, clubs and related	3.39

someone aged 30–44. Comparison of Tables 4 and 3 showing the items of greatest expenditure shows that by the age of 30–44, mortgage repayments have replaced rent as the item of greatest aggregate expenditure. This suggests that by 2020, many Generation Y Australians will have achieved their housing dream, as planned (Cassells and Harding 2007).

Table 4 demonstrates that housing is projected to be the key expenditure item for households headed by someone in this age group in 2020. Households headed by 30–44 year olds will be spending an estimated $13.8 billion on the interest component of mortgage repayments and a further $10.2 billion on the principal component—a substantially greater amount that that spent on these items by the younger 15-29 year old households. Rent payments remain a substantial expense for 30–44 year olds in 2020, being the third greatest aggregate expenditure of the age group, at $10 billion. The eighth greatest expenditure item for 30–44 year old households is expenditure on purchase of dwelling excluding mortgage repayments but including outright purchase and deposit on a home. Expenditure on additions and extensions, as well as internal renovation also feature in the ten items of greatest expenditure indicating that this age group are settling into their 'home' and further investing in it.

Purchase of a motor vehicle also features highly for this age group, as does petrol expenditure. Between saving and renovating, households in the 30–44 age group still find time and money to include takeaway food and restaurant meals in their items of highest aggregate expenditure.

The Baby Boomers and households headed by 45–59 year olds in 2020

Demographer Bernard Salt explains that the Baby Boomers have redefined 'old' age—now regarding age 55 to 65 as their 'lifestyle' period (somewhat like a second adolescence), with 'retirement' being pushed back to 65 to 75 years and 'old age' to 75 years onward (2005). For the purposes of this study, the age group representing the Baby Boomers in 2003–04 has been defined 49–59, so many of this age group are still of prime working age. However the influence of retirement, or retirement planning, can perhaps be seen in Table 5, which shows the ten items of greatest aggregate expenditure for households headed by people in this age group in 2020.

Housing remains a major expenditure item for households in this age group. However, the amount of aggregate expenditure is less than for the households headed by 30–44 year olds. Unlike the younger age groups, housing costs are not the greatest expenditure items for this age group. Rent payments remain the seventh greatest expenditure item, although the total amount is less than for the younger age groups, perhaps suggesting fewer households renting. The sixth greatest expenditure

Table 5: The ten items of highest aggregate expenditure for 45–59 year olds in 2020

	Expenditure items	Expenditure ($ billion—2003–04 real dollars)
1	Purchase of motor vehicle (other than motor cycle)	10.29
2	Mortgage repayments—interest component (selected dwelling)	7.21
3	Mortgage repayments—principal component (selected dwelling)	7.09
4	Superannuation and annuities	6.49
5	Petrol	6.34
6	Capital housing costs not elsewhere classified	5.09
7	Rent payments	4.85
8	Fast food and takeaway (not frozen)	4.60
9	Meals in restaurants, hotels, clubs and related	3.84
10	Hospital, medical and dental insurance	3.56

item is 'capital housing costs not elsewhere classified' which includes items such as building materials, bricks, wood, concrete, door and window frames, electricity and gas connection and surveyors fees, indicating that the Baby Boomers are extending and renovating their existing homes, or building their own new ones.

Purchase of a motor vehicle is the item of greatest aggregate expenditure. This purchase could be associated with preparation for retirement, as people upgrade while they still working, or purchase a car for travelling in their retirement. Petrol is also an item of high expenditure for the group.

Superannuation features in the top ten expenditure items for the first time for this age group, perhaps indicating that people are investing more money into their superannuation in preparation for retirement. Households headed by 45–59 year olds are also projected to be spending $3.56 billion on insurance for hospital, medical and dental, indicating the age group's forethought.

As for the younger age groups, restaurant and takeaway meals feature highly in aggregate expenditure.

The Builders and households headed by someone over 60 in 2020

This section looks at the projected expenditure habits in 2020 of households headed by someone over 60—the age group of households that will experience the most significant growth in number between 2003–04 and 2020. Those born before and during World War II, aged 60 and over in 2003–04 are sometimes called the 'Builders'—referring to their fighting two world wars and experiencing the depression, and the sacrifices and hardships they went through that shaped society and the economy for coming generations. For this reason, this larger group is discussed together, although the distinction between households headed by someone aged 60–74 and over 75 is maintained in order to capture the different expenditure patterns of these two groups.

Similarly to the Baby Boomers aged 45–59, some of the 'war generation' (aged 60–74) would still be preparing for retirement, or be newly retired, although more of this older age group are likely to be retired than still working. The influence of retirement is visible in the ten greatest items of expenditure for this group in 2020. First, the aggregate amounts of expenditure are relatively small in comparison to the households headed by people of 'prime' working age. Purchase of motor vehicle is the item of greatest expenditure for this age group, followed by petrol. This could indicate that retirees are investing in a new car to travel in, or, as purchase of a vehicle is an item of high expenditure for all age groups in this study, it could suggest that Australians continue to upgrade their cars throughout their life.

Overseas holiday packages are the item of the ninth greatest expenditure, showing that travelling

Table 6: The ten items of highest aggregate expenditure for 60–74 year olds in 2020

	Expenditure items	**Expenditure ($ billion—2003–04 real dollars)**
1	Purchase of motor vehicle (other than motor cycle)	4.93
2	Petrol	3.19
3	Hospital, medical and dental insurance	2.52
4	Meals in restaurants, hotels, clubs and related	2.44
5	Superannuation and annuities	2.29
6	Rent payments	2.29
7	Electricity (selected dwelling)	2.12
8	Fixed telephone account	2.06
9	Airfare inclusive package tours—overseas (4 nights or more)	1.86
10	Fast food and takeaway (not frozen)	1.66

is a key focus for the age group. Restaurant and takeaway meals are also high expenditure items for this age group—as for all age groups in the study. Hospital, medical and dental insurance is the third greatest expenditure item for households headed by 60–74 year olds, although the aggregate amount spent on it is less than for 45–59 year olds for whom it was the item of tenth greatest expenditure. Expenditure on superannuation and annuities, although a high expenditure item for households where the head is aged 60–74, is lower than that of the 45–59 age group also. This is not surprising, as most people in this older age group would either be retired or working less and thus not contributing as much to super. However, its appearance in the top ten expenditure items suggests it is important to this group as they prepare for retirement.

Rent payments still feature as a key expenditure item for this age group, although mortgage repayments and other items associated with home ownership do not feature in the ten greatest items. This suggests that while rent remains a major expenditure item for households that do not own their home, many of those that do have finished paying it off by the age of 60–74. Electricity and fixed telephone accounts enter the ten greatest items for the first time. This could perhaps reflect increased time spent in the home by retirees or that these living expenses take a up a greater proportion of total expenditure as people have less income and less demand for many other types of goods and services.

The most significant point about Table 7—showing the ten items of greatest expenditure for households headed by someone over 75 in 2020—is that the aggregate dollar amounts are relatively very small compared with the aggregate expenditure amounts of the younger age groups. As discussed, while the number of households headed by someone over 75 is expected to grow significantly by 2020, this group's share of expenditure in 2020 will not increase proportionately, because older households generally have lower incomes and less demand for many types of goods and services. Table 7 indicates that at this age, as expenditure decreases, general living costs begin to dominate expenditure.

Notably, expenditure on home ownership re-enters the ten greatest expenditure items for households headed by someone aged over 75 in 2020. The item of greatest expenditure for the group is 'Purchase of selected dwelling or other property (excluding mortgage repayments, but including outright purchase, deposit, net of sales)'. As mortgage repayments do not feature as a high expenditure item for this age group, this could indicate that by the age of 75 many Australians own their homes, and thus if they purchase a new home they may not need a mortgage due to the downsizing occurring among this age cohort. Households headed by someone over 75 will also be spending an estimated $0.85 billion in rent, indicating that while many older Australians will own their homes, a significant proportion will still be renting.

Hospital and nursing home charges are a major expense for households headed by an over 75 year old—at an estimated $0.62 billion in 2020. The

Table 7: The ten items of highest aggregate expenditure for households headed by someone over 75 in 2020

	Expenditure items	Expenditure ($ billion—2003–04 real dollars)
1	Purchase of selected dwelling or other property (excluding mortgage repayments but including outright purchase, deposit, net of sales)	1.71
2	Purchase of motor vehicle (other than motor cycle)	1.31
3	Hospital, medical and dental insurance	0.99
4	Electricity (selected dwelling)	0.92
5	Rent payments	0.85
6	Fixed telephone account	0.82
7	Petrol	0.75
8	Meals in restaurants, hotels, clubs and related	0.70
9	Hospital and nursing home charges	0.62
10	Vehicle servicing (including parts and labour)	0.50

nursing home component of this aggregated figure probably represents the expenditure on living in a nursing home for households where a partner remains at home, as the HIES only captures households in private dwellings. For this reason, the expenditure of a family where the single remaining member or both members of a couple live in a nursing home or aged hostel is not captured in the survey. Thus the estimated figure of $0.62 billion aggregate expenditure on nursing home and hospital charges in 2020 for households where the head is over 75 probably significantly underestimates the total expenditure of *all* over-75 year olds in Australia on nursing home and hostel accommodation. Hospital, medical and dental insurance is also a key expenditure item for the older age group.

Transport is another key expenditure item for the over-75 age group, with purchase of vehicle, petrol and vehicle servicing featuring in the ten items of greatest aggregate expenditure in 2020. Electricity and fixed telephone accounts are also significant expenditure items for the age group, indicating that these 'everyday costs' make up a greater proportion of expenditure as older households begin to spend less on other items.

While restaurant meals remain a high expenditure item for households headed by someone over 75, takeaway food does not feature for the first time for this age group, perhaps suggesting that, with more leisure time, older Australians would prefer to eat out.

Conclusions

It is projected that by 2020 Australia's population will be significantly different, with the greatest proportion of households headed by someone over 60 years of age. The types of households estimated to experience the greatest growth are singles and couples where the head is aged over 65 and other households without children, while younger households and those with children are expected to decrease in their share of the population.

This transition towards an older population will have dramatic impacts on the relative growth rates of different types of goods and services over coming years. While older households will increase their share of the population, their share of total expenditure increases at a lesser rate (as older households tend to have lower incomes and less demand for many types of goods and services). The variation in projected rates of expenditure growth—across the age of the household head and across different types of goods and services—shows that products relying on younger markets will experience lower growth rates than in high demand from older consumers.

Comparison of the ten items of greatest aggregate expenditure shows that although goods associated especially with older households are those projected to experience high growth by 2020, housing and motor vehicle costs dominate aggregate expenditure across age groups, although the dollar amounts change significantly with age. For households headed by someone aged 15–29, rent is the key housing cost. In the years from 30 to 60, home ownership expenditure, including mortgage repayments, becomes the greatest aggregate housing cost and expenditure on renovation of the home also becomes significant. The significance of housing costs is reduced in the years over 60, although the highest aggregate item for households headed by someone over 75 is cost related to the outright purchase of or deposit on a home, although this is a much smaller dollar amount than for younger age groups. Takeaway and restaurant meals feature significantly across age groups. In the years over 60 'everyday' living expenses, such as fixed telephone accounts and electricity become high aggregate expenditure items, perhaps indicating that people are spending more time in the home after retirement, or simply that these items represent a greater proportion of expenditure as older households spend less.

This study captures the spending habits of each generational group in 2003–04 and projects what people in the same age groups will be spending in 2020 (assuming that the proportion of income spent on each item remains the same). The study does not attempt to capture trends that may result from technological change, or those that will characterise the next generation. As life expectancy increases, it is possible that Australians might begin to retire later in life—and thus older households might have increased spending power. Either way, the growth of households headed by older people will have substantial impact on the goods and services experiencing high growth and relative decline over the next 13 years in the Australian economy. The key for business over this period of transition will be to maintain demand from the younger customer base (who will be spending greater aggregate amounts), while tailoring products to suit older consumers also (in order to capture the significant growth predicted to come from this market).

References

Australian Bureau of Statistics (ABS) (2006). *Household Income and Expenditure Survey 2003–04* Confidentialised Unit Record File. ABS, Canberra.

Australian Bureau of Statistics (2006). *Population Projections, Australia*. Cat no. 3222.0. ABS, Canberra, <http://www.abs.gov.au/AUSSTATS/abs@.nsf/ProductsbyCatalogue/5A9C0859C5F50C30CA25718C0015182F?OpenDocument>.

Australian Prudential Regulation Authority (APRA) (2005). *Annual Superannuation Bulletin*. APRA, Sydney.

Bell P (2002). *Weighting and Standard Error Estimation for ABS Household Surveys*. ABS Methodology Advisory Committee, Canberra, <http://www.ausstats.abs.gov.au/ausstats/subscriber.nsf/0/02784B0B80A4CD18CA25713000781B40/$File/1352055029_Jul2000.pdf>.

Cassells R and Harding A (2007). Generation whY. *AMP.NATSEM Income and Wealth Report* **Issue 17, July**. <http://www.amp.com.au/group/3column/0,2449,CH5273%255FNI74070%255FSI3,00.html>.

Coupland D (1997). *Generation X: Tales for an Accelerated Culture*. St Martin's Press, New York.

Harding A, Payne A and Vu QN (2006). Tomorrow's consumers. *AMP.NATSEM Income and Wealth Report* **Issue 15, December**. <http://www.amp.com.au/group/3column/0,2449,CH37252%255FSI3,00.html>.

Harding A, King A and Kelly S (2002). Incomes and assets of older Australians: Trends and policy implications. *Agenda* **9(1)**, 3–18.

Harding A and Percival R (1997). *Who Smokes Now? Changing Patterns of Expenditure on Tobacco Products in Australia, 1974-75 to 1993-94*. Discussion Paper No. 24, National Centre for Social and Economic Modelling, Canberra.

Harding A and Robinson M (1997). *Forecasting the Characteristics of Consumers in 2010*. Discussion Paper No. 40, National Centre for Social and Economic Modelling, Canberra.

Harding A, Kelly S and Bill A (2003). Generation Xcluded: Income and Wealth of Generation X. *AMP.NATSEM Income and Wealth Report* **Issue 6**. <http://www.amp.com.au/group/3column/0,2449,CH7497%255FSI3,00.html>.

Huntley R (2006). *The World According to Y: Inside the New Adult Generation*. Allen and Unwin, Sydney.

Kelly S, Bolton T and Harding A (2005). May the labour force be with you: changing face of the Australian labour force 1985–2005'. *AMP.*

NATSEM Income and Wealth Report **Issue 12**. <http://www.amp.com.au/group/3column/0,2449,CH27105%255FNI136044%255FSI3,00.html>.

Productivity Commission (2005). *Economic Implications of an Ageing Australia, Overview*. Productivity Commission, Canberra. <www.pc.gov.au>.

Salt B (2005). 'The Big Shift: How Demographic Change is Shaping Attitudes Towards Retirement". Paper presented to *The Australian Financial Review* Ageing Population Summit 2005, Sydney, 27 September, AFR, Sydney.

Treasury (2002). *Budget Paper No. 5, Intergenerational Report 2002–03*. Commonwealth of Australia, Canberra, <http://www.budget.gov.au/2002-03/bp5/html/index.html>.

Treasury (2007). *Intergenerational Report 2007*. Commonwealth of Australia, Canberra, <www.treasury.gov.au/igr>.

SECTION 3

Demographic and Settlement Transitions

Australia made its transition to a predominantly urban society 125 years ahead of the world as a whole and its level of *urbanisation* continues to increase, albeit slowly (83% urban in 1966 to 88% in 2006; Hugo). Forecasts of future urban growth (McDonald) indicate that Australia's major cities will continue to act as the nation's key *demographic absorbers*; they also represent the key magnets for international and domestic tourism (Dwyer *et al.*; Newton).

No new urban centres are expected to emerge in Australia's settlement system over the coming decades; instead, the following trends are likely:

- continued population increases in the larger urban centres
- Australia's four *mega-metropolitan regions* are expected to accommodate 80% of the projected population increase over the next decade
- continued increase in the *inner city population* of the state capitals; but *outer suburban and peri-urban growth* to dominate in the absence of metropolitan planning capable of generating sufficient new housing in established suburbs
- continued growth of *coastal 'string cities'* (linked to sea-and-tree amenity) as well as *inland 'sponge cities'* (the provincial cities and towns)
- decline in small town rural Australia.

All these trends have associated with them major environmental, housing, transport and, possibly, cultural challenges, which are linked in part to Australia's high levels of *international migration*. New migrants favour the large cities, in particular Sydney and Melbourne, which are becoming more Asian in character (McDonald), while Brisbane and Perth remain more British. (The nation 'currently imports' over half of its annual rate of population growth, which runs at double the OECD average. Our capacity to absorb a greater volume of international migration from populations in developing countries displaced as a result of possible future climate-induced change is currently unknown.) Population redistribution via internal migration will continue as the key source of growth in the sun-belt capitals and the sea-change regions. The challenge of achieving sustainable development in the 21st century will be won or lost in urban Australia. Critical here will be developing an appropriate governance model for planning and managing the 21st century mega-metropolitan region.

PWN

9

Demographic Transitions

Peter McDonald

The five Australian cities with populations in excess of one million people all developed as administrative capitals of colonies of the British Empire and, following the creation of the Federation of Australia, as capital cities of the five largest states. Each has had a dominant role in their respective states because of their administrative function. While their administrative role has not been the central determinant of their historical population growth, it has constrained the capacity of other Australian cities to compete with these five cities. Only two reasonably sized new cities, the Gold Coast and Canberra, emerged in Australia in the 20th century, with Canberra also as the nation's administrative capital. Both cities have current populations under 500,000. The growth of cities in the United States looks very different. Many US cities that were non-existent or tiny settlements at the beginning of the 20th century became multimillion cities by the end of that century. Using estimates from the US Bureau of the Census (2007), some examples include: Los Angeles (Metropolitan Statistical Area population in 2006 = 13.0 million); Dallas (6.0 million); Houston (5.5 million); Miami (5.5 million); Phoenix (4.0 million); Portland (2.1 million); and Las Vegas (1.8 million).

The Australian experience has been much more like the history of settlement patterns and growth of large cities in Canada. Both countries have vast tracts of almost empty land and populations concentrated in narrow bands of settlement—the east coast in Australia's case and along the US border in Canada. There are also considerable similarities in the economies of the two countries. If we wish to predict the future settlement pattern of Australia as it adds its next 10 million people, we should look at what happened in Canada as its population increased from 21 million to 31 million. No substantial new cities emerged and growth was absorbed largely within the existing cities, with the growth being somewhat greater in smaller cities (like Hamilton) that were close to the large cities. This is also likely to be the future scenario in Australia. The future population growth of Australian cities is considered later in the chapter. First, however, recent historical growth is reviewed.

Population growth in Australia's five major cities: 1977–2006

Over the past 30 years, the annual levels of growth in the five largest cities have been volatile, most particularly in Sydney (Figure 1).

From 1977 to 2006, the annual increment to Sydney's population ranged from a high of 62,500 in 1988 to a low of 20,800 in 1990, with a 30-year average of 37,200 per annum. Sydney's population growth increased sharply in the second half of the 1980s, but then collapsed immediately. The period of growth and the subsequent collapse were associated with changing levels of international migration

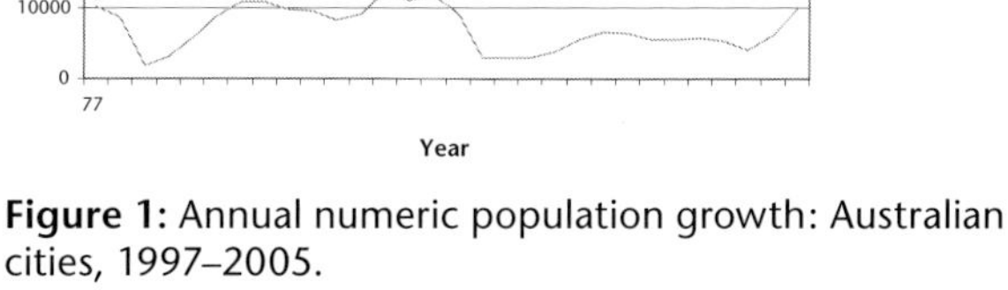

Figure 1: Annual numeric population growth: Australian cities, 1997–2005.

Table 1: Percentage decomposition of the numeric growth of Australia's cities from 1987 to 2004.

	Percentage of growth due to given component, 1987–2004				
Component of population change	**Sydney**	**Melbourne**	**Brisbane**	**Adelaide**	**Perth**
Natural increase	69	61	36	67	39
Net internal migration	−64	−32	35	−22	3
Net international migration	95	71	29	55	58
TOTAL	100	100	100	100	100

Source: Author's calculations from ABS data.

to Australia. Sydney's net gain from international migration reached almost 60,000 in 1989, but had fallen to just 12,000 by 1993. Within eight years, the net gain from international migration for Sydney had again peaked at around 56,000 in 2001, but, once more, fell back rapidly to its 30-year average of around 37,000. The Australian Bureau of Statistics estimated in 1998 that about 39% of all of Australia's net international migration gain accrued to Sydney (ABS 1998), but this estimate has been reduced downwards more recently as other states have become more active in the recruitment of international migrants than has been the case in New South Wales.

Thus, in historical terms, the story of Sydney's growth has been tied up with trends in international migration to Australia. When international migration rose to a peak, Sydney's growth also rose to a peak, but the peak growth was short-lived as each international migration peak was followed closely by a sharp fall in Sydney's growth. The importance of international migration to Sydney's growth is indicated in Table 1, which shows that, over the period, 1987–2004, the natural increase in the population of Sydney was roughly balanced by net out-migration from Sydney to other parts of Australia. This means that in this 18-year period, Sydney's population would have remained the same without international migration. However, there are potential flaws in this logic first because natural increase includes the natural increase of the international migrants and, secondly, and more importantly, there is evidence that out-migration to other parts of Australia from Sydney is associated with the level of net international migration gain to Sydney. From 1987 to 2004, the correlation between the level of international migration to Sydney in a given year and Sydney's net internal migration in the following year was a very high, at −0.76 (Table 2). It has been suggested that an increase in international migration to Sydney is followed by an increase in net out-migration from Sydney because of the impact of increased international migration on the Sydney housing market (Flood *et al.* 1991). The increased demand for housing pushes up housing prices and longer-term Sydney residents seeking a change in lifestyle sell up and realise their housing investment. Those leaving Sydney tend to move to the coast or to South-East Queensland where housing prices have been lower than in Sydney. Additionally, young, longer term residents of Sydney may give up on their objective to buy in the Sydney housing market and move north. Irrespective of whether or not the mechanism is the housing market, there is no question that the driving force of demographic change in Sydney has been international migration to Australia.

Melbourne's population growth has also been very much tied up with the level of international migration (Table 1). In 1998, the Australian Bureau of Statistics estimated that 22% of all net international migration to Australia accrued to Melbourne (ABS 1998). Since that time, this percentage has been increasing, while Sydney's percentage has fallen. However, net out-migration from Melbourne to other parts of Australia has not been as prominent as it has been for Sydney. In sharp contrast with Sydney, there is no relationship between net international migration in one year and net internal migration the following year for Melbourne. If there has been a housing market mechanism associated with population movements in Sydney, there is no sign of the same result occurring in Melbourne, nor in any other major city in Australia (Table 2).

Table 2: Correlation between net international migration in a given year and net internal migration in the following year, Australian cities, 1987–2004.

City	Correlation coefficient
Sydney	−0.76
Melbourne	+0.01
Brisbane	+0.03
Adelaide	+0.38
Perth	−0.12

Source: Author's calculations from ABS data

The story of population growth for Brisbane is very different to that of Sydney and Melbourne. Unlike all the other Australian cities, Brisbane has had a large positive net internal migration from other parts of Australia. In relative terms, international migration has been a much smaller component of its growth than has been the case in other cities. In 1998, it was estimated to receive about 9% of Australia's international migrants in net terms (ABS 1998). The components of population growth for Adelaide were similar to those for Sydney and Melbourne over the period 1987 to 2004. However, the migration movements (international and internal) have been relatively small compared with the other two cities. Finally, Perth's pattern of growth lies between that of Sydney, Melbourne and Adelaide on one hand and Brisbane on the other. Net internal migration has had almost no bearing on Perth's growth and international migration has been prominent in its growth.

Rates of population growth provide another perspective on population dynamics in the Australian cities (Figure 2). The chart of rates of growth commences in 1992–93 when international migration to Australia was running at a relatively low level because of the downturn in the Australian economy in the early 1990s. In 1992–93, the total population growth in Sydney, Melbourne and Adelaide was lower than natural increase, meaning that all three cities were losing population in net terms through migration. While these cities displayed the expected impact of recession on population growth, the rate of growth in Brisbane in the same year was very high at about 2.5% per annum—roughly three times the level of natural increase. Almost all of its net positive migration was due to internal migration indicating that, in this period of recession, Brisbane received large numbers of 'refugees' from the southern depressed states. Subsequently, the rates of growth in Sydney and Melbourne rose reaching peak levels in the second half of the 1990s. By 2001, Sydney and Melbourne both had growth rates of 1.5%. Interestingly, as the rates rose in the two large southern States, the rate of growth in Queensland fell. This was driven by a considerable decline in the net gain to Brisbane from internal migration.

In the years after 2001, the situation changed again as Sydney's growth rate fell back to around 0.6% and Brisbane's rate rose again to around 2.5%. In these years, house prices rose very considerably in Sydney and so its share of Australia's international migration fell. The New South Wales Government, while not being specific in policy terms, also attempted to limit the population growth of Sydney. These patterns provide another indication of an interaction between the growth of Sydney and Brisbane. In the years 1993 to 2005, the correlation between the rates of population growth of Sydney and Brisbane was a high, at −0.60. It appears that Brisbane's growth rate is partly a function of what is happening in Sydney. While Melbourne's growth also fell after 2001, it has remained noticeably higher than Sydney's growth rate in the early years of the 21st century.

Melbourne in these years increased its share of international migration to Australia and the Victorian Government introduced a specific policy to increase the population growth of the state. International migration policy in these years shifted somewhat towards a new regional approach where migrants were directed towards regions that wanted

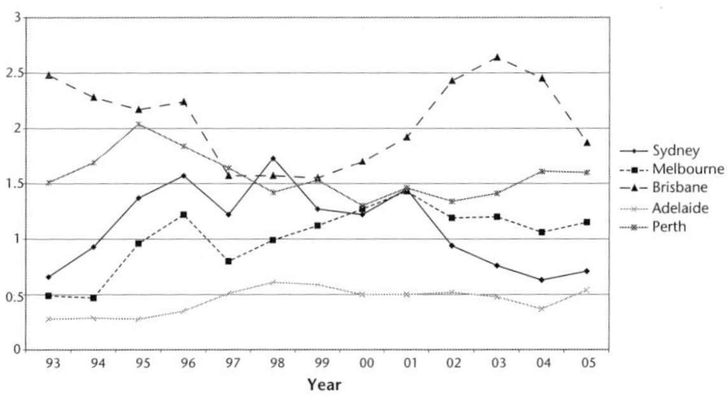

Figure 2: Annual rates of population growth (%): Australian cities, 1993–2005.

their populations to grow. Melbourne was defined as a region wanting more growth while Sydney (with Newcastle and Wollongong) was excluded from this scheme. In 2001, New South Wales received 42.0% of all international migration in net terms. This had fallen to 31.7% by 2006. In contrast, Victoria in 2001 received 21.7% of net international migration compared with 28.4% in 2006.

As the growth rates of the three largest cities swung up and down, the growth rates of Adelaide and Perth remained relatively flat throughout this period—Adelaide at a low level of 0.5% and Perth at a relatively high level of 1.5%. However, most recently, the regional international migration schemes have been successful in directing more immigrants in the direction of South Australia (almost exclusively to Adelaide). Net international migration to the state rose from 3,300 in 2001 to 11,150 in 2006. The growth of the City of Adelaide, at a little over 1%, is now at its highest level for over two decades.

As the final word on city growth, underlining volatility is the recent increase in growth in the city of Sydney— and a change of direction on the part of the Government of New South Wales to be more positive to immigration from overseas. Sydney's growth is once more approaching 1% per annum. Will this lead to a fall in Brisbane's growth? The answer from Figure 2 and data from the Australian Bureau of Statistics for 2006 appears to be 'yes'.

In contrast with overall population growth rates, the rates of fertility and mortality of the five largest cities do not change much from year to year, with the movements for cities usually being very consistent with the movements at the national level. The fertility rates in Sydney, Brisbane and Perth tend to be very similar (1.72–1.75 births per woman in 2005). Fertility is lower in Melbourne and Adelaide (1.64 births per woman in 2005). Mortality, expressed in terms of expectation of life at birth, is very close across all the Australian cities, with only Adelaide having a slightly lower expectation of life. In general, if we were to project the population of these cities, the changes in fertility and mortality used in the projections for each city would mirror the projected changes at the national level.

The main conclusion to be drawn from the above discussion is that the growth rates of Australian cities, particularly the three largest cities, tend to be very volatile. The variations in the growth rates of Sydney and Melbourne are associated mainly with changes in the level of international migration to Australia, which, in turn, reflects the state of the Australian economy and the policies of the two state governments to population growth. Periods of high levels of international migration to Sydney seem to be associated—with a lag of one year—with an increase in out-migration from Sydney to other parts of Australia. At times of low growth in Sydney, the growth is not only low because of low levels of international migration but also because of out-migration to other parts of Australia, including Brisbane. Brisbane's growth then tends to move in the opposite direction to Sydney's growth. This volatility, and some uncertainty surrounding its explanation, complicates the projection of the future populations of the three largest cities. In contrast, the growth rate for Perth has been relatively stable with internal migration playing a relatively minor role. Until very recently, Adelaide's growth had also been low and stable, but new policy initiatives have helped to push up the growth rate of that city. It is too early to conclude whether or not Adelaide's higher growth will be sustained.

Projections of the populations of the five major cities

The projection of the populations of Australia's cities is a difficult exercise because of the volatility of the migration movements. Since the late 1970s, Australia' fertility rate has shifted in a narrow range between 1.73 and 1.91. For this reason, and based on an assessment of the recent upward trend in Australia's fertility rate, the projections made here assume that the 2005 fertility rates in each city will remain constant into the future. The total fertility rates of the Australian cities in 2005 were: Sydney 1.74 births per woman; Melbourne 1.64; Brisbane 1.75; Adelaide 1.64; and Perth 1.73. Thus the variation across cities was also small. Mortality rates are presumed to continue to fall at a rate similar to their falls over the past 30 years and the same (national) mortality schedules are applied to each city.

With standard assumptions about fertility and mortality, the city projections are made with a range of migration scenarios. An underlying assumption, however, is that net international migration for Australia, presently at a high 160,000 per annum, will continue at a high level into the future, even increasing across time. This assumption is based on the logic that there will be a very strong demand for labour in Australia long into the future that will not be able to be met from the domestic supply of labour (McDonald and Withers 2007). Also, it is presumed that most of Australia's international migration will go to these five major cities and that, while some other places will grow very rapidly (see below), Australia will maintain roughly the same broad settlement pattern into the future as it has now. No new cities will emerge in Australia although the population concentrations around Brisbane will tend to coalesce with Brisbane into a very large sprawling South-East Queensland metropolis that, if taken as one, will be the largest city in Australia. The other examples where regional cities will grow in association with the large city are Melbourne–Geelong–Ballarat–Bendigo and Sydney–Newcastle–Wollongong.

Sydney is the most difficult of all cities to project because of the wide fluctuations in migration to and from Sydney. In the past 20 years, its net annual migration has been as high as 38,000 and as low as –2000, with an average of 13,000 per annum. There seems to be strong mechanism for out-migration to other parts of Australia from Sydney to balance international migration. With a higher level of international migration to Australia and increases in housing prices in other markets, it might be expected that Sydney's migration will run, on average, somewhere between about 20,000 and 30,000 per annum. At 20,000 per annum, Sydney's population would reach 5.76 million by 2051; at 30,000 per annum, it would reach 6.35 million (from 4.28 million in 2006). The range of these estimates, based on fairly small shifts in net migration, once more underlines the difficulty in projecting Sydney's population. An increase of over two million people in Sydney's population and improved transport infrastructure could see a flow-over of some of these people to Newcastle and Wollongong.

Melbourne has also had a wide range of annual net migration in the past 20 years, with a minimum of zero, a maximum of 30,000 and an average of 13,000—the same as Sydney's average. In recent years, international migration has shifted more in the direction of Melbourne, rather than Sydney, and the expectation would be that this shift will hold in future years because the expansion of Melbourne is not so constrained geographically and recent Victorian Governments have been more in favour of migration than their equivalents in New South Wales. Accordingly, the projections here assume that net migration to Melbourne will range between about 25,000 per annum and 35,000. Under the lower assumption, Melbourne's population would increase to 5.26 million in 2051 and to 5.83 million under the higher assumption (from 3.74 million in 2006).

Brisbane's annual net migration has always been strongly positive in the past 20 years. Its lowest level of annual net migration was 13,000—equivalent to the averages for Sydney and Melbourne. The highest annual net migration to Brisbane in the past 20 years was 33,000, and the average was 20,000. This movement is driven more by internal migration than by international migration, although in recent years direct international migration to Brisbane has been increasing. It is assumed here that Brisbane's net migration will lie between a low of 25,000 per annum and a high of 35,000 per annum—the same assumption as was made for Melbourne. The high level of these assumptions is based on the continuation of a mechanism by which Brisbane's population grows as a result of international migration to Sydney, but now combined with higher direct international migration to Brisbane. Among immigrants who had entered since 2001 and were still living in Australia at the time of the 2006 Census, 10.8% were living in Brisbane. In comparison, for those entering before 2001, 8.2% were living in Brisbane. The population of Brisbane would rise to 3.46 million by 2051 on the lower assumption and to 4.05 million on the higher assumption (from 1.82 million in 2006).

Adelaide's net annual migration has averaged only 2000 over the past 20 years, with the higher numbers being earlier in this period. In recent years, the South Australian Government has undertaken a major campaign to increase the state's population and this has met with some success in relation to increases in net international migration. Net international migration to Adelaide is now as high as it has ever been. On the other hand, like Sydney, increases in international migration have recently become more associated with increases in out-migration from Adelaide to other parts of Australia and so the net balance of migration has remained relatively low. This is a situation that can be expected to continue. However, if Adelaide's annual net migration were to continue at its average of the past 20 years (2000), its population in 2051 would be roughly the same as it is now, 1.15 million. Furthermore, its population would have aged considerably. Given the high population growth elsewhere and a strong Australian economy, this result seems unlikely. It is more likely that Adelaide will have a somewhat higher level of annual net migration. At 5000 per annum net migration, Adelaide's population would rise to 1.39 million by 2051.

Over the past 20 years, Perth's annual net migration has averaged 14,000, which is higher than that of Sydney and Melbourne. Net internal migration to Perth is relatively small. This means that Perth's net migration tends to be associated with net international migration to Australia. As Australia's net international migration is high at present and assumed to increase in the future, and as the Western Australian economy is booming and can be predicted to continue to boom, relatively high future annual net migration can be assumed for Perth. Here, a low of 20,000 per annum and a high of 30,000 per annum is assumed. At 20,000 per annum, Perth's population would rise to 2.75 million by 2051 and at 30,000 per annum to 3.33 million (from 1.52 million in 2006).

Overall, the results of these projections suggest relatively sizeable future growth for four of the five major Australian cities. This implies a major focus on planning for growth, especially in relation to infrastructure and the environment. The future economy of Australia will be more and more dependent upon human capital, especially high technologically-skilled workers. These workers will be heavily concentrated in these five major cities. This is the central reason that they will maintain their dominance.

The second-tier Australian cities

Seven other Australian cities have populations in excess of 200,000 persons. In the further discussion in this chapter, these seven cities are included along with the five major cities. The City of Darwin is also added given its significance. The boundary definitions of these additional cities are shown in Table 3 along with their 2006 populations and their 2001–06 population growth rates.

It is evident that the rates of population growth in these cities range widely. Of the five factors influencing population growth (birth rates, death rates, age structure of the population, internal migration and international migration), only age structures and internal migration contribute significantly to the wide range of the growth rates of the populations of these cities. The high population growth rates of the two South-East Queensland cities are

Table 3: Second-tier Australian cities: definitions, 2006 population and 2001–06 population growth rates

City	Definition		Population 30 June 2006	Average annual population growth rate, 2001–06 (%)
	Statistical division	Statistical subdivisions		
Newcastle	NSW 10	05	517,500	0.8
Wollongong	NSW 15	05	278,000	0.5
Gold Coast–Tweed	Qld 07 NSW 20	All 05	580,100	3.2
Sunshine Coast	Qld 09	All	295,100	3.4
Geelong–Barwon	Vic 10	05 and 10	228,800	1.0
Hobart	Tas 05	All	205,600	0.8
Canberra–Queanbeyan	ACT NSW 45	All 05	381,700	0.8
Darwin	NT 05	All	114,400	1.3

Source: Australian Bureau of Statistics Estimated Resident Population data.

mainly attributable to very high levels of internal migration. Darwin and Canberra have low rates of internal migration, but young populations, and this keeps their growth rates higher than they would be otherwise. Note the Queanbeyan component of Canberra–Queanbeyan has a high growth rate (around 2.3%) driven mainly by the availability of cheaper housing outside the ACT border than within it.

The demographic, social and economic characteristics of Australian cities in 2006

The remainder of the chapter uses data from the 2006 Census of Population and Housing to describe the characteristics of the 12 largest Australian cities and the City of Darwin. The analysis compares each of the 13 cities with the rest of Australia. The aim is to try to establish differences and similarities in character between the various cities to gain a greater understanding of their driving forces.

Ages of the people

The age distributions of Sydney, Melbourne and Perth are remarkably similar, while Brisbane is somewhat younger and Adelaide somewhat older than these three. With the exceptions of Canberra and Darwin—both of which are notably younger than all of the other cities—the second-tier cities are older than the first tier. This is mainly because second-tier cities have a lower percentage of their populations in the working ages of 15–64. In terms of age, the rest of Australia looks much like the older second-tier cities, although it has a higher percentage of children and the lowest percentage in the working ages. Broadly, however, places across Australia, at this level of disaggregation, do not vary greatly from each other in terms of the ages of their populations.

Indigenous population

The percentages of the total population represented by Indigenous Australians vary significantly across the Australian cities, from 10.7% in the case of Darwin to 0.3% in the case of Hobart (Table 5). The percentages are low in Melbourne and Geelong, but high in Newcastle. As expected, the Indigenous percentage is much higher in the rest of Australia than in the first- and second-tier cities, with the exception of Darwin. Despite the relatively low percentages that the Indigenous populations represent in the cities, almost 40% of Indigenous Australians live there simply because of the size of the cities.

Table 4: The age of the populations of Australia's cities, 2006

Place	Median age	Aged <15 (%)	Aged 15–64 (%)	Aged 65+ (%)
Sydney	35.3	19.2	68.9	12.0
Melbourne	36.0	18.5	69.0	12.5
Brisbane	35.0	20.0	69.1	10.9
Adelaide	38.5	17.6	67.4	15.0
Perth	36.0	19.1	69.2	11.7
Newcastle	38.0	19.5	65.3	15.2
Wollongong	37.0	19.8	65.5	14.7
Geelong–Barwon	39.5	19.3	64.8	15.9
Gold Coast–Tweed	39.4	18.3	66.9	14.8
Sunshine Coast	41.0	19.3	64.6	16.1
Hobart	38.0	19.0	66.7	14.2
Canberra–Queanbeyan	35.2	19.1	71.5	9.4
Darwin	32.2	22.0	72.8	5.2
Rest of Australia	38.0	21.1	64.6	14.3

Source: Medians are from the 2006 Census. Other characteristics are from the Australian Bureau of Statistics Estimated Resident Population data

Table 5: The Indigenous status of the populations of Australia's cities, 2006

Place	Indigenous (%)
Sydney	1.1
Melbourne	0.4
Brisbane	1.8
Adelaide	1.2
Perth	1.6
Newcastle	2.6
Wollongong	1.9
Geelong–Barwon	0.7
Gold Coast–Tweed	1.5
Sunshine Coast	1.3
Hobart	0.3
Canberra–Queanbeyan	1.4
Darwin	10.7
Rest of Australia	5.6

Source: 2006 Census

Countries of birth

Sydney, Melbourne and Perth have been, and continue to be, the major places of settlement of immigrants to Australia. In 2006, about one in three people living in Sydney, Melbourne and Perth were born overseas. At the next level are a range of cities where about 25% of the population were born overseas. This group includes Brisbane, Adelaide, Gold Coast–Tweed and Canberra-Queanbeyan. Then follow Wollongong, Geelong–Barwon, the Sunshine Coast and Darwin at around 20% born overseas. Hobart (13%) and Newcastle (10%) have the lowest percentages born overseas of all of the Australian cities.

A remarkable feature of immigration to Sydney and Melbourne is that the range of immigrant groups living in each city is very wide, with no one group being dominant. In Sydney, the main countries of birth of immigrants are the United Kingdom (4.6% of the population), China (2.9%), New Zealand (2.1%) and Vietnam (1.6%). These four countries are also in the top five countries of birth of immigrants in Melbourne, but in the case of Melbourne, Italy (2.2%) is second only to the United Kingdom (4.7%) as a source of population. Two large immigrant groups—Indian-born and Vietnam-born—make up similar proportions in both cities. Despite these similarities in ethnicity, it is the differences in ethnicity between the two largest cities that seem to mark their character. Melbourne has a southern European character and the proportion of the population that is Italian- and Greek-born in Melbourne is about twice the proportion of Sydney. On the other hand, Chinese, Filipino, Lebanese and Pacific Islanders (including Maori) are much larger components of the population in Sydney than in Melbourne—marking Sydney's distinctiveness.

Immigrants in Perth have a very different composition to those in Sydney and Melbourne, with a heavy concentration of persons from English-speaking countries. A remarkable 12.7% of Perth's population was born in the United Kingdom, 2.6% in New Zealand and 1.4% in South Africa—meaning that one in six Perth residents was from these countries. Indian-born immigrants outnumber China-born in Perth by two to one—the exact opposite of the situation in Sydney. The largest immigrant groups in Perth, therefore, might be described as groups that are less conspicuously different from the Australian population of British origin. Speculatively, those immigrants from South Africa and the United Kingdom may have been escaping ethnic situations that they feel less comfortable with in their countries of origin in search of an ethnically more homogeneous environment.

The countries that stand out in migration to the three Queensland cities are the United Kingdom and New Zealand, with New Zealanders being much more prominent than is the case for Western Australia. The UK and New Zealand account for 10.2% of the population of Brisbane, 14.8% of Gold Coast–Tweed and 12.4% of the Sunshine Coast. Immigrants from Asian countries are relatively rare in Gold Coast–Tweed and the Sunshine Coast.

For Adelaide and Wollongong, the birthplace composition reflects past movements from the United Kingdom and Southern Europe. For Darwin, the Greek-born and the Philippines-born are relatively prominent, while Canberra–Queanbeyan has a wide representation across most of the groups shown in the table. For the rest of Australia, the only groups of any size are the United Kingdom-born and the New Zealand-born.

The latest available information on the countries of birth of international migrants is the data on

Table 6: Countries of birth of the populations of Australia's cities, 2006

Place	Country of birth, percentage in given country										
	Australia	**China**	**Fiji**	**Greece**	**India**	**Italy**	**Lebanon**	**New Zealand**	**Philippines**	**South Africa**	**United Kingdom**	**Vietnam**
Sydney	65.5	2.88	0.71	0.84	1.40	1.17	1.44	2.14	1.37	0.75	4.62	1.64
Melbourne	69.0	1.64	0.22	1.56	1.52	2.21	0.44	1.57	0.73	0.52	4.68	1.73
Brisbane	76.8	0.69	0.41	0.17	0.46	0.41	0.05	4.42	0.60	0.78	5.77	0.72
Adelaide	74.9	0.73	0.08	0.94	0.59	2.00	0.14	0.82	0.45	0.36	9.29	0.99
Perth	66.3	0.57	0.04	0.21	1.05	1.40	0.06	2.58	0.39	1.40	12.74	0.75
Newcastle	89.9	0.21	0.07	0.16	0.16	0.32	0.02	1.05	0.27	0.22	3.49	0.06
Wollongong	78.5	0.65	0.08	0.41	0.21	1.51	0.25	0.95	0.39	0.23	6.73	0.27
Geelong–Barwon	83.9	0.25	0.03	0.27	0.21	0.99	0.05	0.89	0.34	0.24	5.39	0.13
Gold Coast–Tweed	74.0	0.42	0.14	0.12	0.26	0.35	0.05	7.45	0.56	0.81	7.34	0.08
Sunshine Coast	80.7	0.10	0.09	0.05	0.13	0.21	0.01	4.74	0.26	0.59	7.64	0.03
Hobart	87.2	0.36	0.07	0.24	0.22	0.34	0.02	0.85	0.21	0.28	5.11	0.05
Canberra–Queanbeyan	77.7	1.05	0.21	0.39	0.82	0.34	0.11	1.31	0.53	0.33	5.16	0.70
Darwin	79.5	0.32	0.13	0.94	0.51	0.76	0.02	2.30	1.55	0.27	4.61	0.49
Rest of Australia	89.3	0.11	0.06	0.09	0.17	0.43	0.02	1.44	0.28	0.27	4.19	0.05

Source: 2006 Census

Table 7: Religions of the population of Australia's cities, 2006

Place	Christian	Muslim	Buddhist	Hindu	No religion
Sydney	71.5	4.4	4.2	1.9	15.8
Melbourne	66.3	3.2	4.0	1.3	22.5
Brisbane	73.6	0.9	2.0	0.7	21.3
Adelaide	67.5	1.0	2.2	0.5	27.1
Perth	68.0	1.7	2.5	0.6	25.2
Newcastle	81.3	0.3	0.6	0.2	16.7
Wollongong	80.1	1.4	1.0	0.3	16.2
Geelong–Barwon	72.9	0.6	0.8	0.2	24.2
Gold Coast–Tweed	74.3	0.6	1.3	0.3	21.9
Sunshine Coast	71.4	0.1	1.0	0.1	25.8
Hobart	72.6	0.3	0.8	0.3	24.6
Canberra–Queanbeyan	68.5	1.4	2.3	1.0	25.1
Darwin	64.0	1.0	2.6	0.5	30.2
Rest of Australia	76.5	0.3	0.6	0.1	21.3

Source: 2006 Census

settler arrivals to the states for the 2006–07 financial year (Department of Immigration and Citizenship 2007). A very high proportion of the immigrants to the states go to the cities shown in Table 6, so these figures are indicative of movements to the cities. The largest single movement of any country of birth to any State was the 11,278 New Zealanders who settled in Queensland; the second largest was the 7286 people from the United Kingdom who settled in Western Australia. Third place went to the 5450 China-born persons who settled in New South Wales followed closely by the 5250 India-born also settling in New South Wales. The India-born was the largest group settling in Victoria (4826). This suggests that the 'Asianisation' of Sydney and Melbourne continues to progress along with the 'Britishisation' of the three South-East Queensland cities and Perth.

Religions of the people

The migration flows described above explain city differences in religion. Muslims, Buddhists and Hindus are much more prominent in Sydney and Melbourne than in any of the other cities. These three religions are followed by 10.5% of Sydney's population and 8.5% of Melbourne's population. Outside of the 13 cities in Table 7, only 1% of the population followed these religions.

The percentage with no religion is much lower in Sydney than in the other four major cities—once more marking Sydney as different from other cities. Indeed, only Newcastle and Wollongong come close to Sydney in this regard. Darwin is the least religious of the cities, followed by Adelaide—the city of churches.

Family life

Sydney also emerges as conservative in relation to family measures. Of the 13 cities, it has the highest percentage of 15–24 year-olds still living with their parents, the second lowest percentage of 15–34 year-olds living in group households (after Geelong), the second lowest percentage of lone parents among women aged 35–44 (after Canberra) and the lowest percentage of men aged 35–44 living alone. Lone parenthood for women and living alone for men aged 35–44 years is prominent in Adelaide and Hobart—both cities with relatively low economic growth. The Gold Coast and Sunshine Coast also have high proportions of lone parents. Overall, excluding Canberra and Darwin, the incidence of lone parents is higher in the second-tier cities than in the first tier.

Brisbane stands out as having the lowest proportion of 15–24 year-olds living with their

Table 8: Selected family indicators for Australian cities, 2006

Place	Living with parents, percentage of all persons aged 15–24	Living in group household, percentage of all persons aged 15–34	Living in a couple relationship, percentage of all persons aged 35–44	Lone parents, percentage of all females aged 35–44	Living alone, percentage of all males aged 35–44
Sydney	70.0	7.6	73.3	13.3	10.5
Melbourne	68.3	9.1	73.5	13.4	11.2
Brisbane	58.9	9.9	73.4	13.7	10.8
Adelaide	64.2	8.1	71.0	15.2	14.9
Perth	63.7	7.5	74.1	13.7	12.9
Newcastle	64.7	6.9	72.4	15.7	12.0
Wollongong	68.8	6.9	73.8	14.7	11.6
Geelong–Barwon	65.8	7.1	73.4	15.2	12.7
Gold Coast–Tweed	59.0	10.1	71.0	16.2	11.5
Sunshine Coast	63.3	7.7	73.1	15.7	10.9
Hobart	59.8	9.1	70.8	15.6	13.7
Canberra–Queanbeyan	61.5	10.7	73.8	13.0	12.4
Darwin	51.5	8.6	68.0	13.9	15.1
Rest of Australia	59.1	5.7	73.3	13.1	13.0

Source: 2006 Census

parents. This may be related to in-migration of people of this age from other parts of Australia. Canberra tops the scale for young people living in group households. Again, however, the broad conclusion is that variations in family indicators across the cities are not very wide. This is underlined by the percentages of persons aged 35–44 living as couples. With the exception of the somewhat lower percentage for Darwin, this indicator is very much the same across the cities of Australia.

Relationship formation and commencement of child-bearing

Continuing the theme of conservatism in Sydney, women in Sydney appear to marry earlier than women in the other major cities. Only Sydney's neighbours, Wollongong and Newcastle, had earlier ages at marriage (low percentage never married in age group 25–34) than Sydney. More strikingly, women aged 20–34 in couple relationships are much less likely in Sydney to be in a de facto relationship, and more likely to be married, than in any other city. Wollongong, once again, is next in line to Sydney on this measure. Melbourne also has a low percentage of de facto couples. The fact that the largest and most cosmopolitan city in Australia consistently scores highly on measures of conservative behaviour is very interesting. It is likely that part of this result derives from Sydney's ethnic composition, including the ethnic origins of the second generation. De facto living is much less common among those of Southern European and Asian descent than it is for other descent groups. This would also help to explain the low incidence of de facto relationships in Wollongong and Melbourne as well. However, in the case of Sydney, given the size of difference with other cities, there is probably more to the explanation than ethnicity.

Darwin has the highest percentages living de facto and the second lowest percentage ever married at age group 25–29. This can be expected from the high Indigenous proportion of Darwin's population. However, what is not expected is the late

Table 9: Selected relationship indicators for Australian cities, 2006

Place	Percentage never married, females aged 25–34	Percentage de facto among those in couple relationships, females aged 20–34	Percentage with no births, females aged 25–29
Sydney	45.6	26.7	69.6
Melbourne	49.5	31.0	74.2
Brisbane	47.4	36.6	64.0
Adelaide	48.8	36.3	65.9
Perth	48.0	37.3	66.1
Newcastle	44.9	35.4	53.2
Wollongong	41.2	29.0	56.2
Geelong–Barwon	47.7	36.0	63.0
Gold Coast–Tweed	49.4	40.4	61.4
Sunshine Coast	46.0	38.9	55.6
Hobart	52.9	43.1	58.9
Canberra–Queanbeyan	49.3	36.0	71.7
Darwin	52.1	45.5	58.5
Rest of Australia	44.5	38.1	46.9

Source: 2006 Census

marriage and high percentage of de facto couples in ethnically homogeneous Hobart. From Table 8, Hobart was also observed to have a high percentage of female lone parents and men aged 35–44 living alone and a low percentage of 15–24 year-olds living with their parents. Hobart as liberal and Sydney as conservative—these are not the standard views of Australian cities.

The final column of Table 9 provides an indicator of the timing of the first birth. The first birth occurs latest in Melbourne followed by Canberra. Second-generation Southern Europeans probably contribute to the late child-bearing in Melbourne and this may also help to explain the relatively late child-bearing in Sydney. However, the three cities with the latest child-bearing—Melbourne, Canberra and Sydney—are all cities where the pursuit of careers by young women is likely to be strongest and where education levels are highest. As expected, child-bearing is earliest outside of the 13 cities, but also in Newcastle, Wollongong and the Sunshine Coast.

Education levels

All of the Australian cities considered in this chapter have universities and, as such, they are all centres of education. However, some are more educated than others. Using the education indicators for young people shown in Table 10, Canberra consistently emerges as the highest educated of Australia's cities, but Melbourne is a close second followed by Sydney and Brisbane. Two of the smaller cities—Wollongong and Hobart—have relatively high percentages of 15–24 year olds being in university or in any form of education. In regard to 15–24 year olds currently in any form of education, the lowest results apply to Darwin, Sunshine Coast and Gold Coast–Tweed. However, based on the percentage of 25–34 year-olds who have completed Year 12, the industrial cities of Wollongong and Newcastle have the lowest levels. In general, with the exception of Canberra, education levels are higher for younger people in the first-tier cities.

Income levels

The median weekly gross income of men and women aged 35–44 years is used as a measure of income earning capacity in the cities of Australia. This age group is chosen because career paths are usually established by this age and because, for both sexes, labour force participation rates are relatively high at these ages. The hierarchy of relative

Table 10: Selected education indicators for Australian cities, 2006

Place	Persons aged 15–24		Percentage completed Year 12, persons aged 25–34
	Percentage in any form of education	Percentage at university	
Sydney	53.3	18.8	75.9
Melbourne	56.2	21.2	76.6
Brisbane	48.2	19.7	74.5
Adelaide	51.6	18.8	67.4
Perth	48.5	19.1	67.7
Newcastle	49.6	14.7	55.7
Wollongong	52.9	16.9	59.9
Geelong–Barwon	53.7	14.1	65.1
Gold Coast–Tweed	42.5	13.1	64.9
Sunshine Coast	43.5	9.2	64.5
Hobart	51.4	16.9	60.8
Canberra–Queanbeyan	56.2	21.8	83.1
Darwin	35.5	9.0	62.8
Rest of Australia	42.9	7.2	54.8

Source: 2006 Census

income levels does not exactly match the hierarchy of relative education levels shown in Table 10. For both men and women, income levels are considerably higher in Canberra than in any other city. For men, the next highest income levels are in the boom town of Perth followed by Darwin. For women, incomes in Darwin are second only to Canberra. Both Canberra and Darwin have relatively high percentages of their workforces employed in professional or paraprofessional jobs in the public sector and this no doubt explains their high income levels. More basically, Canberra and Darwin also have much higher labour force participation rates for women than the other cities. The male workforce in Perth has a high percentage of its workforce associated in one way or another with the booming mining, industry including many very high paid miners who commute by air from their homes in Perth to the mining sites.

After Canberra, Perth and Darwin, for men, the next highest cities in order of income level are Sydney, Wollongong and Brisbane. Sydney and Brisbane can have been expected to have high incomes because of the nature of their workforces, but the relatively high placement of Wollongong is more difficult to explain. Women also have relatively high incomes in Sydney and Brisbane, but in Wollongong are almost on the bottom of the income scale. Incomes for men are lowest in the Sunshine Coast, Gold Coast, Adelaide and Hobart. Workers on the Sunshine and Gold Coasts tend mainly to be in lower paid service industries that support the tourism industry and the retired populations that live in these areas. The low incomes in Adelaide and Hobart are associated with the low economic growth rates of these cities.

If the incomes of men and women are combined to provide an indication of couple incomes, the Canberra couple aged 35–44 was earning a median income of around \$105,000 per annum and the Darwin couple around \$87,000. This compares with around \$80,000 for couples in Sydney, Brisbane and Perth, \$75,000 for couples in Melbourne and \$70,000 for couples in Adelaide. The other interesting feature of Table 11 is the income ratio between the sexes, the median income for women as a percentage of the median income for men. The ratio is high in Canberra and Darwin, probably because of the relatively high labour force participation rates of women in these cities. The ratio is low in the industrial cities of Geelong and Wollongong, but also, interestingly, in Perth. Perth may

Table 11: Median gross income for males and females aged 35–44, Australian cities, 2006

Place	Median weekly gross income, ages 35–44		
	Males ($)	Females ($)	Females/Males (%)
Sydney	999	532	53
Melbourne	944	489	52
Brisbane	969	535	55
Adelaide	847	489	58
Perth	1059	485	46
Newcastle	894	449	50
Wollongong	983	429	44
Geelong–Barwon	910	421	46
Gold Coast–Tweed	850	484	57
Sunshine Coast	799	461	58
Hobart	835	502	60
Canberra–Queanbeyan	1209	795	66
Darwin	1020	647	63
Rest of Australia	782	436	56

Source: 2006 Census

well be still operating in a mode where women are less likely to work because their husbands are earning high incomes.

Conclusion

Australian cities have some remarkable similarities, but there are emerging differences. In demographic terms, fertility and mortality rates do not vary much across the various cities—both first- and second-tier cities. With the exception of the somewhat older age structure of Adelaide, the first-tier cities also have very similar age distributions. In general, the second-tier cities also have very similar age structures, which are somewhat older than the first-tier cities, except for Adelaide. Two cities—Darwin and Canberra—stand out as having younger populations and higher percentages in the working ages.

The essential demographic difference between the cities is their patterns of internal and international migration. Sydney's growth is determined by the level of international migration, but there is a strong negative association of international migration and internal migration. As international migration to Sydney increases, out-migration from Sydney to other parts of Australia also increases with a one-year lag. In recent years, largely as a result of state government policy, Sydney has been taking a smaller share of Australia's immigrants and its growth has slowed. However, labour shortages may be reversing this policy direction. Brisbane's has grown mainly through internal migration from other parts of Australia, although it has increased its share of international migrants, especially from New Zealand, in recent times. Its population growth rate seems to move in concert with Sydney's population growth, but in the reverse direction—Brisbane's growth rises as Sydney's growth falls, and vice versa. From the very low growth rates of the early 1990s, Melbourne's growth has risen to levels above 1% per annum. The Victorian Government has been promoting population growth in Melbourne and that city's share of Australia's international migration has increased. Perth growth has been very stable over a long period at a relatively high level. Its growth difference stems almost exclusively from international migration. Finally, Adelaide's growth has been low because of a low share of international migration and negative internal migration. More recently, through aggressive government policy, it has increased its numbers of international immigrants, but this has been associated with an increase in movement out of the city to other parts of Australia.

No major shift in Australian settlement patterns is expected in the next 50 years, although some cities, such as the Gold Coast and the Sunshine Coast, will grow faster than others. The projections made for the chapter suggest that the ranges of population sizes in 2051 will be: Sydney 5.76–6.35 million; Melbourne 5.26–5.83 million; Brisbane 3.46–4.05 million; Adelaide 1.39 million; and Perth 2.75–3.33 million. If the urban regions in South-East Queensland are considered as one unit, this will be the largest city in Australia by 2051. My expectation, based on the likely impact of labour demand upon immigration levels, is that the 2051 populations of Australia's cities will be near the top end of these estimates. This implies major infrastructure planning issues.

Analysis of 2006 Census data also indicates some emerging differences between the 13 first- and second-tier cities generated by international migration. International migration to Sydney and Melbourne is very Asian in character, while international migration to Brisbane and Perth is very British in character. These are strong trends at present and, if they were to continue, the broad cultural similarity of the Australian cities may diverge somewhat. This divergence will be driven more by the strengthening 'Britishness' of Brisbane and Perth than by the more cosmopolitan characters of Sydney and Melbourne. Sydney in particular stands out as taking on more conservative characteristics, such as a low percentage with no religion, high percentages of non-Christian religions, children staying at home longer with their parents, earlier marriage, low percentages in de facto relationships and lower percentages of middle aged people living alone. This is somewhat at variance with the popular image of Sydney as liberal and trendy. No doubt what is being observed here is a wider diversity of cultures in than in any other Australian city. Despite the cessation of migration from Southern Europe, Melbourne maintains its Southern European characteristics through its second generation.

In human capital terms, Canberra stands out as having by far the highest education and income levels. This is especially the case for the income levels of women. Canberra women aged 35–44 years have almost the same average income level as men of the same age in Hobart and Adelaide, despite having lower labour force participation. Following Canberra, male income levels are highest in Perth and Darwin, although these two cities do not score highly in terms of education levels. In education terms, Melbourne rates second to Canberra, closely followed by Sydney, but Sydney income levels are higher than Melbourne income levels. The rapidly growing coastal cities, Gold Coast and Sunshine Coast, have relatively low income levels for middle aged male workers and relatively low education levels for young people.

References

Australian Bureau of Statistics (1998). *Population Projections, 1997 to 2051*, ABS Catalogue No. 3222.0, Australian Bureau of Statistics, Canberra.

Department of Immigration and Citizenship (2007). *Settler Arrivals, 2006–2007*. Australian Government, Canberra.

Flood J, Maher C, Newton P and Roy J (1991). *The Determinants of Internal Migration In Australia*, CSIRO, Division of Building, Construction and Engineering, Melbourne.

McDonald P and Withers G (2007). 'Population and Australia's future labour force', policy Discussion Paper. Academy of the Social Sciences in Australia, Canberra.

United States Census Bureau (2007). *Metropolitan and Micropolitan Statistical Areas*. <http://www.census.gov/population/www/estimates/metroarea.html> Accessed on 7 October 2007.

10

Future Urban Population and Settlement Transitions

Graeme Hugo

Three-quarters of a century ago, Griffith Taylor (1947) declared that the broad structure of Australia's population distribution had been set before the end of the 19th century and, although the population has subsequently increased five times, his judgement has been borne out (Rowland 1982; Hugo 2003). However, within these broad lineaments imposed by massive physical constraints, change has occurred in Australia's settlement system—between 1947 and 2006 the proportion of the population living in urban areas increased from 76.1 to 88.0%, the proportion in mainland state capitals increased from 51.0 to 60.2% and the proportion living within 50 km from the coast increased from 79.9% in 1991 to 85.3% in 2006. The settlement system is too often seen as a static backdrop against which other social, economic political and cultural forces are played out, but in fact it is constantly changing. Moreover, where people live *is* important. It influences their day-to-day lives in a myriad of ways and, in turn, shapes the nature and extent of their impact on the physical environment. As the 21st century progresses, the forces impinging on the settlement system in Australia are changing and the settlement system is as dynamic as it has ever been and further and new transitions can be anticipated.

The present chapter seeks first of all to provide a more comprehensive conceptualisation of the Australian settlement system. It is argued that the conventional rural–urban dichotomy fails to capture the growing complexity and dynamism of the way in which Australians are distributing themselves across the nation. The next section seeks to describe the contemporary Australian settlement system and its dynamics. It also is important to examine the changing mix of internal migration, international migration and natural increase in the growth of the various elements in the settlement system. In particular, the changing role of international migration in shaping the growth of settlements is examined. Such elements as the increasing scale of temporary migration and the State Specific and Regional Migration Scheme are discussed. The focus then moves to a consideration of some of the drivers of contemporary and future change in the Australian urban system. The forces examined are, firstly, climate change and the increasing mismatch in Australia between rainfall run-off and that of the population. The ageing of baby boomers as they enter retirement will also be an important exogenous force shaping the settlement system over the next two decades. Changes in the structure of Australia's economy, such as the increasing significance of mining, will also be examined as will social change processes. There will then be a consideration of the possible impact of these emergent forces in shaping the Australian settlement system of the next two decades. Finally, the role for policy in the settlement system will be addressed and its challenge of developing interventions that will enhance efforts to move toward sustainability.

Conceptualising the Australian settlement system

Conventionally, the Australian settlement system has been differentiated between urban and rural areas. The former are defined as population clusters of 1000 or more, while rural areas constitute the residual. The distinction within these categories varies by size of urban places and between rural localities (clusters of 200 to 999 persons) and other rural areas. The population of these areas has generally been defined as the 'usually resident population' (ABS 2006a). While these definitions have never adequately captured the space–time complexity of Australian settlement systems, they are becoming less appropriate with each passing decade (Champion and Hugo 2004). What are these elements of increasing complexity in the settlement system?

Firstly, from a spatial perspective, the urban/rural dichotomy has become less useful as new forms of settlement emerge that are transitional between quintessentially urban and rural places. Moreover, there is increasing recognition that the extent of 'urbanness' and 'ruralness' represent only one of the spatial dimensions of the settlement system. Coombes and Raybould (2001) have identified two other elements within the system that need to be recognised. One of these relates to the density of population—the significance of which has been established in Australian non-metropolitan areas by Smailes *et al.* (2002) although this has not been investigated in urban areas. The other essential dimension for classifying places within nations relates to the degree of accessibility (or remoteness). In the Australian context, the ARIA system developed by the National Centre for Social Applications in Geographical Information Systems (GISCA) (Bamford *et al.* 1999) produced an index for each location in non-metropolitan Australia in terms of its road distance from the nearest places in five levels of the urban hierarchy. This was later taken up by the ABS to divide Australia into five zones:

1. **Highly accessible**—locations with relatively unrestricted accessibility to a wide range of goods and services and opportunities for social interaction.
2. **Accessible**—locations with some restrictions in terms of accessibility to some goods, services and opportunities for social interaction.
3. **Moderately accessible**—locations with significantly restricted accessibility to goods, services and opportunities for social interaction.
4. **Remote**—locations with very restricted accessibility to goods, services and opportunities for social interaction.
5. **Very remote**—locationally disadvantaged, with very little accessibility to goods, services and opportunities for social interaction.

GISCA has subsequently developed a version of ARIA for application within large Australian metropolitan areas (GISCA 2002).

It is also evident that there are other key differences between Australian settlements that go beyond the three dimensions considered above. There are, for example, distinctive differences between non-metropolitan coastal settlements and their inland counterparts regardless of whether they are urban or rural. Some (for example, Holmes 1994; Hugo and Bell 1998) have argued that there is an increasing dichotomy emerging in non-metropolitan Australia between coastal and inland areas. Moreover, there are clear distinctions between the pastoral rangelands of the interior and the dry farming areas of the wheat–sheep belt in relation to settlement patterns. One useful classification of Australian settlement systems, which takes some of these elements into account, has been developed by ABARE (Garnaut *et al.* 2001) and includes the following categories:

1. **Metropolitan**—capital cities.
2. **Regional cities**—non-capital cities with more than 100,000 persons.
3. **Populated coastal**—173 Statistical local areas (SLAs) in densely populated areas within 80 km of coast.
4. **Populated inland**—338 SLAs inland from coast but not in remote and very remote ARIA categories.
5. **Remote**—136 SLAs in remote and very remote ARIA categories.

With respect to identifying the populations of the various spatially defined categories in the settlement system, it is becoming increasingly the case that the conventional use of the usually resident population has limitations (Hugo 2007a). The fact Australians can, and do, have their residence in one type of settlement and work in another type has long been the case, as was demonstrated in relation to rural–urban commuting (Holmes 1970) and urban–rural town farming (Williams 1970) more than four decades ago. However, the mobility of Australians has increased exponentially and this has increased their capacity to move readily between settlement types on a regular basis. More importantly, it means that the real population of particular settlements can vary greatly between day and night, weekdays and weekends and seasonally. The phenomena of weekly commuting between capital cities, fly-in–fly-out work arrangements in mining communities, grey nomads, increasing second home ownership, the winter movement from south to northern Australia, tourist movements, student migrations, and so on, all produce volatility in the real population of places at different times. The increasing scale and complexity of this temporary migration has been demonstrated by Bell and Ward (2000) but it was evident as long ago as in the data collected at the 1981 Australian population census (Hugo 1986). The ABS has plans to test questions on multiple residences at the 2011 census enumeration and this is a timely recognition of the fact that Australians increasingly divide their lives between multiple locations and that the distribution of 'usual residents' is only one of Australia's population geographies and for many communities it is not the most useful one for planning.

The changing Australian settlement system

Several of the conventional definitions of the Australian settlement system show a deceptive stability when intercensal changes in the distribution of the national population are examined. Figure 1, for example, shows the percentage of the national population living in the main settlement categories over the last century. It will be noted that there is a pattern of an increasing proportion of the population living in urban areas—especially with more than 100,000 residents—up to 1971, but thereafter a degree of stability, although the pattern of increasing urbanisation appears to have resumed in the last decade, albeit at a slower rate than the earlier post-war years. This apparent stability in the system, however, masks a considerable degree of change as will be demonstrated below.

Firstly, although the proportion of Australians whose usual residence is within cities with more than 100,000 residents has increased from only 63 to 67% between 1986 and 2006, the spatial configuration of that population distribution has changed significantly under the impact of two opposing forces. On the one hand, there has been a strong pattern of revival of growth of the populations of the central core and inner suburbs of the nation's major metropolitan areas. This is demonstrated in Figures 2 and 3, which show the most recent intercensal population growth in statistical local areas in Sydney and Melbourne.

A number of interesting elements have been involved in this reversal of the doughnut pattern of population change that prevailed in Australian

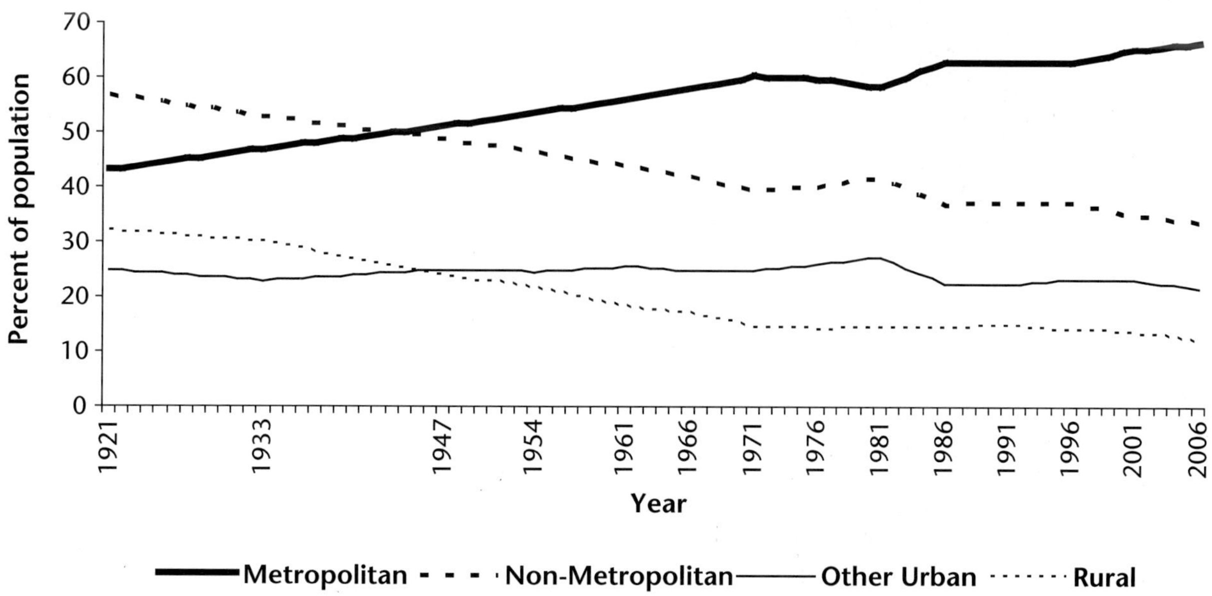

Figure 1: Australia—changing distribution of the population between urban and rural sectors, 1921 to 2006 (Source: Australian Censuses, 1921–2006).

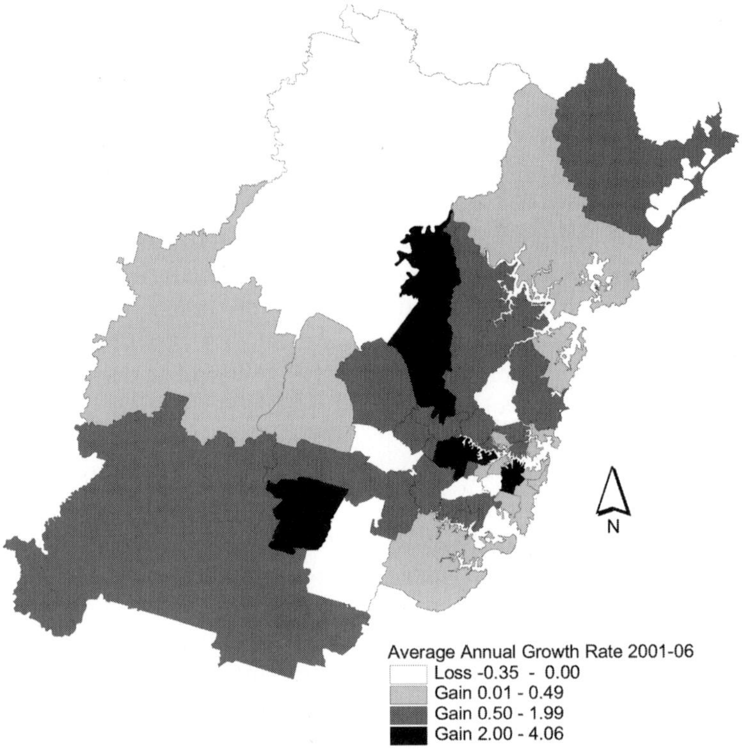

Figure 2: Sydney Statistical Division—average annual growth rate by LGA, 2001–2006 (Source: ABS, 2006 Census).

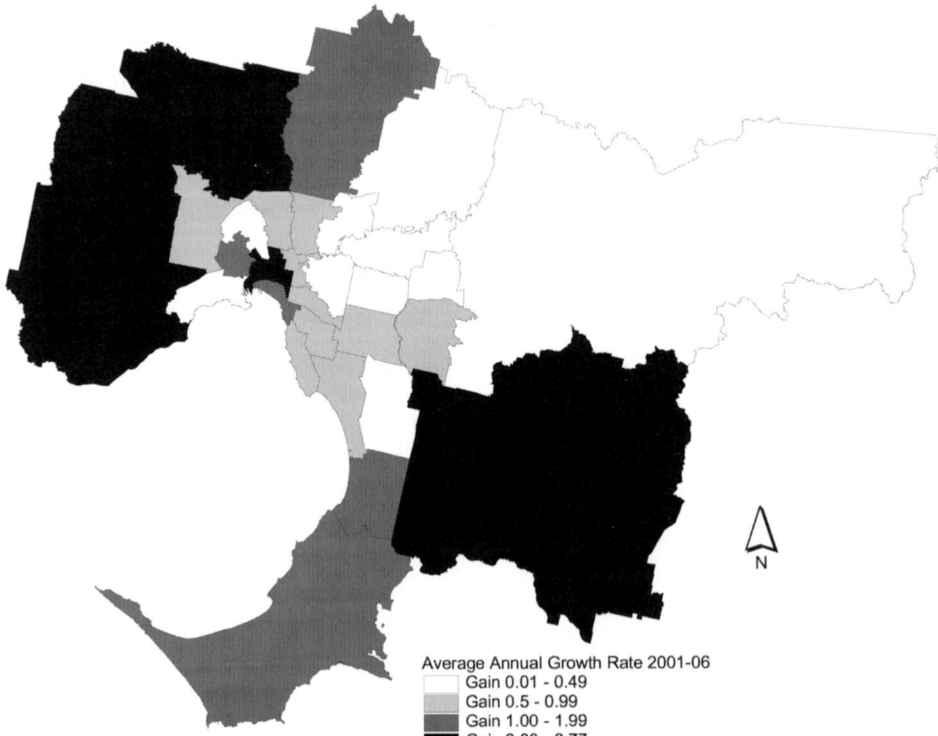

Figure 3: Melbourne Statistical Division—average annual growth rate by LGA, 2001–2006 (Source: ABS, 2006 Census).

cities during the first four post-war decades. These include:

- The processes of gentrification, which has involved several subgroups of Australians expressing a preference for inner city living. This involves not only young professionals, but also increasingly baby boomers who have entered the empty nest of the life cycle and sold their outer suburban family homes and chosen to move to inner or older seaside suburbs in cities such as Sydney, Melbourne, Perth, Brisbane and Adelaide.
- The increasing number of temporary migrants to Australia (Khoo *et al.* 2003; Hugo 2006), a disproportionate share of whom choose to live in inner areas while in Australia. Especially prominent here are the 210,690 (in 2006) overseas students who concentrate in inner areas. In the City of Adelaide, for example, where the 2001–06 intercensal population growth was 5.24% (the second highest in the state), more than a quarter of dwellings are occupied by students, most of them foreign.
- The activities of local government and private developers to increase the number of new housing units in inner areas, which is building on, and reinforcing, the growing preference among some groups for inner city living. Urban consolidation has become a popular, though contested, policy of local governments in Australian cities.
- Structural change in the economy, which has rendered significant tracts of inner city land previously occupied by industry redundant and provided opportunities for the development of housing.
- Technological obsolescence of high rise commercial buildings in the capital city CBDs, which are converted to residential apartments

While the centripetal forces encouraging inner city growth have arisen in recent decades, the centrifugal forces that drove the rapid spread of low-density outer suburbs in Australian cities during the post-war era have continued. This is apparent in Figures 2 and 3, but the rapid growth of population on the fringes of the built-up areas of Australian cities is only one element of this. There has been an extension of the commuting zone outside the continuous urban fabric of these cities, encompassing dormitory communities of many kinds. The improvement of transport infrastructure has facilitated this, as well as the increasing numbers of Australians choosing to have their home in a semi-rural situation while continuing to work within metropolitan areas. There can be little doubt that Australia's metropolitan areas are extending their areas of influence. It has been suggested (Hugo *et al.* 2003) that Australia needs to follow many other nations and recognise that there exists a distinct form of peri-urban zone in which there are mixed urban and rural land uses, but the influence of the major metropolitan area remains substantial. Certainly these zones present a number of distinctive issues for planners and challenges to environmental sustainability.

There has also been a shift in the trajectory of growth of Australia's five major cities as demonstrated in Figures 4 and 5, which show not only the overall size of the major capitals, but also some important shifts in the *rates* of population growth. Melbourne and Adelaide were the fastest growing in the early post-war years as major centres of industrial expansion. With the structural change of the 1970s, however, they lost out to Sydney, which enhanced its status as the nation's global city (Connel 2000). However, in recent years Melbourne has resumed a faster growth rate. The continued rise of Brisbane, both in terms of total population and growth rates is strongly apparent and the growth of the south-eastern Queensland conurbation that incorporates not only Brisbane but also the Gold and Sunshine Coasts (2006 population of 438,261 and 201,806) is a major element in Australia's emerging settlement system. Perth's consistent pattern of growth since the 1960s has been strongly influenced by the state's long-lasting mining boom. The extent to which the 'commuting fields' of Newcastle–Sydney–Wollongong overlap and constitute a single megalopolitan region—and Melbourne overlaps with its satellites, especially Geelong—is increasingly important and perhaps there needs to be new thinking about these metropolitan regions in the settlement system (refer to the following chapter).

Moving to an examination of trends in population change in Australia's 'other urban' centres of less than 100,000 inhabitants, Table 1 shows the

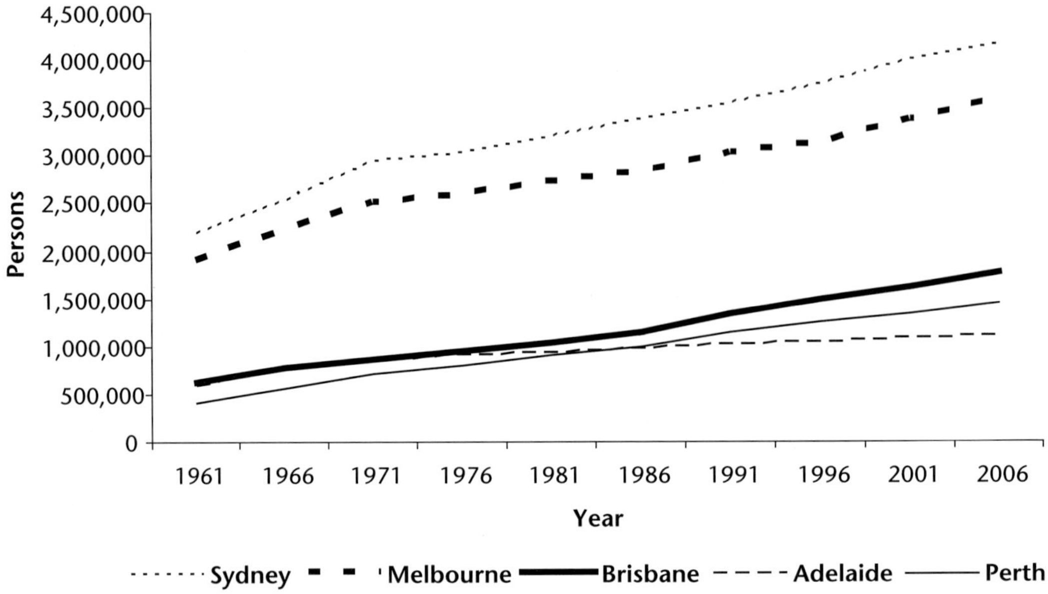

Figure 4: Major Australian capital cities—population, 1961 to 2006 (Source: Australian Censuses, 1961 to 2006)

pattern of growth in various size categories of urban centres. Of course, this is influenced by the intercensal movement of centres up, and to a lesser extent, down, the hierarchy of size categories. Nevertheless a pattern of strong growth is evident in several size categories. However, such data can be misleading with single-size categories containing both declining and rapidly growing centres.

This is evident in Figure 6, which depicts all 'other urban' places and rural localities in the nation and indicates whether they are growing or declining. Some clear patterns are evident:

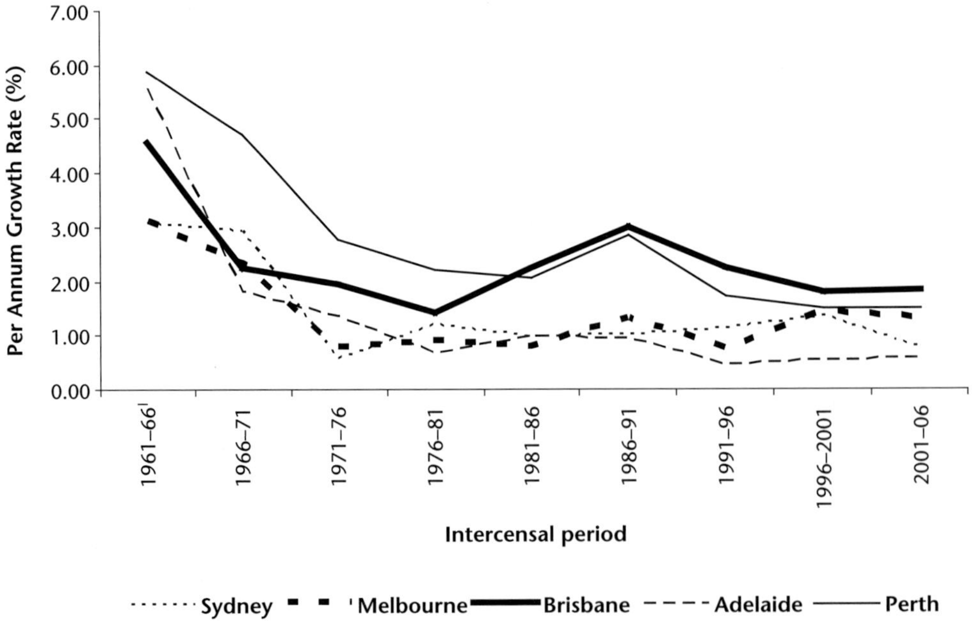

Figure 5: Major Australian capital cities—average annual intercensal population growth rates (%), 1961–66 to 2001–06 (Source: Australian Censuses, 1961–2006).

Table 1: Distribution of population by settlement size, 1966, 1986, 1996 and 2006

	Number of urban centres				Percentage of population			
	1966	1986	1996	2006	1966	1986	1996	2006
500,000 and over	5	5	5	5	56.0	54.5	53.1	55.6
100,000–499,999	4	7	8	10	5.4	8.2	9.2	11.5
75,000–99,999	1	1	2	4	0.8	0.6	1.0	1.6
50,000–74,999	5	7	6	8	2.4	2.8	2.1	2.5
25,000–49,999	5	11	25	23	1.5	2.3	4.6	3.9
20,000–24,999	11	22	17	10	2.1	3.1	2.1	1.1
15,000–19,999	17	17	14	19	2.5	1.9	1.3	1.7
10,000–14,999	19	29	37	34	2.0	2.2	2.5	2.1
5000–9999	61	83	86	91	3.8	3.8	3.5	3.2
2500–4999	103	127	165	172	3.1	2.7	3.2	3.0
2000–2499	50	71	64	62	1.0	1.0	0.8	0.7
1000–1999	178	252	312	297	2.2	2.2	2.5	2.1
Total urban	459	632	741	732	82.9	85.4	86.0	88.0
Total rural					16.9	14.5	14.0	12.0
Total population[a]					100.0	100.0	100.0	100.0
Total number ('000)					11,599	15,602	17,892	19,855

[a] Includes migratory population.

Source: Rowland 1982; ABS Censuses of 1986, 1996 and 2006

- Most striking is the concentration of rapid growth centres along the eastern, south-eastern and south-western coasts. This is clear evidence of what has become known as the sea-change phenomenon (Burnley and Murphy 2004).
- Equally in evidence is the clustering of growth centres within, and slightly beyond, the commuting zones of the major metropolitan centres. This reflects the increasing extent of commuting and lateral extension of metropolitan influence discussed above.
- Inland growth centres are less in evidence, but include centres in favoured localities such as mountains and riverscapes where the 'tree change' phenomenon is increasing. Also, larger provincial centres, which have taken over the roles and functions of smaller centres within their extending hinterlands, are also growing. Furthermore, the mining boom associated with China's seemingly insatiable demand for coal and minerals has seen growth in communities across the northern states and South Australia. Some remote indigenous communities in the northern states have also grown.
- The wheat–sheep belt and much of the rangelands pastoral area stand out as areas in which the majority of centres are static or declining. Similarly, some older mining centres in remote areas are declining.

Hence there is considerable diversity in the pattern of population growth (or decline) across non-metropolitan Australia. Moreover, there is diversity in the form that the population growth takes. Along the coasts, for example, there has been the emergence of 'string cities' in which previously discrete coastal centres are becoming linked by more-or-less continuous housing development between them.

A different perspective on the dynamics of the settlement system can be gained if we examine change within the various accessibility/remoteness zones. Table 2 shows the increasing concentration of the Australian population in the most accessible zones over the last decade.

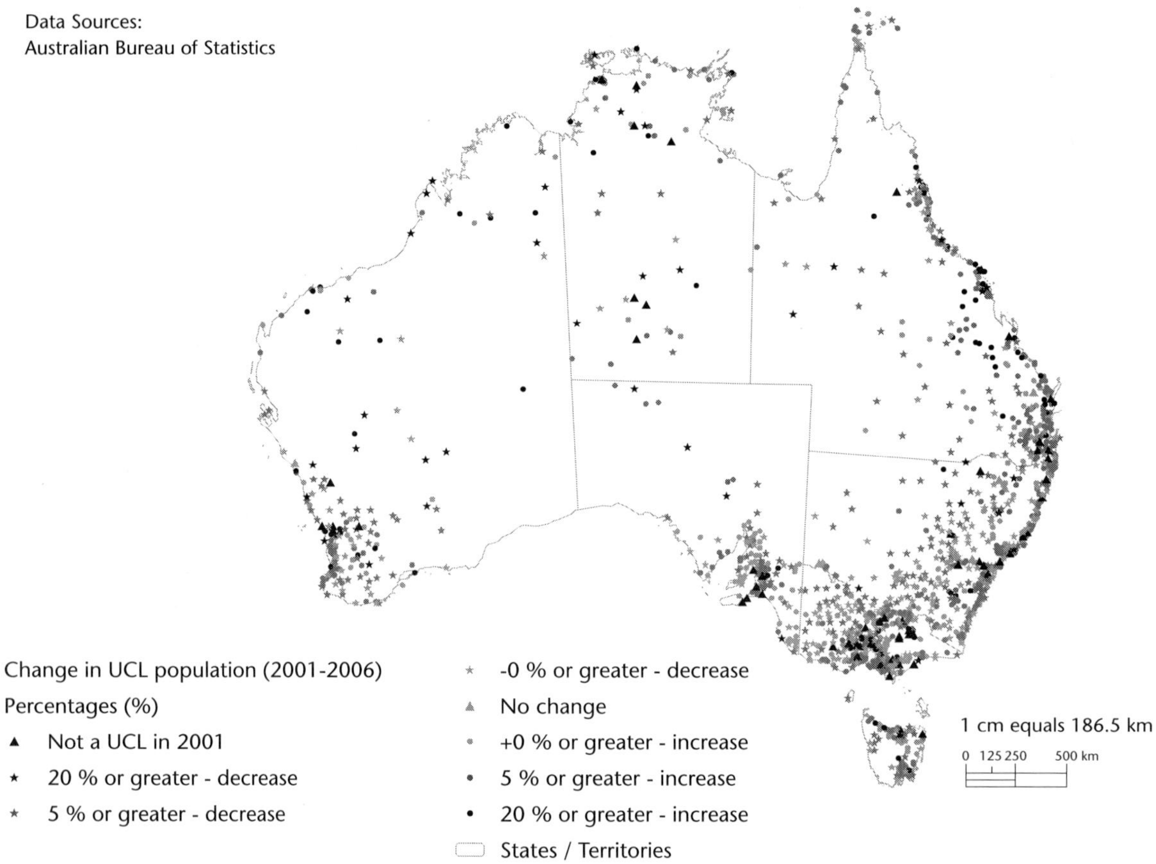

Figure 6: Australia—population change in country towns, 2001–2006 (Source: Australian Censuses of 2001 and 2006).

We can summarise the contemporary dynamics of Australian settlement systems as follows:

- Australia's population is one of the most concentrated in the world. At the 2006 census 66.3% of the national population lived on 0.12% of the national space occupied by major urban areas.
- Australia's population is increasingly concentrated in coastal areas with the percentage of the national population living with 50 km of the coast increasing from 79.9% in 1991 to 85.3% in 2006.
- Australia has one of the most urbanised of national populations with the proportion of the population living in urban areas increasing from 76.1% in 1947 to 85.1% in 1991 and 88.0% in 2006.
- Within these urban areas, there is an increasing concentration in the largest cities, with the proportion in the 1 million plus cities of Sydney, Melbourne, Perth, Brisbane and Adelaide increasing from 49.7% in 1947 to 59.7 in 1991 and 60.2 in 2006.

Demographic processes shaping the Australian settlement system

Changes in population distribution can occur through a number of demographic processes:

- **natural increases** (excess of births over deaths) varying between settlements and regions
- **internal migration** whereby Australian residents leave some areas and concentrate in others
- **international migration** whereby arrivals from overseas concentrate in particular areas and adopt a settlement pattern different from that of the established population.

Table 2: Australia: Persons by remoteness area, 1996–2006

	1996	2001	2006
		Population	
Major cities of Australia	11,698,118	12,921,901	13,558,142
Inner regional Australia	3,670,124	3,613,686	3,926,989
Outer regional Australia	1,974,234	1,905,962	1,879,461
Remote Australia	339,785	337,357	297,391
Very remote Australia	198,949	193,439	151,097
Total	17,881,210	18,972,345	19,813,080
		Growth rate (%) pa	
		1996–2001	2001–2006
Major cities of Australia		10.5	4.9
Inner regional Australia		−1.5	8.7
Outer regional Australia		−3.5	−1.4
Remote Australia		−0.7	−11.8
Very remote Australia		−2.8	−21.9
Total		6.1	4.4

Source: ABS Censuses

All three elements have played a role in producing varying levels of population growth in different parts of Australia and shifts in population distribution. There are variations in age structure, fertility and mortality between areas, which influence regional population growth—although these effects are less substantial than migration.

Fertility is significantly lower in large metropolitan areas than outside the cities (Hugo 2005). In 2006, fertility differences between capital cities and rest of state varied between 7% for Queensland and 28% for South Australia (ABS 2007a). For mortality, the differences are generally smaller with lower levels in major urban areas (Hugo 2005). In 2006, differences ranged between 1.7% for Queensland and 14% for Victoria. These variations are influenced partly by the variations between areas in the composition of the population and the fact that mortality and fertility differentials exist between various groups in Australian society (Jain and McDonald 1997). For example, across Australian cities there are substantial differences in levels of fertility—outer suburban areas tend to have higher levels of fertility than inner and middle suburbs, even after standardising for differences in age structure (Faulkner 2005). Similarly, it is possible to identify differences in mortality between particular parts of the country. Areas with a high proportion of their population of indigenous origin, for example, have significantly higher mortality than elsewhere.

While differences in natural increase levels between settlement types exist in Australia, the major demographic forces causing differences in population growth between areas relate to migration. First, it is important to stress that international migration not only affects the size and rate of growth of the national population but also influences population distribution. In 2006, 24% of the total national population was born overseas and another 26% were Australia-born but had a parent born overseas. Overseas migrants to Australia have not settled across Australia in proportion to the existing distribution of the population, but have concentrated in particular parts of the country and thereby influenced the distribution of the population.

One of the most distinctive features of post-war immigration to Australia has been the tendency for migrants to settle in the largest urban areas. Over the period 1947–2006, the number of Australia-born persons living in cities with 100,000 or more inhabitants more than doubled so that, in 2006, 61% lived in such cities (Table 3). On the other hand, the overseas-born population in the largest

The sea-change phenomenon is sometimes characterised as being dominated by retirees, but it is evident that, while this group are important, it is predominantly a young family's movement. In addition, the sea change is often depicted as a 'stampede from the city', but, although former city dwellers are part of the inflow, people from other non-metropolitan areas are also very significant.

- Another significant settlement zone experiencing net in-migration is the area around Australia's major urban centres, beyond the built up fabric of the city including, and beyond, the daily commuting zone. Figure 11 shows the profile for the Outer Adelaide Statistical Division and it shows a pattern similar to coastal areas, with net gains in all groups but the young adults who move out to the large cities. This zone includes the areas known as 'tree-change' areas which are alternative inland locations within a reasonable distance from large cities that are attracting similar types of people as the 'sea-change' areas.
- Further inland, net migration losses are common across much of the wheat–sheep and rangeland zones. As Holmes (1994) has pointed out, there is an increasing dichotomy between growing coastal and declining inland Australia. In the inland non-metropolitan zone, the phenomenon of 'sponge cities' (Salt 1996), whereby increasing centralisation in large centres is drawing services and shoppers away from smaller centres, has resulted in growth of those centres. However, in many rural areas and country towns, populations are generally stable or in decline. The exceptions are mining areas, tourist centres, some indigenous inland settlements and sponge cities'. A typical inland net migration profile is shown in Figure 12 for the Murrumbidgee Statistical Division. Here there is stability in all ages except young adults, where there is a heavy net loss, especially among young women. Drought has exacerbated out-migration from areas such as Murrumbidgee.

Australia is one of the most mobile societies in the world with 41.4% of the population having shifted house between the 2001 and 2006 population censuses. This gives a consistent dynamism to the Australian urban system, but it does not necessarily lead to a massive redistribution of population. Two key characteristics of Australian internal migration need to be noted. The bulk of the mobility is over short distances and there are strong patterns of reciprocating flows. Nevertheless, internal migration not only has significant demographic effects in Australia's settlement system, but it is also an important social and economic process.

Forces for change in Australia's settlement system

Analysis of Australia's population centroid—the 'centre of gravity' of the distribution of the population—has it located around 54 kilometres east of the town of Ivanhoe (ABS 2007c). However, it has moved less than 100 km during the previous century. Superficially, this would indicate that Australia's settlement system has a high degree of stability. However, this is far from the case. The system is in a constant state of change and is highly dynamic. This will continue and in this section we attempt to identify some of the processes that are likely to impinge on the Australian settlement system over the next decade.

The first issue to be considered is climate change. Australia's settlement system has been strongly shaped by the environment and its limitations, and climate has been an important part of this (Taylor 1947). There are a number of different elements of the climate change process influencing Australia, as is evident in Figure 13, and these undoubtedly will have an impact on the future distribution of Australia's population. Most attention has been focussed on the decline in rainfall and surface run-off that is being observed across South-eastern Australia. Figure 14 shows the below average and above average rainfall areas in 2006 and these roughly corresponds to the areas that are experiencing a long-term decline and increase in run off. At the 2001 population census, the area of long-term decline was home to 89.1% (17 million persons) of the national population. There is a longstanding mismatch in Australia in the distributions of run-off and population as is evident in Table 6 and, two decades ago, Pittock and Nix (1986) commented presciently …

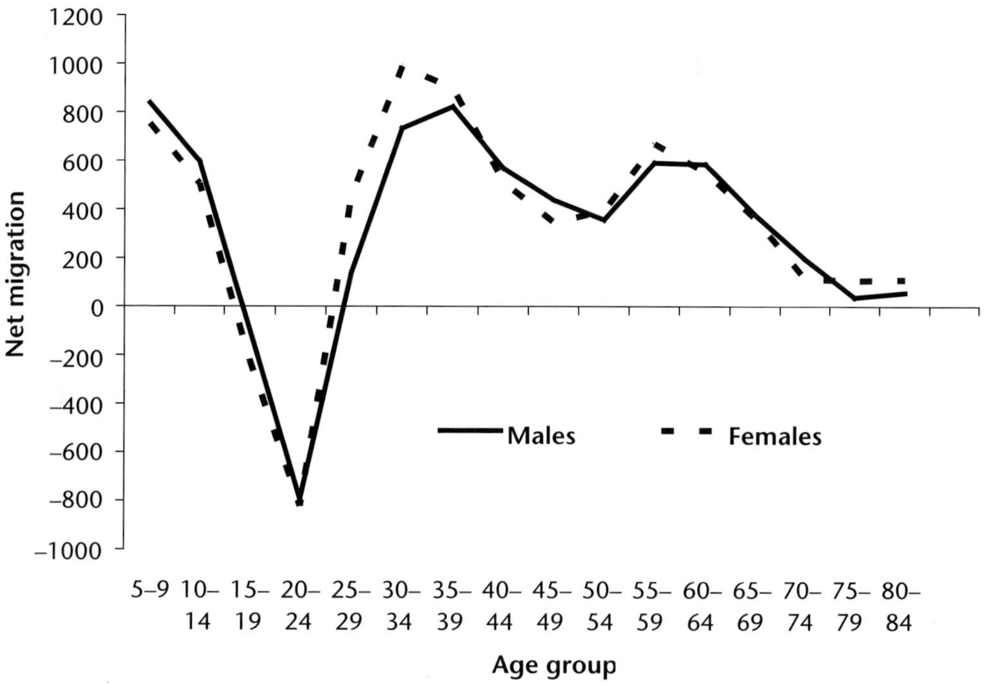

Figure 11: South Australia: Outer Adelaide Statistical Division net migration profile, 2001–2006

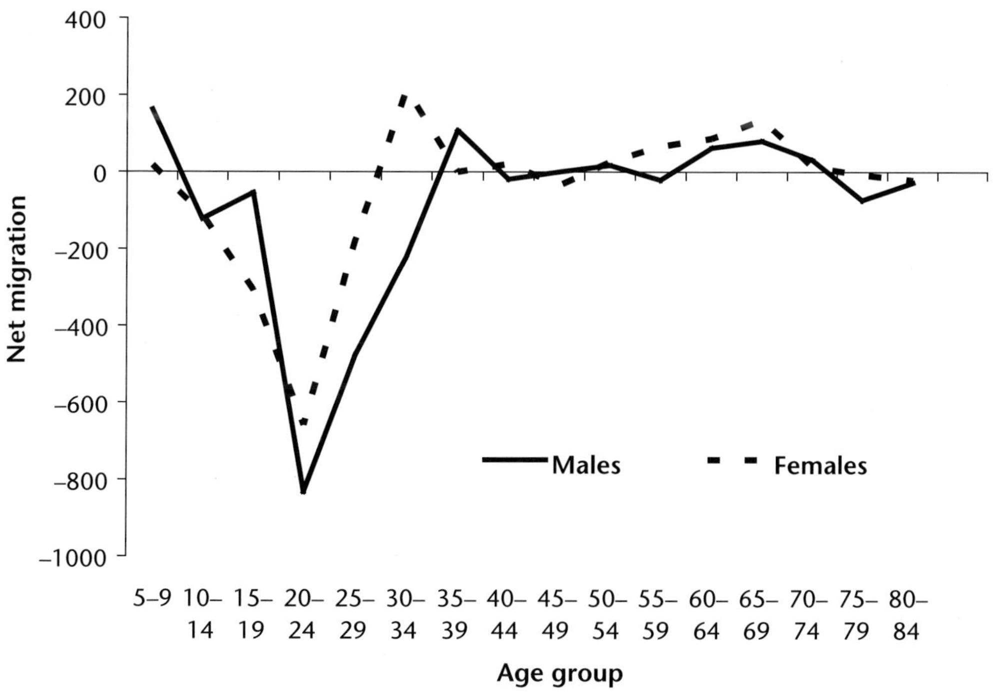

Figure 12: New South Wales: Murrumbidgee Statistical Division Net Migration Profile, 2001–2006 (Source: ABS, 2006 Census).

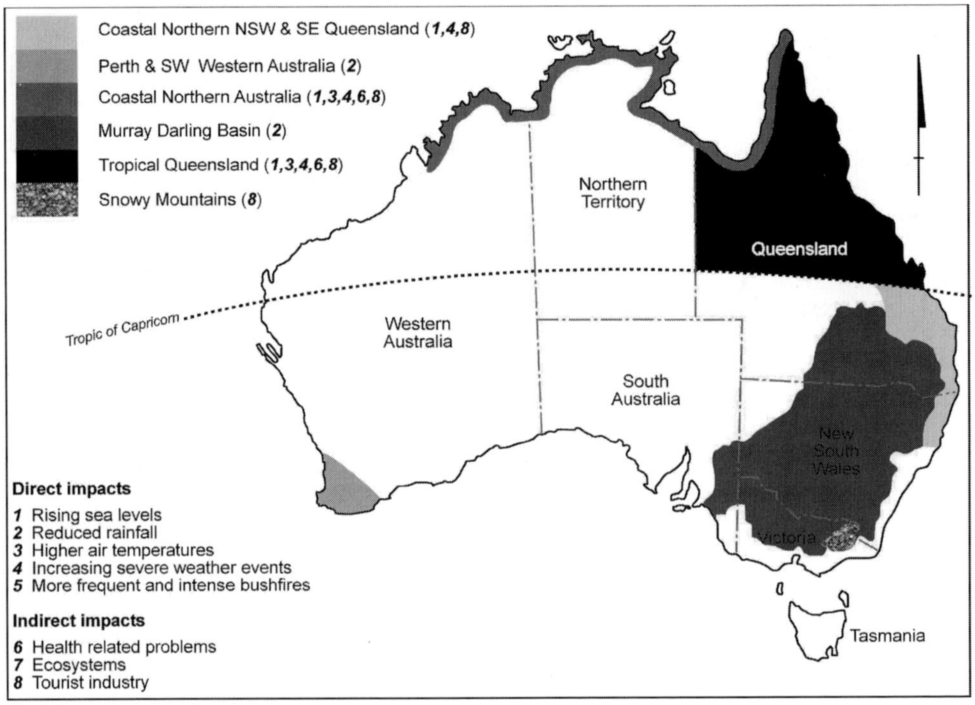

Figure 13: Climate change impact hotspots in Australia (Source: Climate Action Network 2006).

By far the largest volumes of uncommitted water are in northern Australia and Western Tasmania. In the most heavily populated regions of south-western and south-eastern Australia surface waters are committed to a high degree and the consequences of climate change are potentially most serious.

This clearly has implications for future patterns of population settlement in Australia. It is unlikely to lead to a major redistribution of the current residents of the south-eastern part of Australia to the north and to Tasmania where there are large water resources. There are however some possible impacts:

- While the current population can currently be supported in the south-east, there will need to be much better conservation and management of the water available in that region.
- Consideration needs to be given to the fact that a part of the projected population growth—from 21 million in 2007 to 24 million in 2021 and 26 million in 2031 (ABS, 2006b)—will need to settle in areas with better water supply.
- In Australia, more than two-thirds of water usage is by agriculture (ABS 2006c) so it would be seem likely that part of the irrigation-based agriculture in the south-east could be transferred to better-watered areas in the north

A second element of change that may impinge upon population settlement patterns in Australia is the passage of the baby boomer generation, who make up around 28% of the population, into the older age groups. The exit of this very large group from the workforce will have major demographic and economic impacts in Australia (Costello 2002, 2004; Department of Treasury 2007). Traditionally, older Australians have been the least residentially mobile of all age groups (Bell and Hugo 2000); however, the baby-boom cohort differ in many respects from earlier generations of Aus-

Table 6: The mismatch between water and population

	Far North Australia (%)	Southern Australia (%)
Population	2	82
Potentially arable land	4	65
Annual mean surface run-off	52	27

Source: Nix 1988

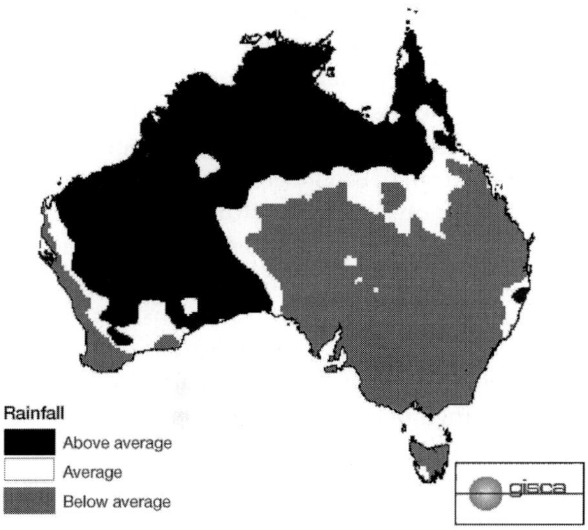

Figure 14: Australian rainfall 2006 (Source: Australian Government, Bureau of Meteorology).

tralians entering retirement. Moreover, they are, on average, a relatively well-off group and they may respond to retirement by moving house more than previous generations. Even if they only move to the same extent as the current generation of older Australians, they will have a significant impact because there are so many more of them. While ageing in place has been the dominant pattern in the past, it may be that baby boomers will be more prone to making more of the following types of moves:

- **Sea change:** the current generation of older Australians have moved more to coastal communities than any previous group of retirees. It is likely that the baby boomers will move in significant numbers to non-metropolitan coastal communities.
- **Trading down:** increasing number of older people on achieving 'empty nester' status are selling the family home in low-density suburbs in major cities and moving to inner-city, higher-density housing or to similar housing in coastal suburbs. Again, this may be a more common pattern among baby boomers.
- **Grey nomads:** many retirees are opting to spend extended periods travelling around Australia.

There are such large numbers of baby boomers that their settlement patterns will have a substantial impact on the overall distribution of population. Little is known of their intentions in respect of their retirement housing, but this could give a significant boost to already significant migration to non-metropolitan coastal communities and to growth of the inner-city populations of large cities.

A third area of potential change to the settlement system relates to the role of mining in the Australian economy. Over a long period, mining activity has been significant for population growth in remote parts of Australia. The current seemingly insatiable demand for minerals from China is seeing an unprecedented expansion of mining activity in non-metropolitan mainly remote, Australia. This has the potential for changing dramatically the settlement patterns in those areas. However, much depends on the extent to which fly-in–fly-out systems are used and the extent to which there is value-added processing of minerals as opposed to the export of ore.

Future patterns of international migration will also have an influence. It was shown above that international migration has been the main engine behind the growth of Australia's large cities, especially Sydney and, to a lesser extent, Melbourne. However, in recent years the introduction of the State Specific and Regional Migration Programs (SSRM) has undoubtedly diverted some migrants away from Sydney and toward slower growing centres such as Adelaide and, to a lesser extent, regional areas. Labour shortages are appearing in many rural areas and it is likely that there will be increased diversion of international migrants into these areas.

One of the most striking changes in settlement patterns over the last two decades has been the rapid growth of 'sea-change' coastal, non-metropolitan communities. So significant has this been that it has been suggested by some (for example, Salt 2004) to represent a third distinctive major settlement type (after rural and urban) in Australia. Growth in these areas does not appear to be slowing and continued growth of settlement along the eastern, south- eastern and south-western coasts of the country is leading to a distinctive pattern of ecological settlements linking fairly discrete coastal urban centres. There are important environmental implications of such settlement. Coastal environments are often ecologically fragile—they contain much of the nation's scarce quality agricultural land, they often have limited water resources, they are vulnerable to pollution and people living in these areas are often heavy users of fossil fuels.

Just as 'sea-change' areas are experiencing the pressures of rapid rates of population growth, many inland communities are facing the opposite problems of population decline and erosion of their economic and social vitality and indeed sustainability. Impacts of increasing personal mobility have been exacerbated by programs to centralise services by governments, banks and other service providers, and by prolonged drought. Problems experienced by those communities relate to loss of services, erosion of social capital and neglect of the environment. There is a need for new thinking, which may involve governments being location leaders in keeping education, health and other key services in rural communities, cooperation between proximate communities in developing housing and services, levering off economic activity, which diversifies the economic base of local communities.

Conclusion

Where people live is important. It influences their behaviour and wellbeing in a myriad of ways. However, where they live also has local regional and national impacts as well, and is a significant element in achieving environmental, social and economic sustainability. The lineaments of Australia's settlement pattern have been shaped over the period of European settlement by environmental constraints and the interplay of economic and social forces. While Australia's population distribution in macro terms conforms to the broad patterns of coastal concentration established more than a century ago, a number of dynamic elements are altering the population distribution within this matrix. While the distribution is unlikely to undergo a massive change over the next few decades, there are likely to be important changes within this structure, which are going to influence the lives of many Australians. Population distribution can be influenced by policy, both directly and indirectly, and policy makers have to be aware both that they can affect population distribution and that many of their actions actually do influence population distribution even though this may not be their primary intention. Equally, it needs to be realised that settlement policy can contribute substantially to the achievement of national goals of environmental sustainability, economic prosperity and equity in access to services and opportunities.

The Australian economy and society are undergoing transitions, with structural change, climate change, globalisation, the effect of the revolution in information technology and the ageing of the population. The implications for the distribution of population are unclear and need to be investigated. There has never been greater potential for analysing changing population distribution patterns and their implications. The development of spatial information system methodology and technology, the revolution in the capacity of computers and the explosion of data availability at the small area level, not only from the ABS but also from administrative data sources, provide us with an unprecedented opportunity to investigate and answer many of the pressing questions regarding population distribution—and in a timeframe that can better assist decision makers.

A number of important forces will impinge on Australia's population distribution in the future. Demographically, the scale and composition of international migration, the ageing of the population (especially the baby boom generation) and changing patterns of family and household formation will have an influence. Internal migration will be shaped by shifts in regional economic structure, changes in tastes and preferences for different lifestyles, political developments and the advance of information technology. It is crucial that these developments be monitored and their implications for environmental sustainability, equity, economic development and the wellbeing of all groups of Australians are established.

References

Australian Bureau of Statistics (ABS) (1990). *Population Growth and Distribution in Australia*, Cat. No. 2504.0, ABS, Canberra.

ABS (2006a). *Census Dictionary Australia 2006 (Reissue)*, Cat. No. 2901.0, ABS, Canberra.

ABS (2006b). *Population Projections Australia, 2001 to 2101 Reissue*, Cat. No. 3222.0, ABS, Canberra.

ABS (2006c). *Water Account Australia 2004–05*, Cat. No. 4610.0, ABS, Canberra.

ABS (2007a). *Births Australia*, Cat. No. 3301.0, ABS, Canberra.

ABS (2007b). *Deaths Australia*, Cat. No. 3302.0, ABS, Canberra.

ABS (2007c). *Australian Demographic Statistics-December Quarter 2006*, Cat. No. 3101.0, ABS, Canberra.

Bamford EJ, Dunne L, Taylor, DS, Symon, BG, Hugo, GJ and Wilkinson, D (1999). *Accessibility/Remoteness Index of Australia (ARIA)*, Department of Health and Aged Care Occasional Papers Series No. 6, Commonwealth Department of Health and Aged Care, Canberra.

Bell M and Hugo GJ (2000). *Internal Migration in Australia 1991–96: Overview and the Overseas-born*. Australian Government Publishing Service, Canberra.

Bell M and Ward GJ (2000). Comparing temporary mobility with permanent migration, *Tourism Geographies* **2**, 87–107.

Borrie WD (1954). *Italians and Germans in Australia. A Study of Assimilation*. Cheshire, Melbourne.

Burnley I and Murphy P (Eds) (2004). *Sea Change: Movement from Metropolitan to Arcadian Australia*, University of New South Wales Press, Sydney.

Champion A and Hugo GJ (Eds) (2004). *New Forms of Urbanisation: Beyond the Urban-Rural Dichotomy*. Ashgate, Aldershot, UK.

Connel J (Ed.) (2000). *Sydney: The Emergence of a World City*, Oxford University Press, Melbourne.

Coombes M and Raybould S (2001). Public policy and population distribution: developing appropriate indicators of settlement patterns, *Environment and Planning C: Government and Policy* **19(2)**, 223–248.

Costello P (2002). *Intergenerational Report 2002–03, 2002–03 Budget Paper No. 5*. Commonwealth of Australia, Canberra.

Costello P (2004). 'Australia's demographic challenges'. Discussion Paper, 25 February. The Treasury, Canberra.

Department of Treasury (2007). *Intergenerational Report 2007*, Commonwealth of Australia, Canberra.

Faulkner D (2005). Spatial dynamics of fertility in South Australia, 1976 to 1996. Unpublished PhD. Thesis, Department of Geographical and Environmental Studies, University of Adelaide.

GISCA (National Centre for Social Applications in GIS) (2002). 'Metro ARIA - index of metropolitan accessibility: a pilot application consulting report for Centrelink'. National Centre for Social Applications of Geographical Information Systems, University of Adelaide.

Holmes JH (1970). External commuting: the Maitland experience, *Australian Geographical Studies*, **8(1)**, 133–148.

Holmes JH (1994). Coast versus inland: two different Queenslands, *Australian Geographical Studies* **32(2)**, 17–82.

Hugo G (1986). *Australia's Changing Population: Trends and Implications*. Oxford University Press, Melbourne.

Hugo GJ (1989). Australia: the spatial concentration of the turnaround. In *Counterurbanization: The Changing Pace and Nature of Population Deconcentration* (Ed. AG Champion). Edward Arnold, London.

Hugo GJ (1994). The turnaround in Australia: some first observations from the 1991 Census. *Australian Geographer* **25(1)**, 1–17.

Hugo GJ (2003). Changing patterns of population distribution. In *The Transformation of Australia's Population: 1970–2030* (Eds SE Khoo and P McDonald), pp. 185–218 University of New South Wales Press, Sydney.

Hugo GJ (2005). The state of rural populations. In *Sustainability and Change in Rural Australia* (Eds C Cocklin and J Dibden) pp. 56–79. University of New South Wales Press, Sydney.

Hugo G (2006). Temporary migration and the labour market in Australia. *Australian Geographer* **37(2)**, 211–231.

Hugo G (2007a). Space, place, population and census analysis in Australia. *Australian Geographer*, **38(3)**, 335–357.

Hugo G (2007b). Immigrant settlement outside of Australia's capital cities, Revised Paper submitted to *Population, Space and Place*, November 2007.

Hugo GJ and Bell M (1998). The hypothesis of eelfare-led migration to rural areas: the Australian case. In *Migration Into Rural Areas-Theories and Issues* (Eds P Boyle and K Halfacree) pp. 107–133. John Wiley and Sons, Chichester, UK.

Jain SK (1992). Recent trends in mortality in Australia: an analysis of the causes of death through the application of life table techniques. *Journal of the Australian Population Association* **9(1)**, 1–24.

Jain SK and P McDonald (1997). Fertility of Australian birth cohorts: components and differentials. *Journal of the Australian Population Association* **14(1)**, 31–46.

Khoo SE, Voigt-Graf C, Hugo G and McDonald P (2003). Temporary skilled migration to Australia: the 457 visa sub-class, *People and Place* **11(4)**, 27–40.

Maher C and McKay J (1986). '1981 Internal migration study final report: internal migration in Australia'. Department of Immigration and Ethnic Affairs, Canberra.

Nix N (1988). Australia's renewable resources, in *How Many More Australians? The Resource and Environmental Conflicts* (Eds LH Day and DT Rowland) pp. 65–76. Longman Cheshire, Melbourne.

Pittock AB and Nix HA (1986). The effect of changing climate on Australian biomass production – a preliminary study. *Climatic Change* **8**, 243–255.

Rowland DT (1982). Population Growth and Distribution, in *Population of Australia*, Volume 1, ESCAP Country Monograph Series No. 9. pp. 10–22. United Nations, New York.

Salt B (1996). *National Report on Population Growth Ranking in Australia & New Zealand*, 8th edn, Coopers & Lybrand Consultants, Melbourne.

Salt B (2004). *The Big Shift: Who We Are and Where We Are Headed.* Hardie Grant Books, Melbourne.

Smailes PJ, Argent N, Griffin TLC and Mason GL (2002). *Rural Community Social Area Identification*, National Centre for Social Applications of Geographical Information Systems, University of Adelaide.

Taylor TG (1947). *Australia: A Study of Warm Environments and Their Effect on British Settlement.* Methuen, London.

Williams M (1970). Town farming in the Mallee lands of South Australia and Victoria, *Australian Geographical Studies* **8(1)**, 173–191.

11

Metropolitan Evolution

Peter Newton

At the outset of the 21st century, the overall structure of Australia's settlement pattern is judged to be relatively fixed:

- Australia is a relatively 'empty' continent, with approximately 0.2% of the land mass accommodating three-quarters of the population (Hugo 2003).
- We are a nation of coastal dwellers, with 85% of the population living within 50 kilometres of the coastline (ABS 2003).
- We are a highly urbanised society with 87% of the population living in urban centres of 1000 or more people (ABS 2003).
- There is high metropolitan primacy, with the capital cities of the seven states and territories accommodating two-thirds of the total population (ABS 2005).

Within this spatial framework, however, there are a number of key shifts that are occurring.

Recent population and settlement trends

'The Bush': Rural and remote

The persistent long-term movement of labour out of agriculture presents continuing threats to the social and economic viability of rural service centres. Figure 1 dramatically illustrates the difference in population fortunes between the 'bleeding bush' and the 'booming beach' (Salt 2004). Over the past two decades there has been a decline in number of farmers (averaging 2.2% per annum) as well as an underlying trend of few younger people entering agriculture as a vocation. Greater farm losses have been off-set by increased dependence on off-farm income. Beer and Keane (2000) clearly modelled the vicious cycle of rural settlement dynamics as follows:

employment loss → services loss → disincentive for in-migration → disincentive for business → employment and population loss ...

and this is a scenario broadly evident across Australia, especially for the smaller rural service centres.

NIER analyses (Shepherd 2003) indicate how widespread population loss has been across rural municipalities between 1991 and 2001 (Table 1). Sixty-three per cent of all rural municipalities lost population during this period—a major contrast to all other regional groupings—with the exception of the resource-based centres. The larger rural regional and provincial centres in most states, however, continued to gain population (ABS 2005).

Australian Natural Resources Audit (http://audit.ea.gov.au/aura) forecasts a decline of between 30 and 55% in farmer numbers to 2020 and a further increase in median farmer age, which will peak in 2011. Regions experiencing greatest threat are the dryland (wheat, sheep) farming regions (Hugo 2002)—regions identified as highly vulnerable under future climate change scenarios of reduced rainfall (Fisher 2005).

There are also questions being raised (Productivity Commission 2007) as to the viability of the small remote indigenous settlements due to their size, high levels of welfare dependency and poor access to services. This is a major dilemma for social policy given the national cultural significance of these communities.

By way of contrast, the mining boom of the past decade has continued to support the existence of

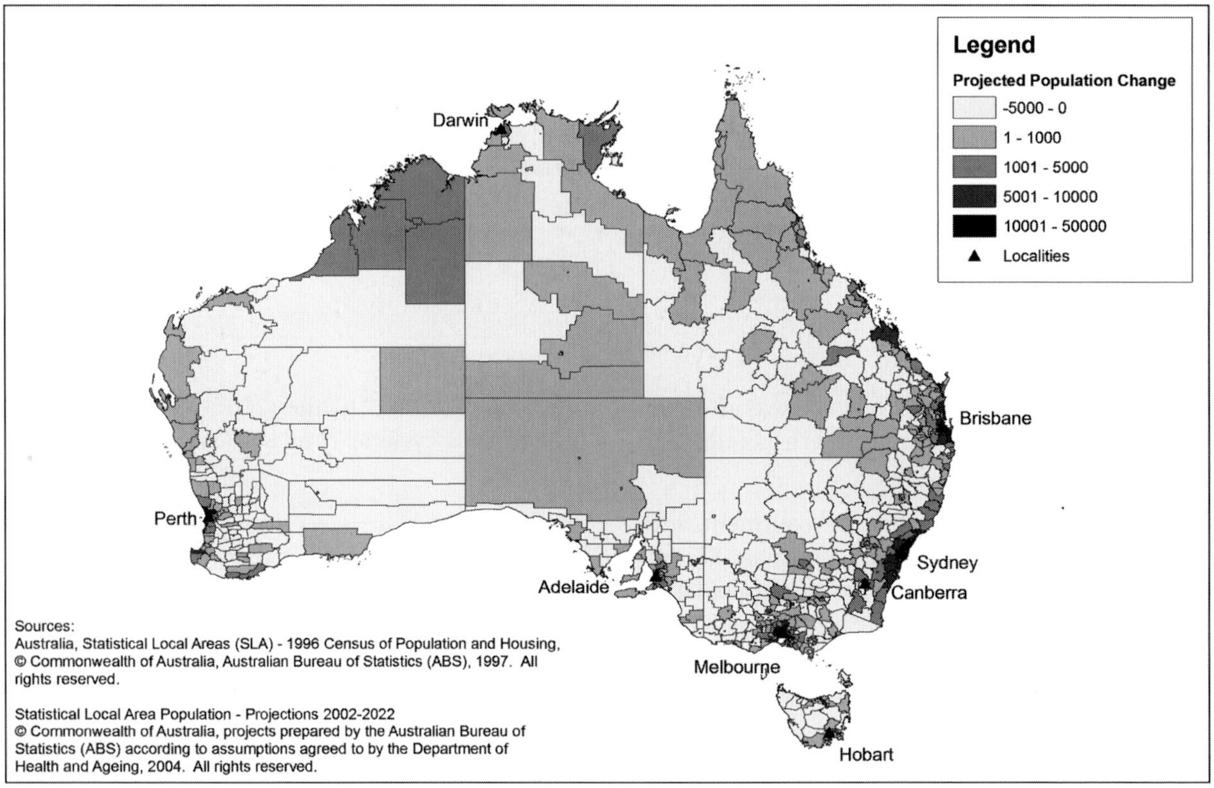

Figure 1: Population Change, Australia: 2002–2011 (Source: Commissioned from ABS for the 2006 State of Environment Report).

Table 1: Changes in population by region, 1991–2001

Region	LGAsc (No.)	Population (No.)	Decline (%)	Population (No.)	Increase (%)
Core metro	63	8	13	55	87
Dispersed metro	59	0	0	59	100
Lifestyle	41	0	0	41	100
Production zone	67	13	19	54	81
Resource based	102	48	47	54	53
Rural	295	187	63	108	37

Notes:
1. Core metro includes inner city regions of capital cities
2. Dispersed metro covers the suburban areas of cities plus regional cities
3. Lifestyle regions are the coastal areas
4. Production zones are areas of heavy industry and manufacturing
5. Resource zones are linked to mining or forestry regions
6. Rural areas are those dominated by agriculture

Source: Shepherd 2003.

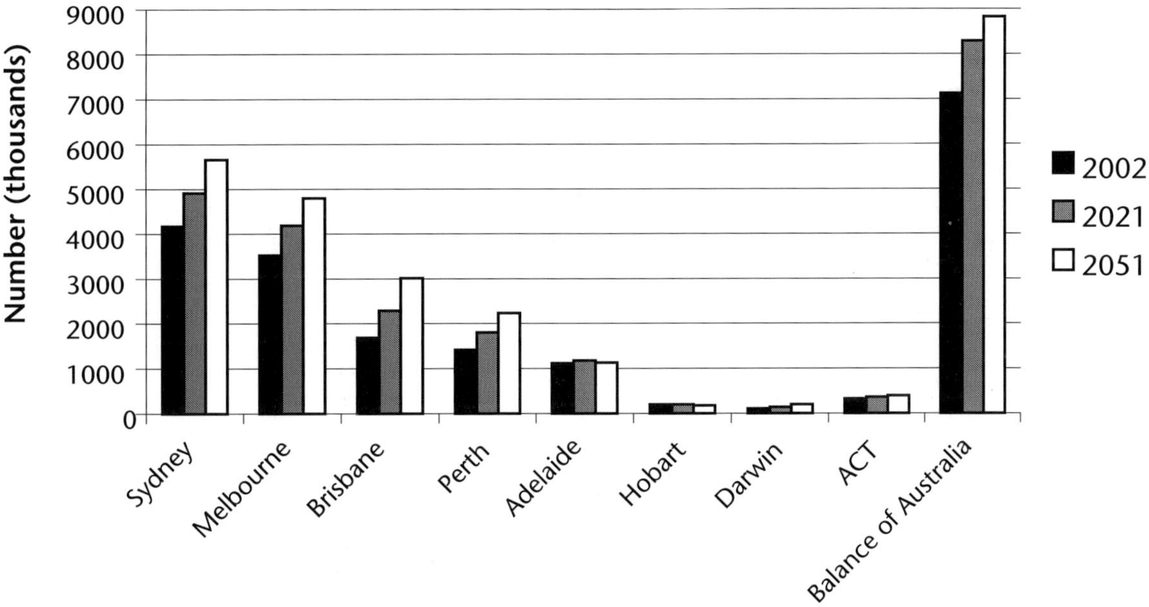

Figure 2: Population projections to 2051 by capital city (Source: Commissioned from ABS for the 2006 State of Environment Report).

Australia's remote resource towns, which remain for the most part linked to single company mining operations (Brealey *et al.* 1988).

The 'Capital Cities' and the rise and rise of the mega-metro regions

Australia's seven capital cities were home to almost two-thirds (64%) of the nation's population of 20.6 million in 2006 and accounted for over 60% of population growth in that year (ABS 2007). Sydney, Melbourne, Brisbane and Perth dominate currently and are expected to continue to do so into the future (Figure 2).

With the continuing growth of freeways and high-speed telecommunications that enable the substitution of communication via internet for daily face-to-face contact for certain occupations and industries, the peri-urban regions now extend 100km beyond the centre of the major state capital cities. It is this zone—which includes city water catchments—where there are a key set of environmental pressures (such as biodiversity and water quality) and land use conflicts (such as agriculture versus urban).

With continued growth of the major capital cities, their peri-metropolitan extensions are beginning to merge in an increasingly functional and spatial sense with their surrounding second-tier urban centres and provincial regional centres, forming what have been termed mega-cities (Forster 2004) or mega-metropolitan regions (O'Connor 1994) and are in line with the growth of mega-cities globally (Hall 2001). Four principal mega-metro (MM) regions are emerging in Australia:

- Sydney Mega-Metro (extending from Newcastle to Wollongong)
- Port Phillip Region (extending from the Bellarine to the Mornington peninsula)
- South-East Queensland (extending from the Sunshine Coast to the Gold Coast and Coolangatta–Tweed–Byron)
- Perth Corridor (extending from Wanneroo to Mandurah)

This represents the most significant settlement transition (Figure 3) in Australia since the auto-city transition, which began mid-20th century. The transition to mega-metro systems is supported by a platform of technologies that allow new peri-urban development without the necessary extension of increasingly costly centralised water and energy infrastructures as well as those that can radically transform space-time relations

Settlement Transitions	Walking City	Transit City	Auto City	Mega-metropolitan Region
Energy Transitions	Human & Animal Muscle Power; Fuel from Timber	Fuel from Coal; Steam Porter	Liquid Fossil Fuels; Internal Combustion Engine	Distributed, Renewable Energy; Hydrogen Economy
Communication Transitions	Face -to-Face Communication; Print	Analog; Centralised Telecomms	Digital (mixed mode); Mobile Communications	High Bandwidth, Wireless, Internet; Telepresence
Industrial Transitions	Agrarian to Industrial Economy	Industrial to Services Economy	Services to Information Economy	Emergence of Green Economy
Water & Wastewater Transitions	Settlement located near fresh water supplies; waste to landfill, pits, receiving waters	Diverted water; linear system; centralised and separate water and sewerage systems; discharge of waste to environment		Integrated Urban Water System; City as Catchment
Built Environment Transitions	Increasing building heights and densities possible through materials and engineering revolutions			
Transport Transitions	Horse-drawn Carriages	Train, Tram	Automobile	Fast Rail (eg. Maglev); High Speed Freeways

Figure 3: Key transitions related to human settlement (Source: Newton 2006).

via high speed ground transport and broadband communications.

In 2001 the populations resident in each of these regions were: 4.9 million for Sydney MM; 3.8 million for Port Phillip MM; 2.4 million for South-East Queensland and 1.5 million for the Perth Corridor. The population added to these regions between 1996 and 2001 were 290,000, 211,000, 237,000 and 108,000, respectively (Figure 4), with percentage rates of change at, or well above, the national average (Figure 5).

These mega-metro regions have become Australia's principal demographic absorbers and are expected to accommodate 80% of the projected pop-

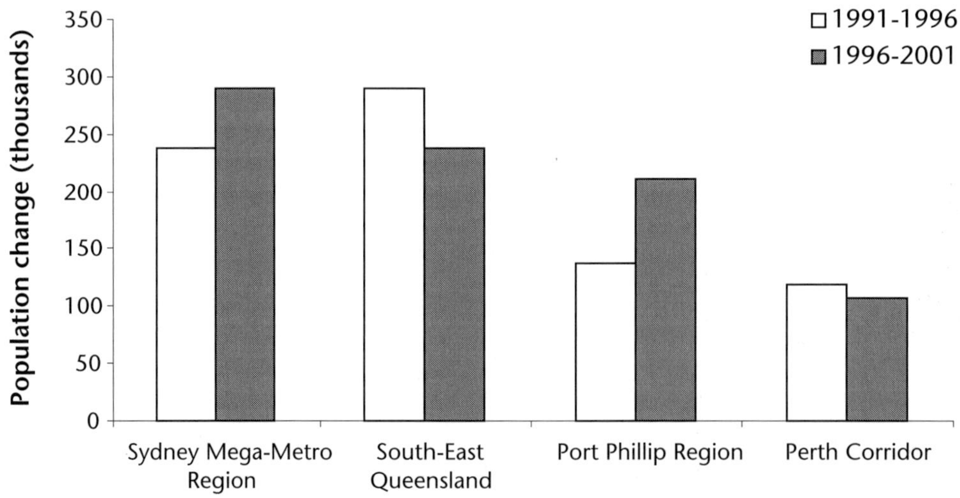

Figure 4: Population change—mega-metro regions (Source: see Newton (2006) for data sources and a detailed definition of the mega-metro regions).

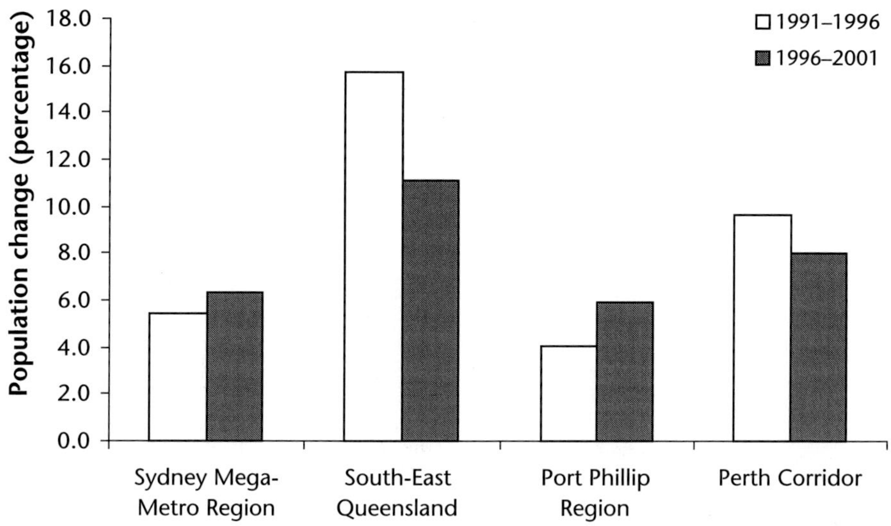

Figure 5: Population change (percentage)—mega-metro regions (Source: Newton 2006).

ulation increase to Australia of 1.9 million by 2011 (Table 2). They also constitute the principal engines of the Australian economy—attracting employment across all sectors of industry (Figure 6).

Australia currently imports half of its population growth each year—one of the key reasons why the nation's annual rate of population growth is high by OECD standards.

Permanent arrivals, which tend to fluctuate in a band ranging from 80,000 to 100,000 per year (although in 2006–7 net overseas migration was 177,600 people, representing 56% of national growth), are significant from a human settlements perspective in that they tend to gravitate to the capital cities—especially Sydney and Melbourne (Table 3). This has in recent years put significant pressure on Sydney and Melbourne housing markets and has led to calls by the New South Wales government for significant cuts in international migration (McDonald and Temple 2003).

A response by federal and state governments has been to devise policies to encourage immigration to rural and regional Australia via new visa classes that give special consideration to those applicants (ranging from humanitarian and refugee to employee-sponsored skilled migrants) prepared to locate in urban centres other than the major capitals. Data for the most recent period would suggest that the policy is having an effect: in 2004, two-thirds of immigrants to Australia settled in capital cities compared with figures of 80% during the 1990s. Notwithstanding, Sydney and Melbourne combined continue to attract half of all permanent immigrants to Australia.

The most striking immigration change over the course of the past decade has been in relation to

Table 2: Population forecasts to 2011—Australia and the mega-metropolitan regions

Mega-metro region	2002 Population	2011 Population	Population change	Population change (%)
All Australia	19,660,189	21,521,416	1,861,227	9.47%
Perth corridor	1,475,667	1,676,906	201,239	13.64%
Port Phillip	3,831,731	4,202,811	371,080	9.68%
South-East Queensland	2,436,464	2,926,671	490,207	20.12%
Sydney	4,940,474	5,351,541	411,067	8.32%

Source: ABS analysis commissioned for 2006 Australian SoE Report

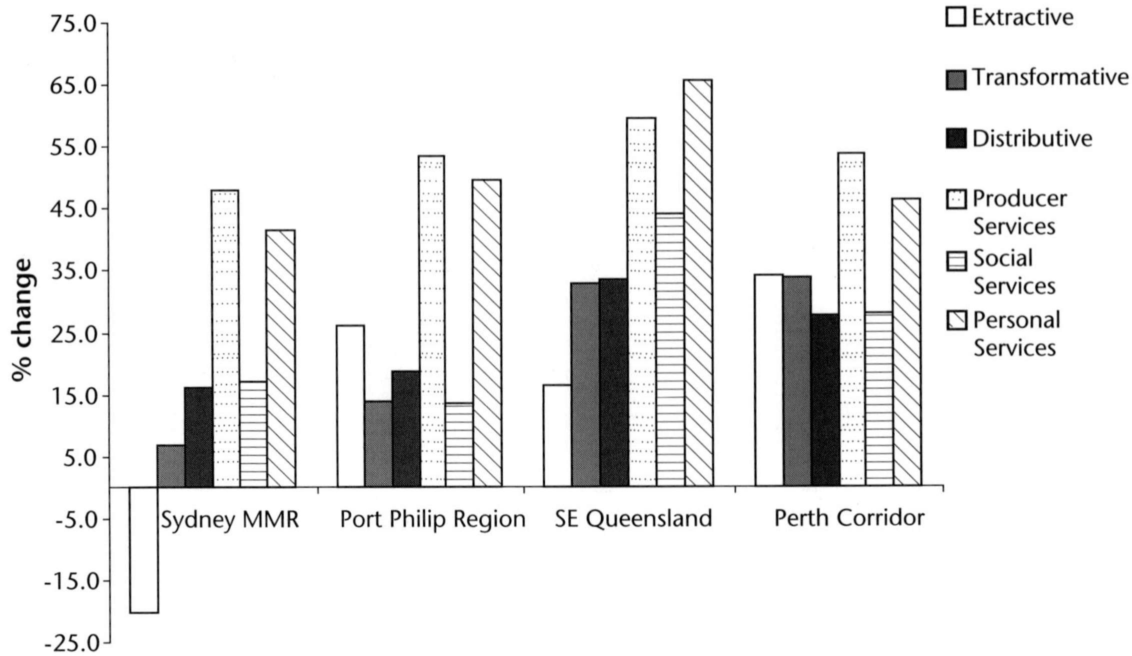

Figure 6: Percentage change in employment in four mega-metro regions by industry sector using Singlemans classification (Source: see Newton 2006 for definition of Singleman's classification and a full allocation of industry classes).

long-term visitor arrivals—or what Hugo (1996) termed 'temporary migration'—driven by two key categories: students and temporary business entrants (the 457 visa—see Khoo *et al.* 2003). The long-term visitor arrivals have increased from 62,000 in 1993–04 to 191,000 in 2003–04—a rate of increase of 20% per year. In both of these categories, destinations would seem to favour the cities—the educational institutions that attract foreign students are primarily located there as are the high-skilled jobs (McDonald and Temple 2004). Initial locations chosen by immigrants are significant, because surveys (Richardson *et al.* 2002) suggest that once they arrive in Australia, migrants tend to stay in the place of first settlement.

Australia's mega-metro regions are also the primary destinations for both international and domestic tourism. In 2003–04, Australia received more than five million international visitors and Australians themselves made over 85 million visits to places within Australia. Volumes of visitors of this magnitude—both international and domestic—bring with them massive benefits to local economies and employment (estimated at up to $50 billion per year; Productivity Commission 2005), but, depending on the destinations chosen

Table 3: Destination of immigrants[1] to Australia 1996, 2001, 2004

	1996		2001		2004	
Destination	**No**	**%**	**No**	**%**	**No**	**%**
All capital cities[2]	69614	78.3	63774	78.1	64585	67.1
Total immigrants	88882	100	81652	100	96232	100

Notes: 1. Excludes immigrants from New Zealand. For data and commentary on New Zealand migration to Australia between 1979 and 2003 see Bedford *et al.* (2003).

2. Total of all capital city statistical divisions.

Source: Department of Immigration and Multicultural and Indigenous Affairs 2005.

Table 4: International tourism by destination, 2004

		Percentage of Total
Capital city region	Sydney region	24.3
	Melbourne region	12.0
	South-East Queensland	16.5
	Adelaide region	3.0
	Perth region	5.0
	Hobart region	0.7
	Darwin region	1.2
	Canberra	1.3
Total Capital City Regions		64.0
Coastal Region	NSW	3.6
	Victoria	2.0
	Queensland	13.8
	South Australia	0.8
	Western Australia	1.6
	Tasmania	1.2
	Northern Territory	2.0
Total coastal regions		25.0
Inland region		5.0
Outback region		6.0
Australia Total		**100.0**

Note: Total international tourists to Australia in 2004 numbered 11.3 million

Source: Derived by author from Tourism Australia spreadsheet (March 2005)

for holidaying, an additional layer of population pressure and costs on infrastructure, services and the environment.

International and domestic visitor survey data collected annually by Tourism Australia across all tourism regions for those travelling for leisure provide insights that run counter to popular conception: that Australia relies on its unique flora, fauna and landscapes—its natural assets—as the key drawcard for international inbound tourism. Tourists' revealed preferences —where they actually visit while in Australia—suggest that it is Australia's cities which are the principal magnets (Table 4). Two-thirds of all international tourism is to Australia's capital city regions. Following the cities, the next most significant 'environment' is the tropical and subtropical east coast, which runs from Byron Bay to Cairns and includes the rainforest and reef (17%); with the Outback attracting about 6% of tourists (mostly to the Northern Territory).

For domestic tourism, the capital city regions continue to dominate as destinations, capturing 50% of total visitor nights (Table 5). A further 25% of domestic tourism is to areas that are contiguous with the city regions—encompassing the mega-metro region. Continued sprawl of Australia's major cities, which test and punch holes in the urban growth boundaries established by state planning agencies in an attempt to contain low-density development at the fringe, has the capacity to destroy the landscape attractions that draw urban tourists out of the city centres.

Useful as they may be as demographic absorbers and economic engines (see Willis and Fry 2001), mega-metro regions, by virtue of their sheer size, create complex and multi-faceted entities on scales never experienced before. Key challenges relate to urban governance—developing urban management and planning practices appropriate to multi-municipal, poly-centric mega-metropolitan regions; and

Table 5: Domestic tourism by destination, 2004

		Percentage of total
Capital city region	Sydney region	12.1
	Melbourne region	9.6
	South-East Queensland	15.2
	Adelaide region	4.1
	Perth region	4.0
	Hobart region	1.3
	Darwin region	1.1
	Canberra	1.7
Total capital city regions		49.1
Coastal region	NSW	10.1
	Victoria	3.6
	Queensland	8.8
	South Australia	1.9
	Western Australia	4.5
	Tasmania	2.1
	Northern Territory	0.2
Total coastal regions		31.2
Inland region		15.6
Outback region		4.1
Australia total		**100.0**

Note: Based on visitor nights. Number of visitor nights for Australian domestic tourism numbered 294 million in 2004
Source: Derived by author from Tourism Australia spreadsheet (March 2005)

the fact that these mega-metro regions occupy bio-regions of multiple catchments and major ecosystem function, requiring more sophisticated systems modelling between environmental and economic domains. They also provide opportunities for developing a new class of resource-based industries—based on eco-industrial clusters—capable of delivering more sustainable urban regions (see Batten *et al.*, this volume).

Suburbanisation dominates re-urbanisation

Shifting our focus to changing patterns of settlement within Australia's major cities we can see that the re-urbanisation of inner city capitals continues, but, in terms of absolute population growth, it is dwarfed by sub-urbanisation. Between 1991 and 2001 Australia's five leading capital cities added 76,000 residents to their inner cities and 1.24 million to their suburbs (6% versus 94%, respectively, of total population change).

Until the 1990s, most of the inner suburbs of Australia's cities were associated with consistent population loss. The process of re-urbanisation identified in the mid 1990s (Newman *et al.* 1996) has subsequently intensified as a result of the supply of inner city dwellings meeting demand from a group of consumers that include students, single- and dual-person households at both ends of the family life cycle, but primarily in the 20–30 year bracket (DoI 1998). Table 6 and Figure 7 illustrate the high rates of growth in percentage terms for the inner suburbs, but the absolute levels of growth remain modest, except for the east coast capitals.

Since 2001 inner city growth has continued. During 2003–04, for example, Perth City municipality recorded the fastest growth rate of all municipalities in Australia, with a 12.1% increase,

Table 6: Population change (percentage) in selected capital cities inner city areas versus metropolitan areas 1991–1996 and 1996–2001

	Number		Percentage	
	1991–1996	1996–2001	1991–1996	1996–2001
Inner Sydney	20 513	25 307	15.9	16.9
Sydney SD	208 281	247 136	5.67	6.4
Inner Melbourne	5 057	10 957	14.6	27.6
Melbourne SD	127 702	188 347	4.1	5.7
Inner Brisbane	2 139	7 556	10.0	32.1
Brisbane SD	161 996	130 431	11.9	8.6
Inner Adelaide	1 509	458	13.3	3.6
Adelaide SD*	21 276	29 549	2.0	2.7
Inner Perth	894	2 208	19.0	39.4
Perth SD	106 330	97 910	8.9	7.6

Source: Newton 2006, including definition of inner city regions.

while the population of the adjacent local government area of Subiaco increased by 3.6%. Melbourne's City municipality recorded an annual growth rate of 6.4%, the population of Adelaide's City municipality increased by 4.6%, and Sydney's equivalent increased by 3.6%. Elsewhere in Australia, other inner-city areas to experienced high levels of growth include Brisbane's inner municipalities, Darwin's Stuart Park, and the Canberra municipalities of Turner and Braddon.

There is emerging debate, however, as to the long-term sustainability of these recent levels of growth. It is viewed by some as a response to long-pent-up demand for inner-city apartment living, but not a recipe for transplanting en masse to suburban (activity) centres of Sydney and Melbourne under current 'compact city' state planning strategies (Salt 2004; Birrell *et al.* 2005):

[inner city living is] fine for a component of the community, but it is not as pivotal to grass roots Australian values as is the quest for space, for independence, for privacy and for low-density living… (Salt 2004, p. 6)

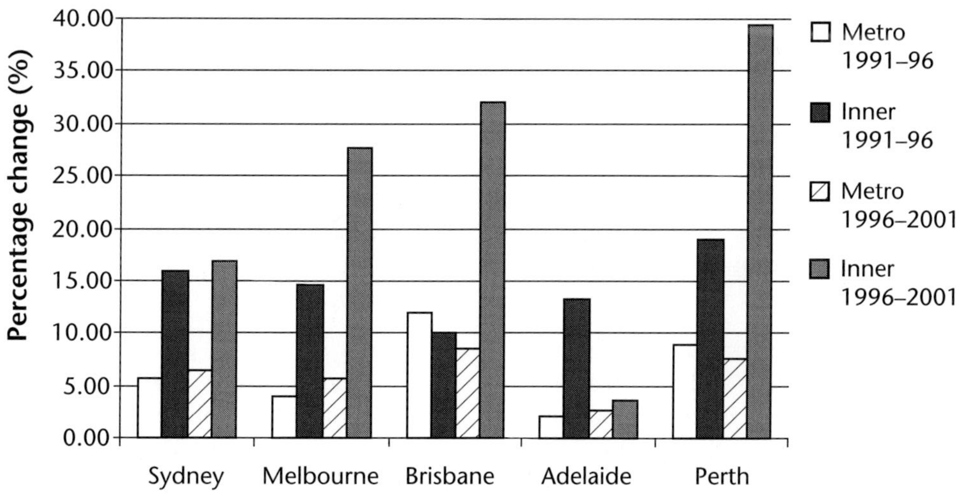

Figure 7: Population change (percentage) in selected capital cities, inner city area versus metropolitan areas 1991–1996 and 1996–2001 (Source: see Newton 2006 for data sources and definitions of the inner city regions).

It is the middle and outer suburbs that will continue to accommodate Australia's urban population growth—the question is, at what density?

Large increases in population continue to occur in many outer Local Government Areas (LGAs) of capital cities. In Sydney, the LGAs of Blacktown, Baulkham Hills and Liverpool experienced large growth (up 5500, 3500 and 2300 people, respectively), while in Melbourne the fringe LGAs of Casey, Wyndham and Melton increased by 8700, 8400 and 5900 people, respectively. Recent studies by Birrell *et al.* (2005) indicate that outer suburbia has now become dominated by the trade-up market of middle income, working-class couples with children—to the exclusion of first home buyers who are now finding it increasingly difficult to gain entry to Australia's capital city housing markets.

These data, and the research by O'Connor (1999), indicate that it is primarily the suburbs that offer households opportunities of matching both housing and jobs. However, transport indicators suggest that this match is far from optimal—as revealed by continued high growth vehicle kilometers traveled (VKT), which have continued to exceed rates of population growth over the past 10–15 years (Newton 2001, 2006).

Growth of coastal high-amenity regions

Next to Australia's cities and mega-metro areas, the regions to experience the most sustained growth over the past 20 years have been the high-amenity coastal regions (the so-called 'sea change' areas), which for the most part extend far enough inland to pick up many of the high-amenity 'hill change/tree change' areas as well (Burnley and Murphy 2004).

The drivers for non-metropolitan coastal development include:

- The attractions of high environmental amenity areas associated with beach and bush that serve resident quality of life preferences (Burnley and Murphy 2004; Salt 2004) and the attraction of space: 'Space is the amenity that people leave the cities in search of and it is the feature that tourists come to Australia to enjoy' (Hamilton 2002, p. 4). Lifestyle-related reasons were given by over 40% of those moving to archetypal coastal centres (Stimson *et al.* 1996).
- Employment (O'Connor 2001). Recent analyses indicate that coastal growth is not predominantly a retirement or baby-boomer-led trend as the popular press often portrays. ABS (2004) data for 2001 indicates that 80% of new residents to coastal growth regions were aged under 50. Unemployment rates, however, are generally higher (18%) and among youth much higher (23%) than the national level (less than 5%).
- Housing affordability continues to be an attraction of the housing markets in the coastal non-metro regions (especially in comparison with the capital cities).

O'Connor's (2001) analysis of coastal population growth between 1976 and 2000 separates out that component located within metropolitan areas—as opposed to other coastal regions outside of the cities. What that data revealed was that the coastal municipalities of cities experienced a three-fold increase in population 1976 to 2000 compared with an increase of two-thirds in the non-metropolitan coastal regions. When more recent data is analysed similar trends are evident (Table 7), although they are not as striking. For the period 1996–2004, coastal Australia, including those parts within capital cities, gained over 440,000 compared with 362,000 for non-metropolitan coastal regions. In this respect *coastal city* living emerges as the most sought after urban environment in Australia.

The impacts of rapid coastal growth have led to recent remarks that 'we are loving our coasts to death' (*The Age* 15 January 2005, p. 1). A combination of rapid urban growth in permanent residents and the impacts of tourism that can see populations of some coastal communities double in size during peak holiday periods (Shepherd 2005) has resulted in:

- a loss of natural and cultural heritage, with previously 'sleepy' towns that for decades had been primarily holiday destinations now transformed into suburbs by the sea
- environmental demands that result in inadequate water supply in many centres as well as loss of native vegetation and local pollution

Table 7: Changes in estimated resident population of coastal areas in Australia, 1996–2004

Coastal areas (as defined by Statistical Local Areas)[a]	Area[b] (000 km^2)	Estimated resident population (000s)			Average annual population Change (%[c])	
		1996	2001	2004	1996–2001	2001–04
All coastal areas	2163.1	7482.0	7971.7	8283.5	1.3	1.3
Coastal areas excluding capital cities	2149.2	2977.7	3193.4	3339.0	1.4	1.5
Coastal areas within capital cities	13.8	4504.4	4776.3	4943.9	6.1	3.5
Australia	7705.3	18 310.7	19 413.2	20 111.3	1.2	1.2

Notes: a coastal areas all Statistical Local Areas (SLAs) with a boundary adjoining the sea, including those boundaries adjoining harbours, and rivers such as Leichhardt (A) in Sydney Harbour and South Perth (C) on the Swan River in Perth. Note that many SLAs extend inland for large distances (for example, East Pilbara Shire in Western Australia has a coastline of roughly 80 kilometres and an area of over 350,000 square kilometres)

b based on 2004 Australian Standard Geographic classification (ASGC) boundaries c. average annual growth rate.

Source: Beeton *et al.* 2006

- a significant increase in demand for infrastructure (roads, water and sewerage), education, community services, and so on.

With most towns reliant on a resident population rate base, the extra layer of transient population represents a fiscal burden that is proving difficult to service.

Environmental implications of contemporary patterns of urban development

The completion of a State of Environment Report for Australia (Beeton *et al.* 2006), together with specialist theme commentaries on human settlements, atmosphere, biodiversity, coasts, inland waters, land, natural and cultural heritage (www.environment.gov.au/soe), provides opportunity for a high-level assessment of the impact that Australia's population and settlement—in particular its cities—are having on the environment.

In summary, both the 2001 (Newton *et al.* 2001) and 2006 (Newton 2006) reports on Human Settlement identify levels of resource consumption across all headline indicators that are running at unsustainable levels equivalent to three- to four-planet living—and increasing (ecological footprint calculations for Australia's major cities tend to be of the order of 7–8 hectares per person (Turner and Foran, this volume). The planning and development paradigm that has operated since permanent settlement began in Australia over 200 years ago has been one premised on an abundant and cheap supply of land, energy and materials—similar to that of North America, where Rees and Roseland (1991, p. 17) argue that:

> *... cities were built using technologies which assumed that abundant and cheap energy and land would always be available. Communities therefore grew quickly and inefficiently, and became dependent on lengthy distribution systems. Cheap energy influenced the construction of our spacious homes and buildings, fostered our addiction to the automobile and increased the separation of our workplaces from our homes.*

Headline indicators clearly show that Australia has followed this path of urban development and rank among the world leaders in relation to:

- per capita water use at 115 KL/year (2000–01; 3.2% per year increase over 1996–67 level).
- per capita energy use at 266 GJ/year (2003–04; with a forecast increase of 2.2% per year to 2019)
- per capita waste generation of over 1 tonne/year
- per capita mobility by car of 8000 km/year (and where vehicle kilometres travelled in capital cities are forecast to be one-third higher by 2020 than in 2002).

- per capita CO_2 generation of 27.5 tonnes/year
- floor area of new dwellings, which has increased at a rate of 2.2% per year over the 10 year period to 2003–04 (where the average was 239 square metres).

In all instances, annual rates of change in per capita consumption are above the rate of Australia's population change (which is 1.2%/year or twice the OECD average); in combination, these twin factors are driving growth in Australia's total consumption.

Vulnerabilities to this style of urban development and living are now becoming more evident in the face of global warming and climate change (and its implication for cities), peak oil, lack of a safe yield for the water supply supporting most cities, urban traffic congestion and air pollution linked to continuing increases in VKT and halos of coastal pollution surrounding the major coastal urban settlements. The need to make the transition to more sustainable forms of urban development has never been more evident.

Pathways towards more sustainable cities

The significant problems we currently have cannot be solved by the same level of thinking with which we created them. Albert Einstein (attributed).

Change is required in three key areas that are capable of delivering transformational change to the sustainability performance of Australia's cities:

- planning and design
- technological innovation
- human behaviour.

Each of these is examined in the context of 3-horizon thinking and action, which effectively identifies the different time horizons required for widespread diffusion of a particular innovative policy, process, product or behaviour (Newton 2007). Horizon 1 initiatives include those that are currently available, have demonstrable benefit and yet are not widely implemented. Horizon 2 innovations exist as a model, prototype product or system, scoped process or similar, but require some challenging testing, extension, real-world application and case-study, in order to develop a convincing 'business case' for wider application. Horizon 3 innovations tend to be more radical—based on concepts, technologies and human behaviours that are vastly different from those currently operating and where there are currently major barriers (of a technical, social, corporate or governance nature) to be overcome, and where widespread implementation is unlikely inside 20 years or so. The 3-horizons for technology innovation, planning and design and human behaviour change are outlined in Figure 8.

Planning and design

In Australia, city planning and design is a responsibility of state governments. Until the release of the most recent round of metropolitan planning strategies for the nation's mega-metro regions (summarised in Buxton 2006), there had been little attempt to rein in low-density sprawl and to attempt better integration of mixed land use development with public transit at higher residential densities. Australia's recent attempts at more compact cities are aligned to the 'smart-growth' policies that have emerged from the USA (see Falconer and Giles-Corti, this volume), but the jury is still out on the triple-bottom-line success of this new model. The principal benefits of more compact, higher density cities would seem to lie most clearly with environmental gains. These can be summarised as follows:

1. ***Energy:*** reductions of between 20 and 60% in operating energy use in medium- and high-density housing versus detached housing have been documented by Miller and Ambrose (2005) and Newton *et al.* (2000), although for high-rise apartments, energy use for services (such as lifts) and poor design for cross-ventilation currently negate energy gains in this category of high-density development (Pears 2005). More compact cities can expect to deliver transport energy savings of the order of 40% compared with their low-density counterparts due to greater use of public transport (Newton 1997; Newton and Manins 1999).
2. ***Materials:*** medium density buildings embody approximately half the material intensity of detached housing stock (Deilmann 2001; Moriarity 2002).

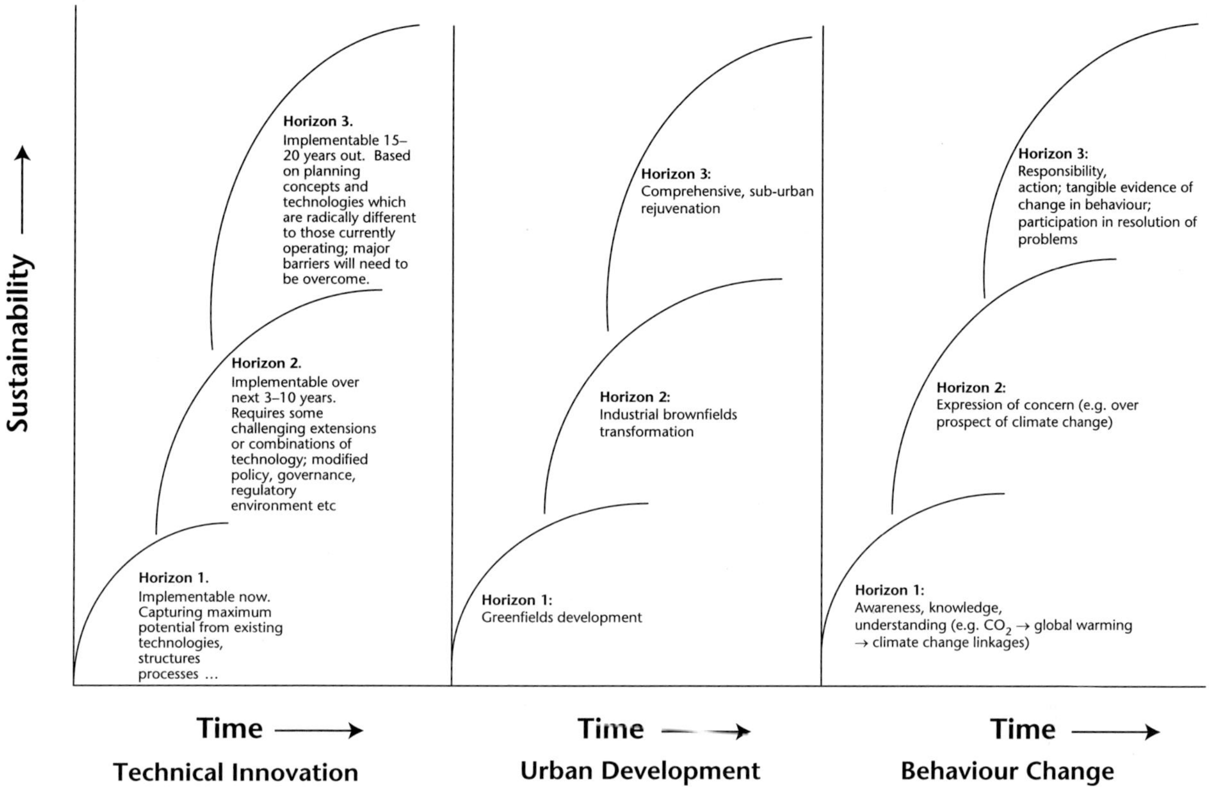

Figure 8: 3 Horizons of urban transitions in technology, planning and behaviour

3. ***Water:*** water consumption in higher density residential developments is approximately one-third to one-half less than in low-density developments—due primarily to less outdoor water use (Moriarty 2002; Tomalty and Alexander 2007).
4. ***Waste:*** per capita generation of municipal solid wastes in high-density cities is approximately half that of low-density cities with comparable GDP (Matsunga and Themelis 2002).
5. ***Land:*** higher density development reduces demand for conversion of land from green space or agricultural use (Rees 1996).

However, the benefits of higher density compact city development are being questioned in relation to social wellbeing:

1. ***Liveability:*** both infill development (2–3 residential units replacing a single detached unit) or higher density projects are associated with perceived loss of amenity, privacy, neighbourhood character, green space and increased social isolation of residents (Searle 2003; Valence *et al.* 2005;) an emergence of 'Save Our Suburbs' movements (Lewis 1999) and a dramatic increase in number of submissions to state government administrative appeals tribunals who are being asked to adjudicate on development proposals unsatisfactorily resolved at municipal level (Newton *et al.* 2001). Enhanced heat-island effects and restricted air flow (for building ventilation) are also evident with higher density development (Coutts *et al.* 2005; Miller and Ambrose 2005).
2. ***Affordability:*** there has been a significant escalation of residential property prices in Australia's largest cities over the past decade and an associated decline in housing affordability (Figure 9). Critics of smart-growth policies have attributed this to the land rationing that is linked to cities with designated growth boundaries (Cox 2006). While there is a supply-side explanation (UDIA 2006)—linked to a reduction in supply of allotments and rate of completions of new housing—

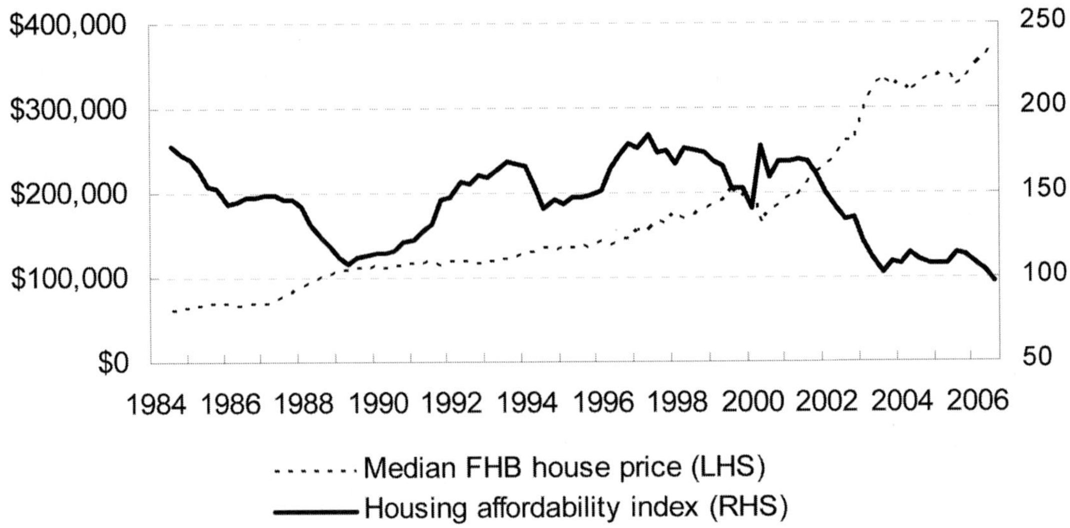

Figure 9: Median house prices and housing affordability index (Source: Yates and Milligan 2007).

it is not the full story. A low interest rate environment, a ready supply of finance for investment in residential property for owner-occupancy and investment, a buoyant economy for more than a decade and increased demand relative to supply of additional new stock(a significant share of recent building activity has been directed to redevelopment and additions/ improvements of existing property) are major contributing factors.

While there are some preliminary studies (such as Sturm and Cohen 2003) that have identified city versus suburban differentials in population health, which are linked to lower rates of car use and greater pedestrian activity among inner city residents, there is a much broader spectrum of research underway in relation to human health—urban form/neighbourhood design inter-relationships (see Falconer and Giles-Corti, this volume).

The principal Horizon 3 challenge for urban planners and designers is to transform the performance of the bulk of our cities that are already developed and in operation—the established suburbs. The barriers to innovation and development are much lower in greenfield and brownfield settings, where pathways to sustainable urban development are more evident (www.brownfields.com). The challenge for established suburbs is regeneration on a scale beyond that of current 'infill', which is proving sub-optimal from an environmental, social and economic perspective. Regeneration on a precinct scale will enable higher quality design of mixed use, higher density development and provide a prospect for absorbing the forecast growth of population within Australia's mega-metro regions in a more sustainable manner.

Technology-based innovation and settlement transition

The evolution of settlement through history has been fundamentally linked with the emergence of transformative technologies capable of radically reconfiguring economic, social and spatial relationships (refer again to Figure 3)—in a world where 'the natural environment' had not been a major constraint on urban development during most of the 20th century. With the environmental sustainability of future settlement systems now clearly on the radar screens of private and public organisations alike, increased attention is now being given to identifying those technologies capable of assisting cities to make the transition to a more sustainable future. It has been argued elsewhere that Horizon 3 technological innovation is required to ultimately achieve sustainable cities (Newton 2007, Figure 8). The technology-based platform for this transition exists, but remains to be implemented:

- A solar–hydrogen economy that services the energy requirements of both mobile (transport) and stationary (buildings) end uses—with zero air pollution and zero greenhouse gas emissions (100% renewable), together with a portfolio of other renewable sources of energy.
- An integrated urban water management system that uses treated stormwater and wastewater as substitutes for a significant portion of potable water supplies that are currently diverted from environmental flows and used in a manner that is not optimal economically or environmentally. Currently, stormwater and wastewater volumes in Australia's largest cities are estimated to be 1.5 times that of freshwater supplies (Mitchell 2005); again a portfolio of options for delivering sustainable supplies of urban water are available and awaiting implementation.
- A system of virtual design and construction of the built environment that enables automated sustainability performance assessment of buildings and infrastructure prior to assembly or construction (Seo *et al.* 2007)—maximising their eco-efficiency performance and optimising their contribution to population health and productivity.
- Hybrid buildings that integrate an energy-efficient shell, energy-efficient appliances and renewable, distributed micro-generation technologies for on-site energy generation—delivering the basis for carbon-neutral building (Newton and Pham 2007).
- A communications system that delivers high-bandwidth mobile computing, which enables any configuration of organisational and spatial development within the mega-metro region and beyond, but currently favouring a model of centralisation (of information, knowledge and non-routine producer services activities) with decentralisation (of data processing and routine services) (Newton 1995).
- A system of fast-rail (400+ km/hr) that provides primary links within polycentric mega-metropolitan regions, as well as between regions (Newton *et al.* 1996; Hall and Pain 2006): for moving people and knowledge and re-defining provincial centres as suburbs of the major cities.
- A further transformation of industries that enables the emergence of a green sector of the economy: either as more sustainable extensions of existing businesses (such as mining, agriculture, manufacturing and services) or new businesses that emerge to provide new renewable energy products (Stern 2006), or new eco-industrial clusters that transform present-day waste streams into future resource streams (Graedel and Allenby 2003; Batten *et al.*, this volume).

The Horizon 1 and 2 innovations that provide the bridge to a Horizon 3 future are outlined elsewhere (Newton 2007).

Changing human behaviour

The unsustainable pattern of urban consumption documented earlier represents one of Australia's greatest 21st century challenges. In part, it is 'built into' our cities in terms of how our buildings and infrastructure have been designed and the manner in which planners have allowed our cities to be shaped. But a significant element of consumption behaviour is household and lifestyle based—and is likely to prove the most intractable to change.

Total consumption is the product of total population and per capita consuption. On all counts there are rapid rates of increase. Despite several 'population summits' (Vizard *et al.* 2002) and multiple publications that attempt a quantification of Australia's 'carrying capacity' (Foran and Poldy 2002), there is no agreement on what the nation's optimal population should be. Under current federal immigration policy, Australia is importing more than half of its population growth each year—which explains why its growth rate of 1.2+% per year is twice that of the OECD.

Per capita consumption—or what Hamilton and Dennis (2005) have termed 'affluenza'—is a key national sustainability challenge. As the incomes and material wellbeing of Australians have increased over the past decade, their *concern* for the environment has diminished (Table 8). Reference again to Figure 8 reveals that *increasing concern* among the population is key to transitioning to more sustainable behaviour change that becomes reflected in the tangible *actions* of individuals and households. A necessary, but not sufficient condition. Whether the nation's drought and recent

Table 8: Percentage of people concerned about environmental problems 1992–2004

	Capital city	Balance of state/ territory	Total state/ territory
1992	75.2	74.1	74.8
1994	70.3	66.6	68.9
1996	69.2	67.1	68.4
1998[#]	63.2	62.8	63.1
2001	61.7	63.1	62.2
2004	57.6	56.4	57.1

[#]Revised data based on March 1998 Monthly Population Survey. Previous data were based on discontinued Population Survey Monitor

Source: Australian Bureau of Statistics 2004a.

pronouncements about how climate change will impact Australia will cause some shift in level of concern is as yet unclear. But transitioning the attitudes and behaviour of Australians to those required to live sustainably within our national and global contexts is another Horizon 3 challenge. Evidence suggests that as a nation we are still operating in Horizon 1.

References

Australian Bureau of Statistics (ABS) (2003). *Census of Population and Housing: Population Growth and Distribution, 2001*. Cat. no. 2035.0. ABS, Canberra.

Australian Bureau of Statistics (2004a). Environmental Issues: People's Views and Practices. Cat. no. 4602.0. ABS, Canberra.

Australian Bureau of Statistics (2004b). Seachange – new residents in coastal areas. In *Australian Social Trends*, ABS, Canberra.

Australian Bureau of Statistics (2005). *Regional Population Growth, Australia and New Zealand*. Cat. no. 3218.0. ABS, Canberra.

Australian Bureau of Statistics (2007). *Regional Population Growth, Australia, 2005–06*. Cat. no. 3218.0. ABS, Canberra.

Bedford R, Ho E and Hugo G (2003). Trans-tasman Migration in Context: Recent Flows of New Zealanders Revisited, *People and Place*, **13(4)**, 53–62.

Beer A and Keane R (2000). Population decline and service provision in regional Australia: a South Australian case study, *People and Place* **8(2)**, 69–76.

Beeton, RJS, Buckley K, Jones G, Morgan D, Reichelt R and Trewin D (2006). Australia State of the Environment 2006, Department of Environment and Heritage, Canberra.

Birrell B, O'Connor K, Rapson V and Healy E (2005). *Melbourne 2030, Planning Rhetoric versus Urban Reality*. Monash University Press, Melbourne.

Burnley I and Murphy P (2004). *Sea Change, Movement from Metropolitan to Arcadian Australia*. University of NSW Press, Sydney.

Buxton M (2006). 'Urban form and urban efficiency'. Paper prepared for the 2006 Australian State of the Environment Committee, Department of the Environment and Heritage, Canberra.

Coutts AM, Beringer J and Tapper N (2006). Investigating the climatic impacts of urban Planning strategies through the use of regional climate modelling; a case study of Melbourne, Australia, *International Journal of Climatology* (in press).

Cox W (2006). *Smart Growth Not So Smart?* Demographia web site <www.demographia.com>.

Cox W (2006). *War on the Dream: How Anti-Sprawl Policy Threatens the Quality of Life*. Demographia web site. <www.demographia.com>.

Deilmann C, Müller B, Banse J, Effenberger K–H, Gruhler K, Neumann I and Schiller G (2001). Nachhaltige Entwicklung des Wohnungsbestandes in Sächsischen Grob-und Mittelstädten Entwicklungss Zenarien Ausgewahlter Wohngebiete unter Ressourcen-und nutzungs orientierten Aspekten, final report <www.ioer.de/nawo> .

DoI (Department of Infrastructure) (1998). *From Doughnut City to Café Society.* DoI, Melbourne.

Fisher B (2005). 'Predicting the Impacts of Climate Change on Agriculture. A Case Study of the Australian Broadacre Industry'. Presentation to ABARE Outlook Conference (unpublished).

Foran B and Poldy F (2003). Between a rock and a hard place: Australia's population options to 2050 and beyond, *People and Place* **11(1)**, 1–15.

Forster C (2004). *Australian Cities.* Oxford University press, Sydney.

Graedel T and Allenby (2003). *Industrial Ecology.* 2nd edn. Prentice Hall, New Jersey.

Hall P (2001). *Global City-Regions in the 21st Century,* Oxford University Press, Oxford.

Hall P and Pain K (2006). *The Polycentric Metropolis.* Earthscan, London

Hamilton C (2002). Population Growth and Environmental Quality: Are They Compatible? *People and Placc* **10(2)**, 1–5.

Hamilton C and Denniss R (2005). *Affluenza: When Too Much is Never Enough.* Allen & Unwin, Sydney.

Hugo G (2003). Australia's Future Population: Where Will They Live? In *Australia's Population Challenge* (Eds S Vizard, H Martin and T Watts). Penguin, Melbourne.

Hugo G (2002). Population Trends in South Australia, *People and Place* **10(1)**, 26–42.

Hugo G (1996). Global mobility. In *Population Shift. Mobility and Change in Australia* (Eds P Newton and M Bell) AGPS, Canberra.

Khoo S–E and McDonald P (Eds) (2003). The Transformation of Australia's Population: 1970 – 2030, UNSW Press, Sydney.

Lewis M (1999). *Suburban Backlash, The Battle for the World's Most Liveable City.* Bloomings Books, Melbourne.

Matsunaga K and Themelis N (2002). *Effects of Affluence and Population Density on Waste Generation and Disposal of Municipal Solid Wastes.* <www.columbia.edu/cu/wtert/waste-affluence-paper.pdf>.

McDonald P and Temple J (2003). International migration and the growth of households in Sydney. *People and Place* **11(4)**, 63–74.

Miller A and Ambrose M (2005). *Sustainable Subdivisions and Energy Efficient Design.* CRC for Construction Innovation, Brisbane.

Mitchell G (2005). Personal Communication (Dr G Mitchell, Monash University).

Moriarty P (2002). Environmental sustainability of large Australian cities. *Urban Policy and Research* **20(3)**, 233–244.

Newman P Birrell B Holmes D Mathers C Newton P Oakley G O'Connor A Walker B Spessa A and Tait D.(1996) Human Settlements In *Australian State of the Environment 1996.* CSIRO Publishing, Melbourne.

Newton PW (1995).Changing Places? Households, Firms and Urban Hierarchies in the Information Age. In *Cities in Competition.Productive and Sustainable Cities for the 21st Century.* (Eds J Brotchie, M Batty, E Blakeley, P Hall and P Newton). Longman, Sydney.

Newton PW (Ed.) (1997). *Re-Shaping Cities for a More Sustainable Future – Exploring the Link Between Urban Form, Air Quality, Energy and Greenhouse Gas Emissions.* Research Monograph 6. AHURI, Melbourne.

Newton PW (2006). 'Human settlements'. Theme Commentary prepared for the 2006 Australian State of Environment Committee, Department of the Environment and Heritage, Canberra. (www.environment.gov.au/soe).

Newton PW (2007). Horizon 3 Planning: meshing liveability with sustainability, *Environment and Planning B. Planning and Design* **34** 571–575.

Newton PW and Manins P (1999). Cities and Air Pollution. In *East-West Perspectives on 21st Century Urban Development.* (Eds J Brotchie, P Newton, P Hall and J Dickey). Ashgate, Aldershot.

Newton PW and Pham L (2007).The Challenges of Environmental Sustainability Assessment.In *Proceedings of Building Australia's Future Conference,* Gold Coast, Queensland, 24–26 September.

Newton PW, Brotchie JF and Gipps PG (1996). 'Cities in transition: Changing economic and technological processes and Australia's settlement system'. Technical Paper, Department of the Environment, Canberra.

Newton PW, Tucker SN and Ambrose M (2000). Housing form, energy use and greenhouse gas emissions. In *2000 Achieving Sustainable Urban Form*. (Eds K Williams, E Burton and M Jenks) pp. 74–78. E &FN Spon, London.

Newton PW, Baum S, Bhatia K, Brown S, Cameron S, Foran B, Grant T, Mak S, Memmott P, Mitchell VG, Neate K, Smith N, Stimson R, Pears A, Tucker S and Yencken D (2001). 'Human settlements, Australia state of the environment report 2001'. (Theme Report). CSIRO Publishing on behalf of the Department of Environment and Heritage, Canberra. (www.environment.gov.au/soe).

O'Connor K (1994). Mega-metropolitan areas in Australia. *People and Place* **2(1)**, 21–24.

O'Connor K (1999). Misunderstanding modern suburban development. *People and Place* **7(4),** 21–25.

O'Connor K (2001). Coastal development: just a little shift in Australia's geography? *People and Place* **9(4)**, 49–56.

Pears A (2005). Does higher density really reduce household energy requirements? It Depends… *Urban Policy and Research* **23(3)**, 367–369

Productivity Commission (2005). 'Assistance to Tourism'. Commission Research Paper. Productivity Commission, Melbourne.

Productivity Commission (2007). *Overcoming Indigenous Disadvantage*. Productivity Commission, Melbourne.

Rees W E and Roseland M (1991). Sustainable communities: planning for the 21st Century. *Plan Canada* **31**, 15–25.

Richardson S, Miller-Lewis L, Ngo P and Ilsley D (2002). 'The settlement experiences of new migrants'. Report prepared for the Department of Immigration and Multicultural and Indigenous Affairs, Canberra.

Salt B (2004). *The Big Shift*. 2nd edition. Hardie Grant Books, Sydney.

Searle G (2003). 'The limits to urban consolidation'. Issues Paper 14, Urban Frontiers Program. University of Western Sydney.

Seo S, Tucker S and Newton PW (2007). Automated material selection and environmental assessment in the context of 3D building modelling, *Journal of Green Building* (in press).

Shepherd C (2003). The risk/reward trade-off that is shrinking the population of rural Australia. In *Australia's Population Challenge* (Eds S Vizard, SH Martin and T Watts) pp. 160–165. Penguin, Melbourne.

Stern Review (2006). *The Economics of Climate Change*. HM Treasury, London.

Stimson RJ, Minnery J, Kakamba A and Moon B (1996). *Sunbelt Migration Decisions. A Study of the Gold Coast*. Bureau of Immigration, Multicultural and Population Research, Melbourne.

Sturm R and Cohen D (2004). Suburban sprawl and physical and mental health. *Public Health* **118(7)**, 488–496.

Tomalty R and Alexander D (2006). Urban form and the environment: the centrality of land use planning to sustainability. *Journal of Industrial Ecology* (in press).

Urban Development Institute of Australia (UDIA) (2006). 'State of the Land Report. The National Land Supply Study'. <www.udia.com.au>.

Vallance S, Perkins SC and Moore K (2005). The results of making a city more compact: neighbours' interpretation of urban infill. *Environment and Planning* **B32(5),** 715–733.

Vizard S Martin H and Watts T. (Eds) (2002). *Australia's Population Challenge*. Penguin, Melbourne.

Williams K, Burton E and Jenks M (Eds) (2000). *Achieving Sustainable Urban Form*. E &FN Spon, London.

Willis A-M, and Fry T (2001). Sydney (UN) LTD, *People and Place* **9(2)**, 40–48.

Yates J and Milligan V (2007) *Housing Affordability. A 21st Century Problem*, Australian Housing and Urban Research Institute, Melbourne.

SECTION 4

Transitioning to Resilient Cities

The resilience of Australia's social and human capital, natural capital and built capital will be significantly tested as a result of the combination of driving forces and pressures outlined in Section 3. Key among these is *climate change*. Australia—as a dry country with a fragile environment—stands to suffer more from climate change than any other developed country based on the regional scenarios outlined by Hennessy. *Mitigation strategies* to address climate change form the basis of Section 5. *Adaptation strategies* are inherently spatial and require planning by region and by type of threat and form the basis of Section 4. The urban development undertaken to date has been influenced by historically expected extreme events so beloved of actuaries and the insurance industry. *Urban development risks* are now being revisited in the light of information emerging from climate change research in the following domains:

Sea-level Rise (Church *et al.*). Much of Australia's coastal urban development has occurred during a period of stable sea levels. With IPCC projections indicating a rise in sea level of 18–79 cm between 1990 and 2095, significant adaptation will be required in coastal settlements, especially in relation to the *combination of sea-level rise with extreme events*, such as storm surges and estuarine flooding, as well as coastal erosion.

Flood (Abbs). Flooding is likely to increase due to *increases in extreme daily rainfall events*, even though rainfall averages may remain much the same. Resilience is enhanced by governments incorporating knowledge from flood models that incorporate the latest climate change–rainfall scenarios into their urban planning schemes, risk-based premiums for flood insurance and a high level of preparedness by emergency services.

Drought and High Temperatures (Howden and Crimp). It is the impact of the *frequency of extremely hot days* (Australia's cities could be 3.4°C hotter and have up to 84 more days over 35°C by 2050) on peak demand for water and electricity—rather than the predicted increase in mean temperature and precipitation—that will create greatest stress on urban water and electricity supplies in the future. Integrated water and distributed energy systems can significantly alleviate peak demand in next generation utility networks.

Bushfire (Leicester and Handmer). There is an *increasing risk of megafires*, which are associated with greater risk to loss of life and property. Urban resilience can be enhanced by appropriate design of houses, planning that avoids housing closer than 500 metres from a bushland interface, and adoption of 'stay or leave early' strategies by residents.

To date, however, the increasing cost of natural disasters in Australia can be ascribed, according to Crompton *et al.*, to changing societal factors—where, how and what we build, and changing economic value—rather than anthropogenic climate change. Notwithstanding, the need to embody the precautionary principle in mitigation and adaptation strategies for future urban development remains evident.

Biosecurity represents a continuing threat to the resilience of Australia's economic and population health and biodiversity (Cook *et al.*). In a globalised world, there is no prospect of zero breaches of national biosecurity; instead, what is required is a biosecurity system that can *manage the risk* and minimise the negative effects of some breach (a similar logic to that behind planning for the impact of an engineered terrorist event).

Biodiversity loss can be directly linked to urbanisation as well as globalisation processes, especially given the absence of mechanisms in the city planning process for effectively evaluating the biodiversity and conservation value of 'green space' (McDonnell and Holland). Consequently, the loss of indigenous plants and mammals has been immense and is set to continue, given the private ownership of fringe land and the pressures of the property development industry on governments.

Levels of *social and human capital,* including *population health* (Baum and Woolcock; Timmins *et al.*), are at the heart of community resilience and the extent to which different communities can accommodate change—whether associated with extreme natural events or economic tsunamis linked to globalisation and technological change (see Stilwell and Jordan, Section 2). In 21st century Australia there is evidence of significant regional disparities in economic and social wellbeing that warrants a range of people-and-place-based policies capable of redressing the balance.

PWN

12

Sea-level Rise

John Church, Neil White, John Hunter, Kathleen McInnes, Peter Cowell and Siobhan O'Farrell

Most Australians live near the sea, with all of the state capitals being located on the coast. The modern 'sea change' phenomena is seeing even more people moving to the coast, leading to rapidly growing populations in such areas as the Gold and Sunshine Coasts in South-East Queensland, and the northern New South Wales coastal region. Chen and McAneney (2006) estimate that about half of Australia's population lives within 7 km of the coast, with as many as 30%, or about six million people, within 2 km of the coast. They also estimate about 6% of Australian addresses are situated within 3 km of shorelines in areas with elevations below 5 m above mean sea level.

The coastal margins are intimately affected by the level of the sea, which varies over a wide range of time scales. Sea level is influenced by ocean waves, twice-daily and daily tides, meteorological fluctuations over days to weeks, and seasonal, inter-annual, decadal and multi-decadal variations. Superimposed on these variations, there has been an overall rise at a rate of just less than 2 mm per year in global mean sea level during the 20th century. This rise was caused predominantly by rising global temperatures (Bindoff *et al.* 2007; IPCC 2007). Coastal vulnerability is affected by an increase in ocean volume causing a rise in globally averaged sea level, redistribution of ocean mass associated with changing climate and by vertical movement of the land. There are large-scale vertical motions of the Earth's crust associated with past and ongoing changes of surface loads on the earth (glacial isostatic adjustment) and local tectonic motions (particularly subsidence) in some coastal regions. The combination of these effects causes the sea level to rise relative to the land ('relative sea-level rise').

The impacts of sea-level rise include coastal inundation and erosion, higher waves at the coast, and salt water intrusion into estuaries, wetlands and aquifers. In some regions, beaches and coastal wetlands become squeezed between the hard barrier of coastal developments and the landward migration of the coastline. If not protected in some way, the natural environment may be lost. Virtually all of these impacts are felt most acutely during extreme high sea-level events and storms (storm surges) when substantial inundation of low lying areas and rapid coastal erosion can occur. Much of Australia's (and the world's) coastal development over decades and centuries has occurred during a period of stable sea levels, with little concern for any potential changes in future sea levels. With sea levels now beginning to rise as one component of anthropogenic climate change, both coastal infrastructure and some of Australia's iconic coastal environments are threatened.

Global mean sea-level change

Historical rates of global sea-level change

Globally, the sea level has varied by more than 120 metres over glacial/interglacial cycles. At the time of the last interglacial, the sea level is thought to have been 4 to 6 metres above present day level, with a significant contribution coming from a partial deglaciation of Greenland (Stirling *et al.* 1998; Overpeck *et al.* 2006). At that time, temperatures over Greenland are estimated to have been several degrees warmer than today's values (Otto-Bleisner *et al.* 2006), and perhaps similar to temperatures projected for the end of the 21st century (IPCC 2007). For roughly the 100,000 years following the last interglacial, sea level fell by over 120 metres as major ice sheets were formed in northern America, Europe and Asia and the Antarctic ice sheet grew. From the last glacial maximum about 20 thousand years ago until about 7000 years ago, sea level rose by over 120 metres at an average rate of about 10 mm per year (1 metre per century), with peak rates as high as 40 mm per year (4 metres per century) (Fairbanks 1989; Lambeck *et al.* 2002). Over the last 7000 years, the rate of sea-level rise has been substantially slower (Lambeck 2002). The height of ancient Roman fish tanks built about 2000 years ago indicate there has been little net change in sea level between then and the start of the 19th century (Lambeck *et al.* 2004).

Sediment cores collected from coastal salt marshes in the western and eastern North Atlantic Ocean (Donnelly *et al.* 2004; Gehrels *et al.* 2005; Gehrels *et al.* 2006) indicate an increase in the rate of sea-level rise in the 19th to early 20th century, consistent with the few available long tide-gauge records, the longest of which date back to about 1700 (Woodworth 1999). (This technique has recently been applied on the east coast of Tasmania to determine a relative sea-level curve for the past few centuries, but the results are not yet available.)

Since 1993, there have been high-quality satellite-altimeter observations of sea levels over most of the globe (about 66°N to 66°S), allowing accurate estimates of both globally averaged and regional sea-level change. Global correlation patterns (called *empirical orthogonal functions*) estimated from the satellite altimeter record have been combined with coastal and island tide-gauge data (corrected for glacial isostatic adjustment) to estimate globally averaged sea levels since 1870 (Church and White 2006). The results show that, from 1870 to the present, global sea level has risen by about 20 cm, at an average rate of 1.7 mm per year during the 20th century, with an increase in the rate of rise over this period (Figure 1). Jevrejeva *et al.* (2006) and Holgate and Woodworth (2004) used quite different techniques to analyse historical tide-gauge data and found quite similar historical rates of sea-level rise. For the modern satellite period (1993 to present), sea level has been rising more rapidly at an average rate of a little greater than 3 mm per year (see, for example, Beckley *et al.* 2007). Note that these rates of increase are an order of magnitude faster than the average rate of rise over the previous several thousand years.

Australian historical mean sea-level change

The oldest surviving Australian sea-level observations were made at Port Arthur, Tasmania from 1841 to 1842 (Hunter *et al.* 2003). They indicate that, from 1841 to 2002, local sea level rose about 14 cm relative to the land, or about 17 cm allowing for vertical land motion. This value is consistent with the estimate of global sea levels from 1870 to 2001 (Figure 1).

Monthly averaged sea-level variations (Woodworth and Player 2003) for the period 1920 to 2000 for Australian locations (Church *et al.* 2006) were compared with the sea levels estimated from a reconstruction of sea levels (Figure 2) on a global 1° by 1° grid using the technique of Church *et al.* (2004). Note that both the observed and reconstructed sea levels are *relative* sea levels; that is, they are the height of the sea surface relative to coastal benchmarks as required for understanding the local impacts of sea-level rise.

The observed inter-annual sea-level variability is strongest at locations along the north-western and western Australian coast and southern Australian coast. This variability is clearly related to El Niño-Southern Oscillation (ENSO) events and is transmitted through the Indonesian Archipelago from the equatorial Pacific Ocean and then anticlockwise around Australia, getting weaker as it progresses. The inter-annual variability on the east Australian coast is generally smaller in magnitude.

The observed records also show a long-term trend of rising sea levels. This trend is well reproduced in the reconstructed sea levels, with two exceptions. Firstly, the Gulf St Vincent site is a composite record from three sites near Adelaide (Port Adelaide Inner Harbour, Port Adelaide Outer Harbour and Port Stanvac). Compaction of sediments with a subsequent sinking of the coastal tide gauge is known to be occurring in this region leading to a larger relative sea-level rise (Belperio 1993). This larger local sea-level rise is not reproduced in the reconstructions. Secondly, the short section of data around 1970 at Brisbane appears anomalous compared with the reconstruction and other sea-level time series, and is suspected to be associated with a change in the datum of the tide gauge

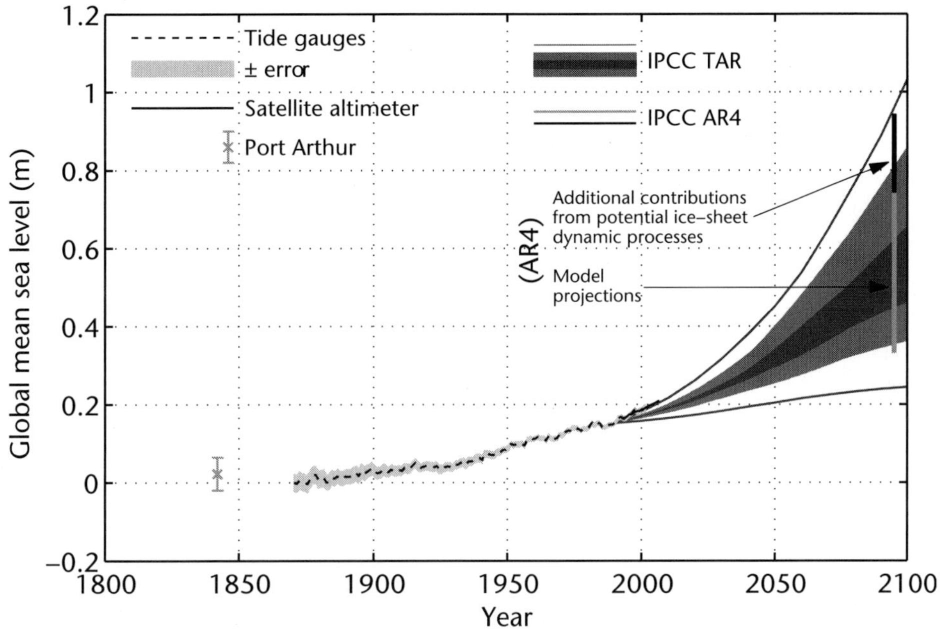

Figure 1: Sea levels from 1800 to 2100. The dotted line from 1870 to 2006 indicates the global average sea level and the black line from 1993 to 2006 indicates the modern satellite altimeter record. The curves from 1990 to 2100 are the IPCC 2001 projections and the bars at 2095 are the IPCC 2007 projections. The estimated local sea level (compared with present day local sea level and corrected for land motion) at the Port Arthur (Tasmania, Australia) benchmark in 1840 (Hunter *et al.* 2003) is shown in grey. The error bars on these estimates are one standard deviation.

between the two segments of data (John Broadbent, personal communication).

Also shown in Figure 2 is an estimate of the Australian-average relative sea-level change (excluding the anomalous Gulf St Vincent record). When examined over the complete 80 year period, these trends show a rise in relative sea level of about 1.2 mm per year. There are suggestions in both the Australian mean time series and in a number of the individual records (for example, Fremantle) that the rate of sea-level rise was at a minimum from the mid-1970s to the mid-1990s. This minimum in the rate of rise occurs during the period of more frequent, persistent and intense ENSO events, as evidenced by the Southern Oscillation Index since the mid-1970s (Folland *et al.* 2001). It is also consistent with global analyses for the period 1950 to 2000, which show a global minimum in sea-level rise in the western equatorial Pacific and in the eastern Indian Ocean (Church *et al.* 2004) over this period. It is important to note that the global analysis gives a global increase in ocean volume equivalent to a sea-level rise of 1.8 mm per year over the period 1950 to 2000 despite these regional areas of lower rates of sea-level rise. The longer records (Fremantle and Fort Denison), the reconstructed sea levels and the Australian Baseline Sea Level Monitoring Array (http://www.bom.gov.au/oceanography/projects/abslmp/abslmp.shtml) suggest that this period of low rate of rise has now passed.

The modern satellite-altimeter data indicate that, since 1993, the rate of rise has not been uniform around the globe, with significantly larger rates of rise in the western Pacific and eastern Indian Ocean and as a consequence around much of the Australian coastline, particularly the coast of Western Australia. This is consistent with the larger rates of relative sea-level rise recorded in the tide-gauge records in the last decade or so, as discussed above, and is a result of both the variable (time and space) pattern of sea-level rise and also the increase in the rate of global-averaged sea-level rise. The large rates of rise since 1993 in the western Pacific/eastern Indian Oceans are again a signature of climate variability – there was a period of sustained El Niño-like conditions and low sea levels in the western Pacific at the start of the altimeter record.

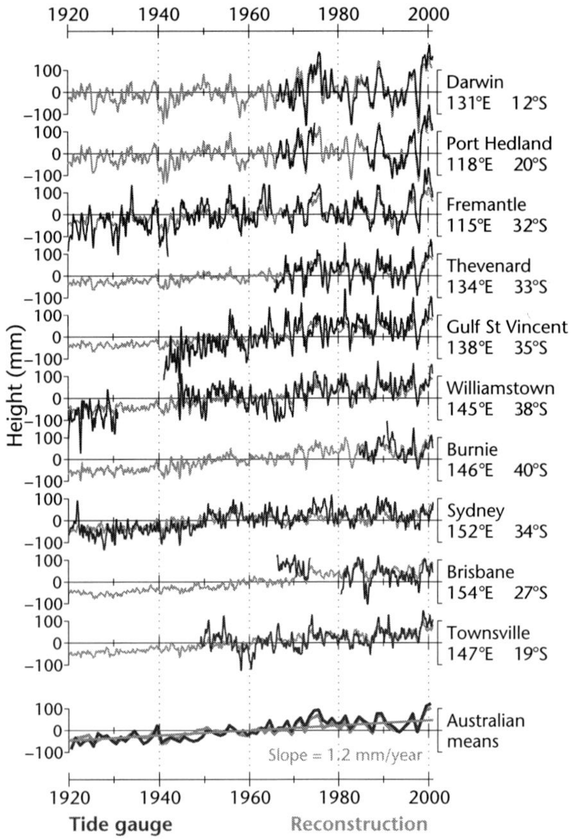

Figure 2: Observed (with coastal tide gauges) and reconstructed sea levels for the period 1920 to 2000. The observed tide-gauge records are monthly average values and can be obtained from the PSMSL website. The Gulf St Vincent site is a composite record from three sites. The bottom panel shows the average observed and reconstructed sea-level record for all Australian sites (with the exception of the Gulf St Vincent site) around the Australian coastline. (Source: after Church *et al.* 2006.)

Projections of global averaged sea-level rise

The two major reasons for sea-level rise are expansion of ocean waters as they warm (and an associated decrease in ocean density) and an increase in the ocean mass, principally from land-based sources of ice (glaciers and ice caps, and the ice sheets of Greenland and Antarctica). Global warming from increasing greenhouse gas concentrations is a significant driver of both contributions. The amount of thermal expansion is non-uniform due to the influence of ocean currents and spatial variations in ocean warming.

The Intergovernmental Panel on Climate Change (IPCC) provides the most authoritative information on projected sea-level change. The IPCC Third Assessment Report (TAR) projected a global averaged sea-level rise of 9 to 88 cm between 1990 and 2100 using the full range of IPCC greenhouse gas scenarios, a range of climate models and an additional uncertainty for land-ice changes (Church *et al.* 2001). The IPCC's Fourth Assessment Report (AR4; IPCC, 2007) used a larger range of models to project a sea-level rise of 18 to 59 cm (with 90% confidence limits) over the period from 1980–1999 to 2090–2099 (Meehl *et al.* 2007). This rise in sea level is mainly a result of thermal expansion of the upper ocean and from the melting of glaciers and ice caps. In these projections, the ice sheets of Greenland and Antarctica together contribute little to sea-level rise during the 21st century. Increased snow fall on Antarctica partially offsets positive sea-level contributions from other components. However, it is recognised that current ice sheet models are incomplete and do not allow for a rapid dynamic response of the ice sheets, as indicated in some recent observations. To allow for ice sheet uncertainties, IPCC AR4 increased the upper limit by 10 to 20 cm and stated that 'larger values cannot be excluded, but understanding of these effects is too limited to assess their likelihood or provide a best estimate or an upper bound for sea-level rise.' The end result is that the upper values given in the IPCC TAR and AR4 projections (Figure 1) are similar.

From the start of the IPCC projections in 1990 to 2006, observed sea level has been rising more rapidly than the central range of the IPCC model projections and near the upper end of the total range of the projections (Rahmstorf *et al.* 2007), suggesting that one or more of the model contributions to sea-level rise is underestimated.

Sea-level rise over the early 21st century has largely been determined by past emissions of greenhouse gases. However, sea-level projections closer to, and beyond, 2100 are critically dependent on future greenhouse gas emissions, with both ocean thermal expansion and the ice sheets potentially contributing metres over centuries for higher emission scenarios. There is increasing concern about the longer term contributions of the ice sheets. For example, for the Greenland Ice Sheet, a sustained global average temperature increase relative to pre-industrial values of greater than 3.1°C (with a 90-percentile range of 1.9°C to 4.6°C) would lead to surface melt exceeding precipitation, resulting in an ongoing wastage of the Greenland Ice Sheet for centuries and millennia (Gregory and Huybrechts 2006), consistent with sea levels in the last interglacial being several metres higher than today's value. This temperature increase could potentially be crossed late in the 21st century if effective greenhouse gas mitigation measures are not adopted. A second cause for concern is associated with the poorly understood dynamic responses of the Greenland and West Antarctic Ice Sheets, which could lead to a significantly more rapid rate of sea-level rise than from surface melting alone.

Projections of mean sea-level rise in the Australian region

Sea-level rise during the 21st century is not expected to be uniform around the globe. This is a result of changing atmospheric conditions (in particular, surface winds) and ocean currents (Lowe and Gregory 2006). The oceans surrounding Australia are influenced by a number of climate variability patterns: for example, the El Niño Southern Oscillation (ENSO), the Southern Annular Mode (SAM) and the Indian Ocean Dipole. These climate variations overlay the global mean sea-level rise, resulting in significant regional variability in the magnitude and trend of sea-level rise in the oceans surrounding Australia. For example, ENSO results in sea-level variability in the tropical Indian and Pacific Oceans, while the SAM is a major driver of sea-level variability in the Southern and mid-latitude Indian and Pacific Oceans. The impact of climate variability results in a regionally complex pattern of sea-level rise and variability in the Indian and Pacific Oceans.

The strongest spatial signatures of sea-level rise projections from 16 atmosphere-ocean models, used for the IPCC AR4, are a minimum in the Southern Ocean south of the Antarctic Circumpolar Current, and a maximum in the Arctic Ocean (Meehl *et al.* 2007). The next strongest features are maxima in sea-level rise at latitudes of about 30° to 40°N in the Pacific and to a lesser extent the Atlantic Ocean, and at about 40° to 50°S in all of the southern hemisphere oceans, at the poleward extremities of the subtropical gyres. This band of maximum sea-level rise in the southern hemisphere is connected to an above average sea-level rise (about 0.2 m above the global average) off the east coast of Australia.

For the Australian region, the spatial pattern of sea-level rise by 2070 was investigated for a number of the climate models. The patterns of thermal expansion for 17 of the AR4 models are shown in Figure 3 for the A1B (mid-range) emissions scenario (CSIRO and Bureau of Meteorology 2007). Each model's globally averaged thermal expansion has been subtracted so that the diagrams represent the regional variation of sea level either above or below the global average sea-level rise projection. Although there is wide variation in the amount of projected sea-level rise, there are also similarities in some regions. For example, in 13 of the 17 models, the thermal expansion along the east coast of Australia south of 30°S is positive relative to the global average values, leading to an additional rise of around 10 cm above the global average change. This feature is attributed to a strengthening of the East Australian Current, which results in the movement of warmer water along the east coast and is caused by a southward shift in wind patterns in the 21st century simulations in the models. More work is required before there is confidence in these regional patterns of sea-level rise.

Extreme events in the Australian Region

Historic extreme events

Sea-level records show high and low extremes related to both tides and meteorological effects—the latter being generally referred to as 'surges'. Australia experiences a wide variation of tidal types and magnitudes. In western and north-eastern Australia, the tides occur approximately once per day (they are 'diurnal') while in north-western, eastern and south-eastern Australia, and in the Great Australian Bight and Bass Strait, they occur twice per day (they are 'semidiurnal'). Elsewhere the tides are 'mixed' and have both diurnal and semidiurnal qualities. Figure 4 (Bureau of Meteorology 2004) shows the tidal range (defined here as twice the sum of the amplitude of the M_2, S_2, K_1 and O_1 tidal constituents) around Australia,

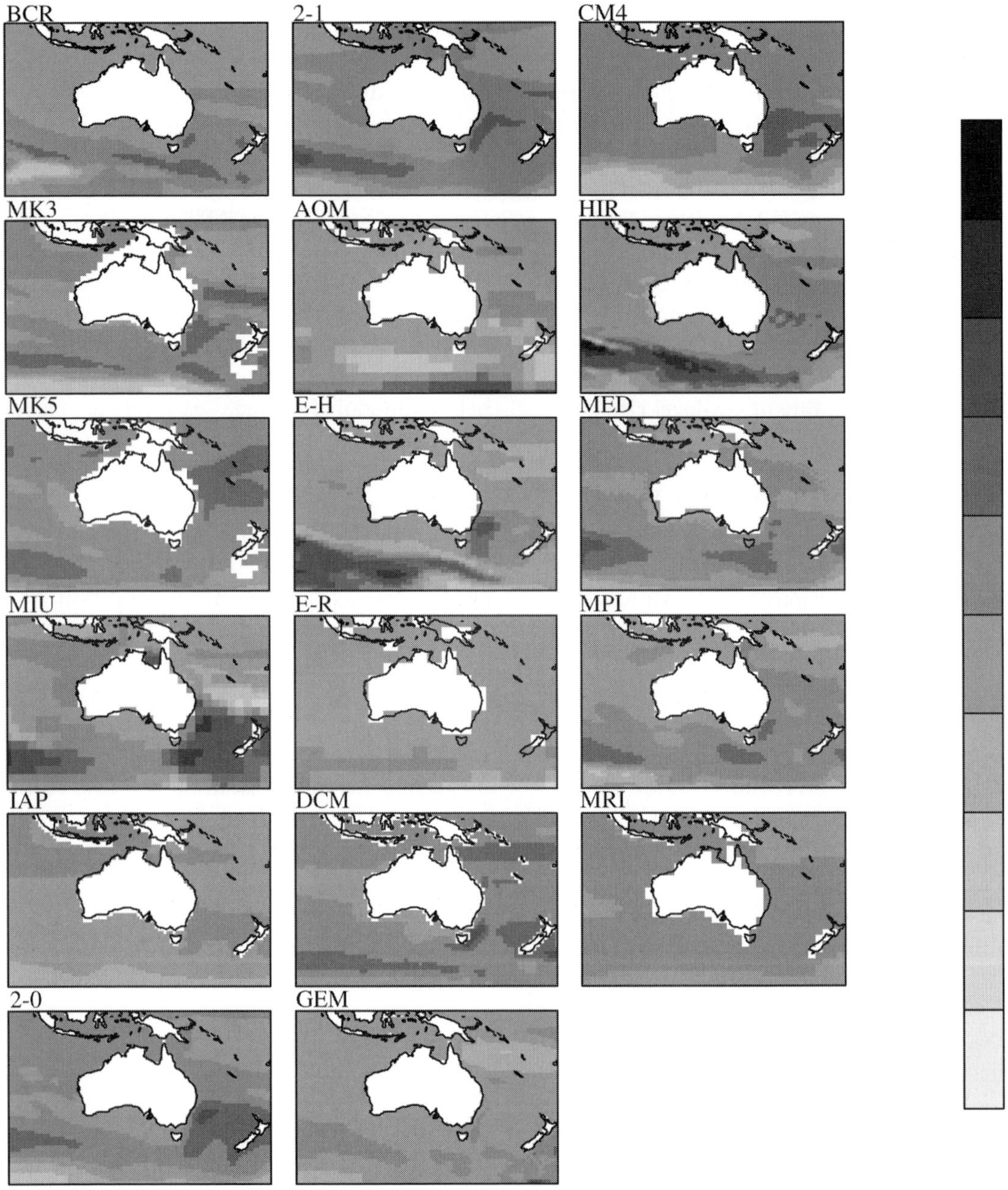

Figure 3: Projected regional contribution of thermal expansion by 2070 to sea level relative to the globally averaged value for each model showing the high degree of variability between models, but the general indication that rises on the east coast of Australia may be higher by about 10 cm than the global average rise though to 2070. Calculations are based on the sub-set of AR4 models that provided sea level data to PCMDI. Units are metres. Simulations used IPCC Scenario SRES A1B.

indicating that the largest tides occur in northern Australia (maximum range about 11 m) and in Bass Strait (maximum range about 3 m). The weakest tides occur off south-western Australia with ranges as small as 0.7 m.

Sea-level rise is often experienced through its effect on extreme levels: high (flooding) sea-level events getting higher, and extremes of a given height becoming more frequent. Figure 5 shows the effect of sea-level rise during the 20th century on

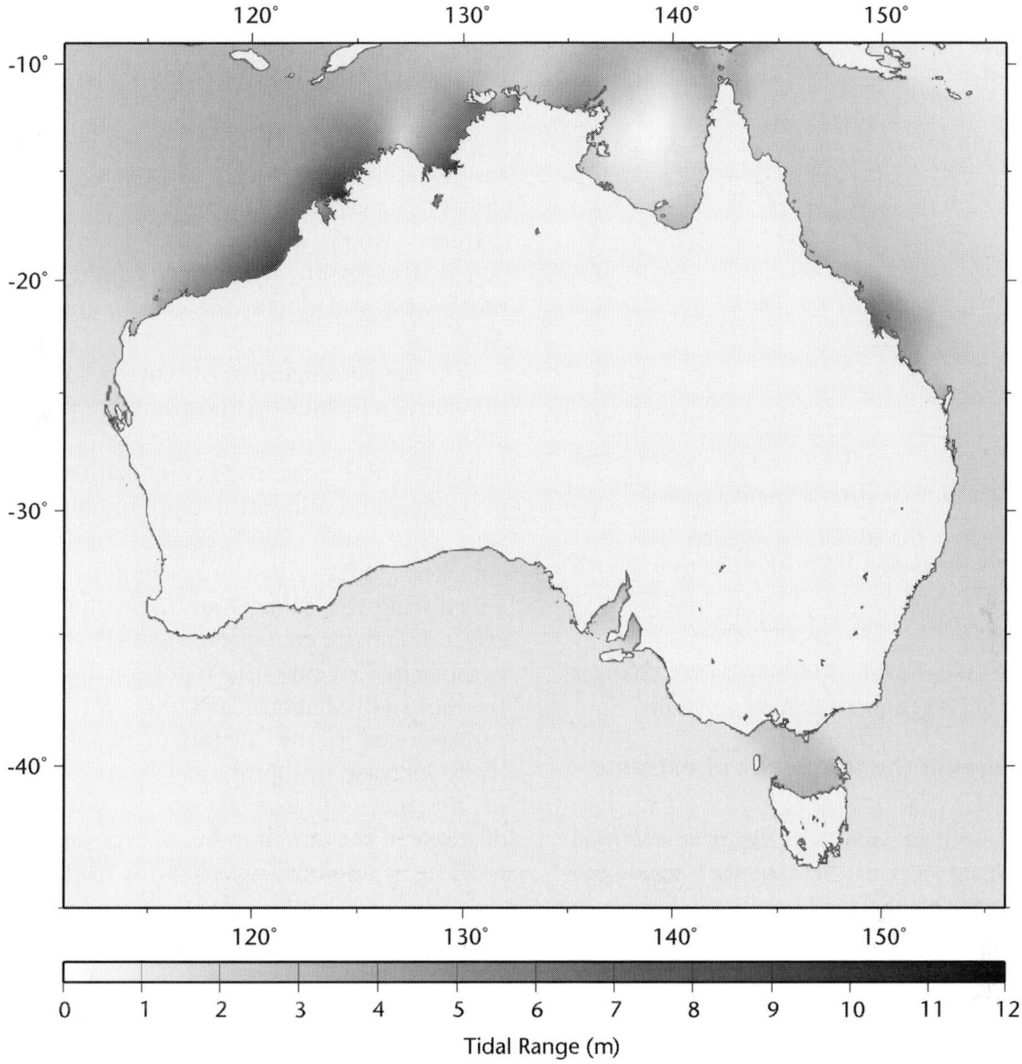

Figure 4: Tidal range around Australia (Source: Bureau of Meteorology 2004).

extremes observed at Fremantle, Western Australia (after Church *et al.* 2006). The curves indicate the levels of extremes with a certain average recurrence interval (defined as the average time between exceedance events of a given height), for data prior to 1950 (black) and after 1950 (grey) (the data starts in 1897 and ends in 2004). The post-1950 curve has been displaced upwards by about 0.1 m, which represents the overall sea-level rise between the two subsets of data. This displacement caused extremes that occurred on annual to decadal timescales to increase their frequency of occurrence by a factor of about three during the 20th century. A similar factor was found for Fort Denison, Sydney, which also has an approximately century-long record (Church *et al.* 2006).

Sea-level extremes may also change as a result of a change in the variability of sea level about the mean. For Fremantle (and for Fort Denison, Sydney), this effect has so far been of secondary importance—the dominant change in extremes being due to the rise in mean sea level. This is not, and will not, necessarily be the case in all regions. For example, in the North Sea, where large storm surges are experienced, an increase in winds can have a larger impact than mean sea-level change on the extreme height of storm surges, at least in the near term. Of course, for larger rises in sea level, the

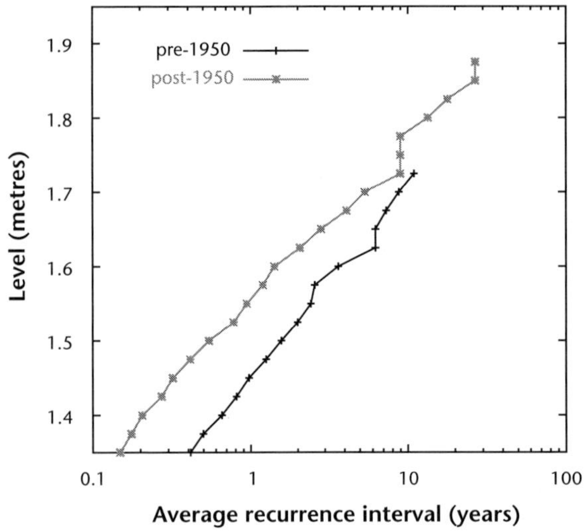

Figure 5: Changes in the average recurrence intervals for Fremantle for the period 1897–2004 (Source: Church *et al.* 2006).

change in the mean value dominates over changes in the height of surges about this mean value.

Future changes in the frequency of extreme events

The severity and frequency of extreme sea-level events in the future will increase with rising sea levels and may also change as a result of changes in the frequency and intensity of the meteorological drivers of storm surge. It may be seen from either curve in Figure 5 that the sea level varies approximately logarithmically with the average recurrence interval, indicating that the extremes approximately follow a Gumbel distribution. The increase in the frequency of extreme high levels, caused by a future increase in mean sea level, may be estimated from the slope of such curves. If a sea-level rise of h increases the frequency of occurrence by a factor r, then, a sea-level rise of H increases the frequency of occurrence by a factor $r^{H/h}$ (this is a consequence of the form of Gumbel distribution), which can become very large, even for modest increases in sea level. Figure 6 shows the estimated increase in the frequency of occurrence of extreme high levels, caused by a sea-level rise of 0.1 m, for the 29 Australian tidal records that are longer than 30 years. This multiplying factor has a range of 1.8 – 5.8 and a mean of 3.1, which is broadly consistent with the 20th century observations for Fremantle and Fort Denison, Sydney (Church *et al.* 2006). For a mid-range 21st century rise of mean sea level of 0.5 m (see above), the mean multiplying factor for Australia would therefore be $3.2^{0.5/0.1}$ or 336, indicating that events which now happen every 10 years would happen about every 10 days in 2100. Figure 6 shows that even larger increases in the frequency of extremes would occur around Sydney, Brisbane and in Bass Strait, with smaller increases off the north-west and north-east coasts, and off South Australia.

The meteorological drivers of sea-level extremes in Australia include cyclones in the tropics north of 30°S, westerly winds associated with cold fronts along the south coast and a combination of cyclones of tropical and mid-latitude origin on the east and west coasts south of 30°S. Storm surges occur more commonly during the warmer months in the north and during the colder months in the south. The effect of climate change on these systems has been summarised recently in a report by the CSIRO and the Bureau of Meteorology (2007). Studies of tropical cyclones in the Australian region indicate a likely increase in the number of tropical cyclones in the more intense categories, but a possible decrease in the total number of cyclones. They also indicate a poleward extension of tropical cyclone

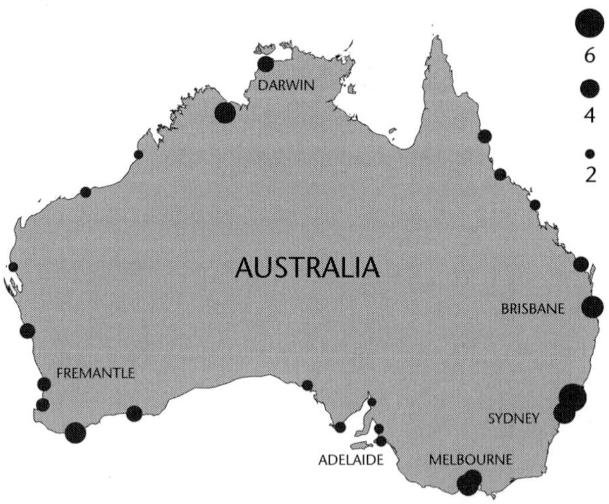

Figure 6: Estimated multiplying factor for the increase in the frequency of occurrence of high sea-level events (indicated by the diameters of the discs), caused by a sea-level rise of 0.1 m.

tracks. For example, Leslie *et al.* (2007) found a poleward shift of over 2° in latitude of the genesis region of tropical cyclones and of the tropical cyclone tracks. Abbs *et al.* (2006) found a poleward shift of 0.7° in the tropical cyclone genesis region across Australia and of 3° in the decay location of cyclones on the east coast.

The projected large-scale change to wind patterns that will affect the incidence of storm surges in mid-latitude regions is one of a poleward shift of the mid-latitude westerlies, bringing weaker winds to the southern Australian coast to the west of 140°E during the colder months. However, increasing wind speeds are more likely over the Victorian, Tasmanian and southern NSW coasts. This is likely to be associated with a southward extension of the East Australian Current, which brings warmer water further southward along the east coast and enables a greater intensification of storm systems in this region. In a climate model simulation in which this pattern of EAC intensification occurs, it was found that east coast low pressure systems with central pressures of 990 hPa or deeper were 60% more frequent in 2070 (McInnes *et al.* 2007b).

There have been two broad approaches for evaluating storm-tide return levels under current and future climates in Australia: one that is used in tropical regions and another that is used in the mid-latitude regions. The basis for these approaches lies with the frequency of occurrence of the storm-surge generating weather systems and whether the associated storm surges have been well captured by the tide-gauge network. In the tropics, the infrequency of tropical cyclones together with their small spatial scale (of order 100 km or less), and hence the small spatial scale of the storm surge, means that the observational record of storm-surge occurrence is poor. This precludes an estimation of the storm-surge recurrence intervals from observational records. Instead, an approach in which a hydrodynamic model is used to simulate the cyclone-induced storm surges from synthetically generated tropical cyclone wind and pressure fields have typically been used (for example, McInnes *et al.* 2003; Hardy *et al.* 2004). The population of synthetic cyclones is derived from the characteristics of the observed cyclones in the region of interest. On the south coast of Australia, the methods developed to evaluate storm-surge return periods exploit the frequency and coherence of storm-surge events along the southern coastline. Previous studies have shown that these events are coherent over thousands of kilometres and are a response of coastal-trapped waves to the atmospheric forcing (Church and Freeland 1987). Therefore methods to quantify storm-tide return periods in this region have involved using a hydrodynamic model to simulate an observed population of extreme sea-level events identified in tide-gauge records (McInnes *et al.* 2007a). In both approaches, statistical methods are used to estimate storm-surge recurrence intervals from the population of modelled events.

The effect of climate change on storm-surge recurrence intervals has been investigated in several studies around Australia. The tropical studies of McInnes *et al.* (2003) for Cairns in northern Queensland and Hardy *et al.* (2004) for a number of locations along the Queensland coast considered an increase in mean sea level (0.25 m and 0.3 m, respectively) and a 10% increase in the intensity of all cyclones. Hardy *et al.* also considered a southward shift of cyclone tracks of 1.3 degrees (about 140 km). They found that sea-level rise provided the largest contribution to the future increase in extreme sea levels, with cyclone intensity changes and the southward movement of tracks having a negligible impact on the height of the 1 in 100 year storm surge. On the other hand, McInnes *et al.* found that an increase in the intensity of tropical cyclones increased the 1 in 100 year storm surge by around 0.3 m. A key difference in the two approaches is in the way tropical cyclones were selected for the hydrodynamic modelling. Hardy *et al.* use an autoregressive model whose coefficients were derived from observed tropical cyclones over the Coral Sea from 1969 to 2001. McInnes *et al.* used extreme value statistics fitted to cyclones dating back to 1908 to evaluate the cyclone probabilities and this approach includes more intense cyclones than have been observed in the last few decades (Church *et al.* 2006).

Along the eastern Victorian coast, the study of McInnes *et al.* (2007a) evaluated the possible change to storm surge recurrence intervals by considering the range of wind speed change in 13 climate models over Bass Strait. Although the changes in wind speed were found to increase in the majority of

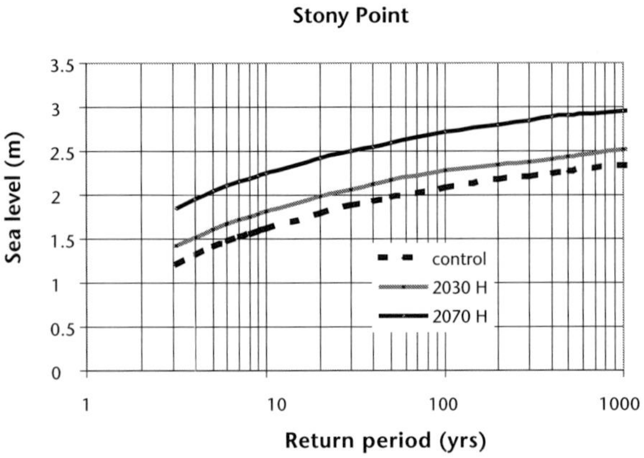

Figure 7: Storm-tide return-period curves for Stony Point under the high climate-change scenarios for 2030 and 2070. Note that a high scenario comprises a high wind-speed scenario combined with a high sea-level rise scenario. The 1 in 100 year storm-tide height in 2070 is around 0.7 m higher than the control value. Of this increase, around 75 to 80% is due to sea-level rise, while the remainder is due to wind speed increase. The frequency of the extreme events increases in the two scenarios shown also with the control climate 1-in-100-year event being exceeded on average once every 30 years in 2030 and once every 5 years in 2070.

models, the response of the various models ranged from a decrease of 5% to an increase of 10% by 2070. These low and high wind-speed changes were applied uniformly to the wind fields for each modelled storm-surge event. For a storm-surge height of 0.75 m—which represents roughly a 1 in 100 year storm surge along much of the eastern Victorian open coast—the wind speed changes induced water-level changes of −0.075 m to +0.15 m (that is, twice the percentage change in wind speed). The changes in extreme sea levels from potential wind-speed changes are somewhat smaller than those that can occur from mean sea-level rise, which may be as high as 0.49 m by 2070 in these simulations. An example of the return periods of extreme sea level under current and future conditions is given in Figure 7 for Stony Point on the Victorian coast.

Coastal erosion around Australia

Factors Influencing Coastal Erosion

Geomorphic and stratigraphic evidence from field and model investigations in Australia and elsewhere, summarised in Roy *et al.* (1994), clearly show massive coastal change of tens of metres in response to sea-level variations during the last 10,000 years of the Holocene Epoch. Confidence is gained from these studies that advanced numerical models are capable of reproducing observed geomorphic responses to the complete range of factors responsible for coastal change, including effects of sea-level rise (Cowell *et al.* 1995, 2003; Stolper *et al.* 2005). However, despite the historical increase in mean sea level around the Australian coast during the 20th century, other factors causing coastal erosion have to date masked any local response to sea-level trends and it has not been possible to unambiguously attribute an erosion signal to sea-level rise, in the presence of other anthropogenic activities.

Geomorphic impacts can be characterised by changes in location of shorelines. Shoreline change has particular significance to society through land loss and effects of erosion on property and infrastructure. Shoreline movements are a manifestation of horizontal and vertical translations in the *shore face* due to changes in sea level, sediment availability or geometry of the shore face (Cowell *et al.* 1995). Erosion, deposition and sediment transport rates generally decrease in deeper water due to the weakening of wave motions with water depth. (The shore face comprises the beach and bed of any coastal water body, extending from the landward limit of sedimentary processes associated with wave run-up, out to the offshore limit of sediment transport; Cowell *et al.* 1999.)

The tendency for the coast to erode or accrete depends on the combined effects of four factors:

1. a gradual rise in mean sea level, which causes a landward recession of the coast wherever it comprises loose sediments or friable bedrock (Roy *et al.* 1994)
2. changes in the magnitude of transient storm-erosion events (Zhang *et al.* 2004; McInnes *et al.* 2007b)
3. supply and loss of sediments from nearby sources and sinks (List *et al.* 1997; Stive and Wang 2003)
4. realignment of shorelines due to changes in directional statistics of wave fields (Ranasinghe *et al.* 2004; Goodwin *et al.* 2006), which can be expected if wave refraction patterns are

altered by sea-level rise or there are other changes in climate.

The first of these mechanisms may be simply understood by consideration of the transport of sediment by incoming ocean waves. Under equilibrium conditions, the shape of the shore face is governed by a balance between onshore/offshore transport of sediment by surface waves. Under a given level of wave activity, the shore face takes on a particular offshore profile. If the sea level rises, the onshore/offshore transport changes until the shore face returns to its original profile and equilibrium is re-established under conditions of raised sea level. However, in this new equilibrium, the shoreline has been moved upwards and onshore resulting in recession of the shoreline. This mechanism was formalised in a very simple way by Bruun (1962; see following Section) and later re-formulated to allow for changes in profile equilibrium through time (for example, Cowell *et al.* 2006).

Generally, the most common experience of acute coastal erosion in Australia has been associated with transient erosion due to storm events involving high waves and abnormally high water levels, especially when storm surges coincide with spring tides. These conditions occurred in south-east Australia during the strong La Niña of May–June, 1974, and the resulting storm erosion that caused destruction of houses at Narrabeen and Bilgola on the Sydney coast has become a benchmark in coastal-hazard planning.

While dramatic property losses during such storm events capture the attention of coastal managers and the public alike, extreme storms that have recurred historically do not represent an increased threat. Nor are they *necessarily* indicative of more serious chronic erosion problems. Coasts recover from these transient events through the return of sand to beaches and dunes following its temporary displacement offshore during storms. Only with an increase in magnitude of erosion events will related storm impacts penetrate further inland. Historically, Australia has generally been free of chronic coastal erosion problems, except where such problems were triggered by coastal engineering. For the most part, property and infrastructure losses associated with extreme storms to date reflect the inappropriate development of assets within dynamic coastal environments that were mistakenly thought to be stable under existing and historical conditions.

At present, the strength of non-sea-level rise factors driving erosion, and uncertainty in quantifying these drivers, generally makes detection of the sea-level rise response signal problematic. Almost all the chronic coastal-erosion hotspots around Australia lie in regions where alongshore sand transport is a dominant process and the erosion is largely due to disturbances in the alongshore transport gradient or the existence of local sinks that cause sand loss from the littoral environment. In most cases, the disturbance to littoral transport has been induced through coastal engineering works.

The possibility of shore-face sand supply to beaches has further obscured the geomorphic signal of sea-level rise over recent decades (Figure 8a). Evidence for sand supply to beaches from the lower shore face is surprisingly ubiquitous around much of the Australian coast (Figure 8b) and is likely to be common on most coasts with shore faces comprising a concave up cross-shore profile due to onshore skewness in the amplitudes of water velocities under shoaling waves. For example, results from geological investigations in south-east Australia (Cowell *et al.* 1995) and sediment-transport modelling on the central Netherlands coast indicates that sand supply from the lower shore face fully compensates for the effects of significant relative sea-level rise (Beets *et al.* 1992). Strong evidence of shore-face sand supply also exists in field data from the Columbia River littoral cell in the US north-west Pacific (Cowell *et al.* 2001).

Historical coastal erosion

One of the best known examples of coastal erosion is the result of the Tweed River entrance training walls interrupting the alongshore sand feed from northern NSW to the Gold Coast beaches (Boak *et al.* 2001). Coastal erosion occurred immediately following initial construction of the training walls in the early 20th century, and again in the late 1960s when the walls were lengthened. The latter phase occurred after massive tourist development on the Gold Coast, so the ensuing erosion attracted much more attention. The publicity was not enough to prevent similar costly mistakes being made elsewhere, such as in Victoria at Portland (Gourlay

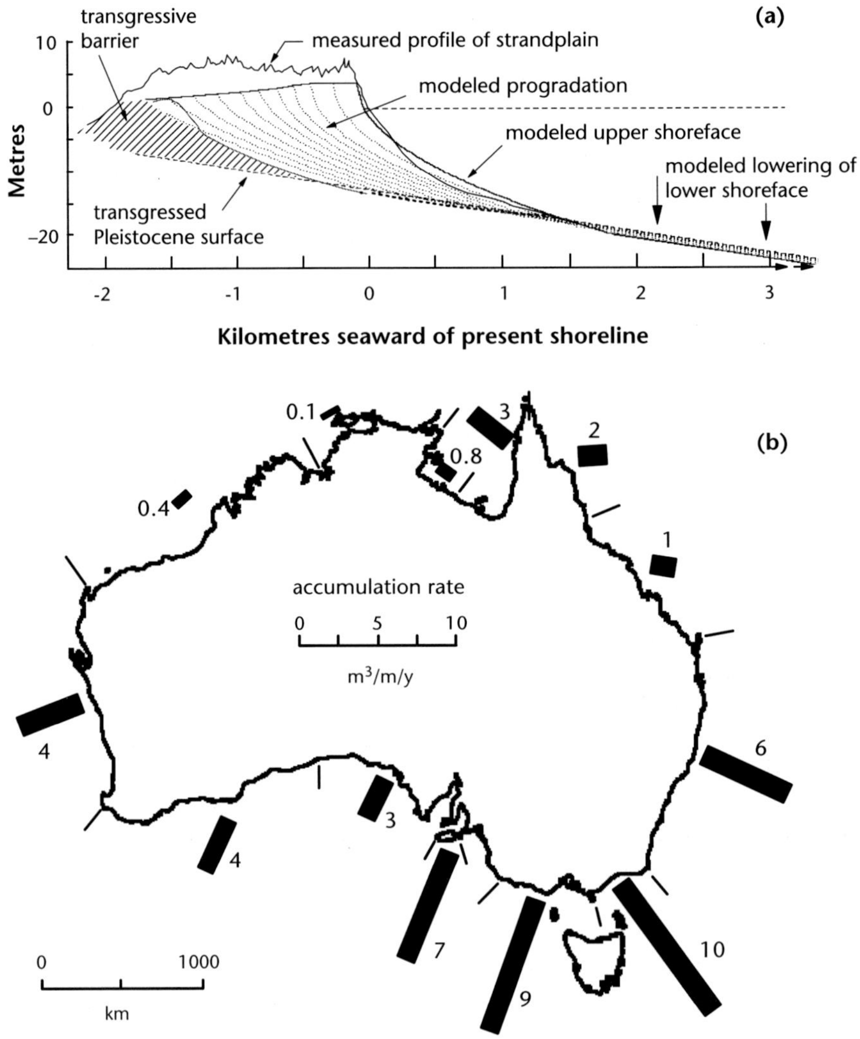

Figure 8: Sand supply for barrier progradation from the shore face: (a) modelled and measured supply from Tuncurry, south-east Australia (Source: adapted from Cowell *et al.* 2003); (b) volume estimates for segments of the Australian coast (Source: Short 2003).

1996) and Port Fairy (BMT-WBM 2007), and in Western Australia along down-drift beaches flanking Fremantle Harbour (Searle and Semeniuk 1985), the Peel Inlet at Mandurah, and the Port Geographe marina entrance near Busselton. Such training-wall installations interrupt alongshore sand transport, thereby inducing positive alongshore transport gradients on down-drift beaches, causing systemic erosion. The walls also tend to promote sequestration of sand by estuarine sinks, causing sand losses from adjacent beaches.

A chronic-beach erosion history along Adelaide's coast is thought to have a sea-level rise component. However, significant uncertainties in estimates of alongshore sand transport (Coastal Engineering Solutions 2004) make it difficult to evaluate the contribution of sea-level rise. As with erosion hotspots elsewhere, erosion may be as much to do with development having encroached too close to the shoreline, thus interrupting littoral transport, making it difficult to identify the response to the sea-level rise.

Similar coastal management issues play a significant role in beach erosion at Byron Bay in NSW. Construction of a car park and its revetments on the active beach in the 1960s caused a realignment of the down-drift coast, contributing to coastal erosion. The coastal realignment problem is compounded by fluctuations in the rate of sand bypassing Cape Byron. These fluctuations occur with ENSO and longer interdecadal climate variations that affect the wave regime, causing fluctuations in coastal alignments (Ranasinghe *et al.* 2004; Goodwin *et al.* 2006). Coastal realignments due to engineering works and climate fluctuations have been especially significant when the transient erosion they cause is compounded by severe storm-erosion events. In 1974, these combined effects caused the destruction and permanent abandonment of Sheltering Palms—a village immediately north of the Brunswick River entrance-training walls on the Byron Bay coast.

On the central and south coast of NSW, the highly embayed coastline features and prominent headlands extending into deep water means that leakage of sand to adjacent beaches is negligible (Roy and Thom 1981). Under such conditions, the closed littoral sediment budgets eliminate the main source of contamination in the potential geomorphic signal in beach response to sea-level rise. The resulting opportunity for detection of this signal is, however, somewhat diminished due to sand contributions to beaches from the lower shore face. The Collaroy–Narrabeen Beach in Sydney probably represents the greatest exposure to imminent property loss due to coastal erosion at present in Australia (Hennecke *et al.* 2004). Here again, the problem relates to coastal development of land subdivided almost 100 years ago on what is now known to be the historically active beach zone. The available evidence indicates that the beach has been historically stable in terms of mean-trend changes, and that erosion episodes have been transient throughout the historical past (Thom 1974).

Projected coastal erosion

While local disturbances and changing wave conditions have been the dominant contribution to detectable, systematic coastal change at erosion hotspots on the Australian coast to date, sea-level rise and changes in storms will become significant and begin to dominate over several decades, particularly if the upper range of sea-level projections and intensification of storms eventuate. Current erosion hotspots will probably be the first places to suffer serious impact from erosion induced by increased sea level.

Coastal recession tendencies due to sea-level rise are the result not only of inundation but also systematic patterns of bed erosion and sedimentation (see above). These geomorphic effects can be expected in all marine littoral environments, including wave-dominated and protected oceanic settings, as well as in estuaries, coastal lakes and wetlands. Stive (2004) demonstrated from consideration of shore-face translation and sediment-mass continuity that the rate of shoreline retreat, c_R, can be expressed as

$$c_R\,(h_* + h_d) = \frac{\partial S}{\partial t} L_* + \frac{\partial Q_y}{\partial y} - Q_x - V \qquad (1)$$

where $\frac{\partial S}{\partial t}$ is the rate of sea-level rise; L_* and h_* are the distance and water depth defining the offshore limits of the active zone; h_d is the height of the dune or friable cliff; Q_x is the time-averaged shoreward sediment flux into the active zone from the lower shore face; Q_y is the alongshore sediment flux integrated over L_*; V is a local sediment source or sink (such as a delta or estuary); and x and y are across-shore and alongshore dimensions. The *active zone* is the upper-shore-face region in which rates of fluctuation in bed elevation, due to cycles of erosion and deposition, are larger than the rate of sea-level rise.

Equation 1 is general, especially if the source/sink term, V, incorporates any changes in the maximum transient-erosion volume due to storms. Such changes can be expected if systematic variation occurs in storm magnitudes and frequencies. For the special case, in which sources, sinks, net littoral transport and shore-face supply of sediments can all be ignored, equation 1 reduces to the widely applied *Bruun Rule* (Bruun 1962):

$$c_R = \frac{\partial S}{\partial t}\left(\frac{L_*}{h_* + h_d}\right) \qquad (2)$$

Equation 2 is widely applied because of its practical simplicity, whereas equation 1 demands use of numerical models (Cowell *et al.* 2006). Unfortunately, the special-case conditions underpinning equation 2 seldom arise in nature, so application of the Bruun Rule as the sole explanation for observed or projected changes is generally invalid. One of the few field studies to have validated equation 2

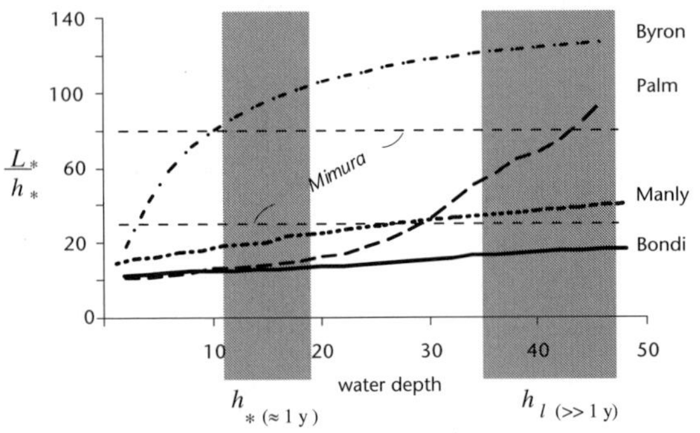

Figure 9: Variability of shoreface slope in equation 2 (L_*/h_*), for five iconic beaches in south-east Australia, in relation to water depths indicative of limiting depths of the active zone (h_*), and the slowly evolving lower shoreface (h_l) on the annual time scale (Hallermeier 1981): h_* is the limiting depth of detectable changes in bed elevation, and h_l characterises the limiting depth of significant sand movement. Although Hallermeier's h_* has been used often to calibrate application of the Bruun Rule, Hallermeier recommended that h_l is more appropriate in place of h_* in equation 2 because significant sand movement during a typical year makes consequential bed-level changes possible on the multi-decadal time scale.

demonstrated that applicability existed along coastal tracts free of tidal-inlet effects and littoral transport gradients, which comprised more than 30% of the US coastline studied from South Carolina to Long Island NY (Zhang *et al.* 2004). Another study, in which validity of equation 2 was evident, involved dominance by strong relative sea-level rise (Mimura and Nobuoka 1995). More generally, under conditions of weak sea-level rise, the other terms in equation 1 tend to dominate coastal change (Roy *et al.* 1994).

These field studies also demonstrate the weakness of the rule-of-thumb that c_R is on the order of 50 to 100 times the rate of sea-level rise. For example, variability in the reciprocal slope term characterising the shore face, L_*/h_*, is illustrated in Figure 9 and demonstrates the effect that choice of h_* has on estimates obtained via the Bruun Rule. Taking uncertainty about choice of h_* into account gives erosion factors (L_*/h_*) ranging from 44 to 64 for Manly, 29 to 99 for Palm Beach and only 28 to 39 for Bondi Beach (Figure 9). These factors correspond poorly to the 50–100 range proposed for coasts elsewhere (for example, Mimura and Nobuoka 1995). On this basis, for example, the response of Manly Beach to an increased sea level by 2100 of 0.28–0.79 m (see above) would involve an expected horizontal erosion distance of between 12 and 51 m. This range of estimates does not, however, take into account uncertainties at Manly related to future changes in shore-face geometry, effects of rock reefs, and a seawall. Application of more appropriate simulation modelling to deal with these uncertainties provides a probabilistic basis for projecting coastal erosion at Manly in response to sea-level rise and possible changes in the wave climate (Figure 10).

The simulations upon which the Manly recession risk estimates were based take into account the additional factors included in equation 1, but excluded from equation 2. For example, the possibility of sand supply from the shore face (Q_x in eq.1) introduces the possibility that the beach may undergo time averaged accretion rather than recession despite sea-level rise, although the probability that this will occur is estimated to be less than 10% (Figure 10). Because Sydney beaches, including Manly, are not subject to significant alongshore transport of sand, projections of coastal recession do not involve effects of disturbances in littoral sediment budgets. In coastal regions subject to these effects, described above in relation to the historical erosion hotspots, erosion risk is much greater than illustrated in Figure 9.

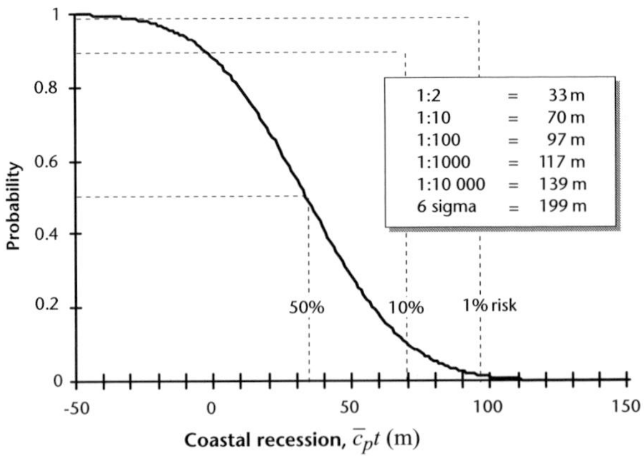

Figure 10: Probability of shoreline recession by year 2100 due to climate-change impacts at Manly Beach, Sydney, based on simulations in which model variables were sampled from probability distributions for each of 2000 repetitions. Negative recession values imply time averaged accretion of the coast (Source: based on Cowell *et al.* 2006).

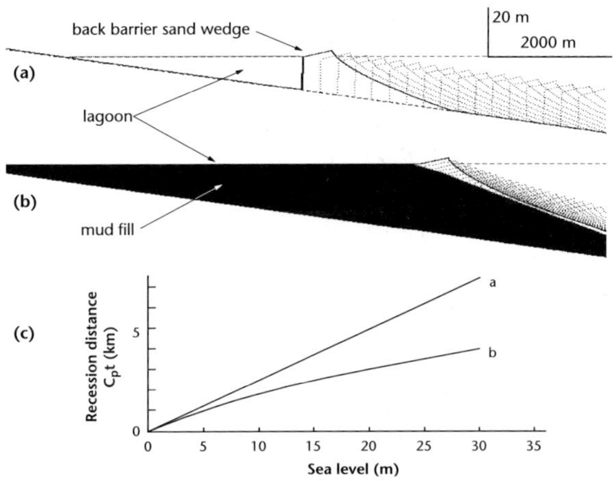

Figure 11: Moderating effects on rate of coastal recession due to deposition of fine sediments in a barrier estuary under constant rates of sea-level rise: (a) barrier rollover with no estuarine mud deposition, for which the rate of recession is identical to the rate of simple inundation on a sloping plane; (b) effect of mud accumulating vertically at the same rate as sea-level rise; and (c) comparative rates of horizontal recession of the shore face (Source: adapted from Cowell *et al.* 2003).

For the next few decades, the role of estuaries, including river mouths and lagoons, as sinks for beach sand is unlikely to be significant. Exceptions will exist in the vicinity of entrances that are artificially deepened for navigation. If sea levels continue to rise in subsequent centuries, the resulting deepening of estuaries and tidal inlets will tend to cause a progressive loss of sand to estuarine sinks from adjacent beaches (Figure 11).

Deposition of sediments other than marine sand (including fine sediments, such as estuarine muds and bay-head (fluvial) delta sands; Roy, 1994) within estuaries and coastal wetlands due to either natural processes or artificial fill plays a significant role in reducing the potential sink capacity of estuaries (Figure 11). Such sedimentation reduces the sink volume and thus limits sand demand exerted on the adjacent coast.

Conclusion

A substantial body of evidence indicates sea level has risen at a significantly larger rate during the 20th century than over the previous several millennia. This is true both globally and around Australia, with evidence coming from geological data, in situ instrumental data and, most recently, satellite-altimeter data. This rise will continue during the 21st century and beyond because of the long time scales associated with the deep ocean and the ice sheets. The IPCC projections indicate a global-averaged sea-level rise of 18 to 79 cm from 1990 to 2095, with recent observations indicating a rise closer to the upper end of the range. In addition, regional variations from this global averaged rise are expected. The projected sea-level rise will require significant adaptation.

Rising sea levels will be felt most acutely through extreme events. The rise in mean sea level has already led to an increase in the frequency of high sea-level events of a given magnitude, by a factor of about three on both the east and west coasts of Australia. The logarithmic relationship between the height and return period of high sea-level extreme events means that by late in the 21st century, events that now occur once every 10 years could be occurring once every few weeks and the present one-in-a-hundred-year event could occur several times a year. In addition, the most severe high sea-level events will have a greater impact. Again, these changes will require significant adaptation and planning for many regions around the Australian coastline.

Urgent and effective mitigation is required if the world is to significantly decrease the risk of multi-metre sea-level rise from ice-sheet contributions. Paleogeomorphic data indicate that, when temperatures were similar to those we can expect in 2100 if efforts to mitigate greenhouse gas emissions fail, the sea level was several metres higher than today's. We presently have insufficient understanding of the response of ice sheets to both atmospheric and oceanic warming to determine whether such a rise would occur over decades to centuries through a rapid dynamic ice-sheet response, or would occur over millennia from surface melting alone. These rises would have major impacts on virtually all regions of the Australian coastline.

As well as inundation from extreme events, coastal erosion will be an important issue. To date, erosion hot spots have occurred as a result of coastal developments on active dune areas. While it is likely that sea-level rise has not yet produced a detectable impact on coastal erosion, it will become a dominant contributor during the 21st century, with the existing hot spots the first (and probably the most severely) regions to be affected. As sea levels continue to rise, coastal erosion will be felt more broadly along the Australian coastline.

Adaptation strategies will be required and include (Church *et al.* 2007):

- **accommodation** through forward planning and appropriate use of low-lying coastal regions (for example, to ensure escape and emergency routes are available for future flooding events, and to increase the resilience of coastal developments and communities)
- **protection** via 'hard' measures such as sea walls (for example, 10 million people who live below sea level in the Netherlands are protected by dykes and levees) and 'soft' measures such as increased beach nourishment
- **(planned) retreat** through planning instruments, such as implementation of no-build areas or building setbacks for areas susceptible to flooding and erosion.

Adaptation plans must not only consider modern urban development, but also allow for the protection of historical sites, sensitive environmental areas and ecosystems. Developing management policies that simultaneously tackle these potentially conflicting goals presents a major challenge. However, with proactive planning, we can substantially lessen the impact of 21st century sea-level rise.

References

Abbs DJS Aryal S, Campbell E, McGregor J, Nguyen K, Palmer M, Rafter T, Watterson I and Bates B (2006). 'Projections of extreme rainfall and cyclones'. A report to the Australian Greenhouse Office, Canberra, Australia, 97 pp. Bureau of Meteorology, Canberra.

Beckley BD, Lemoine FG, Lutchke SB, Ray RD and Zelensky NP (2007). A reassessment of global and regional mean sea level trends from TOPEX and Jason-1 altimetry based on revised reference frame and orbits. *Geophysical Research Letters* **34**, L14608 <doi:10.1029/2007GL030002>.

Beets DJ, van der Valk L and Stive MJF (1992). Holocene evolution of the coast of Holland. *Marine Geology* **103**, 423–443.

Belperio AP (1993). Land subsidence and sea level rise in the Port Adelaide estuary: implications for monitoring the greenhouse effect. *Australian Journal of Earth Sciences* **40**, 359–368.

Bindoff NL, Willebrand J, Artale V, Cazenave A, Gregory J, Gulev S, Hanawa, K, Le Quéré C, Levitus S, Nojiri Y, Shum CK, Talley LD and Unnikrishnan A (2007). Observations: Oceanic Climate Change and Sea Level. In: *Climate Change 2007: The Physical Science Basis. Contribution of Working Group I to the Fourth Assessment Report of the Intergovernmental Panel on Climate Change* (Eds S Solomon, D Qin, M Manning, Z Chen, M Marquis KB Averyt, M Tignor and HL Miller). Cambridge University Press, Cambridge, UK and New York.

BMT-WBM (2007). 'Port Fairy East Beach coastal erosion engineering and feasibility study'. Report prepared for Moyne Shire Council BMT WBM Pty Ltd, Brisbane.

Boak L, Jackson LA, McGrath J and Brosnan M (2001). Proceedings 15th Australasian Coastal and Ocean Engineering Conference. In: *An Overview of Gold Coast Coastal Management 1960–2001.* The Institution of Engineers, Australia, Barton ACT.

Bruun P (1962). Sea-level rise as a cause of shore erosion. *Journal of Waterways Harbors Division* **88**, 117–130.

Bureau of Meteorology (2004). 'Modelling of tides and tsunamis'. National Tidal Centre, Bureau of Meteorology, Australia.

Chen K and McAneney J (2006). High-resolution estimates of Australia's coastal population. *Geophysical Research Letters* **33**, L16601. <doi:10.1029/2006GL026981>.

Church JA and Freeland HJ (1987). The energy source for the coastal-trapped waves in the Australian Coastal Experiment region. *Journal of Physical Oceanography* **17(3)**, 289–300.

Church JA and White NJ (2006). A 20th century acceleration in global sea-level rise. *Geophysical Research Letters* **33**, L01602. <doi:10.1029/2005GL024826>

Church JA, Gregory JM, Huybrechts P, Kuhn M, Lambeck K, Nhuan MT, Qin D and Woodworth PL (2001). Changes in sea level. In *Climate Change 2001: The Scientific Basis. Contribution of Working Group 1 to the Third Assessment Report of the Intergovernmental Panel on Climate Change* (Eds. JT Houghton, Y

Ding DJ Griggs, M Noguer PJ van der Linden, X Dai, K Maskell and CA Johnson) Cambridge University Press, Cambridge, UK and New York.

Church JA, White NJ, Coleman R, Lambeck K and Mitrovica JX (2004). Estimates of the regional distribution of sea level rise over the 1950–2000 period. *Journal of Climate* **17**, 2609–2625.

Church JA, Hunter JR, McInnes KL and White NJ (2006). Sea-level rise around the Australian coastline and the changing frequency of extreme events. *Australian Meteorological Magazine* **55**, 253–260.

Church JA, Nicholls R, Hay JE and Gornitz V (2007). Ice and Sea-level Change, In *Global Outlook for Ice and Snow.* (Ed. J Eamer) United Nations Environment Programme, New York.

Coastal Engineering Solutions (2004). 'Coastal processes study of Adelaide beaches'. Report for Department for Environment and Heritage, Natural and Cultural Heritage Coastal Protection Branch, Adelaide.

Cowell PJ, Roy PS and Jones RA (1995). Simulation of large-scale coastal change using a morphological behaviour model. *Marine Geology* **126**, 45–61.

Cowell PJ, Stive MJF, Roy PS, Kaminsky GM, Buijsman MC, Thom BG and Wright LD (2001). Shoreface sand supply to beaches. In *Proceedings of the 27th International Conference on Coastal Engineering.* pp. 2495–2508. American Society of Civil Engineers, Sydney.

Cowell PJ, Stive MJF, Niedoroda AW, Swift DJP, De Vriend HJ, Buijsman MC, Nicholls RJ, Roy PS, Kaminsky GM, Cleveringa J, Reed CW and de Boer PL (2003). The coastal-tract (Part 2): Applications of aggregated modelling to lower-order coastal change. *Journal of Coastal Research* **19(4)**, 828–848.

Cowell PJ, Thom BG, Jones RA, Everts CH and Simanovic D (2006). Management of uncertainty in predicting climate-change impacts on beaches. *Journal of Coastal Research* **22(1)**, 232–245.

CSIRO and Bureau of Meteorology (2007). 'Climate change in Australia'. Technical report. <www.climatechangeinaustralia.gov.au>.

Donnelly JP, Cleary P, Newby P and Ettinger R (2004). Coupling instrumental and geological records of sea-level change: Evidence from southern New England of an increase in the rate of sea-level rise in the late 19th century. *Geophysical Research Letters* **31**, L05203. <doi: 10.1029/2003GL018933>.

Fairbanks RG (1989). A 17,000-year glacio-eustatic sea level record: influence of glacial melting rates on the Younger Dryas event and deep ocean circulation. *Nature* **342**, 637–642.

Folland CK, Karl TR, Christy JR, Clarke RA, Gruza GV, Jouzel J, Mann ME, Oerlemans J, Salinger MJ and Wang S-W (2001). Observed climate-variability and change. In *Climate Change 2001: The Scientific Basis. Contribution of Working Group 1 to the Third Assessment Report of the Intergovernmental Panel on Climate Change* (Eds JT Houghton, Y Ding, DJ Griggs, M Noguer, PJ van der Linden, X Dai, K Maskell and CA Johnson), Cambridge University Press, Cambridge, UK and New York.

Gehrels WR, Kirby JR, Prokoph A, Newnham RM, Achterberg EP, Evans H, Black S and Scott DB (2005). Onset of recent rapid sea-level rise in the western Atlantic Ocean. *Quaternary Science Reviews* **24**, 2083–2100.

Gehrels WR, Marshall WA, Gehrels MJ, Larsen G, Kirby JR, Eiríksson J, Heinemeier J and Shimmield T (2006). Rapid sea-level rise in the North Atlantic Ocean since the first half of the nineteenth century. *The Holocene* **16(7)**, 949–965.

Goodwin ID, Stables MA and Olley JM (2006). Wave climate, sand budget and shoreline alignment evolution of the Iluka–Woody Bay sand barrier, northern New South Wales, Australia, since 3000 yr BP. *Marine Geology* **226**, 127–144.

Gourlay MR (1996). History of coastal engineering in Australia. In *History and Heritage of Coastal Engineering* (Ed. NC Kraus) pp. 1–88. American Society of Civil Engineers, New York.

Gregory JM and Huybrechts P (2006). Ice-sheet contributions to future sea-level change. *Phil. Trans. R. Soc.* A **364**, 1709–1731.

Hallermeier RJ (1981). A profile zonation for seasonal sand beaches from wave climate. *Coastal Engineering* **4**, 253–277.

Hardy T, Mason L and Astorquia A (2004). 'Queensland climate change and community vulnerability to tropical cyclones: ocean hazards assessment – Stage 3', Report to Queensland Government, 56 pp. <www.longpaddock.qld.gov.au/ClimateChanges/pub/>

Hennecke WG, Greve CA, Cowell PJ and Thom BG (2004). GIS-based coastal behaviour modelling and simulation of potential land and property loss: implications of sea-level rise at Collaroy/Narrabeen Beach, Sydney (Australia). *Coastal Management* **32**, 449–470.

Holgate SJ and Woodworth PL (2004). Evidence for enhanced coastal sea level rise during the 1990s. *Geophysical Research Letters* **31**, L07305. <doi:10.1029/2004GL019626>.

Hunter J, Coleman R and Pugh D (2003). The sea level at Port Arthur, Tasmania, from 1841 to the present. *Geophysical Research Letters* **30**, 7, 1401. <doi:10.1029/2002GL016813>.

IPCC (2007). *Climate Change 2007: The Physical Science Basis. Contribution of Working Group 1 to the Fourth Assessment Report of the Intergovernmental Panel on Climate Change* (Eds S Solomon, D Qin, M Manning, Z Chen, M Marquis, KB Averyt, M Tignor and HL Miller). Cambridge University Press, Cambridge, UK and New York.

Jevrejeva S, Grinsted A, Moore JC and Holgate S (2006). Nonlinear trends and multiyear cycles in sea level records. *Journal of Geophysical Research* **111**, C09012. <doi:10.1029/2005JC003229>.

Lambeck K (2002). Sea level change from mid Holocene to recent time: an Australian example with global implications. In *Ice Sheets, Sea Level and the Dynamic Earth*, Geodynamics Series 29, pp. 33–50. American Geophysical Union, Washington DC.

Lambeck K, Yokoyama Y and Purcell T (2002). Into and out of the Last Glacial Maximum: sea-level change during Oxygen Isotope Stages 3 and 2. *Quaternary Science Reviews* **21(1–3)**, 343–360.

Lambeck K, Anzidei M, Antonioli F, Benini A and Esposito A (2004). Sea level in Roman time in the Central Mediterranean and implications for recent change. *Earth and Planetary Science Letters* **224**, 563–575.

Leslie LM, Karoly DJ, Leplastrier M and Buckley BW (2007). Variability of tropical cyclones over the southwest Pacific Ocean using a high resolution climate model. *Meteorology and Atmospheric Physics (Special Issue on Tropical Cyclones)*, in press.

List JH, Sallenger AH, Hansen ME and Jaffe BE (1997). Accelerated relative sea-level rise and rapid coastal erosion: testing a causal relationship for the Louisiana barrier islands. *Marine Geology*, **140(3–4)**, 347–365.

Lowe JA and Gregory JM (2006). Understanding projections of sea level rise in a Hadley Centre coupled climate model. *Journal of Geophysical Research* **111**, C11014. <doi:10.1029/2005JC003421>.

McInnes KL, Walsh KJE, Hubbert GD and Beer T (2003). Impact of sea-level rise and storm surges on a coastal community. *Natural Hazards* **30(2)**, 187–207.

McInnes KL, Hubbert GD, Macadam I and O'Grady JG (2007a). Assessing the impact of climate change on storm surges in southern Australia. *MODSIM 07* conference, Christchurch, New Zealand, 10–13 December.

McInnes KL, Macadam I, Abbs DJ, O'Farrell SP, O'Grady J and Ranasinghe R (2007b). Projected changes in climatological forcing conditions for coastal erosion in NSW. *Coasts & Ports 2007* conference, Melbourne, Victoria, 18–20 July.

Meehl GA, Stocker TF, Collins WD, Friedlingstein P, Gaye AT, Gregory JM, Kitoh A, Knutti R, Murphy JM, Noda A, Raper SCB, Watterson IG, Weaver AJ and Zhao ZC (2007). Global climate projections. In *Climate Change 2007: The Physical Science Basis. Contribution of Working Group 1 to the Fourth Assessment Report of the Intergovernmental Panel on Climate Change* (Eds S Solomon, D Qin, M Manning, Z Chen, M Marquis, KB Averyt, M Tignor and HL Miller). Cambridge University Press, Cambridge, UK and New York.

Mimura N and Nobuoka H (1995). Verification of Bruun Rule for the estimate of shoreline retreat caused by sea-level rise. In *Coastal Dynamics 95* (Eds WR Dally and RB Zeidler) pp. 607–616. American Society of Civil Engineers, New York.

Otto-Bleisner BL, Marshall SJ, Overpeck JT, Miller GH, Hu A and CAPE Last Interglacial Project members (2006). Simulating Arctic climate warmth and icefield retreat in the last interglaciation. *Science* **311**, 1751–1753.

Overpeck JT, Otto-Bliesner BL, Miller GH, Muchs DR, Alley RB and Kiehl JT (2006). Paleoclimatic evidence for future ice-sheet instability and rapid sea-level rise. *Science* **311**, 1747–1750.

Rahmstorf S, Cazenave A, Church JA, Hansen JE, Keeling RF, Parker DE and Somerville RCJ (2007). Recent climate observations compared to projections. *Science* **316,** 709. <doi:10.1126/science.1136843>.

Ranasinghe R, McLoughlin R, Short A and Symonds G (2004). The Southern Oscillation Index, wave climate, and beach rotation. *Marine Geology* **204**, 273–287.

Roy PS (1994). Holocene estuary evolution - Stratigraphic studies from southeastern Australia. In *Incised-Valley Systems: Origin and Sedimentary Sequences* (Eds RW Dalrymple, BA Zaitlin and R Boyd). pp. 241–263. SEPM (Society for Sedimentary Geology), Tulsa, Oklahoma.

Roy PS and Thom BG (1981). Late Quaternary marine deposition in New South Wales and southern Queensland – an evolutionary model. *Journal of the Geological Society of Australia* **28**, 417–89.

Roy PS, Cowell PJ, Ferland MA and Thom BG (1994). Wave dominated coasts. In *Coastal Evolution: Late Quaternary Shoreline Morphodynamics* (Eds RWG Carter and CD Woodroffe) pp. 121–186 Cambridge University Press, Cambridge, UK.

Searle P and Semeniuk V (1985). The natural sectors of the inner Rottnest Shelf coast adjoining the Swan Coastal Plain. *Journal of the Royal Society of Western Australia* **67**, 116–136.

Short AD (2003). A survey of Australian beaches. In *Coastal Sediments 2003, Proceedings of the International Conference, Clearwater Beach, Florida* (Eds RA Davis, A Sallenger and P Howd). World Scientific Publishing, River Edge NJ.

Stirling CH, Esat TM, Lambeck K and McCulloch MT (1998). Timing and duration of the Last Interglacial: evidence for a restricted interval of widespread coral reef growth, *Earth Planetary Science Letters* **160**, 745–762.

Stive MJF (2004). How important is global warming for coastal erosion? *Climatic Change* **64(1–2)**, 27–39.

Stive MJF and Wang ZB (2003). Morphodynamic modeling of tidal basins and coastal inlets. In *Advances in Coastal Modeling,* (Ed. C Lakhan) pp. 367–392. Elsevier, Amsterdam.

Stolper D, List JH and Thieler ER (2005). Simulating the evolution of coastal morphology and stratigraphy with a new morphological-behaviour model (GEOMBEST), *Marine Geology* **218 (1–4)**, 17–36.

Thom BG (1974). Coastal erosion in Eastern Australia. *Search* **5**, 198–209.

Woodworth PL (1999). High waters at Liverpool since 1768: the UK's longest sea level record. *Geophysical Research Letters* **26(11)**, 1589–1592.

Woodworth PL and Player R (2003). The Permanent Service for Mean Sea Level: an update to the 21st century. *Journal of Coastal Research* **19(2)**, 287–295.

Zhang K, Douglas BC and Leatherman SP (2004). Global warming and coastal erosion. *Climatic Change* **64**, 41–58.

13

Flood

Deborah Abbs

Australia has one of the most variable climates in the world, with this variability appearing in our folklore through references such as 'droughts and flooding rains' in Dorothea Mackellar's 1918 poem *My Country*. In Australia's early years of colonisation, flood hazards played a major role because of lack of knowledge of the magnitude of infrequent floods and the consequent occupation of unsuitable low lying areas. Adaptation to, and risk management of, flooding commenced early in the 19th century when Governor Lachlan Macquarie provided allotments on higher ground for the dwellings of settlers farming the lower floodplains of the Nepean and Hawkesbury rivers. These 'Macquarie towns' were intended to be above the levels that floods could reach. Unfortunately, this was not the case and later in the century a number of towns relocated from flood-prone sites to locations on higher ground. Farmers began to construct levees around cropland, while some towns, especially those in western New South Wales, developed the habit of building earth levees when floods were approaching.

Despite these measures, floods can still have a major impact on society. During significant floods, lives can be lost and the cost of damage to homes, businesses and infrastructure can run into hundreds of millions of dollars. These costs may be further exacerbated by lost production and intangible costs, such as health effects and household disruption. Total economic costs of natural disasters can be considerably greater than insured losses—the total economic losses due to flood are estimated to be ten times greater than insurance costs for flood (Bureau of Transport Economics 2001). Between 1967 and 1999, floods were the costliest natural disaster type in Australia—accounting for almost 30% (see Table 1) of the total cost of all natural disasters.

Queensland and New South Wales account for the largest proportion of floods, with each state bearing approximately 40% of the total economic costs that floods generate nationally and the majority of deaths that they cause.

Extreme rainfall and flooding is frequently defined in terms of an 'average exceedance probability' (AEP). AEP is a statistical benchmark used for flood comparison and is defined as the probability of a flood event of a given magnitude being equalled or exceeded in any one year (Institution of Engineers 1987). It is estimated that at present approximately 170,000 residential properties in Australia are susceptible to flooding by a 1% AEP event (Leigh and Gissing 2006). Loss due to flooding varies widely from year to year and is dependent on a number of factors such as the severity of the flood and its location. The most costly year for floods was 1974, when flooding affected a large area of southern Queensland and northern New South Wales, including Brisbane, Ipswich and the Gold Coast. The total cost from this single event is estimated to have been $2.9 billion (BTE 2001). Widespread flooding also occurred in 1990, affecting the region extending from central western Queensland and western NSW to south-east Victoria. This event is estimated to have cost $1 billion and inundated an area larger than Germany.

However, floods should also be seen as natural processes that recharge underground aquifers, deposit silt and nourishment on flood plains that help maintain agricultural and natural ecosystems, and replenish water storages for our towns and cities.

Table 1: Average cost of Australian natural disasters by state and territory for the period 1967–1999 (excluding death and injury costs)

	Average annual cost ($million)						
State	**Flood**	**Severe Storms**	**Cyclones**	**Earthquakes**	**Bushfires**	**Landslide**	**Total**
NSW	128.4	195.8	0.5	141.2	16.8	1.2	484.1
QLD	111.7	37.3	89.8	0.0	0.4	0.0	239.2
NT	8.1	0.0	134.2	0.3	0.0	0.0	142.6
VIC	38.5	22.8	0.0	0.0	32.4	0.0	93.6
WA	2.6	11.1	41.6	3.0	4.5	0.0	62.7
SA	18.1	16.2	0.0	0.0	11.9	0.0	46.2
TAS	6.7	1.1	0.0	0.0	11.2	0.0	18.9
ACT	0.0	0.1	0.0	0.0	0.0	0.0	0.2
Total	314.0	284.4	266.2	144.5	77.2	1.2	1087.5
Proportion of total (%)	28.9	26.2	24.5	13.3	7.1	0.1	100.0

Note: Figures may not add to totals due to rounding; costs are in 1999 dollars.
Source: BTE 2001.

Chapman (1994) defines flooding as 'water where it is not wanted'. The magnitude and frequency of floods depend on the amount of rainfall and its intensity, but they are also significantly affected by prevailing conditions, such as soil moisture and ground cover, and the ability of the watercourse to convey the run-off. Not all floods are caused by heavy rainfall—in coastal areas, inundation may occur because of a storm surge or a high tide coinciding with higher than normal river levels.

Climatological aspects of flooding

The flooding discussed in this chapter is primarily caused by heavy rainfall. Heavy rainfall occurs where the rate of rainfall is high for an extended period, which is caused by moisture-laden air being lifted and cooled to the point of condensation (Doswell 1996). This requires a source of moisture, a transport process to bring the moisture to the location in question and a lifting mechanism to cool the air. In the Australian context, the moisture sources are the surrounding oceans, especially the warm waters of the tropics, but also the Tasman Sea, the southern Indian Ocean and the Southern Ocean. The moist air is usually transported by winds associated with the synoptic weather systems affecting the region. The lifting mechanism causes moist air to rise and cool, so that the water vapour condenses to form clouds. The moist air can be lifted by meteorological features, such as fronts and convection, or through topographic features, such as mountain ranges and land–sea contrasts. The final requirement for heavy rain is an atmosphere that is buoyant, so that once lift is triggered the air will continue to rise to produce deep, moist convective clouds.

Seasonality and the weather systems that cause flooding

Australian rainfall has distinct seasonal regimes (Figure 1) caused by the north–south movement of the sub-tropical high-pressure belt. The tropical north experiences a summer-dominant pattern with distinct wet and dry seasons. During the wet summer season, the monsoon moves in over northern Australia and moist north-westerly winds bring humid conditions and frequent showers and thunderstorms to the region.

Tropical cyclones and monsoon depressions are the dominant cause of flood-producing rains in summer or early autumn in the north of the continent. These systems can extend or move south bringing heavy rain and flooding to central, southern and eastern Australia. Recently, there has been concern that the relative frequency of very strong tropical cyclones may be increasing (Emanuel 2005;

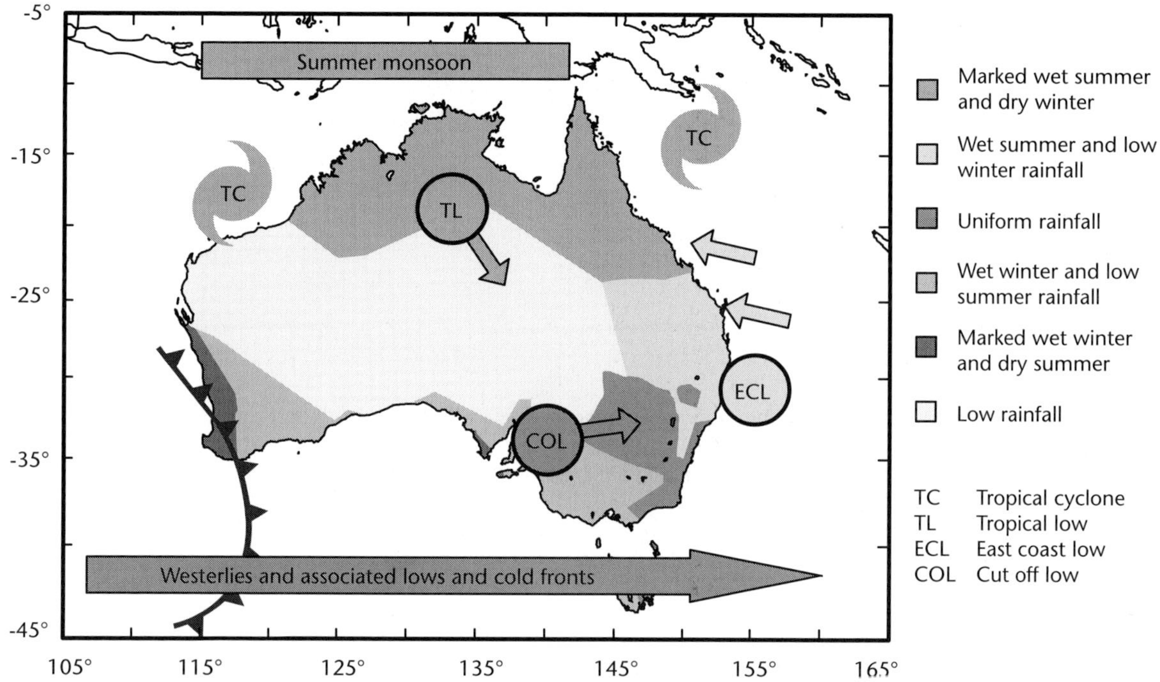

Figure 1: Schematic map showing major seasonal rainfall zones and major synoptic influences on rainfall in Australia (Source: adapted from http://www.bom.gov.au/lam/climate/levelthree/ausclim/zones.htm).

Webster *et al.* 2005; Hoyos *et al.* 2006), although the quality of the historical tropical cyclone data on which these studies relied has been questioned (McBride *et al.* 2006). Analysis of the Australian tropical cyclone database indicates substantial increases in detected tropical cyclone numbers with the advent of weather radar in the late 1950s, although there has been an apparent decrease in numbers of eastern Australian tropical cyclones since the 1970s, largely due to increasing numbers of El Niño events (Nicholls *et al.* 1998). Data also show an increase in the number of more intense cyclones (Kuleshov 2003; Harper and Callaghan 2006) and a reduction in the number of weak and moderate systems.

In the south of the continent, most rainfall occurs in the winter and spring when the subtropical high-pressure belt moves north. At this time, southern Australia comes under the influence of westerly winds and rain-bearing cold fronts.

Rainfall along the coastline of sub-tropical eastern Australia is primarily caused by the influence of the easterly trade winds. Rainfall is relatively high and concentrated in the summer months in Queensland, but is more evenly distributed throughout the year in New South Wales. The Great Dividing Range acts to produce sufficient uplift to produce rainfall and, along the coastal ranges, heavy falls can be produced by sustained, deep easterly airflow. Flood-producing rainfall can be produced when this moisture laden air interacts with a pool of cooler-than-average air at higher levels in the atmosphere.

Flooding along the south-eastern coast—stretching from approximately the Tropic of Capricorn to Gippsland—is frequently caused by a weather phenomenon known as an east coast low (ECL). These are intense low-pressure systems that form at, or close to, the coast. They are characterised by heavy rainfall and often accompanied by strong winds, sometimes of hurricane force. They are a major cause of flood-producing rains, with Hopkins and Holland (1997) estimating that over 16% of all coastal heavy rainfall between 20° and 40°S is directly related to east coast lows. ECLs are a primary cause of flooding in the Sydney Metropolitan area (Speer and Geerts 1994). ECLs occur primarily in autumn, winter and spring in regions of warm sea surface temperature (SST) anomaly. The coastal mountain ranges provide uplift and a mechanism for the northward deflection of the

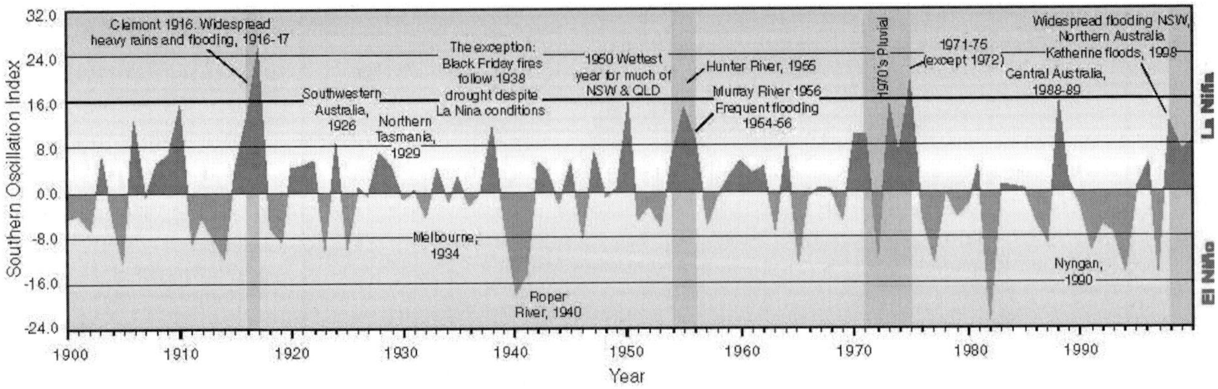

Figure 2: Time-line showing the relationship between major Australian flood episodes and the Southern Oscillation Index. *Note:* Extended periods of high SOI in 1916–17, the mid-1950s, and the early to mid-1970s, were periods of widespread, frequent flooding. (Source: Bureau of Meteorology 2004).

low-level easterly flow. A series of east coast lows was responsible for the flooding experienced in Newcastle and the Hunter Valley and East Gippsland during June 2007.

Localised flash flooding occurs in most parts of Australia. Flash flooding is mainly caused by slow moving thunderstorms or to a succession of thunderstorms that affect the same location.

Longer term variability of Australian flooding

The variability of floods is strongly correlated with the variability in rainfall. Total rainfall is correlated with the frequency and intensity of extreme rainfall, suggesting that extreme events are more frequent and intense during years with high rainfall (Haylock and Nicholls 2000). This variability in Australian rainfall is due to variations in the main rain-bearing weather systems that affect Australia. These weather systems are influenced on multi-year timescales by changes in large-scale patterns including the El Niño-Southern Oscillation (ENSO), the Indian Ocean Dipole (IOD) and the Southern Annular Mode (SAM).

The pattern of rainfall variability associated with ENSO is the dominant influence on the frequency and intensity of floods in eastern Australia. The Southern Oscillation Index (SOI) is used as an indicator of El Niño or La Niña conditions; during El Niño conditions the SOI is persistently negative and vice-versa for La Niña conditions. In the La Niña phase of ENSO the waters of the central and equatorial Pacific cool to below average temperatures and the waters to the north and east of Australia warm. Increased moisture availability from the seas surrounding northern Australia and increased moisture transport provided by the enhanced trade wind regime leads to significantly higher rainfall over northern and eastern Australia, with a strengthened Australian summer monsoon and a greater likelihood of tropical cyclones off the north-east coast. There is a strong tendency for ECLs to occur after El Niño years and, in particular, between large swings of negative to positive SOI values (Hopkins and Holland, 1997). Consequently, there is a much greater likelihood of floods to occur during La Niña years (Figure 2), although it should be noted that floods can occur during El Niño years. Instrumental and paleoclimate records show variations in the frequency and intensity of ENSO over time (see, for example, Allan *et al.*1996; Power *et al.* 1999; van Oldenborgh and Burgers 2005; Tudhope *et al.* 2001; McGregor and Gagan 2004; Shulmeister and Lees 1995) with an increase in the frequency of El Niño events in recent years (IPCC 2001).

The second mode of variability—the Indian Ocean Dipole—affects oceanic and atmospheric circulations in the Indian Ocean. During the negative phase of the IOD, sea-surface temperatures close to Indonesia are higher than average and the winds across the Indian Ocean tend to become more westerly, resulting in enhanced rainfall over Australia.

The third major pattern of rainfall variability is the Southern Annular Mode (SAM) (Thompson and Wallace 2000; Cai *et al.* 2003; Marshall 2003). The SAM is associated with the movement north or south of the mid-latitude westerly winds and associated low-pressure systems and cold fronts that bring rain to southern Australia (Yin 2005; Bengtsson *et al.* 2006; McInnes *et al.* 2007a). During the positive phase of the SAM, the circumpolar westerlies increase in strength and the circumpolar vortex contracts resulting in reduced winter rainfall along the southern coastal regions of Australia (Cai and Cowan 2006; Hendon *et al.* 2007; Meneghini *et al.* 2007). In recent decades, the SAM has increasingly spent more time in its positive phase, resulting in decreased potential for storm formation over southern Australia, a decrease in the intensity of low-pressure systems (Lim and Simmonds 2007) and a southward deflection of some storms (Frederiksen and Frederiksen 2007). Each of these factors contributes to the trend indicating reduced winter-time rainfall over southern Australia. It is thought that the current trend is mainly due to Antarctic ozone depletion (Cai 2006; Miller *et al.* 2006; Cai and Cowan 2007).

Australian rainfall also exhibits variability on inter- and multi-decadal timescales. Much of this variability is attributed to the Inter-decadal-Pacific Oscillation (IPO). The IPO is a 15–30 year pattern of Pacific climate variability, similar to ENSO, that appears to modulate the impact of ENSO on Australia. Negative phases of the IPO correspond to higher regional sea surface temperatures (SST) and increased tropical cyclone formation in the region.

Impacts of climate change

Since the early 1970s scientists and policy makers have been aware of the possibility of human activities affecting the composition of the global atmosphere. By the mid 1980s, atmospheric observations had confirmed that the atmosphere was changing and over the following two decades there has been a major international research effort to establish the nature of the changes and the causes. Since the late 1980s the scientific findings on atmospheric change and the potential for human based changes in the Earth's climate have been reviewed and summarised by the Intergovernmental Panel on Climate Change (IPCC). The most recent report, the Fourth Assessment Report on the Physical Science Basis of climate change (IPCC 2007) had the following among its key conclusions:

- *'Global atmospheric concentrations of carbon dioxide, methane and nitrous oxide have increased markedly as a result of human activities since 1750 and now far exceed pre-industrial values determined from ice cores spanning many thousands of years'*
- *'Warming of the climate system is unequivocal'*
- *'Most of the observed increase in globally averaged temperatures since the mid-20th century is very likely due to the observed increase in anthropogenic greenhouse gas concentrations ... Discernible human influences now extend to other aspects of climate, including ocean warming, continental average temperatures, temperature extremes and wind patterns'*
- *'Continued greenhouse gas emissions at or above current rates would cause further warming and induce many changes in the global climate system during the 21st century that would very likely be larger than those observed during the 20th century.'*

Climate changes projections are based upon the outputs from global and regional climate models that account for possible changes in the emissions of key greenhouse gases and aerosols. The emissions are those due to human activities, such as energy generation, transport, agriculture, land clearing, industrial processes and waste. To provide a basis for estimating future climate change, the IPCC prepared 40 greenhouse-gas- and sulphate-aerosol-emission scenarios for the 21st century that combine a variety of assumptions about demographic, economic and technological driving forces likely to influence future emissions. Carbon cycle models are used to convert these emissions into atmospheric concentrations, allowing for various processes involving the land, ocean and atmosphere. Increasing concentrations of greenhouse gases affect the radiative balance of the Earth. This balance determines the Earth's average temperature. The SRES greenhouse gas concentrations are converted to a radiative forcing of the climate system.

Climate model equations are based on well-established laws of physics, such as conservation of mass, energy and momentum and have been tested against observations. This provides a major source of confidence in the use of models for climate projection. The other basis for confidence comes from

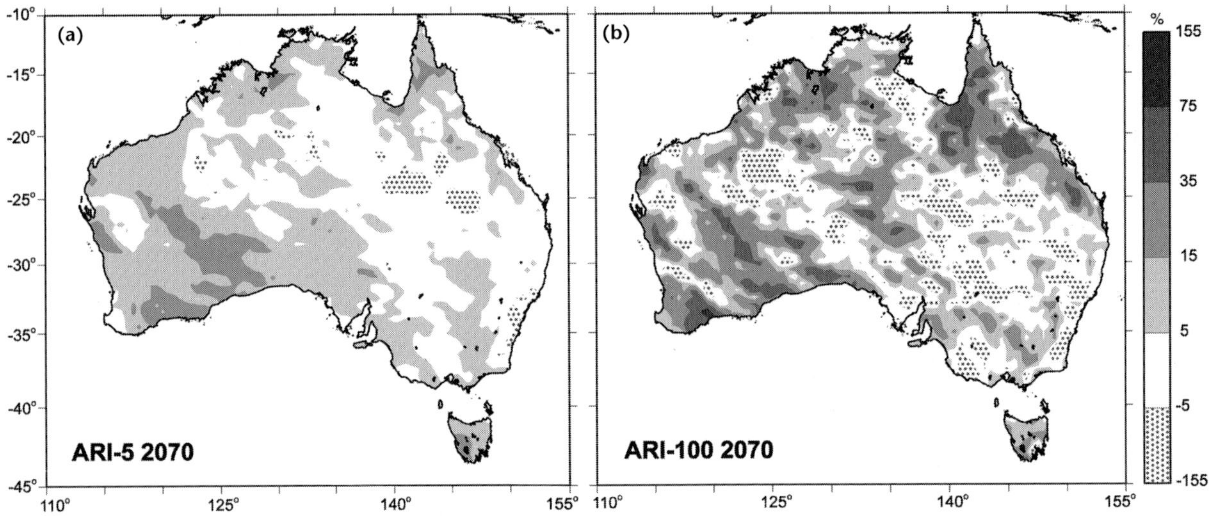

Figure 3: Projected median change for 2070 of (a) the 20% AEP, and (b) the 1% AEP 24-hour rainfall based on outputs from CSIRO climate change simulation. *Note:* The projected change is relative to the present climate.

the ability of models to represent current and past climates, as well as observed climate changes. The most important limitation of models is that a number of important physical processes occur at scales too small to be explicitly resolved by the model, and therefore these have to be represented in approximate form as they interact with the larger scales. Differences in the representation of such processes are the principal cause of differences in the magnitude and patterns of climate change found in different models.

Rainfall changes are not directly forced by rising greenhouse gases, but a warmer atmosphere can hold more water vapour, and hence produce heavier rainfall. Rainfall over some parts of Australia may change little, but elsewhere either decreases or increases may occur due to small differences in the circulation and other processes, discussed above. Projected rainfall changes for Australia, based on the mid-range A1B scenario, can be summarised as a decrease in annual average rainfall for most of the continent, especially along the southern fringe. The 'top end' is likely to experience little change in annual average rainfall.

However, the projected change in extreme rainfall and the weather events that produce them are of interest for flood applications. The general consensus is that increases in extreme daily rainfall are likely even though average rainfall may remain more or less the same. Recent extreme rainfall projections for various regions are that the intensity of the 5% AEP daily rainfall event is likely to increase by:

- up to 10% in parts of South Australia by the year 2030 (McInnes *et al.* 2002),
- 5 to 70% by the year 2050 in Victoria (Whetton *et al.* 2002),
- up to 25% in northern Queensland by 2050 (Walsh *et al.* 2001),
- and up to 30% by the year 2040 in South-East Queensland (Abbs 2004).

In NSW, the intensity of the 2.5% AEP is projected to increases by 5 to 15% by 2070 (Hennessy *et al.* 2004).

Widespread increases in extreme rainfall for 2070 are projected (Figure 3), especially for the south-western and northern parts of the continent. The less frequent, more-extreme events are projected to increase in intensity more than moderate events. These projections are based on the outputs from five separate CSIRO climate change simulations that have been performed using relatively coarse-resolution global and regional climate models.

Statistical and dynamical downscaling techniques are used to translate the projections obtained from coarse-resolution climate models to the catchment and rain-gauge scale required by hydrologists and planners. Global and regional climate models

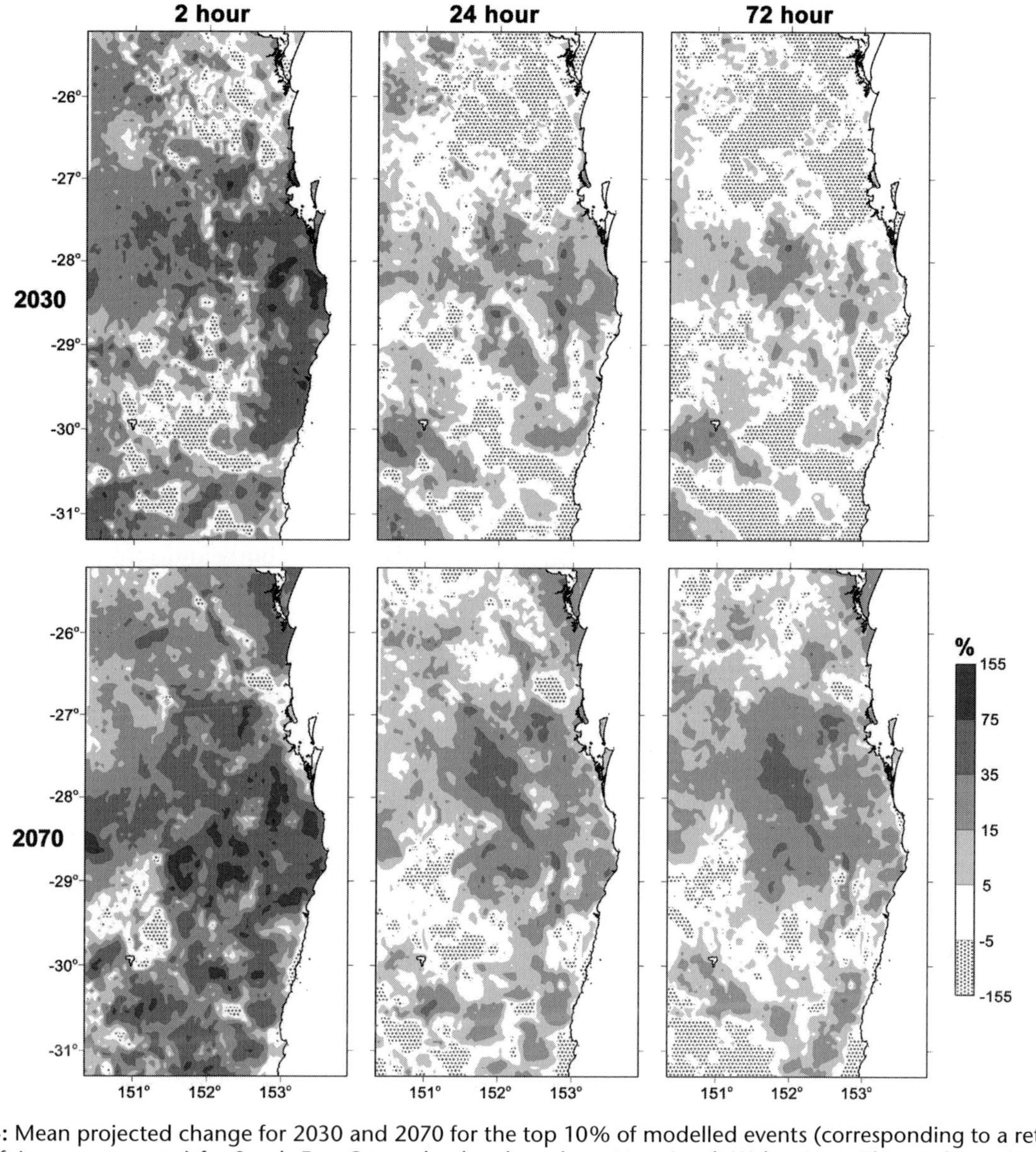

Figure 4: Mean projected change for 2030 and 2070 for the top 10% of modelled events (corresponding to a return period of 4 years or greater) for South-East Queensland and northern New South Wales. *Note:* The results are based on a composite of projected changes from two sets of downscaled model simulations using a model with a grid spacing of 5 km. Results are for extreme rainfall events of 2, 24 and 72 hours (vertical columns). The projected change is relative to the present climate.

show a broad qualitative agreement with the downscaled results. However, downscaling is able to identify important localised regions of projected increases in rainfall intensity not captured by the coarser models (Figure 4). Using these techniques, Abbs *et al.* (2007) identified localised regions of projected significant rainfall increases occurring in the headwaters of the Brisbane and Albert–Logan River systems. Their results also show that with time, the spatial extent of the area of the region affected by increases in extreme rainfall enlarges and the maxima increase in magnitude to values greater than a 40% *average* increase in the intensity of the most extreme events. The short-duration (2-hour) rainfall events show much larger projected changes with the largest changes in the high terrain inland from the Gold Coast. These regions of rainfall increase also increase in spatial extent in the future. The 2-hour rainfall events have average increases in intensity of more than 70% in these regions. These increases in extreme rainfall are likely to result in significant increases in flooding for the rivers of northern New South Wales and south-eastern Queensland.

An understanding of likely changes in the main rain-bearing systems that affect a region can also provide guidance of the changes in flood risk in the region.

Projected changes in tropical cyclones are subject to the sources of uncertainty inherent in climate change projections. These include the future climate-forcing scenario, model dynamics and physics, errors in modelled tropical cyclone climatology and regional patterns and magnitude of change for various fields and climate patterns such as ENSO. The IPCC (2001) concluded that on the basis of climate modelling:

... there is some evidence that regional frequencies of tropical cyclones may change, but none that their locations will change. There is also evidence that the peak intensity may increase by 5% to 10% and precipitation rates may increase by 20% to 30%. There is a need for much more work in this area to provide more robust results.

Since that time, a number of studies have examined the results from medium and high-resolution global climate models (GCMs). The majority of these indicate a consistent signal of fewer tropical cyclones (TCs) globally in a warmer climate. However, there are significant regional variations and these vary between models. Substantial disagreement remains between climate models concerning future changes in TC intensity, although the highest resolution models show evidence of an increase in TC intensity in a warmer world.

Of the studies that have been undertaken specifically for the Australian region, Walsh *et al.* (2004) and Leslie *et al.* (2007) find no significant change in tropical cyclone numbers off the east coast of Australia. Abbs (unpublished) finds a significant decrease in tropical cyclone numbers for the Australian region. Off the coastline of Western Australia, TC numbers are simulated to decrease by 44% while off the east Australian coastline decreases of 9% are simulated. Each of these studies finds a marked increase in the frequency of severe tropical cyclones. This implies a significant increase in rainfall intensity associated with tropical cyclones and the tropical lows that result from such cyclones as they move inland. Both Abbs and Leslie *et al.* (2007) predict that east Australian TCs will be longer lived. In contrast, Abbs finds a decrease in the number of long-lived cyclones for the Western Australian coast.

Despite these recent studies, there still remains considerable uncertainty concerning the effect of climate change on tropical cyclones. This is due in part to the fact that tropical cyclone activity and intensity is strongly influenced by natural climate variability, such as El Niño Southern Oscillation (ENSO), the Madden-Julian Oscillation (MJO) and the Interdecadal Pacific Oscillation (IPO). The effect of climate change on these natural modes of variability is still poorly understood. There are also limitations in the capacity of the climate models to represent the physical processes associated with cyclone formation and movement.

Less information is available for projected changes in the climatology of east coast lows. Abbs and McInnes (2004), Hennessy *et al.* (2004) and McInnes *et al.* (2007b) discuss the ability of various climate models to represent the weather systems conducive to extreme rainfall and winds for locations along the New South Wales and southern Queensland coastlines. Abbs and McInnes (2004) find an increase in the frequency of the weather systems that produce both extreme winds and extreme rainfall in a single regional climate model. However, McInnes *et al.* (2007b) found that projected changes in the frequency of east coast low events are sensitive both to the threshold used to define intensity and to the choice of model. For example, they found that one regional climate model produced a decrease of 60% in the frequency of lows of 990 hPa or deeper by 2070, while a second model produced an increase of 40% in the frequency of these lows.

Changes in Australian average rainfall and rainfall extremes will be affected by changes in the longer term patterns of variability such as ENSO and SAM. Both tropical cyclones and east coast lows are affected by the ENSO cycle, and a tendency for flood events to occur more commonly during the La Niña phase of ENSO has been commented upon earlier. ENSO also has a major influence on the global climate and consequently has been the focus of many studies investigating the effect of global warming. Despite these efforts, the effect of global warming on ENSO is not clear, leading the IPCC (2007) to conclude that:

... there is no consistent indication at this time of discernable changes in ENSO amplitude or frequency in the 21st century.

Power (pers. comm.) analysed outputs from the CSIRO global climate model to examine the ENSO response into the 21st century. His results suggest that in south-east Australia El Niño events will tend to become drier and La Niña events will tend to become wetter, even if Pacific variability linked to ENSO does not increase.

All climate models exhibit a trend in the SAM towards its positive phase when forced with increasing greenhouse gases. SAM is projected to keep trending positive over the 21st century (Arblaster and Meehl 2006; Miller *et al.* 2006) even though the ozone hole is projected to recover within the next 50 years. Changes in SAM will alter storm tracks and synoptic weather systems in the mid-latitudes (Hope *et al.* 2006; Frederiksen and Frederiksen 2007) and affect rainfall over southern Australia.

In many coastal regions, flooding may be due to, or exacerbated by, saltwater inundation arising from a storm tide. A storm tide is the combination of the normal astronomical tide and a storm surge. Storm surges accompany tropical cyclones as they come ashore, but they may also be formed by intense low-pressure systems in non-tropical areas or sustained winds blowing along the eastern coastline in the mid-latitudes. Pounding waves also generated by the powerful winds can add to the impact of a storm tide. This combined with sea-level rise would result in higher storm surge during storm events and a greater area flooded.

A number of other factors also contribute to the increased threat of climate change. For example, more than 80% of Australia's population resides within 50 km of the coast, with 25% of Australia's population growth between 1991 and 1996 occurring within 3 km of the coastline. Thus, there are an increasing number of communities exposed to extreme events such as tropical cyclones, storm surges and flooding of coastal rivers. These socio-economic and settlement trends are projected to continue, and thus Australia's vulnerability to extreme events will continue to increase.

Vulnerability

In a climate change context, vulnerability refers to a system's ability to experience adverse consequences from climate change (Smit *et al.* 1999). For example, the range of climate conditions that different systems can cope with before experiencing adverse effects varies significantly. A major factor affecting a system's vulnerability is its adaptive capacity—its ability to manage or adjust to change. Human settlements, particularly those in developed countries, are generally resilient to daily and seasonal variability in the climate system. However, changes in the frequency of climate extremes, such as severe weather events and floods, will require an adaptive response to minimise the exposure of communities to loss of life, property and business.

Risk can be seen as the interaction between a hazard (in this case, flood), the elements (people, houses, infrastructure, industries) exposed to the hazard and the degree to which those elements are vulnerable to an adverse impact. Climate, and climate change, therefore represents a changing source of risk that needs to be considered in the decision-making process to enable robust and informed decisions on adaptive strategies for the management of flood risk. Flood risk analysis and management in Australia is usually undertaken at the local government level using flood risk assessment approaches such as those outlined in the New South Wales Floodplain Development Manual (NSW Government 2005).

Climate projections based on the outputs from climate models provide a useful basis on which to test the sensitivity of different systems to the possible scale of change. These projections are not forecasts, but are based on scenarios of feasible future greenhouse gas concentrations that combine a variety of assumptions about demographic, economic and technological driving forces likely to influence future emissions. Most impact studies use these projections as one component of a 'what-if' experiment designed to assess the additional risk that a particular level of climate change may pose to an urban region, agriculture, biodiversity, or the economy more generally.

Urban areas and infrastructure have been built to accepted risk limits based on the historically expected return frequency of severe winds, heavy precipitation events and so on. Below these thresholds, relatively light damage to property and human health and life is expected, but at levels above the thresholds, damage, injury and death can accelerate in a non-linear manner. An examination of insurance data shows that more severe flood events cause escalating damage costs (The Australian Climate Group 2004).

Climate projections at a national to regional scale provide information to governments, industry and communities regarding event-risk, but for assessment of future flood risk, projections of changes in extreme rainfall events are required (Hennessy *et al.* 2004; Abbs and McInnes 2004). In order to develop meaningful estimates of the risk of such climatic events, low-resolution climate model simulations of average changes in climatic conditions are not sufficient. Instead, high-resolution modelling through statistical or dynamical downscaling and nested modelling techniques is often required to realistically simulate such events (Abbs *et al.* 2006).

Decision-makers are now using results from these studies. For example, the impacts of increasing rainfall intensities by between 10 and 40% would have a significant impact on the current design events (McLuckie *et al.* 2006) because the current 100-year event could occur every 50 years as a result of an increase in rainfall intensity of just over 10%. These are indicative impacts only—impacts for specific catchments are dependent on the physical characteristics of the floodplain. A 30% increase in rainfall intensity reduces the recurrence interval to just 20 years. The likely impacts of increases in extreme rainfall are more regular damaging flood events with potentially devastating economic and social outcomes.

The Gold Coast City has extensive flood plains and a large population that is exposed to the threat of flooding. The major exposure is in the Nerang River floodplain, which contains almost 60,000 dwellings, 40% of which are flood prone. A GIS-based flood damage model has been developed by the Gold Coast City Council to assess the economic benefits of proposed flood mitigation options and to assess proposed floodplain developments (Betts 2002). Flood damage models estimate flood damage to building structures based on the depth of over-floor flooding, duration of the inundation, velocity of the floodwaters and the types of building material. The Council is now using this model, coupled with CSIRO projections of changes in extreme rainfall, to identify possible impacts of climate change on flood risk in the Nerang catchment (Rahman 2007). Their study found that for a 20% increase in rainfall, the number of properties affected by the 100-year flood event increased from the current 4500 to 7000 residential properties. The direct damage to properties from floodwaters entering the residences increased from \$160M to \$230M. A similar study for flooding of the Albert–Logan catchment (Abbs *et al.* unpublished) found that damage due to flooding more than doubled when a 20% increase in rainfall was combined with sea-level rise of 20 cm or more. Awareness of the changed flood risk enables the council to implement new flood mitigation strategies, respond through land-use planning, minimise future risk in emergencies and to engage with and educate the community.

The devastation caused to New Orleans by Hurricane Katrina provided a clear example of some of the possible consequences—to the developed world—of climate change through the impact of extreme events. These include large loss of life, displacement of millions of people from their homes, loss of water, sewerage and electricity services and knock-on effects to the economy through damage and clean-up costs and the closure of businesses. Katrina also exposed the lack of preparedness of the emergency services for an extreme event of such magnitude. Similar impacts are experienced in coastal regions of India and Bangladesh, but these regions typically do not receive the same levels of media interest as created by Katrina. Climate change is projected to increase the energy of tropical cyclones and so the return frequency of severe storms such as cyclone Tracy could reduce significantly with an associated increase in exposure of tropical population centres to increased flooding and damage from extreme winds. By 2050, sea level may rise 0.1 to 0.4 metres and tropical cyclone intensity around Cairns in northern Queensland could increase by up to 20%. This would increase the flood level associated with a 1-in-100 year storm tide in Cairns from the present height of 2.3 to 2.6 metres to 2.7 to 3.0 metres

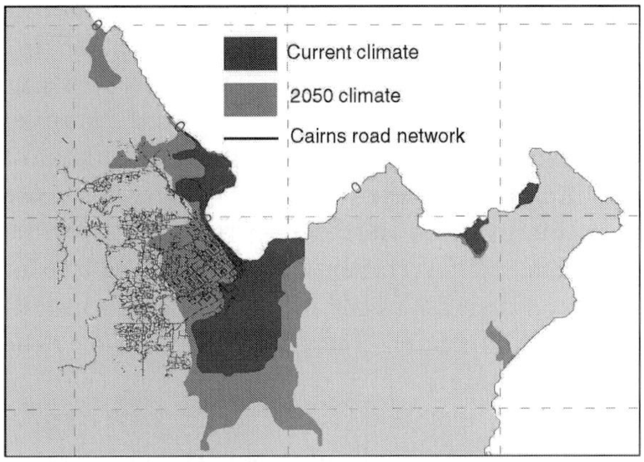

Figure 5: Average area of flooding produced by the top 5% of modelled storm surges (corresponding to a return period of 100 years or greater), in the present climate and the enhanced greenhouse climate of 2050 for Cairns. *Note:* Dark shading indicates flooded area for a 1% AEP storm surge in the present climate and dark shading that in the climate of 2050. The Cairn road network is also shown (Source: Courtesy Kathleen McInnes, CSIRO Marine and Atmospheric Research).

(Figure 5). The extent of flooding implied by the simulated storm surge heights under the present climate indicates that much of the township of Cairns remains above the flood line with the inundation confined to Admiralty Island and the wetland to the south-east of the township. When climate change is factored in, flooding occurs over an area about twice that historically affected, and includes much of the Cairns CBD (McInnes *et al.* 2003). These simulations do not factor in changes in extreme rainfall or riverine flood.

It is unlikely that Australia will experience the level of tropical cyclone devastation caused by Katrina due to differences in the physical exposure of our coastal communities and the scale of the vulnerability of the population (Handmer 2006). However, inundation due to both riverine flooding and storm tide are real risks in northern population centres such as Townsville, Mackay and Hervey Bay. The rapidly expanding Gold Coast canal estates are just above sea level and are vulnerable to minor storm surges and riverine flooding. Studies such as those above suggest a possible need for changes in zoning, building regulations and evacuation procedures if the exposure of Australian communities to flood risk is not to be significantly increased in the future.

Adaptation

Adaptation is a risk-management strategy involving actions to reduce the negative impacts of climate change and take advantage of new opportunities that may arise. The types of adaptation measures adopted will vary from region to region. Adaptation measures are generally implemented at the local level and reduce risk from the bottom up. Some level of climate change is inevitable, irrespective of emission reduction strategies, due to the action of greenhouse gases already accumulated in the atmosphere. Many infrastructure and housing projects that are currently being planned or built are expected to have a lifetime of at least 50 years. By then the impacts of climate change will be evident and thus these developments should now factor in possible increases in flood risk due to increases in extreme rainfall as a consequence of climate change:

An adaptation strategy will aim to increase the resilience of human and natural systems to possible changes in climate conditions where this is likely to be feasible and cost effective, and takes account of the social dimensions of distributing losses. It is a framework for managing future climate risk. It offers the potential of reducing future economic, social, and environmental costs as well as protecting life. An adaptation strategy, to be effective, must result in climate risk being considered as a normal part of decision-making, allowing governments, businesses and individuals to reflect their risk preferences just as they would for other risk assessments. (Allen Consulting 2005)

Consideration of climate change based upon the best available information is a fundamental part of informed flood-risk management and thus needs to be incorporated into management plans as they are developed or reviewed, as do strategic risk-management decisions in relation to future developments and decisions on mitigation works. There are a number of ways in which additional adaptive capacity can be built into these decisions (McLuckie *et al.* 2006):

- Considering the potential for new floodways to form under various climate change scenarios. This may have an impact on future land use as it may identify areas that need to be set aside or developed with the knowledge that they may perform as floodways in the future. Examples include areas which may have an impact upon emergency response, areas that need to be set aside to offset the impacts of urbanization on flood behaviour (e.g. detention basins) and areas where there is a continuing hazard from flooding.

- Considering the potential for significant changes in flood hazard in particular areas. Future developments should be directed to areas compatible with the changing hazard with the result that more vulnerable developments are steered to less exposed areas. For example, some commercial and industrial activities may be suitable for areas that are not suitable for residential developments due to likely flood impacts.
- Considering the potential for changes in flood behaviour to impact upon the management options appropriate for a particular location. Managing flood risk for existing developments presents significantly more challenges than management for future developments. Some options for managing flood risk in existing developments include designing or upgrading levees to account for climate change, voluntary house raising or voluntary purchases of dwellings in hazardous parts of the floodplain. Emergency response planning should be reviewed to ensure that evacuations routes and critical facilities such as hospitals remain above the projected levels of the larger scale floods.
- Allowing additional levels of 'freeboard' in flood planning levels to account for climate change when setting fill levels and floor level controls for new developments. Current flood planning typically requires the minimum floor level of dwelling to be the 1% AEP event plus a freeboard varying between 30 and 50 cm. Freeboard is usually expressed as height above the flood standard and is used to compensate for factors such as wave action, localised hydraulic effects and uncertainties in the flood design levels.
- The decision making process should be flexible to enable future works to allow for climate change.

Adaptation to climate change depends on costs and benefits, lifetimes of the structures and the acceptability of redesigned structures to the community. New infrastructure design and planning approaches should be developed in the context of communities and their needs. Conflicts may arise between investors and developers with short time horizons and government bodies that need to plan on longer time scales and who bear the responsibility for planning and emergency measures over the coming decades.

The insurance and finance sectors have a growing concern about climate change. Flooding is responsible for significant damage, but, unlike most weather-related disasters, flood insurance is not offered in a comprehensive way. In Australia, 10% at most of potential flood victims are estimated to be covered by flood insurance (Smith and Handmer 2002). There is a widening gap between insured and economic losses and the taxpayer will ultimately pay this difference in losses. Many people, particularly in areas that are subject to regular flooding, such as Wollongong, Coffs Harbour and Katherine, are making demands on governments, local councils and their insurance companies to provide cover for flood damage. Flood insurance programs are hampered by the availability of information relating to the probability distributions of floods and especially the damage caused at the individual property level. This is now being tackled by products such as FloodAUS (Risk Frontiers 2003), but the development of better tools to more accurately assess the risk of flood should allow comprehensive flood insurance to be offered. This may have a positive impact on urban planning in response to climate change but there may be downsides if property values are adversely affected through disclosure of increased hazards and more expensive insurance cover. Another adaptation measure in the insurance sector is to reward proactive approaches to reducing climate related risk by decreasing premiums and by increasing premiums so that poor risk management is penalised.

Concluding remarks

Climate change is happening and its effects on our day-to-day lives will often be felt by changes in the frequency and intensity of severe weather events and the floods caused by these events. The regions most vulnerable to flooding are the rapidly growing urban regions of Australia's east coast. These regions currently experience flooding caused by tropical cyclones and east coast lows. Climate change projections suggest that, in the future, tropical cyclones will become more severe and extreme rainfall events along this coastline are likely to be more intense.

We cannot ignore climate change and its potential for significant impact—large increases in potential flood damage and risk to life. Studies are required on adaptation and mitigation options, to ascertain their public acceptability, costs, benefits, side effects and limits. These studies need to take into account human behaviour and socioeconomic impacts. If these issues are to be tackled by decision-makers, there needs to be strong interaction with regional stakeholders to ensure that socioeconomic considerations, ecosystem impacts, adaptation options and regional climate change knowledge are integrated in order to provide the best adaptation solution for each region.

References

Abbs DJ (2004). A high resolution modelling study of the effect of climate change on the intensity of extreme rainfall events. In *Staying Afloat: Floodplain Management Authorities of NSW 44th Annual Conference: Conference Proceedings*. pp. 17–24. Coffs Harbour, New South Wales.

Abbs D, Aryal S, Campbell E, McGregor J, Nguyen K, Palmer M, Rafter A, Watterson I and Bates B (2006). 'Projections of extreme rainfall and cyclones'. Final report to the Australian Greenhouse Office, July 2006. CSIRO,

Abbs DJ and McInnes KL (2004). 'The impact of climate change on extreme rainfall and coastal sea levels over south-east Queensland. Part 1, Analysis of extreme rainfall and wind events in a GCM – A project undertaken for the Gold Coast City Council'. CSIRO Atmospheric Research.

Abbs D, McInnes K, and Rafter A (2007). 'The impact of climate change on extreme rainfall and coastal sea levels over southeast Queensland. Part 2: A High-Resolution Modelling Study of the Effect of Climate Change on the Intensity of Extreme Rainfall Events'. Prepared for the Gold Coast City Council. CSIRO Marine and Atmospheric Research,

Allan RJ, Lindesay JA and Parker DE (1996). *El Niño Southern Oscillation and Climatic Variability*. CSIRO Publishing, Melbourne.

Allen Consulting (2005). 'Climate Change Risk and Vulnerability, Promoting an efficient adaptation response in Australia'. Report to the Australian Greenhouse Office, Department of the Environment and Heritage, Commonwealth of Australia, Canberra. <http://www.greenhouse.gov.au/impacts/publications/pubs/issues-paper.pdf>.

Arblaster JM and Meehl GA (2006). Contributions of external forcings to Southern Annular Mode trends. *Journal of Climate* **19**, 2896–2905.

Australian Climate Group (2004). Climate Change: Solutions for Australia. World Wildlife Fund, Sydney. <http://wwf.org.au/publications/acg_solutions/>.

Betts H (2002). Flood damage analysis using GIS at Gold Coast City Council. *Australian Journal of Emergency Management* **17**, 33–37.

Bengtsson LK, Hodges I and Roeckner E (2006). Storm tracks and climate change. *Journal of Climate* **19**, 3518–3543.

Bureau of Meteorology (2004). *Drought, Dust and Deluge: a Century of Climatic Extremes in Australia*. BOM, Melbourne.

Bureau of Transport Economics (2001). 'Economic costs of natural disasters in Australia'. BTE Report 103. Commonwealth of Australia, Canberra.

Cai WJ, Whetton PH and Karoly DJ (2003). The response of the Antarctic Oscillation to increasing and stabilized atmospheric CO_2. *Journal of Climate* **16** 1525–1538.

Cai, WJ and Cowan T (2006). SAM and regional rainfall in IPCC AR4 models: Can anthropogenic forcing account for southwest Western Australian winter rainfall reduction? *Geophysical Research Letters* **33**, L24708. <doi: 10.1029/2006GL028037>.

Cai WJ (2006). Antarctic ozone depletion causes an intensification of the Southern Ocean supergyre circulation. *Geophysical Research Letters* **33**, L03712 <doi: 10.1029/2005GL024911>.

Cai WJ and Cowan T (2007). Trends in Southern Hemisphere circulation in IPCC AR4 models over 1950–99: Ozone-depletion vs. greenhouse forcing. *Journal of Climate* **20**, 681–693.

Chapman D (1994). *Natural Hazards.* Oxford University Press, Melbourne.

Doswell CA III, Brooks HE and Maddox RA (1996). Flash flood forecasting: an ingredients-based methodology. *Weather and Forecasting* **11**, 560–581.

Emanuel K (2005). Increasing destructiveness of tropical cyclones over the past 30 years. *Nature* **436** 686–688.

Frederiksen JS and Frederiksen C (2007). Interdecadal changes in Southern Hemisphere winter storm track modes, *Tellus*, submitted.

Handmer J (2006) American exceptionalism or universal lesson? The implications of Hurricane Katrina for Australia. *The Australian Journal of Emergency Management* **21**, 29–42.

Harper BA and Callaghan J (2006). 'On the importance of reviewing historical tropical cyclone intensities'. Presented at the 27th Conference on Hurricanes and Tropical Meteorology, Monterey, California, April 24–28. American Meteorological Society, Boston, Massachussetts.

Haylock M and Nicholls N (2000). Trends in extreme rainfall indices for an updated high quality data set for Australia, 1910-1998. *International Journal of Climatology* **20**, 1533–1541.

Hendon HH, Thompson DWJ and Wheeler MC (2007) Australian rainfall and surface temperature variations associated with the Southern Hemisphere Annular Mode. *Journal of Climate* **20**, 2452–2467.

Hennessy KJ, McInnes KL, Abbs DJ, Jones RN, Bathols JM, Suppiah R, Ricketts JR, Rafter AS, Collins D and Jones D (2004). 'Climate change in New South Wales. Part 2, Projected changes in climate extremes'. Consultancy report for the New South Wales Greenhouse Office. Aspendale: CSIRO Atmospheric Research.

Hope PK, Drosdowsky W and Nicholls N (2006). Shifts in the synoptic systems influencing southwest Western Australia. *Climate Dynamics* **26**, 751–764, <doi:10.1007/s00382-006-0115-y>.

Hopkins LC and Holland GJ (1997). Australian heavy-rain days and associated east coast cyclones: 1958–92. *Journal of Climate* **10**, 621–635.

Hoyos CD, Agudelo PA, Webster PJ and Curry JA (2006). Deconvolution of the factors contributing to the increase in global hurricane intensity. *ScienceExpress.* <doi:10.1126/science.1123560>.

Institution of Engineers (1987) *Australian Rainfall and Runoff: a guide to flood estimation. Volumes 1 and 2*, Institution of Engineers, Canberra.

IPCC (2001). *Climate Change 2001: The Scientific Basis.* Contribution of Working Group I to the Third Assessment Report of the Intergovernmental Panel on Climate Change (IPCC). Cambridge University Press, Cambridge, UK.

IPCC (2007). *Climate Change 2007: The Physical Science Basis. Summary for Policymakers.* Contribution of Working Group I to the Third Assessment Report of the Intergovernmental Panel on Climate Change. IPCC Secretariat, Geneva, Switzerland.

Kuleshov YA (2003). *Tropical Cyclone Climatology for the Southern Hemisphere. Part 1: Spatial and Temporal Profiles of Tropical Cyclones in the Southern Hemisphere.* pp. 1–22. National Climate Centre, Australian Bureau of Meteorology, Canberra.

Leigh R and Gissing A (2006). 'How many floodprone properties are there in Australia?' 46th NSW Floodplain Management Authorities Conference, 28 February–2 March 2006. Lismore, NSW.

Leslie LM, Karoly DJ, Leplastrier M and Buckley BW (2007). Variability of tropical cyclones over the Southwest Pacific Ocean using a high resolution climate model. *Meteorology and Atmospheric Physics* (Special Issue on Tropical Cyclones) (in press).

Lim E-P and Simmonds I (2007). Southern Hemisphere winter extratropical cyclone characteristics and vertical organization observed with the ERA-40 reanalysis data in 1979–2001. *Journal of Climate* **20**, 2675–2690.

Marshall GJ (2003). Trends in the Southern Annular Mode from observations and reanalyses. *Journal of Climate* **16**, 4134–4143.

McBride J, Emanuel K, Knutson T, Landsea C, Holland G, Willoughby H, Chan J, Lam C-Y, Heming J and Kepert J (2006). Statement on tropical cyclones and climate change. WMO/CAS Tropical Meteorology Research Program, by participants of the 6th WMO International Workshop on Tropical Cyclones (IWTC-VI), Costa Rica,World Meteorological Organisation, November 2006. < http://www.wmo.ch/pages/prog/arep/tmrp/documents/iwtc_statement.pdf>

McGregor HV and Gagan MK (2004). Western Pacific coral delta-^{18}O records of anomalous Holocene variability in the El Nino-Southern Oscillation. *Geophysical Research Letters* **31**, L11204.

McInnes KL, Suppiah R, Whetton PH, Hennessy KJ and Jones RN (2002). 'Climate change in South Australia: report on assessment of climate change, impacts and possible adaptation strategies relevant to South Australia'. CSIRO Atmospheric Research, Hobart.

McInnes KL, Walsh KJE, Hubbert GD and Beer T (2003). Impact of sea-level rise and storm surges on a coastal community. *Natural Hazards* **30**, 187–207.

McInnes KL, McBride JL and Katzfey JJ (2007a). Winter storm tracks in a global climate model simulation. *International Journal of Climatology*, submitted.

McInnes KL, Abbs DJ, O'Farrell SP, Macadam I and O'Grady J (2007b). *Coastal Zone Climate Change Impact Event Definition: A Project Undertaken for the NSW Department of Natural Resources.* CSIRO Marine and Atmospheric Research.

McLuckie D, Lord D and Gibbs J (2006). 'Climate change – the future is uncertain'. 46th NSW Floodplain Management Authorities Conference, Lismore, 28 February-2 March, 2006.

Meneghini B, Simmonds I and Smith IN (2007). Association between Australian rainfall and the Southern Annular Mode. *International Journal of Climatology* **27**, 109–121.

Miller RL, Schmidt GA and Shindell DT (2006). Forced annular variations in the 20th century of IPCC FAR models. *Journal of Geophysocal Research* **111**, D18101.

NSW Government (2005). *Floodplain Development Manual: the Management of Flood Liable Land.* DIPNR 05_020. Department of Infrastructure, Planning and Natural Resources, Sydney.

Nicholls N, Landsea C and Gill J (1998). Recent trends in Australian region tropical cyclone activity. *Meteorology and Atmospheric Physics* **65**, 197–205.

Power S, Casey T, Folland C, Colman A and Mehta V (1999). Inter-decadal modulation of the impact of ENSO on Australia. *Climate Dynamics* **15**, 319–324.

Rahman K (2007). 'Flood management initiatives in the Gold Coast'. Presentation at the Coastal Cities Natural Disasters – Sydney Conference. 20–21 February 2007. <http://www.conferenceworld.com.au/resources/other/7%20Khondker%20Rahman%20ppt.pdf>.

Shulmeister J and Lees BG (1995). Pollen evidence from tropical Australia for the onset of and ENSO-dominated climate at c. 4000 BP. *The Holocene* **5**, 10–18.

Risk Frontiers (2003) *Quarterly Newsletter* **3**, 2.

Smit B, Burton I, Klein RJT and Street R (1999). The science of adaptation: a framework for assessment. *Mitigation and Adaptation Strategies for Global Change* **4**, 199–213.

Smith D and Handmer J (Eds) (2002). *Residential Flood Insurance: The Implications for Floodplain Management Policy.* Water Research Foundation of Australia, Canberra.

Speer M and Geerts B (1994). A synoptic-mesoalpha-scale climatology of flash floods in the Sydney metropolitan area. *Australian Meteorological Magazine* **43**, 87–103.

Thompson DWJ and Wallace JM (2000). Annular modes in the extratropical circulation. Part I: Month-to-month variability. *Journal of Climate* **13**, 1000–1016.

Tudhope AW, Chilcott CP, McCulloch MT, Cook ER, Chappell J, Ellam RM, Lea DW, Lough JM and Shimmield GB (2001). Variability in the El Niño-Southern Oscillation through a glacial-interglacial cycle. *Science* **291**, 1511–1517.

van Oldenborgh GJ and Burgers G (2005). Searching for decadal variations in ENSO precipitation teleconenctions. *Geophysical Research Letters* **32**, <doi:10.1029/2005GL023110>.

Walsh KJE, Hennessy KJ, Jones RN, McInnes KL, Page CM, Pittock AB, Suppiah, R and Whetton PH (2001). *Climate Change in Queensland Under Enhanced Greenhouse Conditions: Third*

Annual Report, 1999–2000. CSIRO Atmospheric Research, Hobart.

Walsh KJE, Nguyen KC and McGregor JL (2004). Finer-resolution regional climate model simulations of the impact of climate change on tropical cyclones near Australia. *Climate Dynamics* **22**, 47–56.

Webster PJ, Holland GJ, Curry JA and Chang H-R (2005). Changes in tropical cyclone number, duration, and intensity in a warming environment. *Science* **309**, 1844–1846.

Whetton PH, Suppiah R, McInnes KL, Hennessy KJ and Jones RN (2002). *Climate Change in Victoria : High Resolution Regional Assessment of Climate Change Impacts.* Victorian Department of Natural Resources and Environment, East Melbourne.

Yin JH (2005). A consistent poleward shift of the storm tracks in simulations of 21st century climate. *Geophysical Research Letters* **32**, L18701, <doi:10.1029/2005GL023684>.

14

Drought and High Temperature Extremes: Effects on Water and Electricity Demands

Mark Howden and Steve Crimp

Climate has a significant role to play in terms of the infrastructure, design and functioning of our cities and the activities, comfort and lifestyles of the occupants. This can be seen in many ways such as:

- the location and form of buildings
- the frequency and severity of weather-related damage to infrastructure
- the impact of air pollution, heat waves and cold spells on human health, energy and water use
- the types of gardens preferred
- the dominant sporting activities and their timing,
- tourism potential
- migration towards warmer regions such as south-east Queensland.

This existing sensitivity to climate provides a strong qualitative basis for suggesting that future climate changes, driven by increased atmospheric greenhouse gas concentrations (IPCC 2007), will have increasing impacts on our cities.

Two of the critical issues facing our cities are the implications of change in climate extremes on urban water use and on energy demand as a result of growth in atmospheric greenhouse gas concentrations (Hennessy *et al.* 2007). The important impacts of climate change on sea level, flooding, bushfires and biosecurity are dealt with in other chapters in this book, while human health has been comprehensively covered by McMichael *et al.* (2006). In this chapter, following an analysis of past climate changes and prospective future climate changes in Australia, we will present quantitative analyses of likely impacts of climate change on urban water use and energy demand. This will provide a context for evaluating their significance over the coming decades and to indicate key areas where further analysis could improve risk assessment.

Changes in climate factors

The main climate variables that are likely to be important in terms of their impact on electricity and urban water demand are maximum and minimum temperature temperatures and rainfall (Zhou *et al.* 2000; Howden and Crimp 2001). In particular, there are concerns about changes in extreme temperatures and in drought conditions (Kundzewicz *et al.* 2007). We will summarise briefly here the changes in these climate factors that have been observed over the past decades and the anticipated changes over forthcoming decades—drawing on the recent IPCC Fourth Assessment Report (Hennessy *et al.* 2007).

Australia has been getting warmer. Since 1910, the average maximum temperature has risen by 0.7°C and the minimum temperature by 1.1°C, with much of this change occurring since 1950 (Alexander *et al.* 2007). To place this into context—in terms of average Australian temperatures—the difference between a cool year and a hot year in the same decade is about 1.2°C (for example, a 1.5°C difference between 1989 and 1998). There is increasing evidence of a human 'fingerprint' in recent temperature trends with increases very likely to have been in response to continued growth in atmospheric greenhouse gas concentrations arising from human activity (for example, Karoly and Braganza 2005).

As well as changes in average temperatures, there have been changes in temperature extremes. From 1957 to 2004, on average across Australia, there has been an increase in hot days (35°C or more) of 0.10 days/year, an increase in hot nights (20°C or more) of 0.18 nights/year, a decrease in cold days (15°C or less) of 0.14 days/year and a decrease in cold nights (5°C or less) of 0.15 nights/year (Alexander *et al.* 2007).

Rainfall has changed too, with these changes differing between regions. Since 1950 summer monsoon rainfall has increased in the north-western region of Australia, while southern and eastern Australia have become drier (Smith 2004b). While there have been limited studies attributing the causes of decreased rainfall in the east (Crimp and Day 2003; Syktus *et al.* 2005), rainfall decreases in the south and south-west are probably due to a combination of increased atmospheric greenhouse gas concentrations, natural climate variability and land use change (IOCI 2002; SEACI 2006). Increased rainfall in the north-west during the period 1910 to 2006 may be in response to anthropogenic factors, including increased aerosol pollution in Asia and changes in cyclone activity in response to warming in the Pacific and Indian oceans (Cai *et al.* 2003; Nicholls 2006).

Extreme rainfall events have changed in some regions too. Heavy rainfall has increased in north-western and central Australia and over the western tablelands of New South Wales (NSW) since 1950, but decreased in the south-east, south-west and central east-coast (Gallant *et al.* 2007). Similarly, increases in extreme rainfall events (that is, a 1-in-100 year return period) have increased in South-East Queensland since 1950 (Abbs 2004). The changes in the frequency and intensity of most extreme temperature and rainfall events are rising faster than the changes in mean conditions (Alexander *et al.* 2007) implying an increase in the variability of climate. Additionally, droughts have become hotter since about 1973 because temperatures are higher for a given rainfall deficiency (Nicholls 2004).

Climate change projections for Australia—based on global circulation models driven by scenarios of greenhouse gas emissions and other climate drivers—suggest that for Australia's major cities, average temperatures (relative to 1990) will increase by 0.1 to 1.3°C by the year 2020, by 0.3 to 3.4°C by 2050 and by 0.4 to 6.7°C by 2080. This translates to 1 to 32 more days/year over 35°C by 2020 and 3 to 84 more by 2050, with 1 to 16 fewer days/year below 0°C by 2020 and 2 to 32 fewer by 2050 (Suppiah *et al.* 2007). A tendency for decreased annual rainfall is likely over most of southern and sub-tropical Australia, with a tendency for increases in Tasmania, central Northern Territory and northern New South Wales (Hennessy *et al.* 2007). In response to declining rainfall, run-off is also likely to decrease in southern and eastern Australia.

Increases in extreme daily rainfall are also likely. For example, the intensity of the 1-in-20 year daily-rainfall event is likely to increase by up to 10% in parts of South Australia by the year 2030 (McInnes *et al.* 2002), by 5 to 70% by the year 2050 in Victoria (Whetton *et al.* 2002), by up to 25% in northern Queensland by 2050 (Walsh *et al.* 2001) and by up to 30% by the year 2040 in South-East Queensland (Abbs 2004). In New South Wales, the intensity of the 1-in-40 year event is projected to increase by 5 to 15% by 2070 (Hennessy *et al.* 2004). Increases in episodic extreme rainfall events are likely to lead to a higher proportion of rainfall from the events running off the soil surface, thus increasing episodic dam filling, but at the expense of an increased sediment load, which can restrict the use of water until it has cleared.

The dams that supply the majority of water demand of Australian cities are likely to be substantially affected by changes in stream flows, which are projected to fall substantially across most regions. For example, annual stream flow in the Murray–Darling Basin (in which Canberra and Adelaide are located) is likely to fall by 10–25% by 2050 and by 16–48% by 2100 (Beare and Heaney 2002). There may also be reductions in water quality (for example, 5–8% in the Murrumbidgee River by 2050) via increases in salt concentrations. Similarly in south-west Western Australia analyses suggest run-off reductions of around 14% in the period 2021–2050 relative to 1961–1990 (Chiew *et al.* 2003) with a range of –25 to +10% (Chiew and McMahon 2002). In Victoria, stream flow is likely to decline 7–35% (Howe *et al.* 2005) or by 0–45% (Jones and Durack 2005) by 2070. There is likely to be a reduction in annual run-off of 6–8% in most of eastern Australia, where Sydney will be affected, but the indicators for the Brisbane region are more mixed with changes of ±15% (Hennessy *et al.* 2007). Darwin and Hobart may be little affected given the expectation of increased rainfall in current scenarios. Around Perth, there have already been reductions in rainfall and stream flow that have been linked to climate change influences (Power *et al.* 2005).

While the likely changes in cyclone characteristics are covered in another chapter in this book, they are likely to impact on the issues being discussed in this section via the effects on catchment recharge and replenishment of dams and increased risk of large sediment loads.

The frequency of severe tropical cyclones (Categories 3, 4 and 5) on the east Australian coast is projected to increase by 22% for the mid-range greenhouse gas emission scenarios from 2000 to 2050, with a 200 km southward shift in the cyclone genesis region, leading to greater exposure in South-East Queensland and north-east NSW (Leslie *et al.* 2007). For scenarios with approximately twice the current atmospheric CO_2 concentrations, there would be a 56% increase in the number of tropical cyclones expected over north-eastern Australia from model simulations (Walsh *et al.* 2004). In contrast to the increased storm-related risk in northern cities, there are projected to be decreases in hail frequency for Melbourne (Niall and Walsh 2005).

Potential evaporation (or evaporative demand) is likely to increase with climate change (Jones 2004) and this, combined with anticipated reductions in rainfall, suggests that up to 20% more droughts will occur over most of Australia by 2030 (Burke *et al.* 2006 Mpelasoka *et al.* 2007). The effects of climate change on drought frequency are likely to be modulated by phases of the ENSO and IPO in the future as they have been in the past (Power *et al.* 2006).

Water demand and supply

The flow rates of Australian rivers are strongly linked to the nature and frequency of rainfall events. The highly variable climate leads to much higher variability in flow rates than other continents, with the ratio between high annual flows and low annual flows being around 1.9 to 2.4 for the Rhine, Yangtze and Nile, but 15.5 for the Murray and 54.3 for the Hunter (Chartres and Williams 2006). This high variability in stream flow results in substantial risk to the reliable supply of water to our cities and for other uses. In response to high variability in stream flow, the construction of large dams has been predominant in Australia. This substantial infrastructure investment has previously ensured that water restrictions in Australian cities were infrequent and of relatively short duration in accord with design expectations. However, water supplies to most Australian cities are being increasingly challenged as a result of the long-term dry conditions being experienced, combined with increasing population and demand for water. For example, last year Australia's population grew at a record net increase of 307,000 (ABS 2007). These factors have resulted in all capital cities, except for Darwin, being currently under some level of water restrictions even though per capita demand has fallen. In several cases, such as in Canberra and Melbourne, permanent water restrictions have been put in place as well as significant demand-management programs. Projections of drier and hotter conditions clearly may impose further difficulties in providing reliable water supplies for Australian cities. This has fostered several studies assessing the implications of climate change for catchment hydrology and water supply. For example, in Western Australia, the increased risk of drought under global warming was integrated with projections of population growth, economic development and social norms in relation to water use, resulting in the construction of a major new dam and proposed development of new desalination facilities and other water sources (Power *et al.* 2005).

While many of the existing studies have examined the potential impacts of climate change on stream flow, few formal studies have examined the likely impacts of climate change on urban water demand, even though current experience of water restrictions has demonstrated sensitivity to climate conditions. Consequently, the effects of climate variations are not well quantified with these effects being confounded by changes in other factors such as strong growth in populations in many urban centres, changes in urban design (that is, smaller housing lots with smaller gardens, more medium to high density housing), uptake of water-saving technologies such as dual flush toilets, changes in water pricing schemes and water restrictions imposed during droughts (for example, ABS 2005).

Key uses of water that are likely to vary with climate include watering of lawns, parks and gardens and swimming pools. Approximately 44% of household water is used for such outdoor purposes (ABS 2005) although this proportion varies between cities. For example, outdoor use is estimated to be 40% in Sydney (Maheshwari 2006), 44% in Canberra (ACTEWAGL 2007), 56% in Perth (Loh and Coughlan 2003) and 35% in Melbourne (WRSC 2002). In addition to this component of water demand that is responsive to climate, there is an underlying demand from industrial and domestic usage that is unrelated to climate.

The following sections describe a new method for analysing the effect of climate and climate change on city-scale water demand, illustrating this with analyses for Canberra.

Data and Methods

Urban outdoor water demand can perhaps be most simply described as a function of evaporation: evaporation is the key process that removes water from lawns, swimming pools and other moist urban surfaces. In Canberra, evaporation explains about 74% of the variance in daily water use from 1968 to 2000, with this being highly non-linear (unpublished data). However, there are limitations in such an approach for analysing long-term data sets and for climate change assessments:

- it cannot easily allow for changes in the water use behaviour that have been recorded over the past decades
- it cannot allow climate change assessments where rainfall may change independently of temperature or evaporation
- typical climate change scenarios provide changes in temperature and this does not linearly translate to alterations in evaporation.

While complex models exist to simulate urban water demand (Zhou *et al.* 2000) and water balance (Mitchell *et al.* 2001) limited data and process understanding suggest the utility of a simple modelling approach that:

- has a clear representation of the underlying physical processes
- is a parsimonious model of these processes using a very restricted set of parameters
- has parameters that can be established on only a few years of data so that the model can be tested against the maximum number of years of independent data
- is largely driven by climate, but also allows changes in aggregate consumer water use (behavioural norms in water use) to be expressed
- can be used to calculate climate-adjusted use
- captures aggregated demand as opposed to sector specific water demand.

For this analysis we have adapted an approach based on the concept of a simple soil moisture budget (Fitzpatrick and Nix 1970; Littleboy and McKeon 1997) driven by climate averaged across the city area, with an explicit behavioural response to simulated soil moisture deficit.

$$D = SM_2 - SM_1 - P + E + R$$

where D is the amount of water simulated to be applied on a given day (i.e. water demand), P is precipitation, E is evapotranspiration (modified Penman-Monteith method; Allen 1996; 1998) and R is the sum of run-off and recharge—with all these components in millimetres. The subscripts refer to daily time steps. There is a fixed, calibrated soil moisture holding capacity parameter. The amount of applied water (D) is calculated as a linear proportion (b) of the soil moisture deficit—the difference between the soil moisture holding capacity and the current soil moisture. A high value of the b parameter represents the behaviour of high water use norms and vice versa. Water use varies with day of the week (e.g. Zhou *et al.* 2000), but to keep the model simple we aggregate the daily calculated demand values to weekly values. The three parameters used for the model are fitted using genetic algorithms (software: Evolver, Palisade Corp) to minimise the least squares difference between modelled demand and recorded demand values for the most recent three years of data. The fitted model was tested against the remainder of the data set not used in the calibration process. In all cases, there were strong linear relationships between simulated and observed demand, with the model explaining about 94% of the variation in use. These relationships were used to scale the simulated demand (D in mm) to the observed per capita demand (in litres: Figure 1).

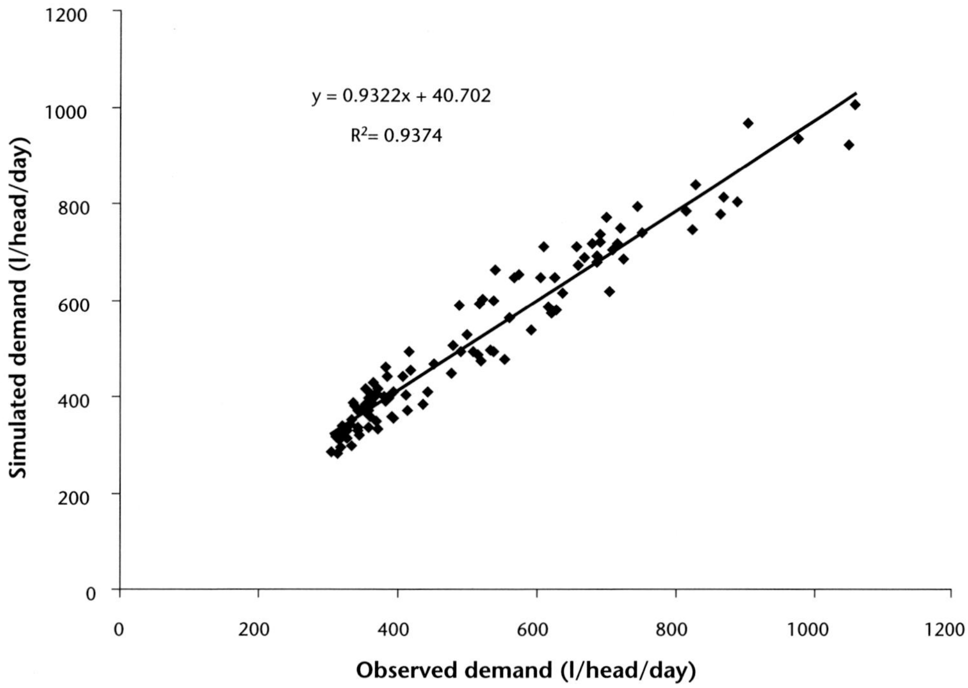

Figure 1: Simulated versus observed demand (litres per head per day) for 1996 to 1997 for Canberra. The line is a least squares linear regression ($P < 0.001$)

In this chapter we use this model for the city of Canberra for the period 1968 to 2000. This period preceded the long-term water restrictions in the ACT that have been in place since the summer of 2002–3. All climate data were retrieved from the SILO Datadrill (Jeffrey *et al.* 2001) and aggregated for a region covering the Canberra urban area (Howden and Crimp 2001) to restrict the impact of localised climatic events.

Climate-adjusted water use was assessed by calculating the slope of the regression between observed demand and demand simulated on the basis of climate: a higher slope indicates greater use of water for a given climate stress and vice versa (Figures 2 and 3). The intercept of the regression reflects water use when there is no demand arising from climate factors (Figure 2). This is a measure of industrial and internal domestic use only. By calculating these values for each year, we can follow changes in behavioural responses due to pricing policies and regulations without the confounding effect of variations in climate between years. For example, a reduction in water use following introduction of a particular policy or technology may have little to do with such changes, but instead have

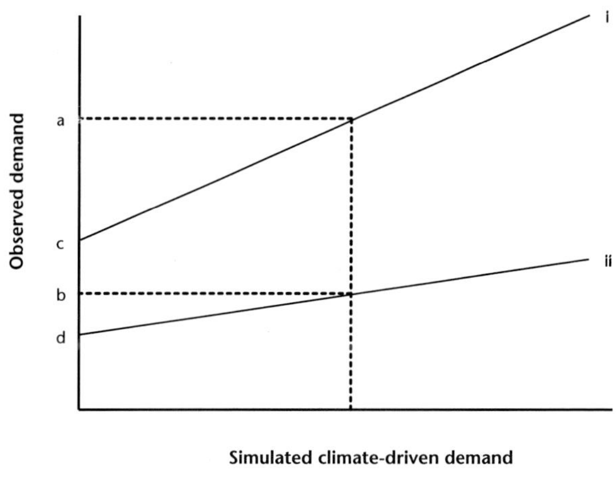

Figure 2: A conceptual diagram illustrating how climate adjusted water use was calculated from the Canberra water demand model. The slope of line (i) is greater than line (ii) showing that for a given increase in climate-driven demand, there is a greater increase in observed demand. For situations when the climate-driven demand is effectively zero (such as in the middle of winter during a rainy period) the intercept of the regressions (c and d) indicate the non-climate demand component. The combination of the greater intercept and slope of line (i) means that for a given climate demand there will be greater useage (a) than for the alternative option (b).

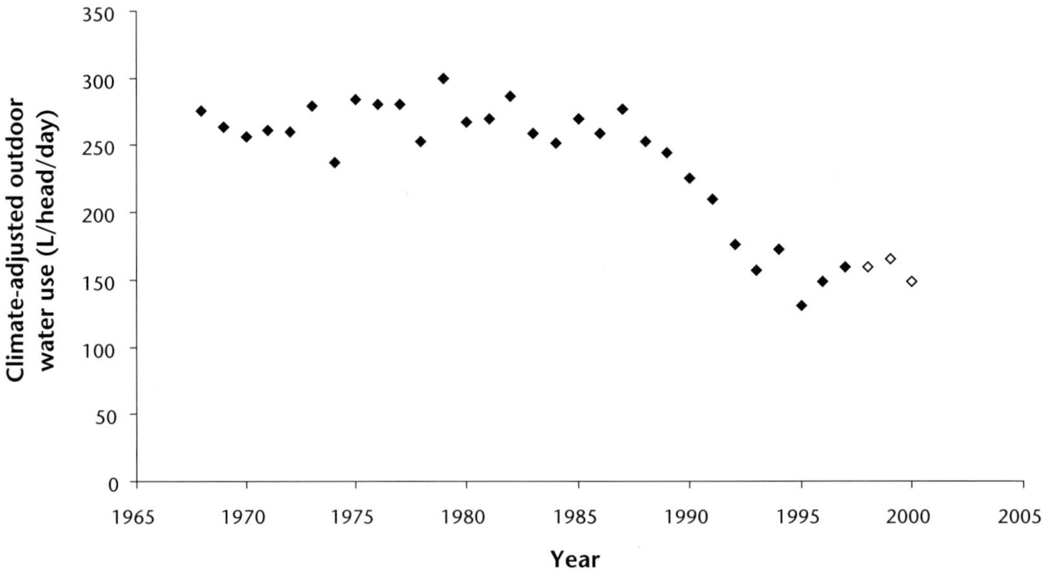

Figure 3: Climate adjusted outdoor water use as estimated by the Canberra water-demand model for each year from 1968 to 2000. The unfilled data points for the last three years are to indicate that this was the calibration data period for the model.

been due to a series of years with above average rainfall. Comparison of the slope and intercept of the regression removes this climate effect. However, in this study we do not attempt to attribute changes in climate-adjusted demand to specific policies.

Climate change impacts on water demand were calculated by three methods. The first two methods use the model described above, but with two different ways of representing climate change. Firstly with climate change scenarios which increased the mean temperature by 1°C increments up to 6°C and had proportional changes in the daily rainfall in 10% increments from −40 to +40% to encompass the full range of prospective change for the year 2070 (CSIRO 2007). These scenarios as used here explicitly do not include changes in the probability distribution of daily temperatures. If the increases in mean temperature involves a shift in the probability distribution towards higher temperatures as well as a change in the mean (as has happened for hot years in the historical record (Figure 4), then this may have disproportionately large impacts on water demand. Hence, this method of implementing climate changes may be conservative. Consequently, for a second approach we sourced daily climate output from the CSIRO Mk 3.5 Global Climate Model (GCM). These GCM models aim to represent the dynamics of the atmosphere-land and ocean and so may incorporate more effectively changes in temperature probability distributions. The GCM data was for three periods: historical baseline climate from 1961 to 2000, the period 2045 to 2065 and the period 2081 to 2100. However, as there remains significant variation between GCMs in terms of their results, we have also analysed climate change impacts on demand using a third method that can include changes in probability distributions. This used an empirical analysis where we assess simulated demand data against historical temperature and rainfall variation to provide relationships (Figure 5). These multiple regressions of temperature, rainfall and demand were extrapolated: this allows in a coarse way for some changes in such probability distributions (as per earlier discussion and Figure 4). We use simulated water demand in these analyses so as to remove the substantial changes in user behaviour that have occurred over time (Figure 3). We present the results of the first and third analytical methods as a response surface of percentage changes in annual demand across the range of temperature and rainfall changes. The second method (model with GCM input) relates to specific combinations of rainfall and temperature change and

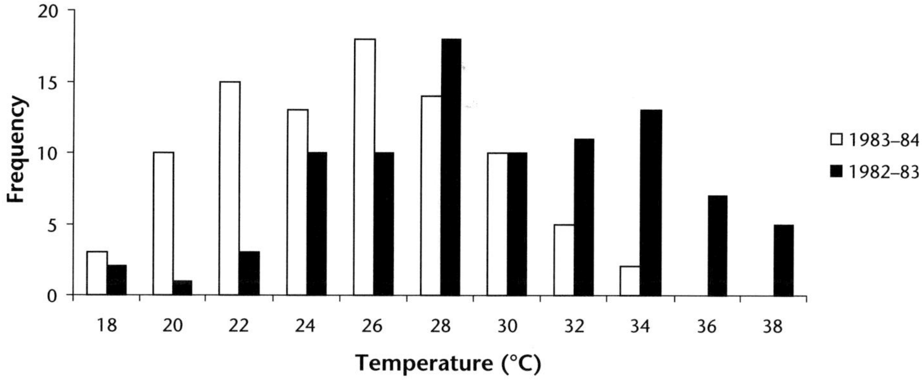

Figure 4: Histogram of temperatures (°C) for a cool year (1983–84) and a hot year (1982–83) in summer in Canberra, illustrating that in hot years there is both an increase in the mean temperature and in the probability distribution of temperature, particularly the relative frequency of very hot days.

these are compared with the equivalent change scenarios in the alternative methods.

We did not simulate the effects of elevated atmospheric CO_2 on evapotranspiration as there appears to be an almost complete trade-off between reduced transpiration per unit leaf area and increased leaf area resulting in little change in water use (Samarakoon and Gifford 1996).

Results

The water demand model performed consistently well over the test period, with the model explaining in excess of 90% ($P < 0.001$) of the variance in weekly demand in most years.

Climate-adjusted demand for outside use was relatively stable at around 250L/head/day from 1968 to 1988 (Figure 3). This was followed by a decline over the following decade to about 150 L/head/day. Non-climate-related demand has remained fairly stable at around 250L/head/day over the period of analysis, with some periods when this has been lower (data not shown). Outdoor use as estimated by this model accounted for about 37.5% of total demand in recent years.

For scenarios of climate change using the first analysis method, increasing temperatures increased simulated demand but this was only modest—being about 2% per degree Celsius when no rainfall change was simulated (Figure 6a). For the harshest scenario, with 40% reduction in rainfall and temperature increase of 6°C, demand was calculated to increase by about 20% (Figure 6a). For mid-range results from GCMs and using emission scenarios matching the current emissions trajectory (for example, Rahmstorf *et al.* 2007), indicative increases in demand are about 8%.

The simulated increases in demand when using the GCM climate results were consistently above

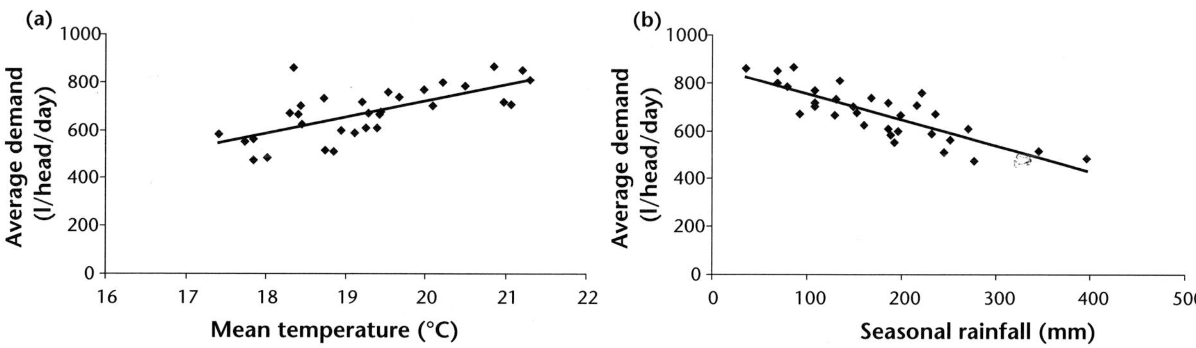

Figure 5: Relationship between (a) mean temperature (°C) and (b) seasonal rainfall (mm) and simulated water demand for the summer season in Canberra from 1968 to 2000. The lines are least-squares linear regression.

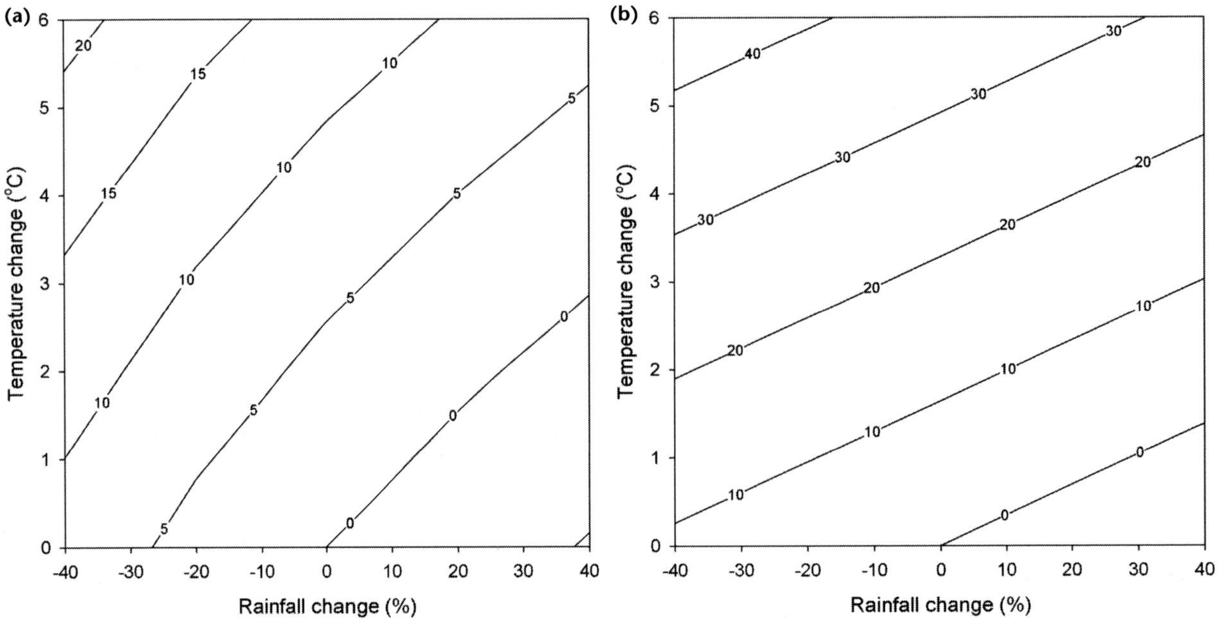

Figure 6: A response surface of change in annual average demand for changes in rainfall (± 40%) and temperature (0 to 6°C) for (a) the Canberra water demand model with simple changes to mean rainfall and temperature and (b) for a multiple regression model of historical temperature, rainfall and demand extrapolated to cover changes in rainfall (± 40%) and temperature (0 to 6°C). This latter approach incorporates increases in frequency of very hot days, as discussed in the text.

(about double) those found when the model used climate change based on mean changes to the historical climate record (Table 1). In contrast, the model results when the GCM climate was used were consistently below (about 30%) those found when a multiple regression approach based on past climate and demand relationships was used to explore future climate impacts. For the period around 2050, using this representation of future climate, demand may increase by around 14% (Table 1)

Higher sensitivities to climate change are indicated using the third analysis method (multiple regression), with increases in demand of about 6% per degree Celsius (Figure 6b). Using this method, under the harshest climate scenario, demand was calculated to increase by 45%. Mid-range scenarios for 2050 indicate increased demand of about 20% (Figure 6b). Inspection of the data showed that this sensitivity was closely related to the frequency of high water demand weeks ($R^2 = 0.87$, $P < 0.001$) with the shift in the frequency of very hot conditions in years such as 1982–83 driving demand upwards (data not shown).

Discussion

Climate change is likely to have significant implications for water demand in Canberra and for other cities in similar situations. Mid-range changes for the year 2050 are for between 8 and 20% increase in per capita consumption, although more significant increases (of up to 45%) are possible if the more extreme climate change scenarios eventuate.

Importantly, these increases in per capita demand will be in addition to substantial possible reductions in dam inflow of around 10 to 35% (Bates *et al.* 2003) by 2050 as well as substantial increases in the population of Canberra. For example, the population may increase by up to 24% by 2021 and 68% by 2051 when compared with 2004 population levels (ABS 2006). The combined effect of changes in demand, supply and population is likely to make the provision of reliable water supplies to Canberra increasingly challenging. For example, using the above figures and assuming no major changes in water supply or demand management, Canberra residents in 2050 may have less than half (44%) of the pre-year 2000 per capita

Table 1: Changes in simulated future water demand (%) for Canberra

Year	Emissions scenario	Rainfall change (°C)	Temperature change (%)	Demand change: model with mean change (%)	Demand change: model with GCM (%)	Demand change: regression (%)
2046–2065	A2	–16	2.65	8.1	14.5	19.5
	B1	–8	2.17	5.6	12.4	14.9
2081–2100	A2	–16	4.68	12.6	21.1	31.9
	B1	–1	2.52	5.1	10.6	15.6

Note: Simulations are for two different periods (2045–2065 and 2081–2100) and for two emissions scenarios (A2—a high emissions scenario and B1—a low emissions scenario) with mean temperature and rainfall changes associated with these. The demand changes shown are for the three different methods of estimating demand change: the demand model with either mean changes in climate or climate changes from the GCM and a multiple regression approach extrapolating past demand as a function of temperature and rainfall. The baseline is the period 1968 to 2000.

water availability and furthermore 20% more water may be needed to provide the same level of service that they have historically experienced. Similar situations are likely to apply to other cities in southern Australia. The confluence of these three factors will, in the absence of major adaptive responses, result in substantial reduction in the services that this water provides (for example, lawns and gardens that add amenity and real estate value, urban cooling, recreational options, tourism activities such as Floriade). The reductions in these services that Australian cities have previously experienced in past 'droughts' are thus much more likely in the future. In particular, the reduction in urban cooling from reduced urban evapotranspiration from lower outdoor water use could be quite significant for Canberra (Cleugh *et al.* 2005), increasing temperatures and increasing energy use in the warm months, feeding back to further climate change via greenhouse gas emissions (see next section). In response to these potential risks, the ACT Government has just announced major infrastructure and policy changes. These include expansion of the Cotter Dam from 4GL to 78GL, a pipeline to increase transfer of water from the Murrumbidgee to Googong Dam, a water purification plant and demand reduction programs (ACT 2007).

There are many prospective options to manage the challenge of reduced per capita water availability. These include improved water demand-management programs to increase end-use efficiency, such as ultra-low flush toilets and other more water-efficient appliances (e.g. Sarac *et al.* 2003, ABS 2005). These programs can be delivered through a variety of mechanisms such as subsidised retrofitting or discount incentives. Other options involve increased water recycling at the household or municipal levels or to increase water capture at the household level via rainwater tanks or downpipe diverters. Many of these have been addressed already in the water savings plans of the ACT and other Australian cities (ACT 2007). There are also options to increase the access of urban users to irrigation water previously allocated to farmers. This may be through improved functioning of a water market or via policies that aim to increase the efficiency of farm use and redistribute part of the saved water to urban users. Some of these options raise issues of equity and sustainability of the farming communities thus affected. Exploration of groundwater supplies and the construction of desalination plants are also being undertaken. However, in all cases, there is a need to invest in new infrastructure and knowledge—diverting investment away from other options.

There appears to have already been a substantial amount of adaptive response to reduce both indoor and outdoor water consumption as shown in this study of Canberra as well as in other capital cities (ABS 2005). This raises the prospect of a situation where many of the easy options to increase efficiency or reduce discretionary use have already been taken and are now implemented as a matter of course, so further reductions in water useage may be more difficult and costly (Inman 2007).

The analyses here demonstrate that the impact of climate change on urban water demand is heavily dependent on the representation of the way the climate changes. If climate change involves only a shift in mean temperature but not disproportionate increases in the frequency of very hot conditions, then the impact on demand may be modest (for example, a 7% increase by 2050). However, if the warming is expressed as a shift towards increasing the frequency of very hot days, then the impacts may be much more significant. Using climate data from a GCM, rather than just changing mean climate, increased the demand to 14% higher than baseline for the same temperature increase and rainfall decrease and using the same water demand model. The doubling of the impact of climate change is due to the increase in frequency of very hot days. Consequently, significant attention has to be paid to the representation of future climate in these types of analyses. Consideration also needs to be given to the analysis method. The simple approach of multiple regression of climate variables against demand may significantly overstate the sensitivity of demand to climate change. Clearly, further analyses of these methodological issues could improve the confidence of urban water resource managers in the capacity to undertake climate change analyses for this sector.

It is important to note that this analysis was undertaken during a period (1968 to 2000) when there were no long-term water restrictions. The model based on un-restricted use is thus likely to over-estimate water consumption at the present time (2007) when there are severe restrictions on outdoor water use. However, the model is designed to be able to incorporate this change in behaviour through change in the demand response term (the *b* coefficient) using more recent water demand data and information on the period and level of water restrictions. Additional climate change scenarios could then be generated either by assuming different levels of water use restrictions or, preferably, by integrating both the demand-side and supply-side (that is, dam fill) aspects of climate change.

Lastly, this model provides a new way of providing climate-adjusted demand to allow more effective policy analysis. For example, it shows very clearly the reduction in demand since 1987, although it does not attempt to operate at the fine-scale that enables attribution of such changes to specific technologies, infrastructure alteration, behavioural changes or policies such as water pricing. However, it can provide independent confirmation of previous studies that have measured outdoor useage as a proportion of total use. In this case, we calculate that outdoor use is about 38% of total water use, similar to previous figures for Canberra of 44% (ACTEWAGL 2007).

Electricity demand

Electricity demand is sensitive to climate, with both average and peak demand being affected (Segal *et al.* 1992; Sailor 2001; Howden and Crimp 2001; Pardo *et al.* 2002; Mirasgedis *et al.* 2006; Thatcher 2007). The response of consumers to low temperatures is typically to increase the use of heating, while high temperatures increase the use of air conditioning and evaporative cooling. The change in demand with climate conditions is region-specific, with the average change varying in both magnitude and direction, often with warmer temperatures decreasing demand in cold climates, but increasing demand in warmer climates. The following sections describe an analysis of electricity demand as a function of climate and potential climate change.

Data and methods

There are several possible approaches to analysing the effects of climate on electricity demand including variations of the heating and cooling degree day concepts (e.g. Howden and Crimp 2001; Sailor 2001; Thatcher 2007), neural networks (Chow and Leung 1996; Khotanzad *et al.* 1996) and Bayesian analysis (Pezzulli *et al.* 2006). The choice of model needs to be a balance between model performance and complexity and the robustness of the representation of future climates. Given that climate change scenarios remain highly uncertain, we have previously adopted and successfully used the heating and cooling degree approach (HDD and CDD; Howden and Crimp 2001). Essentially this describes electricity demand as a linear function of the difference between mean daily temperature and a baseline or threshold temperature value: as temperature increases above this threshold there is an increasing tendency to use energy for cooling and vice versa for heating. We enhanced the basic approach to include a temperature-response threshold selected using least squares optimisation and demonstrated that this varied between cities, with it being higher in warmer climates—probably reflecting adaptation in both the building stock and in behaviour such as the amount of clothing worn. We have here modified the analysis method of Howden and Crimp (2001) and apply it to explore the implications of increased temperature scenarios for Adelaide. The key modifications here are:

- to include two temperature-response thresholds—a lower one and an upper one so as to better describe the responses to extreme temperatures, with both thresholds being quantified by least squares optimisation. Implicitly, the intervening temperatures are interpreted as having no particular climate influence. This essentially describes a 'comfort range', rather than a single 'comfort temperature': an approach that intuitively seems more likely to represent expressed behaviour;
- a daily model where each day of the week is treated separately to allow for different patterns of energy usage so as to more adequately represent peak demand, with the daily results aggregated to weekly totals for ease of reporting
- a regression term to allow for changed electricity consumption associated with economic growth, changes in industry and population growth.

The model was calibrated for the period 1999 to 2002 using NEMMCO demand data and city-area averaged climate data (Jeffrey *et al.* 2001) as described in Howden and Crimp (2001). The Christmas and Easter periods were excluded from the model.

We then interrogated this model with simple climate scenarios based on increasing the mean temperature by increments of 1°C up to 6°C (the maximum temperature scenario increase for Adelaide for the year 2070; Suppiah *et al.* 2007). As discussed in the earlier section on water demand, this approach does not include the possibility that temperature may change in variance as well as mean and, in particular, that there may be a change in the probability distribution of temperatures with higher frequencies of very hot days. These scenarios could be explored by using daily output from global climate models as this becomes more readily available. For each temperature scenario, we calculated the percentage change in electricity demand for the mean and, to investigate extreme or peak demands, also the 95th and 99th percentile demand values. These are the 1-in-20 and 1-in-100 values of electricity demand: that is, the 99th percentile demand value would be exceeded on average only once in 100 days.

Results

The revised model was effective at describing electricity demand for Adelaide when tested against an independent data set (2003), with predicted weekly demand strongly related to observed demand (r^2 = 0.84; Figure 7) with no apparent bias. This is a substantial improvement over the original model which explained significantly less electricity demand variance in Adelaide (72%)

Scenarios of higher temperatures increased mean demand and peak demands (Table 2; Fig 8). Mean demand increased by 0.4, 2.7 and 6.6% for temperature increases of 2, 4 and 6°C, respectively. The peak demand as estimated by the 95th and 99th percentile values were significantly more sensitive to scenarios of elevated temperature with 5.9, 12.0 and 18.0% increases in the 95th percentile demand values, with temperature increases of 2, 4 and 6°C, respectively (Table 2).

Table 2: Change in mean, 95th and 99th percentile electricity demand for Adelaide (%) with increasing temperatures (°C).

	Temperature increase (°C)					
Change in demand (%)	1	2	3	4	5	6
99th percentile	2.6	5.3	8.0	10.9	13.6	16.3
95th percentile	2.9	5.9	8.9	12.0	15.0	18.0
Mean	0.0	0.4	1.3	2.7	4.5	6.6

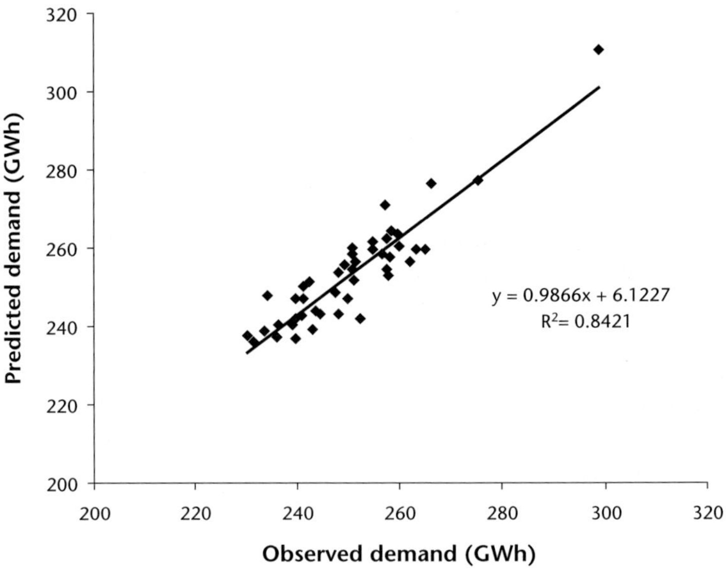

Figure 7: Predicted versus observed electricity demand (GWh) for Adelaide for 1999 to 2002.

Discussion

Peak electricity demand in Adelaide was more sensitive to temperature increases than mean demand, as found by Howden and Crimp (2001), with increases of up to 18% with temperature increases of 6°C. This was similar to the increases found for other Australian cities, such as Melbourne, Sydney and Brisbane, by Howden and Crimp (2001). Some of these cities are already having difficulty in meeting peak demands with current generating and transmission capacity. This study suggests that these difficulties will increase over the forthcoming decades and, along with underlying economic growth, will require installation of new capacity. About 10% of the existing asset levels may be required to allow for climate-related increases in peak demand alone by 2030 (PB Associates 2007). Additionally, the recent drought has shown the vulnerability of hydroelectric and coal-fired power generation to water shortages. The congruence of

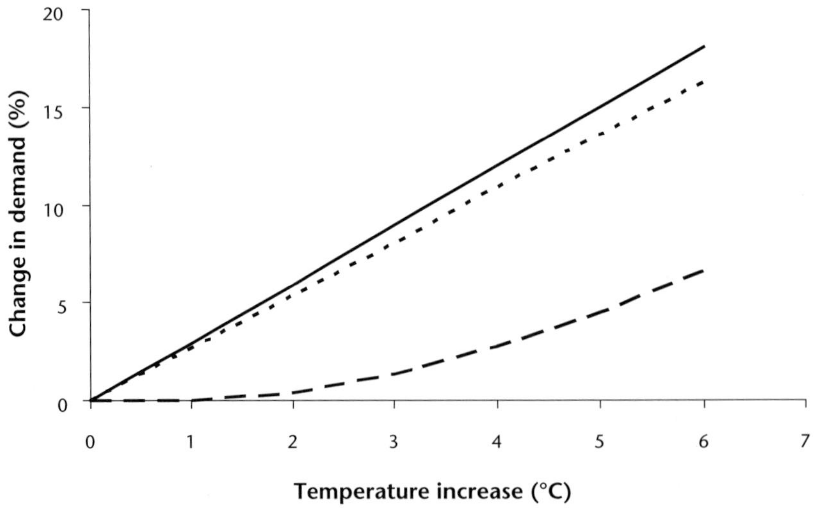

Figure 8: Change in electricity demand (%) for the mean (dashed line), 95th percentile (solid line) and 99th percentile (dotted line) with increased temperature.

these shortages with anticipated future water shortage (see earlier section) suggests that additional peak-load capacity will increasingly have to be sourced from non-hydro and non-coal sources, such as gas turbines, wind, solar and hot dry rocks—adding considerable expense. The cumulative economic implications could be significant. Furthermore, the limited expected expansion of renewable energy sources and the current uncertainty about nuclear options means that the majority of the increase in average demand suggested in all cities under the warmer scenarios is likely to be satisfied by fossil-fuel combustion (gas and coal, assuming that water is not a limiting factor). This will further raise national emissions, increasing the difficulty of Australia meeting any emission targets—such as those proposed under the Kyoto Protocol—and will feed back to create further climate change.

Climate change is also likely to affect energy infrastructure in Australian through impacts of severe weather events on wind-power stations, electricity transmission and distribution networks, oil and gas product storage and transport facilities, and off-shore oil and gas production (see Hennessy *et al.* 2007). There are also likely to be costs and damages that can be avoided by adaptation and mitigation. An assessment of potential risks for Australia (Hennessy *et al.* 2007) found:

- increased peak and average temperatures are likely to reduce electricity generation efficiency, transmission line capacity, transformer capacity and the life of switchgear and other components
- if climate changes gradually, both the generation utilities and the equipment manufacturers are likely to have enough time to adjust their standards and specifications
- vulnerability to the above impacts is low, but there is medium vulnerability to a decline in water supply for large-scale coal, hydro and gas turbine power generation.

There is a potential change in the mix of electricity-generating options too, both as a function of climate changes and of policies such as renewable energy targets. The vulnerability to climate change of different mixes of electricity generation and delivery systems has yet to be evaluated.

There appears to be a significant capacity to adapt electricity demand to prevailing climate conditions as the threshold response values for heating and cooling days calculated by Howden and Crimp (2001) varied systematically between cities. There was a consistent reduction in these thresholds with transitions to the colder locations, suggesting that the inhabitants find progressively lower temperatures tolerable (that is, the temperature at which heaters are turned on is lower in southern cities than in the north). Similarly, there was a systematic decline in a threshold that included humidity as well as temperature, indicating that the tolerance to hot, humid conditions was greater in the sub-tropical locations than in the more temperate locations further south (that is, the conditions of heat and humidity become more severe in the subtropics before air conditioning is turned on compared with the southern locations). Possible causes of this adaptation could be the design of infrastructure, behaviour (such as the clothing worn) and physiological adjustment. If these adaptations are transferable between cities, then at least some of the negative impacts of climate change on electricity demand could be offset. Policies which encourage such transitions could be worth pursuing.

The coefficients of the CDD and HDD terms of the regression were similar in value in both this study and in Howden and Crimp (2001). This contrasts with the US where the coefficients for CDD are always considerably greater than those for HDD (Sailor and Munoz 1997). This difference may be due to Australians having lower air conditioner usage, greater use of alternative cooling systems (such as evaporative coolers and fans) or greater tolerance of discomfort from hot and humid weather. Again, policies that encourage increased efficiency of cooling options including housing design would be worthwhile and the value of these is likely to increase as temperatures increase. Tackling the implications of, say, different pricing policies or changes in household or industry behaviours, and how these may interact with climate changes, would require a different and more complex economic modelling approach than the one taken here as the aim of this study was only to identify the vulnerability of Australia's cities to climate change.

Conclusion

There is growing evidence that the climate is changing and that human activities are highly likely to be contributing to these changes. Existing sensitivities to climate variations indicate that there are a large number of potential impacts of climate change on Australia's cities. Two of these key areas of sensitivity are in relation to water use and electricity use. We have undertaken case study analyses for these two issues—assessing them for Canberra and Adelaide, respectively.

We developed a new method of assessing the impacts of climate on city-wide water use. Climate change is likely to have significant impacts on water demand over the next several decades, potentially increasing demand by between 8 and 20% to deliver the same services. These increases in water demand in conjunction with likely reductions in supply resulting from projections of lower rainfall, combined with a large growth in population, mean that water availability per capita and the level of services derived from that water may decline markedly. There are a range of potential adaptation responses to this challenge, ranging from implementation of more efficient water-use technologies in households to large-scale infrastructure change and policy responses. It is unclear how effective some of these changes are likely to be given that many of the more easily implemented options and reductions in discretionary water use may have already occurred. It is also unclear what the implications of such responses are to the economy as a whole.

The magnitude of the impact of climate change on city-wide water demand was surprisingly dependent on the way in which future climates were represented, with a factor of two difference between different approaches. Where climate change scenarios were implemented with no change in the climate variability, changes in water demand were only half of those where there was alteration in the variability (that is, increased frequency of very hot days), even though the average changes in temperature and precipitation were identical. Enhanced confidence in climate change analyses of urban water issues requires attention to this, and other, methodology.

Electricity demand is also sensitive to increasing temperatures, particularly in relation to peak demand. Generating capacity across Australia is already being increasingly challenged in summer due to the higher temperatures, larger populations and increasing penetration of the market by air conditioners. Climate change is also likely to increase the vulnerability of the electricity system to a range of disturbances and to limit availability of key resources such as water for coal-fired power stations. The likely combination of increased peak demand, increased vulnerability from reduced water availability and other factors, as well as possible emissions targets, is likely to result in evaluation of a more resilient supply network that includes smaller, non-hydro and non-coal sources such as gas turbines, wind, solar and hot dry rocks. These options may be considerably more expensive.

The prospects are for significant increases in temperature in Australia's cities, which have potential impacts on quality of life and health. Outdoor water use can significantly reduce urban temperatures via evapotranspiration, and with projections of reduced water availability for such purposes, there will be a reduction in urban cooling from green space. In addition, the trend towards drier conditions is likely to result in increased local temperatures as well as general global warming. For example, in Canberra, summers with rainfall 40% less than the average have temperatures that are approximately 0.4°C above the norm. Furthermore, pressures for urban consolidation—in response to many factors including water availability—will increase afternoon air temperatures by about 1°C due to the urban heat-island effect (Cleugh *et al.* 2005). Unfortunately, if higher urban temperatures result in increased use of air-conditioners, this will create more greenhouse emissions, feeding back to the underlying problem. Similarly, responses such as dam construction, operation of desalination plants and other water-treatment works—or the long-distance transport of water—are all highly energy demanding, causing correspondingly higher levels of greenhouse gas emissions. Furthermore, some of these options may also add to peak energy demands, adding to the challenges of providing adequate generating capacity. Clearly, there is an increasing need to include several different aspects of climate change into an effective planning framework that includes other large-scale drivers such as population growth.

Nevertheless, there are potential positive aspects of this issue such as increasing the focus on developing more resource-efficient cities and hastening the improvement of building design, materials and construction so that buildings require less cooling and heating, consume less water and have smaller greenhouse emissions 'footprints'. The prospects of large-scale impacts as outlined here could well make some of these decisions more palatable than they would otherwise have been, thus providing opportunities for researchers to engage with decision-makers to forge more sustainable pathways for our cities as we make the transition to a future with a changing and uncertain climate.

References

Abbs DJ (2004). A high-resolution modelling study of the effect of climate change on the intensity of extreme rainfall events. In *Proceedings Staying Afloat: Floodplain Management Authorities of NSW 44th Annual Conference.* pp. 17–24. Floodplain Management Authorities of NSW, Coffs Harbour, New South Wales.

Alexander L, Hope P, Collins D, Trewin B, Lynch A and Nicholls N (2006). Trends in Australia's climate means and extremes, a global context. *Australian Meteorological Magazine* **56**, 1–18.

Allen RG (1996). Assessing integrity of weather data for use in reference evapotranspiration estimation. *ASCE Irrigation and drainage Engineering Journal* **122(2)**, 97–106.

Allen RG, Pereira LS, Raes D and Smith M (1998). 'Crop evapotranspiration. Guidelines for computing crop water requirements'. Irrigation and Drainage Paper No. 56, Food and Agriculture Organization, Rome.

ABS (Australian Bureau of Statistics) (2005). *Household Water Use and Effects of the Drought. Australian Economic Indicators.* Cat. no. 1350.0, ABS, Canberra.

ABS (2006). *Population Projections, Australia, 2004 to 2101.* Cat. no. 3222.0, ABS, Canberra.

ABS (2007). *Australian Demographic Statistics'.* Cat. no. 3101.0, ABS, Canberra.

ACTEWAGL (2007). <http://www.actewagl.com.au/education/water/SavingWater/InTheGarden.aspx>.

ACT (2007). *ACT and Region's Future Water Supply.* <www.thinkwater.act.gov.au/permanent_measures/water_supply_security.shtml>

Bates BC, Charles SP, Chiew F, Harle K, Howden SM, Kirby M, Peel M, Suppiah R, Siriwardena L, Viney N and Whetton P (2003). 'Climate change projections and the effects on water yield and water demand for the Australian Capital Territory'. Report from CSIRO to ACTEW, 116 pp. ACT Electricity and Water, Canberra

Beare S and A Heaney (2002). 'Climate change and water resources in the Murray Darling Basin, Australia. impacts and adaptation'. Conference Paper 02.11. Australian Bureau of Agricultural and Resource Economics, Canberra, Australia.<http://www.abarepublications.com/product.asp?prodid=12389>.

Burke EJ, Brown SJ and Christidis N (2006). Modelling the recent evolution of global drought and projections for the 21st century with the Hadley Centre climate model. *Journal of Hydrometeorology* **7(5)**, 1113–1125.

Cai WJ, Whetton PH and Karoly DJ (2003). The response of the Antarctic Oscillation to increasing and stabilized atmospheric CO_2. *Journal of Climate* **16**, 1525–1538.

Chartres C and Williams J (2006). Can Australia overcome its water scarcity problems? *Journal of Developments in Sustainable Agriculture* **1**, 17–24.

Chiew FHS and McMahon TA (2002). Modelling the impacts of climate change on Australian stream-flow. *Hydrological Processes* **16,** 1235–1245.

Chiew FHS, Whetton PH, McMahon TA and Pittock AB (1995). Simulation of the impacts of climate change on runoff and soil moisture in Australian catchments. *Journal of Hydrology* **167,** 121–147

Chiew FHS, Harrold TI, Siriwardenena L, Jones RN and Srikanthan R (2003). Simulation of climate change impact on runoff using rainfall scenarios that consider daily patterns of change from GCMs. In *Proceedings of the International Congress on Modelling and Simulation.* (Ed. DA Post) pp. 154–159. Modelling and Simulation Society of Australia and New Zealand MODSIM 2003. Townsville.

Chow TWS and Leung CT (1996). Neural network based short-term load forecasting using weather compensation. *IEEE Transmission on Power System* **11(4)**, 1736–1742.

Cleugh HA, Bui E, Simon D, Xu J and Mitchell VG (2005). The impact of suburban design on water use and microclimate. In *Proceedings of the International Congress on Modelling and Simulation*. pp. 2019–2025. Modelling and Simulation Society of Australia and New Zealand, Canberra.

Crimp SJ and Day KA (2003). Evaluation of multi-decadal variability in rainfall in Queensland using indices of El Niño–Southern Oscillation and inter-decadal variability. In *Proceedings of the National Drought Forum: Science for Drought*. pp. 106–15. Queensland Department of Natural Resources, Brisbane.

CSIRO (2005). *Melbourne Water Climate Change Study, Implications of Potential Climate Change*. Melbourne Water Resources, Melbourne.

Fitzpatrick EA and Nix HA (1970). The climatic factor in Australian grassland ecology. In *Australian Grasslands* (Ed. RM Moore) pp. 1–15. Australian National University Press, Canberra.

Gallant A, Hennessy K and Risbey J (2007). A re-examination of trends in rainfall indices for six Australian regions. *Australian Meteorological Magazine* (in press).

Hennessy KJ, McInnes KL, Abbs DJ, Jones RN, Bathols JM, Suppiah R, Ricketts JR, Rafter AS, Collins D and Jones D (2004). 'Climate change in New South Wales. Part 2, Projected changes in climate extremes'. 79 pp. Consultancy report for the New South Wales Greenhouse Office, CSIRO Atmospheric Research, Canberra. < http://www.cmar.csiro.au/e-print/open/hennessy_2004c.pdf>.

Hennessy K, Fitzharris B, Bates BC, Harvey N, Howden SM, Hughes L, Salinger J and Warrick R (2007). 'Australia and New Zealand. Climate change 2007: impacts, adaptation and vulnerability'. Contribution of Working Group II to the Fourth Assessment Report of the Intergovernmental Panel on Climate Change (Eds. ML Parry, OF Canziani, JP Palutikof, PJ van der Linden and CE Hanson) pp. 507–540. Cambridge University Press, Cambridge, UK.

Howden SM and Crimp S (2001). Effect of climate and climate change on electricity demand in Australia. In *Integrating models for natural resources management across disciplines, issues and scales*. (Eds. F Ghassemi, P Whetton, R Little and M Littleboy) pp. 655–660. Modelling and Simulation Society of Australia and New Zealand, Canberra, Australia.

Howe C, Jones RN, Maheepala S and Rhodes B (2005). *Implications of Potential Climate Change for Melbourne's Water Resources*. CSIRO Urban Water, CSIRO Atmospheric Research, and Melbourne Water, Canberra and Melbourne.

Indian Ocean Climate Initiative (IOCI) (2002). 'Climate variability and change in south west Western Australia'. IOCI Panel for Department of Environment, Water and Catchment Protection, Perth. <http://www.ioci.org.au/publications/pdf/IOCI_TechnicalReport02.pdf>.

Inman M (2007). The Water Efficient City. (this volume)

IPCC (2007). *Climate Change 2007: The Physical Science Basis. Contribution of Working Group I to the Fourth Assessment Report of the Intergovernmental Panel on Climate Change*. (Eds S Solomon, D Qin, M Manning, Z Chen, M Marquis, KB Averyt, M Tignor and HL Miller). Cambridge University Press, Cambridge, UK and NY.

Jeffrey SJ, Carter JO, Moodie KM and Beswick AR (2001) Using spatial interpolation to construct a comprehensive archive of Australian climate data. *Environmental Modelling and Software*, **16(4)**, 309–330.

Jones RN (2004). 'Applying a Climate Scenario Generator to Water Resource Policy Issues: Final Report'. RIRDC Project No. CSM–5A. CSIRO Atmospheric Research, Hassall and Associates Pty Ltd, NSW Department of Infrastructure, Planning and Natural Resources, Sydney.

Jones R and Durack P (2005). 'Estimating the impacts of climate change on Victoria's runoff using a hydrological sensitivity model'. Consultancy report for the Victorian Department of Sustainability and Environment, Melbourne. <http://www.greenhouse.vic.gov.au/CSIRO%20Report%20–%20Runoff.pdf>.

Karoly DJ and Braganza K (2005a). Attribution of recent temperature changes in the Australian region. *Journal of Climate* **18**(3), 457–464.

Khotanzad A, Davis M, Abaye A and Maratukulam D (1996). An artificial neural network hourly temperature forecast with applications in load forecasting. *IEEE Trans. on Power Systems* **11(2)**, 870–876.

Kundzewicz ZW, Mata LJ, Arnell NW, Döll P, Kabat P, Jiménez B, Miller KA, Oki T, Sen Z and Shiklomanov IA (2007). Freshwater resources and their management. In *Climate Change 2007: Impacts, Adaptation and Vulnerability. Contribution of Working Group II to the Fourth Assessment Report of the Intergovernmental Panel on Climate Change.* (Eds ML Parry, OF Canziani, JP Palutikof, PJ van derLinden and CE Hanson), pp. 173–210. Cambridge University Press, Cambridge, UK.

Leslie LM, Karoly DJ, Leplastrier M and Buckley BW (2007). Variability of tropical cyclones over the southwest Pacific Ocean using a high resolution climate model. *Meteorology and Atmospheric Physics* **97(1–4)**, 171–180.

Littleboy M and McKeon GM (1997). 'Subroutine GRASP: grass production model. Documentation of the Marcoola version of subroutine GRASP. Appendix 2 of Evaluating the risks of pasture and land degradation in native pasture in Queensland'. Final Project Report for Rural Industries and Research Development Corporation, Canberra.

Loh M and Coghlan P (2003). *Domestic Water Use Study: Perth, Western Australia 1998–2001.* Water Corporation, Perth.

Maheshwari B (2006). 'The efficiency and audit of residential irrigation systems in the Sydney metropolitan area. Technical Report No. 01/06. CRC for Irrigation Futures, Darling Heights, Queensland.

McInnes KL, Suppiah R, Whetton PH, Hennessy KJ and Jones RN (2002). 'Climate change in South Australia: report on assessment of climate change, impacts and possible adaptation strategies relevant to South Australia'. CSIRO Atmospheric Research, Aspendale. <http://www.cmar.csiro.au/e–print/open/mcinnes_2003a.pdf>.

McMichael AJ, Woodruff R and Hales S (2006). Climate change and human health: present and future risks. *The Lancet* **367,** 859–869.

Melbourne Water (2005). *Water Supply–demand Strategy for Melbourne 2006–2055.* Melbourne Water. <http://www.southeastwater.com.au/SiteCollectionDocuments/Corporate%20reports/WSDS.pdf>.

Mirasgedis S, Sarafidis Y, Georgopoulou E, Lalas D, Moschovits M, Karagiannis F and Papakonstantinou D (2006). Models for mid-term electricity demand forecasting incorporating weather influences. *Energy* **31**, 208–227.

Mitchell VG, Mein RG and McMahon TA (2001). Modelling the urban water cycle. *Environmental Modelling and Software* **16,** 615–629.

Mpelasoka F, Hennessy KJ, Jones R and Bathols J (2007). Comparison of suitable drought indices for climate change impacts assessment over Australia towards resource management. *International Journal of Climatolology* (in press).

Niall S and Walsh K (2005). The impact of climate change on hailstorms in southeastern Australia. *International Journal of Climatology* **25**, 1933–1952.

Nicholls N (2004). The changing nature of Australian droughts. *Climatic Change* **63**, 323–336.

Nicholls N (2006). Detecting and attributing Australian climate change: a review. *Australian Meteorological Magazine* **55**, 199–211.

Pardo A, Meneu V and Valor E (2002) Temperature and seasonality influences on Spanish electricity load. *Energy Economics* **24**, 55–70.

PB Associates (2007) 'Assessment of the vulnerability of Australia's energy infrastructure to the impacts of climate change'. Issues Paper, 158292A–REP–002. Report to the Australian Greenhouse Office, Canberra.

Pezzulli S, Frederic P, Majithia S, Sabbagh S, Black E, Sutton R and Stephenson D (2006). The seasonal forecast of electricity demand: a hierarchical Bayesian model with climatological weather generator. *Applied Stochastic Models in Business and Industry* **22**, 113–125.

Power S, Haylock M, Colman R and Wang X (2006). The predictability of interdecadal changes in ENSO and ENSO teleconnections. *Journal of Climate* **8**, 2161–2180.

Power S, Sadler B and Nicholls N (2005). The influence of climate science on water management in Western Australia: Lessons for climate scientists. *Bulletin of the American Meteorological Society* **86**, 839–844.

Rahmstorf S, Cazenave A, Church J, Hansen JE, Keeling RF, Parker DE and Somerville RCJ

(2007). Recent climate observations compared to projections. *Science,* **316** (5825), 709.

Sailor DJ and Munoz J.R (1997). Sensitivity of electricity and natural gas consumption to climate in the USA – methodology and results for eight states. *Energy* **22**, 987–998.

Sailor DJ (2001). Relating residential and commercial sector electricity loads to climate – evaluating state level sensitivities and vulnerabilities. *Energy* **26**, 645–657.

Samarakoon AB and Gifford RM (1996) Elevated CO2 effects on water use and growth of maize an wet and drying soil. *Australian Journal of Plant Physiology* **23**, 53–62.

Sarac K, Day D and White S (2003). What are we saving anyway? The results of three water demand management programs in NSW. *Water Science and Technology: Water Supply* **3**, 215–222.

Segal M, Shafer H, Mandel M, Alpert P and Balmor V (1992). Climatic–related evaluations of the summer peak–hours' electric load in Israel. *Journal of Applied Meteorology* **31**, 1492–1498.

South-Eastern Australia Climate Initiative (2006). 'Six–monthly milestone report'. Murray Darling Basin Commission, Canberra. <http://www.mdbc.gov.au/subs/seaci/whatsnew_sreports.html>

Smith IN (2004). Trends in Australian rainfall – are they unusual? *Australian Meteorological Magazine,* **53(3)**, 163–173.

Suppiah R, Hennessy KJ, Whetton PH, McInnes K, Macadam I, Bathols J and Ricketts J (2007). Australian climate change projections derived from simulations performed for the IPCC 4th assessment report. *Australian Meteorological Magazine* (in press).

Syktus J, McKeon G, Watterson I and Walsh K (2005). 'What is causing the decline in coastal rainfall in eastern Australia? The impact of stratospheric ozone depletion and CO_2 on the Southern Annular Mode and regional climate changes: Implications for water resources in Australia. In *Hydrometeorological Applications of Weather and Climate Modelling.* Extended abstracts of presentations at the 17th annual BMRC Modelling Workshop, 3–6 October 2005. BMRC Research Report No. 111. Bureau of Meteorology Research Centre, Melbourne.

Thatcher M (2007). Modelling changes to electricity demand load duration curves as a consequence of predicted climate change for Australia. *Energy* **32**, 1647–1659.

Walsh K, Hennessy K, Jones R, McInnes KL, Page CM, Pittock AB, Suppiah R and Whetton P (2001). 'Climate change in Queensland under enhanced greenhouse conditions – third annual report, 1999–2000'. CSIRO Atmospheric Research consultancy report for the Queensland Government, Aspendale. <http://www.cmar.csiro.au/e–print/open/walsh_2001a.pdf>

Walsh KJE, Nguyen KC and McGregor JL (2004). Finer–resolution regional climate model simulations of the impact of climate change on tropical cyclones near Australia. *Climate Dynamics* **22**, 47–56.

WRSC (2002). '21st Century Melbourne: a Water Smart city. Strategy directions report, Water Resources Strategy Committee for the Melbourne Area'. Victorian Government, Melbourne.

Whetton PH, Suppiah R, McInnes KL, Hennessy KJ and Jones RN (2002). 'Climate change in Victoria: high resolution regional assessment of climate change impacts'. CSIRO consultancy report for Department of Natural Resources and Environment, Victoria. <http://www.greenhouse.vic.gov.au/climatechange.pdf>

Zhou SL, McMahon TA, Walton A and Lewis J (2000). Forecasting daily urban water demand: a case study of Melbourne. *Journal of Hydrology* **236**, 153–164.

15

Bushfire

Robert Leicester and John Handmer

Living in the interface between rural and urban areas is one of the favoured lifestyle choices of Australians. It has the potential to combine the enjoyment that the bush has to offer, combined with the social, cultural, work opportunities and amenities to be obtained in an urban setting. However, associated with this desirable lifestyle, there is a fear of bushfire attack, which is regularly reinforced by dramatic reports on some of the disasters of past summers, particularly where loss of life has been incurred. Table 1 gives a summary of some of the more devastating fires that have occurred during the past 70 years. To the insurance industry, the annual property damage arising from wildfires is small in comparison with that of other natural hazards (Walker 2003). But for residents, the trauma arising from the risk of death and loss of property can be considerable.

Using data obtained by Chen and McAneney (2004) and McAneney (2005, and *pers. comm.*) it is estimated that currently within Australia there are roughly 250,000 houses within 50 m of an urban–bushland interface, and this interface edge is roughly 10,000–15,000 km in length. On average, the probability that a major wildfire will occur at any particular interface within a 50 year period is about 2%. This is a sufficient risk that it needs to be taken seriously, yet it is too infrequent for local residents to learn to cope with the hazard through personal experience.

Commencing with the early work by George Barrow in 1944, field investigations on wildfire events have clearly shown that houses do not explode when attacked by fire (Barrow 1944; Wilson and Ferguson 1986; Ramsay *et al.* 1996; Gledhill 2003; Leonard and Bowditch 2003; Tolhurst and Howlett 2003; Blanchi *et al.* 2006b). In fact, if they catch fire at all, they most usually commence burning from the inside due to the ingress of embers. These embers can often be extinguished by using a limited amount of human action. By contrast, the danger to humans arises

Table 1: Data on some iconic wildfire events in Australia (after Blanchi *et al.* 2006b, Emergency Management Australia 2007).

Year	State	Iconic name	Houses lost*	Deaths*
1939	Vic.	Black Friday	1300	71
1961	WA	Dwellingup	160	0
1967	Tas	Black Tuesday	1293	62
1983	Vic., SA	Ash Wednesday	1894	75
1994	NSW	Sydney (Como/ Janali)	202	4
2003	ACT	Canberra (Duffy)	519	4
2006	SA	Wangary	50	9

* Note: Within the published literature, there are variations on some of these statistics.

from exposure to the radiant heat of a wildfire. Fortunately the shell of any house provides an effective shelter against such radiant heat.

This combination of factors gives rise to a symbiotic relationship in which people protect houses and houses protect people. It has also led to the AFAC (Australasian Fire Authorities Council) policy of 'Prepare, stay and defend (your property) or leave early' (AFAC 2005; Bushfire CRC and AFAC 2006). Handmer and Tibbits (2005) summarise the historical evidence for this approach. The following will examine the application and potential of fully implementing this policy within Australia.

Wildfire basics

The rate of spread of a wildfire depends on weather and fuel factors. The danger associated with weather can be assessed in terms of an FFDI index (Forest Fire Danger Index) which is a function of temperature, wind speed, relative humidity and fuel moisture (Noble *et al.* 1980). The fuel is essentially characterised in terms of the weight of fine fuel less than 6 mm diameter. On days with an FFDI greater than 50, extremely dangerous fires can develop, even when there is aerial suppression support (Plucinski *et al.* 2007). Under such conditions, once developed, even a fire travelling as slow as 2.0 km/hr through 8-year old fuels cannot be stopped (Koperberg 2003).

Most of the highly destructive iconic fires of Australia have occurred for an FFDI greater than 100 (Leonard and Blanchi, *pers. comm.*). Typically for this index to occur the air temperature is likely to exceed 40°C, the wind-speed could be greater than 50 km/hr and the relative humidity may be less than 10%. These represent extreme weather conditions.

Another aspect that needs to be appreciated is the effect of the heat radiated from a fire. If a flame is approximated as a semicircular disc of fire, with the centre at the level of an observer, then at a distance of 1.0 times the flame height away from the fire, the radiated heat is about 30 kw/m^2, regardless of the height of the flame. At a distance of 5.0 times the flame height the radiation is about 2 kw/m^2.

The significance of radiant heat levels is that they can be used to estimate safe separation distances from potential sources of flames. For example, depending on their detail and whether they are in contact with glowing embers, wooden objects will ignite and windows can fail at radiant heat levels in the range 10–40 kw/m^2 (Cohen 1995; Bowditch *et al.* 2006). People can operate safely outdoors only if the radiant heat level there is less than 2 kw/m^2.

Obviously, in developing any safety and protection plan, it is important to be able to assess the potential size of flames from the burning bushland and also the size of the flames that may arise from the ignition of any combustible objects within the interface zone, such as adjacent houses, sheds, fences and trees. A flame may be 50% higher than a burning object. So, a forest fire could well exceed 10 m in height, in which event the safe outdoor environment would need to be more than 50 m away.

The above factors have been taken into consideration in the development of the AFAC policy known as 'Prepare, stay and defend (your property) or leave early' (Gledhill 2003; AFAC 2005; Bushfire CRC and AFAC 2006), frequently abbreviated in terminology as the 'Stay-or-Go' policy. Key features of this policy are as follows:

- Home owners are responsible for making their properties defendable. To this end, fire agencies will assist in the drafting of land management regulations, such as the CFA (2007) 'Building in a Wildfire Management Overlay'.
- Fire agencies cannot always protect every home.
- Home owners who plan to stay to defend their homes should seek refuge inside against radiant heat as the main fire front passes.
- People who plan to leave should do so early and in response to an agreed trigger. Mass evacuations must be avoided.
- Residents who leave should be allowed to return as soon as possible after the fire has passed so that they can extinguish minor fires in and around their homes

The reason why fire agencies cannot always defend all homes is because of the large number of houses that can be simultaneously involved in a wildfire attack. For example, Gill (2006) has estimated that the ACT Emergency Services Authority has a fleet of only about 50 fire suppression vehicles available for the entire city of Canberra, whereas

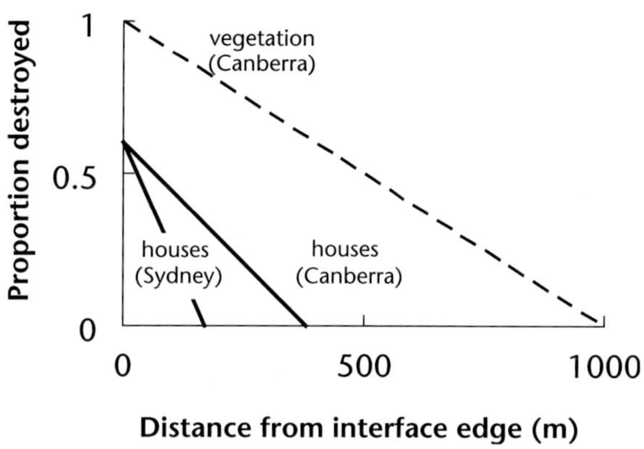

Figure 1: Proportion of houses and vegetation destroyed by wildfire (Source: Chen and McAneney 2005)

during the Canberra fires of 2003 several thousand houses were threatened and more than 500 houses were burnt to destruction.

Protection of property

In considering the defence of property, three key pieces of information are required. The first is the rapid attenuation of the impact of wildfires as they penetrate an interface edge. An example of this is illustrated in Figure 1 that shows the data analysed for two of the iconic fires listed in Table 1. The figure illustrates that near the interface edge we would expect about 50–60% of the houses and 100% of the vegetation to be destroyed; however, it would be rare for any house further than 500 m from the interface edge to be lost. In another study of about 8 million addresses of capital cities, Chen and McAneney (2005) evaluated statistics to provide a rough measure of city houses at risk from a wildfire attack. They found that about 4% of addresses were closer than 50 m from a bushland interface, and about 15% of all addresses were closer than 400 m to the interface.

Second—based on data reported by Blanchi *et al.* (2006a) for some iconic fires—the following would appear to be the typical proportions for the main modes of destruction for houses located at about 50 m from the bushland interface:

- not destroyed: 55%
- destroyed by radiant heat: 5%
- destroyed by ember entry: 30%
- destroyed by ember burn through: 10%.

Third, Blanchi *et al.* (2006a) have reported that the following are typical values for the survival rates of this category of houses:

- no human intervention: 30%
- occupant or neighbour present: 90%.

These three pieces of information lead to the following conclusions. They indicate that about 85% of the destruction of houses has been associated with ember attacks, and that these attacks rarely penetrate further than 500 m from the interface edge, although the destruction of vegetation may go twice that distance. The entry of embers causes a house to burn from the inside and so an alert person can be highly effective in preventing the destruction of a dwelling. It was found that for past fires typically there had been a human presence at the time of arrival of the fire front for about 50% of houses at risk (Blanchi *et al.* 2006a).

The protection of a house from radiant heat (emanating either from the bushland vegetation or from items within the interface zone) is best achieved by the use of appropriate separation distances as discussed earlier, or by the use of shields such as earth mounds, steel fences or window shutters.

Based on studies of the destruction of houses, the Australian Standard AS 3959, 'Construction of Buildings in Bushfire-prone Areas' (Standards Australia 1999) has been developed to provide design guides for increasing the resistance of buildings to wildfire attack. For example, the use of mesh with a minimum aperture dimension of 2 mm is specified for all external envelope openings in order to prevent the entry of embers.

All States in Australia now have model building regulations for providing defendable space around homes. A notable feature of these regulations is that the latest wildfire science has been used in developing them. An example of this is the Victorian Country Fire Authority 'Building in a Wildfire Management Overlay' (CFA 2007). These regulations specify in detail the requirements for setting up two sets of concentric vegetation management zones around each house, together with water supply and vehicle access requirements.

Risks to residents who decide to stay

The primary risks to people within the interface zone during a wildfire attack arise from the associated radiant heat and smoke. As mentioned earlier, houses provide an excellent protection from the radiant heat of a wildfire. Usually a flame front will take only a few minutes to pass a house, whereas the shower of embers could continue for several hours.

Hence, once a wildfire has passed, occupants should actively search for and eliminate any signs of ignitions arising from ember attack. In doing so, the occupants need to be aware that radiant heat will also emanate from any burning objects. When outdoors, they should stay sufficiently far from burning objects that the radiant heat acting on them is lower than the safe working level of 2 kw/m^2. In addition it should be mentioned that deaths have occurred when occupants have remained in buildings that (unknown to them) had been ignited.

The health risks arising from inhaling bushfire smoke have been discussed by Reisen and Brown (2006a, 2006b). For healthy persons, the health risks from smoke are minor unless the smoke hovers for many days. The most noticeable short-term effect may be an increase of asthmatic attacks due to the heavy concentration of particulate matter in smoke. Closing all windows and doors is ineffective except for slightly reducing peak values of the concentrations of particulate matter. The use of half-face respirators with a 10-fold protection factor is reasonably effective. An equally effective (and more comfortable) strategy is to shut the doors and windows and to use air-cleaning devices that can work at a rate that achieves four air changes per hour.

The practical problems for residents who decide to stay and defend their properties during a wildfire attack have been described by Gill (2006). Apart from problems associated with visibility and the lack of knowledge of the progress of a fire, Gill discusses the difficulties arising from the failure of equipment and water supplies, the value of long warning times and the variable competencies of people. In particular he questions whether an inexperienced person would be able to relate factors such as vegetation density, ground slope and wind velocity to assess the speed of travel of an approaching flame front.

Risks to residents who decide to leave at the last minute

Studies such as that by Krusel and Petris (1992), have shown that nearly all wildfire related deaths occur away from houses. For those caught out in the open, there is a high risk of death by radiant heat. For those who are in vehicles, the poor visibility caused by smoke creates a high risk of fatal crashes. Even for those who decide stop and seek shelter in a parked vehicle, the risks are high.

For those who decide to flee in vehicles and happen to be caught by a wildfire, advice on the best course of action to undertake is given in a publication by the Australasian Fire Authorities Council 'Interim Position Paper on Guidance for People in Vehicles during Bushfires' (AFAC 2007). In a recent experimental study on the efficacy of sheltering in a car (Sargeant *et al.* 2007), it was found that a person sheltering under a blanket in the foot well of a car could survive a simulated attack of 20 kw/m^2 for a period of about 10 minutes before the car became untenable. In all cases, untenable conditions first arose from exposure to the toxic gases associated with the burning of interior furnishings following a breach in the outer shell of the car.

For people who decide to leave their homes, it is important that they appreciate not only the risks involved if caught in a fire, but also the dangers associated with the uncertainties of the location and progress of a fire front. There is still considerable debate on the optimum time to flee, but that time is certainly at least many hours before the expected arrival of a fire front.

Implementing the 'Prepare, stay and defend (your property) or leave early' policy

Experience in implementation

As indicated earlier, the single most important factor in the protection of lives and property rests with the actions of the residents of the interface zone, and that these actions are probably most usefully guided through application of the AFAC 'Stay-or-Go' strategy.

While this approach is based on sound evidence and historical experience, full implementation is not easy for a range of factors (Tibbits and Whittaker 2007). These include the reality that there is often not time to make the decision to leave early, and people are faced with having to stay, even if unprepared, or take the risks of fleeing at the last minute. More fundamentally, many people are simply unprepared mentally to face the full force of a wildfire event. Even if they had previously decided to stay, they may find the proximity of a wildfire overwhelming and leave at the last minute.

There is often a significant lack of commitment to staying and defending a property in the sense that many of those who stay will also have a 'contingency plan'. For example, the car might be packed with valuables and ready to 'dash for safety'. Many people assert that they will wait and see how the fire develops and leave if it gets too much. In particular it seems that women and children are likely to leave in this way, or to flee at the last moment. Unfortunately, this is far from safe and is generally the highest risk action.

People also leave at the last minute, or during the passage of the fire front, because they run out of water for extinguishing embers, have a minor accident or feel overcome by smoke. They may also be urged to leave by family members or emergency service personnel who are unfamiliar with the 'Stay-or-Go' strategy. There are also concerns—generally unfounded—about legal liabilities for fire and emergency service workers.

Fieldwork following the October 2006 Hobart interface fire showed that the presence of fire agency personnel was very effective in instilling confidence in householders remaining with their properties.

Education needs

The main aim of the 'Stay-or-Go' policy is to prevent last minute evacuation on the grounds that this represents the single most dangerous action. The policy is now officially titled 'Prepare, stay and defend (your property) or leave early' and was recently reworded so as to give increased emphasis on the dangers of late evacuation.

One major gap concerns mental preparedness. People who stay emphasise that in addition to the arrival of embers, smoke and heat, the noise of the fire itself can be terrifying. Standard approaches can help with household physical preparedness, but these do not seem to help much with psychological preparedness. Ways are needed to tackle this important gap. In particular, people need to be confident that their houses will offer them protection from the fire and also that they have the capacity to defend their properties.

Education methods

Clearly, improved public understanding of the basis for the 'Stay-or-Go' policy, and what it means in practice, is needed for fuller implementation. All Australian fire agencies undertake extensive public education programs. Much effort also goes into intensive engagement through 'Fire Ready', 'Fire Wise' and the NSW 'Community Fire Units', among other approaches. These approaches attempt to move away from reliance on pamphlets and websites to direct face-to-face communication. Lowe and Haynes (2007) have examined the Community Fire Units, which are street-based groups with some fire-fighting equipment to enable them to defend their homes until the fire services arrive. The idea has been very popular with several thousands of people now involved on the Sydney urban–bushfire interface (Bushfire CRC 2007).

As each fire season starts there is a more immediate need for advice from fire agencies. This is generally achieved the same way as set out above through the various fire agency community engagement programs, brochures, printed media articles, radio advertisements and increasingly through TV documentary programs. Much of the Australian media promote the 'Stay-or-Go' policy and the government broadcaster, the ABC, has even developed scenarios and documentaries on the approach as part of its role as official carrier of emergency service material. How well these education programs work is uncertain, although there is now a concerted effort to establish this (Elsworth *et al.* 2007; Paton and Wright 2007).

Although such efforts have had significant impacts in terms of awareness of the policy, there are questions about the effectiveness of using relatively standard public education approaches in the very rapidly expanding and evolving urban interface.

Communication during a wildfire

The ABC has been established as the primary broadcaster during emergencies, and has taken a keen interest in fire-related research and policy. This appears to have worked very well for the broadcaster and certainly for the affected communities. Following the 2003 Canberra fire, praise for the local ABC radio station was universal, as the station had dedicated itself totally to the fire coverage. Local stations in other parts of Australia have since followed suit to similar commendation.

Almost all wildfire-related information, including near-real-time updates, is now on fire agency websites. Efforts are made to make these accessible and useful for those at risk, and it seems reasonable to assume that many people now rely on the web for much of their information. However the web may experience overload and failures due to power outages during a wildfire. The same applies to landline phone systems.

Increasingly, personal warnings via text messaging have become available on a commercial basis. Typically these bypass official channels, sending messages directly to subscribers wherever they are. This trend is part of the world of unofficial or informal warnings. When considering wildfire-related communications, it is important to appreciate that much of this will be outside official channels. It will consist of people's personal networks and unofficial media commentary, and increasingly such sources as international media, blogs and other alternate warning systems.

Legal aspects

The legal aspects relating to community safety have been discussed by Loh (2007). At first sight they may appear complicated due to the changing nature of common law and the range of relevant fire and emergency service legislation over the different state and territory jurisdictions. However, in the last few years there has been increased regulation of liability by statute, and in many states there are statutory immunities aimed at ensuring that individuals are not liable for performing their duties. However, some gaps in legal protection still remain.

To be legally protected when implementing the 'Stay-or-Go' policy, volunteers and career officers must act in a way that is consistent with the policy. In light of this, agencies or their members would be less likely to be found liable of negligence if they do not evacuate residents rather than they would if they insisted on undertaking last-minute evacuations. The latter is in direct contravention of the 'Stay-or-Go' policy and may put the evacuees in greater risk than they would have been if they had been left in their homes.

Future

If we look to the future, there are predictions that there will be an increase in the frequency of days of high fire risk and, as a consequence, an increase in the frequency of occurrences of high-intensity wildfires (Lucas *et al.* 2007). This has the potential to cause an increase in the annual losses due to wildfires.

However if there is the political and community will, it would be quite feasible to apply our current knowledge to modify and/or develop urban–bushland interface zones in such a way that communities can live there in comparative safety and even with reduced property loss. To do this protection strategies must be developed on the assumption that attacks by high-intensity wildfires could occur at any specified interface and that communities should develop protection strategies that do not rely on fire services to protect them when these attacks happen.

An example of such a protection strategy would be one based on the fact that the frequency of destruction of property is small for dwellings that are more than 500 m from a high-intensity wildfire front. Thus the risk of destruction can be minimised if all land within 500 m of houses is subjected to land management and town planning controls that are designed to minimise the severity of potential ember attacks and there is ready access to and from any fires that may occur within the interface zone.

The second feature of a protection strategy would be to specify house construction (design, material selection and house maintenance) that is resistant to radiant heat and ember attack. This would require some new education and regulation for the building industry, and perhaps some associated independent assessment and certification procedures applied to building construction, but otherwise the policy should not be difficult to implement. For the homes of persons who are highly risk averse, special features could be added. These may include internal and external sprinkler systems, and perhaps even smoke-resistant rooms.

Finally, the most important feature of a protection strategy would be the education of the community on the correct behaviour in the event of a high-intensity wildfire attack. In particular, the potentially lethal effects of radiant heat need to be appreciated. The current 'Prepare, stay and defend (your property) or leave early' policy is an effective framework on which to develop a suitable behaviour model. However, to be successful, it will be necessary to develop methods to prepare communities to be psychologically prepared for the experience of a real fire. Procedures for providing people with all relevant information before, during and after a wildfire attack will also be needed.

The above suggestions represent extreme mitigation measures and for many cases may not be justified solely in terms of cost effectiveness. Consequently, the actual level of protection attained in the future will depend more on the risk averseness of our society. This applies to both new and existing construction.

The current progress of research related to wildfires and the enthusiastic uptake of it by fire agencies, all indicate that there is a high probability that within the foreseeable future Australia will develop rural–urban interface zones that are both enjoyable and safe places for living.

References

AFAC (Australasian Fire Authorities Council) (2005). 'Position paper on bushfires and community safety'. AFAC, Melbourne <www.afac.com.au>.

AFAC (2007). 'Interim position paper on guidance for people in vehicles during bushfires'. AFAC <www.afac.com.au>.

Barrow GJ (1945). A survey of houses affected in the Beaumaris fire. *Journal of the Council for Scientific and Industrial Research* **18(1)** February, pp. 22–27.

Blanchi RM, Leonard JE, Leicester RH (2006a). Bushfire risk at the rural-urban interface. In *Proceedings of Bushfire Conference 2006, Life in a fire-prone environment: Translating science into practice.* Brisbane, Australia, June 6–9. Paper 303, CD-ROM, Griffith University, Brisbane and SECQ Fire and Biodiversity Consortium, Brisbane.

Blanchi RM, Leonard JE and Leicester RH (2006b). Lessons learnt from post bushfire surveys at the urban interface in Australia. Paper presented at the *International Conference on Forest Fire Research,* November 27–30, Figueira da Foz, Portugal, <www.bushfirecrc.com>.

Bowditch PA, Sargeant AJ, Leonard JE and Macindoe L (2006). *Window and Glazing Exposure to Laboratory-simulated Bushfires.* Bushfire CRC, Melbourne.

Bushfire CRC (2007). *Resilience at the Urban Interface. A Study of New South Wales Fire Brigades Community Fire Units.* Bushfire CRC, Melbourne.

Bushfire CRC and AFAC (2006). 'The stay and defend your property or go early policy. Fire Note'. Bushfire Cooperative Research Centre and Australasian Fire Authorities Council.

CFA (Country Fire Authority) (2004). *Living in the Bush. Bushfire Plan Survival Workbook.* CFA, Victoria.

CFA (2007). Building in a wildfire management overlay. Applicant's Kit 2007. Country Fire Authority, Melbourne.

Chen K and McAneney J (2004). Quantifying bushfire penetration into urban areas in Australia. *Geophysical Research Letters.* **31**(L12212), 4 pages.

Chen K and McAneney J (2005). The bushfire threat in urban areas. *Australian Science* **January–February**, 14–16.

Cohen JD (1995). Structure ignition model (SIAM) USDA Forest Service'. General Technical Report PSW-GTR-158. pp. 85–92. Pacific Southwest Southern Forest Fire Laboratory. Albany, California.

Emergency Management Australia (2007). *Disasters Database.* <www.ema.gov.au/ema/emadisasters.nsf/>

Gledhill JB (2003). Community self-reliance during bushfires: The case for staying at home. In *Proceedings of 3rd International Wildland Fire Conference.* Sydney, 3–6 October. Paper No. 349. CD-ROM, ISBN 1 877040 15 0, Australasian Fire Authorities Council, Melbourne.

Gill AM (2006). Fire, science and society at the rural-urban Interface. In *Proceedings of Bushfire Conference 2006, Life in a Fire-prone Environment: Translating Science into Practice.* June 6–9, Brisbane. Paper No. 301, CD-ROM, Griffith University, Brisbane and SECQ Fire and Biodiversity Consortium.

Handmer J and Tibbits A (2005). Is staying at home the safest option during bushfires? Historical evidence for an Australian approach. *Environmental Hazards* **6(2)**, 81–91.

Koperberg P (2003). In *Proceedings of 3rd International Wildland Fire Conference*. Sydney, October 3–6, Paper No. 332, CD-ROM, ISBN 1 877040 15 0, Australasian Fire Authorities Council, Melbourne.

Krusel N and Petris SN (1992). 'A study of civilian deaths in 1983 Ash Wednesday bushfires Victoria, Australia'. Occasional Paper No 1. CFA, Melbourne.

Leonard JE and Blanchi R (2007). 'CSIRO sustainable ecosystems,' (pers. comm.).

Leonard JE and Bowditch PA (2003). Findings of studies of houses damaged by bushfires in Australia. In *Proceedings of 3rd International Wildland Fire Conference*. Sydney, October 3–6. Paper No. 342, CD-ROM, ISBN 1 877040 15 0, Australasian Fire Authorities Council, Melbourne.

Loh E (2007). Don't get burnt by the law: the legal implications of the 'prepare, stay and defend or leave early' policy. In *Community Bushfire Safety*. (Eds J Handmer and K Haynes). CSIRO Publishing, Melbourne.

Lowe T and Haynes K (2008). Community fire units. In *Community Bushfire Safety*. (Eds J Handmer and K Haynes). CSIRO Publishing, Melbourne.

Lucas C, Hennesey K, Mills G and Bathols J (2007). 'Bushfire weather in South East Australia: Recent trends and projected climate change impacts'. Consultancy Report prepared for the Bushfire CRC and the Climate Institute of Australia. September. Bushfire CRC, Melbourne.

McAneney KJ (2005). Australian bushfire: quantifying and pricing the risk of residential properties. In *Proceedings of Symposium on Planning for Natural Hazards. How Can We Mitigate the Impacts?* University of Wollongong, NSW, February 2–5, University of Wollongong.

Noble IR, Bary GAV and Gill AM (1980). McArthur's fire-danger meter expressed as equations. *Australian Journal of Ecology* **5**, 201–203.

Plucinski M, Gould J, McCarthy G and Hollis J (2007). The effectiveness and efficiency of aerial firefighting in Australia. Part 1. Technical Report Number A0701. Bushfire Cooperative Research Centre, Melbourne Australia.

Ramsay GC, McArthur NA and Dowling VP (1996) Building in a fire-prone environment: Research on building survival in two major bushfires. *Proceedings of Linnaeus Society. NSW* **116**, 133–140.

Reisen F and Brown SK (2006a) Impact of prescribed fire smoke on downwind communities. In *Proceedings of Bushfire Conference 2006: Life in a fire-prone environment: Translating science into practice*. Brisbane. Paper No. 208, CD-ROM, Griffith University, Brisbane and SECQ Fire and Biodiversity Consortium.

Reisen F and Brown SK (2006b). Implications for community health from exposure to bushfire air toxics. *Environmental Chemistry* **3**, 235–243.

Sargeant AJ, Brown SK, Leonard J and Blanchi R (2007). Civilian passenger burnover experimentation. Paper presented at *Joint AFAC/Bushfire CRC Conference*. Hobart. September 19–21. Bushfire CRC, Melbourne.

Standards Australia (1999). AS 3959 Construction of Buildings in Bushfire-prone Areas. Standards Australia, Sydney.

Tibbits A and Whittaker J (2007). Stay and defend or leave early: experiences of the policy during the 2003 Victorian bushfires. *Environmental Hazards* **7** (in press).

Tolhurst KG and Howlett KA (2003) House ignition likelihood index – A hazard assessment method for land managers in the wildland-urban interface. In *Proceedings of 3rd International Wildland Fire Conference*. Sydney, October 3–6. Paper No. 89, CD-ROM, Australasian Fire Authorities Council, Melbourne.

Walker GR (2003). An analysis of natural catastrophe insurance risk in Australia, based on insurance loss records. In *Proceedings of Australian Disaster Conference*. Canberra, Australia, September 10–12 <walker ADC2003 GRW Paper 1.pdf>.

Wilson AG and Ferguson IS (1986). Predicting the probability of house survival during bushfires. *Journal of Environmental Management* **23**, 259–270.

16

Biodiversity

Mark McDonnell and Kelly Holland

The preservation, restoration and ecologically sound management of urban biodiversity are crucial to the maintenance of global and regional biodiversity. Throughout the world, the number and size of urban areas is increasing. It has been predicted that by the year 2025 over 60% of the world's population will be living in urban settlements (World Resources Institute 1996). Unfortunately, most of our ecological understanding of the world has been obtained from studying natural areas specifically selected to minimise the presence of humans and thus we lack the information and tools to effectively maintain biodiversity and ecological processes in cities and towns.

The preservation of a diversity of indigenous plants and animals in cities and towns contributes to the quality of life of city dwellers. In addition to providing valuable ecosystem services, such as cleaning the air, ameliorating climate and reducing impacts of potential pests and pathogens, and so on (Daily *et al.* 1997; Bolund and Hunhammar 1999; Williams *et al.* 2001), biodiversity satisfies the human need to have contact with other organisms. This relationship between humans and nature is a serious topic of scientific study and has been referred to as 'biophilia' by E. O. Wilson (Wilson 1984; Kellert and Wilson 1993).

Typically, the preservation of biodiversity in urban environments requires the preservation of open space and remnant plant communities, but gardens, roadside and riparian strips can also provide valuable habitats for indigenous organisms (Scott *et al.* 2002). More work needs to be done to enhance the role of the 'matrix' of urban and suburban areas in preserving local biodiversity to supplement the loss of remnant patches in the landscape. Although patches of indigenous vegetation located in cities and towns may not be considered of high regional conservation significance, they are vital living laboratories for environmental education that provide many urban dwellers their only opportunity to experience and learn about the region's natural heritage (McDonnell 2007).

The biodiversity of cities and towns in Australia, and indeed across the globe, is under threat due to a range of human activities, including urbanisation. The conservation, management and restoration of biodiversity present a formidable challenge in cities and towns around the world.

Australian's capital cities

Australia's capital cities were founded in the late 1700s and early 1800s (Scott 1916). Melbourne has grown from 10,000 inhabitants in 1840 to over 3 million residents today (Harvey 1982; State of Victoria 2002). In the 1960s, Melbourne covered some 97,851 ha, but by 1990 it had spread to cover over 202,600 ha. Sydney's population was 100,000 in the mid 19th century, but now hosts over 4 million people (NSW Department of Planning 2005). Much of this spread resulted in the conversion of farmland and remnant vegetation to suburban land use. By 2051 it is projected that the populations of Sydney and Melbourne will have grown to 5.6 and 5.0 million, respectively—increases of approximately 1.4 million people per city from 2004. A similar increase is predicted in Brisbane, which will grow from 1.8 million people in 2004 to 3.4 million in 2051. Perth is expected to grow from 1.5 to 2.5 million people in the same time period (Australian Bureau of Statistics 2004). Recent projections have anticipated the need for an additional 620,000 dwellings over 30 years in Melbourne

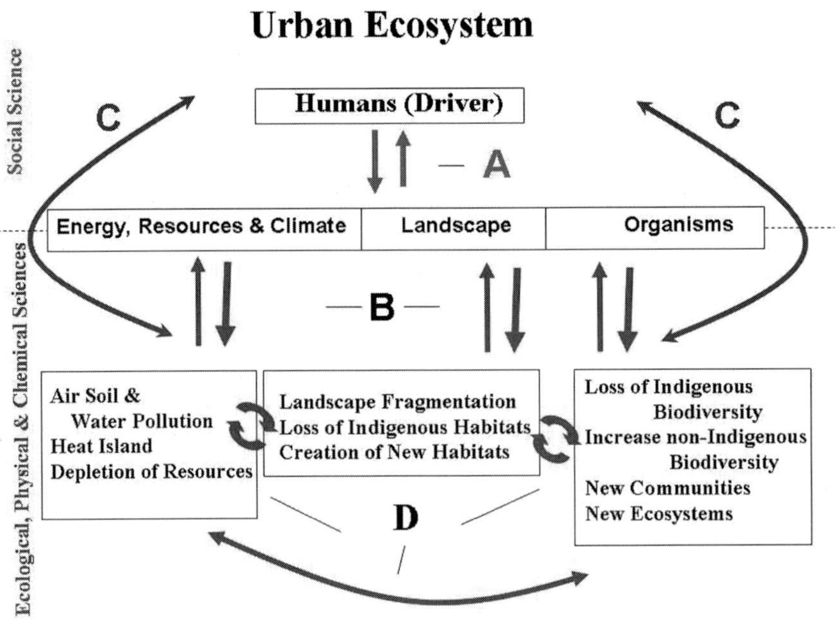

Figure 1: A hierarchical representation of urban ecosystems, which are driven by human activities. Humans primarily alter three components of urban ecosystems: energy, resources and climate; landscape structure; and the presence of other organisms. Arrows in the diagram represent potential interactions between components within and between the levels of organisation. Currently, the interactions labelled D and B are fairly well understood, but interactions A and C are not. To understand these types of interactions will require more trans-disciplinary research involving ecologists, physicists, chemists and social scientists.

(State of Victoria 2002), 528,000 in Brisbane (Larsen 2000), 375,000 in the Perth area (State of Western Australia 2004), 150,000 in Adelaide (Larsen 2000) and 640,000 dwellings over 25 years in Sydney (NSW Department of Planning 2005). Even though planning processes are being put in place to control this new growth, it will still most likely be in the form of urban sprawl. The rapid urbanisation of Australian cities has undoubtedly had a significant impact on the distribution and abundance of indigenous plants and animals in these areas.

Impacts of urbanisation

Cities are complex ecosystems in which humans are the primary drivers of changes to the structure of the system and the interactions between components within the systems (Figure 1). As the density of human populations increases during the building of cities and towns, there are many changes to the surface of the land, the quality of the air, soil and water and the diversity of organisms (Young 2000). The process of urban development invariably results in an increase in impermeable surfaces, the number of cars, energy use, pollution and the number of non-indigenous species, while concurrently reducing green space, indigenous species, remnant habitats and the quality of the air, soil and water. The impacts of these alterations to the environment on biodiversity caused by urbanisation are summarised in Table 1. The general impacts of air, soil and water pollution associated with urban development are fairly well documented in the literature (Freedman 1989; Graedel and Crutzen 1989; Paul and Meyer 2001). In this chapter we will discuss in more detail the impacts on biodiversity associated with changes in disturbance regimes, heat island effects, light pollution, noise pollution, loss of native habitats, nitrogen deposition and non-indigenous species invasions.

Disturbance regimes

The important role disturbance regimes play in maintaining biodiversity and ecosystem processes was highlighted over 20 years ago (Pickett and

Table 1: Major impacts on biodiversity caused by the creation of cities and towns.

Impact	Effect
Air, soil and water pollution	Loss of indigenous plants and animals
Change in disturbance regimes	Loss of indigenous plants and animals
Heat island	Changes in climate Alters the competitive interactions Encourage non-indigenous species
Light pollution	Loss of abundance and diversity of night flying insects
Loss of native habitat	Habitat destruction Landscape fragmentation Community sanitisation
Nitrogen deposition	Alters competitive plant interactions encouraging the invasion of non-indigenous species.
Non-indigenous species invasion	Loss of indigenous plants and animals

White 1985). In Australia, one of the major influences on the composition and structure of indigenous plant and animal communities is the frequency and magnitude of fires (Bradstock *et al.* 2002). As areas become urbanised, residents of cities and towns alter natural disturbance regimes by suppressing fire. This suppression of fire in grassland and heathland ecosystems has well- documented detrimental effects on the composition and structure of plant communities. Indigenous plants that have adapted to fire regimes are commonly lost when fires are suppressed for long periods of time (for example, more than 5 years) (Fox and Fox 1986; Lunt and Morgan 2002). Similarly, regular frequent prescribed burning which does not match the natural fire regimes can also reduce biodiversity in Australian eucalypt forests (Gill and Williams 1996). The suppression of fire in remnant grasslands and heathlands has reduced the number and abundance of several indigenous plants (Lunt and Morgan 1999; Phelan 2000; Williams *et al.* 2006). Urbanisation also results in additional disturbance. For example, in a study of reserves in Perth, Stenhouse (2004a) found that reserves closer to the Central Business District were more subject to trampling and rubbish dumping, and contained a higher abundance of exotic species, bare patches and erosion. Other reported disturbances caused by human activity in reserves include damage from children's play or vandalism, firewood collection and soil compaction and erosion caused by trail bikes and cars (Stenhouse 2004b).

Heat-island effects

The creation of heat islands is becoming increasingly common in cities across the globe (Landsberg 1981; Oke 1995). This effect manifests itself as warmer nocturnal temperatures compared with the cooler surrounding landscape. It occurs due to a set of complex interactions related to the increase in concrete and pavement used to build cities, generation of heat through wasted energy and air pollution. Typically, the buildings and roads that make up cities store heat during the day and release this heat throughout the night resulting in temperature differences between the city centres and the surrounding landscape of between 1 and 2°C (Landsberg 1981; Oke *et al.* 1999). This change in climate, although it appears small, can have a variety of impacts on organism and ecosystem processes. The urban heat-island effect may be even more significant in light of predicted global climate change (see Hughes 2003). Higher temperatures could:

- reduce the risk of cold damage or death to organisms sensitive to cold temperatures
- alter the competitive interactions between organisms (for example, favour non-indigenous species)
- alter the cycling of nutrients (McDonnell *et al.* 1997).

The average annual maximum temperature for Brisbane was 22.5°C in 2002. This is projected to increase to 26.5 ± 0.7°C by 2030 and 28.5 ± 2.2°C by 2070 (CSIRO and AGO 2002). In more temperate regions such as Launceston, Tasmania, average annual maximum temperature was 16.9°C in 2002, expected to increase to 17.8 ± 0.6°C by 2030 and 19.5 ± 1.8°C by 2070 (CSIRO and AGO 2002). Minimum temperatures are also increasing. Melbourne's inner city weather station recorded an average of 12.2 frost days per year from 1856 to 1950. Between 1951 and 2000 this had decreased to 3.1 days, with only five frost days recorded between

1990 and 2000 (Parris and Hazell 2005). This rise in winter minimum temperatures has been one of the contributing factors in the establishment of a new grey-headed flying-fox colony in Melbourne during the early 1980s (Parris and Hazell 2005; van der Ree *et al.* 2006) and has probably affected the current distribution of other indigenous and non-indigenous organisms in urban areas. It has been suggested that plans to limit urban sprawl and increase density, such as *Melbourne 2030*, will increase the urban heat-island effect (Coutts *et al.* 2004); however, some studies suggest that low-density development has a greater impact on urban climate (Stone and Rodgers 2001). Techniques for ameliorating the urban heat-island effect include retaining and replanting vegetation, 'greenroofs' (vegetation on rooftops), and the use of reflective or light-coloured surfaces (Sailor 1998; Shufro 2005; Synnefa *et al.* 2007).

Light pollution

Artificial lights are often associated with civic pride and identity (see, for example, City of Melbourne 2002; City of Sydney 2005) and are used in advertising. This has led to an addition of decorative and feature illumination in cities, in addition to functional lighting. Excess lighting leads to pollution including sky glow, light trespass and glare (Rich and Longcore 2006). Although the effect of light pollution on astronomic observations is relatively well known (Riegel 1973), the impact of light pollution from cities and towns has only recently be identified as a major threat to biodiversity (Eisenbeis 2002 and *pers. comm.*). Night lighting can alter organism survival, mating and breeding success, competitive interactions and landscape connectivity. Many native Australian animals are nocturnal and lights change their environment considerably. Diurnal and other species may also have changed rest and activity periods, causing tiredness and reduced survival. Plant physiology may also be impacted (Briggs 2006).

Night lights are especially threatening to populations of night flying insects, such as moths, beetles and flies. These organisms are attracted to lights at night and it has been estimated that this results in the death of one-third of attracted insects (Eisenbeis 2006). Studies of moth populations in Germany along a gradient of urbanisation found that in the city centre there were only 120 species of moth while at the edge of town there were 650 species (Reichholf 1989). The direct loss of insects in urban environments due to excessive lighting could also have cascading effects within the ecosystem, which may result in the loss of other species that feed on these organisms. Other well documented examples of light attraction include disorientation of sea turtle hatchlings (Salmon 2006) and migrating birds (Gauthreaux and Belser 2006). Many nocturnal species avoid lit areas, which therefore reduces the amount of habitat available to them and restricts their movement, foraging and mating opportunities. For example, many mammal species will restrict their foraging under artificial light or moonlight, which is thought to be a response to increased predation risk (Beier 2006). Lights can delay or disrupt migration in some fish species (Nightingale *et al.* 2006). Additional impacts of light on wildlife may include permanent or temporary vision loss, and changes to breeding and other behaviour. Further information on the impacts of night lighting on organisms can be found in Rich and Longcore (2006).

There is a growing concern by astronomers, biologists, planners and policy makers about the effect of light pollution on a range of issues and several interest groups have been formed to reduce the amount of light pollution in cities and towns. Appropriate power rating and lighting design can help to reduce light pollution. This involves maintaining sufficient light, while limiting excess. Shielding lights can reduce glare and focus light more efficiently downwards, reducing waste and sky glow (Gent 2007). The use of commercial and feature lighting could also be reduced—possibilities include reduction of upward pointing floodlights, and lighting curfews. Homes and offices can reduce light pollution using presence sensor switches and opaque blinds (Clark 2000). Finally, education about the issue of light pollution is likely to be important for encouraging its reduction. The practice of leaving lights on in office buildings after hours remains widespread and is one example where light pollution could be reduced through simple behavioural change (Clark 2000). For additional information on how to reduce light pollution in your area visit the International Dark-Sky Association website (http://www.darksky.org).

Noise pollution

The large population of cities causes a high concentration of human-generated noise. This comes from many sources, including transport (such as aircraft and traffic), industry (such as construction sites and manufacturing) and residential noise (such as barking dogs, appliances and entertainment systems). There is widespread recognition of the impact of noise pollution on human health and wellbeing (Goines and Hagler 2007). The effects of excess noise on wildlife are less well recognised or studied. Like light pollution, noise represents a potential disturbance to organisms, and may alter physiology or behaviour (Maschke *et al.* 2000; Ditchkoff *et al.* 2006). Animals frequently use noise for communication, but background noise can have a masking effect, thereby potentially disrupting behaviours including finding mates, feeding offspring and alerting others to the presence of predators (Warren *et al.* 2006). There have been few studies of the impact of anthropogenic noise on wildlife, and the effect of noise has rarely been separated from other disturbances, such as pollution or the presence of roads (Stone 2000).

Some bird species in urban areas are known to call at higher minimum frequencies, or make shorter and faster calls, than the same species away from the city, thus avoiding overlap with low-frequency anthropogenic noise (Slabbekoorn and den Boer-Visser 2006). It has been suggested that this behavioural adaptation allows some species to survive in urban areas, while other species are unable to adapt to the acoustic environment (Warren *et al.* 2006). It may require the expenditure of too much energy for some organisms to overcome background sound interference in urban areas. Also, there may be anatomical and physiological limitations to calls that would require evolutionary change to adapt to anthropogenic noise levels—this would require much longer timescales than merely changing behaviour (Katti and Warren 2004). Studies of the impact of roads on birds and frogs have found lower abundance and species richness close to roads (Warren *et al.* 2006). There is some evidence that traffic noise could be a factor. Some bird species will forage close to roads, but not breed there—possibly due to difficulties in parents hearing their chick's begging calls over the traffic noise (Forman *et al.* 2002). Rheindt (2003) found that birds with higher frequency songs had higher abundance near roads. Currently, strategies for minimising noise pollution in cities are overwhelmingly focused around human wellbeing, such as concern about urban planning and the proximity of residential areas to major roads and flight paths (see, for example, Environmental Protection Agency 2005). Further study and recognition of the potential impacts of noise on wildlife may be important to preserving biodiversity in urban areas.

Loss of native habitats

Between 1986 and 1993, about 1600 km^2 of land was converted to suburban land use around Australia's five capital cities (Young 2000). The lost of habitat during the creation of cities and towns is one of the greatest threats to local indigenous biodiversity in urbanising landscapes. As the population of Australian cities grows and urban areas expand, the processes of land clearing and habitat fragmentation continue. Development at the urban fringe is normally in the form of low-density sprawl. By world standards, Australian cities are highly suburbanised and characterised by low density and high dependence on motor vehicles (Forster 2006). Recent development plans for Australia's major cities have focused on limiting sprawl, through urban growth boundaries, increased urban density and development of activity centres (Forster 2006). These plans include *City of Cities* for Sydney (NSW Department of Planning 2005), *Melbourne 2030* (State of Victoria 2002), *Network City* for Perth (State of Western Australia 2004), *Southeast Queensland Regional Plan* 2005–2026 (State of Queensland 2005) and *Planning Strategy for Metropolitan Adelaide* (Building South Australia 2006). There has been some criticism of these strategies. For example, *Melbourne 2030* promotes increased urban density and limited urban expansion through a designated Urban Growth Boundary (UGB) and the use of 'green wedges' (State of Victoria 2002). This plan has been criticised for failing to assess the biodiversity or sustainability of green wedges, or the biodiversity values of planned urban growth corridors (Buxton *et al.* 2006). The implementation of the plan occurs locally and has involved predominantly planning and development interests (Buxton *et al.* 2006). There is concern about development eroding the green wedges and extending past the UGB. For example, in 2005 the Victorian Government made an additional 11,100 ha of land available for development—expanding the UGB by 5% and the area of the growth corridors by 34% (DSE 2005; Buxton *et al.* 2006). There is also considerable resistance to the implementation of these strategies, from groups such as Save our Suburbs in Melbourne, which are concerned with preventing inappropriate developments and preserving the existing character of residential areas (Wood 2006). Consolidation of urban areas and identification of growth boundaries has been blamed for rising house prices (Forster 2006). Dealing with these complexities when implementing urban planning strategies is a major challenge, and one that contributes to the continued loss of remnant native habitats.

In addition to the direct impacts of the destruction of indigenous habitat, there are several indirect impacts resulting from the fragmentation of the landscape. Typically, landscape fragmentation results in an increase in edge environments and a decrease in interior habitats. It also increases the distance between similar habitats, which could negatively impact the survival of small remnant populations of plants and animals. In urban and suburban environments, even when entire habitats are not destroyed, components can be significantly altered. For example, because of societal pressure for clean and tidy living environments, homeowners and park managers tend to sanitise natural remnant patches of vegetation usually resulting in the removal of the ground layer of leaf litter and plants, and any dead or dying plants in the area. These practices reduce the availability of nesting and food resources for many animals. For example, the long-nosed bandicoot (*Perameles nasuta*) in remnant vegetation near Sydney, has been shown to use open, grassy areas for foraging, but requires dense scrub for shelter during the day (Chambers and Dickman 2002). This means that maintaining a mosaic of vegetation types is necessary for its survival. Some bird species require understorey vegetation to persist in urban remnants (Sewell and Catterall 1998). Many Australian animals rely on tree hollows for refuge or breeding purposes (Gibbons and Lindenmayer 2002), and these are threatened by the removal of old or dead trees.

The creation of Australian cities has caused a significant loss of the original vegetation. For example, 28% of the Perth metropolitan region is classified as remnant vegetation (Western Australian Museum 2003). In western Sydney, 13% of the original vegetation is present and considered in good condition (NSW Department of Planning 2005). In Adelaide, just 12% (23,047 ha) of native vegetation remains (SA Urban Forest Biodiversity Program 2001). Two-thirds of the original native bushland was cleared in south-east Queensland over the last 150 years (Sewell and Catterall 1998). In the inner city of Melbourne only 1.66% of the area is still covered in remnant vegetation, while in the outer suburbs nearly 16% of the area of remnant vegetation remains (Figure 2). The loss of veg-

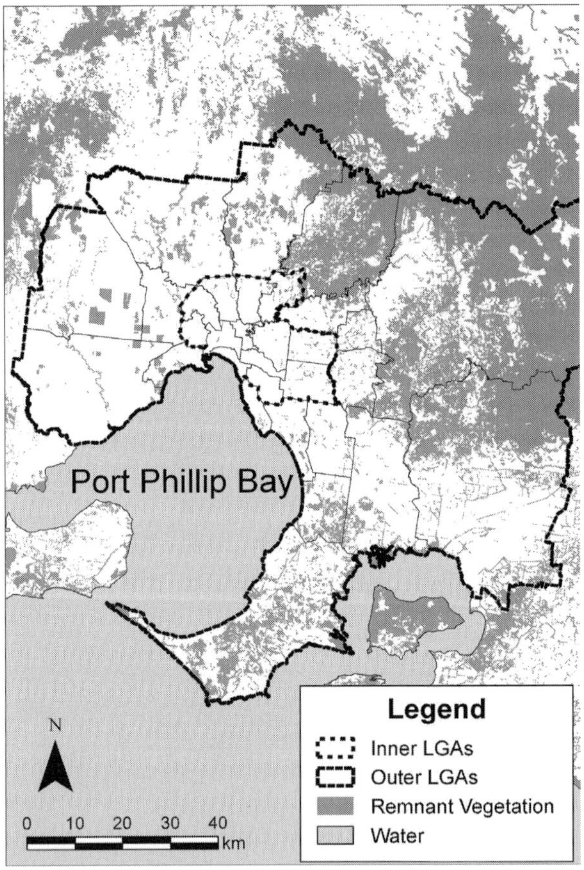

Figure 2: The distribution of remnant vegetation in the greater Melbourne area, Australia. The area within the dotted line is considered inner city; everything outside is referred to as the outer suburbs. Map created by the Australian Research Centre for Urban Ecology and the Victorian Department of Sustainability and Environment.

Table 2: The amount of remnant vegetation remaining in the Greater Melbourne Area separated by inner city and outer suburbs (See Figure 2 for more detail).

Broad vegetation category	Outer suburbs		Inner city	
	No. of patches	Area (ha)	No. of patches	Area (ha)
Coastal	369	2,347	10	10
Forest	7,036	61,897	112	230
Grassland	83	3,949	49	46
Grassy woodland	1,731	7,829	106	239
Heath	1,550	10,516	120	114
Riparian	1,447	6,822	138	257
Wetland	2,145	4,503	104	43
Miscellaneous	182	587	47	41
Total	**14,543**	**98,450**	**686**	**980**

etation in urban areas is a continuing process. Because most of this land is in private ownership, vegetation will be threatened by suburban development as Australian cities grow in the future. In the Perth metropolitan region, 5974 ha of native vegetation was cleared between 1998 and 2004 (Western Australian Government 2007) and half of the vegetation remaining on the Swan Coastal Plain is zoned for future development (Stenhouse 2004b).

A breakdown of the amount of remnant vegetation still present in Melbourne is presented in Table 2. In the outer suburbs patches of forest, heath and riparian vegetation comprise the largest amount of remnant vegetation, while in inner Melbourne forest, grassy woodlands and riparian vegetation are the most abundant. In the inner city, the average size of a remnant patch of vegetation is only 1.43 ha, while in the outer suburbs the size increases to 6.76 ha. Remnant patches in other Australian cities are of similar size (see, for example, SA Urban Forest Biodiversity Program 2001). In a study of the persistence of mammals in Melbourne, van der Ree and McCarthy (2005) found that less than half (26 species) of the 54 species found in the area prior to European settlement were thought to still be present in 2000. Small, ground dwelling mammals were the most significantly affected by urbanisation, with only two of the 15 species originally present still thought to be present.

Loss of native habitat can be minimised during urban planning by reducing low-density sprawl and maintaining green spaces and habitat corridors. However, individual land ownership and local-scale planning make these strategies difficult to implement. The 'matrix' in urban areas can also be enhanced to provide additional habitat and linkages between remnants for a wider variety of species. This includes the use of native plant species in parks, gardens and nature-strips, and retaining leaf litter, dead trees and understorey vegetation. Nest boxes can alleviate the loss of hollow-bearing trees, but are costly long term (Harper 2005; Beyer and Goldingay 2006).

Nitrogen deposition

Due primarily to the burning of fossil fuels, especially through the use of motor vehicles, approximately three times as much nitrogen, in the form of nitrogen oxides and gaseous ammonia, is released into the atmosphere than occurs through natural processes. As a result, NO_2 concentrations are commonly 5–10 times higher in urban than in rural areas (Byntnerowicz *et al.* 1987; Hanson *et al.* 1989). In addition to the increasing human population, the number of vehicles per capita is rising, as well as the number of kilometres travelled, which means that car use is now considered the biggest concern for air pollution in Australian cities (Beeton *et al.* 2006). For the reporting year 2005–2006, it is estimated that 180,000,000 kg (180,000 tonnes) of oxides of nitrogen were released into the atmosphere from the Sydney airshed, 130,000,000 kg from the Melbourne airshed, 94,000,000 kg around Brisbane, 63,000,000 kg from the Perth airshed and

37,000,000 kg from the Adelaide airshed (National Pollution Inventory 2005–2006). In addition to contributing to the creation of smog, some of this nitrogen is transported back to the ground in the form of both wet and dry deposition. In many industrial areas around the world, especially eastern North America and Central Europe, this human-induced input of nitrogen contributes to acidification of the soil and the acidification and eutrophication of lakes and streams. Although the levels of nitrogen deposition in Australia do not result in an acid rain problem, the increased nitrogen in and around cities can result in the fertilisation of the soil as well as lakes and streams. The increased nitrogen inputs can significantly affect many soil–plant and plant–plant interactions (Jeffries and Maron 1997).

Because Australian soils are relatively poor in nutrients, any addition of fertilisers has the potential to increase the competitive advantage of non-indigenous plants over indigenous species (Clements 1983). For example, serrated tussock (*Nassella trichotoma*)—considered a Weed of National Significance—competes more successfully against native grasses in areas with higher soil nitrogen content (Badgery *et al.* 2005). In Melbourne, a positive relationship between soil nitrogen availability and weed invasion has been demonstrated in remnant grassland, woodland and riparian areas (Granger *et al.* 1994; Bidwell *et al.* 2006). Declines in eucalypt remnants have also sometimes been attributed to nitrogen enrichment (in particular, leaves experience higher levels of herbivory due to higher nutrients; Landsberg and Gillieson 1995). Although this has mainly been associated with agricultural areas, continued nitrogen deposition in urban areas may lead to similar dieback. Nutrient release following a fire in Kings Park, Perth, resulted in higher levels of herbivory on eucalypts in burnt area compared with unburnt areas (Radho-Toly *et al.* 2001). Christie and Hochuli (2005) found higher levels of invertebrate herbivory in urban remnants in Sydney compared with continuous forest, possibly due to a combination of nutrient enrichment and loss of insect predators in urban areas.

Non-indigenous species invasions

One of the biggest threats to natural ecosystems around the world is the invasion of non-native species (Adair and Groves 1998). It is an especially acute problem in urban areas due to the abundance of introduced plants and animals. In Australia, it is estimated that there are over 2,800 species of non-native naturalised plants and 81 species of naturalised vertebrates (Department of Environment and Climate Change NSW 2006). In Victoria, some 28% of the flora species are non-indigenous plants, which make up over 65% of the vegetation cover either completely or predominantly. Although the number of naturalised non-native animals is much smaller than the number of introduced plants in Australia, animals such as rabbits, cats, foxes and dogs have had significant negative impacts on native species of plants and animals. Due to the propensity of humans to grow exotic garden plants and to have pets, urban areas will always be exposed to the invasion of non-indigenous species and will typically be the source of invasions into the surrounding landscape. Furthermore, the activities of humans facilitate the spread of these species (Hobbs and Humphries 1995). MacDonald *et al.* (1989) studied several North American nature reserves and found a direct positive correlation between the number of exotic plants present and the number of visitors per year. They suggest that this increase in non-indigenous species is the result of the introduction of non-indigenous plant propagules via the visitors or their vehicles, and an increase in human disturbances resulting in habitat modifications. Both natural and human-induced disturbance of indigenous plant and animal communities can result in reduced competition for the invader as well as increased resource availability, each of which would favour the successful establishment of non-indigenous species.

The introduction and spread of non-indigenous species can be controlled at several levels (Randall *et al.* 2001; The Nature Conservancy 2006). International co-operation and quarantine laws help to limit the import and export of potentially invasive species (AQIS 2007). Business practices to restrict introductions include treatment or control of the release of ballast water from ships (Hillman *et al.* 2004), cleaning shipping containers and packaging (Hallman 2006), and regulation of the exotic pet and plant trades (Nentwig 2006). The replacement of existing non-native species with indigenous and non-invasive garden plants can help limit their introduction to remnant vegetation. Promotion of non-invasive alternatives by nurseries and the media may be important in this process (Randall *et al.* 2001). Approximately 65% of weeds are escaped garden plants, yet many identified weeds continue to be sold by nurseries (Australian State of the Environment 2006). For example, 40% of identified weeds remain available for sale in Queensland, 35% in Western Australia (Groves *et al.* 2005) and 28% in New South Wales (Department of Environment and Climate Change NSW 2006).

Individuals can also limit the spread of non-indigenous species through:

- sensible garden waste disposal—that is, by not dumping in the bush or waterways (Randall *et al.* 2001)
- washing boats or hiking boots when travelling to a new area (UCS 2005)
- responsible pet ownership, such as de-sexing and locking pets up at night, and not releasing unwanted animals or aquarium plants into the wild (Environment Australia 2002).

Conclusion

Maintaining biodiversity in cities and towns is critical to both the ecology of the region and also the health and wellbeing of the human inhabitants. The creation of urban and suburban landscapes typically results in a significant loss of indigenous plant and animal communities. Although many indigenous animals and plants still persist in these human-dominated areas, programs need to be developed to supplement the loss of remnant habitats with the creation of native plantings in the urban matrix. Similarly, land managers and the public must be educated about the need to leave plant litter and branches on the ground, as well as retaining standing dead trees, to provide habitat for many species of mammals and birds. The continued existence of a diversity of organisms in urban and suburban environments is the result of a complex set of ecological, physical and social interactions. In order to maintain biodiversity in cities and towns in the future we must create:

- a solid ecological, social and physical knowledge base on urban environments
- ecologically sound conservation and management practices
- far reaching environmental education programs
- planning laws to ensure genuine attempts at reducing urban sprawl and achievable urban planning visions
- ecologically based policies.

References

Adair RJ and Groves RH (1998). Impact of Environmental Weeds on Biodiversity: a Review and Development of Methodology. Environment Australia, Biodiversity Group, Canberra.

Australian Bureau of Statistics (2004). Population projections, Australia, 2004 to 2101. <http://www.abs.gov.au>.

AQIS (2007). AQIS at a Glance. Australian Government, Australian Quarantine and Inspection Service, Canberra.

Badgery WB, Kemp DR, Michalk DL and King WMCG (2005). Competition for nitrogen between Australian native grasses and the introduced weed *Nassella trichotoma*. *Annals of Botany* **96**, 799–809.

Beeton RJS, Buckley KI, Jones GJ, Morgan D, Reichelt RE and Trewin D (2006 Australian State of the Environment Committee) (2006). 'Australia state of the environment 2006'. Independent report to the Australian Government Minister for the Environment and Heritage, Department of Environment and Heritage, Canberra.

Beier P (2006). Effects of artificial night lighting on terrestrial mammals. In *Ecological Consequences of Artificial Night Lighting.* (Eds C Rich and T Longcore) pp. 19–42. Island Press, Washington DC.

Beyer GL and Goldingay RL (2006). The value of nest boxes in the research and management of Australian hollow-using arboreal marsupials. *Wildlife Research* **33**, 161–174.

Bidwell S, Attiwill PM and Adams MA (2006). Nitrogen availability and weed invasion in a remnant native woodland in urban Melbourne. *Austral Ecology* **31**, 262–270.

Bolund P and Hunhammar S (1999). Ecosystem services in urban areas. *Ecological Economics* **29**, 293–301.

Bradstock RA, Williams JE and Gill AM (2002). *Flammable Australia: the Fire Regimes and Biodiversity of a Continent.* Cambridge University Press, Cambridge, UK.

Briggs WR (2006). Physiology of plant responses to artificial lighting. In *Ecological Consequences of Artificial Night Lighting.* (Eds C Rich and T Longcore) pp. 89–411. Island Press, Washington DC.

Building South Australia (2006). Planning Strategy for Metropolitan Adelaide. Government of South Australia, Adelaide.

Buxton M, Tieman G, Bekessy S, Budge T, Mercer D, Coote M and Morcombe J (2006). *Change and Continuity in Peri-urban Australia. State of the Peri-urban Regions: a Review of the Literature.* RMIT University, Melbourne.

Bytnerowics A, Miller PR, Olszyk DM, Dawson PJ and Fox CA (1987). Gaseous and particulate air pollution in the San Gabriel Mountains of southern California. *Atmospheric Environment* **21**, 1805–1814.

Chambers LK and Dickman CR (2002). Habitat selection of the long-nosed bandicoot, *Perameles nasuta* (Mammalia, Peramelidae), in a patchy urban environment. *Austral Ecology* **27**, 334–342.

Christi FJ and Hochuli DF (2005). Elevated levels of herbivory in urban landscapes: are declines in tree health more than an edge effect? *Ecology and Society* **10**, <http://www.ecologyandsociety.org/vol10/iss1/art10/>.

City of Melbourne (2002). Lighting Strategy. City of Melbourne: Melbourne.

City of Sydney (2005). The City of Sydney Exterior Lighting Strategy. City of Sydney: Sydney.

Clark BAJ (2000). *Outdoor Lighting Principles for Australia in the 21st Century.* Astronomical Society of Victoria Inc. <http://www.gsat.edu.au/astrovic>.

Clements, A (1983). Suburban development and resultant changes in the vegetation of the bushland of the northern Sydney region. *Australian Journal of Ecology* **8**, 307–319.

Coutts AM, Beringer J, Tapper NJ and Cleugh H (2004). Urban planning strategies and climate change: the impact of Melbourne 2030 on local climate. In *Climate and water, proceedings of the 16th Australia New Zealand climate forum.* Lorne. p. 22. ANZCF Conference Committee, Lorne, Victoria.

CSIRO and AGO (2002). *Future Climate Change in Australia.* <http://www.cmar.csiro.au/e-print/open/cechet_2002a.pdf>

Daily GS, Alexander PR, Ehrlich L, Goulder J, Lubchenco PA, Matson H, Mooney A, Postel S, Schneider SH, Tilman D and Woodwell GM (1997). *Ecosystem Services: Benefits Supplied to Human Societies by Natural Ecosystems.* Issues in Ecology No. 2. Ecological Society of America, Washington DC.

Department of Environment and Climate Change NSW (2006). 'New South Wales State of the Environment 2006'. Report of Department of Environment and Conservation, Sydney.

Department of Sustainability and Environment (DSE) (2005). A Plan for Melbourne's Growth Areas. Victorian Department of Sustainability and Environment, Melbourne.

Ditchkoff SS, Saalfeld ST and Gibson CJ (2006). Animal behavior in urban ecosystems: modifications due to human-induced stress. *Urban Ecosystems* **9**, 5–12.

Eisenbeis G (2002). Ecological consequences of light pollution on nocturnal insects. Lecture at Dark Sky: Lucerne, Switzerland. Abstract at <http://www.darksky.ch/ds2002/AbstractsLucerne.pdf>.

Eisenbeis G (2006). Artificial night lighting and insects: attraction of insects to streetlamps in a rural setting in Germany. In *Ecological Consequences of Artificial Night Lighting.* (Eds C Rich and T Longcore) pp. 281–304. Island Press, Washington DC.

Environment Australia (2002) Weeds and Ferals on the Loose! Australian Government, Department of the Environment and Water Resources, Canberra <http://www.environment.gov.au/biodiversity/invasive/publications/loose.html>.

Environmental Protection Agency (2005). Noise Pollution. Queensland Parks and Wildlife Service <http:// www.epa.qld.gov.au>.

Forman RTT, Reineking B and Hersperger AM (2002). Road traffic and nearby grassland bird patterns in a suburbanizing landscape. *Environmental Management* **29**, 782–800.

Forster C (2006). The challenge of change: Australian cities and urban planning in the new millennium. *Geographical Research* **44**, 173–182.

Fox MD and Fox BJ (1986). The effect of fire frequency on the structure and floristic composition of a woodland understorey. *Australian Journal of Ecology* **11**, 77–85.

Freedman B (1989). *Environmental Ecology: the Impacts of Pollution and Other Stresses on Ecosystem Structure and Function*. Academic Press: San Diego, California.

Gauthreaux SA and Belser CG (2006). Effects of artificial night lighting on migrating birds. In *Ecological Consequences of Artificial Night Lighting*. (Eds C Rich and T Longcore) pp. 67–93. Island Press, Washington DC.

Gent, B. (2007). Simple steps to save the night sky. *Astronomy* **35**, 72–73.

Gibbons P and Lindenmayer D (2002). *Tree Hollows and Wildlife Conservation in Australia*. CSIRO Publishing, Melbourne.

Gill AM and Williams JE (1996). Fire regimes and biodiversity: The effects of fragmentation of southeastern Australian eucalypt forests by urbanisation, agriculture and pine plantations. *Forest Ecology and Management* **85**, 261–278.

Goines L and Hagler L (2007). Noise pollution: a modern plague. *Southern Medical Journal* **100**, 287–294.

Graedel TE and Crutzen PJ (1989). The changing atmosphere. *Scientific American* **261**, 58–68.

Granger L, Kasel S and Adams MA (1994). Tree decline in southeastern Australia: nitrate reductase activity and indications of unbalanced nutrition in *Eucalyptus ovata* (Labill.). *Oecologia* **98**, 221–228.

Groves RH, Boden R and Lonsdal WM (2005). 'Jumping the garden fence: invasive garden plants in Australia and their environmental and agricultural impacts'. CSIRO report for World Wildlife Fund Australia, Sydney.

Hallman GJ (2006). Phytosanitary measures to prevent the introduction of invasive species. In *Ecological Studies 193: Biological Invasions*. (Ed. W Nentwig) pp. 367–384. Springer, New York.

Hanson PJ, Rott K, Taylor GE, Gunderson C, Lindberg SE and Ross-Todd BM (1989). NO deposition to elements representative of a forest landscape. *Atmospheric Environment* **23**, 1783–1794.

Harper MJ, McCarthy MA and van der Ree R (2005). The use of nest boxes in urban natural vegetation remnants by vertebrate fauna. *Wildlife Research* **32**, 509–516.

Harvey A (1982). *The Melbourne Book*. Hutchinson Group Pty Ltd, Melbourne.

Hillman S, Hoedt F and Schneider P (2004). 'The Australian pilot project for the treatment of ships' ballast water'. Final report for the Australian Government Department of the Environment and Heritage. National Heritage Trust, Canberra.

Hobbs RJ and Humphries SE (1995). An integrated approach to the ecology and management of plant invasions. *Conservation Biology* **9**, 761–770.

Hughes L (2003). Climate change and Australia: trends, projections and impacts. *Austral Ecology* **28**, 423–443.

Jeffries RL and Maron JL (1997). The embarrassment of riches: atmospheric deposition of nitrogen and community and ecosystem processes. *Trends in Ecology and Evolution* **12**, 74–78.

Katti M and Warren PS (2004). Tits, noise and urban bioacoustics. *Trends in Ecology and Evolution* **19**, 109–110.

Kellert SR and Wilson EO (1995). *The Biophilia Hypothesis*. Island Press, Washington DC.

Landsberg HE (1981). *The Urban Climate*. Academic Press, New York

Landsberg J and Gillieson DS (1995). Regional and local variation in insect herbivory, vegetation and soils of eucalypt associations in contrasted landscape positions along a climatic gradient. *Australian Journal of Ecology* **20**, 299–315.

Larsen C (2000) Population Growth: A Moving Story. Property Council of Australia <http://www.propertyoz.com.au>.

Lunt ID and Morgan JW (1999). Vegetation changes after 10 years of grazing exclusion and intermittent burning in a *Themeda triandra* (Poaceae) grassland reserve in south-eastern Australia. *Australian Journal of Botany* **47**, 537–552.

Lunt ID and Morgan JW (2002). The role of fire regimes in temperate lowland grasslands of south-eastern Australia. In: *Flammable Australia: The fire Regimes and Biodiversity of a Continent*. (Eds RA Bradstock, JE Williams and AM Gill) pp. 177–196. Cambridge University Press, Cambridge, UK.

MacDonald IAW, Loope LL, Usher MB and Hamann O (1989). Wildlife conservation and the invasion of nature reserves by introduced species: a global perspective. In *Biological Invasions: A Global Perspective*. (Eds JA Drake, HA Mooney, F di Castri, H Groves FJ Kruger, M Rejmanek and W Williamson) pp. 215–256. John Wiley and Sons, New York.

Maschke C, Rupp T and Hecht K (2000). The influence of stressors on biochemical reactions – a review of present scientific findings with noise. *International Journal of Hygiene and Environmental Health* **203**, 45–53.

McDonnell MJ (2005). The challenge of preserving biodiversity in urban and suburban environments — the Melbourne example. In *Greening the City: Bringing Biodiversity Back into the Urban Environment*. (Ed. MI Dawson) The Royal New Zealand Institute of Horticulture, Christchurch.

McDonnell MJ (2007). Restoring and managing biodiversity in an urbanizing world filled with tensions. *Ecological Managment and Restoration* **8**, 83–84.

McDonnell MJ, Pickett STA, Groffman P, Bohlen P, Pouyat RV, Zipperer WC, Parmelee RW, Carreiro MM and Medley K (1997). Ecosystem processes along an urban-to-rural gradient. *Urban Ecosystems* **1**, 21–36.

National Pollution Inventory (2005–2006) NPI Location Report. Australian Government, Department of Environment and Water Resources, <http://www.npi.gov.au/index.html>.

The Nature Conservancy (2006). *Invasive Species: Preventing and Containing the Global Spread of Invaders*. The Nature Conservancy: Arlington, Virginia. <http://www.nature.org/initiatives/invasivespecies/>.

Nentwig W (2006). Pathways in animal invasions. In *Ecological Studies 193: Biological Invasions*. (Ed. W Nentwig) pp. 11–27. Springer, New York.

Nightingale B, Longcore T and Simenstad CA (2006). Artificial night lighting and fishes. In *Ecological Consequences of Artificial Night Lighting*. (Eds C Rich and T Longcore) pp. 257–276. Island Press, Washington DC.

NSW Department of Planning (2005). City of Cities: A Plan for Sydney's Future. NSW Department of Planning, Sydney.

Oke TR (1995). The heat island of the urban boundary layer: characteristics, causes and effects. In *Wind Climate in Cities*. (Eds JE Cermack, AG Davenport, EJ Plate and DX Viegas) pp. 81–107. Kluwer Academic, Amsterdam, Netherlands.

Oke TR, Spronken-Smith RA, Jauregui E and Grimmond CSB (1999). The energy balance of central Mexico City during the dry season. *Atmospheric Envrionment* **33**, 3919–2930.

Parris KM and Hazell DL (2005). Biotic effects of climate change in urban environments: the case of the grey-headed flying-fox (*Pteropus poliocephalus*) in Melbourne, Australia. *Biological Conservation* **124**, 267–276.

Paul MJ and Meyer J (2001). Streams in the urban landscape. *Annual Review of Ecology and Systematics* **32**, 333–365.

Phelan G (2000). The effects of urbanisation on Leptospermum myrsinoides heathland in the south-eastern suburbs of Melbourne. BSc (Hons) thesis, The University of Melbourne, Australia.

Radho-Toly S, Majer JD and Yates C (2001). Impact of fire on leaf nutrients, arthropod fauna and herbivory of native and exotic eucalypts in Kings Park, Perth, Western Australia. *Austral Ecology* **26**, 500–506.

Randall JM, Raven PD, Raven PH, Reichard S and White PS (2001). Linking Ecology and Horticulture to Prevent Plant Invasions: Proceedings of the Workshop at the Missouri Botanical Garden (Ed. K Fay). Missouri Botanical Garden, St Louis, Missouri.

Reichholf J (1989). Siedlungsraum – Zur Ökologie von Dorf, Stadt und Straße. Steinbachs Biotopführer. Mosaik, München.

Rheindt FE (2003). The impact of roads on birds: does song frequency play a role in determining susceptibility to noise pollution? *Journal für Ornithologie* **144**, 295–306.

Pickett ST and White PS (1985). *The Ecology of Natural Disturbance and Patch Dynamics.* Academic Press, New York.

Rich C and Longcore T (2006). *Ecological Consequences of Artificial Night Lighting.* Island Press, Washington DC.

Riegel KW (1973). Light Pollution – Outdoor lighting is a growing threat to astronomy. *Science* **179**, 1285–1291.

SA Urban Forest Biodiversity Program (2001).'Conserving Adelaide's Biodiversity: resources'. UFBP, Adelaide, <http://www.urbanforest.on.net/resources.htm>.

Sailor DJ (1998). Simulations of annual degree day impacts of urban vegetative augmentation. *Atmospheric Environment* **32**, 43–52.

Salmon M (2006). Protecting sea turtles from artificial night lighting at Florida's oceanic beaches. In *Ecological Consequences of Artificial Night Lighting.* (Eds C Rich and T Longcore) pp. 141–168. Island Press, Washington DC.

Scott E (1916). *A Short History of Australia.* Oxford University Press, Melbourne.

Scott R, Blake N, Campbell J, Evans D and Williams N (2002). *Indigenous Plants of the Sandbelt: a Gardening Guide for South-eastern Melbourne.* Earthcare St Kilda, Melbourne.

Sewell SR and Catterall CP (1998). Bushland modification and styles of urban development: their effects on birds in south-east Queensland. *Wildlife Research* **25**, 41–63.

Shufro C (2005). Living roofs: green miracles that also cool buildings. E Magazine: the environmental magazine. Earth Action Network, Inc. <http://www.emagazine.com/>.

Slabbekoorn H and den Boer-Visser A (2006). Cities change the songs of birds. *Current Biology* **16**, 2326–2331.

State of Queensland (2005). South East Queensland Regional Plan 2005–2026. Department of Local Government, Planning, Sport and Recreation, Brisbane.

State of Victoria (2002). Melbourne 2030: Planning for Sustainable Growth. Department of Infrastructure, Melbourne.

State of Western Australia (2004). Network City: Community Planning Strategy for Perth and Peel. WA Planning Commission, Perth.

Stenhouse RN (2004a). Fragmentation and internal disturbance of native vegetation reserves in the Perth metropolitan area, Western Australia. *Landscape and Urban Planning* **68**, 389–401.

Stenhouse RN (2004b). Local government conservation and management of native vegetation in urban Australia. *Environmental Management* **34**, 209–222.

Stone E (2000). Separating the noise from the noise: a finding in support of the "niche hypothesis" that birds are influenced by human induced noise in natural habitats. *Anthrozoos* **13**, 225–231.

Stone B, Rodgers MO (2001). Urban form and thermal efficiency: how the design of cities influences the urban heat island effect. Journal of the *American Planning Association* **67**, 186–198.

Syneffa A, Santamouris M, Apostolakis K (2007). On the development, optical properties and thermal performance of cool colored coatings for the urban environment. *Solar Energy* **81**, 488–497.

UCS (2005). What You Can Do... To Prevent Species Invasions. Union of Concerned Scientists, <http://www.ucsusa.org/>.

van der Ree R and McCarthy MA (2005). Inferring persistence of indigenous mammals in response to urbanisation. *Animal Conservation* **8**, 309–319.

van der Ree R, McDonnell MJ, Temby I, Nelson J and Whittingham E (2006). The establishment and dynamics of a recently established urban camp of flying foxes (*Pteropus poliocephalus*) outside their geographic range. *Journal of Zoology* **268**, 177–185.

Warren PS, Katti M, Ermann M and Brazel A (2006). Urban bioacoustics: it's not just noise. *Animal Behavour* **71**, 491–502.

Western Australian Government (2007). 'State of the environment report, Western Australia 2007'. Environmental Protection Authority, Government of Western Australia, Perth. <http://www.soe.wa.gov.au/>.

Western Australian Museum (2003). Welcome to Western Australia: Land and People. Western Australian Museum, Perth, <http://www.museum.wa.gov.au/exhibitions/online/walap/start.html>.

Williams J, Read C, Norton A, Dovers S, Burgman M, Proctor W and Anderson H (2001). Biodiversity. CSIRO Publishing on behalf of the Department of Environment and Heritage, Canberra.

Wilson EO (1984). *Biophilia*. Harvard University Press, Cambridge, Massachusetts.

Wood I (2006). The Implementation and Performance of Melbourne 2030. Save our Suburbs, Victoria, <http://www.saveoursuburbs.org.au/news/M2030Critique.htm>.

World Resources Institute (1996). *World Resources. A Guide to the Global Environment: the Urban Environment.* Oxford University Press, Oxford.

Young A (2000). *Environmental Change in Australia Since 1788.* Oxford University Press, Melbourne.

17

Biosecurity

David Cook, Dave Henry, Andy Sheppard and Mark Lonsdale

How much should we spend on biosecurity (defined here as activities undertaken to protect the environment, the economy and human wellbeing from the impacts of pests, weeds and diseases)? Despite wide recognition of the environmental, economic and social threats and impacts caused by invasive alien species (IAS) (Pimentel *et al.* 2002), there are troubling gaps in biosecurity systems worldwide (Waage *et al.* 2005). This is of particular concern now to all human communities that are well connected to the rest of the world. Increased trade brings with it increased risks of invasive species by creating new entry pathways. Moreover, the relative risk of these pathways is continually changing due to external forces such as advances in transport technologies, the evolution of institutional structures governing domestic and international trade and indirect human-induced factors, such as environmental change.

In planning risk-management strategies to counter the entry and spread of IAS a 'triple-bottom-line' assessment process is needed in which the economic implications of biological invasions are considered, together with environmental and social consequences. Currently, IAS threat assessment typically combines epidemiological models with hypothetical entry scenarios, and estimated costs of eradication within these scenarios (see, for example, Wittwer *et al.* (2005), Bhati and Rees (1996)). To reflect the true diversity of impacts within a region, such assessments must be supplemented with indicators of ecological and socio-economic impact. Such impact assessments have proven difficult enough for economic systems, but eliciting the willingness of societies to pay to protect environmental and social systems is a substantially greater challenge.

In this chapter we explore how biosecurity systems must evolve to build resilience in design and adaptability in response to the increasing threats of IAS to an Australian society under the influence of global change, using examples of biosecurity threats and consequences around economic, environmental and social wellbeing.

What is a biosecurity system?

No physical transfer of goods is without risk, so the social and economic benefits of inter-regional trade are accompanied by potential costs. In an economic context, IAS incursions are termed 'externalities'. This is a particular form of market failure where the full cost of an activity is not incurred by those undertaking it. The potential costs of IAS are not reflected in the price of potentially contaminated imported commodities because importers are not liable for damages incurred, and so make no attempt to recover these costs from consumers. All values are given in Australian dollars unless otherwise stated.

Damage resulting from incursions may affect associated agriculture, but may also include non-market costs to the environment and society. This means the unintended 'victims' of incursions can vary from small through to large business groups, and the threat of IAS potentially involves every member of society.

Once an incursion has occurred, externalities are felt in both market and non-market arenas. *Market effects* refer to impacts on goods and services with known values expressed via a price mechanism. These may be direct or indirect. Direct effects generally result from damage to, or loss of, the commodity or the affected industry, while indirect effects include such things as loss of 'pest-free area' status and consequent loss of export sales. Export bans may extend beyond the area of infection to all regions regardless of the location of the infected site (Wittwer *et al.* 2005), depending on the likelihood of containment.

Non-market effects of IAS incursions include negative environmental, social, cultural or socio-political consequences. These present more difficult analytical challenges where there are no formal prices available to use as a measure of the value of an impact. Environmental consequences may include both annual 'use' values relating to an affected environmental amenity, a rural community or cultural activity and/or 'non-use' values, such as existence, bequest or moral values transferred between generations (Mumford 2001). Also, while market-oriented agricultural processing problems tend to have economic impacts that increase in an obvious fashion (incrementally or exponentially) within production systems, environmental damage may go unnoticed for a long period of time before emerging as an issue (for example, via the media), by which time the damage may be irreversible.

Given the extent and complexity of potential IAS impacts, the notion of the *quarantine continuum* is paramount in understanding the functions of an effective biosecurity system (Nairn *et al.* 1996). Biosecurity policy and systems aim to coordinate activities across a continuum of pre-border, border and post-border quarantine activities to maximise the expected net social benefit of preventing major consequences of breaches on communities, industries, amenities or native flora and fauna. This does not involve the notion of 'zero risk'. No quarantine system in the world can realistically guarantee no-breach barriers with certainty. Instead, the purpose of a biosecurity system is 'risk management', which involves the identification, documentation and implementation of measures to abate the risks of pest and disease entry, establishment and impact.

The task of biosecurity policy making is a difficult one. Notionally, it attempts to ensure a maximum surplus of benefits over costs for any proposed activity (that is, net benefit). Incursion risks decline within increased biosecurity investment up to a point termed the 'optimal level of risk', above which there are diminishing risk mitigation returns on increased investment. Indeed, incursion risk can increase by encouraging illegal activities when mitigation activities become too restrictive or costly (DPIE 1988). Policy makers refer to this optimal level of risk the Appropriate Level of Protection (ALOP), which is defined as: 'where the politically acceptable/defensible benefits produced by any increase in quarantine effort will be insufficient to offset the increased costs'.

Australia's ALOP has been described as 'very conservative' (Kahn *et al.* 1999; Tanner 2001). There is therefore some amount of flexibility in terms of the measures that may be imposed on imported products before they are declared safe to enter the country by Biosecurity Australia (BA). As a Division of the Australian Government's Department of Agriculture, Fisheries and Forestry (DAFF), BA conducts Import Risk Assessments (IRAs) in response to applications from overseas countries wanting to export products to Australia. It then recommends suitable entry conditions in accordance with the World Trade Organization's Agreement on the Application of Sanitary and Phytosanitary Measures, or SPS Agreement (GATT 1994). Once these conditions have been established, a separate DAFF Division—the Australian Quarantine and Inspection Service—has the task of ensuring these import conditions are met.

In setting conditions of product entry, BA refers to relevant standards set by the Codex Alimentarius Commission, the Office International des Epizooties (OIE, the World Organisation for Animal Health) and the International Plant Protection Convention. However, the SPS Agreement allows these international standards to be exceeded where BA's IRAs indicate they are necessary to reduce the risk of invasive alien species incursion to a level consistent with the ALOP. This is not a bilaterally negotiated process and can lead to a loss of shared knowledge and decision-making between trading partners Langlois and Foss (1999). Where information about the probability of invasive species contamination does not exist, there is certainly a need for caution, but when it does, adaptability is a desirable property of quarantine systems.

This is an area where Australia's biosecurity system continues to evolve. It is increasingly recognised that complex, inconsistent IAS problems require an integrated risk-management approach that incorporates interdisciplinary interactions, stakeholder groups and even generations in decision-making processes (Costanza *et al.* 1998). Structures are required that draw together critical parties, relevant experts, authorities and organised interests into a specialised negotiating framework designed to elicit mutually advantageous trade agreements that consider the welfare of trading partners equally alongside that of the domestic economy (Scholz and Stiftel 2005). The emerging field of 'adaptive governance' accepts the need for a collaborative experimental approach acknowledging uncertainty (Costanza *et al.* 1998), in a context where the benefits of a biosecurity strategy must outweigh the potential costs.

Economic impacts

The most widely-cited quantitative estimate of global damage from IASs is given as US$1.5 trillion per annum (Pimentel *et al.* 2002). This aggregate figure is not inclusive of non-market effects and is of little use to governments in making decisions in response to individual IAS outbreaks. Nevertheless, when combined with popular books full of anecdotal 'horror stories', such as Bright (1998) and Baskin (2002), aggregate estimates have proved influential to policy makers. In the United States, for example, Executive Order 13112 of 2000 established an inter-ministerial Invasive Species Council, which channelled tens of millions of dollars of new funding per annum into new governance structures and invasive species management.

The most economically significant biosecurity event in recent times was the 2001 foot and mouth disease (FMD) outbreak in the UK. The incursion and subsequent cull of susceptible livestock served as a catalyst for a great deal of biosecurity research around the world, including projections of likely impacts of this disease if it were to be detected in Australia. FMD is an interesting case study in the sense that the indirect market effects of the disease are far greater than the direct effects on livestock. While rates of infection are typically about 80% among cattle and sheep—and 40% among pigs—yield losses are generally low, and mortality rates are between only 1% (cattle and sheep) and 8% (pigs) among mature animals. Most infected animals make a full recovery and build up immunity to the infecting strain as a consequence (Power and Harris 1973; Harvey 2001).

Despite its minor direct impacts, FMD is listed as a priority disease for World Organisation for Animal Health (OIE) members and, as such, its detection in any member country requires the immediate closure of trade in all potentially contaminated livestock products and the extermination of infected and potentially infected animals. In addition to the cost of culling livestock and compensation for farmers, it was estimated to have cost the UK tourism industry in the order of £2.3 billion in 2001 alone (Blake *et al.* 2003).

Adherence to the OIE rules, rather than an exploration of the benefits and costs of all alternative actions during the 2001 UK FMD outbreak, meant that no ceiling was placed on the costs of the response. Perhaps a more adaptive, case-specific approach to the management of livestock diseases and consideration of the net effects of management actions may have led to a less severe economic consequence. For example, had a vaccination scheme against FMD been implemented, rather than an eradication campaign, the negative cost and revenue implications of the disease for the UK agricultural sector—including export restrictions, estimated to have been in the order of £1.2 billion (Harvey 2001)—would have been diminished. Effects on domestic tourism would have been negligible, as would the effects on domestic consumers of meat and livestock products. In fact, supply disruptions on the UK domestic market due to the foot and mouth disease outbreak were short-lived, and observed price changes were relatively modest. The loss to producers associated with European competition is estimated to have been approximately £50 million (Thompson *et al.* 2002).

In an examination of the possible economic effects of an FMD eradication program in Australia, the Productivity Commission (2002) estimated that the diversion of livestock products from export to domestic markets would decrease prices and create considerable consumer benefits, which offset producer losses. However, they estimate net losses of around $2.5 billion even if eradication were successful. Control costs alone could range from $20 to 420 million, and compensation from $4 to 68 million.

Furthermore, the Australian real Gross Domestic Product would decline by an estimated $8–13 billion over 10 years, which does not include possible effects on the tourism industry as seen in the UK outbreak (due to the lower integration of tourism and agriculture in Australia). When such huge costs are at stake, policy making should perhaps involve the consideration of all management options to ensure the best possible outcome for the home economy, rather than an automatic response dictated by inflexible institutional arrangements.

However, the politics of IAS responses is invariably complicated and difficult to predict when one considers the varied motivations and incentives for rent-seeking behaviour of affected or potentially affected groups. This was amply demonstrated when, in September 1998, a virulent Newcastle Disease (ND) virus emerged on a chicken farm in western Sydney. Further outbreaks were detected on the NSW Central Coast in April 1999 and once again in the Sydney area in early 2000, and an eradication policy ensued. The Mangrove Mountain outbreak was eradicated after the destruction of 1.9 million chickens on 26 farms—as well as hundreds of backyard poultry and aviary birds at an estimated cost of $30 million. A compensation package of $8.5 million was agreed for this regionally-constrained outbreak. However, recovery for the communities was hindered by affected farmer applications for Exceptional Circumstances Financial Assistance being rejected (NSWDOCS 2000). It is interesting to note that, despite almost unlimited supply of funds from government for the initial response, there were fewer funds made available to the community for the recovery phase because the outbreak was not formally declared a 'disaster' under the *State Emergency and Rescue Management Act 1989.* New South Wales Department of Community Services highlighted the impact of the ND outbreak on loss of income (in some cases for more than six months), increased debt levels, lower financial security, decreased land values and a lack of confidence in future investment in the whole region (NSWDOCS 2000). Losses extended across poultry farmers directly affected by the disease, as well as others in the community and local businesses. These flow-on effects are very difficult to quantify.

Flow-on effects are also important in the recent response to an Equine Influenza (EI) incursion in Australia. EI is an acute respiratory disease capable of spreading rapidly through horse populations by direct contact between animals and by the movement of horse-handling equipment. While the number of infected animals can be high, EI is similar to FMD in terms of its direct market impact because the majority of infected horses will make a full recovery after 1 to 3 weeks if well rested. The disease is endemic in continental Europe and North America, although outbreaks tend to be sporadic due to widespread vaccination. Minor epidemics have occurred in other parts of the world, including the United Kingdom, Ireland, South Africa, India, China, Japan, Hong Kong, Singapore, Malaysia and South America.

The horse racing industry is a source of entertainment for millions of Australians, and so the disease has significant potential flow-on effects in addition to direct market effects. Moreover, these flow-on effects are extremely 'visible' to the public because iconic horse racing events such as the Cox Plate, Caulfield Cup and Melbourne Cup are threatened. EI was detected for the first time in Australia in early-August 2007, and by mid-September was known to be present in both NSW and Queensland. On 9th September, the federal government announced a $110 million funding package, which included a range of mechanisms to assist affected groups, including an Equine Workers Hardship Wage Supplement Payment, an Equine Influenza Business Assistance Grant, a Commercial Horse Assistance Payment and Grants For Non-Government, Not-For-Profit Equestrian Organisations (McGauran 2007).

Unlike the ND outbreak in which the Emergency Animal Disease Response Agreement (EADRA) was not in place, EI is listed under this agreement and is subject to response cost sharing (AHA 2002). Cost sharing is defined here as a joint contribution to biosecurity incursion management costs by the affected parties, including the federal government, the states and the relevant industries (see http://www.daffa.gov.au/animal-plant-health/emergency/exercises/minotaur/cost-sharing). The EADRA specifies that in the event of a listed disease incursion one of four cost sharing arrangements for eradication will be activated between industry and the state and federal governments. These four categories are described in Table 1 (AHA

2002), which also lists the diseases to which they are applied under the EADRA. EI is currently classified as a category four disease, indicating a 20% government and 80% applicable industries funding contribution is appropriate should an eradication response to an outbreak be deemed technically and economically feasible (AHA 2002).

Environmental impacts

Cost-sharing arrangements are also in place for plant pest and disease incursions under the Emergency Plant Pest Response Deed (EPPRD), administered by Plant Health Australia (PHA) and detailed in PHA (2001; 2004). However, the development of these arrangements came too late to benefit control efforts for Red Imported Fire Ants (RIFA, *Solenopsis invicta*). The environmental significance of this pest prompted the formation of a separate cost sharing arrangement between the federal and state governments negotiated under the *Environment Protection and Biodiversity Conservation Act 1999* (EPBC Act). Fire ants were discovered in Queensland in February 2001, and are currently restricted to areas around the cities of Brisbane, Ipswich and Logan, and a small portion of Redlands Shire (Commonwealth of Australia 2006). These ants are perhaps the most notorious of a group of ants—referred to as 'tramp ants'—that have been targeted for various control campaigns across Australia. They are known to have severe ecological and social impacts due to their aggressive defensive and predatory behaviour, multiple colony structure and painful sting. In 2002 it was decided that RIFA be subjected to an eradication campaign under the EPBC Act. Now in progress, the campaign is expected to cost $175 million over five years.

As is often the case with environmental IAS, it is difficult to quantitatively determine if this investment in fire ant eradication is appropriate, as many of the benefits generated by control/eradication are intangible. Identifying and capturing non-market values associated with the environment remains a sizeable challenge for economists (Adamowicz 2004), particularly when the benefits of avoiding a given amount of environmental impact may increase logarithmically through time (Waage *et al.* 2005). Precise benefit estimates are difficult to quantify, as there is a tendency for a strategic misrepresentation of level of utility derived from the environment, and demands for specific environmental assets are often embedded within much broader environmental issues (Kristrom and Riera 1996; Whitby 2000). Nevertheless, environmental threats from IAS are recognised under the Convention on Biological Diversity (1991) article 8h, in which it is concluded such threats are 'immense, insidious, increasing and irreversible' (Sandlund *et al.* 1996).

The major IAS group causing environmental impacts are invasive plants, which are considered as second only to land clearing in their harmful impacts on Australia's native flora and fauna. Of the estimated 27,600 plant species introduced (94% for the garden industry) 2,900 have naturalised in Australia 66%, of which escaped from parks and gardens (Groves *et al.* 2005); J. R. Hosking *pers. comm.*). At least 10 new naturalisations are reported each year (Groves and Hosking 1998). Of the 2,190 species that have become invasive (1,980 in an environmental context), 798 cause major problems and 70% are invasive garden plants (Groves *et al.* 2005). Of the known invasive garden plants, an estimated 54% were still on sale in 2005 (Groves *et al.* 2005). In NSW, weeds pose a threat to 44% of the 945 listed threatened species and 89% of endangered ecological communities (Coutts-Smith and Downey 2006). Of the 127 invasive plants identified as threatening NSW biodiversity, 65% were escaped garden plants. Escaped garden plants represented 93% of the species for which a specified invasive plant was identified as the threat. Plant invasions threaten two-thirds of Australia's World Heritage, National Heritage and Ramsar wetland sites and 45% of the threatened species and 71% of the threatened ecosystems listed under the *1999 Environment Protection and Biodiversity Conservation Act* (J.K. Scott *pers. comm.*). Top of the list were lantana, bitou bush, blackberry and kikuyu grass. An estimated 20% of Queensland is affected by rubbervine—an escaped garden plant (Groves *et al.* 2005). Australian gardeners exploits are clearly a primary source of risk to native biodiversity.

While the direct costs to biodiversity can be measured in terms of impacts on threatened species and communities, the indirect costs are more

Table 1: Disease cost-share categorisation in the Government and Livestock Industry Cost Sharing Deed with respect to an emergency animal disease response.

	Description	Diseases
Category 1	*Funded 100% by government* Those diseases that predominantly/ seriously affect human health and/or the environment (depletion of native fauna) but may only have minimal direct consequences to the livestock industries.	rabies Australian lyssaviruses (including bat lyssavirus) Japanese encephalitis Western, Eastern and Venezuelan equine encephalomyelitis Nipah virus
Category 2	*Funded 80% by government and 20% by the applicable industry(s)* Those diseases that have the potential to cause major national socioeconomic consequences through very serious international trade losses, national market disruptions and very severe production losses in the livestock industries that are involved. This category includes diseases that may have slightly lower national socioeconomic consequences, but also have significant public health and/or environmental consequences.	bovine spongiform encephalopathy brucellosis (due to *Brucella abortus*) brucellosis (due to *Brucella melitensis*) Hendra virus foot and mouth disease glanders peste des petits ruminants Rift Valley fever rinderpest screw worm fly sheep pox Tracheal mite Tropilaelaps mite Varroa mite (*Varroa destructor*) – see also Varroa mite Category 4 vesicular stomatitis
Category 3 Moderate Public Benefits	*Funded 50% by government and 50% by the applicable industry(s)* Those diseases that are of moderate public impact that have the potential to cause significant (but generally moderate) national socioeconomic consequences through international trade losses, market disruptions involving two or more states and severe production losses to affected industries, but have minimal or no effect on human health or the environment.	African horse sickness African swine fever anthrax (major outbreaks) avian influenza (highly pathogenic) bluetongue (disease in sheep) bovine tuberculosis due to Mycobacterium bovis contagious bovine pleuropneumonia encephalitides (tick-borne) lumpy skin disease Menangle virus (porcine paramyxovirus) Newcastle disease scrapie Small hive beetle swine vesicular disease trichinellosis vesicular exanthema

	Description	Diseases
Category 4 Predominantly Private Benefits	*Funded 20% by government and 80% by the applicable industry(s)* Those diseases that could be classified as being mainly production loss diseases. While there may be international trade losses and local market disruptions, these would not be of a magnitude that would be expected to significantly affect the national economy. The main beneficiaries of the successful emergency response to an outbreak of such a disease would be the affected livestock industry(s).	Aujeszky's disease Borna disease Braula fly (except in Tasmania) contagious equine metritis dourine east coast fever epizootic lymphangitis equine babesiosis equine encephalosis equine influenza Getah virus haemorrhagic septicaemia heartwater infectious bursal disease (hypervirulent form) Jembrana disease Maedi/visna Nairobi sheep disease porcine Reproductive and Respiratory Syndrome (PRRS) Potomac fever pulmonary adenomatosis sheep scab surra swine influenza Teschen disease Varroa mite (*Varroa jacobsoni*) – see also Varroa mite Category 2 transmissible gastroenteritis Wesselsbron disease

difficult to estimate as they reflect environmental, rural community or cultural amenity values and the moral values of biodiversity bequests transferred between generations. One way to estimate the costs of environmental weeds has been the contingent valuation method, whereby respondents are asked to express their willingness to pay for environmental quality or to accept payment in exchange for a loss of environmental quality. Using this approach Queensland Department of Natural Resources and Mines estimated the cost of environmental weeds in Queensland to be at least $67.2 million per annum (QDNRM 2003).

As with many environmental valuation methods it is somewhat difficult to interpret these findings when respondents expressed preferences for the consequence of weeds (that is, loss of high conservation value vegetation) rather than apportioning blame to weedy species (and therefore preferences for weed management as the means to solve the problem). Moreover, the willingness of respondents to pay to protect the environment cannot be explained without understanding the sociological elements involved in each person's decision-making process. While respondents may not receive tangible benefits from weed control, they may respond positively to suggestions that they might enjoy the 'warm glow' of contributing to environmental conservation strategy (Kahneman and Knetsch 1992; Cook and Proctor 2007). To further complicate matters, it has been shown that people are much more averse to a loss than they are attracted to an equivalent gain (Coursey *et al.* 1987; Kahneman *et al.* 1990; Knetsch and Sinden 1987), so it very much matters whether a questionnaire asks respondents for their willingness to pay or their willingness to accept. The disparity between the two can be reduced with repeated experimentation, but this makes the process of revealing environmental values extremely costly (Portney 1994).

Introduced plants are also the drivers of Australian primary production. Over more than 70 years, Commonwealth Plant Introductions comprised 145,000 accessions of more than 8,200 species that have transformed Australian landscapes, but left a huge legacy of unwanted plants (Cook and Dias 2007). Lonsdale (1994) estimated that of the 463 tropical pasture species introduced between 1947 and 1985, 21 species have been useful and 60 species have become weeds. One such grass, Olive hymenachne, was federally declared as a desirable pasture species for importation and widespread planting in 1988, but by 1994 it was a Weed of National Significance. Of the useful species, 17 have also become weeds—particularly buffel grass and gamba grass, which change native nitrification rates and fire regimes. Such grasses now pose unsurmountable controversy between their recognised economic values to rural pastoral communities, but equally recognised uncontrolled spread and impacts on wet and dry biodiversity rich grasslands of tropical Australia. While such controversy has shut the border for future introductions of pasture species with invasive characteristics and was a motivation towards developing a National Weed Strategy, nothing is being done about these existing controversial plants, and any solution to equitably balancing economic benefits to pastoralists with environmental biodiversity benefits is still a long way off.

Australia now has strong regulatory mechanisms for preventing the importation of exotic plants based on weed risk assessment and permitted species lists. But analytical problems make it difficult to justify rejection when only a small proportion of those proposed are likely to become invasive. Both the benefits and costs of permitting species entry—and the resultant net benefit or cost to society—are important to consider. This raises the problem of adequate representation of future impacts of invasion, particularly in cases where these impacts cannot be reversed. The absence of these future costs in market access decisions have left us with a legacy of naturalised and naturalising exotic flora as 'sleeper' weeds.

Social impacts

The valuation problem is not only encountered in cases of environmental IAS effects, but also when damage to social and cultural amenity is experienced, or social cohesion and activities are affected. These IAS impacts receive little attention in the literature, but are nonetheless often experienced. For example, the decline in viability of certain agricultural industries will affect the functioning of the rural communities that they support. As with environmental amenity, quantitative analyses of social impacts are difficult to undertake. In the same way that an environmental resource may have a cultural existence value or moral value, so too might a rural community. As such, a majority of society may be willing to pay to preserve it, even if they spend most of their time in urban areas and have little social or economic ties to rural communities (Cook and Proctor 2007). Bennett *et al.* (2004) present evidence to this effect in three different regions of rural Australia. Here the maintenance of rural populations is associated with environmental damage mitigation, so it is difficult to draw conclusions about the willingness of society to pay for the preservation of rural communities *per se*.

Damage to communities resulting from IAS can be direct. In a cost-share strategy that mirrors arrangement for fire ant eradication, the federal and state governments have recently embarked on a European house borer (EHB; *Hylotrupes bajulus*) control program after it was discovered on the outskirts of Perth in January 2004. This insect is a destructive pest of seasoned coniferous timber—including pine, fir and spruce—and is therefore capable of causing structural damage to buildings. Residences with untreated radiata pine, southern pine, Douglas fir, hoop pine or bunya pine frames are particularly at risk. Between 2004 and 2006, the Department of Agriculture and Food Western Australia spent $7.1 million on controlling the insect. A national cost-shared control program commenced in January 2007, with government funding for two years—dependent on industry contributing to the costs by January 2008 and eradication being deemed technically feasible.

Ultimately, the direct impact of disease outbreaks and incursions on agricultural communities, and their flow-on effects from economic and environmental impacts, will have consequences for the wellbeing of people and communities. Although these consequences are potentially the most severe of all the impacts, they remain the most difficult to quantify. At the individual and family level, the social impacts could range from strains on family relationships that are normally associated with adverse events and loss through to severe mental disorders. At the community level, the impacts could range from a breakdown of normal community activities in the midst of quarantine and movement restrictions, to the changes in interpersonal relationships affecting the longer term cohesion of the community (NSWDOCS 2000).

The manner in which authorities approach control can have a significant impact on the stress it creates. Decisions are often made about the livestock and property with little or no consultation with farmers, which creates not only feelings of anxiety and grief, but also hopelessness, anger, frustration and injustice that are often directed at individuals and institutions implementing the control strategy. As noted below, these feelings have a flow-on effect on the emergency personnel themselves.

Farmers and their families can also harbour feelings of a sense of personal failure, isolation, loss of reputation and/or the loss of identity and dignity. These effects could be further exacerbated by households splitting up, children being singled out at school (such as bullying) and cancellation of recreational activities.

When the Mangrove Mountain community was surveyed after the Newcastle Disease outbreak, it identified that the development of divisions in the community was a major issue; and without social cohesion communities become unworkable and fail (ENRC 2000; NSWDOCS 2000; Hood and Seedsman 2004). Some small rural communities already face long-term challenges from, for example, the movement of people to regional centres and cities, labour shortages, climate change and extreme weather events, such as drought. An exotic disease outbreak could accentuate the pressure on communities, such as triggering population threshold effects where rural industries operate on the edge of economic viability and where a further small reduction in population makes the provision of certain services unsustainable.

The Productivity Commission (2002) recognised that both individuals and communities can respond very differently to similar adverse circumstances. This makes preparation difficult, except to say that vigilance and responsiveness tailored to individual community needs is essential to create resilience.

Needless to say, the social impact of disease outbreaks will be lessened if the duration and severity are decreased and 'normal' community function can be restored as quickly as possible. Social impacts would entail costs well beyond the length of the outbreak for the individuals and communities concerned and, indirectly, for governments through the health and welfare systems. Trying to build increased resilience into our communities must be a community-driven process, but with full engagement and investment by government.

A further social cost of IAS incursions that is being increasingly recognised is the stress and adverse psychological reactions of emergency services staff to traumatic events and the nature of the work they are expected to perform (Jenner 2007). In responses to livestock diseases, farmers and their families endure enormous emotional losses made worse by psychological human-animal bonds (Hall *et al.* 2004). Long-term psychological difficulties include anxiety, depression, anger or aggression—and even post-traumatic symptoms similar to those seen in the 'victims' of the event. Post-Traumatic Stress Disorder is prevalent in 1–2% of the general population, but has been reported to be as high as 30% in emergency responders and, if left untreated, can continue for years and severely affect on their quality of life. Their personal stress can create further stress within families, which can then compound the problem. Conversely, not all the outcomes of exposures to emergency incidents are negative. Positive aspects include satisfaction of a job well done, performance enhancement, greater appreciation of life and better relationships with others.

Jenner (2007) proposed a number of preventative measures including training, debriefing, counselling and formal recognition of an individual's participation in managing outbreaks. A community's experience and resilience to an emergency animal or plant disease crisis will be dependent upon its unique geographic, economic and social profile, degree of social cohesion, community leadership and history of overcoming crises (Barclay 2005).

Adaptive biosecurity

The complexity and unpredictability of invasive species risk management calls for a flexible, adaptive system with free-flowing, short information pathways, feedback loops and institutional structures capable of embracing necessary changes. The following sections describe some of these.

Post-border cost sharing

Post-border responses are required when species breach the quarantine barrier and may or may not be discovered before they have spread to the extent that eradication is either technically or economically unfeasible. The provision of cost-effective control benefits from a spirit of cooperation between affected parties, and consequently will be difficult because it requires engagement from diverse parties, ranging from agricultural industries to environmental groups, and community groups within society.

Australia is at the forefront of response policy design, with the notion of cost sharing at the centre of eradication response philosophy with respect to invasive alien species of environmental, social and/or cultural significance. As mentioned previously, the Emergency Animal Disease Response Agreement (EADRA; AHA (2002)) and Emergency Plant Pest Response Deed (EPPRD; PHA (2005)) involve 'high-profile' invasive organisms being placed in one of four categories relating to their significance in terms of potential damage to public resources and private industries. These categories dictate an appropriate split of eradication funding between government and private industry. The size, and hence cost, of an eradication campaign is not in the agreement, reflecting the uncertainty around impacts of an invasion on affected parties.

Monitoring

It is increasingly recognised that modern governance should be dispersed across multiple centres of authority, rather than centralised, particularly where multiplicity of parties and jurisdictions is a system feature. Arrangements in Western Australia may serve as a model for cooperative arrangements between industry members. State legislation enables the formation of 'producer committees' to oversee industry activities including the control of invasive species, such as Mediterranean fruit flies (APC 2005). Activities can be funded through fees for service on producers, wholesalers, exporters and retailers, without precluding the capacity to attract public resources and cost sharing when the impacts and benefits of controlling an invasive alien species are broad. Sourcing directly from commercial beneficiaries of control avoids bureaucratic costs of 'public' institutions and reduces reaction time (King and Pitchford 2001)—a critical factor in eradication success (Olson and Roy 2002). Conceivably, spatial diversity within cooperatives may lead to the formation of various opinions with regard to appropriate uses of collective funds, but to a lesser extent than centralised governance structures. Indeed, this is one of the key advantages to multi-layered governance where heterogeneity presides due to relative flexibility. Cost-shared surveillance systems and 'collective-good' based incentive schemes to detect and report intruders as soon as they are identified maximise the chance of eradication and minimise the stress on the biosecurity system. Such cooperatives may form, disband and re-form as required by the demand for governance.

Incentive alignment

The issue of adequate, but not over-compensation for parties affected by species eradications has long been a concern, particularly when time-until-detection inflates the future costs for all parties (Whiting 2003). Decentralising the governance structure and promoting like-minded producer cooperatives can improve motivation to work towards a common goal or standard as there is a natural inclination to minimise disparities between individual and cooperative objectives and to share the burden of an invasion response if and when required. Such activities may lead to other cost-sharing benefits, such as in research and development, technology adoption, adaptive management strategies and development of threat prioritisation techniques. In fact, for the stability and the development potential of the biosecurity structure, functional redundancy represents a reservoir of adaptive responses and enhances evolutionary potential (Rammel and van den Bergh 2003). Having a range of tools and experience to draw upon when the unforeseen occurs can be seen as paramount in the development of sustainable biosecurity governance.

Conclusions

In answering the question posed at the start of the introduction, we can say that finding a desirable level of investment can only be achieved through a process of trial and error, and learning from resource misallocation. An adaptive approach to regulation of incursion risk is needed for effective management. Similarly 'adaptive governance' is required where the new institutions that evolve must be capable of generating long-term, sustainable policy solutions to highly complex problems through effort-coordination across previously independent systems of users, knowledge, authorities, and interests (Scholz and Stiftel 2005).

Like any system involving human interactions with their environment, the robustness of a biosecurity system emerges from three complementary attributes: resilience, adaptability and transformability (Walker *et al.* 2004). The more resilient a system, the greater is its capacity to absorb or to cope with disequilibria. The more adaptable the system, the better its components are able to influence resilience. In the event that circumstances push a system to the limits of resilience, its transformability characteristics will determine if it is possible to configure an entirely new stability landscape with components of the redundant system supplemented with new elements. System longevity is of paramount importance in the case of biosecurity systems and desirable resilience and adaptability properties will avoid the need for a fundamental system transformation.

References

Adamowicz WL (2004). What's it worth? An examination of historical trends and future directions in environmental valuation. *The Australian Journal of Agricultural and Resource Economics* **48**, 419–443.

AHA (Animal Health Australia) (2002). *Government and Livestock Industry Cost Sharing Deed in Respect of Emergency Animal Disease Response.* Animal Health Australia, Canberra.

APC (Agricultural Produce Commission) (2005). 'Annual Report for the Year Ended 30 June 2005'. Agricultural Produce Commission, South Perth.

Barclay E (2005). *Local Community Preparedness for an Emergency Animal Disease Outbreak.* Australian Government Rural Industries Research and Development Corporation, Canberra.

Baskin Y (2002). *A Plague of Rats and Rubber Vines: The Growing Threat of Species Invasions.* Island Press, London

Bennett JM, van Bueren M and Whitten S (2004). Estimating society's willingness to pay to maintain viable rural communities. *The Australian Journal of Agricultural and Resource Economics* **48,** 487–512.

Bhati UN and Rees C (1996). 'Fireblight: A Cost Analysis of Importing Apples From New Zealand'. Report Prepared for the Australian Quarantine and Inspection Service. Australian Bureau of Agricultural and Resource Economics, Canberra.

Blake A, Sinclair MT and Sugiyarto G (2003). Quantifying the impact of foot and mouth disease on tourism and the UK economy. *Tourism Economics* **9**, 449–465.

Bright C (1998). *Life Out of Bounds: Bioinvasion in a Borderless World.* WW Norton & Co., London.

Commonwealth of Australia (2006). *Background Document for the Threat Abatement Plant to Reduce the Impacts of Tramp Ants on Biodiversity in Australia and its Territories.* Department of the Environment and Heritage, Canberra.

Cook DC and Proctor WL (2007). Assessing the threat of exotic plant pests. *Ecological Economics* **63**, 594–604.

Cook GD and Dias L (2007). It was no accident: deliberate plant introductions by Australian government agencies during the 20th century. *Australian Journal of Botany* **54**, 601–625.

Costanza R, Andrade F, Antunes P, van den Belt M, Boersma D, Boesch DF, Catarino F, Hanna S, Limburg K, Low B, Molitor M, Pereira JG, Rayner S, Santos R, Wilson J and Young M (1998). Principles for Sustainable Governance of the Oceans. *Science* **281**, 198–199.

Coursey DL, Hovis JL and Schulze WD (1987). The disparity between willingness to accept and willingness to pay measures of value. *The Quarterly Journal of Economics* **102**, 679–690.

Coutts-Smith A and Downey P (2006). *Impacts of Weeds on Threatened Biodiversity in New South Wales.* Waite Campus, University of Adelaide.

DPIE (Department of Primary Industries and Energy) (1988) 'Australian quarantine requirements for the future'. Report of the Quarantine Review Committee. Department of Primary Industries and Energy, Canberra

ENRC (Environment and Natural Resources Committee) (2000) 'Control of Ovine Johne's Disease in Victoria'. Inquiry Report. Parliament of Victoria, Melbourne.

GATT (General Agreement on Tariffs and Trade Secretariat) (1994) Agreement on the application of sanitary and phytosanitary measures. In *The Results of the Uruguay Round of Multilateral Trade Negotiations: The Legal Texts* pp. 69–84. General Agreement on Tariffs and Trade Secretariat, Geneva

Groves RH, Boden R and Lonsdale WM (2005). 'Jumping the Garden Fence: Invasive garden plants in Australia and their environmental and agricultural impacts'. CSIRO report for World Wildlife Fund Australia, Sydney.

Groves RH and Hosking JR (1998). *Recent incursions of Weeds to Australia 1971–1995.* Waite Campus, University of Adelaide.

Hall MJ, Ng A, Ursano RJ, Holloway H, Fullerton C and Casper J (2004). Psychological impact of the animal-human bond in disaster preparedness and response. *Journal of Psychiatric Practice* **10**, 368–374.

Harvey DR (2001). *What Lessons from Foot And Mouth? A Preliminary Economic Assessment of the 2001 Epidemic.* Department of Agricultural Economics and Food Marketing - University of Newcastle upon Tyne, Newcastle, UK.

Hood B and Seedsman T (2004). Psychosocial investigation of individual and community responses to the experience of ovine Johne's disease in rural Victoria. *Australian Journal of Rural Health* **12**, 54–60.

Jenner M (2007). The psychological impact of responding to agricultural emergencies. *The Australian Journal of Emergency Management* **22**, 25–31.

Kahn SA, Beers PT, Findlay VL, Peebles IR, Durham PJ, Wilson DW and Gerrity SE (1999). *Import Risk Assessment on Non-Viable Salmonids and Non-Salmonid Marine Finfish.* Australian Quarantine and Inspection Service, Canberra.

Kahneman D and Knetsch J (1992). Valuing public goods: the purchase of moral satisfaction. *Journal of Environmental Economics and Management* **22**, 57–70.

Kahneman D, Knetsch JL and Thaler RH (1990). Experimental tests of the endowment effect and the coase theorem. *The Journal of Political Economy* **98**, 1325–1348.

King S and Pitchford R (2001). *Private or Public? A Taxonomy of Optimal Ownership and Management Regimes.* The Australian National University, Canberra.

Knetsch JL and Sinden JA (1987). The persistence of evaluation disparities. *The Quarterly Journal of Economics* **102**, 691–696.

Kristrom B and Riera P (1996). Is the income elasticity of environmental improvement less than one? *Environmental and Resource Economics* **7**, 45–55.

Langlois RN and Foss NJ (1999). Capabilities and governance: the rebirth of production in the theory of economic organization. *Kyklos* **52**, 201-218.

Lonsdale W (1994). Inviting trouble: introduced pasture species in northern Australia. *Australian Journal of Ecology* **19**, 345-354.

McGauran P (2007). Media Release: $110 million Equine Influenza Assistance. Department of Agriculture Fisheries and Forestry, Canberra.

Mumford JD (2001). Environmental risk evaluation in quarantine decision making. In *The Economics of Quarantine and the SPS Agreement.* (Ed. CMK Anderson and D Wilson) pp. 353–383. Centre for International Economic Studies and the Department of Agriculture, Fisheries and Forestry – Australia/Biosecurity Australia, Adelaide.

Nairn ME, Allen PG, Inglis AR and Tanner C (1996). *Australian Quarantine - A Shared Responsibility.* Department of Primary Industries and Energy, Canberra

NSWDOCS (New South Wales Department of Community Services) (2000). *Assessment of the Needs of the Mangrove Mountain Community Following the Outbreak of the Newcastle Chicken Disease in April 1999.* New South Wales Department of Community Services, Sydney.

Olson LJ and Roy S (2002). The economics of controlling a stochastic biological invasion. *American Journal of Agricultural Economics* **84**, 1311–1316.

PHA (Plant Health Australia) (2001). 'Funding and Compensation for Emergency Eradication of Exotic Plant Pests and Diseases: A Discussion Paper'. Plant Health Australia, Canberra.

PHA (2004). *PLANTPLAN: Australian Emergency Plant Pest Response Plan.* Plant Health Australia, Canberra

PHA (2005) *Government and Plant Industry Cost Sharing Deed in Respect of Emergency Plant Pest Responses.* Plant Health Australia, Canberra

Pimentel D, McNair S, Janecka J, Wightman J, Simmonds C, O'Connell C, Wong E, Russel L, Zern J, Aquino T and Tsomondo T (2002). Economic and environmental threats of alien plant, animal and microbe invasions. In *Biological Invasions: Economic and Environmental Costs of Alien Plant, Animal and Microbe Species.* (Ed. D Pimentel) pp. 307–329. CRC Press, London

Portney PR (1994). The contingent valuation debate: why economists should care. *Journal of Economic Perspectives* **8**, 3–17.

Power AP, Harris SA (1973) A cost-benefit evaluation of alternative control policies for foot and mouth disease in Great Britain. *Journal of Agricultural Economics* **24**, 573-596.

Productivity Commission (2002). *Impact of a Foot and Mouth Disease Outbreak on Australia.* AusInfo, Canberra.

QDNRM (Queensland Department of Natural Resources and Mines) (2003). *Economic Assessment of Environmental Weeds in Queensland.* Queensland Department of Natural Resources and Mines, Brisbane.

Rammel C and van den Bergh JCJM (2003). Evolutionary policies for sustainable development: adaptive flexibility and risk minimising. *Ecological Economics* **47**, 121–133.

Sandlund OT, Schei PJ and Viken Ae (1996). The Trondheim Conferences on Biodiversity. In *UN Conference on Alien Species.* Trondheim, Norway. Directorate for Nature Management/Norwegian Institute for Nature Research, Trondheim.

Scholz JT and Stiftel B (2005). Introduction: the challenges of adaptive governance. In *Adaptive Governance and Water Conflict: New Institutions for Collaborative Planning.* (Eds JT Scholz and B Stiftel) p. 300. Resources for the Future Press, Washington, DC

Tanner C (2001). Quarantine reform: Australia's recent experience. In *The Economics of*

Quarantine and the SPS Agreement. (Eds K Anderson, C McRae and D Wilson) pp. 240–264. Centre for International Economic Studies and the Department of Agriculture, Fisheries and Forestry – Australia/Biosecurity Australia, Adelaide.

Thompson D, Muriel P, Russell D, Osborne P, Bromley A, Rowland M, Creigh-Tyte S and Brown C (2002). Economic costs of the foot and mouth disease outbreak in the United Kingdom in 2001. *Revue Science et Technique* **21**, 675–687.

United Nations Environment Programme (1992) Convention on Biological Diversity: Final Text. (United Nations Department of Public Information: New York).

Waage JK, Fraser RW, Mumford JD, Cook DC and Wilby A (2005). *A New Agenda for Biosecurity.* Department for Food, Environment and Rural Affairs, London.

Walker B, Holling CS, Carpenter SR and Kinzig A (2004). Resilience, adaptability and transformability in social-ecological systems. *Ecology and Society* **9**, 5.

Whitby M (2000). Challenges and options for the UK agri-environment: presidential address. *Journal of Agricultural Economics* **51**, 317–332.

Whiting TL (2003). Foreign animal disease outbreaks, the animal welfare implications for Canada: risks apparent from international experience. *Canadian Veterinary Journal* **44**, 805–815.

Wittwer G, McKirdy S and Wilson R (2005). Regional economic impacts of a plant disease incursion using a general equilibrium approach. *The Australian Journal of Agricultural and Resource Economics* **49**, 75–89.

18

Natural Hazards and Property Loss

Ryan Crompton, John McAneney, Keping Chen, Roy Leigh and Laraine Hunter

Worldwide, the costs of natural disasters are increasing (Sigma 2006) and there is a widespread expectation that these will be further aggravated as a consequence of global climate change. This chapter examines that expectation, while reviewing a century of building damage caused by natural hazards in Australia, 93% of which has been caused by meteorological events. We shall demonstrate unequivocally that, at least until now, demographic and societal changes are overwhelmingly responsible for the increasing costs of natural disasters in Australia. We conclude with implications for policy aimed at reducing the vulnerability of the built environment to natural hazards.

We begin by examining Risk Frontiers' database of historical building losses—*PerilAUS*—the most comprehensive record of its kind in Australia. (Risk Frontiers is a research centre devoted to helping insurers better understand and price natural catastrophe risks in the Asia Pacific Region.) The database was compiled by painstaking examination of official records, early newspapers and scientific accounts of nearly 5000 hazard events from 1900 to 1999, and for major events to 2003. For almost 1200 events, it is possible to estimate the number of buildings destroyed, with damaged buildings (including commercial premises) normalised to *residential house equivalents* (RHE) using relative building costs and floor areas for different types of buildings (Blong 2004). One RHE could represent one residential home completely destroyed or 10 homes each experiencing damage to 10% of their replacement value. Bushfires tend to either destroy buildings completely or leave them relatively unscathed, whereas hail generally causes relatively small amounts of damage to individual dwellings spread over a wide area (Shuster *et al.* 2006). In extreme cases, tropical cyclones can inflict significant damage to individual dwellings over very large areas.

Figure 1 shows meteorological hazards dominate building losses in Australia over the relatively short time that records have been kept. In terms of the total number of buildings destroyed (insured and uninsured) between 1900 and 2003, tropical cyclones have been most destructive—accounting for one-third of losses, with floods and bushfires each contributing another 20%, as do thunderstorms when the effects of hail, wind gusts and tornadoes are combined. Earthquakes account for only 6% of losses and the majority of this was caused by a single event—the 1989 Newcastle earthquake. (For long-return period events, such as damaging earthquakes, the historical record in Australia is an inadequate sample on which to judge the future.) The dominance of meteorological hazards suggests that Australia should be more vulnerable than many other countries to changes in global climate. This is an issue we explore in later discussion.

In terms of the geographic spread of building losses, New South Wales and Queensland have been most severely affected (Figure 2). Tropical cyclones dominate losses in Queensland, Northern Territory and Western Australia, whereas bushfires dominate in Victoria, Tasmania and the Australian Capital Territory. New South Wales has been most affected by floods and then thunderstorms, and South Australia by earthquakes. The reader should be aware that risk is not just a function of the hazard, but is the joint outcome of hazards and societal vulnerability to these hazards. This is a feature that has long been recognised by social scientists (for example, Mileti 1999) and by the insurance industry (Walker 1997), but one largely absent from the international

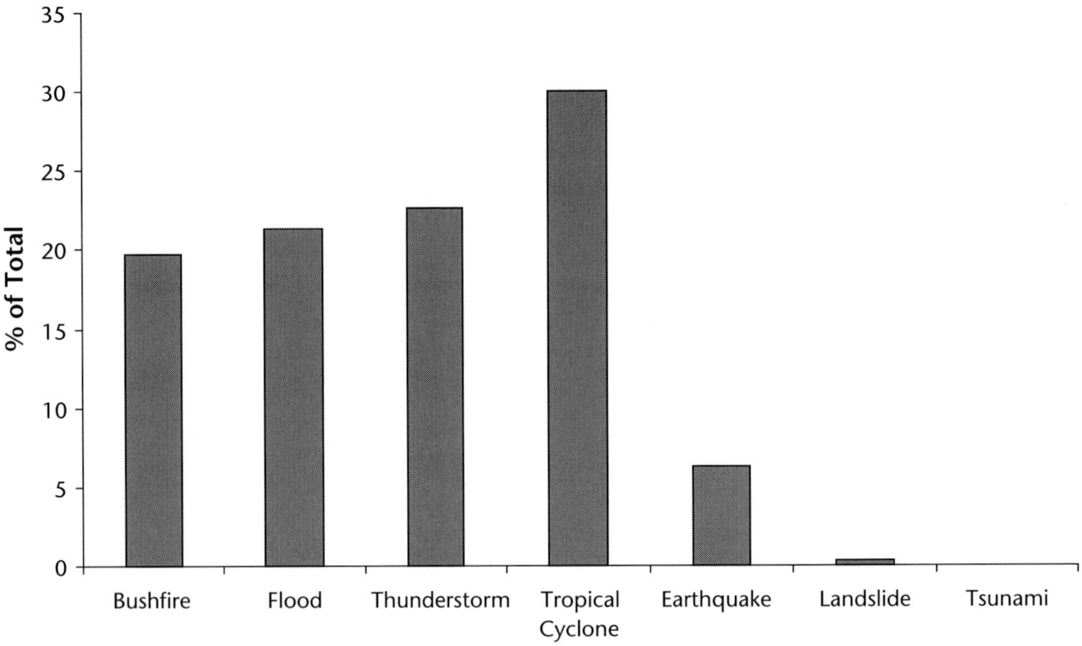

Figure 1: Percentage of the accumulated building damage between 1900 and 2003 attributed to different hazards. Thunderstorm encompasses hail, gust and tornado losses (Source: *PerilAUS* database, Risk Frontiers).

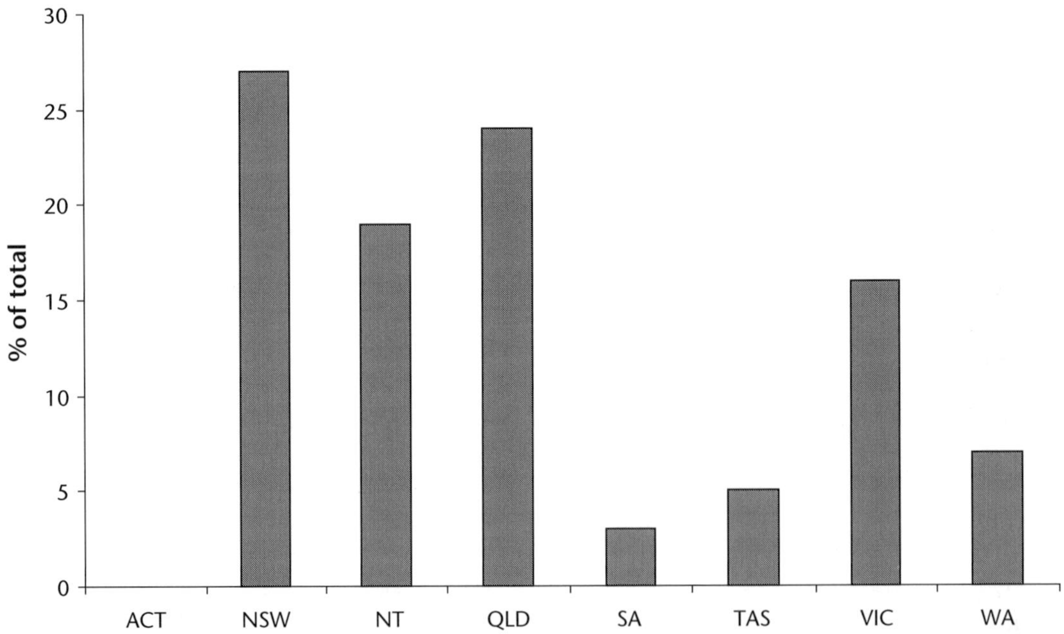

Figure 2: Percentage of the accumulated building damage between 1900 and 1999 by state or territory (Source: *PerilAUS* database, Risk Frontiers).

debate on global climate change, which has been narrowly focused on the control of greenhouse gas emissions (Pielke and Sarewitz 2005).

In the remainder of this chapter we review new information based on work undertaken on the vulnerability of property to riverine flooding, bushfire and coastal inundation. We then consider the results of a recent normalisation of the Insurance Council of Australia's (ICA) Disaster database of insured losses. Our interest here is to see whether any other influencing factor remains once we have adjusted the original losses for changes in dwelling numbers, values and building codes in tropical cyclone-prone locations (Crompton and McAneney 2008). Any residual trend in the normalised losses might then be ascribed to other factors, including climate change caused by humans.

National distribution of properties vulnerable to riverine flooding

An examination of 300 flood-prone areas by Risk Frontiers suggests that there are approximately 170,000 residential properties susceptible to mainstream flooding by an event with an average recurrence interval (ARI) of 100 years. This represents around 2.4% of all properties nationally (Leigh, 2006). This estimate is based primarily on the number of residential buildings liable to overground inundation, although the nature of the data available sometimes meant that alternative measures, such as the number of buildings susceptible to over-floor flooding, were also used. Only residential properties in urban settlements prone to mainstream riverine flooding were considered. Stormwater flooding from overflowing drainage systems and coastal inundation from tsunami and storm surge were excluded (except for storm surge coincident with river flood). All major flood-prone settlements were considered within areas containing approximately 75% of Australia's population.

Over 90% of residential properties expected to be flooded by a 100-year ARI event are located in Queensland, New South Wales and Victoria (Table 1). About 22% of these are located in Brisbane city, Ipswich and the Gold Coast. The ten most flood-prone areas account for 45% of the total number of identified flood-prone properties (Table 2).

Table 1: Number of residential properties susceptible to mainstream riverine flooding by an event with a 100 year ARI.

State	Number of residential properties	Proportion of total
Queensland	62,130	36%
New South Wales	55,677	33%
Victoria	42,376	25%
Western Australia	1,142	1%
South Australia	6,582	4%
Northern Territory	990	1%
Tasmania	723	<0.5%
Australian Capital Territory	0	0%
TOTAL	169,620	100%

Bushfire-prone property

Distance from bushland is not the only variable determining bushfire vulnerability, but it is demonstrably the most important (Chen and McAneney 2004). It is also easily quantifiable. For this reason, Chen and McAneney (2005) calculated separation distances between addresses and adjacent bushland for all addresses in and around major capital cities with the exception of Darwin (Northern Territory), which, despite a history of grassfires, has experienced few building losses.

In order to identify addresses, we employed the Geocoded National Address File (G-NAF) street address database. It contains about 10.9 million addresses with known locations, including up to 12% potential duplicates in its earlier versions. For the time being, it is regarded as Australia's most authoritative geo-located address database.

For bushland maps, we first sourced government and publicly available land-cover databases. Many are now quite out of date and are of coarse spatial resolution. High-resolution imagery is preferable but medium-resolution Landsat 7 ETM+ images (28.5 metre resolution) are more affordable and we used these to classify bushfire-prone vegetation. We focused on identifying large areas of continuous bushland—that is, areas greater than 0.5 square kilometres where large fires could

Table 2: Ten most flood-prone regions in Australia based on a 100 year ARI flood.

Region or city	State	Number of flood-prone residential properties	Proportion of national total
Gold Coast	QLD	20,128	12%
Brisbane and Ipswich	QLD	18,010	10%
Shepparton	VIC	6,572	4%
Melbourne	VIC	6,000	3%
Mackay	QLD	5,924	3%
Brown Hill and Keswick Creeks (Adelaide)	SA	5,000	3%
Hawkesbury–Nepean floodplain (Sydney)	NSW	4,862	3%
Wyong and Tuggerah Lake	NSW	4,651	3%
Grafton	NSW	4,167	2%
Benalla	VIC	3,641	2%

develop and, on occasions, get out of control. We ignored small, scattered and discontinuous areas of vegetation. More recently, we have updated the analysis using the GEODATA TOPO database (Geoscience Australia, 2006).

Given the locations of all addresses and bushland, the calculation of shortest distance between them is straightforward. Five distance ranges were categorised: Group 1 addresses are within 100 metres of bushland and so are most at risk. The other six groups are spaced at intervals of 100 metres to 700 metres—the maximum distance at which we have seen damage (Chen and McAneney 2004).

For the 8,161,680 addresses analysed, about 6% of addresses fall into Group 1 and about 80% are located beyond 700 metres (Figure 3), with negligible exposure to bushfire. The dataset analysed makes up some 75% of national addresses, so that when we extrapolate nationally, and ignore 5% of buildings in Northern Territory and northern Queensland where few building losses have been experienced, we find some 620,000 addresses fall within in the most at-risk category—being either immediately adjacent, or very close, to extensive bushland. After adjusting for duplicates in the early version of G-NAF used, a realistic upper number of most-at risk addresses is about 550,000 nationally.

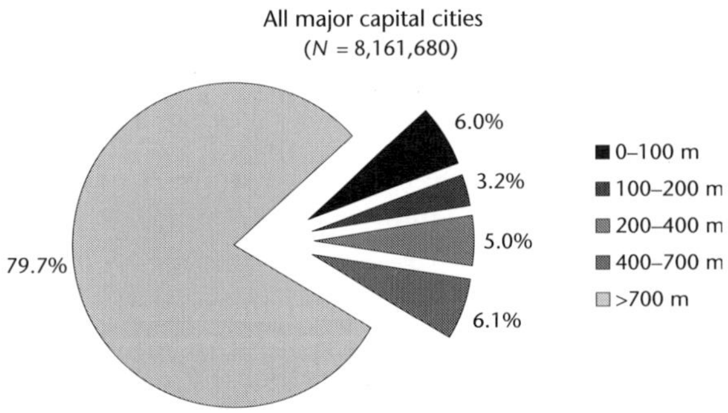

Figure 3: Breakdown of properties by distance category from large areas of bushland.

Coastal vulnerability

Another concern for people, property and infrastructure is coastal flooding caused by a range of ocean-related phenomena, such as storm surge, tsunami and significant sea-level rise, associated with global warming. Using the most recent, fine resolution global and national databases, including a clean version of G-NAF with a total of 9.6 million unique national addresses, Chen and McAneney (2006) quantified the number of addresses in Australia located within 3 kilometres of the coast and having an elevation less than 6 metres above mean sea-level. These thresholds were adopted as national benchmarks for a variety of reasons. When uninterrupted by topography, the 2004 Boxing Day tsunami penetrated coastal regions of Sumatra some 3 to 4 kilometres inland so we assume this is also possible in Australia. The 6 metre elevation threshold arose from considerations of technical accuracy, the fact that storm surges of this magnitude are possible in tropical cyclone-prone areas and as a plausible upper limit on sea level rise in the next 100 or more years (Solomon 2007).

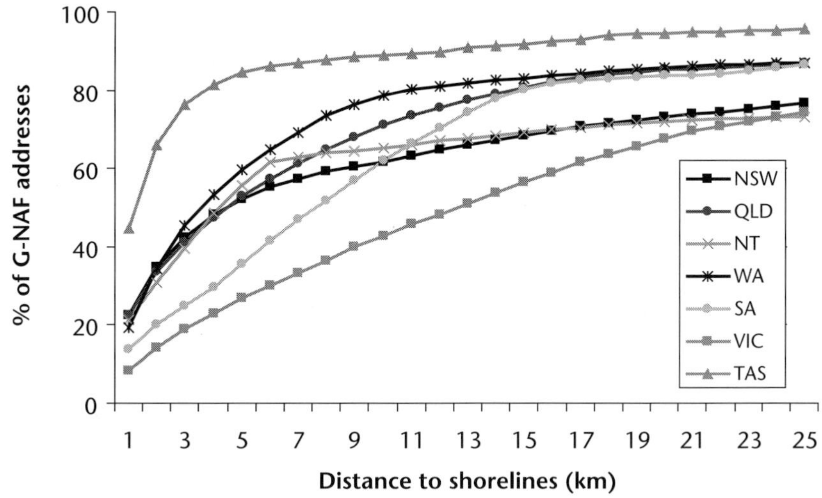

Figure 4: Percentage of coastal addresses by state/ territory and distance to shoreline.

Nationally, 7.4% (about 700,000) of addresses fall within these thresholds and about 50% are located within 7 kilometres of the shoreline (Chen and McAneney, 2006). The proportion of coastal addresses varies significantly by state and territory (Figure 4): in Tasmania more than 40% of addresses are located within 1 kilometre of the coast and 80% are within 4 kilometres, while in Victoria the corresponding proportions are about 9% and 23%. Statistics for other states lie within these extremes.

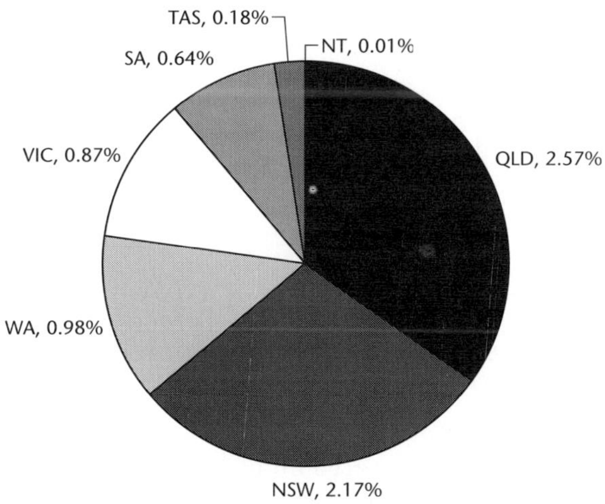

Figure 5: Percentage of coastal vulnerable addresses (distance less than 3 km from the shore and elevation less than 6 m above the mean sea level) by state and territory. Note: Proportions refer to the total number of unique national addresses after removing G-NAF duplicates (9.6 million).

The majority of vulnerable coastal addresses are located on the eastern seaboard—that is, Queensland and New South Wales (Figure 5).

Insured losses – what would they cost today?

The ICA Disaster List is a record of natural hazard events in Australia that have caused significant insured losses since 1967. It contains details of each event including date, areas affected and total insured (industry) cost in 'original' dollars. Although the threshold loss for inclusion in the database has varied over time, most events had losses in excess of AUD10 million. Crompton *et al.* (2005) developed an indexation methodology to account for changes in population, wealth and inflation so the likely impact of past events could be estimated in today's terms —a current (2006) loss. The method uses the increase in the number of dwellings and the nominal (in dollars of the day) dwelling values as surrogates for population, wealth and inflation. This methodology produces very comparable adjustments to that of Pielke and Landsea (1998).

We have already established the importance of tropical cyclones on Australian building losses (Figure 1). Normalising historical losses to current values must also account for improvements in construction standards introduced after Tropical Cyclone Althea devastated Townsville in 1971 and

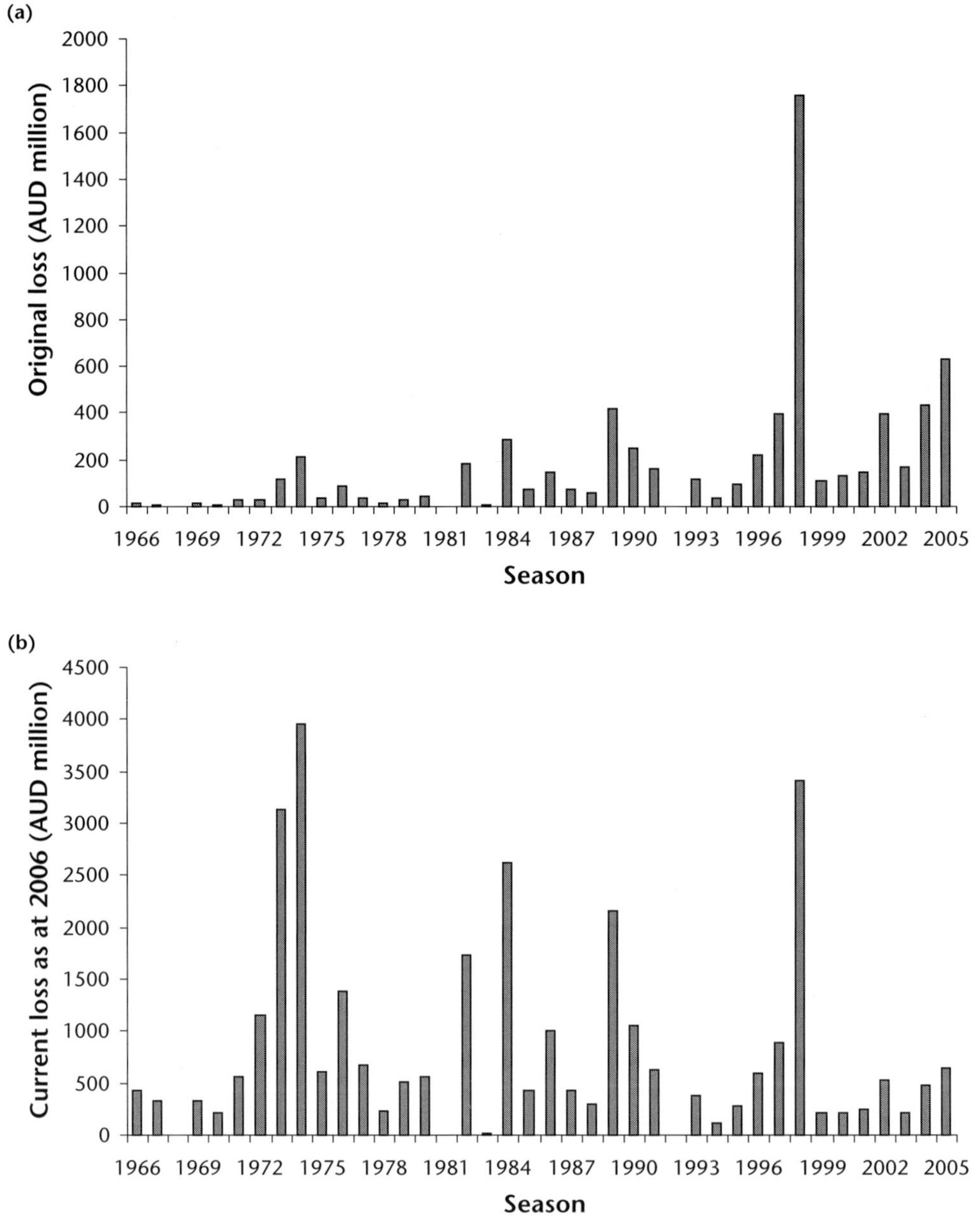

Figure 6: (a) Original annual aggregate insured losses (AUDM) for weather-related events in the ICA Disaster List for 12-month periods beginning 1 July (Crompton and McAneney 2008). (b) Insured losses normalised to 2006 dollars. Tropical cyclone-related losses have been reduced to account for post-1980 improvements in construction standards in tropical cyclone-prone areas following Crompton and McAneney (2008).

Cyclone Tracy destroyed Darwin in 1974. For details of this correction, the reader is referred to Crompton and McAneney (2008).

Figure 6a presents the original losses in the Disaster List with five geological events – four earthquakes and one tsunami – excluded. Annual losses have been calculated for 12-month periods beginning July 1 to take account of the southern hemisphere seasonality of meteorological hazards. The analysis begins at the 1966 season (1966/67) and ends with the 2005 season.

When correctly normalised (Figure 6b), the time series of insured losses exhibits no obvious trend (increase or decrease) over the last four decades. In

Table 3: Ranking of insured event losses indexed in current (2006) dollars. The current loss estimates the loss if the historical event was to reoccur with today's (2006) society.

Rank	Event	Year	Location	State	Original loss (AUD million)	Current loss as at 2006 (AUD million)
1	Earthquake	1989	Newcastle	NSW	862	4300
2	Tropical Cyclone Tracy	1974	Darwin	NT	200	3650
3	Hailstorm	1999	Sydney	NSW	1700	3300
4	Tropical Cyclone Wanda	1974	Brisbane	QLD	68	2090
5	Hailstorm	1985	Brisbane	QLD	180	1710
6	Ash Wednesday bushfires	1983	*Multiple*	VIC/SA	176	1630
7	Hailstorm	1990	Sydney	NSW	319	1470
8	Tropical Cyclone Madge	1973	*Multiple*	QLD/NT/WA	30	1150
9	Hailstorm	1976	Sydney	NSW	40	730
10	Hailstorm	1986	Sydney	NSW	104	710

other words, the increasing cost of insured losses over time is overwhelmingly explained by demographic and societal changes. Contrary to popular belief, there is no discernable evidence that human-induced climate change is significantly impacting building losses in Australia to date. This is an important conclusion and consistent to that reached by Pielke and Landsea (1998) and Pielke *et al.* (2008) in relation to hurricane losses in the US.

Table 3 ranks the top 10 insurance event losses in current dollars (all hazards, not just meteorological) with the Newcastle earthquake and Tropical Cyclone Tracy topping the list with indexed losses of around AUD4 billion. Total direct economic losses (insured plus uninsured) will often be roughly twice these figures. The top 10 indexed losses comprise five distinct classes of extreme event. Indeed if the 1974 Brisbane floods resulting from Tropical Cyclone Wanda were to be re-classified as a flood event (there was little wind damage) then the top six loss events would comprise five different hazards.

Figures 7a and 7b show the breakdown of insured losses and frequency by meteorological hazard category. Twenty-one per cent of loss events (ignoring earthquakes) have been tropical cyclones responsible for 33% of dollar losses—a proportion very similar to that shown in Figure 1. Hailstorms comprise 19% of loss events and have been responsible for 35% of losses in dollar terms. Floods are less prominent than in Figure 1, but are probably under-represented because this hazard has not been uniformly insured and it is not certain whether some flood events were attributed to tropical cyclones.

Figure 7c shows the contribution of the various states and territories to the average annual loss in current (2006) dollars of AUD900 million. New South Wales accounts for nearly half of this figure. This representation is a crude measure of the national distribution of risk. A true metric should also take into consideration the standard deviation or volatility of these losses (McAneney 2005). It is nonetheless reflective of the multi-hazard threats facing states such as New South Wales and Queensland, as well as population density and property values.

Frequency and severity of tropical cyclones that made landfall on the Australian east coast

Previous discussion has examined aggregate losses. In this section, we use data derived from the Bureau of Meteorology to focus on tropical cyclones that have made landfall (www.bom.gov.au). We further restrict the analysis to those crossing the east coast of Australia during the last 45 years with central pressures less than or equal to 995 hPa.

Tropical cyclone activity in the South-Western Pacific region is strongly related to the El Niño–Southern Oscillation (ENSO) phenomenon. Cooler

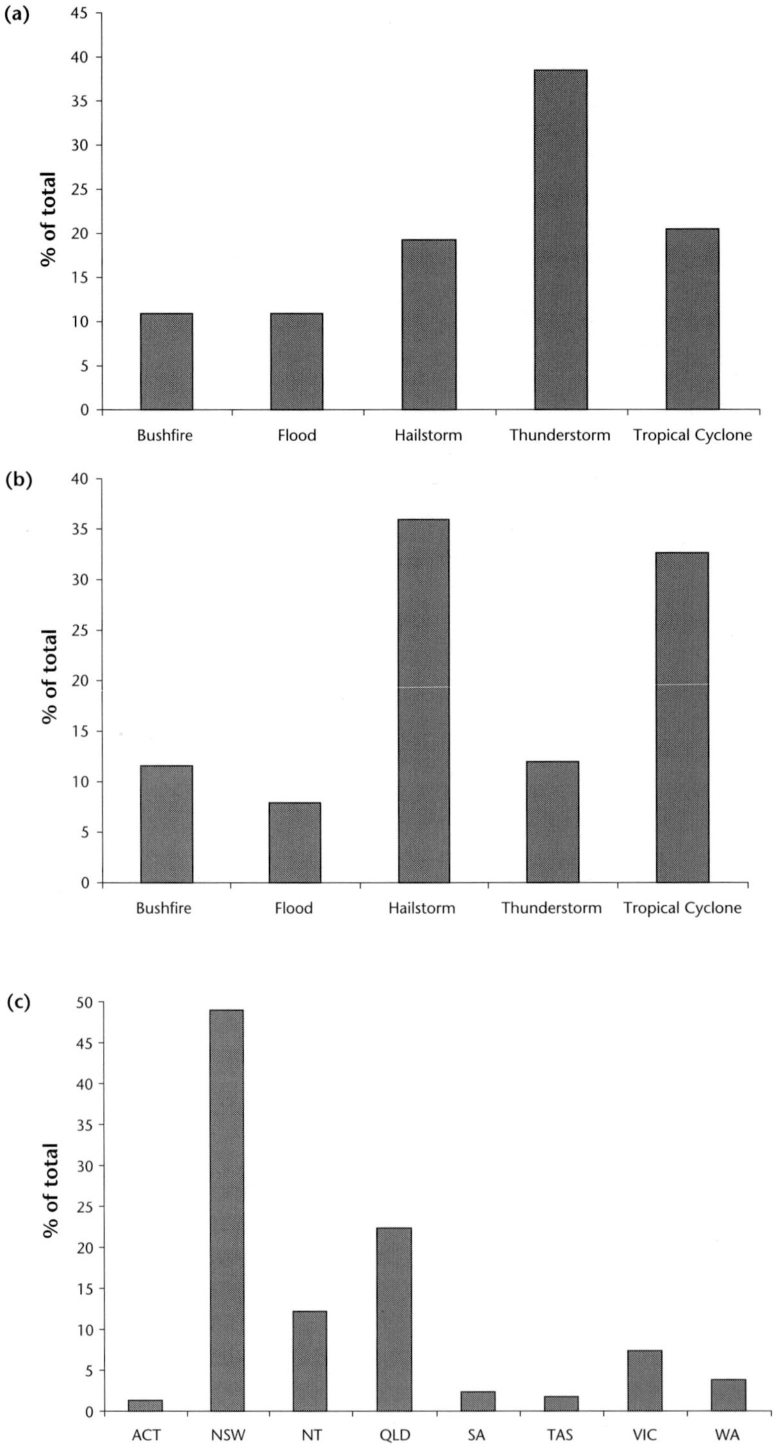

Figure 7: (a) Percentage of the number of weather-related events classified by hazard type in the ICA Disaster List. Event losses from hailstorms have been separated from other forms of severe storms—tornadoes and high winds (data from Crompton and McAneney 2008). (b) Percentage of the total accumulated current losses as at 2006 (AUD million) classified by hazard type in the ICA Disaster List (data from Crompton and McAneney 2008). (c) Distribution by state and territory of the national annual average insured losses in current (2006) dollars. Losses due to geological hazards have been included in this figure.

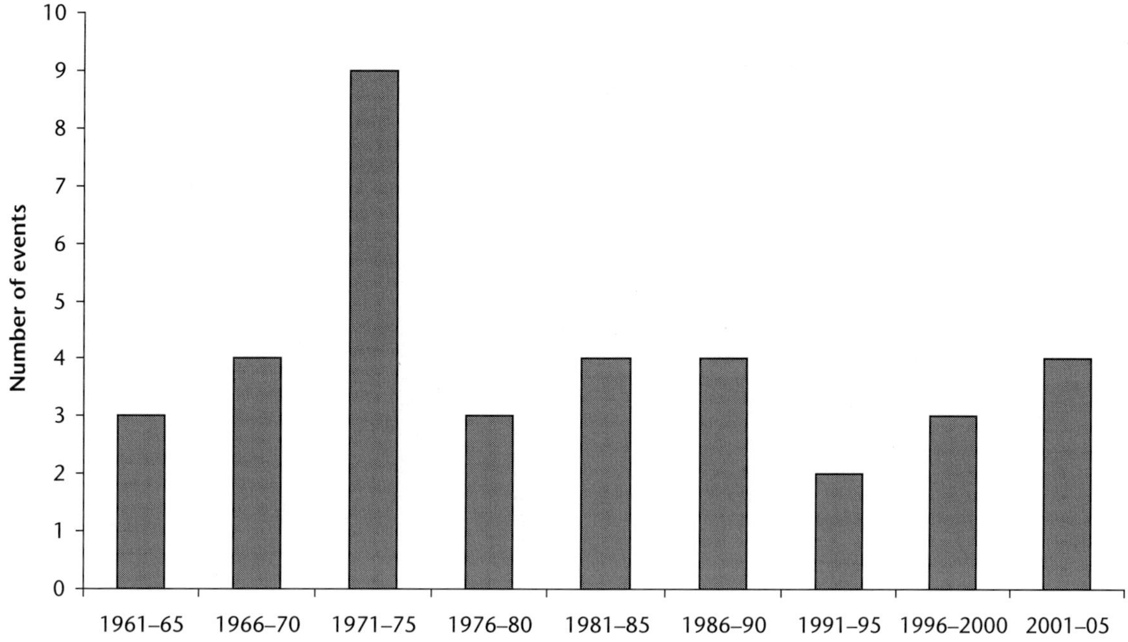

Figure 8: Number of tropical cyclones to cross the east coast (Source: Data from Bureau of Meteorology).

ocean temperatures exist in the Western Pacific and Coral Sea during El Niño episodes and ocean temperatures near the Queensland coast are typically above average during La Niña phases. Consequently, tropical cyclone activity tends to shift further away from the east coast of Queensland and further north during El Niño phases resulting in fewer than average numbers of tropical cyclones making landfall than during La Niña phases.

Figure 8 shows successive five-season period frequencies of tropical cyclones that have crossed the east coast since the 1961 season (1961/62). Within each of the five-season periods there are different numbers of El Niño, La Niña, and Neutral events. With the exception of 1971–75, there have been between two and four tropical cyclones for each period. It comes as no surprise that the 1971–75 period was dominated by La Niña episodes (four out of five seasons).

Figure 8 only tells part of the story. Of more interest is the combination of frequency and intensity. According to Gray (2003), when normalised for coastal population, inflation and wealth per capita, tropical cyclone-spawned damage in the US rises by a factor of four for each successive increase in Saffir-Simpson intensity category. Thus a Category-3 hurricane typically causes about four times the normalised damage upon landfall of a Category-2 hurricane and so on. Figure 9 takes this weighting into account: Category-5 and -4 events have each been assigned an equal weighting of 1/4; Category-3 a weighting of 1/16; Category-2 a value of 1/64 and Category-1 events 1/256. The weighted frequency by intensity is then summed for each 5-year period to obtain a potential destructiveness index, assuming that the same damage relativities between categories hold true under Australian conditions.

By concentrating on the damage potential of the hazard alone, Figure 9 allows for Australia's low population density and the large physical separation distances between population centres. Actual damage arising from individual tropical cyclones will vary widely as a result of the tropical cyclone track in relation to centres of population, terrain and topography, proportions of building construction conforming to improved wind loading codes, wealth per capita of the afflicted population as well as attributes of the cyclone, such as forward speed, storm surge and rainfall.

Again we see no obvious change (increase or decrease) in the potential destructiveness over the past 45 years. The east coast is of particular interest because this is where most building assets are

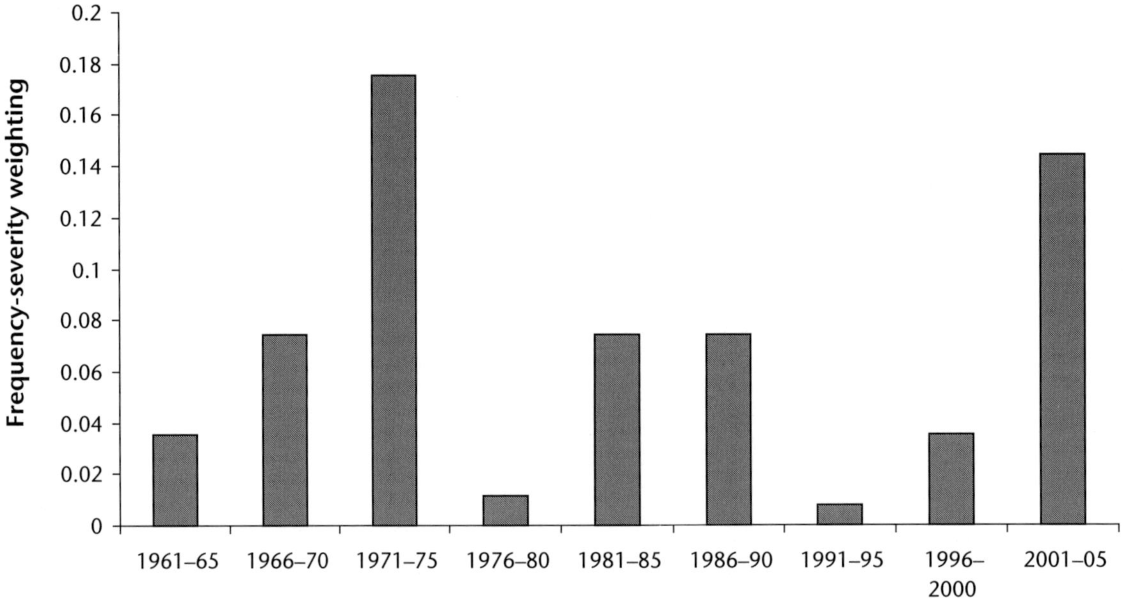

Figure 9: Potential destructiveness index of tropical cyclones that have crossed the east coast of Australia (Source: Crompton *et al.* 2006).

located; very similar results hold for the western and northern coasts of Australia and for the entire coastline. While we acknowledge that the small number of tropical cyclones per 5-year time interval makes it difficult to draw robust conclusions, there is no indication to date that tropical cyclones making landfall are becoming more frequent or have the potential to cause increased losses.

Annual probability of bushfire losses

Many studies on bushfire risk in Australia have concluded that this hazard will increase under high greenhouse gas emission scenarios (for example, Pitman *et al.* 2006). With this in mind, it is interesting to look at the statistics of event losses over the last century. Table 4, drawn from the *PerilAUS* database, lists three relevant statistics calculated between various start years (from 1900 to 1990) and 2003. The first row shows the annual probability of a non-zero building loss anywhere in Australia. The second and third rows list the average annual frequency of having more than 25 and 100 homes destroyed within a single week—in other words, the annual probability of having significant individual event losses. The seven day period is used to help identify particular fire events, rather than accumulated seasonal losses from fires widely dispersed in space and time.

The risk metrics shown in Table 4 are remarkably insensitive to the time period over which they are calculated. At least some home destruction can be expected in approximately 55% of years and the likelihood of losing more than 25 and 100 homes has remained steady at around 40% and 20%, respectively, over the last century. At current average asset values (AUD375,000 for average home and contents), an event loss of 25 homes would come close to AUD10 million—the lower limit for consideration in the ICA Disaster List.

That the probability of building destruction has remained so stable in the face of significant demographic changes and improvements in firefighting capability over the last century is curious. Population alone has increased fourfold over this period, with a much higher proportion now living in capital cities. Life for most of the population is now an urban experience (ABS 2006). To invoke global climate change and argue that increases in temperature and growing populations at the urban–forest interface have been *exactly* compensated for by improved fire management stretches credulity.

The result seems better explained by infrequent very large fires that have occurred about five times in the last 75 years and which have been responsible

Table 4: National bushfire building loss probabilities calculated between the start year and 2003. The first row gives the frequency of any (non-zero) loss, while the second and third rows include only events that have resulted in more than 25 or 100 homes destroyed within a single week. The years 1926, 1939, 1967 and 1983 had significant losses (> 500 homes destroyed) (McAneney *et al.* 2007).

Start Year	1900	1926	1939	1967	1983	1990
Annual probability of a non-zero loss	56%	53%	48%	57%	57%	57%
Annual probability of losing >25 homes in 1 week	39%	40%	42%	41%	38%	36%
Annual probability of losing >100 homes in 1 week	18%	19%	22%	19%	19%	21%

for the majority of building losses. Fires such as these will exhaust the resources of even the most professional fire services. In fact, once fires get beyond a certain intensity, there is very little that can be done except to keep out of the way and hope that the weather changes or that the fire will run out of fuel before it intersects a populated area. Actual losses will then be a function of the physical disposition of dwellings vis-à-vis the advancing fire front and as moderated by the actions of owners and the fire services. Although we concede that bushfire losses are the result of a complicated mix of social and climatic phenomena, one inference is again clear: there is no evidence to suggest that losses are increasing due to human-induced climate change

Policy implications

The collective evidence reviewed here suggests that societal factors—dwelling numbers and values—are the predominant reasons for increasing building losses due to natural disasters in Australia. The role of anthropogenic climate change is not detectable at this time. This being the case, it seems logical that in addition to reducing greenhouse gas emissions, significant investments be made to reduce society's vulnerability to current and future extreme events, irrespective of their possible connection to climate change. We are aware of few policies explicitly developed to help Australian communities adapt to future climate change (Leigh *et al.* 1998). One example of effective adaptation arises from improved construction standards introduced in the 1980s as part of a National Building Code of Australia. These have been mentioned already and, as a result, dramatic reductions in wind-induced losses were observed following Tropical Cyclones Winifred (1986) and Aivu (1989) (Walker 1999) and, most recently, Larry (2006) (Guy Carpenter 2006). The reason that a repeat of the Newcastle earthquake is expected to cause a similar insured loss to a repeat of Cyclone Tracy (Table 3) is a direct result of the fact that no seismic building codes analogous to the wind-loading prescriptions described above exist for the design of residential homes. Seismic risk is simply not treated seriously in this country—a fact that no doubt arises from the relatively low frequency of damaging earthquakes (Figure 1). While the wind code measures were regulated in response to the immediate threat, the resulting benefits will hold true under any future climate scenario.

An increased threat from bushfires under global climate change is often assumed. However, our analyses suggest that while the prevalence of conditions leading to bushfires is likely to increase (Pitman *et al.* 2007), the impact is unlikely to be as dramatic as the combined changes of all of the other factors—increasing population, improvements in fire management and science, and so on—that have so far failed to materially affect the likelihood of bushfires building losses over the last century. This is not to ignore the threat posed by global climate change, but, at least in the case of fire in Australia, the main menace will continue to be extreme fires, such as the 1939 Black Friday, 1967 Hobart, 1983 Ash Wednesday and 2003 Canberra fires, which were each responsible for large losses. The threat to the most at-risk homes on the bushland–urban interface could be diminished by improved planning or enforcement of regulations that restrict where and how people build with respect to distance from the forest. These are political choices.

Conclusions

We conclude that changing societal factors are the principal reasons underlying the increasing cost of natural disasters in this country. In some ways, this is a positive message. While Australia can have little direct control over the worldwide emission of greenhouse gases, it can, if it chooses, control where and how people live and build. In this paper, we demonstrate how relatively straightforward studies can be used to identify homes vulnerable to threats, such as bushfire, riverine flood and coastal flooding, to an accuracy sufficient to enable policy decisions. The success of improved wind-loading codes show what can be done where there is a demonstrated need and political will. Social governance of this kind in relation to other natural hazards would result in immediate improvements in community resilience to both current and future climates.

References

Australian Bureau of Statistics (2006). *2006 Year Book Australia*. Number 88. ABS Catalogue No. 1301.0, Australian Government, Canberra. pp.776.

Blong R (2004). Residential building damage and natural perils: Australian examples and issues. *Building Research & Information* **32(5)**, 379–390.

Chen K and McAneney KJ (2004). Quantifying bushfire penetration into urban areas in Australia. *Geophysical Research Letters* **31**, L12212.

Chen K and McAneney KJ (2005). The bushfire threat in urban areas. *Australasian Science* **26(1)**, 14–16.

Chen K and McAneney KJ (2006). High-resolution estimates of Australian coastal population: with validations of global data on population, shoreline and elevation. *Geophysical Research Letters* 33, L16601.

Crompton RP, McAneney KJ and Leigh R (2005). 'Indexing the Insurance Council of Australia Natural Disaster Event List'. Report prepared for the Insurance Council of Australia. Risk Frontiers, Sydney.

Crompton RP, McAneney KJ and Leigh RJ (2006). Natural disaster losses and climate change: an Australian perspective. In *Proceedings of the Workshop on Climate Change and Disaster Losses – Understanding and Attributing Trends and Projections*. 25–26 May 2006. Hohenkammer, Germany. (http://sciencepolicy.colorado.edu/sparc/research/projects/extreme_events/munich_workshop/crompton.pdf)

Crompton RP and McAneney KJ (2008). Normalised Australian insured losses from meteorological hazards 1967–2006. *Environ. Science & Policy* (in press).

Gray WM (2003). Twentieth century challenges and milestones In *Hurricane! Coping with Disaster* (Ed. R Simpson) American Geophysical Union, Washington DC.

Geoscience Australia (2006). *Geodata Topo 250k Series 3*. CD-ROM. Geocat # 64059. Commonwealth of Australia, Canberra.

Guy Carpenter (2006). *Tropical Cyclone Larry Review*. Instrat, Guy Carpenter and Company, Inc., London. pp. 18.

Leigh RJ, Taplin R and Walker GR (1998). Adaptation of the insurance industry to climate change, and consequent implications. *Australian Journal of Environmental Management* **5(2)**, 81–96.

Leigh R (2006). 'An estimate of the number of residential properties in Australia liable to inundation by an ARI 100-year flood'. Report prepared for the Insurance Council of Australia, Risk Frontiers, Sydney.

McAneney KJ (2005). Australian bushfire: Quantifying and pricing the risk to residential properties. In *Proceedings of the Symposium on Planning for Natural Hazards – How Can We Mitigate the Impacts?* 2–5 February 2005 University of Wollongong (Eds RJ Morrison, S Quin and EA Bryant) pp. 13-22, University of Woolongong.

McAneney KJ, Chen K, Crompton R. and Pitman A (2007). Australian Bushfire Losses: Past, Present and Future. In *Proceedings of 4th International Conference on Wildfire*, May 13-17, Seville, Spain. <http://www.fire.uni-freiburg.de/sevilla-2007/contributions/doc/SESIONES_TEMATICAS/ST5/McAneney_et_al_AUSTRALIA.pdf>

Mileti D (1999). *Disasters by Design*. Joseph Henry Press, Washington, DC.

Pielke RA Jr and Landsea CW (1998). Normalized hurricane damages in the United States: 1925–95. *Weather & Forecasting, American Meteorologocial Society* **13**, 621–631.

Pielke RA Jr and Sarewitz D (2005). Bringing society back into the climate debate. *Population and Environment* **26**, 255–268.

Pielke RA Jr, Gratz J, Landsea CW, Collins D, Saunders MA and Musulin R (2008). Normalised hurricane losses in the United States. *Natural Hazards Review* **9**, 29–42.

Pitman AJ, Narisma GT and McAneney KJ (2007). The impact of climate change on Australian bushfire risk. *Climatic Change* **84**, doi:10.1007/s10584-007-9243-6.

Schuster SS, Blong RJ and McAneney KJ (2006). Relationship between radar-derived hail kinetic energy and damage to insured buildings for severe hailstorms in Eastern Australia. *Atmospheric Research* **81**, 215–235.

Solomon S (Ed.) (2007). *Climate Change 2007: The Scientific Basis.* Contribution of Working Group 1 to the Fourth Assessment Report of the Intergovernmental Panel on Climate Change, Cambridge University Press, New York.

Swiss Reinsurance Company (2006). *Natural catastrophes and man-made disasters 2005: high earthquake casualties, new dimension in windstorm losses.* Sigma No.2, Zurich. pp. 40. http://www.swissre.com/resources/e109a780455c56b897efbf80a45d76a0-Sigma2_2006_e.pdf

Walker GR (1997). Current developments in catastrophe modelling. In *Proceedings of Aon Group Australia Conference: Financial Risk Management for Natural Catastrophes.* (Ed. NR Britton). pp. 17–36. Griffith University Press, Brisbane.

Walker GR (1999). Designing future disasters: An Australian perspective. In *Proceedings of Aon Group Australia Conference: The Changing Risk Landscape: Implications for Insurance Risk Management.* (Ed. NR Britton). pp. 17–24. Southwood Press, Sydney.

19

Community Health

Perri Timmins, Justin Graf and Kuldeep Bhatia

This chapter examines some of the key disparities in the health of people living in urban, regional and remote areas of Australia. Data highlighting trends in several key health indicators across regions are provided from the national mortality and hospital morbidity databases and other available sources. Additional information is drawn from two recent Australian Institute of Health and Welfare (AIHW) reports that have examined the relationship between rural and remote living and health (Knox *et al.* 2005; AIHW 2006c). The final section contains a brief discussion of the possible reasons for the regional health disparities and their implications for future trends in urbanisation and sustainable urban development.

Background

It is often assumed that people living in rural and remote areas live less stressful, more active and generally healthier lives than their overcrowded urban counterparts. When people think of health risks in rural areas, they often focus on injuries and deaths from accidents, such as those involving farming equipment and motor vehicles. However, recent research from around the world points to rural living as a marker for poor health. For some chronic diseases, prevalence and death rates are much higher in rural areas compared with urban areas (Dennis and Pallotta 2001).

There is ample evidence that the health of people living in rural and remote areas of Australia is not only different from those in urban areas—it is by many standards substantially worse. People living in rural and remote areas have, on average, shorter life expectancies than those who live in major cities (AIHW 2005b). They tend to have higher rates of certain health risk factors and chronic diseases than their urban counterparts and are more likely to die in motor vehicle accidents. People living in rural and remote areas are also more likely to rate their dental health as poor or very poor (AIHW 2005a).

Why should people living in rural and remote areas be less healthy than those living in urban areas? Living in a rural setting in itself does not cause poor health; rather, factors often associated with rural or remote living may lead to relatively poor health outcomes. Such factors include socio-economic disadvantage (lower income, less education), geographical isolation (distance from services, poor transport and communication difficulties) and shortage of health-care providers (AIHW 2006b). These factors, in turn, may lead to, or exacerbate, exposure to environmental hazards, unhealthy lifestyles, negative attitudes and social norms towards health, and less access to health-care services.

The setting

About 66% of the Australian resident population live in cities with populations of over 250,000—the remainder live in areas that may be classified as 'regional' or 'remote' (Table 1). The Australian Bureau of Statistics (ABS) Australian Standard Geographical Classification (ASGC) Remoteness Areas structure is used in this chapter for categorising geographical regions in terms of 'remoteness'. The structure has five remoteness categories: major cities, inner regional Australia, outer regional Australia, remote Australia and very remote Australia (ABS 2001). A sixth category covers migratory populations.

Table 1: Distribution of Australian population by ASGC Remoteness area, 2005

ASGC Remoteness area	Per cent
Major cities	66.2
Inner regional Australia	21.2
Outer regional Australia	10.2
Remote Australia	1.6
Very remote Australia	0.9

Source: AIHW analysis of ABS data.

The ASGC Remoteness Area categories are based on Census Collection Districts and scores from the revised Accessibility and Remoteness Index of Australia (ARIA+). ARIA+ was developed in its original form in 1997 by the then Commonwealth Department of Health and Aged Care and the National Key Centre for Social Applications of Geographical Information Systems. ARIA+ relies on road distance to service centres, which are further classified by population size. Therefore, although usually referred to as the five categories, the ASGC Remoteness Areas structure is an ordinal scale, which allows the examination of trends with increasing remoteness.

Mortality across remoteness areas

Mortality data provide a useful window into the health of a community. Cause of death and related information has the advantage of being compiled regularly and, for the most part, reliably (AIHW 2006c). Death rates therefore allow a more precise examination of remoteness effects than, for instance, self-reported survey data. In this section, mortality data are derived from the AIHW National Mortality Database for the years 2003 to 2005 and refer to the usual residence of the deceased rather than the place of death. Trends over time for several causes of death are provided by a recent study that examined inter-regional differences from 1992 to 2003 (AIHW 2006c).

The overall death rates show a consistent increase with increasing remoteness in both sexes. The death rates for all causes in very remote areas are about 70% greater than those in major cities.

At the cause-specific level, much higher death rates in very remote areas compared with those in major cities are noted for injury and poisoning and for diabetes (Table 2). While circulatory diseases and neoplasms are the predominant causes of death in all areas (Figure 1), the death rates in regional

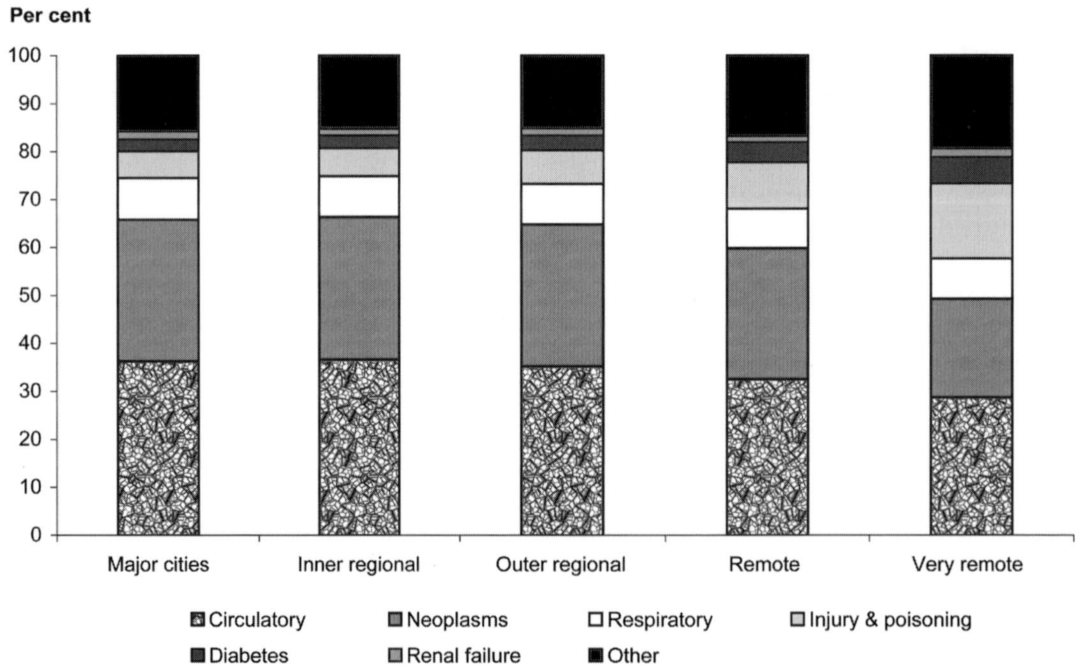

Figure 1: Major causes of death by ASGC Remoteness area, 2003 to 2005 (Source: AIHW National Mortality Database).

Table 2: Standardised mortality ratios (SMRs) for major causes of death by ASGC Remoteness area, 2003 to 2005

Cause of death	Sex	ASGC Remoteness area				
		Major cities	Inner regional	Outer regional	Remote	Very remote
All causes	Males	1.00	1.08	1.15	1.23	1.74
	Females	1.00	1.06	1.08	1.13	1.65
Circulatory diseases	Males	1.00	1.08	1.13	1.20	1.54
	Females	1.00	1.08	1.07	1.07	1.56
Neoplasms	Males	1.00	1.07	1.10	1.03	1.11
	Females	1.00	1.04	1.05	1.04	1.09
Respiratory diseases	Males	1.00	1.05	1.14	1.32	2.03
	Females	1.00	1.02	1.03	1.00	1.57
Injury and poisoning	Males	1.00	1.28	1.52	1.78	3.23
	Females	1.00	1.14	1.30	1.65	2.82
Diabetes	Males	1.00	1.03	1.31	1.84	3.42
	Females	1.00	1.17	1.43	2.30	5.02
Renal failure	Males	1.00	0.94	0.94	0.93	1.84
	Females	1.00	1.04	1.16	1.18	2.88

Notes: 1 SMRs are the ratio of the observed number of persons with the cause of death in the region compared with the expected number of persons if the age-specific prevalence rates in major cities were to apply to the population in that region.

2 Shaded SMRs are not statistically significant, all other SMRs are significantly different compared with major cities ($p < 0.05$).

3 ICD-10 codes: circulatory diseases (I00–I99), neoplasms (C00–D48), respiratory diseases (J00–J99), injury and poisoning (V01–Y98), diabetes (E10–E14) and renal failure (N17–N19).

Source: AIHW National Mortality Database.

and remote areas due to these causes do not rise as dramatically as they do for diabetes or for injury and poisoning.

These variations in death rates should be viewed in the context of trends over time. Death rates have decreased, in both sexes, in all remoteness areas from 1992 to 2003 (AIHW 2006c). The overall rate of decline in regional and remote areas has been similar to that in major cities, although the rate of decline in very remote areas has actually been greater than in major cities. Overall reductions in circulatory diseases and cancer have been the major factors in driving down these rates. However, there has been no significant change in the rates of death due to diabetes in remote or very remote areas, while the rates of suicide has increased significantly in males in very remote areas and in females in inner regional areas (AIHW 2006c).

The variation in death rates across remoteness areas can be viewed in terms of (a) the prevalence of health risk factors, (b) the incidence and prevalence of diseases, and (c) access to (and use of) health-care services.

Prevalence of risk factors and health conditions

Information on the prevalence of risk factors and health conditions in Australia comes mostly from self-report surveys. Population-based self-report surveys usually have relatively small sample sizes outside major cities. Hence, reliable prevalence rates are difficult to estimate and it is often difficult to compare rates across regions. Information is limited by the scope, coverage and design of the surveys.

The ABS National Health Survey (NHS) is a good source of self-reported information about the

Table 3: Standardised prevalence ratios (SPRs) for selected health risk factors by ASGC Remoteness area, 2004-05

		ASGC Remoteness area		
Risk factor	**Sex**	**Major cities**	**Inner regional**	**Other regions**
Current smoker[a] (18+ years)	Males	1.00	1.13	1.26
	Females	1.00	1.25	1.37
Risky alcohol consumption[b] (18+ years)	Males	1.00	1.10	1.27
	Females	1.00	1.05	1.06
Physical inactivity[c] (15+ years)	Males	1.00	1.16	1.39
	Females	1.00	1.02	1.10
Overweight/obese[d] (15+ years)	Males	1.00	1.01	1.09
	Females	1.00	1.08	1.07

[a] Current smokers included all persons, 18 years and over, who were smoking daily, weekly or less than weekly.

[b] Alcohol consumption was recorded for all persons 18 years and over. Risky alcohol consumption was defined as 5 or more Australian standard drinks per day for males and 3 or more Australian standard drinks per day for females.

[c] Physical inactivity included all persons, 15 years and over, who were sedentary (very low or no level of exercise).

[d] Self-reports of height and weight, for all persons aged 15 years and above, were used to determine body mass. Body mass index (BMI) was calculated by dividing the weight (kg) of a person by the square of their height (m). Overweight/obese category includes all persons with a BMI ≥25.

Notes: 1 SPRs are the ratio of the observed number of persons with the risk factor in the region compared with the expected number of persons if the age-specific prevalence rates in major cities were to apply to the population in that region.

2 Shaded SPRs are not statistically significant, all other SPRs are significantly different compared with major cities ($p < 0.05$).

Source: AIHW analysis of the 2004–05 ABS National Health Survey.

presence of several health conditions (such as arthritis, asthma, diabetes, injury, bronchitis/emphysema and psychological distress) and a variety of health risk factors (such as tobacco smoking, risky alcohol consumption, physical inactivity and overweight/obesity). However, the sample size and coverage of the 2004–05 NHS allowed aggregation of data into only three regional categories: major cities, inner regional and other (more remote) regions.

Risk factors

People living in inner regional and other regions of Australia are more likely than those in major cities to report having a variety of health risk factors (Table 3). Increases in prevalence rate with increasing remoteness are observed for a variety of risk factors. Major increases are noted for smoking, risky alcohol consumption and physical inactivity in males and smoking in females. An increase in prevalence of these risk factors is likely to contribute significantly to the incidence of chronic conditions in inner regional and other (more remote) regions in the long term.

Health conditions

The prevalence of bronchitis/emphysema is significantly higher in inner regional and other regions than in major cities. The prevalence rate of bronchitis/emphysema in males in other regions is more than one-and-a-half times that in males in major cities (Table 4).

Diabetes is the only condition that shows a decreased prevalence for males and an increased prevalence for females living in inner regional and other regions, compared with those in major cities. An increase in cancer prevalence is only evident for males in inner regional areas, whereas females are less likely to report cancer in other regions compared with major cities. Reported injury occurrence is significantly lower for males, but higher for females, living in inner regional areas, compared with those in major cities.

An increased prevalence of high or very high levels of psychological distress is noted for males living in other regions compared with their counterparts in major cities. In addition, males living in inner regional areas are more likely to report asthma

Table 4: Standardised prevalence ratios (SPRs) for National Health Priority Areas and selected conditions by ASGC Remoteness area, 2004–05

		ASGC Remoteness area		
Disease/condition	**Sex**	**Major cities**	**Inner regional**	**Other regions**
Poor/fair general health (15+ years)	Males	1.00	1.20	1.35
	Females	1.00	1.08	1.15
Arthritis	Males	1.00	1.30	1.29
	Females	1.00	1.19	1.15
Asthma	Males	1.00	1.18	0.95
	Females	1.00	1.25	1.19
Cancer	Males	1.00	1.39	0.81
	Females	1.00	0.99	0.88
Diabetes	Males	1.00	0.81	0.93
	Females	1.00	1.34	1.20
Heart and circulatory	Males	1.00	1.18	0.95
	Females	1.00	1.25	1.19
Injury[a]	Males	1.00	0.94	0.91
	Females	1.00	1.06	0.95
Psychological distress[b] (18+ years)	Males	1.00	1.05	1.24
	Females	1.00	1.06	1.04
Bronchitis/emphysema	Males	1.00	1.31	1.59
	Females	1.00	1.06	1.15

[a] Includes the most recent event, for which action was taken and an injury was sustained.

[b] Psychological distress was measured using the Kessler Psychological Distress Scale. The scores are grouped into low (K10 score 10–15), moderate (16–21), high (22–29) and very high (30–50). The combined prevalence of high and very high levels of psychological distress was compared between regions.

Notes
1. SPRs are the ratio of the observed number of persons with the disease/condition in the region compared with the expected number of persons if the age-specific prevalence rates in major cities were to apply to the population in that region.
2. Shaded SPRs are not statistically significant, all other SPRs are significantly different compared with major cities ($p < 0.05$).
3. All ages were included in the analysis except where otherwise indicated.
4. Includes all types of arthritis, cancer, diabetes (excluding high sugar levels), and heart and circulatory conditions, included in the survey.

Source: AIHW analysis of the 2004–05 ABS National Health Survey.

and heart and circulatory conditions compared with males in major cities. Females are more likely to report heart and circulatory conditions in other regions than those in major cities.

Use of health-care services across remoteness areas

Close and affordable access to general practitioners, pharmacists, dentists, nurses and hospital facilities is a central aspect of a policy response to ill health. Australians living in regional and remote areas generally have less access to primary health care and specialist health services than those living in major cities (Table 5). Shortages and uneven distributions of health providers and facilities are compounded by access difficulties due to distance, time, cost and transport availability in regional and remote areas. People needing more specialist treatments must travel even further to larger towns

Table 5: Supply of health-care workers (FTEs per 100,000 population) by ASGC Remoteness area, 2004

	ASGC Remoteness area				
Occupation	**Major cities**	**Inner regional**	**Outer regional**	**Remote**	**Very remote**
General practitioners[a]	100	90	79	81	67
Specialists	119	55	33	19	5
Registered nurses	919	832	788	828	867
Enrolled nurses	175	260	282	271	219

[a] Includes all primary care medical practitioners.

Note: Full-time equivalent (FTE) numbers are based on 35-hour working week for nurses and 45-hour working week for GPs and specialists.

Source: AIHW Medical and Nursing Labour Force Surveys.

or cities or wait longer for these services to come to them on a rotation basis.

People living in regional and remote areas tend to use emergency departments as a source of primary care services more so than do people in urban areas (AIHW 2006a). Also, patients might be more likely to be admitted to hospital rather than be treated as outpatients because of long travel distances. These, and issues such as hospital capacity, complicates inter-regional comparisons of health-care-service use. However, statistics on health-service use inherently combine aspects of disease prevalence and severity as well as service availability and access, and should always be interpreted accordingly.

Information on health-service use in this section includes results of a study of general practice activity across remoteness areas from 1998 to 2004, and analysis of data from the AIHW National Hospital Morbidity Database for the years 2003–04 to 2005–06. The general practitioner (GP) data refer to the location of the practice; while the hospital separation data refer to the usual residence of the patient rather than the location of the hospital, except where noted otherwise.

General practice activity

A GP is usually the first point of contact for many health problems in Australia. However, GPs and other health professionals are unevenly distributed in Australia. Unpublished Health Insurance Commission data show that in 2001 the number of people for every GP increased with remoteness: 1090 in major cities to 3886 in very remote areas (Knox *et al.* 2005).

The Bettering the Evaluation and Care of Health (BEACH) remoteness study (conducted between April 1998 and March 2004 with more than 600,000 GP–patient encounters) has found increasing remoteness to be significantly associated with decreases in annual visits by patients and increases in the following measures:

- proportion of encounters with male patients
- proportion of encounters with new patients
- rate of encounters with Indigenous persons
- rate of management of diabetes
- rate of general check-ups
- rate of pathology ordering
- rate of hospital referrals (Knox *et al.* 2005).

Patients visiting GPs in inner regional areas had characteristics and problem management rates distinct from those visiting GPs in major cities or outer regional areas. Compared with the national average, patients visiting GPs in inner regional areas were older, had more chronic problems managed, had fewer new or acute problems managed, and had higher management rates of depression, back complaint, osteoarthritis and coronary heart disease. In inner regional areas, there was a larger number of problems managed at the encounter compared with the national average (152.1 per 100 encounters versus 148.4), but a lower rate of medications prescribed, advised over the counter, or supplied by the GP (68.8 per 100 problems versus 71.4).

Encounters with GPs in very remote areas are also different from the national average and from encounters with GPs in remote areas. For example, very remote areas had the greatest proportion of encounters with male patients (44.8% of encoun-

Table 6: Management rate of problems (per 100 encounters) encountered in general practice, 1998 to 2004

Illness/problem	ASGC Remoteness area					
	Major cities	Inner regional	Outer regional	Remote	Very Remote	Australia
Major illness/injury category						
Respiratory	22.2	**19.5**	**19.4**	**18.3**	19.4	21.4
Musculoskeletal	17.0	18.3	17.5	17.5	15.2	17.3
Circulatory	16.3	18.4	17.0	15.3	14.9	16.7
Psychological	11.4	12.1	10.3	10.2	**8.0**	11.4
Physical injury	7.3	6.9	7.2	7.7	9.0	7.2
Malignant neoplasm	1.7	2.6	2.8	2.7	2.2	2.0
Most frequently managed problems in Australia						
Hypertension	8.9	9.3	8.9	7.4	7.9	8.9
Upper respiratory tract infection	6.6	**4.2**	**4.4**	**3.2**	4.7	5.9
Immunisation/vaccination	5.0	5.0	**3.7**	**3.4**	3.3	4.9
Depression	3.7	4.4	3.7	3.9	**2.5**	3.8
Diabetes	2.8	2.9	3.2	3.9	4.6	2.9
Asthma	2.8	2.8	3.0	3.1	2.2	2.8
Lipid disorder	2.9	2.6	2.6	**2.0**	2.6	2.8
Acute bronchitis/brochiolitis	2.7	2.7	2.9	2.7	2.4	2.7
Back complaint	2.5	3.0	2.7	2.6	2.2	2.6
Osteoarthritis	2.4	2.9	2.6	2.1	1.7	2.5
Female genital check-up/Pap smear	2.1	2.1	1.8	**1.4**	1.4	2.1
General check-up	1.7	2.2	2.6	3.7	3.1	1.9
Prescription	1.9	2.1	1.9	2.0	1.2	1.9

Note: Shaded rates are significantly higher than the national rate, rates in bold type are significantly lower ($p < 0.05$).
Source: Knox *et al.* 2005.

ters), but the smallest proportion of encounters with patients aged 65 years and over. Also, compared with remote areas, there was a larger proportion of encounters with Indigenous persons (27.4% of encounters versus 13.4%), lower management rates for skin cancers and psychological problems (including depression), and higher management rates for urological problems and new lipid disorders.

Rates of problems managed for the major illness and injury categories, and the most frequently managed problems in Australia, tended to vary with remoteness (Table 6). Compared with the national average:

- Respiratory problems are managed significantly more often in major cities.
- Psychological problems are managed significantly less often in very remote areas.
- Malignant neoplasms are managed more often in inner regional, outer regional and remote areas.
- Arthritis is managed more often in inner regional areas.
- Diabetes is managed more often in outer regional, remote and very remote areas, the rate increasing with increasing remoteness.
- The rate of general check-ups increases with increasing remoteness, but the rate of female check-ups decreases with increasing remoteness.

Patients in very remote areas had a higher rate of referrals compared with the national average

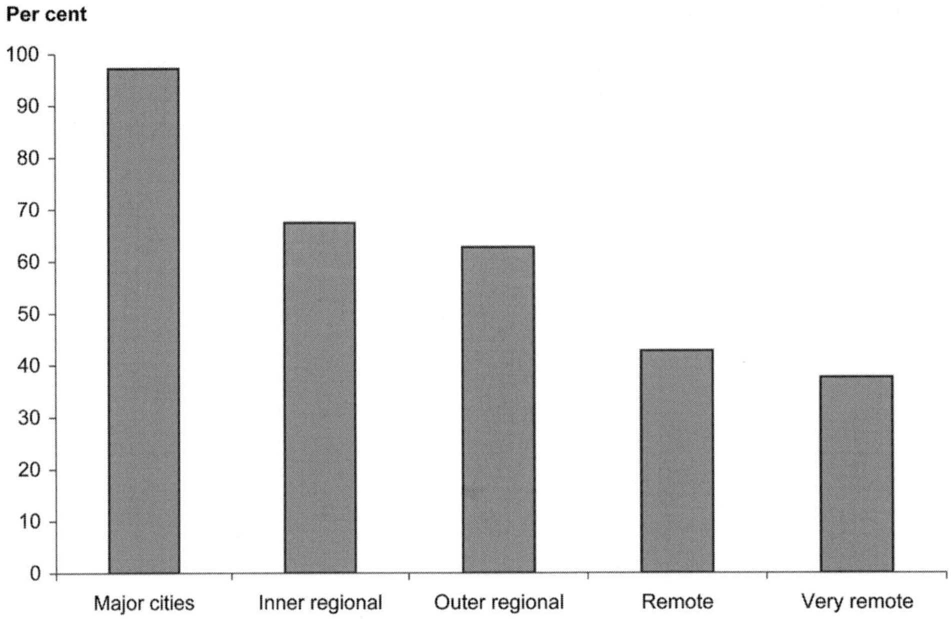

Figure 2: Hospital separations occurring in the same ASGC Remoteness area of the usual residence of the patient, 2005–06 (Source: AIHW National Hospital Morbidity Database).

(10.3 per 100 problems versus 8.0). There was a higher rate of pathology test orders in outer regional and remote areas, and a lower rate in major cities. Referrals to medical specialists were less common in inner regional areas, while referrals to psychiatrists were less common in outer regional areas. Referrals to hospitals were more common in outer regional and remote areas.

Hospital separations

As remoteness increases, people are less likely to be hospitalised in their area of usual residence (Figure 2). That is, whereas 97% of separations involving patients whose usual residence is in major cities occurred in major cities, the corresponding proportion for people whose usual residence is in very remote areas is about 38%. About 12.5% of separations involving people whose usual residence is in very remote areas occurred in major cities.

Hospital separations data from 2003–04 to 2005–06 show strong remoteness effects—in both sexes— for diabetes, respiratory diseases, and injury and poisoning (Table 7). That is, when these conditions were the principal diagnosis, the difference between the hospital separation rate in regional or remote areas and that of people living in major cities increased with increasing remoteness.

Females living in very remote areas with a principal diagnosis of renal failure had the greatest hospital separation rate compared with those living in major cities (about five-and-a-half times the rate for females in major cities). The next greatest difference is noted with females living in very remote areas with a principal diagnosis of diabetes (more than three-and-a-half times the rate for females in major cities).

For males, the greatest difference is noted with those living in very remote areas with a principal diagnosis of diabetes (more than two-and-a-half times the rate for males in major cities).

Intra-urban disparities

The above analyses highlight disparities in health outcomes *between* urban, regional and remote areas. But health outcomes are not uniformly distributed *within* urban areas either. Urban areas also contain or create many risks to health that vary spatially, including lifestyle, availability of energy-rich foods, the urban heat island effect (that is, the tendency of inner cities to be warmer than surrounding areas due to the heat trapping/radiating effect of buildings, roads and pavements), crime, air pollution, overcrowding as well as stress from

Table 7: Standardised separation ratios (SSRs) by ASGC Remoteness area, 2003–04 to 2005–06

Principal diagnosis	Sex	ASGC Remoteness area				
		Major cities	Inner regional	Outer regional	Remote	Very remote
All diagnoses	Males	1.00	0.95	1.00	1.02	1.39
	Females	1.00	0.97	1.04	1.15	1.58
Circulatory diseases	Males	1.00	1.09	1.11	1.02	1.23
	Females	1.00	1.12	1.18	1.22	1.65
Neoplasms	Males	1.00	1.02	0.95	0.76	0.68
	Females	1.00	0.99	0.93	0.74	0.70
Respiratory diseases	Males	1.00	1.10	1.28	1.59	2.13
	Females	1.00	1.14	1.37	1.83	2.60
Injury and poisoning	Males	1.00	1.21	1.45	1.67	2.15
	Females	1.00	1.08	1.24	1.59	2.69
Diabetes	Males	1.00	1.04	1.27	1.43	2.59
	Females	1.00	1.03	1.31	1.55	3.79
Renal failure	Males	1.00	0.66	0.72	0.98	2.44
	Females	1.00	0.75	1.05	2.36	5.42
Mental health problems	Males	1.00	0.88	0.82	0.83	0.85
	Females	1.00	0.74	0.66	0.63	0.67
Musculoskeletal conditions	Males	1.00	1.17	1.25	1.15	1.03
	Females	1.00	1.10	1.12	1.06	1.02

Notes: 1 SSRs are the ratio of the observed number of persons with the principal diagnosis in the region compared with the expected number of persons if the age-specific prevalence rates in major cities were to apply to the population in that region.

2 Shaded SSRs are not statistically significant, all other SSRs are significantly different compared with major cities ($p < 0.05$).

3 ICD-10 codes: circulatory diseases (I00–I99), neoplasms (C00–D48), respiratory diseases (J00–J99), injury and poisoning (S00–T98, with a first reported external cause V01–Y98), diabetes (E10–E14, O24), renal failure (N17–N19, Z49), mental health problems (F00–F99, G30) and musculoskeletal conditions (M00–M99).

Source: AIHW National Hospital Morbidity Database.

work demands, travel, financial pressures and social isolation (Galea and Vlahov 2005; Kjellstrom 2006; Kearns *et al.* 2007). Intra-urban disparities exist in relation to these risks. For example, people living in urban areas of lower socioeconomic status have been found to have higher levels of obesity (King *et al.* 2006) and lower levels of physical activity (Kavanagh *et al.* 2005) than those living in areas of higher socioeconomic advantage, even after taking into account the socioeconomic status of the individual.

These intra-urban disparities notwithstanding, people living in regional and remote areas have demonstrably poorer health. There must be consistently observable reasons for these health disparities.

Reasons for disparities

Disparities in health outcomes within and between populations occur when there is an uneven distribution of the factors that harm, improve or restore health. Besides differences in genetic susceptibilities of populations in different regions, factors that may explain health disparities within and between populations include environmental factors, lifestyle risk factors, social factors, and availability and access to health-care services (Dennis and Pallotta 2001; Galea and Vlahov 2005).

Environmental factors

Environmental factors are hazards within, or created by, natural and built environments. Human settlements are both environments in their own right and a source of pressure on the broader physical environment (Newton *et al.* 1998). Consequently, they generate a range of situations and conditions that affect human health. The size, location, design and demographic composition of a settlement can have an impact on the incidence and prevalence of health problems, such as obesity, cardiovascular disease, cancer, diabetes, mental illness, respiratory diseases and injuries. This is partly because there is variability between settlements in the factors that influence health, such as the supply of fresh food, water quality, pollution levels, crime rates, public safety, social interaction and opportunities for physical activity (Capon and Dixon 2007). It follows that differences between the physical and socio-demographic characteristics within and between urban, regional and remote areas create observable trends and disparities in the health of their inhabitants. For example, as a result of working in rural occupations such as farming and mining, people living in rural areas may be exposed more often to health risks such as herbicides, insecticides, fertilisers, airborne dusts and gases, and diesel exhaust fumes (Dennis and Pallotta 2001). On account of living and working in large cities, exposure is higher to health risks linked to motor vehicle-induced air pollution, especially smog and fine particles (Newton and Manins 1999).

Lifestyle factors

Lifestyle or behavioural factors include modifiable elements such as diet, exercise and tobacco smoking. Health and longevity are greatly influenced by unhealthy behaviours such as smoking tobacco, drinking alcohol more than moderately, not exercising regularly, not maintaining a healthy body weight, not sleeping adequately and not eating regular and nutritious meals (Breslow 2004). Many of these unhealthy behaviours have been found to be more common in regional and remote areas than in major cities (AIHW 2006b). Several of them have been shown in this chapter to be more likely outside of major cities.

Social factors

Social factors include social norms and attitudes. Achieving good health also depends on the motivation to do what needs to be done (a function of personal attitudes and social norms), the ability to do it and the resources to maintain that ability (Breslow 2004). Motivation and personal choice may play a role in what Pearson and Lewis (1998) refer to as rural populations' slow adopter behaviour. That is, they tend to be slower than urban populations to adopt unhealthy behaviours (such as smoking, high-fat diets, and sedentary lifestyles), but are also slower at adopting healthy behaviours. For example, being overweight or obese was at first more common in urban than rural areas; now the opposite appears to be the case (Swinburn 2006).

However, not all slow adopter behaviour is likely to be voluntary—circumstances such as socioeconomic status, distance and lack of free time are also likely to be influential factors. A lower rate of disease screening in remote and very remote areas, for example, might have more to do with less access to adequate health-care facilities as much as, if not more than, a negative attitude towards screening. In the rural American communities they studied, Pearson and Lewis (1998) proposed low socioeconomic status (rural poverty) and poor access to health-care services as explanations for the slow adopter behaviour they observed. In addition to slow adopter behaviour, people living in regional and remote areas may also be exposed to the effects of 'slow provider' behaviour. That is, they may experience delays in the delivery of infrastructure (such as safe roads or adequate sanitation) and public health policy initiatives (such as education about risk factors and promotion of healthy behaviour) that may help to prevent illness and injury.

Availability and access to health-care services

Health-care services include both primary care and specialist care. As well as health risks, urban areas also create and contain factors that improve, restore or maintain health, such as cultural diversity, social support, economic opportunities, high water quality and accessible health-care services. Of these factors, availability of and access to adequate health-care services, particularly specialty services, is likely to be a major practical difference between life in urban, regional and remote areas (Dennis and Pallotta 2001). For example, although disparities in quality of care may exist in urban areas due to wealth inequalities, they are likely to be balanced in comparison with rural residents by (a) greater distances to health-care services for rural residents and (b) health-care providers choosing to work in urban areas for their perceived and actual advantages (Galea and Vlahov 2005).

In short, compared with those living in major cities, people living in regional and remote areas must either be exposed to more harmful environmental factors, lead less healthy lifestyles, have attitudes or norms less conducive to good health, and/or have less access to health-care services leading to significantly worse health. Two other important factors to consider are socioeconomic disadvantage and the distribution and health of Indigenous Australians.

Socioeconomic disadvantage

It is well established that socioeconomically disadvantaged people (such as those with low income or poor education) tend to live shorter and less healthy lives (AIHW 2006a). Having a low income and/or poor education, for example, limits choices and opportunities that may affect health. It is in effect a risk factor that may profoundly influence other risk factors, such as diet and access to health-care services. Socioeconomic disadvantage tends to be greater in regional and remote areas compared with major cities. In 1996, about 55% of people living in very remote areas were classified as being among the most disadvantaged people in Australia (based on national Socioeconomic Indexes for Areas (SEIFA) quartiles) compared with 20% in major cities (AIHW 2006a).

Indigenous Australians

Indigenous Australians generally experience poorer health and die much younger than other Australians (ABS and AIHW 2005). They are more likely than other Australians to be exposed to poor living conditions and to health risk factors such as tobacco smoking and risky alcohol drinking (AIHW 2006c, 2007). Therefore, any examination of regional health disparities must also consider the fact that the proportion of Indigenous Australians within the population increases with increasing remoteness. According to the 2001 Census, about 44% of people in very remote areas are Indigenous, compared with 13% in remote areas, 5% in outer regional areas, 2% in inner regional areas and 1% in major cities. It is well known that overall the mortality rate among Indigenous Australians is much higher than that for other Australians (AIHW 2006a).

Implications

Trends in the health of the Australian population over the past 100 years reveal a transition away from acute, communicable diseases to chronic, non-communicable diseases that require long-term management. Improved understanding of underlying risk factors and advances in medical sciences has contributed to considerable improvement in the prevention and management of many of these diseases. Accompanying these trends has been a shift in the pressures and challenges on the health-care system. The extent of the challenge is observable in the disparities in health outcomes between the service-intense urban and service-stressed regional and remote areas. Besides poorer access to health-care services, these disparities may result from lower socioeconomic status, exposure to relatively harsher environmental and occupational hazards, poorer attention to lifestyle risk factors and higher proportions of Indigenous Australians in regional and remote areas.

Several health outcome and service use disparities have been reported in this chapter. Most of the excess deaths in remote and very remote areas are due to circulatory diseases and injury and poisoning. However, disparities related to diabetes, and to a lesser extent injury and poisoning, appear to have the greatest impact when comparing rates in regional and remote areas with those in major cities.

In regional and remote areas, death rates for diabetes and injury and poisoning provide greatest contrast to those living in major cities; the difference increases with increasing remoteness. For diabetes, the rate of management in general practice and the hospital separation rate also increase with increasing remoteness. Due to the limitation of sufficient data for only three settlement categories, remoteness effects are difficult to reveal from NHS self-reports. Nevertheless, the rate of exposure to risk factors that may lead to diabetes appears to be higher outside of major cities, for both sexes. Interestingly, diabetes in females is reported at a higher rate outside of major cities, but the opposite is noted in males. These findings require further investigation and action.

To various extents, the concentration and accessibility of health services in urban areas reduces the impact of many of the health risks encountered by people living in major cities but this may occur at the expense of those living in regional and remote areas. Similarly, if the incidence of diabetes continues to rise as expected, those living in regional and remote areas, particularly Indigenous Australians, are likely to bear more than their fair share of the consequences.

Acknowledgements

This project was funded by the Australian Institute of Health and Welfare (AIHW) through its Asthma, Arthritis and Environmental Health work program. The authors would like to thank Ms Susan Killion, Senior Executive, Health and Functioning Group, AIHW and Dr Penny Allbon, Director of AIHW for their support in undertaking this project, Mr Andrew Phillips for providing helpful comments, and Ms Faith Margrie for assisting with the preparation of the manuscript.

References

ABS (Australian Bureau of Statistics) (2001). *Australian Standard Geographical Classification (ASGC)*. ABS cat. no. 1216.0. ABS, Canberra.

ABS and AIHW (Australian Institute of Health and Welfare) (2005). *The Health and Welfare of Australia's Aboriginal and Torres Strait Islander Peoples, 2005*. ABS cat. no. 4704.0, AIHW cat. no. AIHW 14. ABS and AIHW, Canberra.

AIHW (2005a). *Oral Health and Access to Dental Care – Rural and Remote Dwellers*. AIHW cat. no. DEN 144. AIHW Dental Statistics and Research Unit, Adelaide.

AIHW (2005b). *Rural, Regional and Remote Health—Indicators of Health*. AIHW cat no. PHE 59. AIHW, Canberra.

AIHW (2006a). *Australia's Health 2006*. AIHW cat. no. AUS 73. AIHW, Canberra.

AIHW (2006b). *Chronic diseases and associated risk factors in Australia 2006*. AIHW cat. no. PHE 81. AIHW, Canberra.

AIHW (2006c). *Rural, Regional and Remote Health – Mortality Trends 1992–2003*. AIHW cat. no. PHE 71. AIHW, Canberra.

AIHW (2007). *Statistics on Drug use in Australia 2006*. AIHW cat. no. PHE 80. AIHW, Canberra.

Breslow L (2004) Perspectives: the third revolution in health. *Annual Review of Public Health* **25**, xiii–xviii.

Capon AG and Dixon JM (2007). Creating healthy, just and eco-sensitive cities. *NSW Public Health Bulletin* **18(3–4)**, 37–40.

Dennis LK and Pallotta SL (2001). Chronic disease in rural health. In *Handbook of Rural Health 2001*. (Eds S Loue and BE Quill) pp. 189–207. Springer Publishing Company, New York, NY.

DHAC (Department of Health and Aged Care) and GISCA (National Key Centre for Social Applications of Geographic Information Systems) (2001). *Measuring Remoteness: Accessibility/Remoteness Index of Australia (ARIA)*. Occasional Papers: New Series No. 14. DHAC, Canberra.

Galea S and Vlahov D (2005). Urban health: evidence, challenges, and directions. *Annual Review of Public Health* **26**, 341–365.

Kavanagh AM, Goller JL, King T, Jolley D, Crawford D and Turrell G (2005). Urban area disadvantage and physical activity: a multilevel study in Melbourne, Australia. *Journal of Epidemiology and Community Health* **59**, 934–940.

Kearns A, Beaty M and Barnett G (2007). A social-ecological perspective on health in urban environments. *NSW Public Health Bulletin* **18(3–4)**, 48–50.

King T, Kavanagh AM, Jolley D, Turrell G and Crawford D (2006). Weight and place: a multilevel cross-sectional survey of area-level social disadvantage and overweight/obesity in Australia. *International Journal of Obesity* **30**, 281–287.

Kjellstrom T (2006). Risks and benefits to health from living in urban environments. In *Program and Abstracts of the Urbanism, Environment and Health Fenner Conference 2006*. pp. 10–12. The National Centre for Epidemiology and Population Health, Canberra.

Knox S, Britt H, Pan Y, Miller GC, Bayram C, Valenti L, Charles J, Henderson J, Ng A and O'Halloran J (2005). *Locality Matters: the Influence of Geography on General Practice Activity in Australia 1998–2004*. AIHW cat. no. GEP 17. AIHW, Canberra.

Newton P, Flood J, Berry M, Bhatia K, Brown S, Cabelli A, Gomboso J, Higgins J, Richardson T and Ritchie V (1998). *Environmental Indicators for the National State of the Environment Reporting—Human Settlements, Australia: State of the Environment (Environmental Indicators Report)*. Environment Australia, Canberra.

Newton P and Manins P (1999). Cities and air pollution. In *East-West Perspectives on 21st century Urban Development* (Eds J Brotchie, P Newton, P Hall and J Dickey) pp. 277–304. Ashgate Publishing, Aldershot, UK.

Pearson TA and Lewis C (1998). Rural epidemiology: insights from a rural population laboratory. *American Journal of Epidemiology* **148**, 949–57.

Swinburn B (2006). Rising obesity and the urban environment: causes and solutions. In *Program and Abstracts of the Urbanism, Environment and Health Fenner Conference 2006*. p. 34. The National Centre for Epidemiology and Population Health, Canberra.

20

Community Social and Human Capital

Scott Baum and Geoff Woolcock

In recent years there has been a resurgence of interest in the nature and significance of communities, their sustainability and what strong communities mean for the people who reside in them. Strong communities are defined as those endowed with social, economic and environmental assets and organisational structures that work towards their sustainable and equitable use (Department for Victorian Communities 2004). Put another way, we can think about 'the extent to which resources and processes within a community maintain and enhance both individual and collective wellbeing in ways consistent with the principles of equity, comprehensiveness, participation, self-reliance and social responsibility' (Black and Hughes 2001, p. 3).

Much contemporary policy has focused on the need to strengthen communities (and families). A key concern in this area is that, in circumstances where the characteristics of a strong community are missing, members of that community have 'less capacity to meet the challenges of economic change and to cope with the pressures that lead to family and social breakdown' (Department of Families and Community Services 1999, p. 4). The notion of a strong community is of critical importance in a period of smaller government and the retreat of the welfare state. Stronger communities (and stronger families) are seen as augmenting neo-liberal policies across a range of systems.

Other concerns centre on how the economic and social transitions underway in contemporary Australia have affected communities. Both within large cities and between smaller regional cities and towns, there is a confronting paradox whereby, despite long periods of social and economic prosperity at aggregate (national-regional) levels, there are diverse and unequitable outcomes at disaggregated community levels. Questions have been continually raised about the extent to which some communities are socially included and some are socially excluded from the broader-based prosperity. An often associated question relates to the extent to which community strength measured in terms of a range of types of 'capital' varies between local communities or across a given region.

How to analyse and benchmark community performance is a long-established concern of regional scientists. In Australia, recent studies have focused on patterns of community opportunity and vulnerability (Baum *et al.* 1999, 2005; Stimson *et al.* 2001), which characterise localities by their level of performance across a wide range of socioeconomic transition measures:

- the changing roles of Australia's metropolitan cities (O'Connor and Stimson 1995)
- patterns of change in regional cities (Beer *et al.* 1994)
- the functional roles of regional urban centres (Beer and Maude 1995; Beer 1999).

Other research has focused on more specific aspects of socioeconomic performance, including: income (Gregory and Hunter 1995); poverty (Fincher and Nieuwenhuysen 1998); and unemployment (Stimson 1997; Stimson *et al.* 1998).

There has also been a gathering interest in the nature and measurement of social and human capital (Onyx and Bullen 1997, 2000; Stone 2001). Importantly, these two 'capitals' are considered as central to understanding community strength and sustainability. Similarly State of Environment reporting addresses stocks of capital including human capital (Newton *et al.* 2001). This chapter provides an oveview of current research, while adding material related to trends in human and social capital and their link to community strength and sustainability.

On human and social capital

Human capital

Human capital is 'the knowledge, skills, competences and other attributes embodied in individuals that are relevant to economic activity (OECD 1998, p.9). At a general level, it focuses on the economic behaviour of individuals, especially the way their accumulation of knowledge and skills enables them to increase their productivity and earnings and, in doing so, to increase the productivity and wealth in their host societies. The underlying assumption is that there are positive returns to investment in human capital and that these returns are felt by individuals as well as by the wider community and society. At the community level, returns to human capital investment are reflected in the capacity and motivation of individuals to contribute to their local community:

it is dependent on their motivation to do so and their ability to do so as measured by their skills and knowledge, their capacity to adjust to changing circumstances, sometimes by acquiring new skills and knowledge, and the management of health and disability. (Black and Hughes 2001, p. 3)

While the strength of a community is largely influenced by the extent of human capital available within that community, it is also influenced by the ways in which human capital is being developed within a community. Local and regional educational systems make an important contribution to the production of human capital. Moreover, while formal human capital is the most readily available indicator of outcomes (for example, the level of education attained), informal human capital which might include skills learned in the workforce are also an important element of a local community's 'stock' of human capital.

Social capital

In contrast to human capital, social capital is a more complicated idea to conceptualise. The concept of social capital has been gaining interest among researchers and policy-makers and a push is also evident from the general community to use social capital as a way to both describe and understand community wellbeing. The origins of social capital as a concept can be traced to the first decade of the 20th century (Hanifan 1916). In the 1980s, Coleman (1988) put the notion of social capital firmly on the intellectual agenda, arguing that it is embodied in the relations among people, and that it can facilitate productive activity, and that it is manifest in level of trust. Woolcock (1998, p. 5) suggests that social capital is a 'broad term encompassing the norms and networks facilitating collective action for mutual benefit'. Putnam (2000, p. 19) who is arguably its leading proponent, suggests that 'social capital refers to connections among individuals—social networks and the norms of reciprocity and trustworthiness that arise from them'.

Two aspects to social capital can be identified in the work of these and other scholars: social structures or social networks; and the norms governing behaviour in these social structures or social networks. A high level of social capital is seen in situations where there are cohesive networks of considerable density and where interactions are governed by norms of trust and reciprocity. Cohesive and strong communities, are therefore characterised by high levels of social capital, whereas weak communities are characterised by low levels of social capital. Social capital's alleged causal affect on the strength of communities is the subject of intense criticism by many scholars (Durlauf 2002; Mowbray 2004). Many draw attention to highly cohesive communities, which can also be socially exclusive communities resistant to change (Portes 1998; Durlauf 1999)—traits often ascribed to excessive levels of 'bonding' social capital.

In discussing formal and informal networks, Putnam (1998) suggests that informal networks include relationships people have with their families, partners, friends and neighbours; whereas formal networks include relationships at work, within community groups and churches and with formal bodies such as business and governments. In this way social capital can be thought of as comprising four broad groupings: informal structures, formal structures, informal norms and formal norms.

Norms governing network behaviour relate to trust, unity and reciprocity. Within formal networks, individuals have what is called particularised trust: a trust that is specific to the individual a person knows (Uslaner 1999; Cox and Caldwell 2000). This is different to the trust people have for strangers since the probity of a stranger cannot be predicted with the same certainty as it can for a person known to the individual. The trust afforded to strangers is by its very nature generalised and so is termed generalised trust (Dasgupta 1998; Putnam 1998; Uslaner 1999). Trust in formal networks, which is referred to as institutional trust, is similarly general because it is not aimed at individuals but rather at institutions and relates to, for example, trust of the government, the police or the church (Giddens 1990; Black and Hughes 2001; Stone 2001).

Closely related to trust are unity and reciprocity. Unity is a feeling of belonging to a network together with the feeling that a two-way relationship exists between the individual and the network. An example of unity within a formal context is when governments include citizens in decision-making, creating a sense of mutual ownership of the task and a two-way relationship between the institution and the individual. Reciprocity concerns the exchanges that take place within a network. Individuals are likely to offer assistance to others if they believe that in the future the assistance will be reciprocated. Volunteering is a good example of reciprocity within a formal context, where individuals may feel that they are 'giving something back' to the community.

For the normative dimension, a high level of social capital is indicated by particularised trust, generalised trust, unity and reciprocity. High levels of social capital are associated with high levels of trust, unity and reciprocity in informal networks, such as family and friendship networks, and in formal networks, such as the community at large, local groups, associations and institutions. Sustaining high levels of social capital is dependent on an ability to intermix networks through open-minded, diverse and overlapping interactions with other networks (Gluckman 1967; Granovetter 1973; Coleman 1988).

Both normative and structural aspects of social capital tap into what are termed bonds, bridges and linkages. Bonds refer to the internal dynamics of a network (measured by size, capacity, homogeneity and openness) and they also focus on normative aspects such as trust, unity and reciprocity, particularly within informal networks. Bridges equate to the density of networks and the capacity of people within one particular network to draw on other networks, both informal and informal. Bridges are characterised by heterogeneity of membership that entail ties that cut across characteristics of social groups such as gender, wealth and religion (Putnam 1998; Krishna and Shrader 1999; Narayan 1999; Woolcock and Narayan 2000). Linkages represent a particular form of bridging social capital—measuring the links people have with authoritative organisations and institutions (Putnam *et al.* 1993; Putnam 2000; Black and Hughes 2001).

Trends on key indicators

Human capital

In Australia there has been significant attention paid to the differentiation across communities of human capital and educational outcomes. The broad trends point to a changing educational base within Australia. Levels of formal education have been increasing steadily over the paste decade or so. The proportion of people undertaking formal education has increased between 1988 and 2005, especially among those aged 19 to 24 years. Similarly, aggregate Year 12 retention rates illustrate a continuing focus on achieving a formal education. Retention rates for males have increased from 53.9 to 69.9% over the period 1988 to 2005, while over the same time the rate for females has risen from 61.8 to 81.0%. Not surprisingly, the level of achieved education—the stock of human capital—has also increased with higher proportions of people holding degrees or higher degrees in 2005 (19.6%) than in previous census years (7.6% in 1988).

These broad numbers hide differentiation that is known to exist at more disaggregated spatial levels. Research by Stimson *et al.* (2004) developed a series of human capital indices and found that, in general, the capital cities have higher average scores on indicators of high formal qualifications than localities in non-metropolitan Australia, although there are significant variations within metropolitan area. However, the picture across regional Australia is not one of universal doom and gloom.

Table 1: Education levels Australia 1988–2001

	1988(a)	1998(a)	2001(b)	2005 (a)
Year 12 Retention rates[1]				
Male	53.9	65.9	68.1	69.9
Female	61.8	77.7	79.1	81.0
Proportion of persons aged 15–64 years who did not complete the highest level of secondary school[2]	43.7	34.1	39.9	31.2
Proportion of persons aged 15–19 years in education[3]		76.9	77.4	76.0
Proportion of persons aged 19–24 years in education[3]	–	32.1	34.8	38.9
Proportion of persons aged 15–64 years with a bachelor degree or higher	7.6	14.3	17.0	19.6

[1] Number of school students in Year 12 expressed as a percentage of their respective cohort at the commencement of secondary schooling.

[2] Data for 1988 are for 15–69 years and hence are not strictly comparable with later data

[3] Persons enrolled in a course of study at an educational institution

Source (a): Australian Bureau of Statistics 2000; 1988 and 1998 data are derived from p. 82.

Source (b): Australian Bureau of Statistics 2002; data for 2001 derived from p. 43.

There are regional centres performing well (such as Armidale in NSW, Townsville in Queensland, which both have regional universities). Other regional centres, such those large cities based on tourism or mining, also have high levels of human capital. There is some evidence that population size does matter. This is seen in the degree to which larger urban areas have more diversified job opportunities and higher levels of human capital than smaller areas. From the research reported by Stimson *et al.* (2004), smaller regional centres and remote communities are the most vulnerable in terms of having low levels of human capital. Moreover, some of the rapidly growing coastal communities, which have in recent times been seen as important new regions of social and economic growth (Salt 2001), have low levels of human capital, suggesting that growth per se is not a panacea for potential problems.

The trends in human capital attainment and community outcomes identified by Stimson *et al.* (2004) are mirrored by the Australian Bureau of Statistic's SEIFA Index of education and occupation—a proxy measure for human capital development of Communities. Figure 1 illustrates that there are marked geographic variations in this index. The index is much higher in some metropolitan sub-regions than in non-metropolitan sub regions. The top 20 sub-regions are all located in metropolitan regions and are over-represented in Sydney (Table 2), reflecting the nature of employ-

Table 2: Top 20 regions, SEIFA index of education and occupation 2001

Region	Index score
Boroondara City (Melbourne)	1180.34
Central Metropolitan (Perth)	1168.73
Lower Northern Sydney	1164.42
Inner Melbourne	1163.53
Eastern Suburbs (Sydney)	1145.85
Central Northern Sydney	1138.17
Inner Sydney	1115.53
Eastern Adelaide	1109.82
Northern Beaches (Sydney)	1097.38
Inner Western Sydney	1095.4
Southern Melbourne	1091.06
Eastern Middle Melbourne	1085.35
Brisbane City	1064.34
St George–Sutherland (Sydney)	1047.38
Northern Middle Melbourne	1033.5
North Metropolitan (Perth)	1018.56
Moreland City (Melbourne)	1014.61
Eastern Outer Melbourne	1012.8
South West Metropolitan (Perth)	1007.02
Southern Adelaide	1006.17

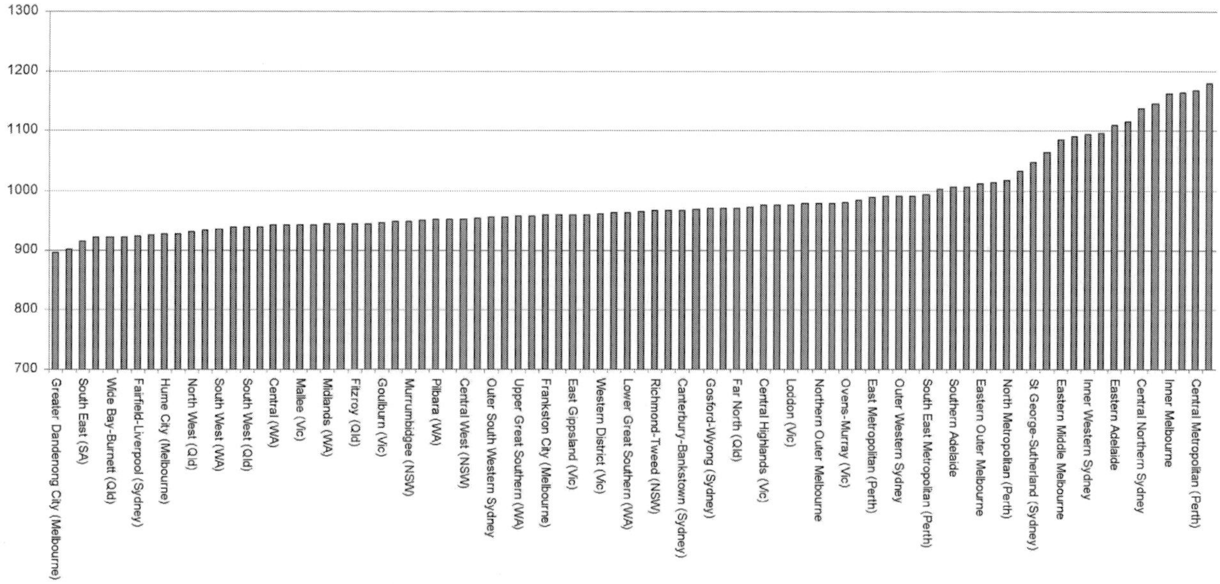

Figure 1: SEIFA index for education and occupation, 2001, metropolitan and non-metropolitan regions

ment and occupational structure of that city and, in particular, the residential location of people employed in new economy and other high-skilled jobs (Baum *et al.* 2005). By way of contrast, the bottom 20 sub-regions represent a mix of metropolitan and non-metropolitan localities (Table 3), with a number of very remote regions (that is, far north New South Wales or northern South Australia) being represented.

Table 3: Bottom 20 regions, SEIFA Index of education and occupation, 2001

Region	**Index score**
Greater Dandenong City (Melbourne)	895.76
Murray Lands (SA)	901.21
South East (SA)	914.68
Northern (SA)	921.33
Wide Bay–Burnett (Qld)	922.62
Far West (NSW)	922.79
Fairfield–Liverpool (Sydney)	923.27
Yorke and Lower North (SA)	926.5
Hume City (Melbourne)	926.7
Northern Adelaide	927.77
North West (Qld)	931.31
Mackay (Qld)	933.01
South West (WA)	935.92
Eyre (SA)	938.6
South West (Qld)	939.37
South Eastern (WA)	939.44
Central (WA)	942.45
Brisbane balance	942.51
Mallee (Vic.)	943.33
Central West (Qld)	943.34

Within metropolitan regions, the distribution of human capital scores differs significantly across communities (see Figures 2 and 3). High scores for Sydney are evident in those communities/suburbs that make up what is referred to as the global arc or global Sydney—centred on the harbour and northern suburbs (mirroring the data in Table 2)—while lower index numbers tend to be concentrated with communities that have traditionally been associated with old economies and disadvantaged neighbourhoods (Baum *et al.* 2005). A similar picture is presented in Melbourne, with higher human capital in the core 'inner east and bayside zones' compared with the outer suburbs.

Social capital

Some significant social capital attributes have been measured that indicate that Australia can claim comparatively high levels of volunteerism and generalised trust. The Commonwealth Bureau of Transport and Regional Economics (BRTE, DATE) recently released data on indicators of social capital

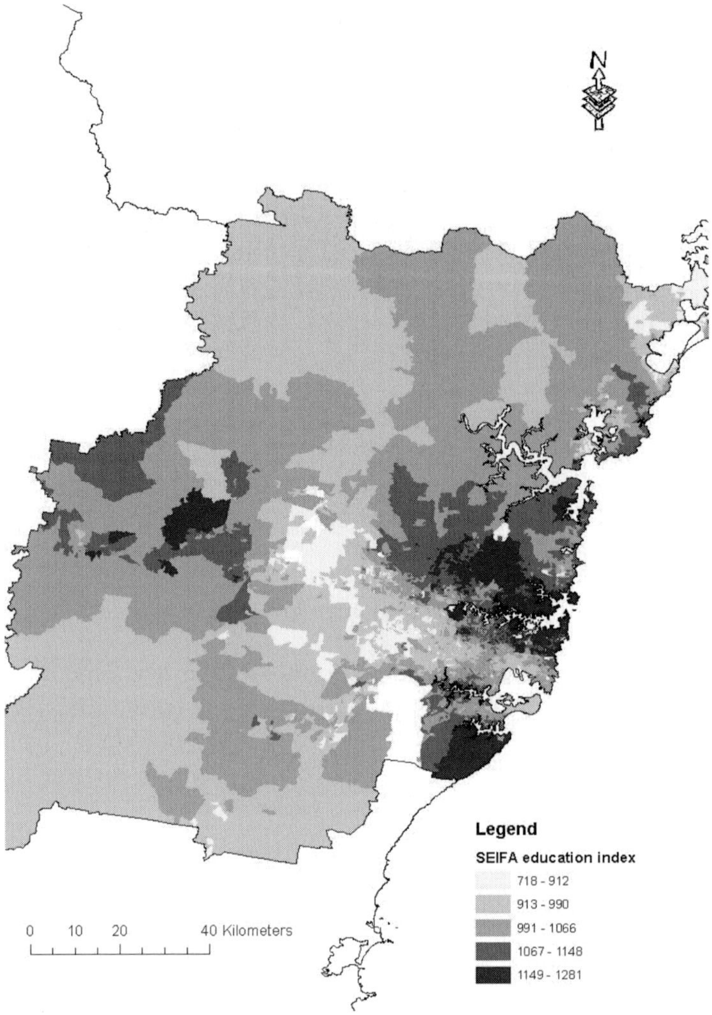

Figure 2: SEIFA index of education and occupation 2001, Sydney

across Australia's regions. The BRTE indices cover the notions of formal and informal structures outlined earlier. Information in Table 4 presents generalised indicators of high versus low social capital. Localities with statistically significant social capital outcomes are labelled as 'H' for above average and 'L' for below average. The first thing to note is that when social capital is viewed as a suite of indicators, there is significant variation in terms of factors that contribute to a community's level of social capital. In some localities, strong as opposed to weak neighbouring is important, while in other places the ability to tap into important social support is more important. Beyond these differences, there is clear spatial variation both across states and between regions. Within states, positive indicators of social capital are more evident in regional and fringe metropolitan areas than in the established metropolitan zones. It is clear that in comparing non-metropolitan and metropolitan localities in New South Wales for example, the level of active membership, volunteering and neighbouring are more likely to be high and above average in non-metropolitan than metropolitan regions. Feelings of community integration are also higher in non-metropolitan regions than metropolitan regions. A similar picture is presented for Victoria and, to some extent, Western Australia, while the patterns are less strong in Queensland and South Australia.

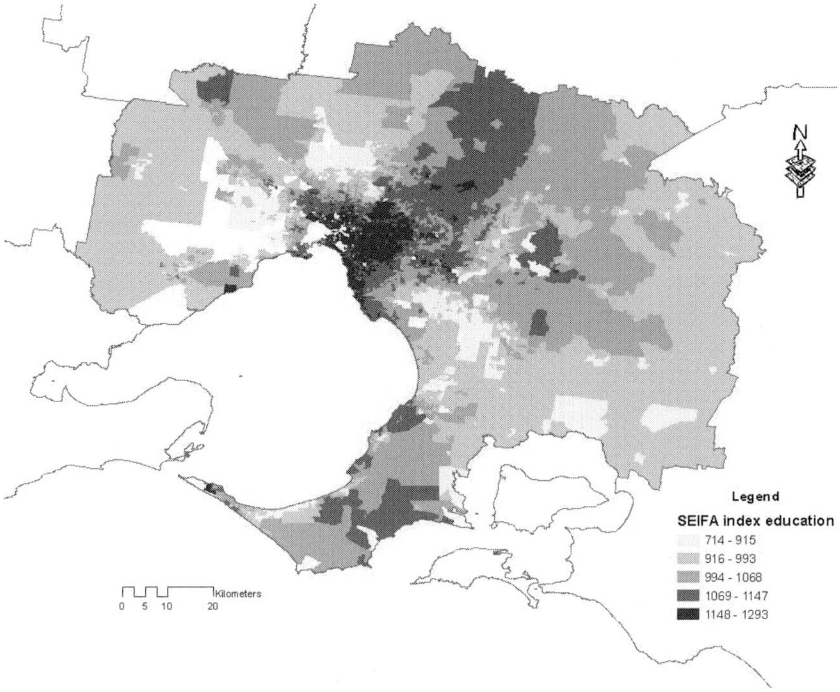

Figure 3: SEIFA index of education and occupation 2001, Melbourne

A more generalised picture is available from BTRE's composite indices of 'general support' and 'community involvement'. Figure 4 represents these indices for regions in metropolitan and non-metropolitan New South Wales, and again indicates that these social capital indicators are higher (above the mean) in non-metropolitan than in metropolitan regions.

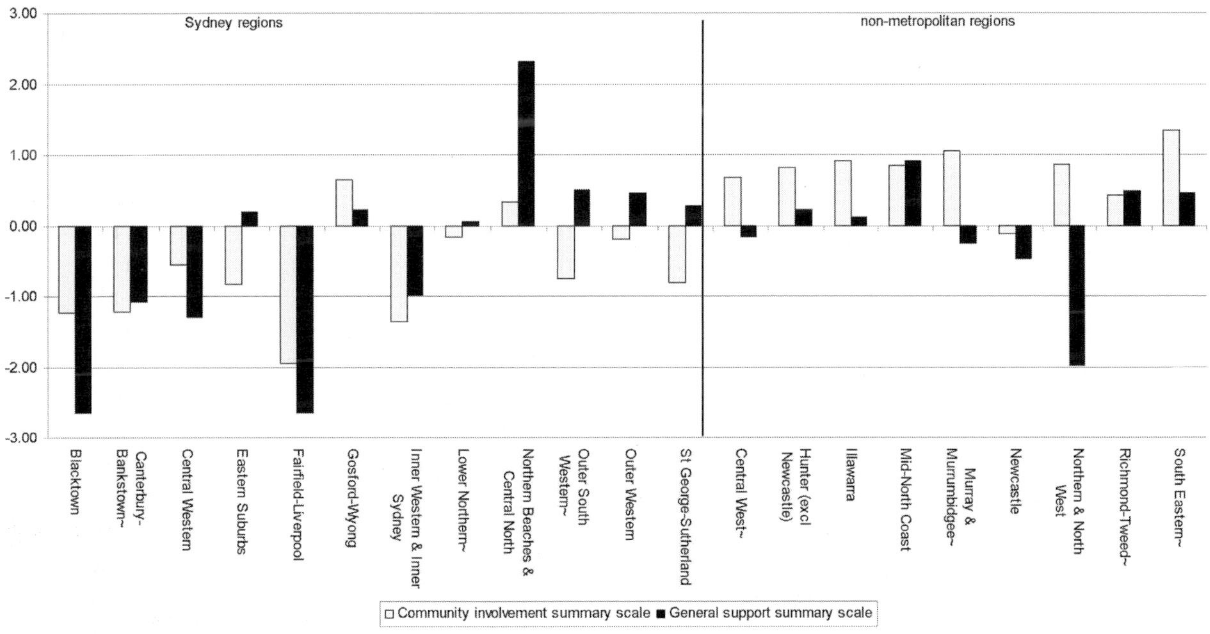

Figure 4: Social capital summary scales, metropolitan and non-metropolitan New South Wales

Table 4: Social Capital Indicators, 2001–02

REGION	How commonly do neighbours help each other out in your neighbourhood?*	Active membership rate (%)	Volunteering rate (%)	Satisfaction with family relationships*	Health barriers to social participation*	Frequency of social contact*	Emotional and general support received from others*	Capacity to raise $2000 in 1 week for emergency*	Integration into the community*	Only get together socially once a month or less with friends or relatives (%)	I often feel very lonely*
SYDNEY											
Blacktown	–	L	–	–	H	–	–	L	–	–	H
Canterbury–Bankstown	–	–	–	–	–	–	–	–	–	–	–
Central Western	–	–	–	–	–	–	L	–	–	–	–
Eastern Suburbs	L	–	–	–	–	–	–	–	–	L	–
Fairfield–Liverpool	–	L	–	–	H	–	–	L	–	–	–
Gosford–Wyong	H	H	–	–	–	–	–	–	–	–	–
Inner Western and Inner Sydney	L	L	L	–	H	–	–	–	L	–	–
Lower Northern	–	–	–	–	–	–	L	H	–	–	–
Northern Beaches and Central North	–	–	–	–	L	–	–	H	–	L	L
Outer South Western	–	–	–	H	–	L	–	–	–	–	–
Outer Western	–	–	–	–	–	–	–	–	–	–	–
St George–Sutherland	–	L	–	–	–	–	–	H	H	–	–
REST OF NSW											
Central West	H	–	–	–	–	–	–	–	H	–	–
Hunter (excl Newcastle)	–	–	–	–	–	–	–	–	H	–	–
Illawarra	H	H	–	–	–	–	–	–	–	–	–
Mid-North Coast	H	–	–	–	–	L	–	–	H	H	–
Murray and Murrumbidgee	–	H	H	–	–	–	–	–	H	–	–
Newcastle	–	–	–	–	–	–	–	–	–	–	–
Northern and North West	H	–	–	–	–	–	–	L	H	–	H

Table 4 (continued)

REGION	How commonly do neighbours help each other out in your neighbourhood?*	Active membership rate (%)	Volunteering rate (%)	Satisfaction with family relationships*	Health barriers to social participation*	Frequency of social contact*	Emotional and general support received from others*	Capacity to raise $2000 in 1 week for emergency*	Integration into the community*	Only get together socially once a month or less with friends or relatives (%)	I often feel very lonely*
Richmond–Tweed	H	–	–	–	–	–	–	–	H	–	–
South Eastern	H	H	–	–	–	–	–	–	H	–	–
MELBOURNE											
Boroondara City	L	–	L	–	L	–	–	H	–	–	–
Eastern Middle	–	–	–	–	L	–	–	H	–	–	–
Eastern Outer	–	–	–	–	–	–	–	–	–	–	–
Greater Dandenong and Frankston	–	–	–	–	–	–	–	L	–	–	–
Inner Melbourne	L	–	–	L	–	H	–	–	L	L	–
Moreland and Northern Middle	–	L	–	–	–	–	–	–	–	–	–
Mornington Peninsula^	–	–	–	–	–	–	–	H	H	–	–
Outer Western and Northern	–	L	–	–	–	–	–	–	–	–	–
South Eastern Outer	–	L	–	–	–	–	–	–	L	–	–
Southern	–	–	–	–	–	–	–	H	–	–	–
Western	L	–	–	–	–	–	–	L	–	–	–
Yarra Ranges Shire Pt A	–	–	–	–	–	–	–	–	–	–	–
REST OF VICTORIA											
Barwon	–	–	–	–	–	–	–	–	–	–	–
Central Highlands^	–	–	–	H	L	–	–	–	H	–	–
Gippsland and East Gippsland	H	H	H	–	–	–	–	–	H	–	H
Goulburn and Ovens-Murray	H	H	H	H	–	–	–	–	H	–	–
Loddon	H	–	–	–	–	–	–	–	–	–	–

Table 4 (continued)

REGION	How commonly do neighbours help each other out in your neighbourhood?*	Active membership rate (%)	Volunteering rate (%)	Satisfaction with family relationships*	Health barriers to social participation*	Frequency of social contact*	Emotional and general support received from others*	Capacity to raise $2000 in 1 week for emergency*	Integration into the community*	Only get together socially once a month or less with friends or relatives (%)	I often feel very lonely*
Mallee-Wimmera	–	–	–	–	–	H	–	–	H	–	–
Western District	–	–	H	–	–	–	–	–	H	–	–
SOUTH EAST QUEENSLAND											
Brisbane City LGA	–	–	L	–	–	–	–	–	L	–	–
Brisbane SD Bal	–	–	–	–	–	–	–	L	L	–	–
Gold Coast LGA	–	–	L	–	–	–	–	–	L	–	–
Logan City LGA	L	–	–	–	–	–	–	L	L	–	–
Moreton SD Bal	H	–	–	–	–	L	–	–	–	H	–
Sunshine Coast	–	L	–	–	–	–	–	L	–	–	–
REST OF QUEENSLAND											
Darling Downs	–	–	–	–	–	–	–	–	–	–	–
Far North	–	–	–	–	–	–	–	–	–	–	–
Fitzroy	–	–	–	–	–	–	–	–	–	–	–
Mackay	–	–	–	–	–	–	–	–	H	–	–
Northern	L	–	–	–	–	–	–	–	–	–	–
Wide Bay-Burnett	–	–	–	–	–	L	–	–	–	H	–
ADELAIDE											
Eastern	L	–	–	–	–	–	–	H	–	–	–
Northern	L	–	–	–	H	–	L	L	L	–	–
Southern	–	–	–	–	–	–	–	–	–	–	–
Western	–	–	L	–	–	–	–	–	–	–	–
REST OF SA											
Outer Adelaide	–	–	–	–	–	–	–	–	H	–	–

Table 4 (continued)

REGION	How commonly do neighbours help each other out in your neighbourhood?*	Active membership rate (%)	Volunteering rate (%)	Satisfaction with family relationships*	Health barriers to social participation*	Frequency of social contact*	Emotional and general support received from others*	Capacity to raise $2000 in 1 week for emergency*	Integration into the community*	Only get together socially once a month or less with friends or relatives (%)	I often feel very lonely*
South East and Murraylands	–	–	–	–	–	–	–	–	–	–	–
Yorke, Northern and Eyre SA	–	H	H	–	–	–	–	L	H	–	–
PERTH											
East Metropolitan	–	–	–	–	–	–	–	–	–	–	–
North Metropolitan	L	–	–	–	–	–	–	–	L	–	–
South East Metropolitan	L	–	L	–	–	–	–	–	L	L	–
South West Metropolitan	–	–	–	–	L	–	–	H	L	–	–
REST OF WA											
Midlands and Central	H	H	–	–	–	–	–	–	H	–	–
South West and Lower Great Southern	H	–	–	–	–	–	–	–	–	–	–

Source: BTRE (2005) Focus on Regions, No. 4, Social Capital. Department of Transport and Regional Services: Canberra

Future problems, vulnerabilities and possible responses for Australia

Building strong sustainable communities is clearly one of the important challenges facing policy makers and planners today. The increasingly differentiated nature of Australian society has resulted in varied patterns of human and social capital, together with other indicators of local community strength. There is ample evidence that low levels of social capital and human capital are associated with a wide range of social malaise and that while these connections are important to understand at the level of the individual they are increasingly important to understand at an aggregate community level. Indeed, the federal government's stronger communities mantra is built on these concerns. In the *stronger families and communities strategy*, it is suggested that in places where the characteristics of a strong and sustainable community are missing or weak:

There is less capacity to meet the challenges of economic change and to scope with the pressures that lead to family and social breakdown. Lack of community leadership and community control over decision making also diminishes community strength. (Black and Hughes 2001, p. 7)

State and local government organisations and NGOs express similar concerns based on a growing body of evidence.

Given this situation, informed policy must focus on ways of building community capacity, strength and sustainability. The highly spatial dimension of the issue presented in this chapter would suggest that policies and programs that focus on local community level intervention may be most appropriate. However, there has long been a debate between people-based versus area-based policy interventions, or what O'Connor *et al.* (2001) refer to as 'place prosperity versus people prosperity'. A focus on place involves dealing directly with local communities when policy is designed and implemented. At regional scale, for example, assistance is often scaled in terms of location-specific attributes—providing higher levels of funding in locations with a higher or lower weighting on specific indicators, such as unemployment. At a more local level, place-based policy might include targeting locations for community development or community regeneration or supporting schemes for increasing the level of connectedness and engagement with community organisations. In contrast, people prosperity is linked to government policies that influence the social and economic outcomes for people, irrespective of where they live. Taxation is perhaps the best example, but policies that increase the capacity of individuals to improve their skills and education are also relevant.

To develop an appropriate mix of place-based and people-based polices requires a sophisticated evidence base and a strong regime of benchmarking and monitoring. The numerous problems inherent in capturing data that depicts local differentiation in community strength have been highlighted by several commissioned reviews (Productivity Commission 2003) and the considerable work of the ABS on social capital conceptualisation and measurement. However, there has been a lack of progress in reaching a national consensus about both *what types* of social capital ought to be measured as well as *how*. Indeed, after several years of intensive research, the ABS has not produced a social capital measurement tool as such, but rather has attempted to identify a number of avenues in which key measures of social capital ought to be included, most notably through the Mapping Australia's Progress (MAP) project.

MAP represents Australia's best attempts to date to incorporate a range of social measures into a system of national accounts, which has been dominated thus far by the economic Gross Domestic Product (GDP) metric. For some time, a number of economists and other social scientists have proposed revisions to the national income and accounts to make them more like national wellbeing accounts—accounts measuring social progress and not just economic progress (Becker 1965; Lancaster 1966; Nordhaus and Tobin 1972; Juster and Stafford 1985; Eckersley 1998; ABS 2004). The current drive in government and business circles to develop 'triple-bottom-line accounts' (economic, social and environmental) reflects similar priorities.

However, the extent to which a community's strengths (including human and social capital) can be measured consistently across all three levels of government as part of a national system of accounts remains problematic in Australia. Other national studies identifying highly location-specific markers of social disadvantage—most notably, those of Vinson (2007)—highlight how such indicators can exacerbate tendencies for vulnerable local communities to be exposed, via media reports on such studies, to discriminatory stigmatisation; and, coincidently, how governments can be reluctant to be accountable to indicators highlighting entrenched social disadvantage Encouraging signs are present in Victoria where a well-founded set of community strength indicators are emerging, supported and resourced by key agencies such as the Department of Victorian Communities, VicHealth and several local governments, together with input from the ABS and universities. The Victorian work has developed from long-term research, originally based at Swinburne University's Institute for Social Research, on international social benchmarks and indicators and their applicability to that jurisdiction (Salvaris *et al.* 2000). They usefully reviewed government-sponsored projects regarded as 'best practice' in a number of areas including national, regional, state, city-wide and local projects to develop social indicators and 'best practice' models of national strategic planning, national social auditing and community participation in policy making, culminating in what is now known as the Community Indicators Victoria project. Similarly encouraging developments are underway in Tasmania, Queensland and South Australia, where the pivotal involvement of the ABS in each jurisdiction adds weight to the ambition that a set of national accounts measuring genuine progress, beyond the traditional and skewed set of economic indicators may be attainable sooner rather than later.

Such a set of national accounts would also be enhanced by the addition of more-sophisticated human capital measures, given the growing pressures on labour markets globally to attract and retain a skilled workforce. However, the increasingly pervasive influence that external forces exert in shaping Australian community life also draws attention to how all levels of government are constrained in their capacity to affect change in our most disadvantaged areas, particularly in relation to increasing social and human capital. Arguably some of the most progressive thinking in strengthening communities to overcome entrenched disadvantage is occurring in remote indigenous communities where in the face of considerable obstacles, there are some hopeful indications that bipartisan and community collaboration will lead to improved community wellbeing. Similarly promising signs are emerging in early childhood studies, where there is growing evidence of a link between social and community capacity and children's developmental outcomes (Daly *et al.* 2006). As Johnson *et al.* (2003) advocate, an enhanced understanding of the relationship between social capital and key societal outcomes, such as economic performance, educational achievement, health, mental health and crime, is a critical step in planning for the development of stronger and more sustainable communities in Australia in the 21st century.

References

Australian Bureau of Statistics (2000). *Australian Social Trends 2000.* Australian Bureua of Statistics, Canberra.

Australian Bureau of Statistics (2002). *Education and Training Indicators Australia 2002.* Australian Bureau of Statistics, Canberra.

Australian Bureau of Statistics (2004). *Measuring Australia's Progress*, Publication 1370.0, Australian Bureau of Statistics, Canberra.

Baum S, Stimson R, O'Connor K, Mullins P and Davis, R (1999). *Community Opportunity and Vulnerability in Australia's Cities and Towns: Characteristics, Patters and Implications*, University of Queensland Press for the Australian Housing and Urban Research Institute, Brisbane.

Baum S, O'Connor K and Stimson R (2005). *Fault Lines Exposed: Advantage and Disadvantage across Australia's Settlement System.* Monash University e-Press, Melbourne.

Becker G (1965). A theory of the allocation of time. *Economic Journal* **75**, 493–517.

Beer A (1999). 'Regional cities within Australia's evolving urban system, 1991–96', Paper presented to the 16th Pacific Regional Science Association conference, Seoul, July.

Beer A, Bolam A and Maude A (1994) *Beyond the Capitals: Urban Growth in Regional Australia.* Australian Government Publishing Service, Canberra.

Beer A and Maude A (1995) Regional cities within the Australian urban system, 1961 to 1991. *Urban Policy and Research* **13**, 135–148.

Black A and Hughes P (2001). 'The identification and analysis of indicators of community strength and outcomes'. Occasional Paper No. 3. Department of Family and Community Services: Canberra.

Coleman JS (1988). Social capital and the creation of human capital. *American Journal of Sociology* **94**, S95–S120.

Cox E and Caldwell P (2000). Making social policy. In *Social Capital and Public Policy in Australia* (Ed. I Winter), pp. 43–73. Australian Institute of Family Studies: Melbourne.

Daly A, McNamara J, Tanton R, Harding A and Yap M (2006). 'Indicators of social exclusion for Australia's children: an analysis by state and age group'. Online conference paper. National Centre for Social and Economic Modelling, Canberra. <http://www.natsem.canberra.edu.au/publication.jsp?titleID=CP0607>

Dasgupta P (1998). Trust as a commodity. In *Trust.* (Ed. D Gambetta) pp. 49–72. Basil Blackford, Oxford.

Department of Families and Community Services (1999). 'Stronger families and communities strategy information paper'.< http://www.facs.gov.au>.

Department for Victorian Communities (2004). *Indicators of community Strength at the Local Government Area Level in Victoria.* Strategic Policy and Research Division of the Department for Victorian Communities, Melbourne.

Durlauf SN (1999). The case 'against' social capital. *Focus* **20**, 1–4.

Durlauf SN (2002). Bowling alone: a review essay. *Journal of Economic Behavior and Organization* **47**, 259–273.

Eckersley R (1998). *Measuring Progress: is Life Getting Better?* CSIRO Publishing, Melbourne.

Fincher R and Nieuwenhuysen J (Eds) (1998). *Australian Poverty: Then and Now.* Melbourne University Press, Melbourne.

Gluckman M (1967). *The Judicial Process Among the Barotse of Northern Rhodesia.* Manchester University Press, Manchester.

Granovetter M (1973). The strength of weak ties. *American Journal of Sociology* **78**, 1360–1380.

Giddens A (1990). *The Consequences of Modernity.* Stanford University Press, Stanford.

Gregory RG and Hunter B (1995). 'The macroeconomy and the growth of ghettos and urban poverty in Australia'. Discussion Paper No. 325, Centre for Economic Policy Research, Australian National University, Canberra.

Hanifan LJ (1916). The rural school community centre. *Annals of the American Academy of Political and Social Science* **67**, 130–138.

Johnson D, Headey B and Jensen B (2003). 'Communities, social capital and public policy: literature review', Melbourne Institute Working Paper No. 26/03, Melbourne Institute of Applied Economic and Social Research, University of Melbourne.

Juster F and Stafford F (1985). *Time, Goods and Well-Being.* Michigan Institute for Social Research. University of Michigan, Ann Arbor.

Krishner A and Shrader E (1999). 'Social capital assessment tool'. Paper presented at the Conference on Social Capital and Poverty Reduction. World Bank, Washington DC.

Lancaster K (1966). A new approach to consumer theory. *Journal of Political Economy* **74**, 132–157.

Mowbray M (2004). *Beyond Community Capacity Building: the Effect of Government on Social Capital.* PASCAL International Observatory. <http://www.obs-pascal.com/resources/mowbray2004.pdf>

Narayan D (1999) Bonds and bridges: social capital and poverty. In World Bank *Poverty Group*, Working Paper, pp. 1–52. World Bank, Washington, DC.

Newton PW, Baum S, Bhatia K, Brown, SK, Cameron AS, Foran B, Grant T, Mak SL, Memmott PC, Mitchell VG, Neate KL, Pears A, Smith N, Stimson RJ, Tucker SN and Yencken D (2001). 'Human Settlements'. Australia State of the Environment Report 2001 (Theme Report), CSIRO Publishing on behalf of the Department of the Environment and Heritage, Canberra.

Nordhaus W and Tobin J (1972). Is Growth Obsolete? Cowles Foundation Discussion Papers 319, Cowles Foundation, Yale University.

O'Connor K and Stimson R (1995). Convergence and divergence of demographic and economic trends. In *Population Shift: Mobility and Com-*

munity Change in Australia. (Eds P. Newton and M. Bell) pp. 108–125. Australian Government Publishing Service, Canberra.

O'Connor K, Stimson R and Daly M (2001). *Australia's Changing Economic Geography: A Society Dividing.* Oxford University Press, Melbourne.

OECD (1998). *Human Capital Investment: An International Comparison.* OECD, Paris.

Onyx J and Bullen P (2000). Measuring social capital in five communities. *The Journal of Applied Behavioural Science* **36**, 23–42.

Portes A (1998). Social capital: its origins and application in modern sociology. *Annual Review of Sociology* **24**, 1–24.

Productivity Commission (2003). 'Social capital: reviewing the concept and its policy implications', Research Paper, AusInfo, Canberra.

Putnam RD (1998). Foreword to the housing policy debate issue on 'social capital': its importance to housing and community development. *Housing Policy Debate* **9**, v–viii.

Putnam RD (2000). *Bowling Alone: The Collapse and Revival of American Community.* Simon and Schuster, New York.

Salt B (2001). *The Big Shift: Welcome to the Third Australian Culture.* Hardie Grant Books, Melbourne.

Salvaris M, Burke T, Pidgeon J and Kelman S (2000). 'Social benchmarks and indicators for Victoria', Consultants Report for the Department of Premier and Cabinet, Victoria. Institute for Social Research, Swinburne University of Technology, Melbourne.

Stimson R (1997). Labour markets in Australia: trends and spatial patterns in the context of global and local processes. In *Pacific Rim Development: Integration and Globalisation in the Asia Pacific Economy.* (Ed. P Rimmer) pp. 151–177. Allen and Unwin, Sydney.

Stimson R Shuaib F and O'Connor K (1998). *Population and employment in Australia: Regional Hot Spots and Cold Spots, 1986 to 1996.* Australian Housing and Urban Research Institute, Melbourne.

Stimson R, Baum S, Mullins P and O'Connor K (2001). Australia's regional cities and towns: Modelling community opportunity and vulnerability. *Australasian Journal of Regional Studies* **7**, 23–62.

Stimson R, Baum S, Mangan J, van Gellecum Y, Shyy T-K and Yigitcanlar T (2004). 'Analysing spatial patterns in the characteristics of work, employment and skills in Australia's capital cities and across its regional cities and towns: modelling community opportunity and vulnerability: main report'. Prepared for the Australian National Training Authority, Melbourne.

Stone W (2001). 'Measuring social capital: towards a theoretically informed measurement framework for researching social capital in family and community life'. Research Paper No. 24. Australian Institute of Family Studies, Canberra.

Uslaner EM (1999). Trust but verify: social capital and moral behaviour. *Social Science Information* **38**, 29–56.

Vinson T (2007) *Dropping Off the Edge: the Distribution of Disadvantage in Australia.* Jesuit Social Services, Richmond, Melbourne.

Woolcock M (1998). Social capital and economic development: toward a theoretical synthesis and policy framework. *Theory and Society* **27**, 151–208.

Woolcock M and Narayan D (2000). Social capital: implications for development theory, research and policy. *World Bank Research Observer* **15**, 225–249.

SECTION 5

Transition to Renewable Energy

'Decarbonising' the economy represents the world's and Australia's largest industrial transformation challenge and opportunity this century. *Energy security* (see Section 2 on peak oil) and *climate change* (see Section 4) are the twin drivers.

A range of initiatives to address these challenges are underway, including clean coal (carbon sequestration) and nuclear—which are both linked to non-renewable forms of energy with a range of environmental issues attached. This section however, focuses on energy efficiency and renewable energy as preferred pathways for developing *a portfolio of energy options* for 21st century Australia.

Graham *et al.* point to the potential for between 15 and 70% improvement in energy efficiency across industry sectors. The way in which buildings (Ambrose) and cities (Newman, Newton, Scheurer, Falconer and Corti) are designed represent avenues for immediate savings in energy use.

Projections of energy use to 2050 by Graham *et al.* point to increasing levels of demand, but identify pathways for more sustainable energy supply. These include:

- *wind* (Coppin) and its potential for integration with gas to overcome intermittency, and offer base-load levels of generation
- *solar* (Fell *et al.*), and its extensive range of emerging technologies, including new building–integrated PV options, hybrid buildings, and so on
- *biofuels* (O'Connell *et al.*), which include both ethanol and biodiesel options, and discussion of the level of competition expected in agriculture between production for food versus fuel
- *hydrogen* (Dicks and Rand), which is capable of application to stationary power generation, portable power and *transportation* (see Lamb), thereby embodying exciting prospects for a future *hydrogen economy.*

The challenges associated with transitioning to sustainable energy are outlined. These include relative costs versus benefits (economic, non-economic), developing new infrastructures for energy supply, integration of centralised with distributed energy (Jones), base versus peak generation capability, energy storage, and the nature of regulatory settings (such as carbon pricing, product labelling of embodied energy, product safety and access to the national grid).

Understanding the *carbon signatures* of alternative buildings, transport systems, food and water systems—to name but a few—should be mandatory in all future urban investment decisions.

PWN

21

Energy Futures

Paul Graham, Luke Reedman and Josephine Cheng

Energy makes a fundamental contribution to life in Australia. It is a basic input into virtually every aspect of personal and business activity. Energy, in some form, is involved in most household activities, such as heating, cooling, cooking, lighting, transport or simply enjoying products and services that require energy. Firms also use energy in virtually all of their activities, whether it is processing and manufacturing materials, transporting goods, heating and cooling premises, providing telecommunication services or powering computers.

This chapter provides an overview of the state of energy in Australia: the resources we have available; the resources and technologies we have chosen to use; the prices we pay; our social attitudes to energy and its relationship to greenhouse gas emissions. Recognising that the issue of greenhouse emissions is providing a strong catalyst for change, the chapter concludes with a general overview of the potential for greater uptake of a variety of new low-emission electricity generation sources in the future.

Energy resources

Australia is well endowed with both fossil and non-fossil energy (Table 1). Australia has vast reserves of coal that are relatively easy to mine and are located close to centres of energy consumption. Its natural gas and uranium resources are substantial, although much of the natural gas is located far from large domestic markets and uranium is not used to produce energy in Australia.

The nation's oil resources are significant but reserves are declining in the absence of new discoveries. While there is the potential for new discoveries to add to the size of this resource, the 'years of availability' statistic, which indicates the number of years until the resource will deplete at current production rates, is not likely to improve because demand is also rising with economic growth.

Compared with our understanding of fossil fuels and uranium, the size of the renewable energy resource is not nearly as well understood. However, many renewable energy technologies, such as geothermal, wave, solar and wind, have enormous potential. This is based on Australia's availability of arable land (biomass), wind speeds (wind power), length of coastline (wave) and the existence of certain geological structures (geothermal). Excluding hydropower, the relatively higher cost of renewable energy has meant the use of these non-fossil-fuel resources has been limited to date. Large-scale hydropower is economically viable but most opportunities have already been exploited. Some potential still exists for smaller-scale hydro plants.

Energy use

Over the past 30 years, energy consumption in Australia has more than doubled to more than 5500 petajoules a year (Figure 1).

Until the early 1990s total energy consumption tended to grow at a rate that closely matched the rate of growth in gross domestic product (GDP). Since then, energy consumption has tended to grow more slowly than GDP (Figure 2). This decline in energy intensity of the Australian economy can be attributed to two main factors—greater energy efficiency, through technological improvements and fuel switching, and the rapid growth of less-intensive sectors, such as services, relative to the more moderate growth of energy-intensive sectors, such as manufacturing and mining (ABARE 2006a).

Table 1: Australian and world energy reserves

	Economic Demonstrated Resources (EDR)[a]	Total Identified Resources (TIR)[b]	Years of availability (EDR)[c]	Years of availability (TIR)[c]	Percentage of world (EDR)	Economic Demonstrated Resources (EDR)
	Australia				World	
Black Coal[d]	39 Gt	107 Gt	100	270	5.3	739 Gt
Brown Coal[d]	37 Gt	194 Gt	550	2,900	23.9	155 Gt
Natural Gas	2,587 Bcm	4,069 Bcm	65	100	1.4	180,000 Bcm
Crude Oil	157 GL	238 GL	7	10	0.1	190,500 GL
Uranium[d]	716 kt[e]	1,191 kt	75	125	36.5	1,962 kt
Biomass[f]	NA	330 PJ/y	Renewable	Renewable	NA	NA
Wind[g]	NA	200 TWh/y	Renewable	Renewable	NA	NA
Solar	NA	Very large	Renewable	Renewable	NA	NA
Wave and tidal	NA	Very large	Renewable	Renewable	NA	NA
Hot dry rocks[h]	NA	2500 EJ	NA	800	NA	NA

Gt: gigatonnes; Bcm: billion cubic metres; GL: gigalitres; kt: kilotonnes, PJ/y: petajoules per year, TWh/y: Terawatt hours per year, EJ: etajoules, NA: Not available.

[a] Economic Demonstrated Resources (EDR) are resources judged to be economically extractable and for which the quantity and quality are computed partly from specific measurements, and partly from extrapolation for a reasonable distance on geological evidence.

[b] Total identified resources (TIR) consist of demonstrated resources (economic and subeconomic) and inferred resources.

[c] Years of availability based on current rates of Australian production.

[d] EDR and total identified resources estimates reflect recoverable resources, not in situ resources.

[e] Recoverable at costs of less than US$80/kg U.

[f] All biomass crops and residues, without competing with food crops but not withstanding soil nutrient maintenance

[g] On shore sites no more than US8c/kWh

[h] Includes resources in granite bodies above 5km and more than 165 degrees Celsius

Sources: Geoscience Australia 2006a, 2006b; Somerville *et al.* 1994; IEA Greenhouse Gas R&D Program 2000; Energy Strategies 2004.

There is significant potential for further improvements in energy efficiency. The Energy Efficiency and Greenhouse Working Group (2003) estimated that energy consumption improvements of between 15 and 35% were achievable under the conservative assumptions that only existing technology was used and that the change would pay for itself within 4 years. Using more optimistic assumptions of new technology becoming available and longer payback periods being acceptable, energy consumption reductions of up to 70% were shown to be economically viable in some sectors (Figure 3).

The three largest uses of energy in Australia are electricity generation, transport and manufacturing processes, which together account for some 75% of Australia's energy consumption (Figure 1).

Once energy has been converted into electricity the residential and commercial services sectors account for around half of electricity consumption. The remainder is consumed by industry, with manufacturing, metals and aluminium smelting being the largest industrial consumers (Figure 4).

Almost 80% of Australia's electricity is consumed in New South Wales, Victoria and Queensland. Western Australia's electricity consumption is around 11%—about half of Victoria's. Electricity consumption in Tasmania and South Australia is around 5% each. Electricity consumption in Northern Territory is around 1% of the total (Figure 5). These figures approximate to each state's respective proportion of the national population.

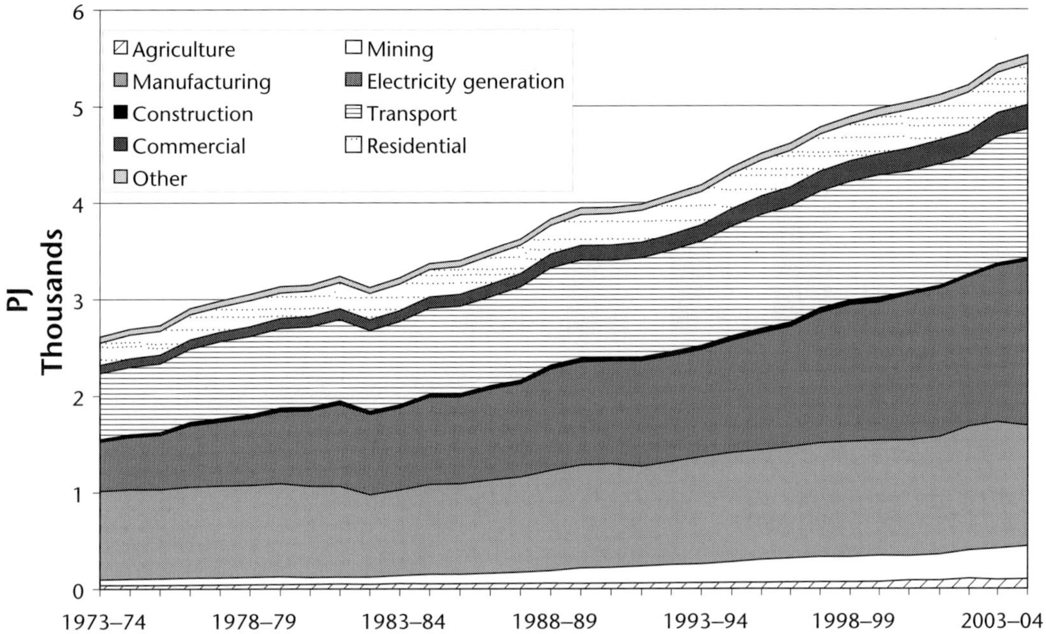

Figure 1: Australian energy consumption by industry (Source: ABARE 2006b).

Road transport accounts for some 77% of the total energy used for transport. Reflecting Australians' desire for personal mobility, the majority of energy used in road transport use is for passenger vehicles (Figure 6).

The transport system is heavily reliant on petroleum-based fuels, which meet more than 97% of Australia's total transport needs. This is in line with world trends. Electricity is used to power some rail transport, natural gas is used for gas pipelines, and biofuels and natural gas play a very minor role in meeting our road transport needs (ABARE 2003).

Fuel shares

Three fossil fuels—coal (brown and black), oil and natural gas—supply 95% of Australia's energy needs (Figure 7).

These fossil fuels fulfil different needs in different markets, and also occupy niche roles in many markets.

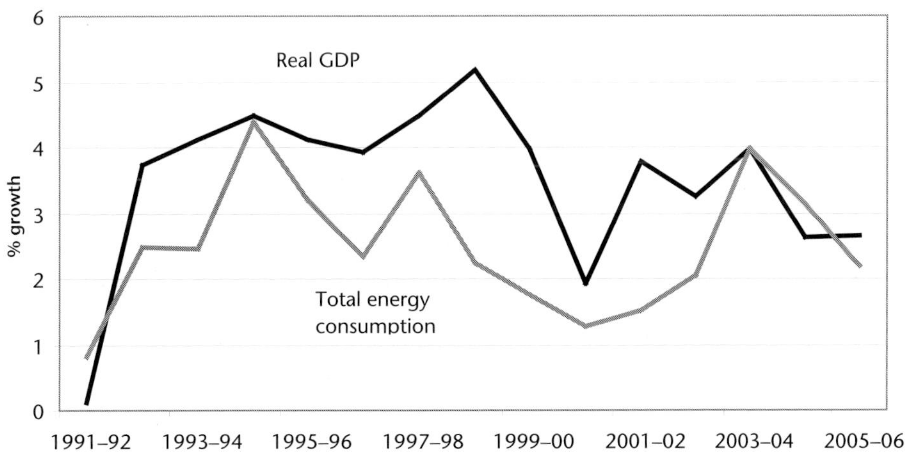

Figure 2: Growth in Australian real GDP and energy consumption (Source: ABARE 2006b, 2006c).

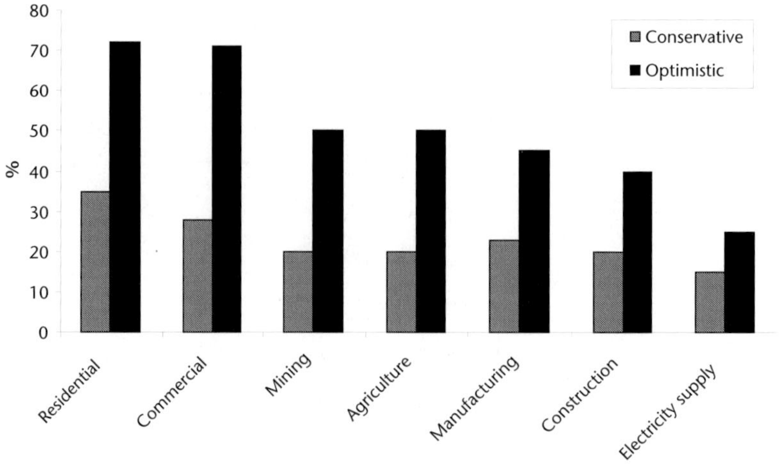

Figure 3: Percentage cost-effective energy consumption reduction potential across sectors (Source: Energy Efficiency and Greenhouse Working Group 2003).

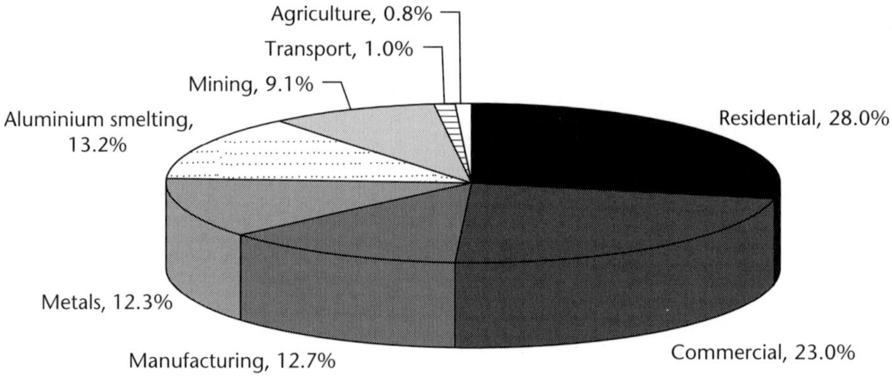

Figure 4: Shares of electricity consumption in Australia by end use categories, 2005–06 (Source: ABARE 2006b).

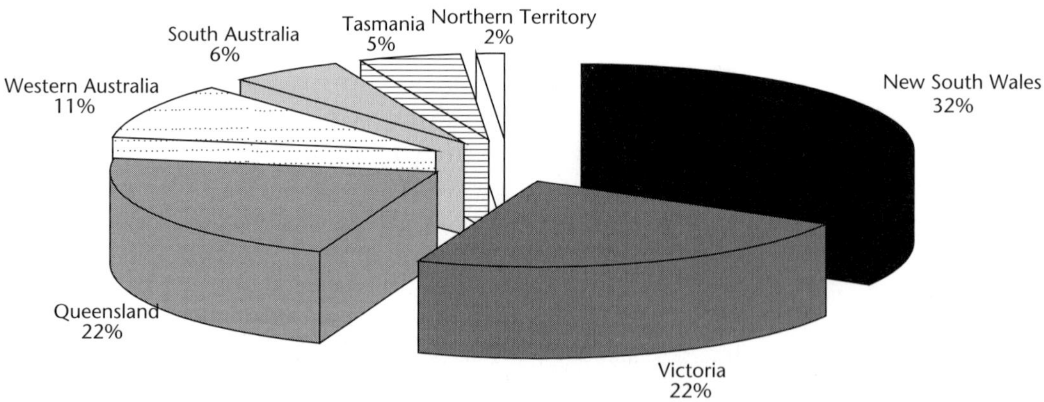

Figure 5: Electricity consumption by state, 2005–06 (Source: ABARE 2006a).

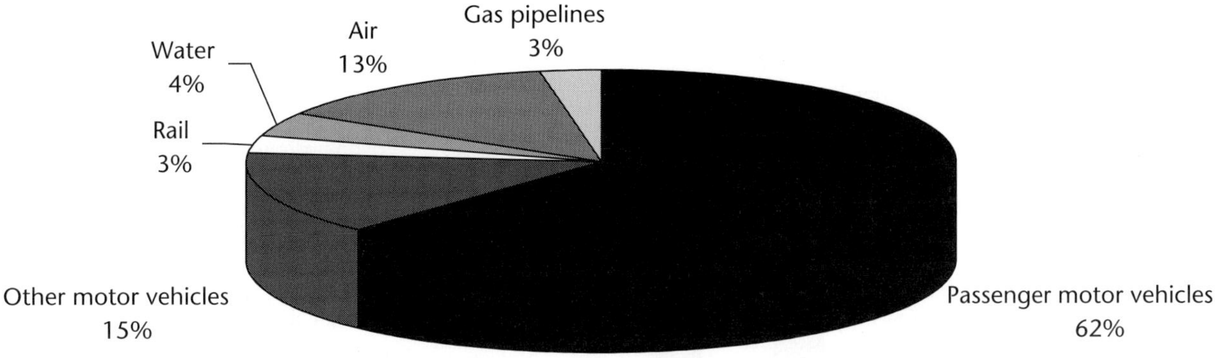

Figure 6: Share of energy consumption in transport by end-use category, 2005–06 (Source: ABARE 2006a).

Australia has large, easily mined deposits of coal close to the major urban centres in the eastern mainland states. The major power stations have been sited close to those coal deposits, thus eliminating much of the cost and inconvenience of moving large tonnages of a bulky material. As a result of its relatively low economic cost (ignoring any cost to the environment, such as greenhouse gas emissions), coal dominates base–load electricity generation in Australia (Figure 8).

Western and South Australia have relatively less coal, but abundant gas. More than half of their electricity is derived from burning gas. The role of natural gas is greater in South Australia, Northern Territory and Western Australia, where access to natural gas, and a comparative lack of coal, makes it more competitive as a source for base-load, as well as mid- and peak-load, electricity generation. Tasmania is unique in that almost all of its electricity is generated from hydroelectric power stations. These states and territories, however, only consume some 20% of electricity generated in Australia so coal remains dominant overall, in total national electricity generation.

Low-emission technologies, including renewables, are currently not cost-effective when compared with conventional technologies. Low-emission technologies would be able to compete if the costs of electricity generation technologies were equated on the basis of their relative emission output—via either a subsidy to low-emission technologies or a cost imposed on high-emission technologies. With the increasing desire of Australians to tackle climate change, it is now highly likely that such subsidies and imposts on high-emission technologies will become a permanent feature of energy markets and therefore we can expect a shift towards these low emission technologies over the longer term. The

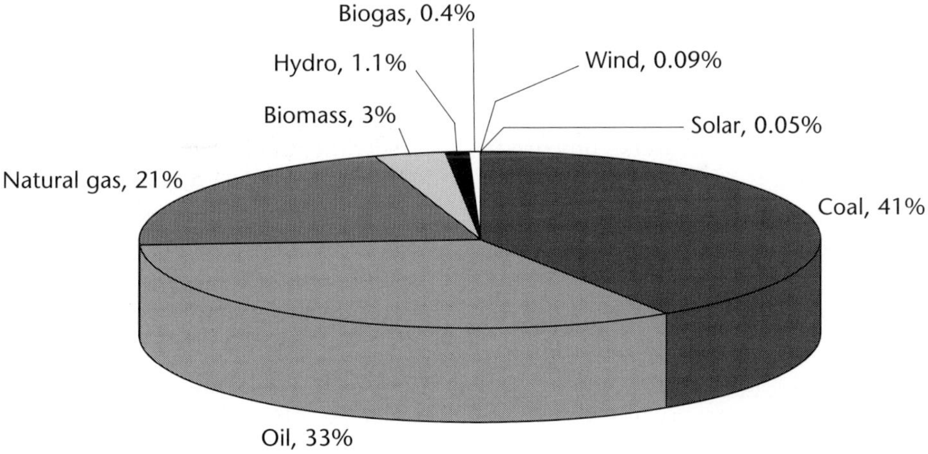

Figure 7: Share of energy consumption by fuels and other energy sources (Source: ABARE 2006a).

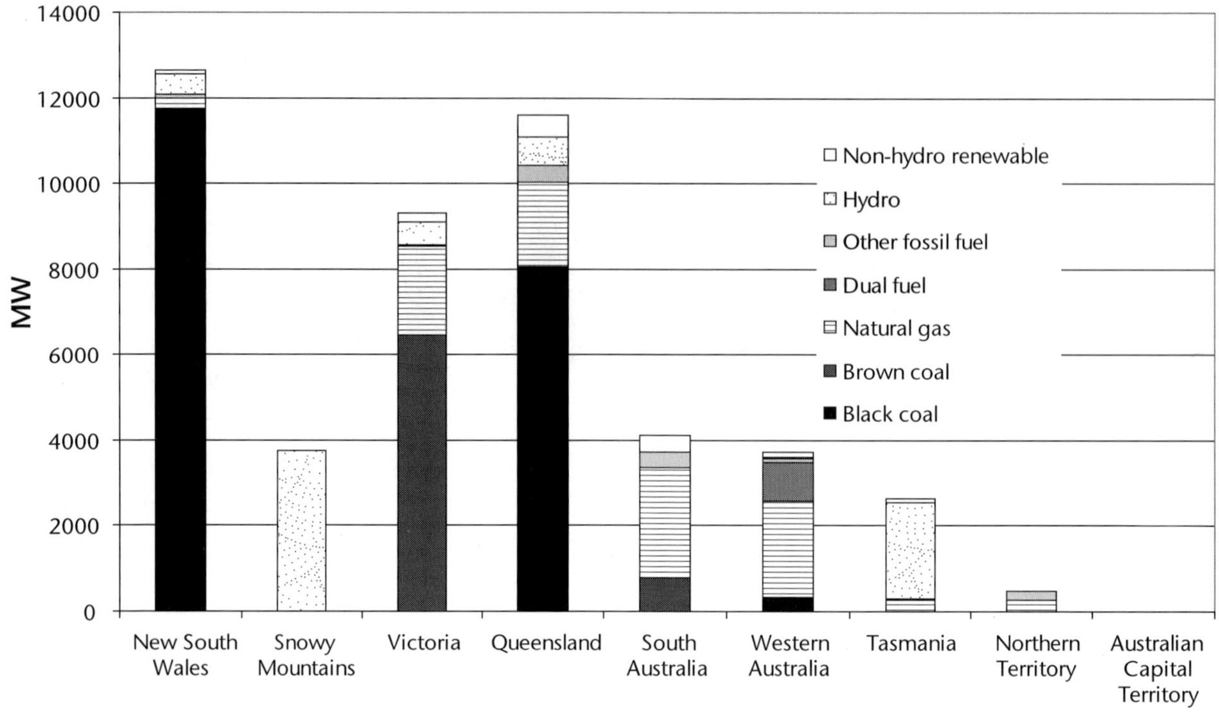

Figure 8: Installed electricity generation capacity by technology category and state (Source: ESAA 2006).

potential uptake of a variety of low-emission electricity generation technology is discussed later in this chapter. Low-emission transport is not covered in such detail, as research exploring the future mix of fuels and technologies lags behind that available for the electricity sector in Australia. Some projections of the uptake of low-emission transport technologies in Australia can be found in CSIRO and ABARE (2006).

Energy prices

Australia has among the lowest electricity prices for large business and industrial users compared with other OECD countries. Australia also has one of the lowest residential electricity prices in the Asian region and prices compare favourably with European countries.

Similarly, gas prices in Australia are among the lowest in the world and are substantially lower than in most OECD countries.

Australia's inexpensive electricity has seen substantial investment in resource processing and other energy-intensive industries. The production of alumina and aluminium, paper, cement, chemicals, metals and minerals processing require large amounts of energy. The international competitiveness of these sectors relies on access to competitively priced, reliable energy.

The operation of the wholesale electricity market and security of the power system in the Australian eastern states is managed by the National Electricity Market Management Company (NEMMCO). Table 2 shows wholesale electricity prices in that market since its inception in 1998.

The retail price that consumers pay for electricity differs from the wholesale price for two reasons. Firstly, the retail price of electricity is significantly higher to cover the costs of transmission and distribution of electricity via the grid. Secondly, the retail price of electricity depends on the type of customer. For example, in Queensland the average residential electricity price is around \$120/MWh (12c/kWh), which is made up of a mixture of night time and daily tariffs. The commercial and rural retail price is similar to this or slightly higher. However, the price paid by very large industrial electricity consumers is much lower at around \$70/MWh (7c/kWh).

Table 2: Average annual wholesale price of electricity in Australian Eastern Regions, $/MWh

Year	NSW	QLD	SA	Snowy	Tas.[a]	Vic.
1998–1999	33.13	51.65	156.02	32.34	–	36.33
1999–2000	28.27	44.11	59.27	27.96	–	26.35
2000–2001	37.69	41.33	56.39	37.06	–	44.57
2001–2002	34.76	35.34	31.61	31.59	–	30.97
2002–2003	32.91	37.79	30.11	29.83	–	27.56
2003–2004	32.37	28.18	34.86	30.80	–	25.38
2004–2005	39.33	28.96	36.07	34.05	190.38	27.62
2005–2006	37.24	28.12	37.76	31.09	56.76	32.47
2006–2007	43.60	39.91	47.14	42.67	47.06	47.07

[a] Tasmania joined the National Electricity Market (NEM) in 2005.
Source: NEMMCO 2007

A consequence of the varying gap between wholesale and retail prices is that for any given change in wholesale electricity prices, retail prices changes are more muted and differ by end-user. Large industrial customers are the most exposed to wholesale electricity price movements.

Retail gas prices are also similarly differentiated by end-users, with industrial users again having the lowest distribution margin as a component of their retail price. For this reason, industrial consumers may be regarded as more vulnerable to changes in wholesale electricity and gas prices.

Petrol prices are a significant part of the overall Australian energy price picture. Between 2000 and 2004 retail petrol prices ranged between 80 and 90 cents per litre. In late 2004, average prices reached A$1 per litre and remained above that level (fluctuating up to A$1.50 per litre) from March 2005. The fluctuations are largely driven by the international price of oil and the exchange rate (because international oil prices are denominated in US dollars). The other components to the petrol price are excise (a constant 38 cents per litre), refining costs, transport costs, retail costs and goods and services tax (GST).

All fuel excise rates will be incrementally reformed, a process which commenced in 2006 and concludes in 2015. Under this system, fuel excise rates will move from a volume-based system to an energy-based system. This makes sense because fuels are purchased for their energy content. The energy content per litre of a fuel is referred to as its energy density and can vary significantly. Excise rates for high-, medium- and low-energy density fuels will be 38, 25 and 17 cents per litre, respectively. Alternative fuels, such as biodiesel, ethanol, compressed natural gas (CNG) and liquefied petroleum gas (LPG), will remain free of excise until 2011. From 2015 they will attract a discounted excise rate of 8.5, 12.5 and 19.1 cents per litre for low-, medium- and high-energy density fuels, respectively. Overall, the tax incentives for taking up alternative fuels will remain, but won't be as great.

Greenhouse gas emissions and energy

Emissions of greenhouse gases come from a variety of sources and locations. While comprehensive and accurate data do not exist on greenhouse emissions for many nations, some general observations are possible.

The first is that the production and use of energy is a large source of emissions. The World Resources Institute estimates that about 60% of global emissions in 2000 came from carbon dioxide emissions from energy use (WRI 2006).

The second is that a small number of nations (or a union of nations, in the case of the European Union) account for a large proportion of global emissions. According to WRI, six nations (United States, European Union, China, Russia, Japan and India) account for nearly half of global emissions and nearly two-thirds of carbon dioxide emissions from energy. The United States alone accounts for more than one-fifth of global energy-related emissions (Figure 9).

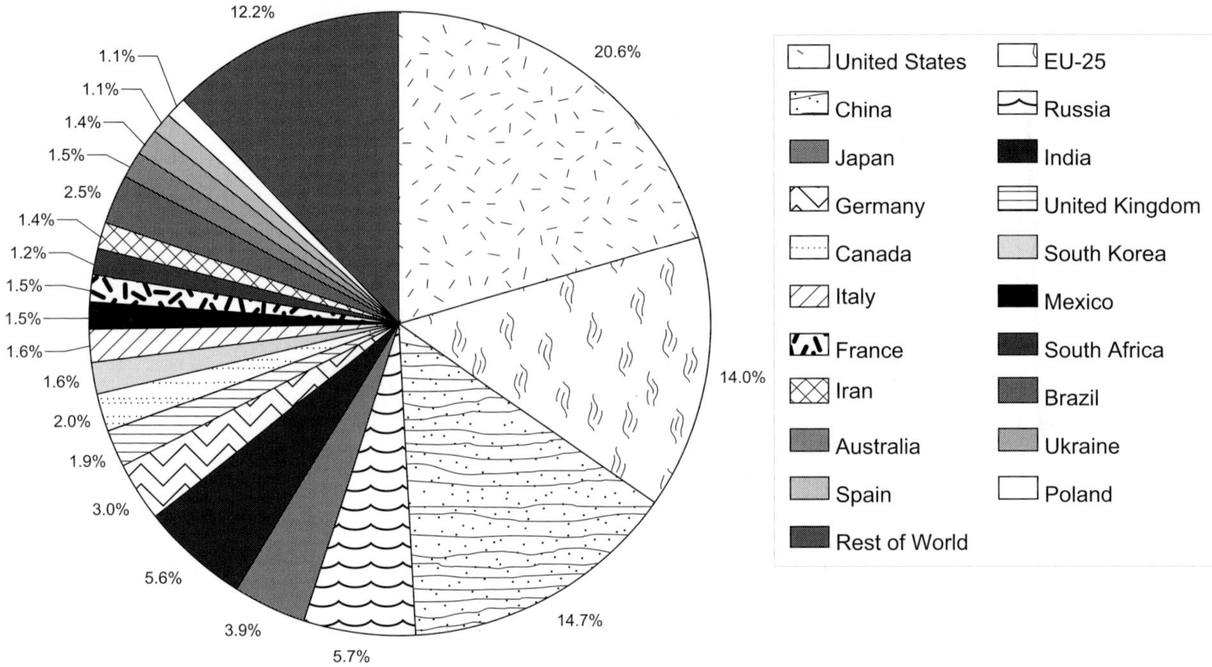

Figure 9: Shares of greenhouse gas emissions (excluding land use change), 2000 (Source: WRI 2006).

Expected economic growth in less-developed countries, such as China and India, will result in emissions from these nations increasing substantially over the next 20 to 30 years. Total emissions from less-developed countries are soon expected to overtake those from industrialised countries.

Australia contributes 1.6% of world greenhouse emissions from energy. Australia is not unique in being a small greenhouse gas emitter. There are around 20 other countries worldwide with emissions less than 3%.

In 2000, Australia was ranked fourth in terms of greenhouse gas emissions per capita, at around 25.6 tCO_2e per capita (WRI, 2006). According to this measure, Australia's ranking is the highest among developed countries, followed by the United States and Canada. The average amongst developed countries is 14.1 tCO_2e per capita. Australia's high ranking in emissions per capita reflects our relatively high proportion of fossil fuels used in energy consumption, high usage of relatively less-efficient private transport and high production of non-ferrous metals per capita (most of which is exported).

Australia's emissions come from a variety of sources (Figure 10). Stationary energy and transport provide the single largest sources—accounting for 69% of the total 565 Mt CO_2e emitted in 2004 (with electricity production accounting for 35% of the stationary energy category and transport energy 14%). Agriculture was the next largest contributor at 16%, with the other 6% of emissions coming from land-use change and forestry, industrial processes and waste. Emissions from non-energy sectors have generally been declining, with total emissions from these sectors falling by 87 Mt from 1990 to 2004. This is due to significant declines in land use, land-use change and forestry emissions. By contrast, energy sector emissions rose by 100 Mt over the same period (Australian Greenhouse Office 2006).

Energy infrastructure

The energy sector is characterised by its connection to long-lived assets. Nuclear and coal-fired power stations have a life of between 40 and 60 years. Motor vehicles have a life of 12 to 20 years. Buildings last for 50 to 100 years. The challenge of such infrastructure is that it requires enormous up-front investments and, to ensure the investment is recouped, it comes with an expectation that the infrastructure will be in place and providing benefits to owners for many decades. This is a problem for key sectors of the economy that, due to climate change concerns, may need to change rapidly.

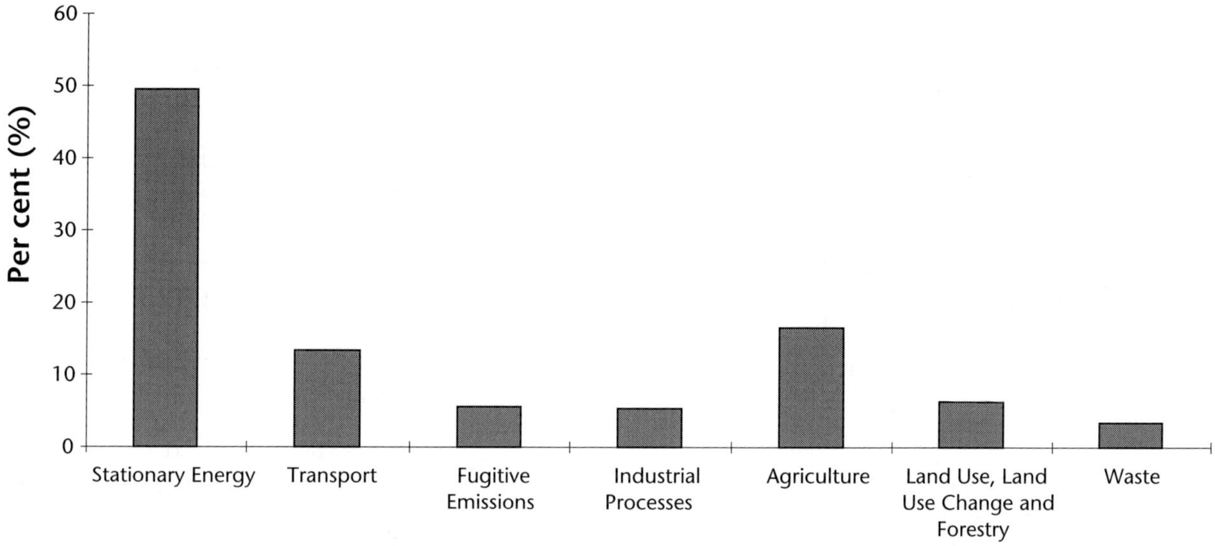

Figure 10: Contribution to net GHG emissions by sector (Kyoto accounting) 2004 (Source: AGO 2006).

While many infrastructure services have been privatised in recent decades (or corporatised, where government ownership is maintained but the entity is given the right to behave like a private company), the bulk of Australia's infrastructure development occurred in the 1950s, 60s and 70s in public ownership. As that infrastructure becomes physically, technologically or socially obsolescent, the decisions about the types of new infrastructure that will replace it will often be taken from a commercial point of view. This is particularly so in the National Electricity Market which is now open to all private investors.

Social attitudes to energy

Australians consume energy for a variety of reasons, including entertainment, comfort (heating, cooling), industrial processes, information management, mobility and even fashion. Apart from a few industrial processes, the energy costs of these activities are often minor when compared with total household or industrial consumption.

In 2006, household electricity and petrol represented around 1.1 and 3.3 per cent of average full time earnings (assuming the earnings support one average house and one car). The general low cost of energy relative to income, as well as the key role that energy plays in the Australian lifestyle, is one of the reasons why demand for energy is often observed by economists as being price inelastic. That is, demand for energy does not change significantly in response to changes in the price of energy.

A literature review of the price elasticity of demand for electricity (for example, ABARE 2003) indicates that for a 10% increase (decrease) in electricity price, demand for electricity is only expected to decrease (increase) by 2.5%. It should be noted, however, that some users of electricity would have much higher price elasticity. For example, industries where electricity is a significant share of total costs of production, such as the minerals sector, would be expected to be more responsive to electricity price changes.

Table 3 shows the contribution of different household activities to energy use and greenhouse gas emissions. In 2005, 43% of households said they considered cost to be the main factor when buying new white goods, 44% nominated the energy star rating as a main consideration, and only 11% stated an environmental factor as their main consideration. Many households indicated that their main reason for installing insulation was to achieve comfort (83%), rather than to save on energy bills (10%) or to use less energy (4%) (ABS 2006).

Many Australians are aware of costs to the environment from the consumption of energy. There are some 725,000 customers of 'green' energy across Australia; that is, electricity derived from resources such as the sun, wind, hydro and waste, which do not emit greenhouse gases (Greenpower Accreditation Program 2007).

Table 3: Household energy use activities and their contribution to greenhouse gas emissions in 2005

Household activities	Energy use (%)	Emissions (%)
Appliances	30.0	53.0
Heating water	27.0	28.0
Cooking	4.0	6.0
Heating and cooling	39.0	14.0
Total	100.0	100.0

Source: ABS 2006

There is also some evidence of changes in car vehicle purchasing preferences in response to high petrol prices. Generally speaking, petrol is considered to be even more price inelastic than electricity. Estimates show that for a 10% increase (decrease) in petrol prices demand for petrol is only expected to decrease (increase) by 1%.

Vehicle size is a major determinant of fuel consumption. During the 1970s, car vehicle purchasing preferences did change significantly in response to the high oil prices being experienced at the time. From 1970 to 1980 the proportion of new vehicles registered in the smaller medium sized car range increased from 14 to 26%. Intermediate medium sized cars also increased from 15 to 35%. At the same time, the share of new registrations of large and larger medium size cars fell (MUARC 1993). However, it should be noted that the trend levelled out, and even reversed by a few percentage points, just after oil (and therefore petrol) prices peaked during the 1980s. It has also been argued that some of the purchases of smaller cars could be accounted for by improvements in income at the time, which meant second cars became affordable for a growing number of households.

Competitiveness of electricity generation technologies

Cost projection methodology

There are several factors that affect projections of future costs of electricity generation technologies in Australia. The three factors which we will consider are:

- technological improvement or 'learning'
- resource constraints or the quality of resources available
- the volatility of generation plant markets.

It is broadly recognised that technological improvement has a close relationship with deployment. The observation was first made in the early part of last century during the study of industrial production of aircraft (Wright 1936). It was found that a reduction in costs of technologies can be observed as a fixed rate for each doubling of cumulative production—this relationship is often called a learning curve.

The learning-curve approach recognises that reductions in the cost of capital or plant are the direct result of learning that occurs through experience ('learning by doing') and economies of scale occur as a technology is adopted. A key implication of this approach is that cost changes can occur at any time as long as there has been a sufficient interval for capacity to be installed and the relevant economic or policy drivers are in place to kick-start adoption.

In relation to Australian deployment, it is assumed that the costs of technology will fall as a function of world technological adoption. To implement this approach, we have collated projections of the future uptake of energy technologies at the global level and used them as our projection of cumulative world production.

The historical learning rate for most technologies has been comprehensively reported elsewhere (see, for example, McDonald and Shrattenholzer 2001). However, for our purposes, we need to know the future learning rate. Future learning rates will change as technologies pass through various stages of their technological development. For example, a technology with a learning rate of 20% for each doubling of cumulative capacity in the last 10 years may have a learning rate of only 15% in the next 20 years as it becomes more mature. As a result, setting a fixed learning rate now, which is based on historical rates, may overestimate future technological change.

To form estimates for future capital costs of our technologies, we used the following approach:

- average learning rates for immature technologies (such as solar-based power) of around 10–15%

- average learning rates for mature technologies (such as conventional coal and gas power) of around 0–5%
- A lower limit will be set on technology costs that is equal to around half their current costs
- the maximum rate of change in a 5-year period is 10% (an exception is made for solar photovoltaics, as technological achievement suggests acceleration above this maximum in the next decade).

In contrast to technological change, resource constraints move the costs in the opposite direction. As the best renewable electricity generation sites are used first, the next electricity generation plant to be commissioned, all other things being equal, will cost more per unit of useful energy output because of a lower capital utilisation rate (due to, for example, lower average wind speed or poorer solar radiation in the next location).

The volatility of generation markets can have a positive or negative effect on generation plant costs. For example, in the years during and following the Asian Economic Crisis, the costs of power plant, particularly gas-fired units, fell significantly as many potential buyers in Asia were forced out of the generation plant market. Currently it appears we have moved in the opposite direction. A surge in demand for new power plants has occurred alongside a period of strong demand growth for metals and other plant input materials (DOE/EIA 2006).

This period of increased prices for generation plant is not expected to be permanent in Australia, but its effects are still being felt. In the United States, at least, the Energy Information Administration is projecting new construction in the power industry to slow for a period, which will take pressure off the market. However, pressure from investment in other countries, such as China, could see the demand premium continue.

Figure 11 shows the projected costs of electricity generation technologies to 2050 combining these three factors—learning, resource constraints and world markets for generation plant. The projections show point estimates over time. However, there is a great deal of uncertainty around each technology cost timeline. In reality, the actual costs of alternate technologies will vary due to different site conditions. Nevertheless the diagram helps us to draw several conclusions:

- All the renewable technologies have higher costs than conventional coal and gas—currently the main source of electricity supply in Australia.
- In the period to 2020, electricity from most of the renewable energy technologies is expected to cost between \$60 and 100/MWh.
- The costs of solar photovoltaics are expected to experience the steepest decline in costs, although from a very high starting point. Solar-hot-water installations are initially more cost effective than large-scale solar thermal plants, but the latter are expected to improve faster in the long term, reflecting their relative immaturity at this time.
- Wind, biomass, landfill gas and tidal electricity generation are also expected to experience significant cost reductions. However, in the case of the first three, the current higher power plant construction costs mean the benefit of technological change does not resume its influence on costs until after 2010. These technologies are in demand worldwide at the present time. Most new renewable electricity generation in the United States is expected to come from biomass and wind (DOE/EIA, 2006).
- In the long term, it is difficult to draw a distinction between the costs of alternative renewable electricity generation technologies. For example, solar photovoltaic electricity, which appears to have higher costs than the others, cannot be ruled out given the uncertainty in the rate of technological change and the potential to sell directly into higher priced retail markets—thus by-passing the grid. Such market opportunities are discussed in the following section.

The width of the uncertainty range is widest for small hydro and wind. In the case of small hydro, this reflects the importance of the specific characteristics of the chosen site. Due to site conditions, the capacity factor or utilisation of capital can vary from 15 to 90%, the size can be roughly between 5 kW to 150 MW and there are a variety of other considerations such as the existing infrastructure, land costs and the quality of the head of water.

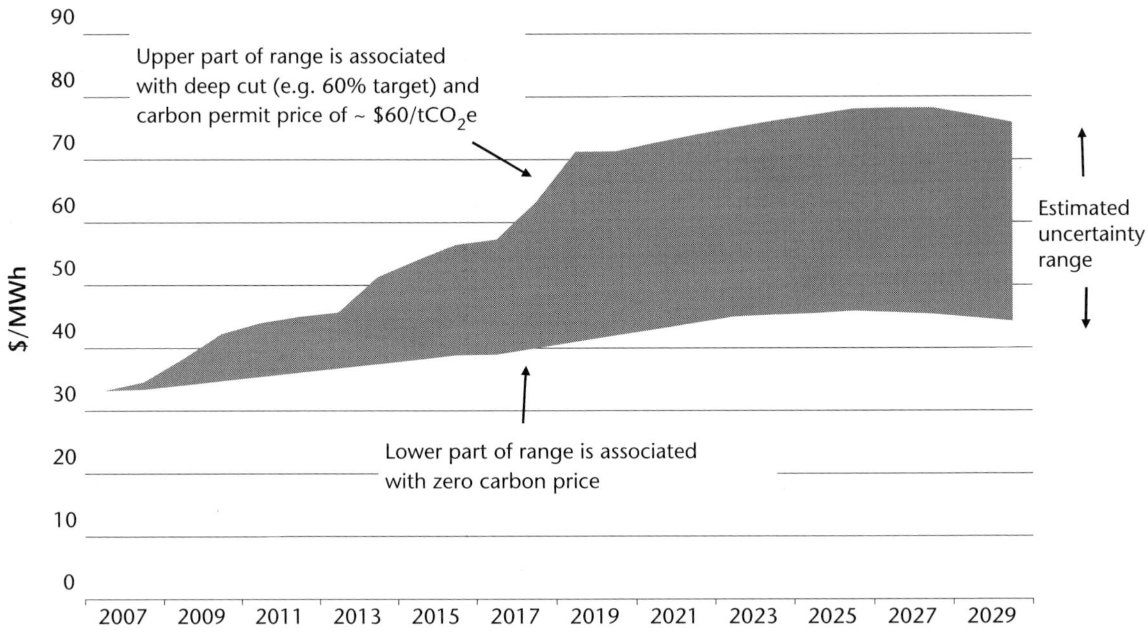

Figure 12: Projected wholesale electricity price range under impact of a carbon price (Source: CSIRO research)

Based on various modelling studies conducted over the past several years (see, for example, Reedman *et al.* 2007, CSIRO and ABARE 2006, Graham *et al.* 2005), it is estimated that the average wholesale price could increase by up to $20/MWh within a decade (Figure 12). The reason for a variety of possible outcomes is because the size of the carbon price depends on the circumstances relating to how the scheme is designed, such as:

- the target and path to that target
- the extent to which other sectors of the economy are involved
- uncertain costs of abatement any constraints on carbon trading, such as price caps
- to whom permits are allocated and how many years ahead they may be purchased, banked or borrowed.

Retail markets

Some technologies operating in the retail market are able to be located on-site with end-users. Retail prices here are much higher than in the NEM. Retail prices are not actually received by the generators of renewable electricity, but are avoided by the presence of on-site generation and, if that on-site generation has lower costs, the end-user's overall energy bill will be reduced.

Retail prices differ by state. For residential customers the headline tariff is around $120–140/MWh (12–14 c/kWh), plus GST. However, some customers will also receive the 'off-peak' rate of around $50–90/MWh (5–9c/KWh) for overnight water heating or other activity that makes up a significant portion of their energy demand. Together with service fees, this result in an average residential retail price of $110–130/MWh (11–13c/kWh).

For business end-users, the tariffs become more complicated and diverse. For relatively low-voltage users, such as farms and small business, the average tariff is generally higher than the residential rate, at around $140–160/MWh (14–16c/kWh). Different rates for different times of day are available, so there will be significant variation around this average depending on the type of business. For very-high-voltage consumers, the energy charge can as low as $40–80/MWh (4–8c/kWh), but, again, this varies widely.

Non-electricity markets

Electricity-generation sources such as natural gas, landfill gas and biomass provide other valuable outputs or benefits besides electricity. In the case of biomass and natural gas, heat can be generated for use in industrial processes. Other useful by-products of biomass include ash and fertilisers. Landfill gas production reduces the total volume of waste that needs to be processed and disposed of. Fees are also paid for the raw materials received and used for landfill gas production.

Another potential benefit, particularly in regional locations, is that local generation avoids network construction and augmentation costs. In the case of remote sites, the cost of network infrastructure per unit of energy delivered is so prohibitive that all power is generated locally (which is referred to as off-grid). In non-remote sites, such as cities, the benefit also exists. The addition of generation near to end-users can often mean that upgrading existing distribution network infrastructure in that area is delayed.

Potential uptake of new electricity generation technologies

Returns to investment in new electricity generation technologies

In this section we discuss whether, ignoring the availability of suitable sites and resources and non-electricity market benefits, a new renewable electricity generator could make positive net returns from installing new plant and selling renewable electricity into either wholesale or retail markets. The relevant information we need to reach a conclusion on this issue is shown in Figure 13. The ranges of costs shown are extracts from the year 2020 taken from the full cost projection paths to 2050 presented in the previous section. The prices received are based on an assessment of the markets available to the technology given its operating configuration and size plus the potential premium available estimated at between \$20 and \$40/MWh based on access to either renewable subsidies or electricity price rises from a potential future greenhouse gas emission trading scheme. A subsidy of \$4000/kW is assumed for solar photovoltaics. This is lower than is currently available, but the subsidy would be expected to decline as the cost of solar photovoltaic units comes down.

With respect to 2020, the following conclusions (from an economic viewpoint only) can be drawn from this data:

- Coal with CO_2 capture and sequestration (Coal+CCS), hot fractured rocks (HFR), natural gas combined cycle (NGCC) and nuclear power are likely to be the most competitive in the base-load wholesale electricity market.
- Any hydroelectric power sites falling within the cost range shown are likely to be economic. Over large areas of the cost range, no subsidy or future increase in electricity prices are required to make positive net returns above long-run marginal costs
- High-quality wind sites can be expected to be economically viable
- Solar thermal will be very close to becoming economically viable in the period around 2015–2020, providing suppliers can access some intermediate- and peak-load wholesale markets.
- Landfill gas is economically viable. Its revenue stream may be a mix of avoided retail electricity bills and returns from sales to the wholesale market. The mix depends on the size of the landfill resource and therefore by how much the electrical output exceeds local needs.
- Where appropriate sites are available, solar hot water will be an economically viable option, with or without subsidies
- Solar photovoltaics (PV) will be economically viable in some retail markets if the subsidy is around \$4000/kW (in real terms).

Non-economic impediments to uptake of energy sources

With the economics of low-emission technologies and resources improving, the various ecological, technical and social constraints to uptake are beginning to come into focus. Table 5 summarises some of these issues. These are discussed in more detail in other chapters. There are many impediments which, if overcome, could radically change Australia's energy future. For example, if hot fractured rock technology proves feasible then it can potentially provide large-scale renewable base-load power, thus overcoming a technical hurdle that has long been considered too great for renewables to achieve.

Similarly, if nuclear power becomes broadly accepted as a power option, it could also provide a share of Australia's future energy needs. Ultimately, a combination of economic and non-economic factors will determine the future mix of low-emission energy sources that Australia will rely on to meet its future energy needs.

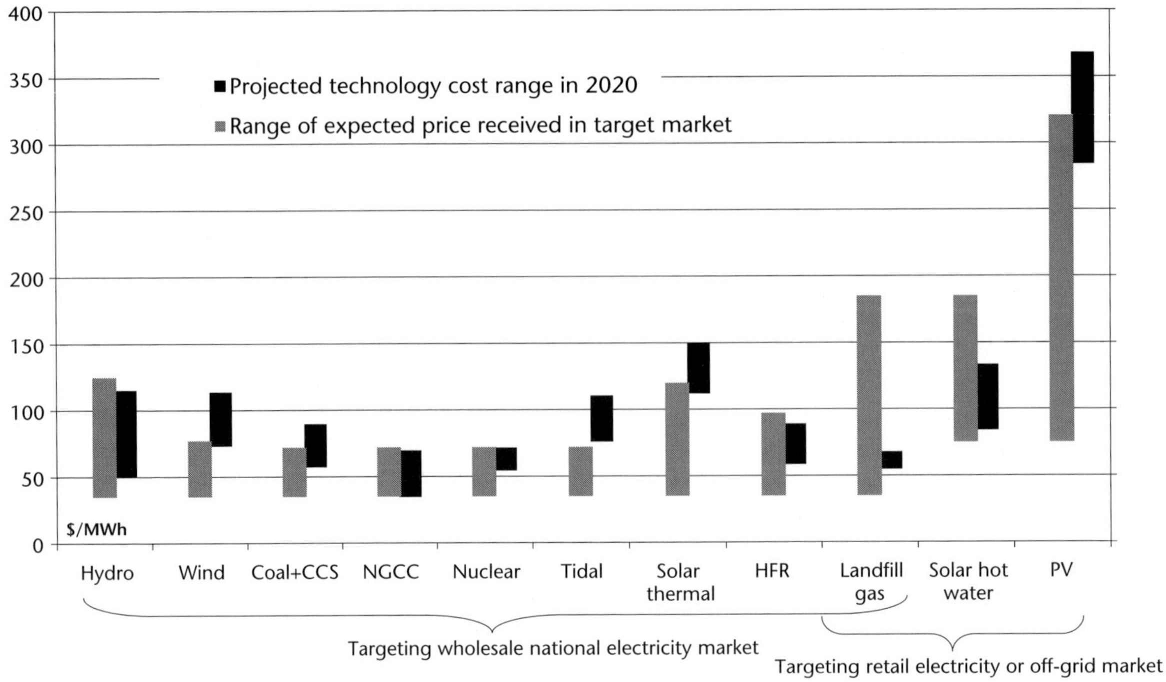

Figure 13: Projected revenue and cost ranges for low-emission electricity generation technologies, 2020 (Source: CSIRO research).

Table 5: Overview of non-cost related impediments to uptake of low-emission energy sources

Technology	Impediments to further uptake
Biomass	Competition with food crops Soil nutrient reduction or, alternatively, increased use of fertiliser Agricultural market is primary focus of fuel suppliers
Solar thermal	Small scale: Conditions for solar-hot-water system uptake are seldom met due to a general aversion to upfront costs, poor knowledge, slow turnover of existing stock of water heaters and inability of owners to capture benefits if not occupying the building Large scale: Intermittent or irregular energy supply, but some positive correlation with peak daily electricity demand
Solar photovoltaics	Intermittent or irregular energy supply, but some positive correlation with peak daily electricity demand
Hydro	Few unexploited sites Intermittent or irregular energy supply in some cases
Wind	Intermittent or irregular energy supply in most cases Impact on bird life Impact on aesthetic quality of landscape
Nuclear	Perceived risk of unsafe plant operation Perceived risk of increased proliferation of nuclear weapons material Perceived risk of unsafe storage of waste Inability of private sector to manage above risks without government intervention and expenditure
Coal or gas with CO_2 sequestration	Yet to be demonstrated at large scale Perceived risk of unintended CO_2 escape from designated geological storage site Damage or sterilisation of nearby underground resources or ecosystems
Hot fractured rocks	Yet to be demonstrated at large scale Trigger for seismic activity, but often in remote locations

References

ABARE (2003). *Australian Energy: National and State Projections to 2019–20.* ABARE eReport 03.10. ABARE, Canberra.

ABARE (2006a). *Australian Energy: National and State Projections to 2029–30.* ABARE eReport 05.9. ABARE, Canberra.

ABARE (2006b). *Australian Commodity Statistics 2006.* ABARE, Canberra.

ABARE (2006c). *Australian Energy Statistics.* ABARE, Canberra. <http://www.aberconomics.com/interactive/energy/index.html>

ABS (Australian Bureau of Statistics) (2006). Australian Social Trends, ABS No. 4102, ABS, Canberra. <http://www.abs.gov.au/AUSTATS/abs@.nsf/DetailsPage/4102.02006?OpenDocument>.

AGO (Australian Greenhouse Office) (2006). *National greenhouse Gas Inventory 2004.* AGO, Canberra. <http://www.greenhouse.gov.au/inventroy/2004/pubs/inventory2004.pdf>.

CSIRO and ABARE 2006. *Modelling Energy Futures Forum scenarios using ESM.* CSIRO, Melbourne. <http://www.csiro.au/resources/pfnr.html>.

DOE/EIA (United States Department of Energy/Energy Information Administration) (2006). *Annual Energy Outlook 2007: With Projections to 2030.* DOE/EIA-0383(2007), Washington DC.

Energy Efficiency and Greenhouse Working Group 2003, 'Towards a National Framework for Energy Efficiency: issues and challenges discussion paper'. <http://www.nfee.gov.au/about_nfee.jsp?xcid=64>.

ESAA (Energy Supply Association of Australia) (2006). *Electricity Gas Australia 2006.* ESAA, Melbourne.

Energy Strategies (2004). *A Clean Energy Future for Australia.* Clean Energy Future Group, http://www.wwf.org.au/publica tions/clean_energy_future_report/.

Geoscience Australia (2006a). *Oil and Gas Resources of Australia 2004.* Geoscience Australia, Canberra.

Geoscience Australia (2006b). *Australia's Identified Mineral Resources 2006.* Geoscience Australia, Canberra.

Graham P, Coombes P, Beer, T, Bouma W and Vincent D (2005). 'Options for electricity generation in Australia (Supplement to technology assessment report 31)'. Technology assessment report 34, Cooperative Research Centre for Coal in Sustainable Development, Brisbane.

Greenpower Accreditation Program (2007). Quarterly Status Report, 1 October–31 December 2007. http://www.greenpower.gov.au/admin/file/content13/c6/Quarterly%2Status%20Report%202007%20Q4%2Final_1.pdf

Greenpower Price Watch (2007). <http://www.greenpowerpricewatch.com.au/block.html>. Accessed 1 May 2007

IEA Greenhouse Gas R&D Program (2000). The potential of wind energy to reduce CO_2 emissions. Report number PH3/24, IEA Greenhouse Gas R&D Program.

McDonald A. and Shrattenholzer L (2001). Learning Rates for Energy Technologies. *Energy Policy* **29**, 255–261.

MUARC (Monash University Accident Research Centre) (1993). 'Vehicle mass, size and safety', Report prepared for Federal Office of Road Safety, Canberra.

National Greenpower Accreditation Program (2007). 'Quarterly Status Report 1 January – 31 March 2007'. <http://greenpower.gov.au/downloads/reportsaudits/2007Q1Report.pdf>.

NEMMCO (2007). Average Price Tables, NEMMCO, Melbourne. <http://www.nemmco.com.au/data/avg_price/averageprice_main.shtm>. Accessed April 2007.

Reedman L, Graham P, Coombes P and Vincent D (2007). 'Impact of carbon price uncertainty on investment in selected electricity generation options'. Technology Assessment Report 59. Cooperative Research Centre for Coal in Sustainable Development, Brisbane

Somerville M, Wyborn D, Chopra PN, Rahman SS, Estrella D and van der Meulen T (1994). *Hot Dry Rocks Feasibility Study.* ERDC Australia, Report 94/243.

WRI (World Resources Institute) (2006). *Navigating the Numbers: Greenhouse Gas Data and International Climate Policy.* WRI, Washington DC.

Wright T (1936). Factors affecting the cost of airplanes. *Journal of the Aeronautical Sciences* **3**, 122.

22

Biofuels

Deborah O'Connell, Tom Beer, Michael Dunlop and David Batten

Demand for energy—for transport fuels as well as stationary energy (electricity)—has grown dramatically throughout the world during the 21st century. Over the last 10 years, debate has accelerated about whether oil—the feedstock underlying the petrochemical and transport fuel industries — may be close to its limit in terms of cheap extraction, and politically and financially secure sources. There is growing evidence that this is the case, although there is still a high degree of uncertainty. Oil prices have risen dramatically since 2002. It is becoming increasingly certain that the atmosphere, land and oceans cannot continue to absorb the greenhouse gases that are produced in using fossil energy without serious consequences.

Consequently, there has been a growing trend worldwide to look for alternative energy sources that are more secure and produce less greenhouse gases (GHG). Impacts can be reduced by improving energy efficiency combined with reducing demand, improving waste management and using 'clean, green' bio-based replacements for the energy and petrochemical industries. There is an emerging recognition that a transition away from oil is required—and biofuels have been proposed as an alternative to increase energy security and to mitigate climate change. These are the first-order drivers of change, but other potential benefits could include opportunities to revitalise rural and regional communities and decrease pollution, thus improving health in cities.

Australia is highly dependent on oil to support our current way of life in cities and in rural areas. The broader subject of future transport has been covered elsewhere in this volume (Lamb, Chapter 28). Transport is the second largest emitter of energy-related GHG—accounting for 14.4% (80.4 Mt) of national greenhouse gas emissions in 2005 (AGO 2006), and this proportion is increasing. In 2004/05 Australia consumed 19.9 GL of petrol and 15.2 GL diesel (ABARE 2006). About 75% of our current transport fuel requirements are derived from local oil and gas sources, but, as current rates of demand increase and without major new discoveries or technological breakthroughs, this fraction is estimated to decline to 45% by 2020 (ABARE 2004). Approximately 85% of Australians live in urban areas. The average trip length of a Sydney dweller was estimated at less than 10 km in 2003 (NSW Department of Planning 2005). Petrol and gaseous fuels are mainly used in urban areas, whereas diesel is used more widely in rural areas (Figure 1).

Providing a significant share of transport energy from *renewable* sources may be achieved by using biofuels and renewable sources of electricity. The leverage for improvements in sustainability of Australia's transport energy system (and therefore many of the key points of intervention to change the system) is strong in cities because of the gross number of kilometres travelled on short trips in urban environments (Figure 1). The developments in vehicle and engine technology that would facilitate a transition to electricity as a source of transport fuel are dealt with in chapter 28. Here, we will focus on biofuels as a renewable and sustainable source of fuels for future transport needs. Biofuels may play a marginal role (2–5% of our total transport needs) or become mainstream (10–20% or more of our transport needs) in Australia's transport energy future. Indeed, it may play a major role—meeting more than 20% of our long-term future transport requirements (Keating *et al.* 2006). This chapter seeks to evaluate the factors to be con-

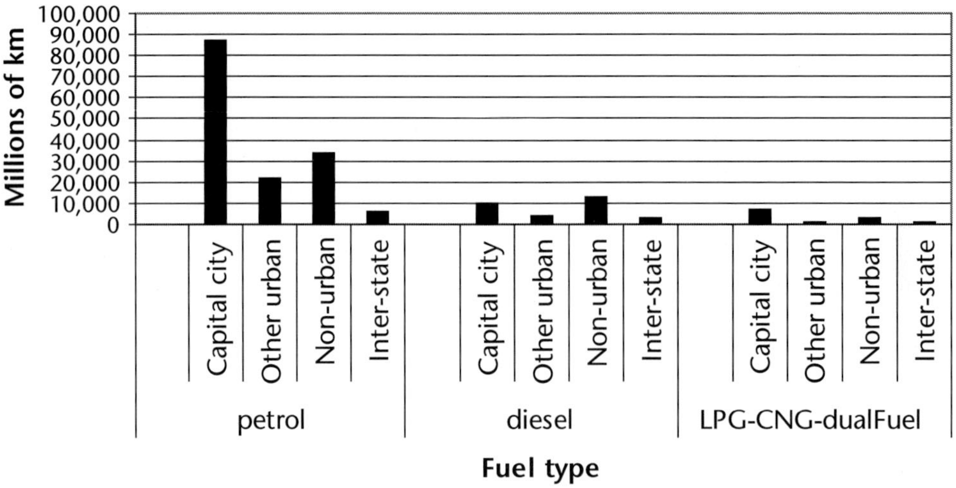

Figure 1: Average kilometres travelled in urban and non-urban environments on various fuel types, from 1998 to 2005 (Source: ABS 2003, Table 7; SMVU is a survey and therefore subject to sampling error).

sidered for biofuels as a major transport energy source, in the light of potential transitional pathways to sustainable futures. It begins with a brief history of biofuels in Australia, provides a snapshot of the current situation, and then focuses on future transitions.

Background – what is biofuel?

There are two main types of transport biofuels in current usage internationally: ethanol and biodiesel. These are known as 'first-generation' biofuels. This term also includes biogas and methanol made from biogas.

The commercial viability of biofuels is determined by the amount and cost of the feedstock, the cost of the conversion technologies, and the price of competitive products such as petroleum. First-generation technologies (for liquid transport fuels) include:

- fermentation and distillation of ethanol from starch and sugar crops
- transesterification of waste cooking oils, tallow and vegetable oil to produce biodiesel
- anaerobic digestion of wet wastes to produce biogas (largely methane).

Second-generation biofuels are made from non-food feedstock. Such technologies are being developed around the world, and may revolutionise the industry. They include conversion of lignocellulose to a range of fuels including ethanol (and other alcohols such as methanol, butanol and propanol), syngas, synthetic diesel, DiMethyl Ether (DME), and, in the long term, hydrogen. Lignocellulose gives plants strength, structure, and resistance to insect and other damage. It is composed of lignin (the 'glue'), celluloses and hemicelluloses (interlocking cell wall components). Breaking down woody and other lignocellulose-like materials into components that can be used for fuel is not simple. Technologies to do so—based on enzyme and bacterial transformation, acid hydrolysis or gasification—are being developed, but are not yet cost-effective.

Third-generation technologies are those where a range of high-value products, including bioproducts (replacements for petrochemical products,) may be produced—with biofuel or bioenergy as one of a suite of co-products (Kamm *et al.* 2006). These technologies are in the early stages of development but offer intriguing prospects for the future.

Where have we been?

The focus of this book is on Australian cities. However, biofuels are in their infancy in this country, and lessons must be drawn from the international arena to understand biophysical, infrastructure and institutional factors and pathways that have led to more significant levels of production or use

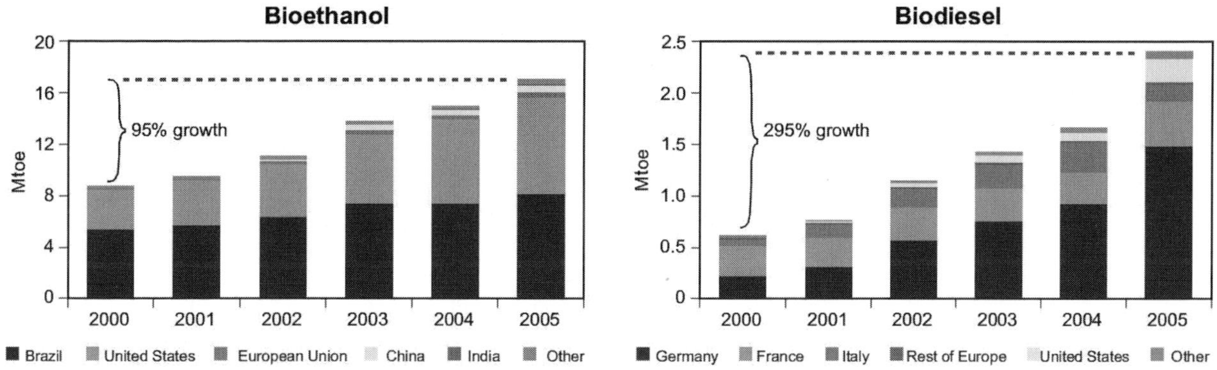

Figure 2: Global ethanol and biodiesel production (Source: IEA 2007).

of biofuels in particular countries. In addition, in an era of globalisation, the impact of increased overseas biofuel production and use has implications for both urban and rural Australia through upward pressure on commodity prices, as well as through potential new markets for 'sustainably certified' Australian biofuel or feedstock.

Early farming systems were relatively self-sufficient in energy and used approximately 20% of the energy produced from crops to feed the draft horses that worked the land. Subsequently, the low prices and the ready availability of fossil fuels enabled both rural and urban areas to import their energy requirements, and thus increase the density of settlement (Smil 2000). The oil shocks of the 1970s led to some research into the production and use of biofuels in Australia (Boardman *et al.* 1980, Howes and Rummery 1980, Stewart *et al.* 1979). However, because this was followed by a period of cheap oil, interest waned off in most countries, except for Brazil.

In the mid-1970s, the government of Brazil launched the National Fuel Alcohol Program (or Proálcool) to increase the share of domestically produced fuels. Passenger cars that could run on ethanol were required, as well as installation of nationwide infrastructure to supply ethanol in all service stations. Supply was guaranteed via strict controls on the planting of sugar cane and the production of both sugar and ethanol. By the late 1980s, ethanol had a larger market share in the transportation sector than petrol. By the mid-1990s, the program was abandoned, because ethanol shortages and low petrol prices led to widespread rejection of ethanol-powered cars. Ethanol retained a significant market share, however, which increased further during the high petrol prices of the last few years. From 1975 to 2002, ethanol replaced around 210 GL of petrol, saving Brazil around US$52 billion (A$65 million). This example shows that the transitional pathway to biofuels becoming mainstream requires integrated and coordinated effort across the entire value chain — from feedstock production, through to conversion to biofuels, distribution infrastructure and utilisation in vehicles. This also shows the active role that governments have played in initiating a change away from the practice of 'business as usual'.

The production and use of biofuels in other countries has increased very rapidly over the last few years. Drivers for the use of biofuels have differed greatly between countries, and between fuels. Current biofuel production is dominated by ethanol in the United States of America (USA) and Brazil, and by biodiesel in the European Union (EU). Figure 2 shows production levels at a global scale. Production of these fuels is predicted to continue to rise steeply.

In countries other than Brazil, ethanol was until fairly recently regarded as a minor player. In the USA ethanol is mooted as an alternative fuel that can help to tackle the issue of energy security. The USA now has major programs to encourage the production and use of ethanol. Biodiesel is strongly supported in the EU where, compared with the USA or Australia, a much higher proportion of passenger cars use diesel. Support of biodiesel was originally part of a package to assist farmers, but, in recent years, policy support has been based on reducing greenhouse gas emissions—as required as part of the EU obligation to the Kyoto protocol.

Where are we now?

Current supply of biofuels

In Australia, there has been a focus on biofuels since 2001 when the Australian Government introduced a target of 350 ML biofuels by the year 2010. A study on the appropriateness of this target (Beer *et al.* 2003) and a report from the Prime Minister's Biofuels Taskforce in 2005 (Biofuels Taskforce 2005) followed. Various states also instigated taskforces recently to investigate the role of biofuels and appropriate policy settings, and there have been numerous government inquiries into the subject. Mid-2007 data indicates that progress towards the 350 ML target has been very slow—but oil companies have recently started to promote and advertise petrol containing ethanol and biodiesel blends.

Australia has a small, but established, commercial biofuels industry. Ethanol in Australia in 2007 is produced by CSR in Queensland from C-molasses, which is derived from sugar cane, and by Manildra in New South Wales from wheat starch. There are also plans to produce ethanol from sorghum. It is estimated that for the 2006/2007 financial year about 75 ML (from the 147 ML existing annual production capacity) of ethanol as a transport fuel will be produced in Australia. This equates to a very small proportion (less than 0.5%) of our current petrol usage (O'Connell *et al.* 2007).

In Australia, the most common feedstock for biodiesel is waste cooking oil, animal tallow or imported palm oil. To date, canola oil has been too expensive to use as a viable biodiesel feedstock. For the 2006/2007 financial year, it is estimated that about 50–75 ML (of the 323 ML existing annual production capacity) of biodiesel as a transport fuel will be produced in Australia—again a very small fraction (less than 0.5%) of our diesel usage (O'Connell *et al.* 2007).

There are plans for production capacity to increase to 1155 ML for ethanol and 1122 ML for biodiesel. These plans are being made in a very uncertain investment environment—influenced by a combination of factors, including government policy conditions, fluctuating oil price, increases in biofuel feedstock prices due to prolonged drought in Australia as well as the influence of increased international biofuel production on grain prices (O'Connell *et al.* 2007). It is therefore not certain how much of the planned capacity will eventuate.

Current demand for biofuels

As a general rule, in Australia, the light vehicle fleet uses petrol and the heavy vehicle fleet uses diesel, so that ethanol and biodiesel have different markets. Both can be used as a fuel in their own right — though there are fewer vehicle alterations required to use biodiesel as a diesel replacement than there are in using ethanol as a petrol replacement.

Ethanol can be mixed with petrol and used in spark ignition vehicles. In Brazil, flexible-fuel vehicles can use any petrol–ethanol mixture. In Australia, legislation limits the maximum allowable ethanol in petrol to 10% and the most common use of ethanol is as E10. At this concentration, all modern spark ignition vehicles (about 60% of the vehicle fleet) should be able to use the fuel. Biodiesel can be used in compression ignition engines as a replacement for diesel fuel, so that, in terms of operability criteria, any blend of biodiesel/petroleum diesel can be used.

The current lack of availability of E10 and B5 to most potential customers throughout the southern and western states is a major barrier to demand growth and sales. Consumer confidence in ethanol (relating to concerns about engine damage) is also a major challenge for the ethanol industry (Biofuels Taskforce 2005). Regional motorists are happier to purchase E10 than city dwellers, with Queenslanders favouring ethanol more than drivers from other states. As part of its broader ethanol-industry-development policies, the Queensland Government has been involved in many initiatives to improve consumer confidence in ethanol blends, including:

- information about the suitability of E10 (consistent with manufacturers' advice) in all motor vehicle registration renewal notices and on a website
- working with vehicle manufacturers to ensure that fuel and engine technologies are optimally compatible
- encouraging service stations in Queensland to sell E10.

Biodiesel does not suffer from the same degree of consumer negativity as ethanol blends. However, confidence can be fragile, and the biodiesel industry will need to ensure that consumers are properly advised on fuel blends.

Future prospects for biofuels

Biofuels will need to demonstrate a stable and substantial role in energy supply if they are to be relevant to major Australian sustainability and energy security strategies. The viable longer term domestic supply, and the gross reduction in greenhouse gas emissions, will therefore determine how significant a role biofuels will ultimately play; that is, a role at the margins (2–5% of future transport), part of the main game (10–20%) or part of the end game (over 20%). Performance against a range of sustainability criteria besides supply and GHG emissions is, however, essential. An understanding of demand is also important in understanding the transitional pathways. These assessments have not yet been substantively made in Australia, but some preliminary estimates are presented here.

Potential domestic supply of biofuels

If all domestic sugar and grain produced in average years were converted to ethanol, a blend of E78 could be supported. This equates to a 50 per cent replacement of 2004–05 petrol energy if a correction is made for energy efficiency—although this could be partially offset by producing vehicles that use the higher octane rating of ethanol. If only the export fraction of Australian crop production in an average year were used, a blend of E50 could be supported. This equates to a 33% replacement of 2004–05 petrol energy (O'Connell *et al.* 2007).

If all domestic waste oil, tallow and oilseed crops produced in average years were converted to biodiesel, a blend of B10 could be supported. This equates to 10% replacement of 2004–05 diesel use as there is a similar energy content. If domestic waste oil was combined with export tallow and oilseeds, a blend of B6 (or 6% of current diesel transport energy) could be supported in average years (O'Connell *et al.* 2007).

These figures give a broad indication of the envelope of upper limits to production in average years. There is, however, a high level of variation between years due to seasonal conditions. This variation has implications not only for the physical limits of production and the impact on fuel security, but also on the cost of production as feedstock is the largest, and most variable, cost component.

Australia may be able to produce the lignocellulose feedstocks to move beyond the limits of first-generation technologies and have a mainstream role in Australia's transport future. The lignocellulose feedstock resources include waste wood material, forest residues, and potentially new large-scale plantings of dryland woody crops such as oil mallee (Wildy *et al.* 2003; Bartle 2006; Hobbs *et al.* 2007). Depending on which sources of feedstocks were used, this could equate to upper limits of between 10 and 140% of current petrol usage (O'Connell *et al.* 2007). The very wide range in these estimates indicates the degree of uncertainty in these data on annual production of biomass, as well as levels of sustainable harvest and economic viability. This is in contrast to the agricultural data, where the range in upper limits is largely due to seasonal variability and export markets.

In terms of physical limits, it is therefore likely that biofuels based on first-generation domestic feedstocks will remain a marginal part of Australia's transport future (2–5% of transport fuel needs). This is because high-input agricultural systems, which are geared towards producing food and animal feed, will make the biofuel feedstock more expensive—especially given the upward pressure on grain and oilseed prices from the international impact of rapidly increasing biofuel production, as well as prolonged drought and impacts of climate change in Australia. Biofuels could, however, move beyond these limits and become mainstream (greater than 10–20%) or even play a major role (greater than 20%) of Australia's transport future if industries develop around second- (or third-) generation technologies. The technical, economic and sustainability potential for this scenario is currently poorly understood.

Future demand for biofuels

Two components determine the aggregate demand for alternative fuels such as ethanol and biodiesel in transport—*intermediate* and *final* demand (O'Connell *et al.* 2007). Intermediate demand consists of the purchasing patterns of intermediate producers, such as oil companies, service stations, farming cooperatives and other intermediaries who process and blend fuels for eventual sale to consumers. Final demand consists of fuel purchases by consumers for use in their own vehicles and fleets.

There are sensitive links between levels of intermediate and final demand. For example, the commercial risks of market entry are high for the oil majors because they are associated with low levels of consumer confidence and thus a lack of consumer demand for the product (O'Connell *et al.* 2007). As seen in the earlier discussion of biofuels in the international arena and in Queensland, however, there are a range of policy and market mechanisms that can be employed by governments to tackle issues of intermediate and final demand. These are important considerations in transitional pathways, but are not of the same strategic importance as issues of security of supply and sustainability, which are harder to resolve.

Sustainability of biofuels

Biofuels (and more broadly bioenergy and bioproducts) must demonstrate their sustainability credentials if they are to gain widespread support as a replacement for the current fossil-based energy system. The capacity to produce feedstock, or fuel that can be certified and sold as 'sustainably produced', may be a key market advantage for Australia in the future. Major challenges exist in estimating the sustainable production of biomass—for current agriculture and forestry, as well as for an emerging industry such as biofuels—over the longer term, when land and water resources in Australia will be increasingly contested for food, fibre, energy production and water yield.

Assessing the sustainability of biofuels (or bioenergy) requires understanding multiple dimensions, including GHG emissions, air quality, energy balance, land and water impacts, biodiversity and socioeconomic factors. The international and Australian discussion abounds with potential benefits of biofuels in terms of lower greenhouse gas emissions, lower air pollutants (and therefore benefits for health in cities) and opportunities for rural and regional communities. The benefits are frequently, however, very specific to the particular production system and value chain, and are not necessarily universal. We focus now on the potential benefits that are of most relevance to sustainable Australian cities, while recognising that a higher level of demand for biofuel in the cities may have significant implications for rural and regional Australia.

GHG emissions

The carbon dioxide (CO_2) emitted during combustion of a biofuel is not counted as a greenhouse gas because it was originally sequestered into plants as they grew. But, to make a biofuel out of plant material, some energy must be provided to grow and transport the crop, convert it to a fuel, and distribute the fuel to retailers before it is combusted in an engine (Figure 3). If this 'external' energy is provided from non-renewable sources (such as fossil fuels), then some GHGs are emitted over the whole life cycle. The emissions profile is therefore extremely sensitive to the particular type of production system. For example, ethanol made from feedstock from a high-input agricultural system (such as irrigated maize that required fertiliser, herbicide, sowing, harvesting, and so on) will have higher emissions over its life cycle compared with the same fuel made from a waste product feedstock (such as wood waste). This has important implications not only for the GHG emissions, but also for the Energy Profit Ratio (EPR), or the Energy Return on Investment (EROI). If the biofuel is using as much, or more, energy to make a litre of biofuel as is contained in that litre of biofuel (EPR close to or less than 1), the net energy gain is minimal. If the input energy is fossil-based (as it is in first-generation systems) then the greenhouse gas emission profile is also poor.

In Australia, first-generation ethanol and biodiesel GHG emissions have been analysed by Beer *et al.* (2003). Greenhouse gas emissions are lower than petrol by a range of 1.7% (from wheat) to 5.1% (from molasses, using co-generated power). Greenhouse gas emissions for biodiesel based on waste vegetable oil range from 89.5% lower for pure biodiesel to 4.18% lower for a 5% biodiesel blend when compared with diesel. Greenhouse gas emissions for biodiesel based on canola range from 15% less for pure biodiesel to 1.5% less for a 5% biodiesel blend when compared with diesel. The influence of the type of feedstock and the proportion of blend is clearly seen in these figures. Improvements in energy efficiency of feedstock production systems

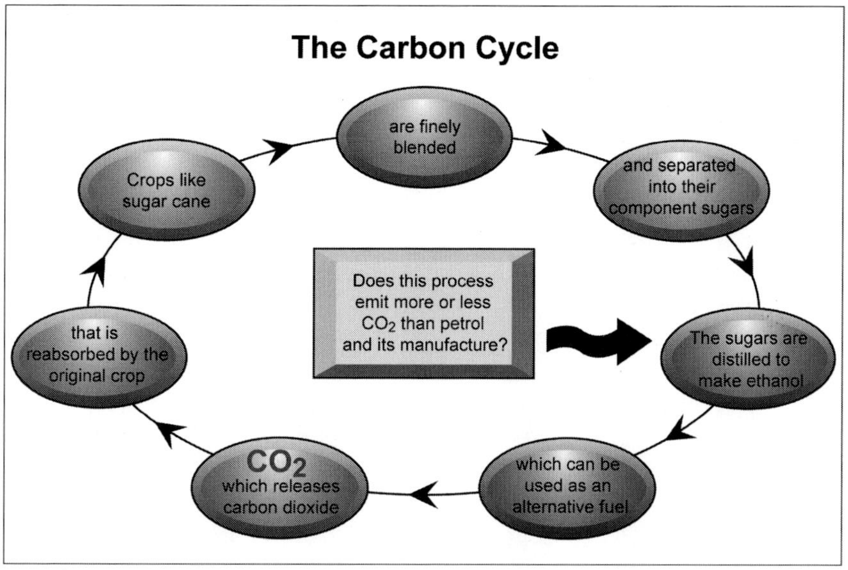

Figure 3: The full fuel cycle of biofuel manufacture indicates that fossil carbon is added during the fuel processing stage. In addition, the fertilisers used during the agricultural stage also emit greenhouse gases.

and transport, as well as in conversion technologies (such as the new molecular sieve technology for ethanol) can further lower greenhouse emissions.

Health

There are only a limited number of Australian studies on motor vehicle air pollutant emissions, and no agreed methods for assessing the associated health risks. Air pollutants are generated by vehicles and the levels depend on the nature and composition of the fuel used, the type and age of the vehicle, the nature of the drive cycle (such as the distance, speed and number of stops and starts), and the tuning of the engine (Beer *et al.* 2001).

There are six categories of pollutants associated with transport—particulate matter, oxides of nitrogen (NOx), carbon monoxide, sulphur dioxide, ozone and lead. All may be emitted from tailpipes (depending on the fuel that is used) except for ozone, which forms from a secondary chemical reaction between NOx and non-methane volatile organic compounds (NMVOCs). Air quality in Australian cities is generally good, but ozone and particle levels remain problematic and exceed standards in some cities (Biofuels Taskforce 2005; Beer *et al.* 2006). Urban air quality is improving due to a number of previous policies to tighten emissions and fuel standards. Sulphur dioxide and lead, having been reduced or eliminated in fuel, are no longer considered to be environmental problems in Australian urban areas. Emissions of major pollution in cities will continue to fall until 2020, when the projected increase in vehicle usage will begin to offset air quality gains (Biofuels Taskforce 2005).

The health impacts of air pollutants include respiratory disease and mortality from particulate matter. Air pollutants can also cause corrosion to buildings and impair visual amenity. The health benefits of using biofuels arise mostly from lower levels of particulate matter emissions (when compared with petrol and diesel).

For biodiesel, all criteria air pollutants except NOx are significantly lower when diesel is replaced with biodiesel. Particulate matter emissions are significantly lower for pure biodiesel (B100) made from tallow, canola or waste oil than for diesel (Figure 4). The benefits of lower particulate matter emissions are greatest for pure biodiesel blends, and lowest in B5 blends where the benefits are negated by the diesel

The benefits of ethanol, particularly in an E10 blend, are less clear. There may be lower particulate emissions from the tailpipe, but there are increased evaporative emissions of smog-forming organic compounds that may have a negative impact on air

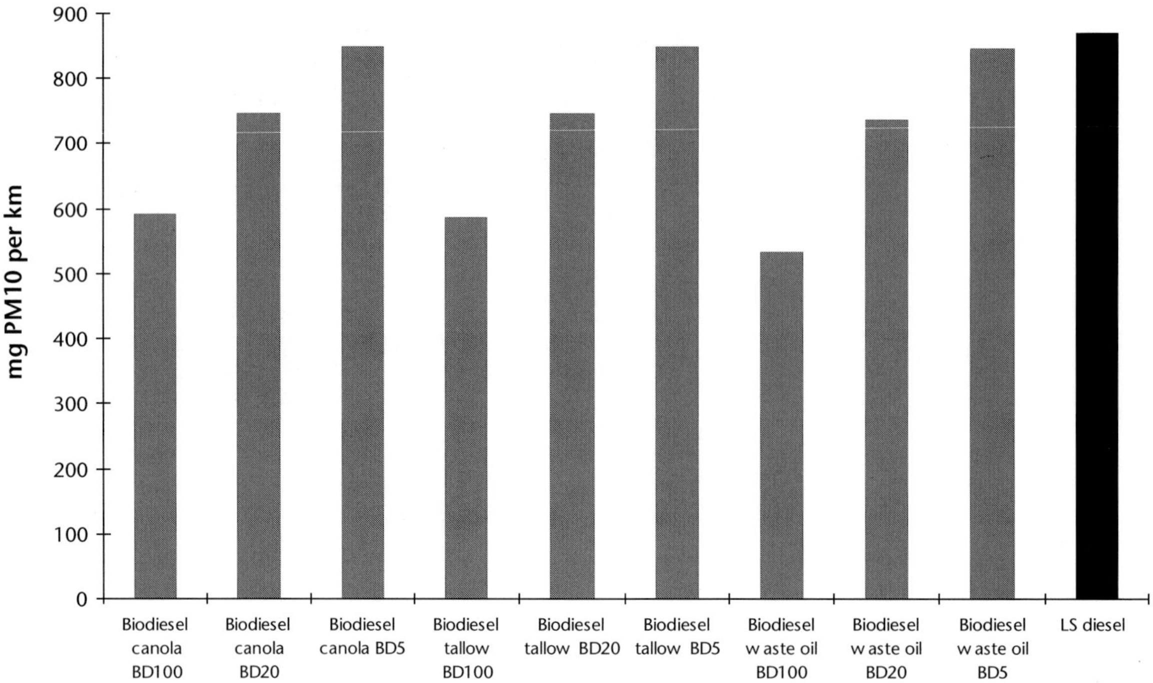

Figure 4: Life-cycle particulate matter (PM10) emissions from low sulphur diesel (LSD) and the use of biodiesel from various feedstocks (Source: Beer *et al.* 2003).

quality. This may lead to worse air quality outcomes in some specific circumstances. The high level of uncertainty in the level of particulate reduction from E10 does not permit a reliable estimate of the health costs and benefits of using E10. Due to the uncertainty in the above estimates, the Department of Environment and Water Resources (DEW) in 2007 commissioned a project led by CSIRO and Orbital Engine Corporation to study the health impacts of E5 and E10.

CSIRO is developing consistent methods to assess such factors as shown in Figure 5 (unpublished). This is a useful information tool, but static information such as this does *not* provide the basis for a sustainable industry— it needs to be balanced with understanding dynamic interactions into the future, potential transition pathways and a broader dialogue in society about sustainability.

Trade-offs between different elements of sustainability

In the above approach, each of the dimensions of sustainability is assessed individually. However, in reality, trade-offs between criteria must be made. For example, there is an international scale trade-off emerging between fuel and food security (UN-Energy 2007). In North America, corn is the dominant ethanol feedstock and corn prices have risen dramatically leading to 'tortilla riots' in Mexico City. There is also international concern at the rapid growth in the palm oil industry due to biodiesel demand. Since the 1990s, the area under palm oil cultivation has increased by about 43%. Clearing rainforests to grow palm oil not only endangers biodiversity and creates social conflict, but releases large amounts of carbon and thus exacerbates the very problem that a move to biodiesel in Europe is seeking to tackle. As at 7 June 2007, the price of Malaysian palm oil (a feedstock for biodiesel) was US$830/tonne—an all–time high (palmoilis.mpob.gov.my). In Indonesia and Malaysia, palm oil is widely used as cooking oil and price increases affect people's ability to feed themselves. Palm oil is also the major feedstock used in the manufacture of soaps and detergents, and the prices of these are also affected.

At a finer scale, trade-offs between individual sustainability criteria can be quantified. For example, in general we can say that measures to reduce greenhouse gases will reduce air pollution (and

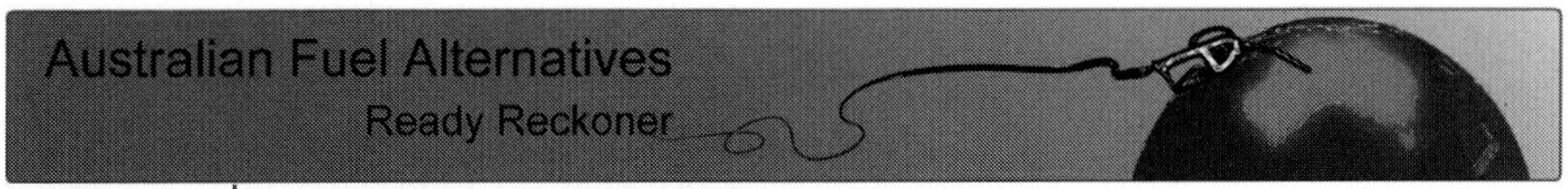

About the project
Fuel pathways
Fuel types
Processes
Feedstocks
Vehicles
Assessment criteria
Fuel report cards
Glossary

Home > Fuel report cards > Ethanol from Wheat in Cars

Print

Report card - Ethanol from Wheat in Cars

Data are based on a mixture of measurements and modelling. These assessed values derive from passenger vehicles (maximum weight 3.5 tonnes) running a wheat-derived E10 (10% ethanol) petrol blend and operating on the European Drive Cycle (EDC) 2003.

Blend

E10

Assessment 15 December 2006

Category	Component	Rating	Reliability of Evidence	Assessed Value (Range)	Units
1. Surety of supply					
1a. Future	% of current transport fuel market which can be supplied	Neutral	Medium	15 (10 - 20)	%
	Likely duration of supply at this annual rate	Advantage	Low	>50	years
	Uncertainty associated with supply (e.g. political, seasonal)	Disadvantage	Low	>30	%
2. Sustainability					
2a. Greenhouse gases	GHG (full life cycle) for cars	Neutral	High	236.09	g CO_2 equiv/km
	Energy Output / Energy Input	Disadvantage	High	0.93	Ratio
2b. Air quality	PM10	Disadvantage	High	10.04	mg/km
	CO	Neutral	High	2.26	g/km
	HC	Advantage	High	0.40	g/km
	NO_x	Advantage	High	0.33	g/km
2c. Land and water quality	Land usage	Neutral	Medium	0.04	ha/GJ
	Water extracted from water supplies	Disadvantage	High	8333	Litres/GJ
	Water uptake by feedstock growing cycle	Disadvantage	High	21920	Litres/GJ
	Impact of cultivation practices (fertiliser and pesticide usage, tillage practices)	**NA**	**Varies**	Varies	Impact/GJ
2d. Biodiversity	Impact on natural environment and habitat	**NA**	**Varies**	Varies	Impact
2e. Social	Impact on quality of life and economic well-being of rural Australia	Neutral	Low	Neutral	Impact
3. Cost of supply					
3a. Supply of feedstock	Cost of extraction and delivery of feedstock materials (car)	Advantage	High	5.2	c/km
3b. Transformation of feedstock	Cost of transformation of fuel and distribution costs (car)	Advantage	High	2.1	c/km
3c. New capital requirements	Processing facilities	Disadvantage	High	Entirely new infrastructure	Infrastructure change
	Distribution	Neutral	Medium	Significant modification to infrastructure	Infrastructure change
	Vehicles	Advantage	High	None to minor mods	Infrastructure change
4. Health and safety					
4a. Production storage and distribution	Refuelling and distribution	Neutral	Low	Similar	Performance relative to petrol
	Accidents and on board mishaps	Neutral	Low	Similar	Performance relative to petrol
5. Consumer acceptance					
5a. Fuel consumption habits	Refuelling and distribution	Advantage	High	No or little change	Scale of change
	Vehicle	Advantage	High	No or little change	Scale of change

>> References and further information (PDF file)

CSIRO This is a prototype website. © CSIRO. A project of the Energy Transformed Flagship

Figure 5: Report card on sustainability of ethanol from wheat (Source: CSIRO 2007).

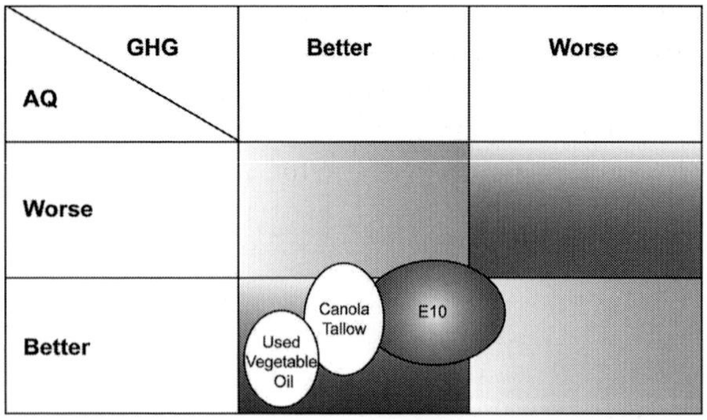

Figure 6: Diagrammatic representation of the uncertainties associated with the policy options of introducing petrol containing 10% ethanol (E10) or using various feedstocks to make biodiesel.

vice versa), but this is not always the case. This can be shown using a four-division quadrant, with air quality and GHG options divided into better and worse outcomes (Beer *et al.* 2005).

The results from studies on ethanol are variable. There are positive GHG and air quality (AQ) benefits (compared to petrol), but in each case there is a 25% probability that the benefits may not materialise. This may be shown diagrammatically, as in Figure 6, by drawing quadrants of possible policy options and placing the 10% ethanol in petrol option (denoted by E10) in a circle that is displaced from the centre such that the displacement is towards the improved GHG and improved AQ quadrants. Using data from Beer *et al.* (2003) similar ellipses for biodiesel (compared to diesel) are illustrated for tallow and canola, and for waste oil. Overall, the options may be expected to be positive, but there is a finite probability of a perverse outcome.

Processes for sustainable transitions

Australia has processes at various levels of government for dealing with sustainability issues in agriculture and forestry, as well as mature processes for environmental and social impact assessment for specific projects. There are also processes in place for reporting on the short-, medium- and long-term economic outlook of agriculture, forestry and energy on a global scale, and, in Australia, at national and state levels. These existing structures or processes are relevant to part of the bioenergy/biofuels debate, but do not cover sustainability issues across the full agriculture, forest industry and energy value chain (Rutovitz and Passey 2004; O'Connell *et al.* 2005).

O'Connell *et al.* (2005) took a local scale ('bottom up') systems view, embedding scientific or economic *content* (for example, economic analysis, life cycle analysis) within a *process* logic. A sequential set of steps and decision points, framed by consultation with 'consenting communities' was defined in order to gain both formal approval, as well as a community 'licence to operate'. This type of approach complements the ('top down') criteria and indicator approaches being developed internationally (Energy Transition Taskforce 2006, UN-Energy 2007). These international approaches seek to specify a universal framework of sustainability requirements, with careful development of the use of biomass for energy, transport and chemicals to avoid competition with food production and deterioration of biodiversity, and maximise positive effects on energy supply, development of agriculture and local development and economic prosperity. Sustainability accreditation, which will rely on 'track and trace' systems, will also be developed to underpin sustainable trade of biomass, biofuels and bioenergy.

The sustainability issues depend, in part, on the ultimate size of the biofuels industry. If the biofuels industry remains marginal in Australia (that is, less than 5% of total transport requirements), then the sustainability issues probably do not differ from those facing our current agricultural systems—which are already substantial. If the biofuels industry were to become mainstream or even a major player in transport, then the sustainability implications will change—with positive or negative impacts depending on how the industry developed. For example the industry could develop to deal with salinity issues in low rainfall areas (such as creating an oil mallee industry), while at the same time diversifying farm incomes and improving biodiversity in the landscape (Bartle 2006). Alternatively, it could lead to intensification of agriculture and a reduction of stream flow in rivers (Stirzaker *et al.* 2002) or removal of too much carbon from agriculture or forest systems (Raison 2006) which would have a negative environmental impact. Transition pathways to a large-scale biofuels industry therefore need to be carefully considered by government, industry and society.

Transitional pathways to the future

The simple quantitative exploration of physical limits discussed in this chapter is inadequate to gain a real understanding of the prospects and implications of a biofuel industry in Australia. The essential physical questions about whether biofuels could solve (a substantial part of) Australia's transport energy sustainability problems cannot yet be answered, because the quantitative work has not yet been done. The following qualitative exploration of scenarios and transitions illustrates the complex analysis required to ensure sustainable transition pathways in a future where biofuels go beyond the margins.

As shown in the introductory section of this chapter, there is scope for substantial leverage in Australian cities for making changes to the transport energy system, because a large fraction of the national fuel demand and usage is currently in the cities. Reducing demand for transport energy in cities could take place through eco-efficient urban design, high-usage public transport systems, and changes to vehicle and engine technologies, such as electric cars (which do not need to travel long distances in cities and are therefore not limited by battery technology). If biofuels do emerge as a major transport option in Australia, however, the factors and interactions to be considered extend way beyond the boundaries of the cities.

The area of land used to grow suitable feedstocks may increase. It is largely unknown — beyond relatively small changes in price—how much additional land could be brought into production for a given change in demand. Farming systems could be optimised for biofuel (as opposed to human food or animal feed), and this would require different agronomic practices, economic production functions and productivity. This, in turn, would add further to the uncertainty about the supply curve—how much could supply increase with a change in price? For example, the high protein content in wheat for use in human food products is not required to produce biofuel. The area of land available for production would increase, and the requirement for nitrogen fertiliser would decrease.

The availability of cost-efficient conversion technology for lignocellulose feedstock could radically alter the demand for first-generation feedstocks and affect the direction of land use changes. For example, increased demand for cereal for ethanol might lead to expansion of crop production into more marginal areas. If, however, second-generation processing were commercial, it may instead lead to a reduction of the area under crop as currently marginal cropland was converted to producing perennial biomass, such as coppiced eucalypts. While these substitutions are likely, a large amount of lignocellulosics could probably be supplied from land that is not currently cropped, reducing trade-offs between bioenergy and food and feed production. However, it is as yet unknown how much lignocellulose feedstock could be supplied sustainably from different regions and plant species.

Cost is clearly a determining factor in the production of feedstocks and biofuels, but there are other considerations including the environmental impacts of feedstock production and biofuel manufacture, distribution and use. Many of these impacts have not yet been quantified—they could change with technology, location and other regional factors. For example, increased feedstock production could have positive impacts on soil condition, regional hydrology, biodiversity, and CO_2 balances—or considerably negative impacts on all counts—depending on the type of feedstock, where it is produced and how it is produced. Hence there is scope for perverse outcomes should incentives for one environmental outcome lead to degradation of others. It also follows that regulation designed to control these impacts might significantly shape the structure of the industry as it develops.

The demand for biofuel will also be affected by other developments in the energy sector. If biomass becomes a significant source of stationary energy (electricity), then this would compete with fuel production, but it might also allow economies of scale to be captured. Current biofuel options are limited to ethanol and biodiesel in internal combustion engines. If other fuels became significant (such as methanol for fuel cell vehicles), there may be greater opportunities for substitution between a range of complementary energy sources (including coal and natural gas) that could all be used to produce the same fuel.

The long-term viability of biofuels will be determined by the transitional pathways that are taken towards whichever 'post-oil' and 'low-carbon' energy systems are developed. Early policy and investment decisions around fuel types, processing and distribution infrastructure, locations of investment, engine types, land use changes and planning for eco-efficient cities could facilitate or retard some transition pathways. For example, biodiesel has many current attractions (such as better CO_2 reductions, more flexible blend choice in the current fleet), but this would contribute little to any transition to a fuel-cell based transport system. Ethanol, in contrast, could provide a logical technological and infrastructural stepping stone to a hydrogen economy. And while ethanol production from first-generation feedstocks may have some negative impacts (such as poor overall CO_2 savings when used as E10 and impacts on other agricultural sectors), it might be a vital part of a transition to a lignocellulose-based biofuel system.

Clearly there is potential for biofuels to move beyond current usage to become mainstream (10–20%) or even a major part (greater than 20 per cent) of our future transport fuel needs. The potential for and impact of biofuels will be affected by medium- to long-term considerations. These include the size, structure and timing of fuel demand, which could be substantially impacted by changes in urban design and function, other demands for biomass, other developments in the energy system, impacts beyond energy production (such as on water and other natural resources), and domestic and global policy regarding CO_2 emissions and energy security. Further analysis of these factors, and their interactions along various transition pathways, is required.

References

ABARE (2006). *ACS Petroleum Tables.* Australian Bureau of Agricultural Research Economics Canberra.

ABARE (2004). *Australian Energy National and State Projections to 2019–20.* Australian Bureau of Agricultural Research Economics, Canberra.

ABS (2003). *Survey of motor vehicle use.* Australian Bureau of Statistics, Canberra.

AGO (2006). *Australia's National Greenhouse Accounts.* Australian Greenhouse Office, Canberra.

Bartle J (2006) New non-food crops and industries for Australian dryland agriculture. *Green Processing* **2006**, 63–68.

Beer T, Grant T, Morgan G, Lapszewicz J, Anyon P, Edwards J, Nelson P, Watson H. and Williams D (2001) *Comparison of Transport Fuels - Final Report to the Australian Greenhouse Office on the Stage 2 study of Life-cycle Emissions Analysis of Alternative Fuels for Heavy Vehicles.* CSIRO, Aspendale, Victoria.

Beer T, Olaru D, Van der Schoot M, Grant T, Keating B, Hatfield Dodds S, Smith C, Azzi M, Potterton P, Mitchell D, Reynolds Q, Winternitz J, Kierce S, Dickson A, Short CLT and Heyhoe E (2003). *Appropriateness of a 350 Million Litre Biofuels Target: Report to the Australian Government Department of Industry, Tourism and Resources*, CSIRO, ABARE, BTRE, Canberra.

Beer T, Grant T and Korn W (2005). *Environmental Sustainability Issues in Relation to Biodiesel*, CSIRO Atmospheric Research, Aspendale, Victoria.

Beer T, Borgas M, Bouma W, Fraser P, Holper P and Torok S (2006). 'Atmosphere', theme commentary prepared for the 2006 Australia State of the Environment Committee, Department of Environment and Heritage, Canberra. <http://www.environment.gov.au/soe/2006/publications/commentaries/atmosphere/index.html>.

Biofuels Taskforce (2005). 'Report of the Biofuels Taskforce to the Prime Minister'. Department of the Prime Minister and Cabinet, Australian Government, Canberra.

Boardman NK, Thring MW, Johnston DR and Swift-Hook DT (1980). Energy from the Biological Conversion of Solar Energy *Phil Trans R. Soc. Lon. Series A, Math and Phys Sci* **295 (1414)**, 477–489.

CSIRO (2007). Alternative Fuels web-based Ready Reckoner tool (in preparation)

Energy Transition Taskforce (2006) *Criteria for sustainable production of biomass*, Energy Transition Task Force, Projectgroep Duurzame Productie van Biomassa, Netherlands.

Hobbs T, Bennell M, Huxtable D, Bartle JR, Neumann C, George N and O'Sullivan W (2007). *FloraSearch Agroforestry Species and Regional*

Industries – Low rainfall farm forestry options for southern Australia, RIRDC, Canberra.

Howes KMW and Rummery RA (Eds) (1980). *Energy and Agriculture: The Impact of Changes in Energy Costs on the Rural Sector of the Australian Economy.* CSIRO, Perth.

IEA (2007). Energy Technology Essentials. Newsletter Jan 2007. <www.iea.org/Textbase/techno/essentials2.pdf>. Accessed 1 July 2007.

Kamm B, Kamm M, Gruber P and Kromus S (2006). Biorefinery systems – an overview. In *Biorefineries – Industrial Processes and Products. Status Quo and Future Directions.* (Eds B Kamm, P Gruber and M Kamm) pp. 3–40. Wiley-VCH Verlag GmbH, Weiheim.

Keating B, O'Connell D, Beer T, Dunlop M, Batten D, O'Connor M, Grant T, Poole M, Miller T and Lamb D (2006). Biofuels from Australian Agriculture – prospects and implications for R&D. In *Proceedings of Australian Institute of Agriculture Science and Technology. Biofuels: Paddock to Pump Symposium.* 19 October 2006. Canberra.

NSW Department of Planning (2005). *Transfigures: 2003 Household Travel Survey and Regional Estimates.* NSW Department of Planning, Sydney.

O'Connell D, Keating B and Glover M (2005). *Sustainability Guide for Bioenergy: a Scoping Study.* Rural Industries Research and Development Corporation, Canberra.

O'Connell D, Batten D, O'Connor M, May B, Raison J, Keating B, Beer T, Braid A, Haritos V, Begley C, Poole M, Poulton P, Graham S, Dunlop M, Grant T, Campbell P and Lamb D (2007). *Biofuels in Australia – Issues and Prospects.* Rural Industries Research and Development Corporation, Canberra.

Raison RJ (2006). Opportunities and Impediments to the Expansion of Forest Bioenergy in Australia. In *Biomass and Bioenergy* (Eds J Richardson, R Bjorheden, O Popescu and CT Smith). **30(12)**, 1021–1024.

Rutovitz J and Passey R (2004). *NSW Bioenergy Handbook,* Mark Ellis & Associates, Sydney.

Smil V (2000). Horse power – The millennium of the horse began with a whimper, but went out with a bang. *Nature* **405**, 125.

Stewart GA, Gartside G, Gifford RM, Nix HA, Rawlins WHM and Siemon JR (1979). *The Potential for Liquid Fuels from Agriculture and Forestry in Australia.* CSIRO, Mentone, Victoria.

Stirzaker R, Vertessy R and Sarre A (2002). *Trees, Water and Salt – An Australian guide to using trees for healthy catchments and productive farms,* Agroforestry Design Guidelines. Rural Industries Research and Development Corporation, Canberra.

UN-Energy (2007). *Sustainable Bioenergy: A framework for decision makers,* United Nations.

Wildy D, Pate J and Bartle J (2003). *Silviculture and Water Use of Short Rotation Mallee crops.* Rural Industries Research and Development Corporation, Canberra.

23

Solar Energy

Chris Fell, Jim Hinkley, Anne Imenes and Wesley Stein

The impact of anthropogenic carbon dioxide emissions on the Earth's climate is now almost beyond question. An overwhelming proportion of these emissions are a direct result of the processes we use to provide convenient sources of energy. If we are to begin to turn this situation around, finding new low- or zero-emission sources of energy will be absolutely critical. Given the enormous existing investment in infrastructure for fossil-fuel sources, such as coal and oil, a key challenge will be how we make the transition from polluting to non-polluting energy sources. Solar energy offers a unique opportunity to make this transition practical, both from an economic perspective and with minimal disturbance to the way we live our lives.

A greater use of solar energy in our energy mix makes tremendous sense for many more reasons than merely its impact on lowering greenhouse gas (GHG) emissions. The use of solar energy can reduce all harmful emissions from our energy processes, making us less dependent on imported fuels. In addition, Australia's expertise in all forms of solar-energy technology could yield benefits from being at the forefront of a burgeoning new industry.

The solar resource

Energy from the sun arrives at the Earth in the form of electromagnetic radiation, which we know as sunlight. It is perhaps staggering that in any given year the amount of energy arriving at the Earth's surface as sunlight represents some ten thousand times the energy the entire human race will consume in that year (although much of that solar energy is used by natural climatic processes). The challenge is to find ways of converting this enormous resource into more familiar forms of energy (energy vectors, such as electricity, heat and fuels) that allow us to integrate solar energy into our lifestyle with a minimum of disruption and cost.

One component of the challenge is that the availability of solar energy varies, depending on the time of day, the time of year and the location of the receiver on Earth. Figure 1 depicts the local variability in insolation, or solar irradiance, which will determine the suitability for deployment of solar technologies. While differences between neighbouring regions are the result of the effect of weather patterns on cloud cover, the most notable variation is the dependence on latitude—there is a region centred on the equator known as the sun belt.

The existence of the sun belt is due to the angle at which the sun's rays strike the Earth's surface. The high latitude regions—the far North and the far South—on average receive sunlight at more oblique angles than the equatorial latitudes (Figure 2). A phenomenon known as the cosine effect means that any radiation striking a surface at an angle will provide less energy per square metre, compared with the same radiation from directly above. The cosine effect is a fundamental parameter for designers of solar collection systems, because it drives not only the latitudinal variations in insolation, but also the seasonal and diurnal (daily) variations in sunlight for any given location. Figure 3a demonstrates the variation in solar irradiance (the rate of energy received per square metre) throughout a clear day, for both winter and summer in Newcastle, Australia. Clearly, the irradiance is greatest in summer and at noon, when the sun is highest in the sky.

While this may appear to suggest that areas outside the sun belt would be much less suited to collecting solar energy, the cosine effect can be

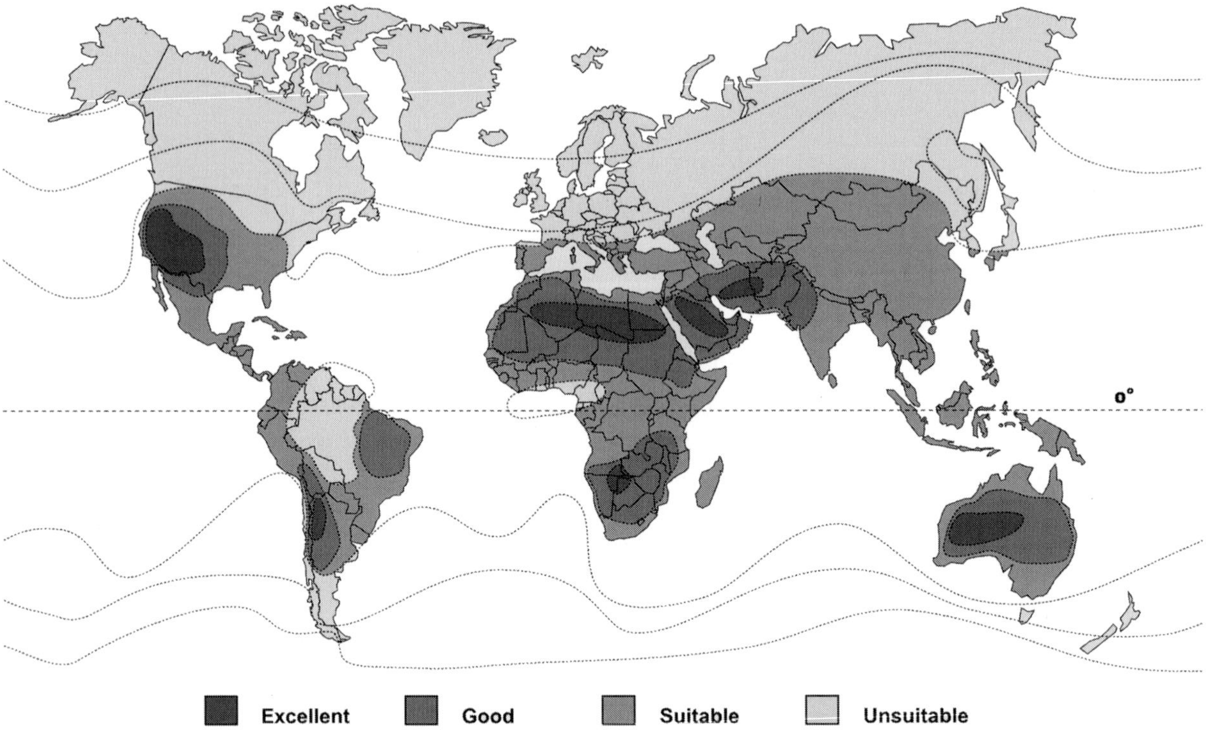

Figure 1: 'The sun belt'—geographic variations in solar irradiance depend on latitude, giving rise to a broad equatorial region highly suitable for the utilisation of solar energy (Source: Solar Millennium AG 2008).

greatly reduced if the solar collector is tilted so that it faces toward the sun. High-latitude sites must still contend with greater atmospheric absorption compared with sites in the sun belt, because the sun's oblique angle means its rays take a longer path through the atmosphere, but this effect is much less significant than the cosine effect. Figure 3b shows the daily solar irradiance variation for the same location, but for a collector that is able to follow the sun on its path through

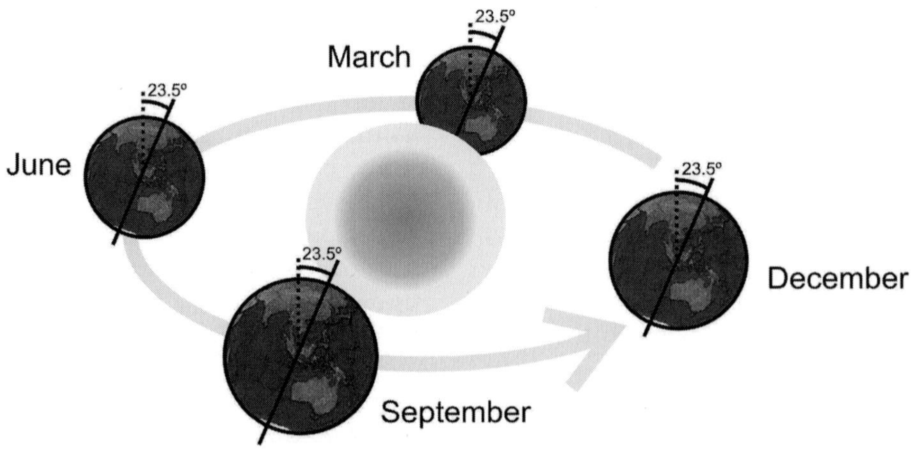

Figure 2: The angle of the Earth's axis remains constant throughout the year, which means that latitudes nearer the poles always receive sunlight at oblique angles.

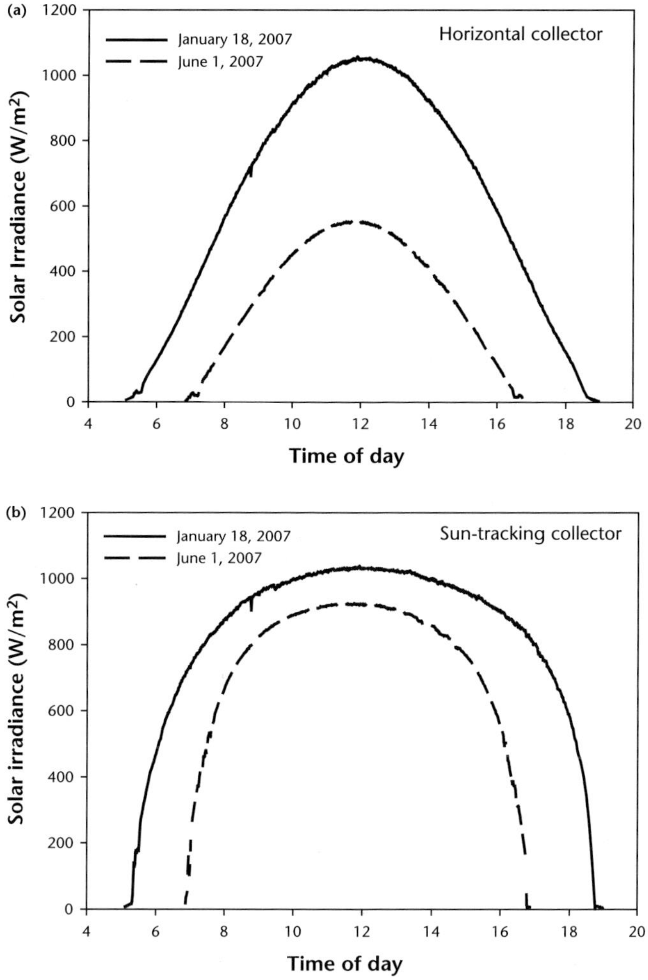

Figure 3: Clear-sky solar irradiance at Newcastle, Australia, comparing the summer and winter diurnal variations for (a) a horizontal solar collector and (b) a solar collector with two-axis tracking. (Source: CSIRO Energy Technology 2008).

the sky. Known as solar tracking, this feature can add to the cost of a system, but offers much more effective collection of solar energy. Because designers must balance the cost and benefit of a solar tracking system, simpler tracking systems that only rotate about one axis are popular, as are installations that do not move but are simply tilted toward the sun. In fact, a quick analysis shows that the simple act of tilting a fixed flat-panel collector by an angle equal to the local latitude provides an extremely cost-effective improvement in the power output of a solar-energy system.

Despite the increased cost, however, some solar collectors depend on solar tracking, because they have a limited field-of-view. A widespread example is the solar concentrator, which magnifies the sun's radiation intensity many times. These mirrors or lenses offer a means to concentrate the solar irradiance onto an expensive solar cell and are also essential in many solar-thermal systems, which can require temperatures of over 1000°C.

Clearly, the choice of an ideal location for solar-energy production involves a number of considerations, such as latitude, the use of solar tracking, and whether a non-tracking system can be angled toward the equator—as opposed to being fixed horizontally, such as a solar pond, or fixed in a vertical position, such as many building-integrated photovoltaics. Perhaps most important, however, is the consideration of local weather and the impact of clouds. Clouds reduce the level of solar insolation in three ways:

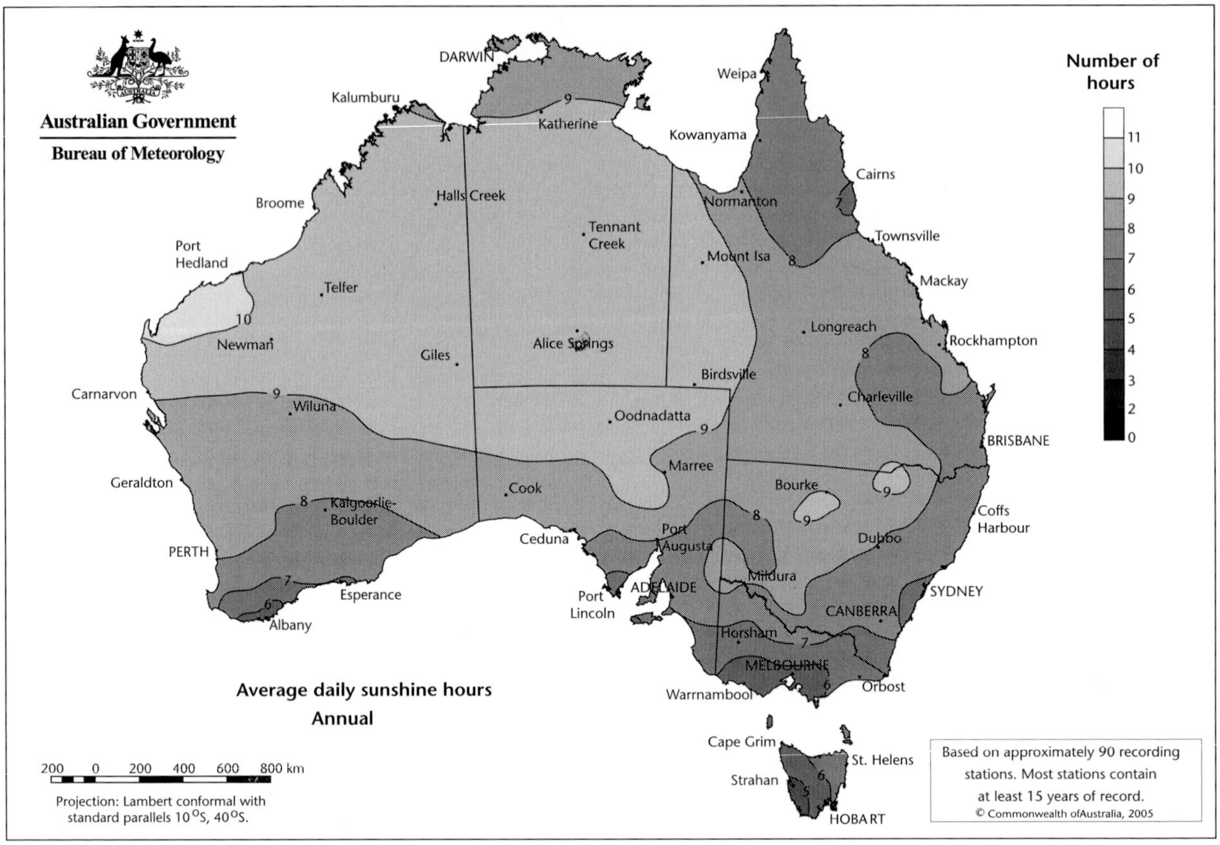

Figure 4: Australian insolation map (Source: Australian Bureau of Meteorology 2000).

- they reflect sunlight back into the upper atmosphere
- they absorb sunlight
- they scatter sunlight sideways.

Although a small amount of scattered sunlight still reaches the Earth and can be collected (we refer to this as diffuse solar radiation), it is at a much lower intensity than direct sunlight and hence provides much less energy.

Due to the combined effects of latitude and relatively low cloud cover, Australia has the highest average solar irradiance of any continent and so is a prime location for solar-energy utilisation. Figure 4 is a map from the Australian Bureau of Meteorology indicating the average number of hours of bright sunshine at 90 recording stations over the past 15 years (Australian Bureau of Meteorology 2000). Much of Australia receives an annual average of more than 9 hours per day. With this much sunshine, a land area of only 50 km × 50 km would accommodate sufficient solar collector area to supply Australia's entire electricity demand.

The ideal site for a solar plant in Australia depends on a number of factors such as proximity to the electricity grid, the type of installation and the particular solar technology being employed. Systems that do not track the sun will need to accommodate the seasonal variation in insolation—as shown in Figure 5 for the major Australian cities. The trend is clearly similar for all cities on the chart, with reduced insolation during the winter months. The latitude effect makes this variation more prominent for cities further from the equator— an issue that must be considered for non-tracking installations. Tracking or fixed-tilt installations will be less susceptible to the summer–winter variation, but must still contend with cloud-cover, which can be seen in this data to particularly affect the city of Darwin between December and March.

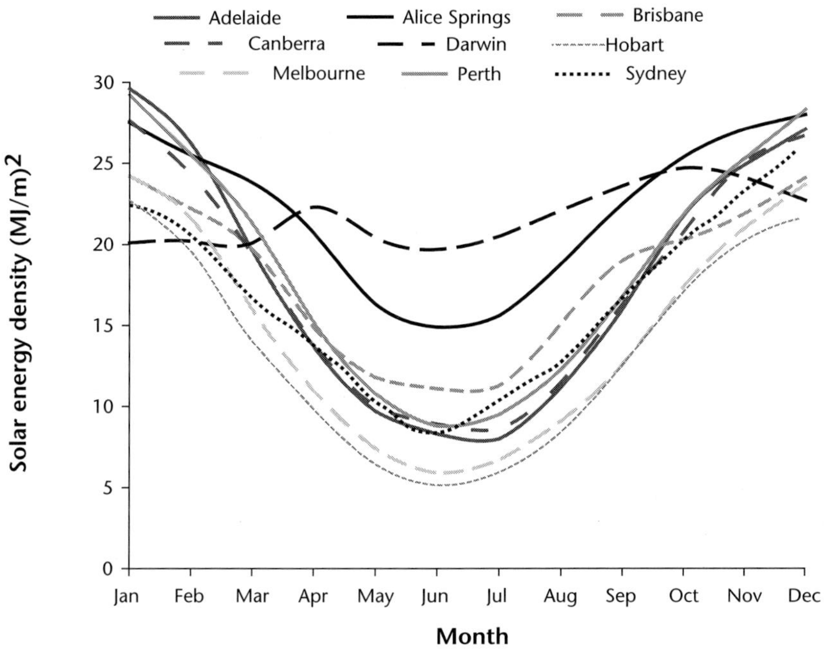

Figure 5: Monthly average of the solar energy received daily on a horizontal surface at Australia's major cities. Note the effect of cloud cover on Darwin during the summer wet season. (Data source: from Morrison and Litvak 1999.)

As an indication, a hypothetical cloudless site would receive about 3500 kWh/m^2/yr direct solar radiation, a very good solar site about 2800 kWh/m^2/yr and Sydney around 1800 kWh/m^2/yr.

While designers must consider geographical and seasonal variations in solar radiation, probably the most significant challenge facing the large-scale introduction of solar energy is its intermittency. No clever engineering can extract solar energy when the sun isn't shining. The cycle of day and night and the unpredictable intermittency of cloud cover mean that methods of storing solar energy will necessarily play an important role in the way the technology is used.

Solar Collectors

Our ability to harness the energy in sunlight depends on the technologies we develop to capture this energy and convert it into one of two familiar forms—either electricity or heat. Solar technologies that produce heat are collectively referred to as solar thermal. We will distinguish these from technologies that directly convert sunlight into electricity, which are known as photovoltaics and are discussed later in the Chapter.

The solar-thermal approach is based on one of the oldest and simplest concepts: radiation from the sun is reflected, focussed, or simply falls directly onto a collector filled with a fluid to collect the heat. Whether the application is to produce hot water at 60°C, electricity generation at 300°C or a thermochemical process at 1000°C, the heating of a fluid represents the fundamental first step for a large number of industrial processes. At present, the vast majority of these processes obtain this heat from the burning of fossil-fuels. Solar-thermal technologies provide a clean, renewable source of heat, without the need to change any downstream components of the process. Given the simplicity of the overall concept, the primary challenge for solar-thermal researchers today is to design ways of collecting solar energy with a minimum of cost. To achieve the high temperatures necessary for industrial processes, most solar-thermal installations employ a concentrator configuration, the specifics of which depend on the application. Concentrators that focus sunlight to a point produce the highest temperatures, and the highest thermodynamic cycle efficiency, but are the most expensive. A list of some of the common configurations is included in Figure 6 and Table 1 and described below.

Table 1: Summary of solar-thermal collectors

Motion	Collector type	Concentration ratio	Nominal maximum operating temperature range (°C) (assuming ambient temperature is 25°C)	Maximum Carnot Efficiency (%)
Stationary	Solar pond	1	90	18
	Unglazed pool heater	1	35	3
	Conventional glazed flat-plate collector (FPC)	1	85	17
	Advanced FPC	1	120	24
	Solar chimney	1	60	11
	Evacuated-tube collector (ETC)	1	150	30
	Compound parabolic collector (CPC)	1–5	200	37
Single-axis tracking	Linear Fresnel reflector (LFR)	10–40	280	46
	Parabolic-trough collector (PTC)	75	400	56
Two-axes tracking	Paraboloidal dish concentrator (PDC)	100–3000	>1000	>77
	Solar tower (central receiver)	100–1500	>1000	>77

(Rabl 1985; Kalogirou 2004)

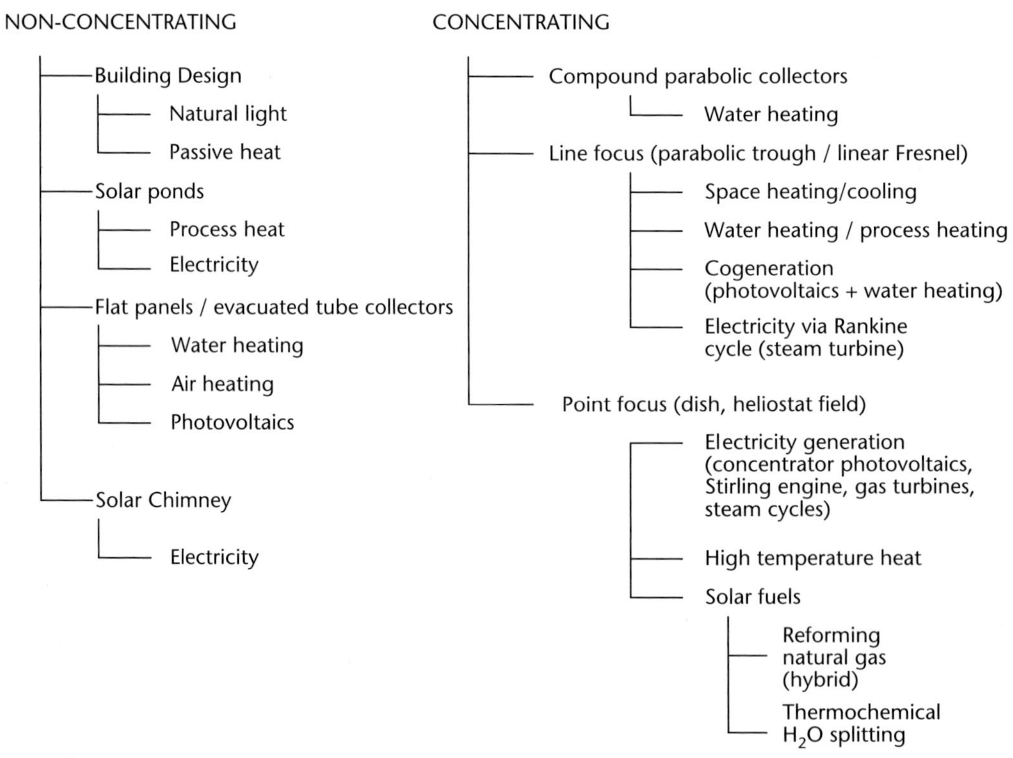

Figure 6: Overview of solar-thermal technologies and applications.

Non-Concentrating Solar Thermal

Solar ponds

Solar ponds use a concentration gradient of salt-laden water to reduce convective thermal losses. When properly constructed, they can achieve temperatures double those of an otherwise similar pond. Heat is removed using heat exchange pipes in the lower (warmer and saltier) part of pond. Australia, with its saline groundwater, is well placed to exploit this technology. A pilot project is in place at Pyramid Salt in Northern Victoria, supplying 60 kW of thermal energy to a flake-salt-drying process.

Flat-plate collectors

Flat-plate collectors (Figure 7) are widely used for solar-hot-water systems and also for air heating. They consist of a flat layered 'sandwich' of materials—a glazing to prevent convective heat loss above an air space, with an absorber mounted on supporting insulation. Some installations are unglazed, such as those for the heating of swimming pool water where the temperature is relatively low. Flat-plate collectors are very effective for producing the moderate temperatures required for household, as well as some industrial and chemical process heating requirements. Their moderate cost makes them competitive with electric water heaters in some regions and they are likely to see even greater market penetration in the near future. Double-glazed versions, which offer higher efficiency, are now emerging.

Evacuated-tube collectors

Evacuated-tube collectors have made great inroads into the hot water market over the last few years

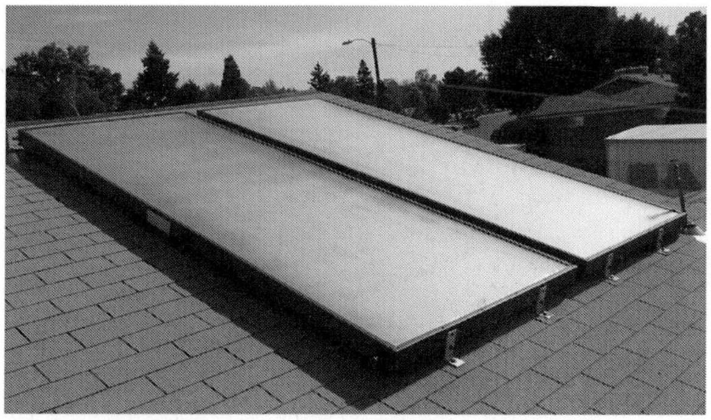

Figure 7: Typical flat-plate solar-hot-water heater (Source: Paul Norton, DOE/NREL Photographic Information Exchange).

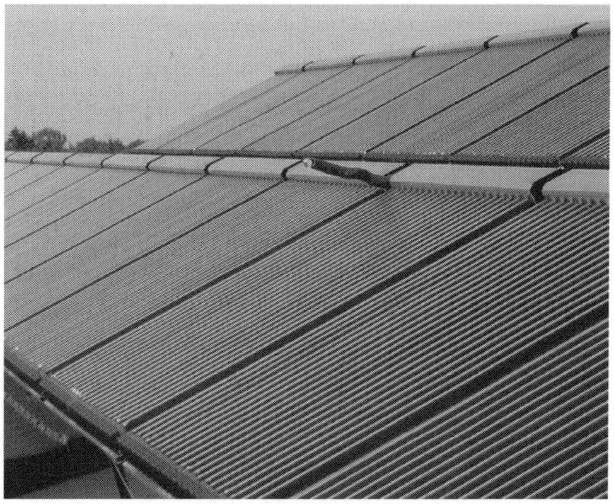

Figure 8: Typical evacuated-tube solar-hot-water heater under testing (Source: Wesley Stein).

(Figure 8). Mass production and take-up in China in particular has caused costs to drop rapidly. Their design uses a vacuum gap to reduce convective losses, and a selective surface to reduce radiation losses. These collectors, used without concentration, are capable of attaining temperatures above those of conventional flat-plate collectors and are particularly suited to cooler climates.

Solar chimneys

The solar chimney is based on the simple concept of hot air rising. A large collection area— operating much like a greenhouse—collects and funnels hot air into the centre, through one or more wind turbines and up a stack. The turbines rotate as a result of the pressure difference created by the updraft or 'stack effect', and generate electricity much like a normal wind-turbine generator. Solar chimneys generally have low solar to electricity efficiencies as a result of the low temperature differences involved, but are relatively easy to construct with no moving parts apart from the wind turbine. A small prototype operated in Spain in the 1980s to prove the concept and a much larger version has since been proposed for Australia.

Concentrating-solar-thermal power

Nearly all of the world's electricity—whether coal, gas or nuclear—is generated by first heating a fluid. Concentrating-solar-thermal power is simply another means of generating a hot fluid that can

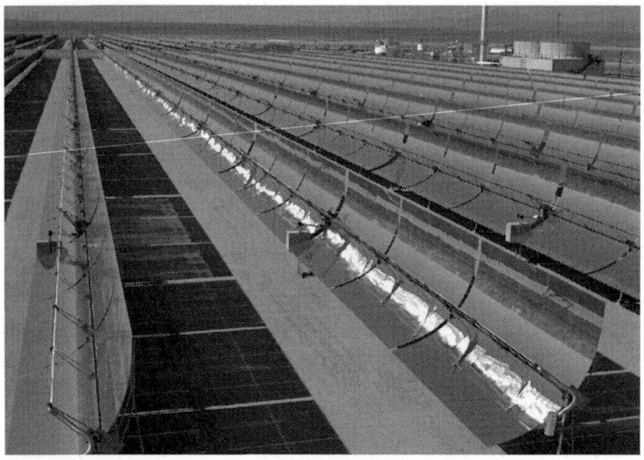

Figure 9: Parabolic trough electricity plants operating in California (Source: Warren Gretz ,DOE/NREL Photographic Information Exchange)

then be used downstream in conventional power generation equipment, such as steam turbines. Some of the advantages of concentrating-solar-thermal power are that:

- it integrates well with conventional thermodynamic cycles and power generation equipment as well as with advanced and emerging technology
- it can be installed in large capacities, yet constructed of modular, repeatable, well-known components
- it integrates readily with thermal storage (the economics of thermal storage are greatly enhanced by increased plant capacity)
- there is generally good *matching* between solar insolation and summer electrical demand. Often the storage required to obtain very high matching is only of the order of several hours.

There are two basic types of concentrating collectors—those which focus the radiation along a line, and those which focus the radiation at a point. It is possible to achieve much higher concentration ratios with point-focus collectors than with linear collectors, although the required optical precision is greater. The challenge is to generate the higher temperatures that provide greater thermodynamic efficiencies, but without the high costs associated with improved optical precision.

Parabolic-trough collectors

The parabolic-trough collector is constructed by forming a sheet of reflective material into a parabolic shape that concentrates incoming sunlight (beam radiation) onto a central linear collector at the focal point of the collector (Figure 9). A single-axis tracking mechanism is used to orientate the collector toward the sun. The collector can be installed in either a north–south orientation or an east–west orientation, depending on the needs of the system. North–south trough installations offer a slightly higher total output over a year, with a summer-oriented peak, while east–west troughs produce less seasonal variation. Parabolic trough installations have been operating on a utility scale in the Mojave Desert in California for nearly two decades, generating a total of 354 MW, with the largest single unit being 80 MW. A number of new installations are currently under construction.

Linear Fresnel reflectors

The linear Fresnel reflector uses a field of long linear mirror strips to concentrate light on a fixed linear receiver (Figure 10). The mirrors are flat or

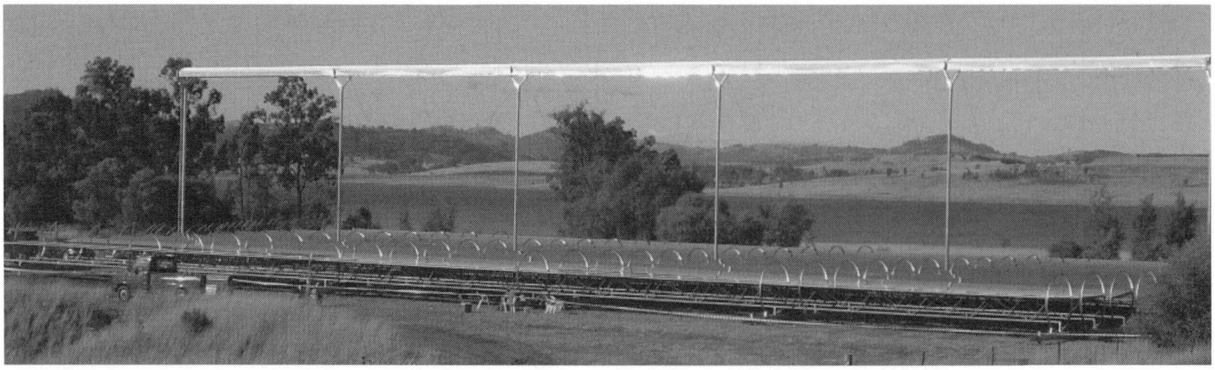

Figure 10: The linear Fresnel solar-thermal prototype at Liddell Power Station, NSW (Source: Ausra, Inc.).

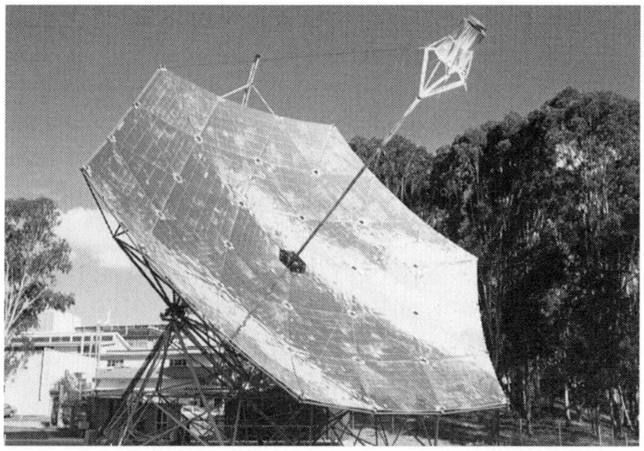

Figure 11: The big solar dish at the Australian National University, Canberra (Source: The Australian National University).

elastically curved, and rotate about the receiver axis to track the sun. A small number of prototypes have been built and are in operation for applications ranging from medium-temperature steam production for power stations to solar-thermal cooling. They offer lower costs on a square-metre basis, but have lower optical efficiency compared with troughs.

Parabolic dish concentrators

The paraboloidal dish concentrator uses a frame in the shape of a parabolic dish to support curved mirrors that concentrate solar radiation onto a receiver at the focal point of the dish (Figure 11). This system requires the sun to be tracked in two axes, but the concentration of sunlight onto a single point can yield a very high concentration ratio, and hence very high temperatures at the receiver. Dishes have been used for pilot-scale natural gas reforming, steam generation, ammonia dissociation, Brayton Cycles and Stirling engines, as well as for concentrating photovoltaics.

Solar towers

Solar towers (also known as central receivers) use a ground-based field of mirrors to focus solar radiation onto a receiver mounted high on a central tower. The computer-controlled mirrors, called heliostats, rotate individually about two axes, maintaining a stationary image of the sun on the target. Solar towers permit very high concentration ratios and are well suited to large-scale implementation for utility power generation. The higher temperatures afforded by solar-tower systems also enable a wide range of applications, such as thermochemical processes and high-temperature steam cycles for more efficient electricity generation. The largest solar electricity plant of this type was recently commissioned near Seville, Spain. In this plant, around 55 MW of solar-generated steam is converted into 11 MW of electrical energy. The plant's design parameters are conservative, ensuring a successful demonstration, and it is widely accepted that similar plants in the future will achieve greater energy-conversion efficiencies by using higher steam temperatures.

At CSIRO's National Solar Energy Centre in Newcastle, work is being undertaken to reduce the cost and increase the range of applications of solar-tower systems. A 200 heliostat prototype solar-tower system, which is capable of achieving more than 1000°C, has been designed and built entirely

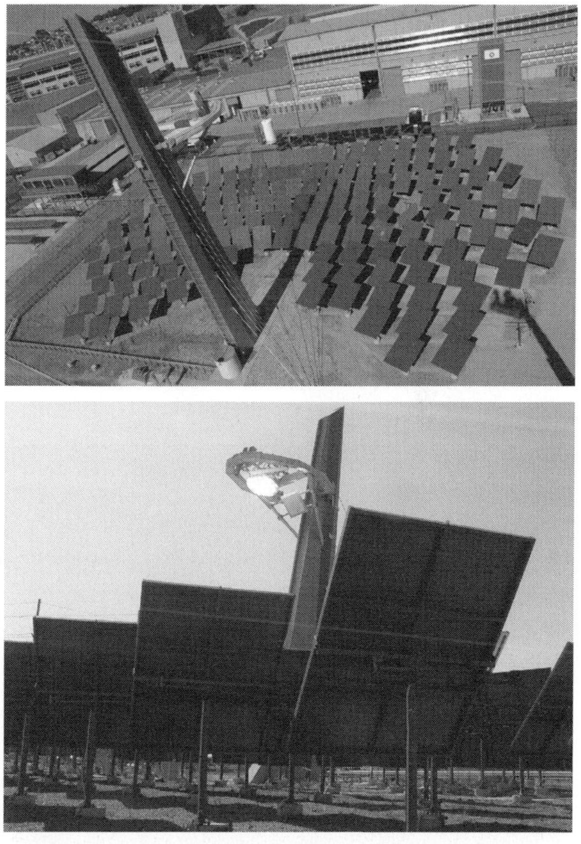

Figure 12: The solar tower facility at the CSIRO National Solar Energy Centre in Newcastle, NSW (Source: Wesley Stein).

from components that are already mass-produced for other industries (Figure 12). This approach aims to achieve cost reductions from both mass production and economies of scale. The use of 'off-the-shelf' components from other industries helps to minimise technical and financial risk during the introduction and commercialisation phase of this technology. This tower is being used to develop SolarGas™ (containing embedded solar energy) as well as for high-temperature steam generation.

Photovoltaics

The photovoltaic cell (solar cell) is a two-terminal electrical device which, when illuminated, produces a DC electrical voltage that is often referred to as a photovoltage. When placed in a circuit, the solar cell produces a photocurrent of some tens of milliamperes per square centimetre of cell area. Although this current is of a usable magnitude, the typical photovoltage is between 0.5 and 1 volt—which is too low for most applications. To produce useful voltages, cells are connected together in series and encapsulated into solar panels—often referred to as modules—with a typical output of around 12 volts. These modules can be further connected in series or parallel into photovoltaic arrays, which generate a higher voltage or current (or both), depending on the demands of the application (Figure 13).

The first real solar cell was fabricated in 1953 and developed 10 milliwatts of electricity, representing an energy conversion efficiency of 5% (Perlin *et al.* 2003). Fifty years later, the global photovoltaics industry shipped nearly 50 million solar modules in 2006 alone, with a combined power generation capacity of nearly 2 gigawatts and generating revenues of over USD 10 billion (SolarBuzz 2007). Despite these figures, photovoltaics still represents less than 0.01% of the world's total energy production today (SolarBuzz 2007). One reason for this is cost. At present, the cost of installed systems is such that the effective price of photovoltaic solar electricity is between 35 and 85 US cents per kilowatt-hour (kWh). The price range is mainly due to the variation in sunshine hours in different locations, but also depends on the conditions of finance, such as the local interest rate. Commercial and industrial systems offer lower prices due to economies of scale, but also have to compete with equivalently cheaper grid electricity (SolarBuzz 2007). For remote applications (away from the electricity grid) this price is competitive with alternative sources of electricity. In order to play a real part in reducing climate change, however, photovoltaics must become cost-competitive with grid-connected electricity sources. In Australia, the retail price of grid electricity from conventional fossil-fuel sources is A$0.10–0.15 per kWh (Electricity Supply Association of Australia 2004). Although this price is expected to rise as carbon costs are imposed, the price of photovoltaic systems still needs to come down before this technology will take a significant share of the global electricity market.

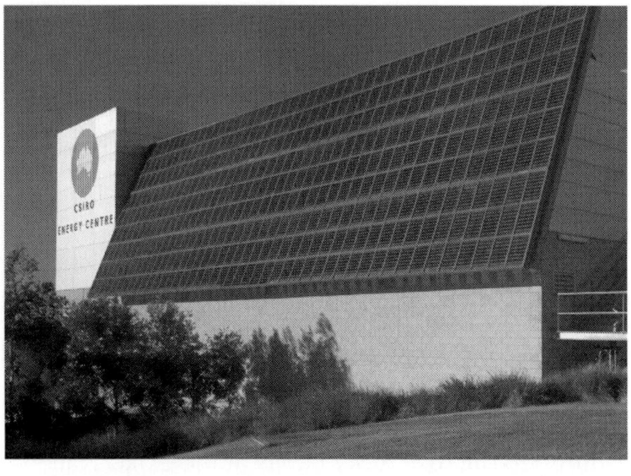

Figure 13: Photovoltaic (solar cell) installation at the CSIRO Energy Technology Centre (Source: Chris Fell).

Heavily influenced by government subsidies, however, global growth in demand for solar cells has been healthy—averaging 38% per annum since 1996 (see Figure 14). Some recent years in particular have experienced very high growth (perhaps due to a jump in public acceptance of climate change), but over the next two decades the industry expansion is expected to return to an average of 25% per year (Rohatgi *et al.* 2003). Combining this projected demand with the historical 'learning curve' illustrated in Figure 15 (manufacturing cost reduces by 20% for every doubling of capacity) suggests the price of photovoltaic electricity will continue to reduce by around 6.6% each year. At this rate, it appears that without a breakthrough development in the technology, it will take all of the next two decades

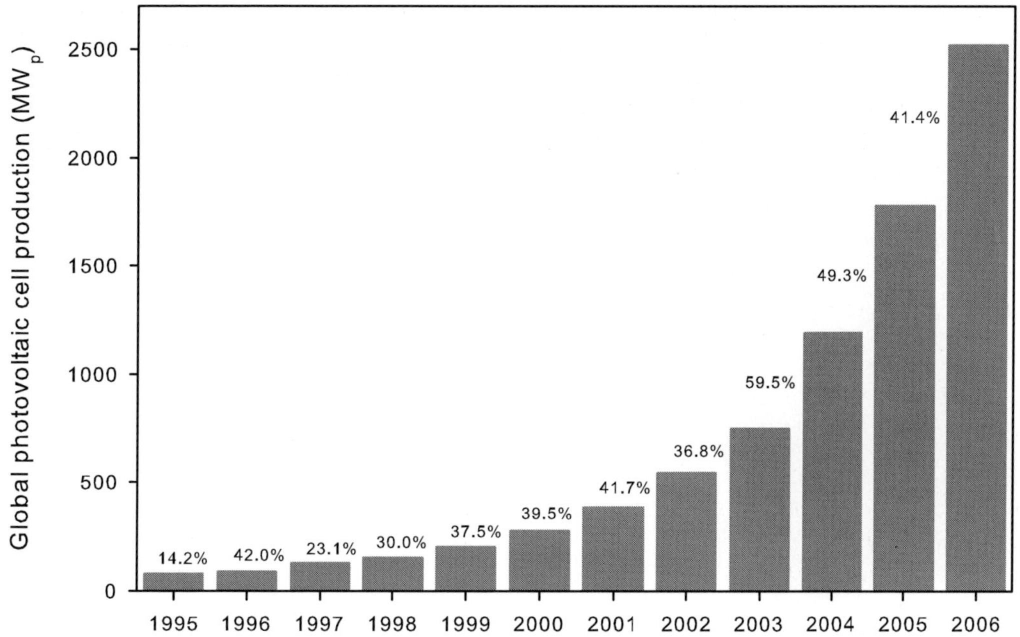

Figure 14: Worldwide photovoltaic cell production (Data source: Bradford and Maycock 2007).

before un-subsidised photovoltaics are genuinely cost-competitive, even in the sunniest locations.

Concerns about the ability of silicon photovoltaics to become cost-competitive with fossil-fuel energy sources are not new. In particular, uncertainty lingers as to when the learning curve will break down, as the manufacturing cost becomes dominated by the cost of the raw materials. Since the 1970s these concerns have driven investment in alternative technologies, with the aim of significantly reducing the manufacturing costs of photovoltaic systems. A common thread to these 'second-generation' technologies is the use of very thin layers of the expensive active materials, supported on inexpensive substrates, leading to a significant reduction in the cost of the raw materials. Manufacturing processes based on thin-film deposition are often also more amenable to

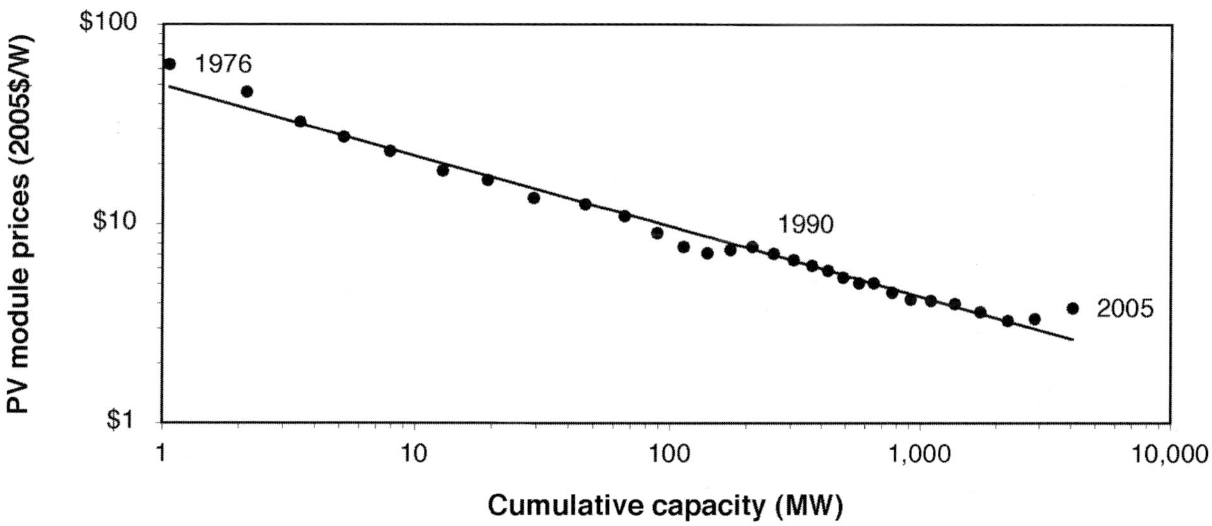

Figure 15: The 'learning curve' for photovoltaics (Source: Gregory Nemet, University of Wisconsin).

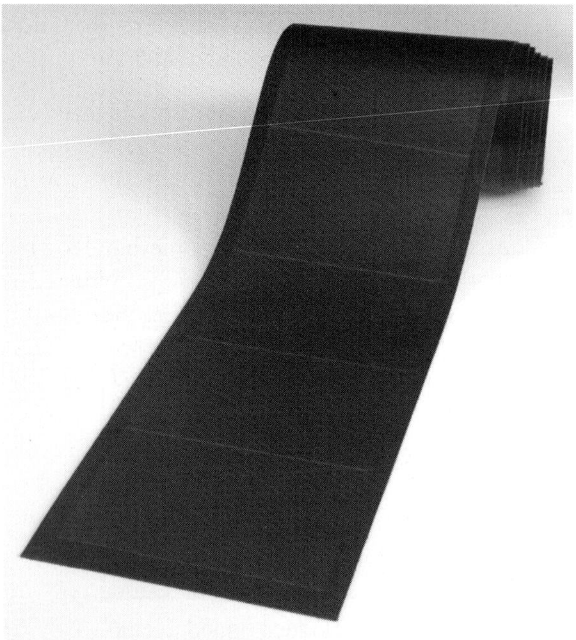

Figure 16: Flexible solar panel (Source: United Solar Ovonic).

automation—offering further opportunities for cost savings.

The most mature examples of second-generation photovoltaics are based on thin films of silicon (either amorphous or crystalline)—leveraging a half-century of expertise in this well-understood material. Thin-film amorphous silicon solar cells on cheap substrates are already being produced on a significant commercial scale, with 100 Megawatts of production in 2006, which represents 5% of the global photovoltaics market (US Department of Energy 2007). This technology is expected to benefit particularly from economies of scale and hence to consistently increase its share of the target market, which is grid-connected, rooftop systems for small, medium and large-scale installations.

Almost as mature as thin-film silicon systems, cadmium telluride (CdTe) thin-film photovoltaics have also reached the full-scale production stage. With typically higher conversion efficiency, this technology currently leads the market for very-large-scale systems used by electricity utilities. As the sandwich structure requires two sheets of glass for every module, CdTe modules don't stand to benefit as much from manufacturing economies of scale, but predicted further improvements in efficiency and production yield will see them stay competitive into the future.

A promising thin-film photovoltaic technology on the near horizon is based on copper–indium–gallium–diselenide (CIGS) and is currently generating much investor interest. A number of large production plants are currently under construction and backers say their factories cost one-third the price of a silicon-cell production plant and can produce devices more than ten times faster. Recently, one of the world's largest solar cell manufacturers sold their entire silicon manufacturing operation in order to focus on CIGS technology, reflecting projections that CIGS could soon provide the lowest electricity cost of all available photovoltaic technologies.

An interesting feature of some thin-film photovoltaics is that they can be manufactured on flexible substrates (Figure 16), expanding market opportunities in the area of building-integrated photovoltaics (BIPV) and creating a new market for photovoltaic power sources for portable electronic devices. A major challenge for manufacturers of flexible solar cells is the development of suitably performing encapsulants, without which the product is vulnerable to degradation due to the ingress of moisture and oxygen from the environment.

Solar technology: into the future

Overview

Despite impressive cost reductions over the last few decades, the cost of solar energy remains higher than traditional fossil-fuel-based energy. Indeed it is unlikely that zero emission energy, such as solar, will ever reach the low costs of conventional coal-fired power seen in Australia over the past decades. Although the primary energy resource is free, solar energy systems are capital intensive. New generations of solar technologies will improve conversion efficiencies, reduce manufacturing costs and improve systems integration to drive market growth and rapid expansion of the solar industry.

The goal is to reduce the cost of solar energy to the point that it becomes competitive against other zero-emission energy technologies. As a near-mid-term strategy, there is strong interest in a transitional approach to deploying solar-thermal systems by building solar/fossil hybrids. An obvious example is where domestic solar hot water is backed up by electricity. At the supply side of the market it makes strong economic sense for concentrated solar-thermal energy to be integrated with existing fossil-fuel power stations so that fossil fuel can be directly displaced, yet allowing the solar component to leverage existing generating infrastructure. There are also thermochemical options that enable solar to reduce the GHG signature of fossil fuels. One of the big issues for the future will be storage. It is well-known that, apart from biomass and geothermal, renewable energy output is not fully dispatchable. Battery storage, which is well suited to photovoltaic output, is covered elsewhere in this publication. The output of a solar-thermal plant can be stored in a variety of ways—thermal storage based on sensible heat, molten salts or using phase-change materials, compressed air energy storage, or thermochemical storage of solar energy in chemical bonds. While hybrid solutions of fossil and solar continue to emerge—and while renewables such as solar are contributing less than 1% to the electricity mix—there is a window of time during which storage technologies can continue to be improved, without detriment to grid stability. Right now, there are certainly economic incentives for storage to enable industrial plants to benefit from supply during peak times of the market. The economic value of storage is increasing as Australia's electricity market moves from a winter-dominated peak to a strong summer peak. However, from a grid stability point of view, the requirement for long-term storage to ensure high capacity factor output, such as base-load power from renewables, will not need to emerge until renewables contribute a much greater proportion of the mix.

There is now therefore a great opportunity to develop and optimise various storage technologies for use with solar. A promising option for concentrating solar thermal is based on thermochemical cycles. These cycles require higher temperatures—of the order of 700–900°C, with some cycles even over1000°C. They do, however, offer the possibility of embodying solar energy in the chemical bonds of a product that may be a gas or liquid. This makes long-term storage and long-distance transport of solar energy much more achievable than would be the case with a simple hot fluid.

Examples of such cycles are already under demonstration in Australia. These include:

- closed-loop storage using ammonia dissociation and synthesis enabling solar energy to be stored for long periods of time with no losses (Lovegrove *et al.* 2004)
- open-loop steam reforming of methane to yield a hydrogen-rich gas that can be combusted in gas turbines, which can be further processed to liquid transport fuels or further enriched to provide pure hydrogen for future fuel cells and hydrogen engines (Stein *et al.* 2007)
- closed-loop version of the methane reforming above such that solar energy can be stored for long periods of time with no losses. This enables solar base-load power to be achieved (Edwards *et al.* 1995).

Markets, costs and investment

Solar radiation is a very diffuse form of energy compared with our conventional fuels. Concentrating solar power (CSP)—where the sun is concentrated to provide high energy fluxes—has attraction in terms of its ability to integrate and thus provide a transitional path with significant scale. The parabolic trough form of concentrated solar power has been in operation in California since the mid 1980s. More than 2 million square metres of mirrors have been concentrating the sun to generate steam for electricity generation for over 20 years in these plants. A study for the World Bank in 2006 concluded that CSP is on the cusp of large-scale uptake in those countries that are actively supporting it. The support is largely strong national policy and incentives that reduce the risks faced by developers. The study recognised the need for a solid policy platform to foster solar-thermal-based electricity generation efforts, as well as strong national programs to help the industry reach a significantly higher capacity for commercial competitiveness (World Bank GEF 2006). Since that study, CSP momentum has increased globally.

Over the next several years it is expected that feed-in laws, renewable portfolio standards and other similar mechanisms will greatly increase solar commercial activities. Mills and Dey (2001) have suggested that growth rates for solar-thermal electric conversion must exceed 25% per annum

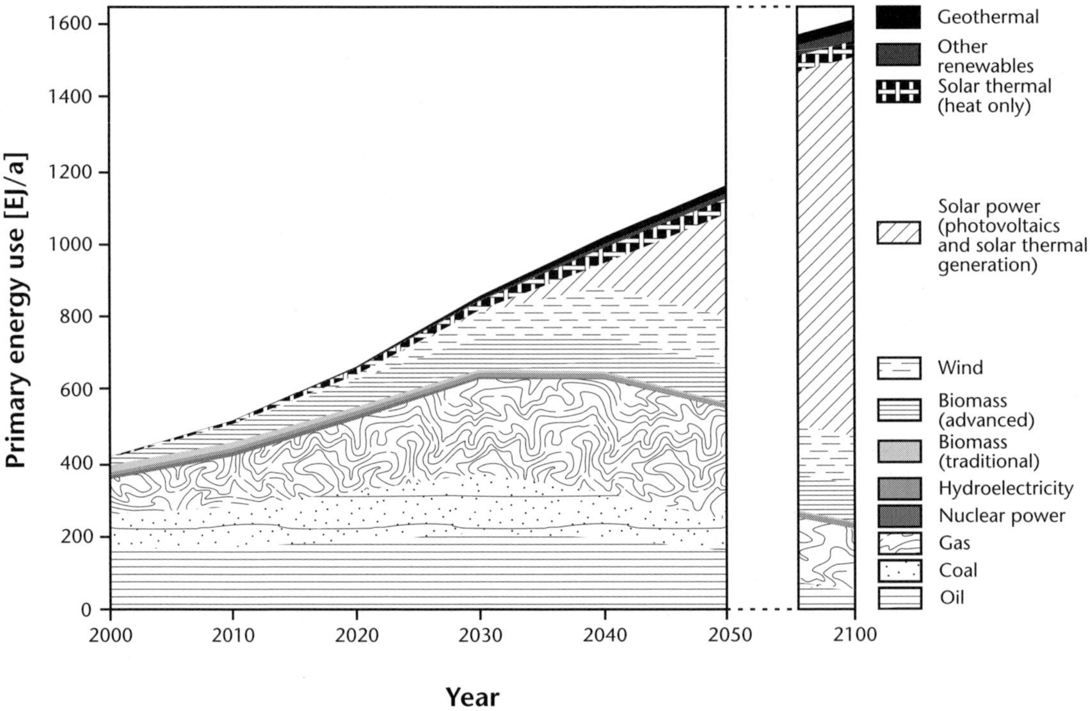

Figure 17: High renewables integration path for the development of solar power technologies up to year 2100. (Source: German Advisory Council on Global Change (WBGU), 2003).

for the next 30 years in order to achieve a dominant market position by 2050. Figure 17 shows a high growth scenario for the development of solar-power technologies up to the year 2100.

The simplest and most viable form of solar-energy technology at present is solar hot water for domestic and commercial use. In many cases, it is competitive now. Given that any energy source is used most effectively when the amount of processing between the sun itself and the point of end use is kept to a minimum, solar hot water makes a great deal of sense, and all efforts should be made to encourage its maximum uptake. However, hot water accounts for about 30% of average household energy consumption, and residential electricity for only about 25% of Australia's electricity demand. Assuming that solar hot water might supply 75% of a household's hot water requirements, and say a maximum of 75% of households take it up (due to shading, inappropriate roof, and so on) domestic solar hot water could at best displace 4–5% of Australia's total electricity demand. While this is important as part of the mix, there is a strong need for other means of producing solar electricity on a large scale. Photovoltaic and solar-thermal electricity are the two solar electricity technologies that can together meet this end. A possible solar energy future would comprise:

- solar hot water
- photovoltaic panels on homes and buildings supplying distributed energy
- small parabolic troughs or advanced evacuated tubes supplying medium temperature heat for industrial, hospital, mining and agricultural processes; and for absorption chillers (such as in shopping centres and commercial buildings)
- large solar-thermal power plants, either in hybrid or stand-alone form, for bulk supply of solar electricity to the grid for distribution to all users
- solar thermochemical power plants supplying solar-enhanced fuels to supplement the petrochemical industry.

Each of the technologies above is at a different stage of commercial maturity. All, however, share major components, such as concentrators, materials and control and operating systems. This offers a relatively smooth transition from technologies of today to more advanced cycles of the future, requiring only to 'tweak' and hone components, rather than develop completely new systems.

Concentrating-solar-thermal power

A large number of solar-thermal trough plants were built between 1984 and 1991, sparked by the 1970s oil crisis and by various tax credits and other incentives in California. The last plant to be installed was an 80 MW unit. During this time, the cost of solar-thermal power decreased from US 48 c/kWh to less than US 15 c/kWh (including incentives). However, this was still not enough to maintain the momentum and, with an oil price decreasing in real terms through the 1990s, investor enthusiasm for new large solar-thermal power plants waned. Though the trough-plant components were performing reliably, with significant reductions in operating and maintenance costs, ongoing cost reductions were expected to come largely from economies of scale. Because new plants were not built, progress along the capital cost reduction curve stagnated. In a market that was again favouring fossil fuel, trough plants of 80 MW were regarded as requiring too much risk capital. In that period, smaller capacity wind turbines of around 1 MW or so emerged—these involved much lower levels of capital risk, so the wind-turbine market grew strongly. Likewise, photovoltaics—though expensive on a per kWh basis—could be installed in kW capacities, which again was much more financially manageable. Now, as greenhouse gas (GHG) issues dominate, and the need for deep cuts is accepted more than ever before, the bulk capacity attributes of solar-thermal power may be its best asset.

Depending on solar radiation levels, the cost of electricity for a one-off plant is currently around A$0.16 to 0.20/kWh, which is presently uncompetitive; however, the CSP roadmap shows that this could drop by 60–70% (in real terms) as deployment increases. There is no fundamental reason why concentrating-solar-thermal technology could not follow a similar cost reduction curve to that experienced by wind energy and, within the lifetime of our existing generating stock, become one of the most cost-effective forms of zero-emission electricity. However, robust and long-term support mechanisms will be required. The US Department of Energy's Solar Energy Technologies Program notes that:

'CSP has been established as one of the most attractive renewable energy options in the U.S. Southwest, with a nominal levelized cost of energy target of US 7–10 c/kWh by 2015 and US 5–7 c/kWh by 2020' (US Department of Energy, 2007).

The most detailed of the recent studies (World Bank GEF 2006) predicts that commercial competitiveness can be achieved with installations of around 42 GW (less than installed wind capacity today), excluding the impact of carbon taxes. This would represent approximately 1% of the global installed power generation capacity and require subsidies of about €12 billion. An allowance for a carbon cost beginning at €7.50 per ton of carbon dioxide (/t CO_2) would decrease this capacity to 22 GW, and reduce required subsidies to €2.5 billion. Cost reductions are expected to come from a combination of larger plants, increased production volume and technological innovation.

The required growth rates for solar-thermal electricity to achieve cost-competitiveness by 2020 are quite moderate compared with that achieved by wind—the fastest growing energy technology in the world. The Global Market Initiative (GMI) is a proposed industry target to reach 5,000 MW of installed CSP by 2015. The major success factors required to create a sustainable solar-thermal electricity market are long-term targets, appropriate financial support, and supporting legislation and regulation. As costs drop over time, the level of mandated support should be correspondingly reduced to encourage competition, while maintaining certainty for investors to ensure roll-out continues through the construction of subsequent plants if stated targets are met (Figure 18). With technological innovations lowering costs and increased market growth leading to new jobs and export opportunities, solar energy can become a major high-technology growth industry that contributes significantly to Australia's economy. An independent review by Sargent and Lundy (2003) states:

'CSP technology is a proven technology for energy production, and it can achieve significant cost reductions, assuming a reasonable deployment of CSP technologies occurs'.

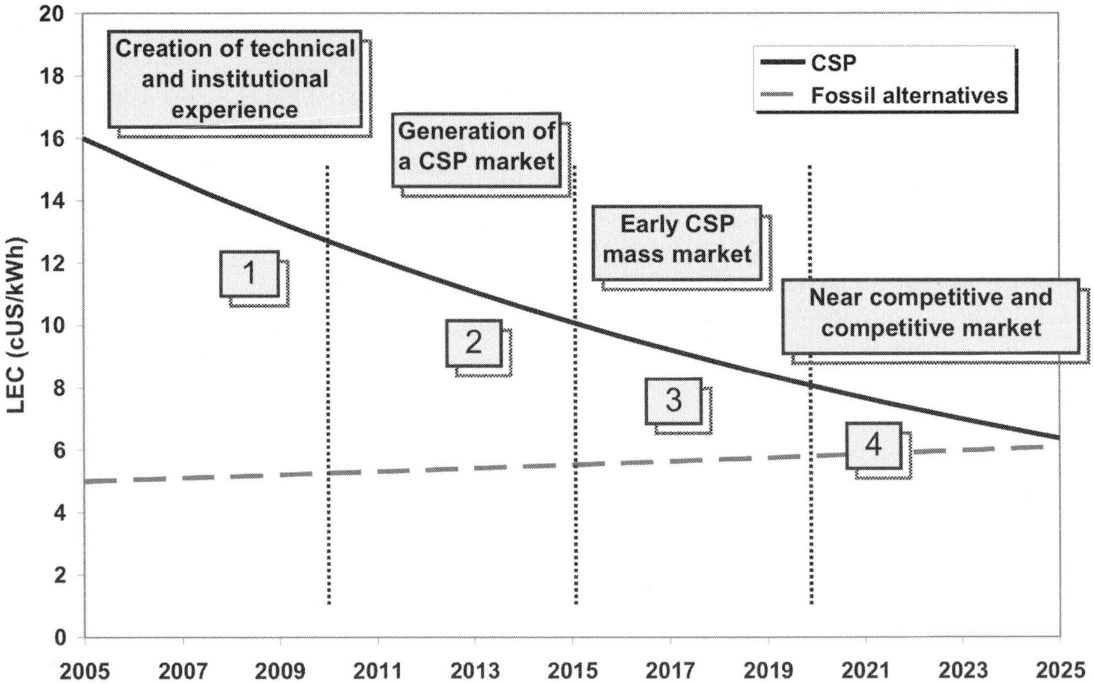

Figure 18: The four market stages of cost reduction for concentrating solar-thermal power (Source: World Bank GEF 2006).

Photovoltaics in the future

Second-generation technologies presently account for around 10% of the global photovoltaics market and it seems inevitable that this share will continue to increase. What, though, of the long-term future for photovoltaic solar energy? It is easy to accept the argument that any technology, sufficiently scaled-up, will evolve to the stage where costs are dominated by those of the constituent materials. In the case of thin-film technologies it will not be the active materials, but the substrate and other encapsulants required to maintain a minimum 20-year operating life, that will govern the cost. Under that scenario, the most cost-effective technology will be the one that can produce electricity with the highest efficiency. A number of pathways have been identified toward a 'third generation' of photovoltaics that could circumvent the theoretical efficiency limit of 33% for a standard solar cell. Tandem solar cells, hot-carrier solar cells and quantum-well solar cells all represent technologies currently at the stage of fundamental exploratory research, but which could one day lead to a revolutionary step-change improvement in the viability of photovoltaics as a source of energy.

An alternative school of thought suggests that the way of the future is not via ultra-high-efficiency, but ultra-low-cost. If photovoltaic devices could be designed from essentially ubiquitous materials, the need for efficiency would play a smaller role in the overall viability. One such device, the dye-sensitised solar cell has reached the pilot manufacturing scale this year. Based on a nanoparticle layer of titanium dioxide—the main ingredient in sunscreen—this technology is also vastly less sensitive to impurities than its competitors, which further reduces the cost of the manufacturing process. Dye-sensitised solar cells have an interesting feature that will probably provide some competitive advantage for building-integrated photovoltaic systems—they are semi-transparent and they can be made in a variety of colours. With current module efficiencies at 6–7%, this important new technology is likely to build a strong presence in the photovoltaics market once some niggling durability problems are solved.

So which is more important, low cost or high efficiency? If the limitations we currently accept were thrown away, what form might the future of photovoltaics take? As long as there are rooftops on which to put solar cells, cost will be the single most important factor that drives our designs. In the clean-energy future, solar cells will likely be made from ultra-thin layers of limitless materials. They will be printed like newspaper, at room temperature, so that the energy required to produce them is quickly paid back. They will be ultra-light, ultra-flexible and offer a range of applications we can now only dream of. One class of materials—organics—could provide all of these features. Organic compounds are molecules based on carbon, and form the basis of all life on Earth. The sheer variety of organic molecular structures leads to an enormous range of material properties, including the semiconducting behaviour that makes photovoltaics possible. When we consider the number of ways these molecules can interact when we blend them, the possibilities increase exponentially. At present, solar cells made from organic materials are in the research phase. A conversion efficiency of 5% has been demonstrated on a laboratory scale, but durability and cost are still a long way from being satisfactory. Improvements will be driven by investment in research, to improve our understanding of the fundamental processes that make these devices work, and to design innovative ways to infiltrate the technology into our daily lives. Whichever technology dominates, it seems inevitable that photovoltaics will one day soon play a key role in the way we generate electricity.

Solar fuels and chemical storage of solar energy

Much of the focus for renewable energy tends to be on electricity production. However, electricity is a low cost bulk-produced commodity with very low margins, and is based on incumbent technology. This makes it very difficult for new technologies to gain a share of the market place. There are numerous non-electricity energy processes that contribute to global GHG emissions and have a higher $/GJ value—transport fuel is one of these. These sectors offer alternative, and possibly higher value, ways for solar to enter the market. Solar-driven thermochemical, electrochemical and photobiological processes hold great interest as the next generation of advanced solar processes.

Inherent in these processes is the ability to store solar energy chemically—that is, in chemical bonds. This effectively provides a form of storage without losses—just as with any of our solid or liquid fuels, the solar energy remains embodied in the chemical structure until released by a downstream process when desired.

Thermochemical energy storage can be accomplished on an open-loop or closed-loop basis. An example of open loop is the solar-enhanced fuel described below. Closed loop systems include those based on the ammonia dissociation/synthesis cycle or the reforming/methanation cycle. Both these closed-loops are examples of pure solar cycles that offer long-term storage and base-load power.

Solar enhanced fuels

A transitional approach for advanced solar-energy cycles is the hybridisation of solar energy and biomass or fossil fuels, which combines the zero GHG emissions of solar with the greatest asset of biomass and fossil fuels: the chemical storage of energy.

One such way is to use concentrated solar energy to reform natural gas. Solar-driven steam reforming has been demonstrated at pilot scale in several research institutes and offers an excellent transitional pathway for a country such as Australia, which is rich in solar energy and natural gas. The product gas embodies 25% solar energy and is a feedstock for the production of hydrogen, industrial chemicals, fertilisers and liquid transport fuels.

An attraction of the solar-enhanced fuel route is that it would offer not only a domestic source of transport fuel—thus reducing reliance on imports—but would also enable Australia to become an exporter of solar energy (in the form of fuel) to countries that have minimal energy resources. Australia's solar energy could for example be transported to Japan embedded in the chemical bonds of an energy-dense liquid fuel.

A number of research institutions are investigating the combination of solar energy with other fossil fuels, which would give solar energy a viable inroad to the market while improving the GHG signature of many of our energy processes. Given that there inevitably has to be a period where fossil fuels and renewable energy co-exist in our energy mix, it makes sense that they do so in a manner which offers greatest efficiency for our economy. If there are cost, thermodynamic and market advantages in a solar-energy process integrating with a fossil fuel, rather than remaining separate, then such technological integration will be an important part of any roadmap for solar energy.

Solar hydrogen

A particular form of solar fuel is hydrogen. As mentioned above, hydrogen can be produced from the solar reforming of natural gas using steam—methane and water are the two highest hydrogen-containing compounds on Earth.

To move to a completely renewable form of hydrogen, however, we need to eliminate the fossil fuel. Extracting hydrogen from water is a highly energy intensive process. For example, direct thermal decomposition requires temperatures of 2500–3000°C—far beyond the realm of engineering materials under such arduous conditions. There are a number of lower temperature routes for splitting water including:

- *thermochemical cycles (800–2000°C).* These processes use reactants that are reduced and oxidised in a cyclic fashion to create a two (or more) step process that is capable of splitting water at much more realistic temperatures than direct thermal decomposition
- *photocatalysis*—the photon-induced excitation of semiconductors in solution can be used to provide the activation energy needed for certain chemical reactions, including splitting water
- *high-temperature electrolysis (500–1000°C).* At such temperatures, the thermal energy available reduces the amount of electrical energy required to drive the water splitting reaction (Licht 2005).
- *a solar beam-splitting system*—which can be used for applications other than hydrogen—where a beam-splitting filter selectively directs the lower-grade thermal energy to heat water and directs the higher-energy photons to a photovoltaic cell. The outcome is a more efficient solar cell due to reduced heat load, plus an additional thermal stream available for other processes. The combination of these electrical and thermal streams can be used to efficiently drive the process of splitting water molecules into hydrogen and oxygen (Lasich 1997).

Conclusions

A cleaner energy future for Australia and the world will require a suite of supply-side and demand-side options. Australia is well endowed in terms of access to large resources of solar energy which are relatively close to existing transmission and distribution infrastructure. There is much land available for harvesting of solar energy and, in any case, the land area required is not huge. There is also a large potential for distributed solar-energy generation to reduce the reliance of the transmission network through production of electricity in urban areas close to the consumer. The wider implementation of mature technologies, such as solar hot water, should also play a role in reducing the amount of extra generation capacity required from non-renewable resources.

Solar technologies that can play a key role in the short term include:

- solar hot water
- solar-thermal electricity technologies and hybrid solar/fossil systems for large capacity power
- solar process heat for smaller urban, commercial and industrial applications
- distributed or building integrated photovoltaics.

In the medium term, thin-film photovoltaics, and more advanced photoelectric applications, solar transport fuels and solar-thermal power with base-load capacity, could be expected to emerge as key technologies. In the longer term, there is the opportunity for a whole new energy model to develop, such as a hydrogen economy. We could also see the list of solar options expand to include solar water splitting and organic photovoltaics.

Solar-thermal power is on the cusp of a large-scale uptake in those countries where incentives are in place. Photovoltaics continue to push the boundaries of performance and cost. There are great opportunities for those countries with sun and solar technology expertise to be a part of a massive new industry.

Solar energy is clean, abundant, and renewable. It is the largest sustainable energy source in the world. It holds tremendous potential in providing benefits to Australia—as well as globally—by diversifying the energy supply, reducing dependence on fossil fuels, improving the quality of the air we breathe, dramatically reducing greenhouse gas emissions and stimulating economies by creating jobs in the manufacture, installation and maintenance of solar energy systems. The uptake and implementation of solar technologies will ease the transition as we necessarily move towards a sustainable society in the 21st century.

References

Australian Bureau of Meteorology (2000). <http://www.bom.gov.au/climate/averages/>

Bradford T and Maycock P (2007). PV Market Update, Renewable Energy World online <www.renewable-energy-world.com>.

CSIRO Energy Technology (2008). Website. <http://www.energy.csiro.au>

Edwards J, Do T, Maitra A, Schuck S and Stein W (1995). The CSIRO/Pacific Power project on using CO2/CH4 reforming for solar energy storage and transmission. *Solar Engineering* **1**, 389–398.

Electricity Supply Association of Australia (2004). Electricity Prices in Australia 2002/2003, Electricity Supply Association of Australia, Sydney.

German Advisory Council on Global Change (WBGU). Website. <http://www.wbgu.de/>

Kalogirou SA (2004). Solar thermal collectors and applications. *Progress in Energy and Combustion Science* **30**, 231–295.

Lasich JB (1997). Production of hydrogen from solar radiation at high efficiency. US Patent 5,658,448.

Licht S (2005). Solar water splitting to generate hydrogen fuel - a photothermal electrochemical analysis. *International Journal of Hydrogen Energy* **30**, 459–470.

Lovegrove K, Luzzi A and Soldiani I (2004). Developing ammonia based thermochemical energy storage for dish power plants. *Solar Energy* **76**, 331–337.

Mills DR and Dey CJ (2001). Development strategies for solar thermal electricity generation. In *Advances in Solar Energy*. Vol. 14. (Eds DY Goswami and KW Boer), pp. 401–423. American Solar Energy Society.

Morrison GL and Litvak A (1999). 'Condensed solar radiation database for Australia'. Report 1/1999 for the Solar Thermal Energy Laboratory, University of New South Wales, Sydney.

Perlin J, Kazmerski L and Moon S (2003). Good as Gold – The Silicon Solar Cell Turns Fifty. Publication for 'Bell Cell' golden anniversary exhibit, National Renewable Energy Laboratory, Golden, Colorado.

Rabl A (1985). *Active Solar Collectors and their Applications*. Oxford University Press, New York.

Rohatgi A, Ristow A and Yelundur V (2003). Cost and technology roadmaps for cost-effective silicon photovoltaics. In *Proceedings of 12th International Workshop on the Physics of Semiconductor Devices (IWPSD-2003)*. Vol. 2. December 16–20, pp. 893–898. New Delhi, India.

Sargent and Lundy LLC Consulting Group (2003). 'Assessment of parabolic trough and power tower solar technology cost and performance forecasts'. Subcontractor report NREL/SR-550-34440, H Price (NREL Technical Monitor), National Renewable Energy Laboratory, Golden, Colorado. Available online at: <http://www.nrel.gov/docs/fy04osti/34440.pdf>

SolarBuzz (2007). Website <http://www.solarbuzz.com/FastFactsIndustry.htm>

Solar Millenium AG (2008). Website <http://www.solarmillennium.com>

Stein W, Benito R and Chensee M (2007). Transport and use of solar energy in hydrogen. *Advances in Applied Ceramics* **106(1–2)**, 2–5(4).

United States Department of Energy (2007). 'National solar technology roadmap – film silicon PV'. Solar Energy Technologies Program Draft Report, available online at <http://www1.eere.energy.gov/solar/solar_america/planning.html>

World Bank GEF (2006). *Assessment of The World Bank GEF Strategy for the Market Development of Concentrating Solar-Thermal Power*. A World Bank Group Global Environment Facility Program Publication, (Eds W Eichhammer, G Morin, S Szewczuk and W Stein). World Bank, Washington DC.

24

Wind Energy

Peter Coppin

Wind energy has been used for centuries but has recently become a major source of low-emissions electricity generation. The familiar Dutch windmills designed for grinding grain have given way to sleek three-bladed giant wind turbines that are highly optimised for generating electricity. The renaissance of wind power began in Europe in the late 1970s and continues unabated. The potential to generate large amounts of electricity directly into the grid has driven the technology towards mass-produced large-scale wind turbines, but there remains a vibrant market for small-scale turbines for individual dwellings or remote-area power supplies. At the time of writing, wind turbines of up to 3 Megawatt capacity are available in production quantities, which stand over 80 metres high at the hub with a three-blade rotor 90 metres in diameter. The total structure is as high as 40 storey building. Each turbine could supply the annual energy need of about 1000 households. Wind power is a true low-emission energy source. The energy 'payback period' is very short—the time taken to generate the energy consumed in manufacturing, installing, operating and decommissioning a wind farm is less than 12 months, compared with an operating life of 20 years (Milborrow 2007). An excellent guide to the technology and all aspects of the industry is available in the form of a guided tour at http://www.windpower.org/en/tour.htm

The global scene

Globally, wind power is a major player in electricity production, with a compound annual growth rate of 20%. At the end of 2006, the world-wide installed capacity of wind-powered generators was 74GW. Although wind power currently produces less than 1% of world-wide electricity use, it accounts for approximately 20% of electricity production in Denmark, 9% in Spain, and 7% in Germany. Global wind-power generation more than quadrupled between 2000 and 2006. Table 1 indicates the installed capacities and new installations for 2006 across the globe. It is interesting to note that these figures are obtained from a very small number of countries with large installed capacities. It is obvious that if more countries were to adopt wind energy the contribution would be many times greater. Forecast growth indicates a further doubling of capacity by 2010 (Figure 1). In 2005 the Global Wind Energy Council issued a blueprint for achieving 12% of global world electricity requirements by 2020, involving a 16-fold increase in installed capacity. With increasing renewable energy targets in many countries, this goal is achievable—China alone has set a target of 30,000 MW. The Global Wind Energy Council also estimates current total worldwide employment in the wind generation industry at more than 150,000 people.

How wind energy is used

Wind energy is currently dominated by generation of electricity direct into local high voltage grids. This involves very large wind turbines, which are arranged in arrays—or 'wind farms'—to reduce the establishment overheads and sited sufficiently close to the electricity grid infrastructure to minimise connection costs. This arrangement uses the potential wind generation most economically (an example of a wind farm in Western Australia is shown in Figure 2). The size of wind farms has been steadily growing, with proposals approaching

Table 1: International rankings of wind power capacity (USDOE 2007)

Cumulative capacity at the end of 2006 (MW)		Capacity added in 2006 (MW)	
Germany	20,652	US	2,454
Spain	11,614	Germany	2,233
US	11,575	India	1,840
India	6,228	Spain	1,587
Denmark	3,101	China	1,334
China	2,588	France	810
Italy	2,118	Canada	776
UK	1,967	UK	631
Portugal	1,716	Portugal	629
France	1,585	Italy	417
Rest of world	11,102	Rest of world	2,305
Total	74,246	Total	15,016

the 1000 MW level. This maximum output is similar in magnitude to a conventional thermal generation station and hence wind farms are rapidly becoming considered as wind 'power stations'.

The wind resource itself varies considerably from place to place, with the annual energy yield from a modern turbine varying approximately in proportion to the square of the mean annual wind speed. The inherent wind resource at well-exposed coastal sites, in shallow coastal waters or on an inland ridge with little vegetation is typically several times that of a sheltered urban location. There is little incentive to have small wind-generating units distributed in urban environments where the wind-energy resource is a fraction of that available in more exposed locations. In addition, the unit cost per installed KW of current small units is several times that of a large turbine.

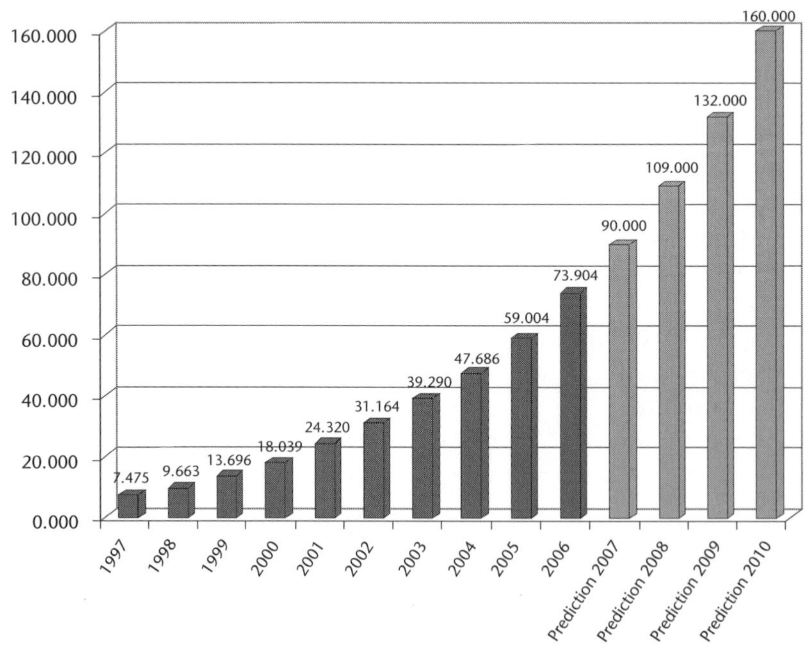

Figure 1: Total world wind-energy installed capacity (MW) with projections for 2007–2010

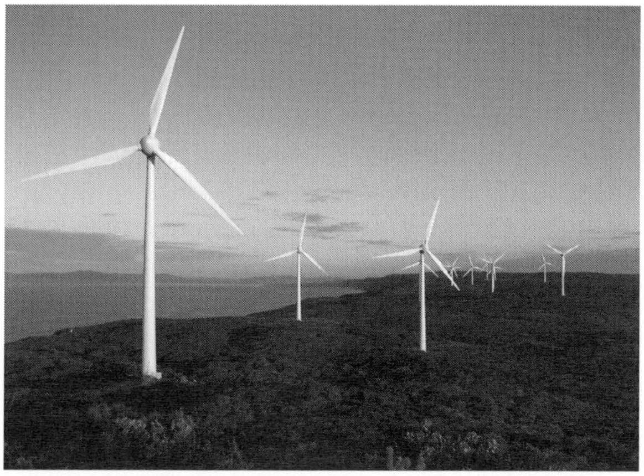

Figure 2: Enercon 1.8 MW wind turbines at the 20 MW Albany wind farm, Western Australia (Source: courtesy David Clarke).

Ironically, even though the trend is towards larger wind farms, wind energy represents a key contributor to distributed generation, given the widespread nature of the wind resource. Small wind turbines (from about 1 kW capacity up to 300kW) are only currently economically viable where the electricity grid is not accessible and individual consumers or remote communities require only a limited amount of energy. However, many individuals have decided to install such devices as a demonstration of the potential of these small devices in local generation.

There are some issues that are inhibiting the widespread uptake of wind energy. In some regions, large installed capacities of wind energy have exposed some of the weaknesses in centralised electricity generation and distribution systems. Current large-scale electricity distribution systems are generally designed to generate electricity at a central location and distribute the energy outward via a grid. Wind energy is an example of distributed generation, which often takes place at locations that may be on the fringes of the grid or on segments that have limited capacity. The inherent variability in the wind resource must also be taken into account when scheduling generation capacity. These issues are discussed later, but it is clear that to successfully exploit wind energy, there must be a transition to market systems and physical infrastructures that are able to assimilate generation sources that are both spatially distributed and variable in nature.

Offshore development

In Europe, the available land area for acceptable sites for wind turbines is approaching saturation, especially with the trend towards large wind farms. There is a concerted effort to enable large-scale wind generation offshore—in both shallow and deeper waters. The disadvantage of higher costs in anchoring the turbines and connecting the power production to the shore is offset by higher wind speeds over water and the ability to build very large wind farms that use bigger and more noisy turbines than would be possible over land. Already wind farms of 160 MW have been commissioned in Danish waters (Figure 3), with wind farms of up to 1000 MW planned (www.offshorewindenergy.org).

The economics of wind power

As the global market has grown, wind power has seen a dramatic fall in cost. A modern wind turbine produces 180 times more electricity annually, and at less than half the cost per unit (kWh), than its equivalent twenty years ago. At good locations in Europe, wind can compete with the cost of both coal- and gas-fired power, with little or no subsidy: *Windpower Monthly* (2006) analyses show that at European sites with an average wind speed of more than 7 metres per second and a capital cost per installed kilowatt of approximately €1000, wind is already cheaper than gas, coal and nuclear power.

The competitiveness of wind power has been further enhanced by the recent rise in the price of fossil fuels—in particular the gas used to fuel power stations. In the United States, this has made wind-generated electricity an increasingly attractive option for power utilities faced with rising costs.

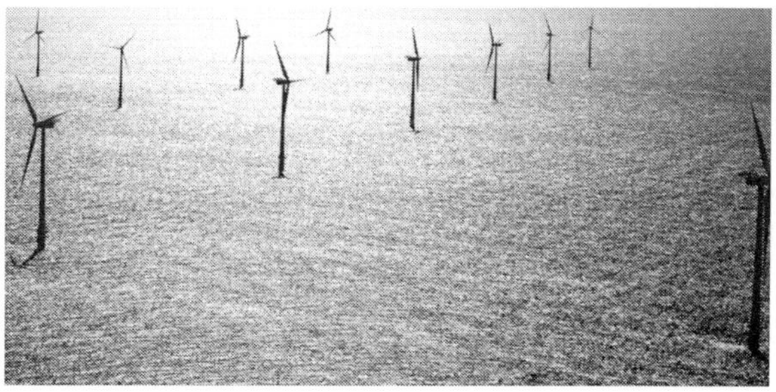

Figure 3: 160 MW Horns Rev offshore wind farm, Denmark (Source: www.hornsrev.dk).

Against the volatility of conventional electricity costs, wind offers an energy source that has no fuel element and is unaffected by world trade issues.

The annual energy yield from the wind at a particular location obviously increases as the average wind speed rises. However, the net cost of the energy varies significantly with the size of the wind farm, connection costs, method of financing, and so on. In Australia, combined wholesale electricity prices and MRET certificate prices (see next section) indicate that projects are viable over about AU$70 per MWh with current turbine technology and wind farm sizes. The locations of these projects also indicate that annual average wind speed above about 8 metres per second is required. In Europe, depending on the country and support mechanism, the wind-speed threshold for viability could be as low as 6 m/s.

However, direct cost comparisons between wind power and other generation technologies are misleading, because they often do not account for the 'external costs' to society and the environment caused by burning fossil fuels or from nuclear generation. These external costs, including the effects of greenhouse gas emissions, air pollution and risk of radiation emissions, are not included in electricity prices. The pan-European study—known as the ExternE project—conducted across all 15 original EU member states, has assessed these costs for a range of fuels. Its latest results showed that wind power has the lowest range of these hidden costs—0.15 to 0.25 € cents/kWh—compared with 2 to 15 € cents/kWh for coal (www.externe.info). The study concluded that the cost of electricity from coal or oil would double—and that from gas increase by 30%—if their external costs associated with the environment and health were taken into account.

The polluting effect of fossil fuels is now taken into account through carbon reduction measures, such as the European Union's emissions trading scheme, which sets a limit on the amount of carbon dioxide which can be emitted by all major industrial enterprises.

The situation in Australia

Australia's history of large-scale wind generation is more recent. Due to the lack of an incentive scheme for large-scale implementation, there was little activity until the federal government introduced its 2% Mandatory Renewable Energy Target (MRET) in 2000. This scheme mandated the generation of 9500 GWh per annum of renewable energy by 2010. The scheme effectively achieved its target four years ahead of its original schedule in 2006. In late 2007 the then newly elected Labour Government issued a new goal of a 20 per cent share for renewable energy in Australia's electricity supply by 2020, and with it a commitment to increasing the MRET from 9500 gigawatt-hours to 45,000 gigawatt-hours in 2020. The expanded measure is to be phased out between 2020 and 2030 as emissions trading matures and prices become sufficient to ensure that an MRET is no longer required to drive deployment of renewable generation technologies.

Under the MRET, Australia's manufacturing industry gained considerable experience. VESTAS established a nacelle assembly plant in Burnie, Tasmania (unfortunately now closed). The nacelle is the enclosure at the top of the support tower containing the gearbox, generator, transformer, yaw and pitch control mechanisms—mostly imported components. Turbine tower manufacturer, Haywards Engineering, also established facilities in Tasmania. In Victoria, Keppel Prince established a manufacturing plant for wind towers and components and until recently, a blade manufacturing plant was based in Portland. South Australia hosts the major steel fabrication company, Air-Ride Technologies. With a likely resurgence in the wind industry, it is certain that Australia will gain significant manufacturing benefits.

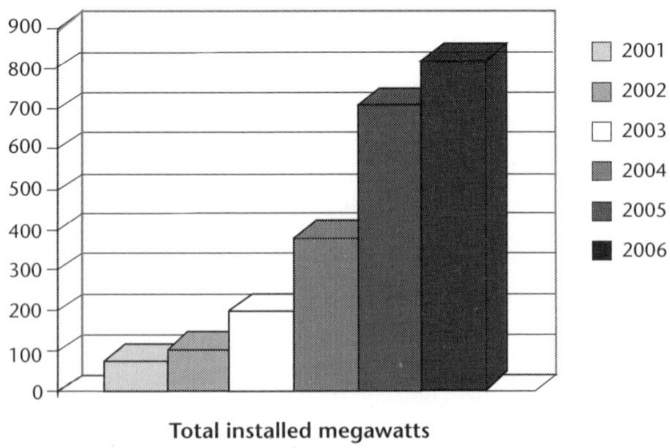

Figure 4: Australian wind-energy capacity—cumulative growth 2001–2006 (Source: Auswind 2007).

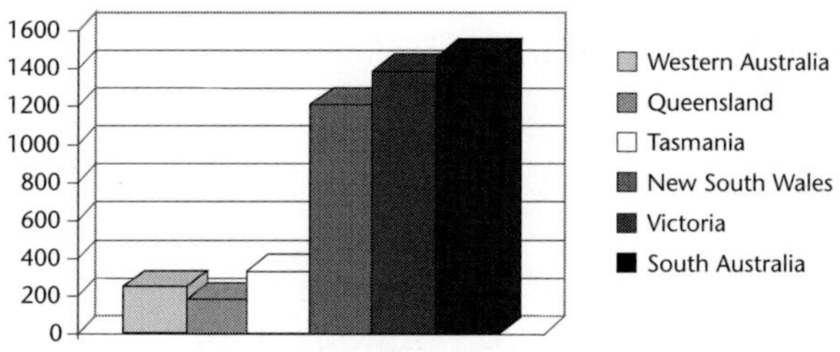

Figure 5: Proposed Australian wind-energy projects by state, 2006 (Source: Auswind 2007).

The wind resource in Australia

Australia has large areas of wind resource, and hence significant wind-energy potential, especially along the southern coasts and adjacent inland mountain areas, such as the Great Divide (Figure 6). The useable area of resource is dependent not only on the wind resource, but also on proximity to power transmission infrastructure and a set of environmental constraints. Although wind farms occupy significant areas, they have a small footprint and allow grazing and cropping up to the base of the turbines. The turbines must be spaced a few hundred metres apart to minimise the interference between units—in some wind directions a turbine may be in the wake of another. With adequate spacing the current turbine technology yields about 20 MW installed capacity per square kilometre.

At least two states—NSW and Victoria—have developed wind atlases, which attempt to overlay all the necessary constraints and reveal areas of useable resource. The Commonwealth Government has included wind energy in its comprehensive Renewable Energy Atlas, now being developed (2008). The amount of land area suitable for generation will ultimately depend on the current economics of the technology. As the economics steadily improve through, for example, technological advances, the greater is the land area potentially available for energy generation. Coppin *et al.* (2003) showed that lowering the economic annual-average wind speed threshold in the Great Divide area of NSW from 8 m/s to 7 m/s increased the land available for consideration by a factor of 20. Figure 7 shows an example of high resolution (100 m) mapping from the *Wind-*

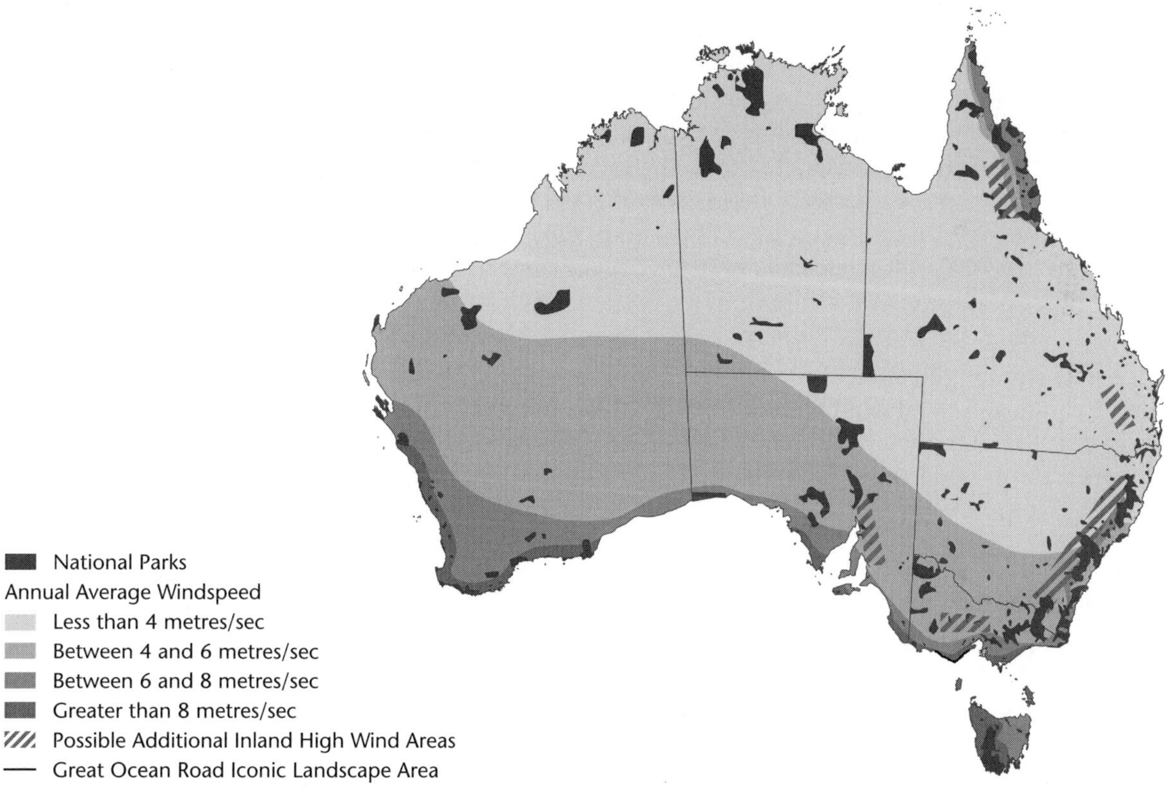

Figure 6: Broad-scale picture of wind resources in Australia

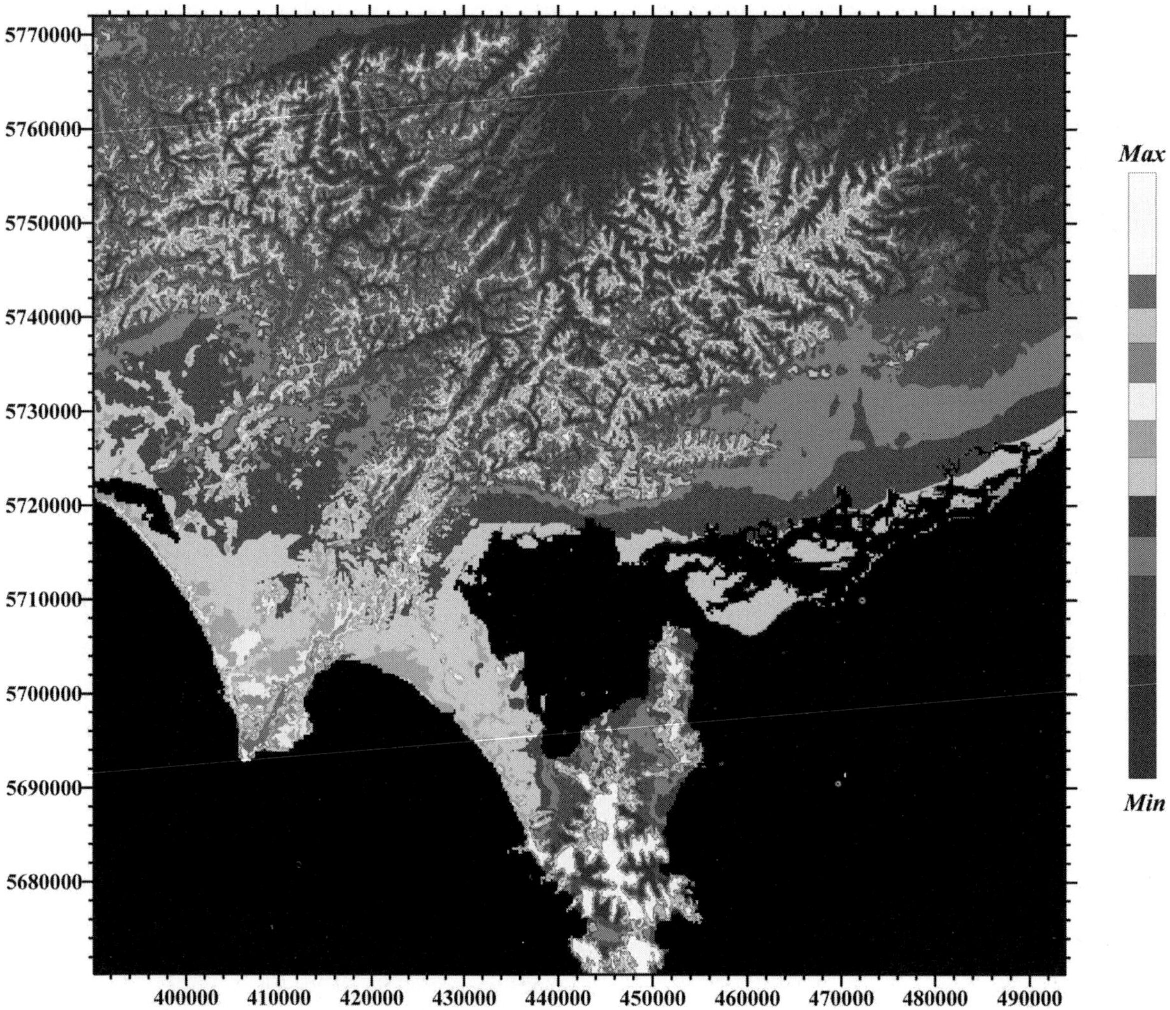

Figure 7: WindScape® predictions of mean annual wind speed (m/s) at 67 m above ground across the Foster region of Gippsland for the year 2000, 100 m resolution estimates (Source: Coppin *et al.* 2003).

Scape® system. This map shows the raw wind resource before any consideration of constraints.

Figure 8 shows an inland wind farm at Challicum Hills, near Ararat in Victoria, which harnesses the local increase in wind speed provided by the hills.

Siting wind farms

The planning requirements for wind farm developments are considerable, and extend well beyond finding a good wind resource. They range from physical site concerns, such as soil integrity for foundations and access road construction, erosion control, and so on, to broader environmental concerns, such as noise propagation, visual impact and a range of ecological factors such as bird impact. In 2002, the Australian Wind Energy Association (Auswind), with funding and support from the Australian Greenhouse Office, developed an extensive *Best Practice Guidelines* (Auswind 2006), which cover the broad range of planning considerations, including community consultation issues.

A key project from Auswind is the *Wind Farms and Landscape Values—National Assessment Methodology* project, jointly undertaken with the Australian Council of National Trusts (ACNT). The project provides a sound and transparent, nationally applicable framework for establishing and assessing potential impacts of wind farms on land-

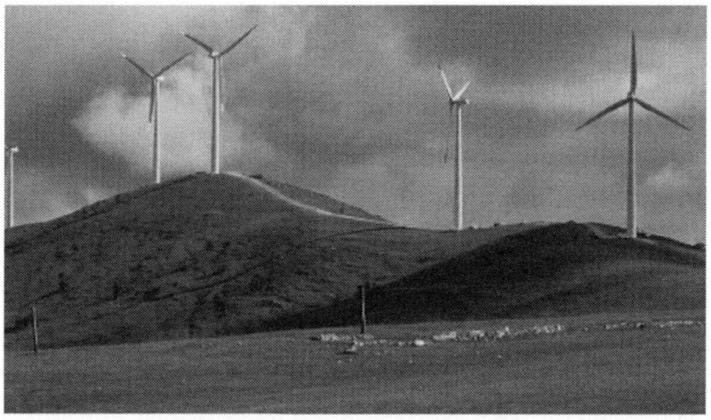

Figure 8: 1.5 MW NEG Micon turbines at the 52.5 MW Challicum Hills Wind Farm, Victoria.

scape values. It will provide site impact assessment and mitigation and community consultation procedures. In particular, the framework highlights the importance of taking account of the values that communities place on landscapes affected by a wind farm development. This project tackles some of the more contentious issues in wind farm development and aims to facilitate appropriately sited and sensitively developed wind-energy projects.

Issues for implementing wind power in the power system

For wind-energy generation to gain an increasing share of the electricity market, several issues need to be tackled: grid adequacy, energy market structures and variability of the generation. To a large extent, the urgency of these issues is proportional to the penetration of wind power, that is, the percentage of power which is derived from wind. Australia is rapidly approaching very high penetration rates at a local level, particularly in South Australia where installed wind capacity will shortly exceed typical overnight minimum demand.

If wind production has a degree of variability, there is heavy reliance on either good interconnection to other parts of the grid or on alternative local generation sources that are able to compensate for the excess or shortfall of production. Even if the variations are a small proportion of demand, the required compensation may not be available if other generation sources are unavailable or are not able to change at a sufficient rate. This principle applies at any geographic scale. Fortunately, the geographic spread of wind farms tends to naturally smooth the aggregated output. However, on occasion, the size of swings in wind generation will test the adequacy of the electricity distribution network.

There are several measures that can be taken to allow increased access of wind power to the grid. The first is the strengthening of transmission infrastructure. The more robust the grid interconnections, (up to a point) the better able the system is to cope with variations in wind-energy production. Denmark is a prime example. With strong interconnections to Sweden, Germany and Norway, Denmark is able to host high wind penetration. Wind power supplies an average of 20% of Denmark's annual energy consumption (in 2006) and the instantaneous contribution from wind regularly exceeds 100%. Excesses or shortfalls are managed through the connections to neighbours, particularly to the hydroelectric power sources in Norway, which can easily be configured to compensate for variations in wind. Where local grids or interconnects are not strong, it may be necessary to reinforce grid infrastructure. The problem then becomes one of equitable sharing of the costs.

A necessary adjunct to grid adequacy is wind-energy production forecasting, which allows grid system operators to give sufficient notice to other generators if large and rapid changes in wind-energy production are anticipated. This is based on high-precision weather forecasting that can predict wind farm production from minutes out to several days ahead (Coppin and Katzfey 2003). The Australian Government has recently funded a state-of-the art wind-energy forecasting system for the National Electricity Market Management Company (NEMMCO), which manages the grid for south-eastern Australia. This system has the capability to cover all large wind farms in Australia, including Western Australia.

Once the knowledge of future wind production volumes and variability is available through forecasting, the problem moves to one of ensuring that sufficient generation is available to respond to changes at the required rates (the rate of change of power is known as the *ramp rate*). The ramp rate achievable by a generator is dependent on the technology employed, with some power sources, such as gas turbines, having a high ramp rate capability and others, such as 'base-load' coal and nuclear power, having a much lower ramp rate capability. Unfortunately ramp rate is often inversely proportional to cost. However, it should be remembered that the required ramp rate depends critically on the aggregation area considered. A region with good grid infrastructure and with a large amount of wind generation spread over a wide area may need much smaller ramp-rate generation backup than a region with weak grid infrastructure and wind generation concentrated in a limited area.

Storage is an obvious answer to the intermittency in wind generation, but, unless the wind generation is accessible to a pumped storage hydro-electric facility, most solutions are presently too expensive (but see later section on future possibilities). For the moment it may be more feasible to tackle the ramp-rate problem directly. One example of an interesting possibility is AGL's Hallett development, north of Adelaide in South Australia, which matches 95 MW of wind generation to 180 MW of gas-fired peaking (rapid response) plant on adjacent sites with a forecast increase to 140 MW of wind and 430 MW of gas-fired plant. This combination could be run as a quasi-base load generator—the gas plant essentially filling in the lulls of its on-site wind turbines and eventually doing the same for other nearby wind farms.

The energy markets need to be aware of the characteristics of wind power, while allowing the wind generators to participate in the market without penalty. Market bidding is made feasible by the use of wind forecasting systems.

Current trends

With wind power now an established technology, what does the immediate future hold? It is clear that the wind-energy industry is heading towards large-scale, grid-connected generation. Obviously there will be a steady development in the turbines themselves, which will allow larger capacity wind farms, and there will be large offshore installations. There will be a steady improvement in basic underlying technology—the size of wind turbines will increase—with 5 MW prototypes being tested (REpower 2007) and 10 MW turbines being developed. However, the greatest gains will be in the spread of wind power to more countries around the globe, which will benefit via technology transfer from the success of the limited number of countries that have significant installed capacities. The focus of much current research and development is solving the issues mentioned above, which are preventing the more widespread penetration of wind-energy generation into the electricity grid.

In Australia, a report prepared for the Australian Greenhouse Office by Outhred (2003), indicates that the National Electricity Market could readily accept an installed wind capacity of over 8000 MW, provided that the progressive development of wind-energy projects is enabled in a dispersed fashion, commercial wind output forecasting is in place, appropriate network management strategies are adopted and interstate connectivity continues to be enhanced. This amount of wind energy would equate to approximately 10% of Australia's electricity needs and is attainable without significant wind-specific modifications to existing electrical infrastructure. Further infrastructure enhancements and other innovations could see this potential figure rise much higher.

With the trend towards large, utility-scale wind-generation facilities, nationally accredited Green Power programs are one of the most cost effective ways of allowing participation in the growth in this technology. Offered by electricity retailers to households and businesses across Australia, these schemes allow individuals—for a small increment on electricity prices—to access the economies of scale of such large renewable energy developments.

The future

While the wind-energy industry is well established and the technology of the wind turbines themselves may be considered mature, there are many examples of innovations that will help to broaden the uptake of wind power.

Large-scale offshore wind

The space required for large wind farms as a result of increased demand will make offshore development increasingly attractive and economic, especially close to densely populated regions, such as those in Western Europe. Both bottom-anchored and floating wind farms have been designed in arrays of up to 1000 MW. Airtricity (the Irish Utility) has released a plan to create a pan European

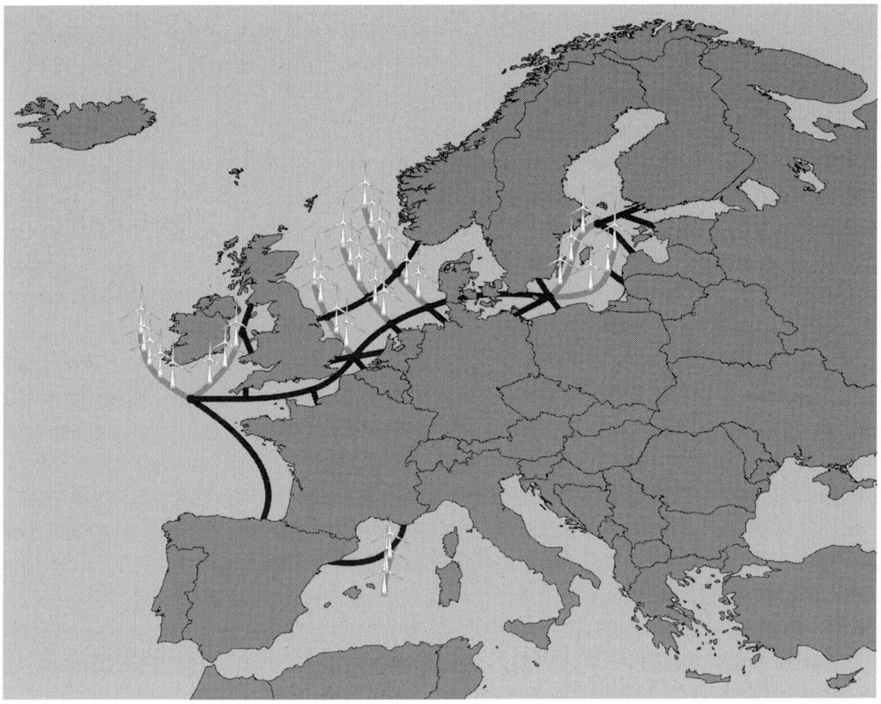

Figure 9: Airtricity Supergrid® proposal for interconnected offshore wind farms (Source: www.airtricity.com)

sub-sea energy grid that will transport energy from offshore wind projects throughout the European region. The Supergrid® (Figure 9) not only harnesses the geographic diversity of the wind resource, but also acts as an interconnector between national markets and thereby helps to create a properly functioning internal market in electricity. The first element of the Supergrid will be a foundation project consisting of a 10GW wind farm located in the North Sea between the UK, Germany and the Netherlands (Figure 10). The technology for this first stage is essentially in place. It will consist of about 2000 5 MW offshore wind turbines located on the seabed in water depths up to 50 metres connected by a high-voltage DC grid.

Storage

Grid limitations are one of the key constraints on the continued growth of wind energy. This is particularly important where substantial wind generation is present—as is often the case—at the extremities of the grid or on relatively small capacity power lines. In these conditions, the variability of the wind power can cause voltage and frequency stability problems, as well as causing excessive peak loads on transmission lines (thus reducing net carrying capacity) and increasing demand for high ramp-rate backup systems.

The variability in wind production covers a broad range of scales from seconds to hours and beyond. Several storage solutions have been trialled to cover the longer-term variations in the wind, including pumped-hydro and compressing air into exhausted underground hydrocarbon wells. NEMMCO has identified that short–term variability (from minutes to less than one hour) can be quite substantial under some conditions where sizeable wind generating facilities are present. Through its Energy Transformed Flagship, CSIRO is developing demonstrations of renewable energy storage systems that provide a solution to the shorter term wind variability problem. Rather than providing bulk storage of the wind farm output at a time-scale of hours—which would be very expensive—they provide smoothing of the output of wind farms at the sub-one hour time-scale.

These systems may use a variety of fast-response storage units, such as Ultrabatteries (traditional lead-acid batteries combined with a super capacitor), in combination with smart control systems

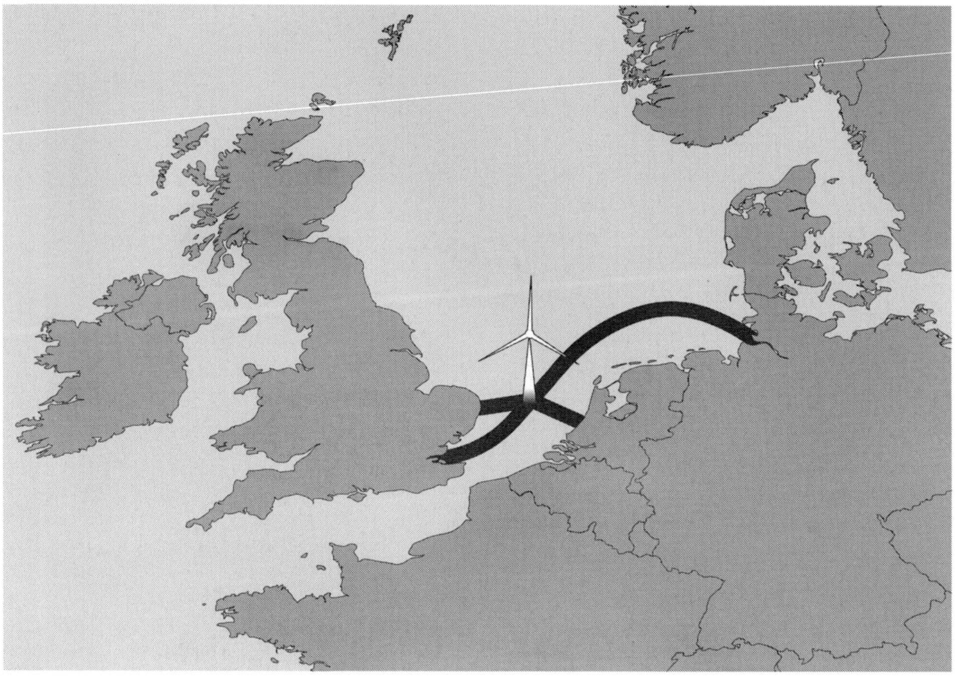

Figure 10: The 10 GW Supergrid foundation project (Source: www.airtricity.com).

that can use wind forecasting to significantly increase their smoothing efficiency. The aim is to achieve a significant reduction in short-term variability from the wind farm (output from a typical large wind farm should see a reduction by 70% in the number of events in which ramp rates exceed NEMMCO error limits). The smoother voltage and power delivery would then result in increased wind-power carrying capacity of local transmission lines and hence increased potential wind penetration on a specific grid. These systems will be equally applicable to remote area power-supply systems, which may include a combination of power sources including wind, solar and diesel. The storage system is able to absorb the variability in the wind or solar production, allowing the diesel system to be run more efficiently at constant load or, at times, not at all.

Urban wind turbines

Unfortunately, urban areas usually experience a lower wind speed that cleared open land, due to shelter from buildings and vegetation and, as a further consequence, the wind also has a higher degree of turbulence. It would obviously be difficult to site current full-scale turbines, which are the size of 40 storey buildings, in urban areas. However, wind energy utilisation in urban environments has seen an enthusiastic level of development, and it is possible to generate wind power economically where local enhancements to wind flow can be found.

One particular series of developments are the vertical-axis wind turbines produced by Windside Wind Turbines in Finland (www.windside.com). These are designed from sailing engineering principles using a turbine rotor that is rotated by two spiral-formed vanes (Figure 11). They are particularly suited to high turbulence environments, such as on the tops of buildings, where conventional horizontal axis wind turbines would not survive and where wind speeds may be greatly enhanced by the wind tunnel effect of building structures—particularly in the roof area of high-rise structures. Aerovironment, Inc. in the USA also offers architectural wind solutions (http://www.avinc.com/wind.asp).

Further into the future, it may be possible to build wind generation into buildings themselves. Research under a European Union funded program 'Wind Energy for the Built Environment' (www.bdsp.com/web) has developed concepts that include using the building itself as a concentrator for the generating elements.

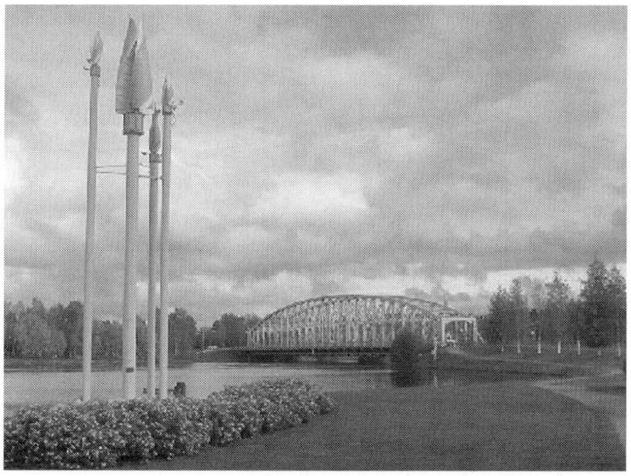

Figure 11: Urban wind turbines suitable for high turbulence environments (Source: www.windside.com).

Figure 12 shows the concept of a high-rise building with integrated wind generation.

Conclusion

With several decades of steady growth, wind energy has become a mainstream source of electrical power that has the enormous benefits of being a renewable energy source. The goal now is to integrate significant numbers of large wind farms into energy delivery networks as power stations, rather than as ad-hoc sources of clean power. The challenges to achieving this lie in providing good forecasting to enable scheduling of the power, judicious augmentation of grid assets and complementary fast-response generation sources, combined with the targeted use of storage technologies. Innovations in wind farm development will continue, including the widespread use of off-shore locations. Eventually we will see wind generators routinely built into our urban landscape.

With good planning, wind energy could provide well in excess of 20% of global energy requirements.

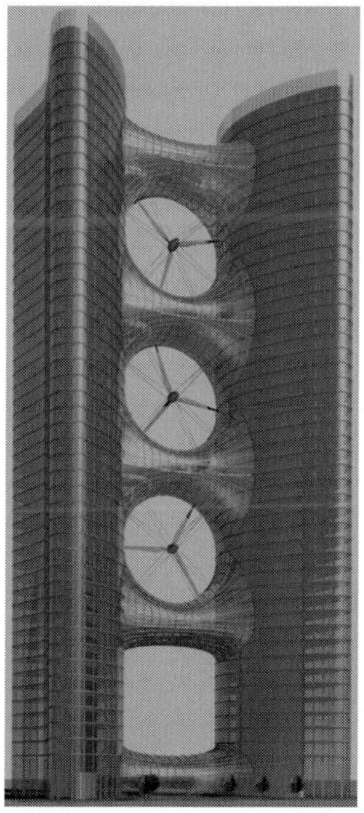

Figure 12: Concept of a high-rise building with integrated wind generation

References

Auswind (2006). *Best Practice Guidelines—For Implementation of Wind Energy Projects in Australia*. Available at: <http://www.auswind.org/bestpractice>

Auswind (2007). Australian Wind Energy Association website. <www.auswind.org>

Coppin PA, Ayotte, KA, and Steggel N (2003). 'Wind resource assessment in Australia – a planners guide, Version 1.1'. Wind Energy Research Unit, CSIRO Land and Water, Report to: Australian Greenhouse Office, October 2003. Available at: <http://www.csiro.com.au/weru>

Coppin PA and Katzfey J (2003). 'The feasibility of wind power production forecasting in the Australian context'. Report to: National Electricity Market Management Company Limited, December 2003. Available at: <http://www.nemmco.com.au/dispatchandpricing/260-0006.pdf>

Milborrow D. (2007). Wind plant earn back emissions in less than a year, *Wind Stats Newsletter*, **20(2)** 1–3.

REpower (2007). Website. <http://www.repower5m.com/index_flash_uk.htm>

Outhred H (2003). 'National wind power study – an estimate of readily accepted wind energy in the national electricity market'. A report prepared on behalf of Unisearch Ltd for the Australian Greenhouse Office. Available at: <http://www.greenhouse.gov.au/renewable/publications/pubs/wind-power.pdf>

Windpower Monthly (2006). Annual analysis of power generation costs. *Windpower Monthly*, Knebel, Denmark, January.

25

Hydrogen Energy

Andrew Dicks and David Rand

Interest in making the transition to a hydrogen economy has arisen from mounting concern over the sustainability of global energy supplies. Among the key drivers are:

1. climate change, ocean surface acidification and air pollution, which imply the need to control and reduce anthropogenic emissions of greenhouse gases, particularly emissions from the combustion of fossil fuels in transportation and thermal power stations;
2. the diminishing reserves of oil and natural gas
3. the need for energy security adapted to each country, such as decreasing the dependence on fossil-fuel imports from regions where there is political or economic instability
4. the expected growth in world population with the ever-increasing aspiration for an improved standard-of-living for all, especially in developing and poor nations.

Hydrogen is being promoted worldwide as a panacea for energy problems in that it may eventually replace, or at least greatly reduce, the reliance on fossil fuels. Although hydrogen is the most abundant element in the universe—the stuff from which stars are made—it does not occur freely on Earth, but is predominantly found in combination with oxygen as water and with carbon as fossil fuels. Chemical, thermal or electrical energy has to be expended to extract hydrogen from these sources. Hydrogen is therefore not a new form of primary energy, but a vector (or carrier) for storing and transporting energy from any one of a myriad of sources to where it may be used. In this respect, it is analogous to electricity—another secondary form of energy. Hydrogen and electricity are complementary: electricity is used for a multitude of applications for which hydrogen is not suitable, whereas hydrogen, unlike electricity, has the attributes of being both a fuel and an energy store. These two energy vectors are, in principle, inter-convertible: electricity may be used to generate hydrogen by the electrolysis of water, while hydrogen may be converted to electricity by means of a fuel cell.

Specifically, hydrogen has the following key attributes:

- it can be derived from fossil and non-fossil sources (renewable or nuclear energy)
- it can serve as an alternative fuel for internal combustion engines
- it is ideal for use in fuel cells for transportation and for distributed energy supply
- it is oxidised cleanly to water, with no emissions of carbon dioxide; when obtained from water using renewables, the fuel cycle is closed and no pollutants are released in the overall process.

The proposal to use hydrogen as a sustainable medium of energy has become known as the 'hydrogen economy'; the overall scheme is illustrated conceptually in Figure 1. The upper part of the diagram is generally referred to as the transitional phase, during which hydrogen is produced from fossil fuels. The lower part relates to the long-term, post fossil-fuel, age when hydrogen will be manufactured from renewable energy sources and used as a storage medium and as a super-clean fuel. Not unexpectedly, the building of a hydrogen economy presents great scientific and technological challenges in production, delivery, storage, conversion and end-use. In addition, there are many policy, regulatory, economic, financial, investment, environmental and safety questions to be tackled.

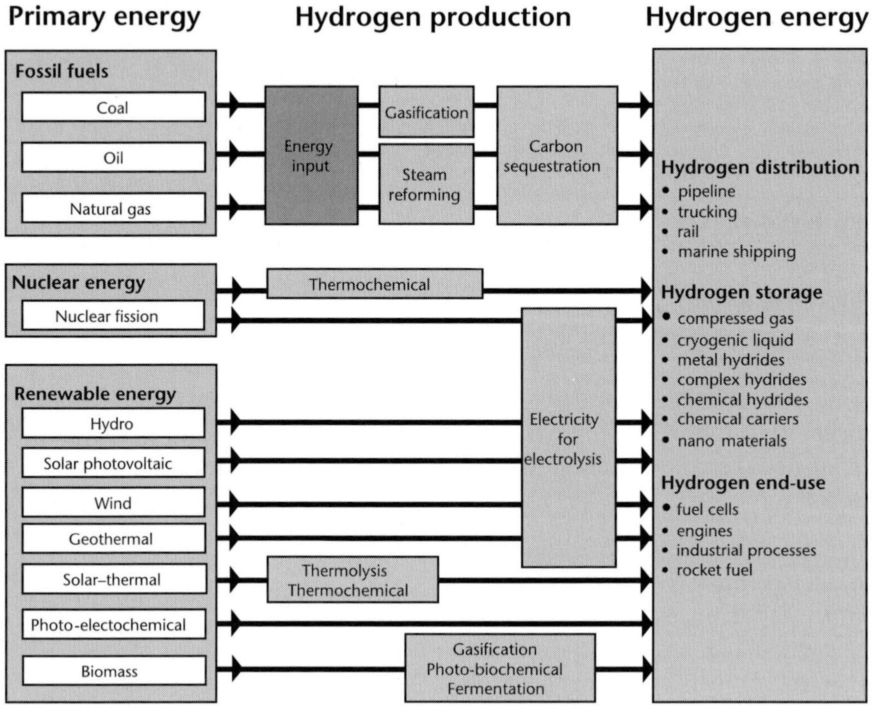

Figure 1: A sustainable hydrogen economy (Source: Rand and Dell 2008).

Hydrogen as a fuel

Hydrogen was first identified as a distinct entity in 1766 by the British scientist Henry Cavendish (1731–1810) after he produced the gas by reacting zinc metal with hydrochloric acid. He called this gas 'inflammable air'. In a paper to the Royal Society in London, Cavendish provided exact measurements of the weight and density of the gas. He also found that water was formed when 'inflammable air' was ignited with a spark in ordinary air, but he continued to hold the belief that, in its production, the gas was released from the metal itself rather than from the acid. It was the French chemist Antoine-Laurent Lavoisier (1743–1794) who first appreciated that water is a compound substance formed from a combination of inflammable air and oxygen. Indeed, Lavoisier gave hydrogen its name by coupling together the Greek words 'hydro' meaning 'water' and 'genes' meaning 'genesis'—hence: the 'maker of water'.

If hydrogen is to be employed as an energy vector and a non-polluting fuel, then we need to take its basic physical properties into account. Hydrogen is a colourless, odourless, tasteless and non-toxic gas. It is the lightest of all molecules, with a density of only 0.0899 kg m^{-3} at normal temperature and pressure—that is, 7% of the density of air. By virtue of its remarkably low density, hydrogen has the best energy-to-weight ratio ('heating value') of any fuel, but its energy-to-volume ratio is poor.

Hydrogen has a wide range of flammability—it has a lower flammability limit in air of 4% by volume and an upper flammability limit of 75% by volume. Because the lower explosive limit of hydrogen in air (13% by volume) is higher than the lower flammability limit, hydrogen generally burns rather than explodes. When mixed with pure oxygen in a 2:1 molecular ratio and ignited, hydrogen detonates violently. The energy required to ignite a hydrogen–air mixture is exceptionally low—just one-fifteenth of that needed to ignite natural gas.

The combination of hydrogen's physical properties therefore has safety implications for its use in bulk. On the positive side, the low density of the gas and its high diffusivity in air means that outdoors leakage of hydrogen soon disperses safely. The situation is quite different in an enclosed space—the wide flammability and explosive ranges of hydrogen in air means that a leakage is very likely to give rise to a fire or explosion. Because hydrogen contains no carbon, it burns with a non-luminous flame that does not radiate heat. Consequently, bystanders are not subject to radiation heat burns. On the other hand, because the flame is almost invisible and does not radiate heat, there is always the possibility of inadvertently straying into its path and being seriously burnt. Clearly, the safety of hydrogen from both a technological and societal perspective will be a key issue if the hydrogen economy is to be taken forward.

Present uses for hydrogen

The world production of hydrogen is around 45–50 Mt per year. Most of this is derived from natural gas by steam reforming; the remainder is obtained principally from oil and coal by partial oxidation processes. The transformation of natural gas and liquid hydrocarbon feedstocks into hydrogen is a straightforward catalytic process, but the route from coal requires an initial step of high-temperature gasification. Only 4% of the hydrogen worldwide is generated by electrolysis—invariably when there are special reasons that make this route economic, such as where there is a surfeit of cheap hydroelectricity, or when the hydrogen is a by-product of the chlor-alkali process for the manufacture of chlorine and caustic soda.

To date, the major consumption of hydrogen has been in the petroleum industry for the refining and upgrading of crude petroleum, and in the chemical industry for the manufacture of ammonia (for example, for fertilisers), methanol, and a variety of organic chemicals. Other important uses are found in the food industry for the hydrogenation of edible plant oils to fats (margarine) and in the plastics industry for making various polymers. Lesser applications occur in the metals, electronics, glass, electric power and space industries. Very little hydrogen is yet used as a fuel.

Hydrogen from fossil fuels and biomass

Reforming of natural gas

Natural gas consists predominantly of methane (CH_4), with small quantities of other gaseous hydrocarbons. It is widely available, easy to handle, and relatively cheap (at present). Compared with other fossil fuels, natural gas is the most desirable feed for making hydrogen—it has the highest hydrogen-to-carbon ratio of all fuels and this serves to minimise the quantity of carbon dioxide produced when it is burnt or reformed to hydrogen.

The steam reforming of natural gas to manufacture hydrogen is a mature industrial process, in which methane is reacted with steam over a nickel-based catalyst at around 900°C and at elevated pressure:

$$CH_4 + H_2O\,(gas) \rightarrow CO + 3H_2 \qquad (1)$$

The resulting mixture is known as 'synthesis gas' (or 'syngas') because it may be used for the preparation of a range of commodities that include various organic chemicals. Syngas may also be converted to liquid fuels by the 'Fischer-Tropsch reaction' in what has become known as the 'gas-to-liquids' (GTL) process. When the object of reforming natural gas is to produce hydrogen, the syngas is further processed using the water-gas shift (WGS) reaction. This converts the carbon monoxide to carbon dioxide by further reaction with steam over a catalyst at a much lower temperature:

$$CO + H_2O\,(gas) \rightarrow CO_2 + H_2 \qquad (2)$$

Summing Reactions 1 and 2 yields:

$$CH_4 + 2H_2O\,(gas) \rightarrow CO_2 + 4H_2 \qquad (3)$$

Thus, each molecule of methane gives rise to four molecules of hydrogen, of which two molecules originate from the steam. In reality, the practical yield of hydrogen is less than this because some of the methane is usually burnt externally to provide the heat needed for reforming (Reaction 1).

Finally, it is necessary to separate the product gas from the carbon dioxide and any other gases that may be present. This is generally accomplished by 'pressure swing adsorption' (PSA)—a well-proven industrial process that yields high-purity hydrogen (up to 99.999%) and recoveries of up to 90%. Other separation processes using ceramic membranes are under development that promise the benefit of being able to run at higher temperatures than PSA and are potentially more elegant and cost-effective at the smaller scale.

Whereas it is likely that natural gas will provide the earliest affordable feedstock for hydrogen production, today's costs are prohibitively high for applications such as transportation while petrol

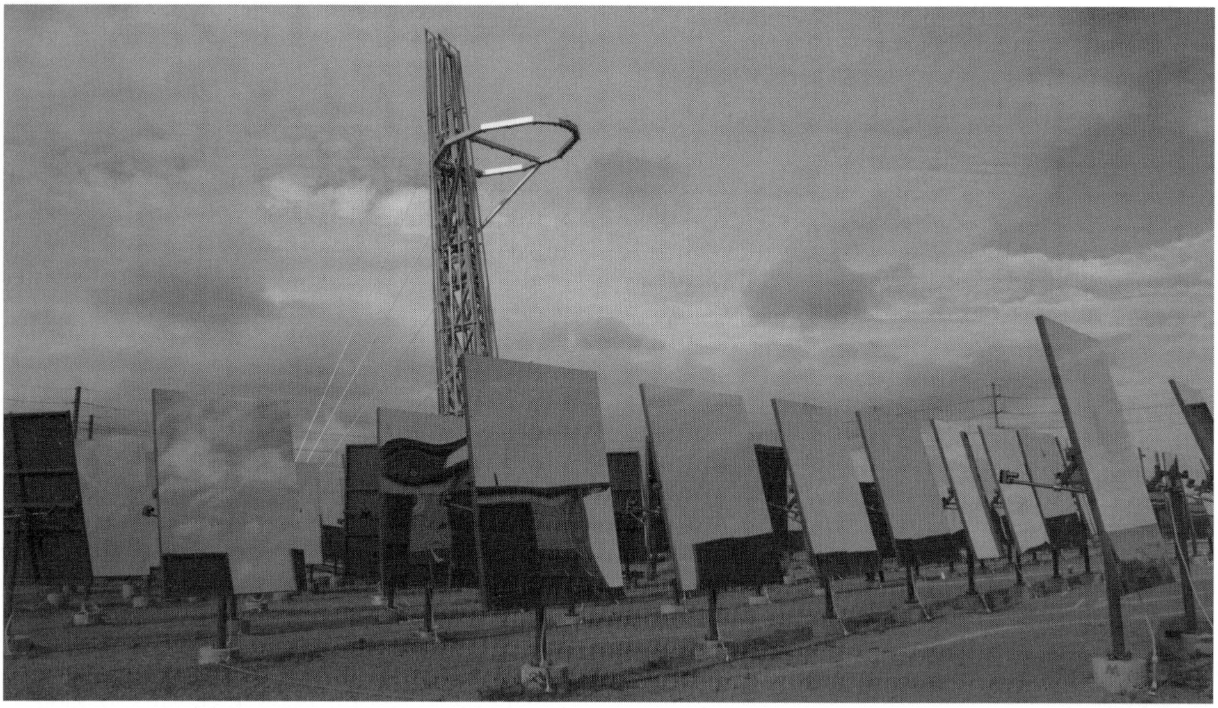

Figure 2: CSIRO solar tower array (Source: CSIRO).

remains still comparatively cheap. Moreover, the fact that processes based on the use of fossil fuels produce carbon dioxide in addition to hydrogen appears to be self-defeating on environmental grounds. Clearly, their future will depend on developing efficient means to capture this greenhouse gas and then store it (so-called 'sequestration')—possibly in underground chambers or in the ocean.

Solar–thermal reforming

The CSIRO is investigating the possible use of solar energy to provide the heat required for the steam reforming of natural gas and other methane-containing gases, such as landfill and coal-bed methane. As the resulting syngas would contain a substantial amount of embodied solar energy (up to 25%), solar–thermal reforming offers the prospects of high thermal efficiencies and greatly reduced emissions of carbon dioxide.

To demonstrate the feasibility of solar–thermal reforming, the CSIRO erected a solar dish of 48 curved mirrors that focused the sun's rays on to a thermal receiver mounted above the dish at its focal point. More recently, decentralised generation is under investigation through the development of more practical 'mini' versions of the solar-tower approach coupled to a small steam reformer; see Figure 2. Apart from the benefit of permitting the generation of hydrogen close to where it is needed, this modular technology is less expensive than a dish and is more flexible in that it allows easier integration of additional units to meet any growth in demand.

Coal gasification

As supplies of natural gas become depleted, gasification of coal is seen as the preferred technology of the future for electricity production, along with renewable forms of energy. This is because coal is plentiful and widely available geographically, the gasification process is more efficient than conventional coal combustion, and the carbon dioxide produced during gasification can be extracted from the syngas before combustion and then stored permanently underground. All these benefits, however, have yet to be practically realised on a commercial scale.

The syngas made from coal, like that produced by the steam reforming of natural gas, may be used to produce a variety of liquid fuels and chemicals via the Fischer-Tropsch process. The longer term goal is to process the syngas further to hydrogen and carbon dioxide by the WGS reaction. The Clean Coal Technology program of the US Department of Energy is motivated by the huge reserves of coal in the USA, as well as by concerns over atmospheric pollution. The ultimate aim is to develop a versatile facility that is capable of producing pure hydrogen (or organic chemicals) and electricity from coal in any desired ratio—a so-called 'poly-generation plant' as shown schematically in Figure 3.

Biomass gasification

In principle, hydrogen may be produced by the gasification of any form of biomass. The most suitable materials are those with water contents in the range of 5 to 30% by weight. These include fast-growing energy crops such as *Miscanthus* ('silver') grass, hemp, poplars and willows, or any forestry or agricultural waste, for example wood chippings, sawdust or bagasse. The gasification is generally conducted at lower pressures and temperatures than for coal, and is limited to mid-size operation due to the heterogeneity of biomass, its localised production, and the relatively high costs incurred in its gathering and transportation.

Wet forms of biomass, such as sewage, liquid manure and silage, may also be decomposed by anaerobic digestion at ambient temperature. The resulting 'biogas' is a mixture of predominantly methane and carbon oxides that may be converted to hydrogen by steam reforming.

Hydrogen from water

Water is a huge storeroom of hydrogen, but energy is required to split this source into its component elements. The water molecule is very stable—decomposition by direct heating only starts to become significant at very high temperatures; namely, about 1% by volume at 2000°C. Moreover, it is difficult to separate the hydrogen and the oxygen at such high temperatures. Some research is under way to produce hydrogen at lower temperatures by employing catalysts for the decomposition

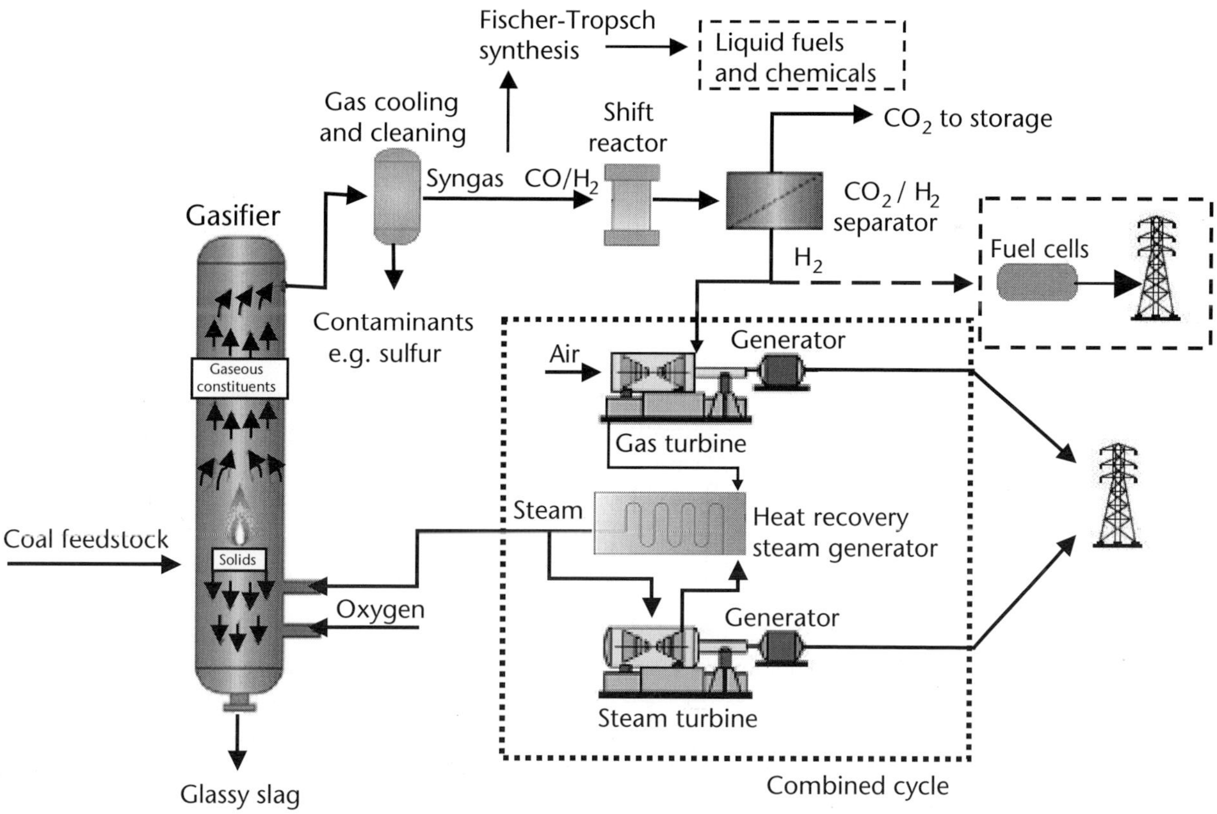

Figure 3: Concept of a poly-generation plant.

reaction together with ceramic membranes for efficient gas separation, but the work is in its very early stages. Given these limitations, the splitting of water must be undertaken indirectly by processes that employ either chemical reactants or other energy forms of high thermodynamic potential (such as electrons or photons). The chemical reduction of water to hydrogen by means of hydrocarbons—in the form of fossil fuels—has been described above. Here, we discuss principally the use of renewable energy, via the medium of electricity or directly as solar radiation, to generate hydrogen from water. The latter option may be accomplished in several ways, as shown schematically in Figure 4.

Electrolysis

The electrolysis of water to generate hydrogen or oxygen is a mature technology, but only a few per cent of world hydrogen is obtained by this method. Electrolysis is extremely energy-intensive: the faster the generation of hydrogen, the greater is the power required per kilogram produced. Large-scale units using alkaline electrolyte typically run at 70 to 75% efficiency, while smaller systems with polymer electrolytes reach 80 to 85%. Steam electrolysers, in which some of the energy required to split water is supplied by heat, may be able to achieve very high efficiencies (over 90%) but they are not yet commercially feasible.

Electrolysers should be well-suited for use with electricity generated from renewable energy sources, such as wind and solar. Conventional electrolysers are designed to operate on grid-quality power, whereas the electricity from these two renewable sources is intermittent and of variable quality. This limitation will necessitate the development of specialised power control and conditioning equipment.

Cost is also a significant issue. Using electricity derived from renewable sources, it is estimated that hydrogen can be obtained at US$ 7–10 per kg; that is, 3–5 times above that from fossil fuels. Even with the mass production of small electrolysers—should it occur—the overall economics would not change much until 'green' electricity becomes cheaper than the conventional alternative.

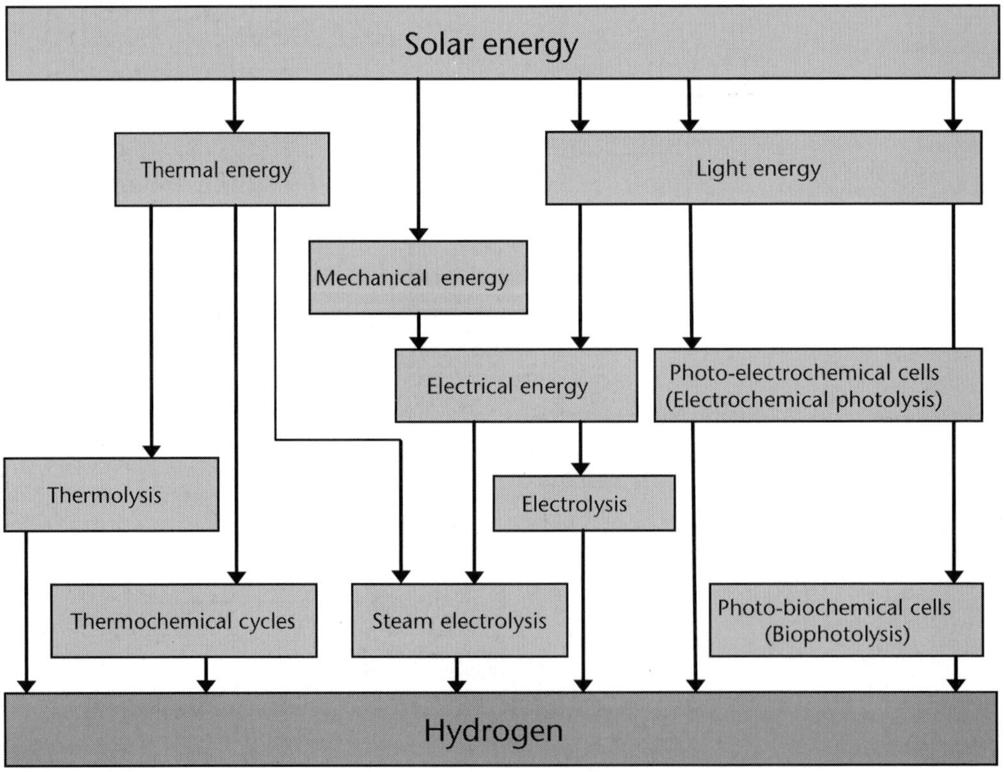

Figure 4: Solar-to-hydrogen conversion options.

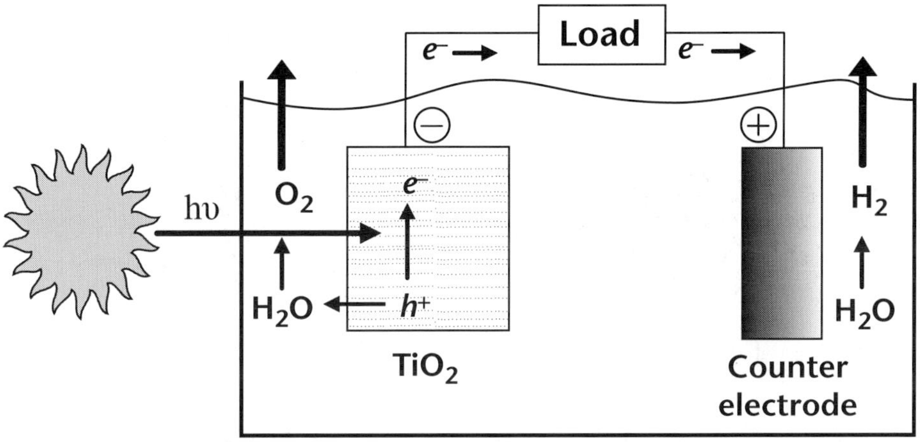

Figure 5: Operating principles of a photo-electrochemical cell for hydrogen production.

Thermochemical cycles

Because of the stability of the water molecule and the very high temperatures required to split it thermally, attempts have been made to accomplish the process indirectly at a more moderate upper temperature. The general idea is to decompose water by reacting it with one or more chemicals that are regenerated via a series of cyclic thermochemical reactions. In this way, the hydrogen and oxygen evolution steps are separated. Clearly, on practical and efficiency grounds, the fewer reactions involved the better. There are several possible cycles and the most widely studied is the sulfur–iodine cycle, which consists of three chemical reactions. The concept of a thermochemical cycle has the attraction that the required heat could be provided by a low-carbon source, such as a high-temperature nuclear reactor or solar heat.

Photo-electrolysis

Photo-electrolysis is a prospective method for the direct production of hydrogen via the harnessing of solar radiation. Light is converted to electrical and chemical energy by using a semiconducting oxide, such as titanium dioxide (TiO_2), to absorb photons. Those photons with energies greater than that of the band-gap generate electron (e^-)–hole (h^+) pairs that become separated by the electric field; see Figure 5. The holes are driven to the surface of the TiO_2 where they oxidise water to oxygen, while the electrons travel round the external circuit to reduce water to hydrogen at a metal counter electrode, such as platinum. By virtue of its relatively low cost, titanium dioxide is most attractive as a photo-electrochemical material, but it does have a somewhat high band-gap energy (~3.2 eV) and therefore absorbs light energy in the ultraviolet rather than in the visible part of the spectrum. Accordingly, present cell efficiencies are only 1 to 2%, which is well below the commercial target of 10%.

Photo-electrochemical reactions may also be used to generate electricity. This is achieved in a dye-sensitised solar cell (DSSC) by means of a subterfuge and involves separating the optical-absorption and charge-generating functions. A dye, which is capable of being photo-excited, is adsorbed on to the surface of porous titanium dioxide where it acts as an electron-transfer sensitiser. To date, such cells have demonstrated sunlight-to-electricity efficiencies of 11% in the laboratory, but only 5% in the field. Nevertheless, development of the DSSC opens up the possibility of improving the efficiency of hydrogen production from photo-electrochemical cells. The two technologies are operated in tandem to ensure absorption of the complementary parts of the solar spectrum; see Figure 6. The front photo-electrochemical cell absorbs high-energy ultraviolet and blue light in sunlight, while longer wavelength light in the green-to-red region of the spectrum passes to, and is absorbed by, the dye-sensitised solar cell. This boosts the energy of the electrons that are then fed back to a hydrogen electrode in the front cell. With such an arrangement, overall photon-to-hydrogen efficiencies of up to 12% have been reported.

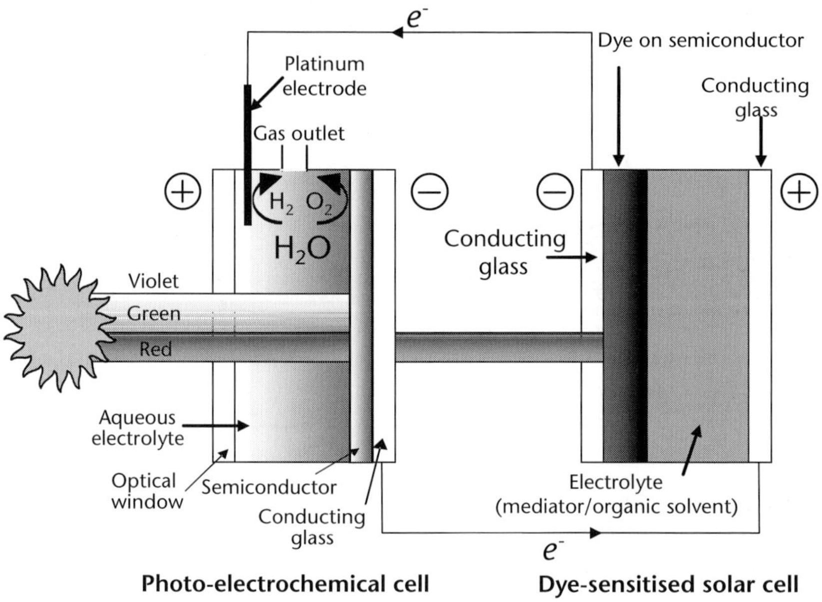

Figure 6: Operating principles of a tandem cell for enhanced hydrogen production.

Biophotolysis

The operation of the tandem cells discussed above bears a close similarity to the processes that take place in natural photosynthesis where there are also two photosystems connected in series. The first step involves splitting water into oxygen and hydrogen, and then hydrogen is used to metabolise carbon dioxide into carbohydrates. There are, however, a few natural bacteria—such as blue-green algae (cyanobacteria)—that are capable of releasing hydrogen freely into the air. This is a photobiochemical process that is often referred to as 'biophotolysis'. In practice, the energy-conversion efficiency of such systems is very low—often less than 1%. To improve performance, two-stage 'indirect biophotolysis' processes are being investigated, namely: photosynthetic fixation of carbon dioxide with concomitant generation of oxygen, followed by dark-anaerobic fermentation with evolution of hydrogen. Such a system, proposed by the Hawaii National Energy Institute, is shown schematically in Figure 7.

Hydrogen distribution and storage

To be useful as a future fuel, hydrogen has to be conveyed to the point of use and stored there until required. The distribution and storage of hydrogen are intricately bound together and depend on both the scale of operations and the intended application. In general, the storage of hydrogen for stationary applications is less demanding than for transportation systems in which there are more severe constraints in terms of acceptable mass and volume, speed of charge–discharge and, for some storage systems, heat dissipation and supply. Finding a satisfactory solution for the on-board containment of hydrogen is one of the major challenges facing the development of fuel-cell vehicles.

Gaseous and liquid hydrogen

In the gaseous state, the most obvious method for distributing hydrogen would appear to be via pipeline. Indeed, this has long been the practice to supply hydrogen in refineries and for industrial chemical processes. The transmission of hydrogen as a universal energy vector is a more difficult proposition because the distances would be much greater and the allowable costs much less. Apart from the huge capital investment that would be required, there are many technical issues with hydrogen transmission, such as materials compatibility if an existing natural gas infrastructure is to be used, pipeline integrity, leak minimisation, and high compression costs. As with natural gas transmission systems, the pipelines themselves would

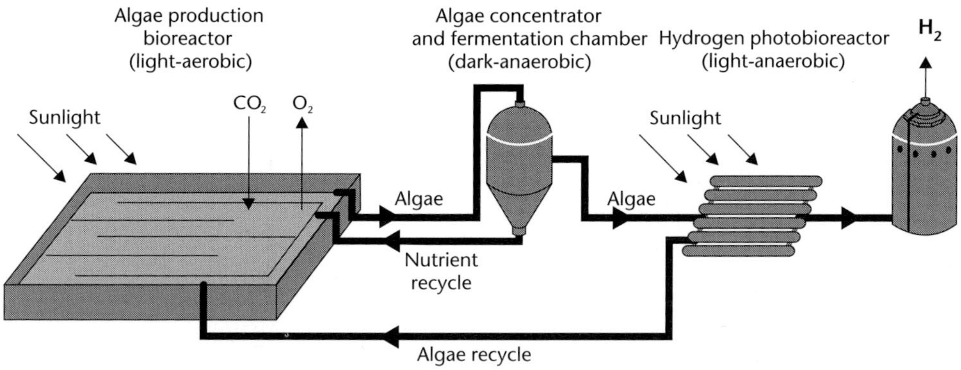

Figure 7: Schematic of developmental system for the biophotolytic production of hydrogen (Source: Adapted from and reproduced with permission from OECD/IEA 2006, Figure 6, p. 13).

provide some degree of storage for hydrogen. Large-scale storage underground in natural or engineered cavities is also a possibility.

On a much smaller scale, steel cylinders are used for storage of gaseous hydrogen at pressures of up to 80 MPa. For portable and mobile applications, however, cylinder weight and volume must obviously be minimised. To some extent this has been achieved by using all-composite vessels composed of carbon-fibre shells with aluminium liners. These can comfortably withstand a pressure of 55 MPa and thereby provide a hydrogen volumetric storage density of 3 to 4 MJ dm^{-3}; see Figure 8a. This approaches the 2010 target (5.4 MJ dm^{-3}) set by the US FreedomCAR Partnership for fuel-cell vehicles, but is far below the 2015 target (9.7 MJ dm^{-3}). Incidentally, the corresponding gravimetric targets (7.2 and 10.8 MJ kg^{-1}) will be even more difficult to reach. In this respect, liquid hydrogen (LH_2) would be more attractive; its density is 850 times greater than that of the gaseous form.

Unfortunately, the energy required to liquefy hydrogen is considerable. For large-scale production, the input is equivalent to around 30% of the energy value of the hydrogen itself, and rises to an even greater percentage for smaller operations. This, together with the high capital cost of the cryogenic plant, is a major disincentive to conveying and storing hydrogen as a liquid. From an automotive-engineering standpoint, there are also practical difficulties in terms of designing, at acceptable cost, LH_2 cryostats for on-board vehicular use that are both appropriate and safe. Nevertheless, some progress has been made towards this goal by the car manufacturer BMW and its hydrogen supplier Linde AG; see Figure 8(b). There are also aspects of vehicle refuelling to be tackled, especially the development of transfer lines for LH_2 that can be operated safely by the general public. Loss through evaporation during fuelling and boil-off on standing would also cause problems with the relatively small tanks that would be typically used in cars.

Metal hydrides

Certain metals and alloys can repeatedly absorb and release hydrogen under moderate pressures and temperatures via the formation of hydrides:

$$M + x/2H_2 \rightleftharpoons MH_x + \text{heat} \tag{4}$$

Heat must be removed during absorption of the hydrogen, but has to be added to effect desorption.

Hydrides have relatively poor mass storage for hydrogen; it ranges from 1 to 2% by weight for materials such as $CaNi_5H_4$ that operate at near-ambient temperatures, up to 3.6% by weight for magnesium-based materials such as Mg_2NiH_4 that function above 300°C. By contrast, hydrides have a relatively good theoretical volumetric performance of 0.1 kg dm^{-3}. Because lower temperature materials would be preferred in most cases, the major research challenges are to develop new alloying techniques that increase the gravimetric density of the hydrides. In addition, operational difficulties associated with the kinetics of heat and hydrogen transfer have yet to be resolved. It has been calculated that if a typical hydride bed capable of holding 5 kg of hydrogen (to provide a vehicle driving range of ~500 km) were to be charged in 3 min (an

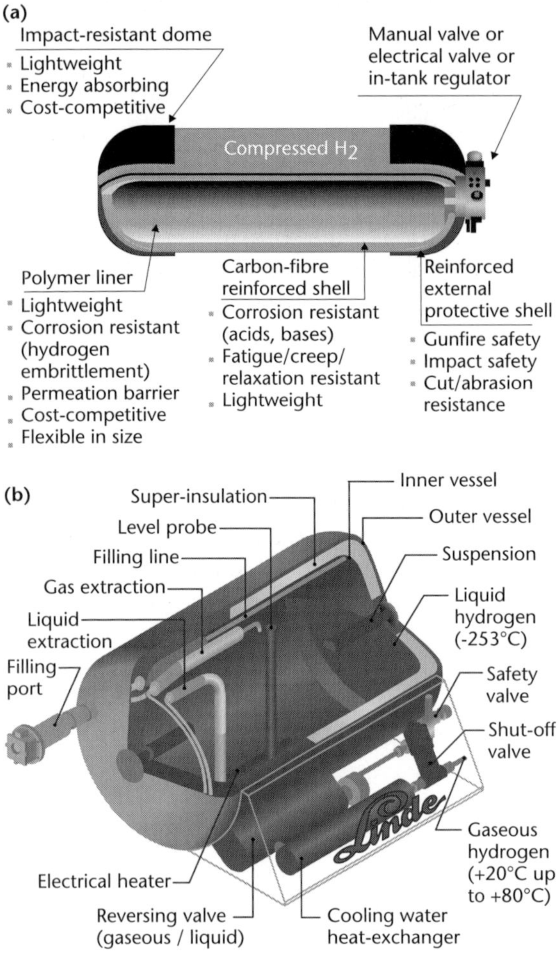

Figure 8: Schematics of vessels for the storage of (a) gaseous hydrogen and (b) liquefied hydrogen on board road vehicles (Source: (a) Reproduced with permission from OECD/IEA 2006, Figure 12, p. 22; (b) redrawn from www.linde.com).

acceptable time for refuelling), it would be necessary to remove heat at a rate of about 500 kW. Clearly, the development of an effective means of heat transfer poses a formidable engineering challenge.

Simple-hydrogen bearing chemicals

Certain organic chemicals contain significant atomic proportions of hydrogen that can be recovered and therefore they may be considered as prospective hydrogen carriers. One of the best known of these is cyclohexane (C_6H_{12}), which can be decomposed catalytically to yield benzene (C_6H_6) and hydrogen:

$$C_6H_{12} \rightarrow C_6H_6 + 3H_2 \quad (5)$$

The hydrogen content is 7.1% by weight. This is an example of a reversible (round-trip) carrier because cyclohexane is manufactured by the hydrogen reduction of benzene over a nickel catalyst at 150–200°C. Methyl cyclohexane (C_7H_{14}) has also been proposed—it dehydrogenates to toluene (C_7H_8), with a yield of 6.1% hydrogen by weight. Liquid ammonia (NH_3), hydrazine hydrate ($N_2H_4{\cdot}H_2O$) and ammonia borane (NH_3BH_3) will theoretically store 17.7, 8.0 and 12% hydrogen by weight, respectively. In practice, however, such materials present many safety and operational difficulties, especially when the hydrogen is intended for use in vehicles.

Complex chemical hydrides

Hydrogen may be stored chemically in the form of ionic salts that are composed of sodium, aluminium or boron, and hydrogen—the so-called 'complex chemical hydrides'. The alanates $Na[AlH_4]$ and $Na_3[AlH_6]$ have been widely studied. Thermal decomposition of $Na[AlH_4]$ to yield hydrogen takes place at temperatures up to 180°C.

Sodium borohydride, $NaBH_4$, is stable up to about 400 °C and is therefore not suitable for providing hydrogen through a thermal-activation process. It does, however, release hydrogen, on reaction with water:

$$NaBH_4 + 2H_2O \rightarrow NaBO_2 + 4H_2 \quad (6)$$

This is an irreversible reaction. Based on the mass of $NaBH_4$, the hydrogen released is 21% by weight—a remarkably high output—but in practice the yield is lowered to around 7% when the weight of the total system is taken into account. Several similar hydrides are being evaluated for their reactivity with water, but they all have the disadvantage that the storage of hydrogen is not rechargeable; the spent solution has to be returned to a processing plant for regeneration of the hydride.

Nanostructured materials

Considerable research is being carried out on materials that have structural elements with dimensions in the nanoscale range and have high specific surface-areas ($m^2\ g^{-1}$). Amongst these materials it has been found that carbon and boron nitride nanostructures, clathrates and metal–organic frameworks can store hydrogen in the molecular state via weak molecular–surface interactions.

Carbon nanotubes were some of the first nanomaterials to be investigated for hydrogen storage and many have been reported with capacities of several per cent by weight that may be further enhanced by various pre-treatments. Nevertheless, there continues to be considerable controversy over such findings because of the difficulty in preparing homogeneous, well-defined, pure and reproducible samples. Moreover, contrary to expectation, significant hydrogen storage in these materials usually requires either high pressure (greater than 10 MPa) or low temperature (at least –100°C).

Hydrogen utilisation: fuel cells

Fuel cells are a key enabling technology for a future hydrogen economy—particularly for electric vehicle propulsion—and therefore are being intensively investigated and developed in many countries. The fuel cell is an electrochemical device for the conversion of chemicals into direct-current electricity. To this extent, it resembles a primary battery. There are, however, some important differences. In a battery, all the chemicals necessary for its operation are normally confined within a sealed container. Thus, the capacity of a battery, measured in ampere-hours, is determined by the quantity of the chemicals that it holds. With a fuel cell, the chemicals are supplied from external reservoirs so that the capacity of the device is limited only by the available supply of reactants.

To demonstrate the principle of the fuel cell at low power output and without regard to cost is remarkably simple and can be done in any school science laboratory. The difficulty lies in developing an engineered, high-power system at acceptable cost that will operate for long periods. The fundamental operation of a fuel cell is illustrated in Figure 9. Fuel is oxidised at the negative electrode with the release of electrons that pass through the external circuit and reduce an oxidant at the positive electrode. The flow of electrons is balanced by a flow of charged ions in the electrolyte. Because a fuel cell functions at a low voltage (that is, well below 1 V), it is customary to build up the voltage to the desired level by electrically connecting cells in series to form a 'stack'.

The first demonstration of a basic fuel cell was by Sir William Grove in 1839, but it was not until the work of FT (Tom) Bacon in the 1940s that a practical system began to emerge. Bacon first developed the alkaline fuel cell while he was at the University of Cambridge (1946–1955). This was adopted by the USA in the 1960s for the Apollo space program and refined later for the Space Shuttle Orbiter. It provided all the electricity, as well as drinking water, when the Space Shuttle was in flight.

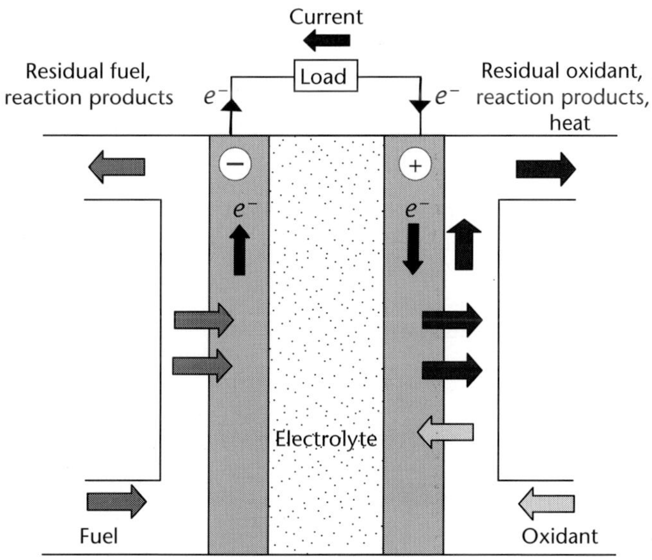

Figure 9: Schematic of fuel cell operation.

Types of fuel cell

There are six major types of fuel cell that can be broadly classified in terms of their temperature of operation:

- low-temperature (50–150°C): alkaline electrolyte (AFC), proton-exchange membrane (PEMFC) and direct methanol (DMFC) fuel cells
- medium-temperature (around 200°C): phosphoric acid fuel cell (PAFC)
- high-temperature (600–1000°C): molten carbonate (MCFC) and solid oxide (SOFC) fuel cells.

Some operational data on each type are given in Table 1.

Apart from the DMFC, which is fuelled by methanol, hydrogen is the preferred fuel because of its high electrochemical activity. The AFC and PEMFC are both demanding in terms of purity of the hydrogen, which must typically contain less than 10 ppm carbon monoxide. By contrast, the PAFC is somewhat more tolerant to carbon monoxide, while the high-temperature designs (MCFC and SOFC) will accept carbon monoxide and a variety of hydrocarbon fuels. All the designs require the fuel to be sulfur-free.

A variety of sub-systems and components are required for a fuel-cell stack to function effectively. The exact composition of this so-called 'balance-of-plant' depends upon the type of fuel cell, the available fuel and its purity, and the desired output of electricity and heat. Typical auxiliary sub-systems may include:

1. fuel processing reactors (such as desulfurisation, steam reformer, shift and partial oxidation reactors)
2. a humidifier
3. fuel and air delivery units
4. power-conditioning equipment; for example, for inverting direct current to alternating current
5. heat- and water-management facilities
6. overall system controls.

The advantages of fuel cells over other generators of electrical energy, such as gas turbines, steam turbines, or internal combustion engines with alternators, are as follows:

- potentially high energy-conversion efficiency over a range of sizes
- greater efficiency when operated at part-load, in contrast to engines
- no moving components apart from auxiliary fans and blowers
- almost silent in operation
- flexibility of fuel supply
- very low exhaust emissions
- pure water emitted when using hydrogen as the fuel
- modular construction and ease of installation.

This impressive list of attributes has provided the incentive for much of the fuel cell research that has taken place in recent years.

Fuel cell applications

The potential uses for fuel cells are extremely diverse and vary greatly in their power demands, from watts to megawatts. Broadly, the applications fall into the three categories: (1) stationary power systems; (2) portable and military devices; and (3) transportation.

Stationary power generation

Among the fuel cells with outputs of 10 kW to several MW that are being developed for the stationary generation of power, the PAFC has until recently been regarded as the prime candidate—at least for units up to 250 kW. A 200 kW system offered by United Technologies is shown in Figure 10a. Concerns over cost and durability of this system are, however, causing a reconsideration of this choice. At the upper end of the size range, it is possible that MCFCs (see Figure 10b) or SOFCs will become more attractive propositions for both commercial and industrial purposes, because of their internal-reforming function and their CHP capability. Again, cost and durability targets present a real challenge. In the case of larger units—that is, greater than 1 MW in output—there is growing interest in hybridisation with gas turbines to take advantage of the high-quality heat for the production of additional electricity.

There are numerous potential mass-markets for smaller stationary fuel-cell systems with power outputs between 0.1 and 10 kW. These include: electronic communications equipment; small-scale

Table 1: Principal types of fuel cell

Fuel-cell technology	Electrolyte	Temperature range (°C)	Electrocatalyst		Fuel	Efficiency[a] (% HHV)	Start-up time (h)
			Positive electrode	Negative electrode			
PAFC	H_3PO_4	150–220	Pt supported on C	Pt supported on C	H_2 (low S, low CO, tolerant to CO_2)	35–45	1–4
AFC	KOH	50–150	NiO, Ag, or Au–Pt	Ni, steel, or Pt–Pd	Pure H_2	45–60	<0.1
PEMFC	Polymer	80–90	Pt supported on C	Pt supported on C	Pure H_2	40–60	<0.1
DMFC	H_2SO_4 Polymer	60–90	Pt supported on C	Pt supported on C Pt–Ru	CH_3OH	35–40	<0.1
MCFC	Li_2CO_3	600–700	Lithiated NiO	Sintered Ni–Cr and Ni–Al alloys	H_2, variety of hydrocarbon fuels (no S)	45–60	5–10
SOFC	Oxygen-ion conductor	700–1000	Sr-doped $LaMnO_3$	Ni- or Co-doped yttria-stabilised zirconia cermet	Impure H_2, variety of hydrocarbon fuels	45–55	1–5

[a] The reported efficiency of a given type of fuel cell varies widely and often no information is provided on whether the higher heating value (HHV) or the lower heating value (LHV) of the fuel is used. The efficiencies that are taken here from the literature should be treated with caution as to their exact meaning and are simply included to provide an approximate comparison of the performance of the respective systems.

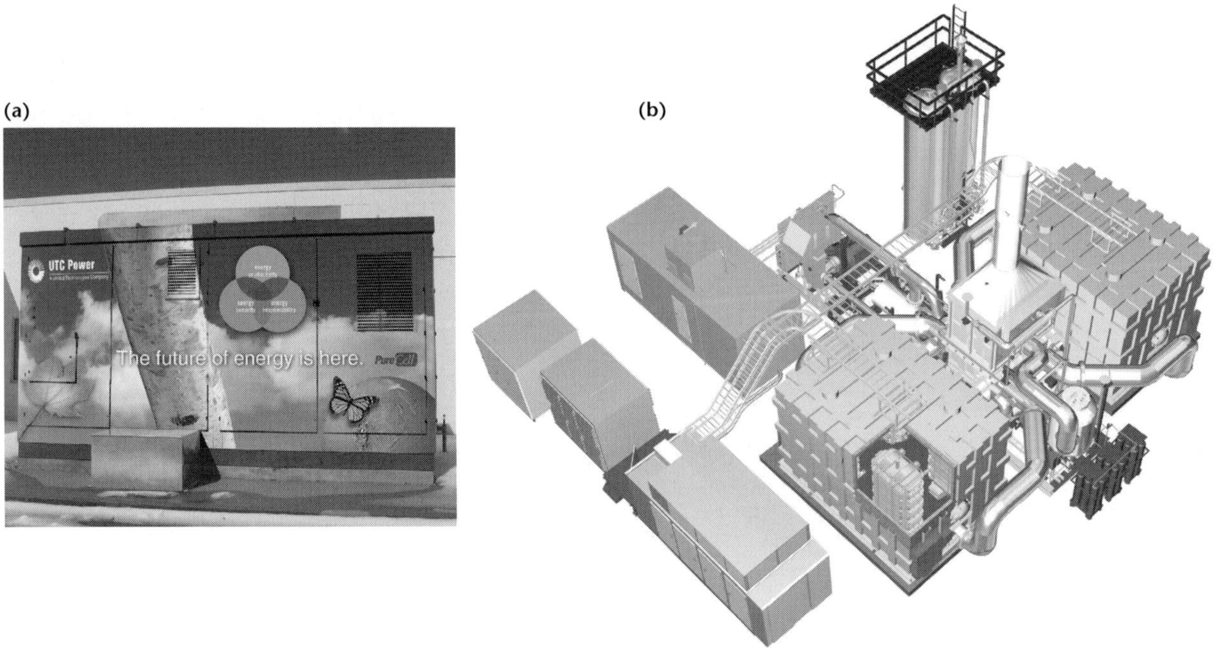

Figure 10: (a) PureCell™ Model 200 Fuel Cell Powerplant (photo courtesy of UTC Power Corporation); (b) 2 MW molten carbonate fuel cell (photo courtesy of FuelCell Energy Inc.).

distributed electricity generation; CHP for individual homes, small apartment blocks or offices; uninterruptible power supplies; and remote-area power supplies.

Portable power

Many companies are engaged in developing fuel cells with power outputs in the range of 1 to 50 W for electronic appliances and communications equipment. As well as their use in mobile phones and laptop computers, there are demands from the military for superior power sources to provide more extensive 'silent watch' operation of surveillance and also communications facilities in the battlefield. Fuel cells are attractive for all these purposes, not only because they are more portable than batteries and have longer life, but also because refuelling is much faster than recharging a battery.

Because these 'miniature' fuel cells are devices for use in high-value markets, the cost of the polymer membranes and the platinum-based electrocatalyst is less significant than for larger-scale applications. Recent examples of prototype devices are shown in Figure 11.

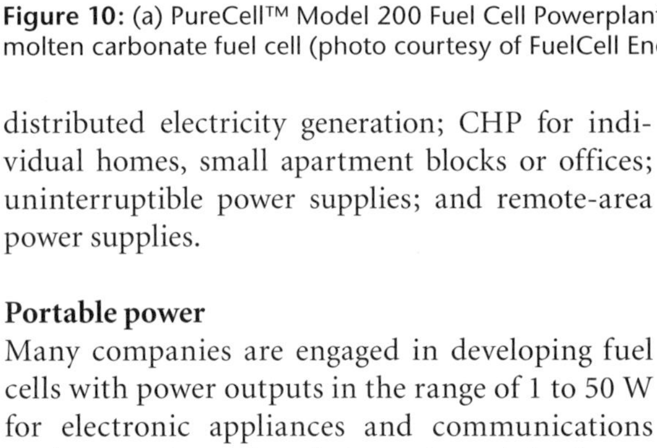

Figure 11: Developmental miniature fuel cells for: (a) mobile phones and (b) laptop computers.

The widespread introduction of miniature fuel cells would obviously have little direct impact on the international energy scene, so their real significance may lie in encouraging developers to address the generic and basic problems associated with fuel cells, and in increasing public awareness and acceptance of such power sources. Many observers believe that volume production of miniature units would be a key technical and economic driver for the entire fuel-cell market.

Transportation

A consensus appears to be emerging among car companies that the *all-electric* vehicle, if it comes into widespread use, will be powered by a fuel cell rather than by a battery. The focus for general transportation applications is the PEMFC, although the specifications that must be met to compete with the internal combustion engine are stringent—particularly regarding the size, performance and overall cost. The progress being made with fuel-cell vehicles is discussed in more detail in the following section.

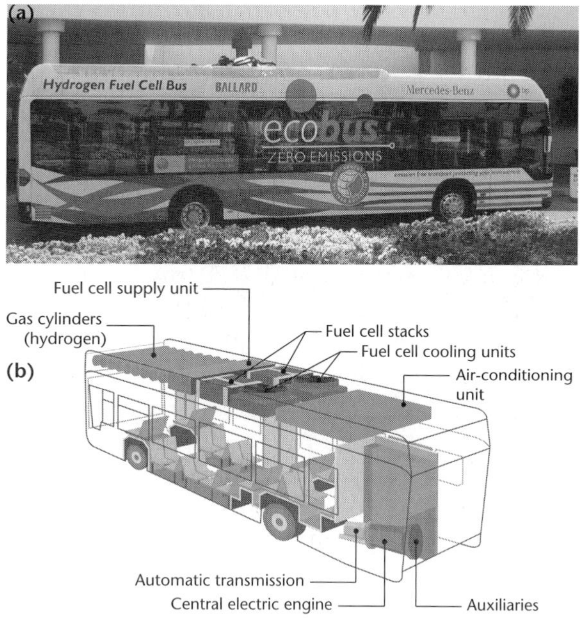

Figure 12: (a) Fuel-cell bus operating under the STEP program in Perth; (b) arrangement of fuel-cell system and on-board gas cylinders.

Hydrogen-fuelled transportation

Internal-combustion-engine vehicles

The combustion characteristics of hydrogen in internal combustion engines are quite different from those of conventional liquid fuels and this necessitates some changes in engine design. In general, hydrogen engines tend to exhibit pre-ignition, backfire and knock, which are phenomena attributable to the low ignition energy and high flame speed of hydrogen compared with petrol. The German engineer Rudolf Erren showed in the 1930s that these malfunctions could be overcome by feeding hydrogen at a slightly increased pressure directly into the combustion chamber, rather than by introducing it with the fuel–air mixture through the carburettor.

BMW has been working with hydrogen engines since 1978. In 2000, the company produced an experimental fleet of 15 dual-fuel cars, each fitted with a 140-litre storage tank that was capable of holding about 8 kg of LH_2 to give the vehicle a cruising range of around 200 km. The cars also had petrol tanks that allowed a further 500 km of travel. In 2001, BMW introduced a 4.4-litre *745h* dual-fuel car that was equipped with an auxiliary PEMFC power unit. The latter functioned independently of the engine to power accessories, such as air-conditioning that could be operated when the engine was shut off. Recently, Ford in the USA has started production of a supercharged 6.8-litre V10 hydrogen engine intended for use in hydrogen-fuelled *E–450* shuttle buses.

Fuel-cell vehicles

The fuel-cell vehicle (FCV) operating on hydrogen, most probably with a PEMFC, is seen by many as the ultimate solution to the increasing energy security and environmental problems that are confronting road transportation. At their present stage of development, PEMFC power systems are far more expensive than internal combustion engines (up to 60 times greater per kW of power produced) and there are serious technical challenges that must be met if costs are to be reduced to the point where FCVs are economically attractive. Above all, there are overriding problems associated with the introduction of a hydrogen fuelling infrastructure. Nevertheless, major investment in FCVs is in place in Europe, Japan and the USA. Emphasis has been on the production of urban buses and private cars—all manufacturers are employing PEMFC systems.

Buses

Urban (transit) buses constitute a prime market for the introduction of fuel cells. In addition to the fact that there is adequate internal space and roof area to accommodate the fuel-cell stacks and the hydrogen storage cylinders, buses have several other positive features with respect to using fuel-cell power. The vehicles usually run on well-defined routes and schedules, are maintained by qualified technicians, and operate from a home depot so that only a single hydrogen-refuelling facility is required to service a fleet. Furthermore, electric designs would neither emit pollutants nor waste energy when buses are stationary in dense traffic.

A notable demonstration of fuel-cell buses has been conducted under an initiative known as the Clean Urban Transport for Europe (CUTE) program that began in November 2001. Twenty-seven buses were built and integrated into the public fleets of Amsterdam, Barcelona, Hamburg, London, Luxembourg, Madrid, Porto, Stockholm, and Stuttgart. Each city received three buses and these were serviced through the construction of a hydrogen-refuelling station at each location. The buses were produced by DaimlerChrysler and were based on the Mercedes-Benz *Citaro* design—a full-size, low-floor vehicle that can accommodate up to 70 passengers. A similar project (ECTOS), also funded by the European Union, was undertaken in Reykjavik, Iceland, between 2001 and 2005. This country is in an unusual position in having available excess geothermal and hydroelectric power that could be used for generating hydrogen for the buses via electrolysis. Three more *Citaro* fuel-cell buses were also trialled in Perth, Western Australia from 2004 to 2007; see Figure 12. The hydrogen fuel for the Australian STEP trial was produced by British Petroleum in its nearby oil refinery at Kwinana.

The CUTE, ECTOS and STEP projects were judged to be highly successful. By October 2005, the buses had carried more than four million passengers and covered almost 1.1 million km. The reliability of the buses was high—with an availability rate of over 90%. Nearly 9000 refuelling operations were carried out safely and collectively dispensed almost 200 t of hydrogen.

Cars

Following work on prototypes commencing in 1999, Honda became the first company to produce a passenger car powered by a fuel cell—the *FCX-V4*—and have it certified for commercial use in both Japan and the USA. The vehicle was supplied to customers in both countries in December 2002. Subsequently, Honda built its own PEMFC stack with improved power density and the capability of start-up at temperatures as low as –20°C, This was introduced into the 2005 Honda *FXC*, which duly received certification for commercial services and is now undergoing further refinement. A working

(a)
(b)
(c)
(d)
(e)

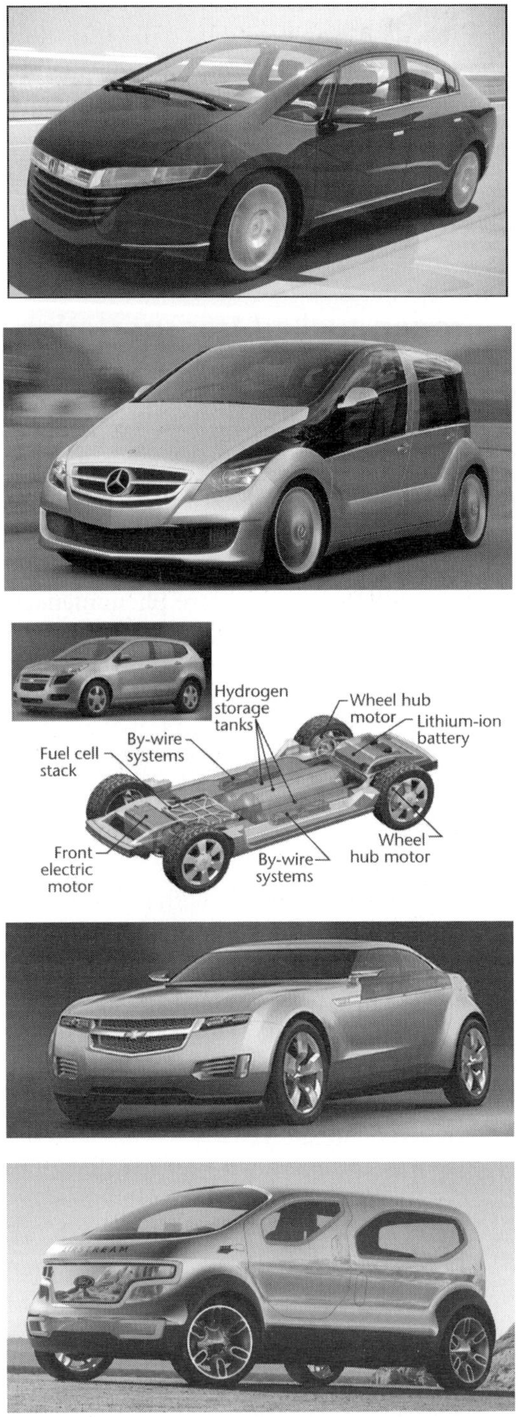

Figure 13: (a) Honda *FCX Concept* (photo copyright Honda); (b) DaimlerChrysler *F 600 Hygenius* (photo courtesy of Mercedes Benz); (c) General Motors *Sequel* and 'skateboard' chassis; (d) General Motors fuel cell variant of the Chevrolet *Volt* (photos copyright General Motors); (e) Ford *Airstream Concept* (photo courtesy of Ford Motor Company of Australia Ltd).

version of the *FCX Concept*, shown in Figure 13(a), is scheduled for limited marketing in Japan and the USA in 2008.

Toyota began producing its vehicle—the *FCHV*—at the same time as Honda. It was a very similar car with comparable performance. The company is now investigating the possibility of incorporating an on-board reformer in the next-generation *FCHV-5* to allow the use of ultra-low sulfur liquid fuel (Clean Hydrogen Fuel®, an evolved form of petrol) that can be supplied by existing petrol pumps.

In Germany, Opel (General Motors) has retrofitted its *Zafira* compact van to produce a fuel-cell car—the *HydroGen3*. This is a somewhat larger car than the 2005 Honda *FCX* or Toyota *FCHV* and differs from these in that it can run on either liquid or compressed hydrogen. In October 2002, DaimlerChysler unveiled its sixth-generation fuel-cell car—the *A-class F-Cell*. Sixty of these cars were manufactured and evaluated by organisations in Europe, Japan, Singapore, and the USA. At present, a research vehicle—the *F 600 Hygenius*—is being used by DaimlerChrysler to develop improved drive-train technology; see Figure 13(b).

In the USA, General Motors presented its latest fuel-cell concept car—the *Sequel*—in January 2005; see Figure 13c. This sporting sedan has a 'skateboard' chassis made of aluminium that integrates a fuel-cell stack, a lithium-ion battery, three carbon-composite hydrogen cylinders, electronics, wheel motors, and suspension components. The vehicle is operated with revolutionary 'drive-by-wire' technology (GM *Hy-wire*®) that allows for steering, acceleration, braking and other systems to be controlled electronically, rather than mechanically, through the use of two handgrips.

Early in 2007, General Motors disclosed plans to build a different version of its series-hybrid—the Chevrolet *Volt*—in which the internal combustion engine will be replaced by a fuel cell stack, so making it a fuel cell and battery (that is, all-electric) hybrid; see Figure 13d. Ford has also chosen to 'build a bridge' from hybrid electric cars to fuel-cell cars. The Ford *Airstream Concept* (introduced in January 2007; see Figure 13e) is powered by a plug-in hybrid power-train that combines a 336-V lithium-ion battery pack with a compact Ballard fuel-cell system as a range extender. Clearly there is scope for much further experimentation of hybrid vehicles based on fuel cells, particularly those with a plug-in facility for battery recharging.

Hydrogen highways

At the end of 2005, there were around 115 hydrogen-refuelling stations across the world, with the majority located in the USA. In Europe, filling stations were established in the nine European cities that participated in the CUTE bus program, and this has been repeated in other countries that have hosted bus trials.

In the USA, the main interest in FCVs is centred on the State of California. Activity is concentrated around the Los Angeles metropolitan area where there is a sizeable cluster of existing or planned refuelling stations. The Governor of California has enunciated his vision for 2010 of 'tens of thousands' of FCVs being commercially available. For this to be a realistic proposition, it will be necessary to have an extensive hydrogen-dispensing network. To achieve this vision, the aim of the California Hydrogen Highway Network Action Plan is to install 150 to 200 refuelling sites throughout the State—that is, approximately one every 30–35 km along the major highways.

Canada has embarked upon a rather more modest program for the implementation of a hydrogen highway. A consortium of organisations will work together to design, build, operate and evaluate a hydrogen-fuelling infrastructure along a corridor between Vancouver and Whistler (170 km to the north), with an extension to Victoria on Vancouver Island. The enterprise is scheduled for completion in time for the 2010 Winter Olympics that will be held in the former two locations.

Proposals have also been made—although at a less-advanced stage—for a hydrogen highway in Europe. Germany's leading hydrogen supplier, Linde, has suggested an initial circular network of around 1800 km running between major German cities, with semi-automated refuelling points every 50 km. This would require some 35 new hydrogen stations, in addition to those already operating in Berlin and Munich. If this demonstration is enacted and proves successful, the plan would then be to implement progressively a Trans-European Hydrogen Highway.

Hydrogen activity in Australia

The vision of a hydrogen economy in Australia can be traced back at least to the 1930s when JF Kemp, a councillor from Brisbane, suggested that vehicles could be powered by hydrogen generated from coal. The Queensland Government ordered construction of a high-pressure plant, but this was never built as the end of World War II signalled a return to relatively cheap oil and petrol.

Interest in hydrogen during the latter half of the 20th century was confined mainly to university researchers, such as HC Watson at Melbourne—who has studied hydrogen-fuelled vehicles since the 1960s—and the CSIRO. In 1992, CSIRO research on solid oxide fuel cells was spun out to Ceramic Fuel Cells Limited. This Melbourne company has subsequently made considerable progress in overcoming the challenges to produce practical, economically-viable and robust SOFC systems. As mentioned above, the CSIRO has also been developing solar–thermal technology for the decentralised generation of hydrogen.

In May 2003, experts from around the world were invited to take part in the conference *The Hydrogen Economy: Challenges and Strategies for Australia* that was held in Broome, Western Australia. This event was part of an initiative by the federal government that cumulated in the publication of Australia's *National Hydrogen Study* in October 2003. In the same year, Australia became one of the first signatories to the International Partnership for the Hydrogen Economy (IPHE), and in 2004 the Department of Industry, Tourism and Resources (DITR) commissioned the CSIRO to prepare a report on *Australian Hydrogen Activity* that was subsequently published in May 2005.

In 2002, the CSIRO set up six 'flagship' projects aimed at focusing research on issues of national importance. The Energy Transformed Flagship is devoted to the development of technologies and systems for the cost-effective reduction of Australia's emissions of greenhouse gases. Specifically, the mission of the project is 'to halve greenhouse gas emissions and double the efficiency of the nation's new energy generation, supply and end use, and to position Australia for a future hydrogen economy.' To emphasise further the role of hydrogen, the Flagship launched a National Hydrogen Materials Alliance in October 2006. This is a research cluster of 11 universities, the Australian Nuclear Science and Technology Organisation (ANSTO), and the CSIRO. The objectives are:

- to develop new materials that improve the efficiency and economics of hydrogen production
- to draw together the significant expertise in Australia in the generation, storage and end use of hydrogen
- to focus this expertise into the most promising research directions.

Australia joined the Hydrogen Implementing Agreement of the International Energy Agency in 2005 and, in June 2007, the DITR issued a tender for the production of a National Hydrogen Road-Map. While many countries developed their vision for hydrogen energy through a period of progressively advancing research, development and demonstration during the latter half of the 20th century, it is evident that Australia has only just begun with a flurry of activities. This can be attributed largely to the lack of drivers for a hydrogen economy in Australia. The country has enjoyed a wealth of indigenous energy resources and does not have the issues of energy security that have forced other developed economies to consider alternative fuels.

The future

Early in the 20th century, the invention of the internal combustion engine had a revolutionary impact on the transportation sector, with a shift from horses and carts to railway locomotives, motorised vehicles and aircraft. The realisation is now dawning that the time is approaching for another Industrial Revolution—based on new forms of energy—which will involve massive disruptions in developed societies. Central to this revolution will be the growing use of natural gas in gas turbines, clean coal technology, the distributed (local) generation of electricity based (in part) on renewables, and the replacement of petrol and diesel engines by other propulsion units and/or fuels.

Although the precise role for hydrogen has yet to be established, there are many reasons why hydrogen energy is stimulating worldwide interest. A particularly desirable feature is its complementarity to electricity as an energy vector and also its potential use as an energy store. Much depends upon future political and economic factors that are difficult to forecast, so hydrogen energy is usually viewed as a long-term option. If oil and gas supplies are restricted or even disrupted, if climate change is as severe as many fear or if there is a breakthrough in lowering the cost of renewable energy then, quite suddenly, a transition to hydrogen energy could become highly desirable. Whatever eventuates, growing reliance on coal, as oil and gas supplies decline, is almost a foregone conclusion. From an environmental perspective, coal is best used through the media of hydrogen and electricity via gasification with sequestration of carbon dioxide. Both the development of hydrogen technology and the establishment of an infrastructure will, however, be neither quick nor easy and this provides the justification for undertaking the basic science and engineering sooner rather than later.

References

ACIL Tasman and Parsons Brinckerhoff (2003). *National Hydrogen Study.* Commonwealth of Australia, Canberra.

Dell RM and Rand DAJ (2004). *Clean Energy.* Royal Society of Chemistry, Cambridge, UK.

Larminie J and Dicks A (2003). *Fuel Cell Systems Explained.* Second Edition. John Wiley and Sons Ltd, Chichester, UK.

OECD/IEA (2006). *Hydrogen Production and Storage: R & D Priorities and Gaps.* IEA, Paris, France.

Rand DAJ and Badwal SPS (2005). *Australian Hydrogen Activity,* Commonwealth of Australia, Canberra.

Rand DAJ and Dell RM (2008). *Hydrogen Energy: Challenges and Prospects.* Royal Society of Chemistry, Cambridge, UK.

26

Distributed Energy Systems

Terry Jones

Energy is an important part of 21st century life and a plentiful supply is essential to the wellbeing of Australian communities. Today, most stationary energy (in the form of electricity) is derived from an abundant supply of coal, which has the significant disadvantage of producing large amounts of CO_2 emissions. Australia's ageing electricity generation and delivery network has lacked investment since deregulation in the 1990s and requires continual upgrade and repair. To cater for a growing population and an ever increasing demand for reliable clean energy, we must face these challenges. What new pathways can we propose to deal with these energy based problems and yet ensure we deliver energy reliably and cost-effectively? What lessons can we learn from the past?

At the beginning of the 20th century electricity started to replace gas for street lighting and domestic lighting. At that time, all electricity was provided by entrepreneurs and councils to local communities. Since then the theory that 'bigger is better' has led to our electricity network being built around large power stations located near to fuel sources (mainly coal) rather than close to consumers. Australia's dependence on large coal-fired power stations and their associated transmission and distribution (T&D) networks has minimised cost of generation, but required significant infrastructure investment to transmit the power from the generation plants to consumers. This is now leading to problems with emissions, network expansion and renewal and the ever-growing base and peak demand. In a world where energy efficiency is critical, we can no longer afford to ignore the losses associated with placing our energy sources such long distances from energy consumers. We should not continue to install a large, costly centralised generating plant that is only fully used a few times a year.

If we are to achieve sustainable growth of our energy supply for urban Australia then we need to examine alternative models of supply to communities across Australia using the latest technology. A new model for sustainable urban development in Australia is needed—distributed energy solutions can fill that need.

Climate scientists tell us that we are already on an accelerating pathway to global warming as a result of our greenhouse emissions. A key challenge is whether we can take decisive, early action to mitigate the effect of this trend by reducing emissions associated with new growth. Distributed energy (DE) comprises a suite of smaller capacity technologies for energy generation located close to the point of consumption and provides an alternative means to take action now, instead of waiting to build large centralised generation capacity using clean-coal technologies that still require 10–15 years to test and prove. DE would also set in place a new standard for sustainable growth that avoids many of the problems of the past.

If we are to create a pathway to sustainable urban development then we have to:

- use our existing infrastructure more efficiently and plan to minimise wasteful network arrangements
- quickly build new generating capacity that is cleaner, more efficient and can supply power with minimal energy loss.

This new pathway to growth will challenge conventional practice and commercial interests. The question is whether the incumbent electricity supply industry can accommodate a major shift in its structure and business model in order to embrace distributed energy and an increased range of consumer choices.

In establishing a case and action plan for sustainable urban development involving DE, we need to examine both the supply side and demand sides of energy provision and consumption. The supply side (generation, transmission and distribution of electric power) is important because today we face problems with emissions, peak power demands, over-capacity of base-load generation and under-capacity to deliver at peak times. On the demand side, consumers face increasing prices and complicated tariffs for energy and a lower certainty of supply. An increased awareness of global warming is encouraging consumers to search for solutions towards low-emission, sustainable living.

The supply side

Current electricity networks and a supply dominated system

In Australia, two drivers have determined the shape of our electricity networks. Firstly the location of natural fossil fuel (primarily coal fields), plus the massive Snowy Mountains Hydro scheme, built by visionaries of the 1940s, has led to a long 'stringy' distribution network down the east cost of the continent. Secondly, much of the nation's population and economic growth is concentrated around the state capitals and major urban centres on the east coast.

While Australia can boast among the longest (44000 KM) transmission networks in the world, the associated losses and inefficiencies of this system can no longer be ignored. Coal-fired generation converts fuel to electricity at between 30 and 40% efficiency. The long transmission and distribution network contributes about another 8% of energy losses. We cannot continue on this pathway of energy growth—we are losing over half of the initial energy resource during generation and transmission. Imagine if we could convert 80% of our fuel into useful electricity and have near zero losses along the way. We could then double our energy capacity overnight.

Levels of investment

Australia's electricity networks are continually growing to feed the needs of the population, and the increasing list of domestic appliances (such as air conditioners and plasma TVs), as well as new urban developments that need to be connected to water, transport, telecommunications and energy infrastructures to make them attractive places to live. This leads to further demands (both base-load and peak) on both the electricity generation and distribution systems. The problem we now face is that the privatisation of the electricity industry in Australia has led to a reduction in asset investment by utilities in search of shareholder value. The capacity that would have been typically created under state ownership has been eroded away, leading to a significant need for new investment in generation of $78 billion to 2050 (Boshier 2007) and T&D infrastructure of approximately $5 billion per year (Curran 2006).

Figure 1 shows the predicted continued growth in energy needs and the retirement of existing power plants creating what we call the future supply–demand gap. The key challenges now are:

- how to fill the supply demand gap with clean energy and meet the needs of growing urban communities for sustainable energy
- finding ways to minimise the investment required (and therefore the cost passed on to the consumer),
- minimising emissions
- improving overall energy efficiency

Distributed energy does not require massive investment in T&D infrastructure because energy is generated close to the consumer. Distributed energy avoids the need for large power stations to be built to cope with peak loads lasting for a few hours each day as DE can be turned on at peak times when necessary. Distributed energy systems are generally cleaner (they produce lower emissions) and more efficient (due to waste heat recovery or use of renewable energy sources) compared with large coal-fired power stations.

International trends

Internationally, the drive towards distributed energy is generally a response to predicted shortages of coal and gas or demands for increased energy security and quality of supply. The USA is actively pursuing DE solutions, driven by recent blackouts and terrorism concerns, in order to make its electricity network more resilient and able to

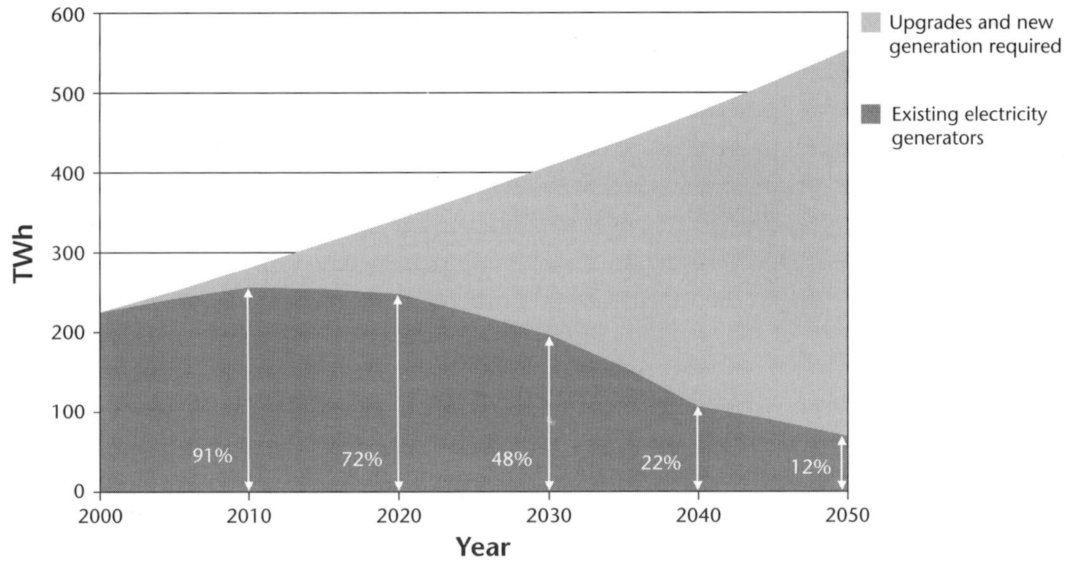

Figure 1: Predicted energy supply and demand in Australia

survive major losses of generation and transmission connections. In Europe, DE is being employed by countries motivated either by concerns over future bulk fuel shortages, alternatives to high-priced energy and, in some cases, the social imperative to reduce emissions. In developing countries DE is also being used to provide power into expanding communities that either have no electricity network connection or where there is simply insufficient generation capacity to meet demand. In all cases, DE provides a low-cost, flexible way to meet community power demands that allows for fast growth, minimises infrastructure cost, and ensures security, peak demand and emissions reductions. DE creates a pathway for future energy planning that will provide a more efficient cleaner electricity supply model and is capable will ensure sustainable future urban development.

The demand side

Towards demand-responsive supply

The growing awareness among the Australian population of global warming, and our own very high level of per capita emissions, represents a new driver for employing low-emission technologies to meet future energy growth. It would be convenient to simply continue building large power stations, but this strategy does not provide triple-bottom-line solutions—allowances for peak-load growth, low emissions and flexible low-cost technologies that can be tailored to the demands of consumers. There is growing acceptance of the need for increasing the use of renewable energy, but this introduces more uncertainty into the supply of electricity as a number of these sources (such as wind or solar photovoltaic) are intermittent and are unlikely to become a dominant part of our energy mix.

How Australia chooses to meet continued consumer load growth for both peak and base demand is a central issue. Energy efficiency and conservation—which are covered in the following Chapter—are key initiatives that are being taken up in Australia. They will have the effect of reducing the base-load requirements from generators, thereby reducing daily emissions. Clearly there is a need to better understand the characteristics of electricity consumption and better match the supply side to consumption. Distributed Energy provides the means to do this by addressing the peak demand, infrastructure cost, asset investment and supply security issue in a more sustainable way than continued expansion of the Centralised Energy model.

What is distributed energy?

Distributed energy is a suite of technologies that aim to reduce the reliance on a centralised supply of energy, reduce emissions and improve energy

Table 1: Domestic energy use

Use of energy	Percentage of energy	Sustainable technology solution
General appliances	29	Standby power reduction
Heating and cooling	26	Solar thermal with gas boost (new technologies)
Hot water	20	Solar thermal with gas boost (existing technologies)
Fridge / freezer	12	Powered by renewables
Lighting	9	Use CFCs
Cooking	4	Gas cookers

Source: *Weekend Australian* 2007

use efficiency. They are all relatively small in capacity (less than 30 MW) and are typically sited close to the point of consumption, thereby avoiding the construction of expensive transmission and distribution assets to get the power from remote locations and the associated energy losses that ensue (see Figure 2).

The technologies range from gas-reciprocating engines and micro turbines to renewable sources of energy, such as solar photovoltaic and wind generators. Waste-heat recovery technologies are important here as they increase the efficiency of many devices. Waste energy can be recovered and reprocessed to supply heating or cooling or be re-used to produce further electricity (the most common form today being co-generation, where waste heat is used in industrial processes and can be used for community heating). Distributed energy uses devices to provide power, heating and cooling that are usually either renewable in nature or highly efficient engines with waste heat recovery (CHP) that double the efficiency of fuel use compared with a conventional coal-fired power station. Distributed Energy devices generally use fuels that have lower greenhouse emissions than coal. The fuel may be natural gas or LPG, which is 20 to 30% cleaner than burning coal or oil. In other cases, the distributed generation unit will be wind or solar based, so there is no fuel combustion to contribute to CO_2 emissions. Around the world, co-generation (simultaneous generation of heat and electricity) is an accepted standard, particularly in bio energy and industrial applications. With further research and development these systems will be available for urban communities as centralised facilities (such as the district heating schemes in Scandinavia) and in residential applications with small, low-cost units that replace conventional boilers. The downside of these devices is the high price, operational noise and physical size, but they are predicted to be in widespread use in urban environments in the next 5 years.

Residential solar power is a well proven technology, but will probably be taken up only when subsidised, due to the high capital costs involved. Progress is being made with integrating solar technology into the building fabric so, assuming an economic price can be achieved, we will see integrated solar panels and films built into environment fabrics in the next 5 years.

Wind generation has some way to go as an acceptable residential power source although it has already been integrated into some communities and small industrial sites successfully. For example, the 4 MW Hepburn Community Wind Park is expected to generate enough energy to power most of the homes in the Daylesford Township. The local community and industry will establish an investment cooperative to finance the wind park (Sustainability Victoria 2007).

Many developers are looking to solar and wind power generation to create an image of being 'clean and green'. The advent of electric vehicles and hybrids will give urban developers the opportunity to use vehicle-based storage technology to further supplement domestic demands. Coupled with renewables DE will potentially give the opportunity for most households to be energy self-sufficient for most of the time—only requiring a grid connection to sell surplus energy back to the grid and for occasional times when self-generation is inadequate.

Another aspect of DE is its ability to manage demand by switching loads on and off—thereby influencing peak demand—at certain times of the day, thus flattening the load curve. This has the

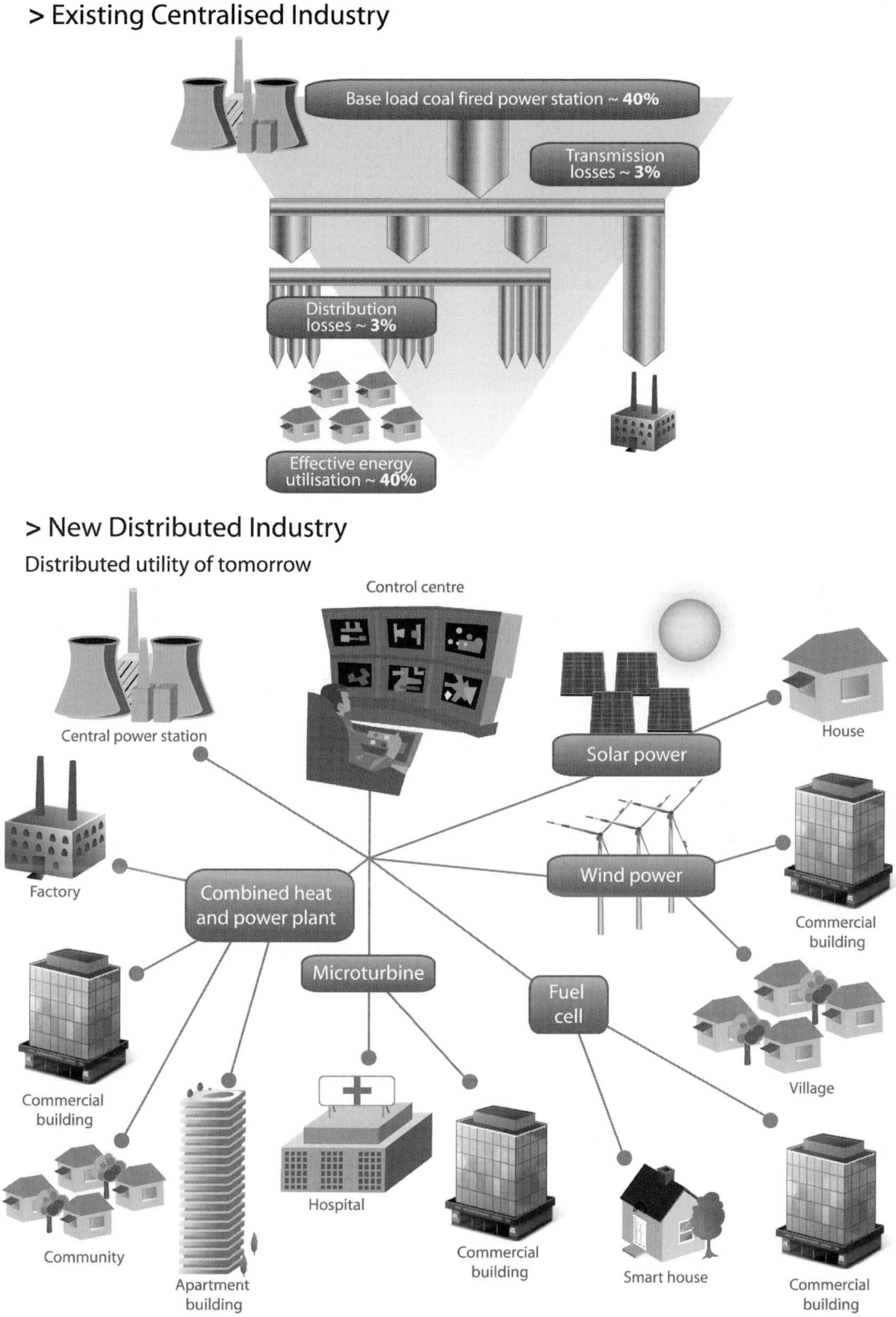

Figure 2: Shift from Centralised to Distributed Energy

effect of reducing the amount of power required from generators and the transmission and distribution network infrastructure at times of peak price (generation shortage) or during network overload situations (network constraints). Due to the small size of these DE units, aggregation and intelligent-control mechanisms are usually required to deliver demand response (peak price response and network overload services) back into the market.

DE is technically feasible today and, assuming supporting financing arrangements and tariffs can be agreed, this platform could change the face of energy supply and demand forever.

Distributed energy transition

To build a pathway for sustainable urban development we need to shift from our supply-dominated energy system to a situation where consumers take greater responsibility for their demand and manage it in more efficient ways. Consumers need to be rewarded for taking this role by receiving capital investment support and energy payments commensurate with the true avoided cost of additional generation, transmission and distribution.

It is clear that the true price of energy (which includes costs to the environment) has to rise in the future—the only question is by how much. The transition to sustainable urban energy development will then be driven by how quickly the real cost of energy is passed on to consumers. The rest of this chapter presents details of how the transition to DE could progress and provides pathways to sustainable urban energy developments.

Energy hubs and community self-sufficiency

Urban environments that integrate mixed-use development (that is, industrial and residential land use) have the greatest opportunities for energy efficiency and self-sufficiency. Industrial processes requiring energy in the form of heat or electricity would drive the establishment of local generation facilities from which the residential population could benefit. Surplus electricity generation would be available and waste heat could be used to provide hot water and drying facilities (with modern non-electrical appliances) It remains to be seen whether cost-effective ways can be found for using waste heat for cooling, but technically this can be demonstrated today. Over the next 10 years the probability of producing a cost-effective cooling device is reasonably high.

Aggregation

Systems and services will become available to aggregate surplus energy from self-generators and offer this with load reductions (Demand Management) to the electricity market, so that aggregators can be rewarded for helping to balance the supply and demand of electricity. The aggregation of small generators and loads, and their orchestration by service suppliers, has yet to be fully demonstrated in Australia, but this is seen as a key element of network design in other countries where CO_2 mitigation and energy supply security are the driving forces.

Comparison of performance of distributed versus centralised energy systems

As discussed earlier, current generation efficiency from coal-fired power stations is only about 30–40% (that is, 60% of the fuel is wasted). The efficiency of our transmission and distribution systems is good, but even so 6–8% losses are considered normal (ABB Inc. 2007).

A financial perspective on this, in Australia, would be that out of the total 196 thousand GWh being supplied into our T&D infrastructure worth $11,000 billion, the T&D losses equate to $660 billion per year. (NEMMCO Annual report 2007, page 17).

Today's overall losses calculation is quite disappointing:

- 60% of energy input is lost in the generators
- 6% more of the remaining 40% is lost through the T&D system, leaving 37% at the point of consumption.
- the average efficiency of domestic appliances is approximately 40%, so that households usefully use only 22% of the energy put into our supply system!

We waste 78% of our energy through this supply chain!

Urban households can be guided towards employing different technology options for home appliances that could halve their emission footprint of typically 14 tonnes CO_2 per year.

Utilisation of DE solutions close to point of consumption would eliminate the 8% T&D losses and by using more-efficient DE devices we could even achieve utilisation of 50% of our fuel energy input.

Residential energy consumption accounts for approximately 27% of total energy used in Australia. This represents one of the smaller segments of energy use in Australia, but nevertheless accounts for over 80% of all energy consumers (NEMMCO 2005) and, due to its 'peaky' demand, requires the creation of large amounts of poorly used capacity in centralised generation. For urban communities to grow sustainably, we need to provide technologies that remove peaks and obviate the need to build further centralised capacity that is needed for only a few hours each year.

The most logical source of local generation in Australia is photovoltaic technology, which is still very expensive and requires government subsidies to encourage its uptake. However, the widespread use of zero emissions technologies based on solar would require re-design of the electricity market to allow intermittent energy sources to be scheduled to address peak demand as opposed to large amounts of capacity being held in reserve in case that demand arises

Small- to medium-scale wind generation is an equally viable intermittent generation option. While proponents of wind power face significant environmental issues, design of alternative turbine mechanisms should minimise the visual and environmental impact. This may reach the point where wind-powered generation is acceptable on the roof of the domestic home.

Using wind and solar as primary power supplies provides a pathway for future sustainable urban energy development, although there will always be the need for a connection to a non-intermittent power source (such as storage or to the grid).

In a future where supply and demand for energy may become increasingly out of balance, due to rising demand and shortfalls of generation, consumers will find that either the price for retail electricity will rise or they will be forced to switch off loads to reduce the likelihood of loss of supply. Consumers will respond in many ways, depending on their technical ability, attitudes to energy use and the environment and the capital they have available to spend on alternative supplies to minimise the effects of these price hikes and power losses. To achieve a more sustainable urban community in the future it is clear that members will need to have options to generate their own power from solar, wind or low-emission engines that are located close to their homes and can be controlled by the consumers themselves. It may well be that they sell this right to the retail utility and in return are rewarded for demand management actions which will tend to flatten their load profile (removing the peaks). Some communities may form their own generation and demand management cooperatives that can respond to price incentives for reducing load and the prospect of energy supply outages as well as being able to minimise the effects of rising prices and enhancing supply security.

Current level of diffusion of DE in Australia

Australia generally lags the rest of the developed world in terms of the percentage of local generation in the overall mix. The BCSE (BCSE 2006) estimates that distributed energy, excluding hydro-and non renewable fuels, accounts for less than 2% of Australia's energy mix. This compares with 7% in the USA and Europe, where there are increasing concerns about fuel supply and where governments are turning to renewable energy and demand management.

From an urban planning and development point of view, this provides an opportunity for property developers to take advantage of current infrastructure funds so that they can install the necessary technologies, avoid paying large charges for connection costs, and thereby avoid continuation of the current centralised supply-dominated system.

It is likely that energy efficiency rating systems, carbon footprint and reliability of energy supply issues will increasingly become selling points for developers who wish to attract consumers to their properties. The differentiating characteristics of these new developments will be that they are environmentally sustainable, highly efficient and providing a measure of control by the consumer

In developing countries, there are also significant opportunities for sustainable urban development along these lines. The current challenges in these countries are rapid urbanisation and the supply of energy infrastructure to support rapid urban growth. Countries such as China could exploit the advantages and benefits of distributed energy as a deferral mechanism until they have time to build large infrastructure transmission and distribution networks. This means that urban

Table 2: Examples of funding sources for distributed energy projects

Low Emissions Technology Demonstration Fund (LETDF)	This fund supports the commercial demonstration of technologies that have the potential to deliver large-scale greenhouse gas emission reductions in the energy sector and will operate from 2005–06 to 2019–20.	$410 million has been allocated out of the announced $500 million in government funding and these projects will involve total investments worth over $3 billion
Renewable Energy Development Initiative (REDI)	This initiative is a competitive merit-based grants program supporting renewable energy innovation and commercialisation and it will operate from 2004–05 to 2010–11.	$100 million in government funding to match private sector funding for each project
Solar Cities	This program is designed to demonstrate how solar power, smart meters, energy efficiency and new approaches to electricity pricing can be combined. The program will run from 2004–05 to 2012–13.	$75 million; will leverage significant private sector funding
Renewable Remote Power Generation Program	This program provides financial support to increase the use of renewable generation technologies in remote parts of Australia that currently rely on fossil fuel for electricity generation. This is typically a 50% rebate of initial capital costs for eligible projects. Funding will be provided until 2012.	$328.5 million

Source: Task Group on Emissions Trading 2007

communities could start with small 'islands' of power supply, which would provide the model for the most economic method of introducing distributed energy into their community. It is highly likely that such urban communities would subsequently decide to continue supplying their own energy into the future.

Barriers and pathways to DE

Recent work by the IEA Demand Response Resources Task XIII (IEA 2006) suggests that there are significant barriers in Australia inhibiting the growth of businesses concerned with DE and demand management, as well as more broadly based community participation in DE. These range from the physical connection issues to all the incumbent requirements for participation in the electricity market which are simply too great for the average consumer to contemplate. These include technical issues, such as the metering and protection requirements and device testing standards that need to be met to connect a DE unit to the grid.

Most of Australia's large cities have significant amounts of emergency stand-by generation that could be deployed in the event of price or demand peaks. These units are currently not permitted to operate, due to EPA environmental pollution regulations despite the availability of simple technical solutions (for example, gas conversion and exhaust gas scrubbing).

There are also significant commercial barriers and interests that are not supportive of distributed energy. Most utilities in Australia have been deregulated, and operate as fully commercial businesses. Retailers are keen to supply more energy at higher prices, as they need to minimise their wholesale trading risk. Network businesses want to maximise their asset base. This will have a direct consequence on consumers, who, when supplied with smart meters, will be exposed to additional costs due to peak pricing tariffs as a means of being persuaded to reduce their load during peak times. If sustainable urban development is to be achieved, then new connection agreements and pricing agreements that encourage consumers to generate their own power and share surpluses with the grid must be drawn up by the utilities and government.

Positive steps towards sustainable urban communities

There are a range of steps that can be taken now to prepare a new infrastructure model for our future urban communities. The following technologies need to be embedded in our urban structures if this new model is to be created and sustained:

- Smart meters that provide valuable feedback to consumers on price signals and usage data in timeframes that can be used to adapt and improve demand patterns.
- Communication infrastructure that can carry telecommunications traffic (such as TV, phone and security), as well as energy metering and demand information, to every resident
- Network connections to be configured in a flexible way, so that local generation can be connected to the grid to feed areas of high demand at peak times
- Local generation capacity (such as solar, wind and micro-engines) needs to be embedded in urban communities to allow residents to choose their energy source and provide flexible supply and demand scheduling to the locality
- Intelligent energy-management systems working need to be deployed at the urban distribution level both to manage demand from individual residences and small businesses and also to manage demand and supply to whole communities. Consumer awareness and choice will become important here.

Specific Australian regulatory issues also need to be tackled if the correct policy framework is to be created supporting low-emission urban communities:

- Renewable and low-emission sources of energy generation need to be easily connectable to the home and grid. New standards need to be put in place and utilities should encourage self-generation
- Feed-in tariffs for surplus energy generated by residential self generators need to be established to offer improved benefits for selling surplus energy
- Standards need to be set for encouraging small local energy and service providers who can bundle services to urban communities (such as electricity, heating and cooling).

A shift in focus from supply-led to consumption-led energy development provides a pathway for sustainable future urban community development that no longer relies on large inefficient centralised systems with generation situated long inefficient distances from the place of demand.

Case studies

There are an increasing number of examples of pioneering applications of distributed energy in urban settings. The following are brief case studies in a residential setting and a commercial building.

Residential case study: GridX, Vision Estate, Glenfield, NSW

MIRVAC and GridX Power launched *Vision Estate in Glenfield*—Australia's first estate that generates three forms of energy from one resource (natural gas). The technology is known as tri-generation and uses natural gas to produce electricity from a 160 kW generator while capturing hot and cold water to re-use in houses for climate control (see www.gridx.com.au; Figure 3).

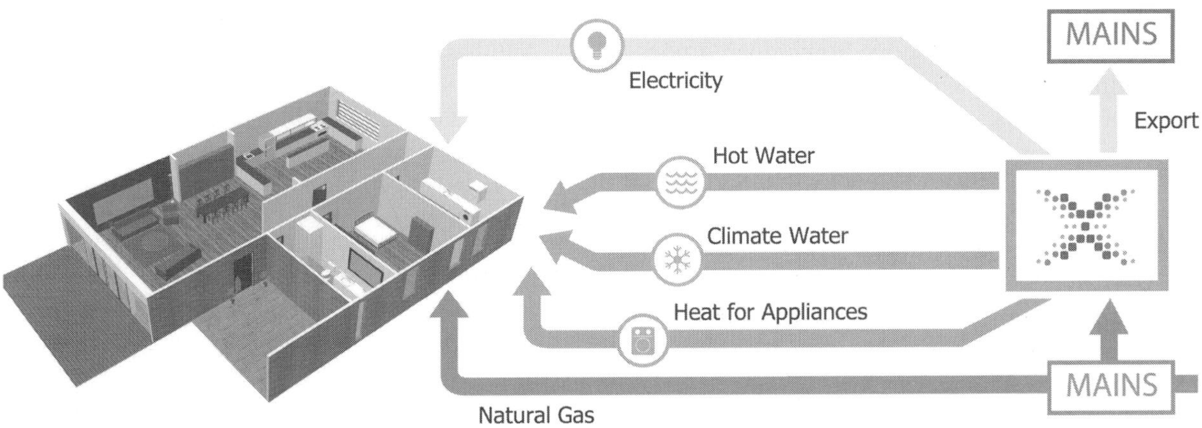

Figure 3: System layout of a GridX house.

	Traditional Grid Power	GridX Power
COST	In most locations in (Australia) delivered grid power cost between 7 and 18 cents/kWh	The GridX Power system with a good fuel cost/availability can be under 10 cents per kWh. ✓
FUEL EFFICIENCY	Delivered grid power is less than 35% efficient depending on the mix of power plants used.	The GridX Power system with CCHP will be 70 - 80% efficient and even more so with the delivered chilling. ✓
RELIABILITY	Grid power is on average 99.97% reliable.	The fully integrated GridX Power system can achieve even higher reliability than the mains grid. ✓
EMISSIONS	Traditionally, Australia relies on brown and black coal power stations which have very high CO_2 emissions.	The GridX Power system relies on natural gas which by comparison emits about 33% less emissions than the grid for the same given output*. ✓
SECURITY	Damage to a few major transmission lines can cause regional blackouts affecting millions of customers.	Severe damage to any portion of the GridX Power system may only affect that section of homes. The GridX Power system is able to operate any portion of its infrastructure independently and in isolation to minimise inconvenience. ✓
CONSTRUCTION CONSTRAINTS	It is difficult to build new transmission lines and substations because of NIMBY ("Not In My Backyard Syndrome") and the high costs involved.	The GridX Power system can be a solution to overloaded and constrained T&D systems. The GridX generators are designed to be out of sight and discretely operated upon the development. ✓

Figure 4: GridX versus traditional grid power.

- By April 2007, 16 houses had been using the technology for 12 months, with stage two of the development expected to be underway by the end of 2007 (another 200 houses)
- Three energy generators are hidden underground and work independently of the main electricity grid
- Excess power is exported to the grid to meet peak demand requirements and greenhouse credits (NGACs) are gained by Integral Energy (GridX's partner). Integral pays GridX for the avoided transmissions tariff
- There was an immediate 33% reduction in greenhouse gas emissions compared with coal-fired power stations
- This is equivalent to taking 864 cars off the road (for full 216 houses) Residents pay 10% less than for conventional electricity (ARF Article, Apr07)
- Hot and cold water are used to supply specially modified appliances (GreenMark), such as clothes driers and dishwashers which have had the electrical elements removed and replaced with fan coils. All houses have ducted air conditioning
- There were 40 blackouts in the Campbelltown area between October 2006 and January 2007, none of which affected the GridX residents
- GridX claims that with good fuel costs and availability, power can cost under 10 c/kWh (compared with 7–18 c/kWh from the grid)
- The project is BASIX and Energy Star Compliant.

A summary of benefits is listed in Figure 4.

Council House 2

Local government and sustainable outcomes – CH2

Fifty-one percent of the world's population live in cities and cities are where the community, business and government interact daily. Sustainable cities need strong local government – local government that can facilitate sustainable outcomes.

The City of Melbourne is committed to a zero emission target by 2020. Areas of focus in the Melbourne greenhouse gas equation include a 50% reduction from building design and 15% from industry efficiency gains; growth of renewable purchases; and doubling the efficiency of non-renewable supply.

The energy efficiency of commercial buildings is a key part of the strategy. The City of Melbourne's own Council House 2 (CH2) project is a lighthouse development that establishes leadership and commitment to the strategy (Figure 5).

Council House 2 (CH2)

CH2 is a visionary new building with the potential to change forever the way the industry approaches ecologically sustainable design. It is a building that is innovative, creative, technologically advanced, environmentally sustainable and financially responsible. CH2 has sustainable technologies incorporated throughout its 10-storey structure.

The total gross floor area of the building is 12,536 m^2, with a typical floor of 1064 m^2. The total building cost is $51.045 million of which $11.3 million has been identified as 'sustainability features'. This leading edge six-star building rated by the Green Building Council of Australia (www.gbca.com.au) will provide considerable environmental savings in areas such as energy and water use as well as deliver a high-quality indoor working environment.

When compared with the City of Melbourne's previous 1970s Council House, CH2 is expected to reduce electricity consumption by 85%, gas consumption by 87%, reduce water mains supply by 72% and produce only 13% of the greenhouse gas emissions (greenhouse gas emissions will be 64% less than a five-star building).

By reducing water and energy consumption, CH2's reliance on public infrastructure will be small. A comprehensive eco audit of the materials used in CH2 assessed all aspects of the manufacture and transportation of materials in relation to their effect on the environment and the occupants of the building.

Although the reduction in energy costs will be substantial, the greatest economic benefit is expected to be in increased productivity, reduced absenteeism and lower staff turnover rates, which cost employers millions of dollars each year. Studies have shown the improved air quality provided by systems such as those employed in CH2, could achieve a 4.9% increase in productivity—in part through reduced sick leave. It is predicted this will save the City of Melbourne up to $1.12 million a year. A business case assessment has estimated the payback period for the 'sustainability features' to be between 8 and 10 years.

Although most of the principles adopted in the building are not new, such as using thermal mass for cooling and using plants to reduce glare, never before

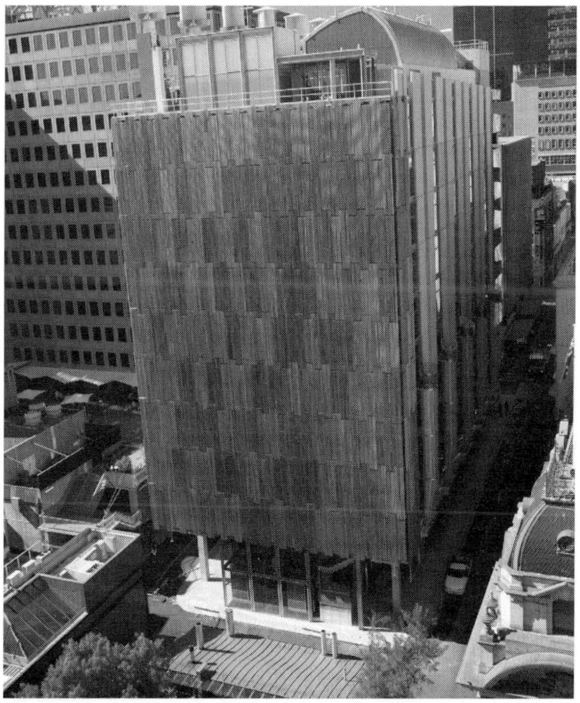

Figure 5: The CH2 building showing recycled timber shutters and six yellow wind turbines on the north side of the roof, which extract air from the offices spaces through ducts on the north façade. The turbines, which were especially designed for CH2, are 3.5 m high and replace electric fans

in Australia have they been used in such a comprehensive, interrelated fashion in an office building.

The City of Melbourne began construction of the new building in 2004 and has occupied the building since October 2006.

How it works

CH2 has been designed to be a holistic and all-inclusive system, with its occupants as participants. The design follows a model that promotes a more interactive role between the city and nature—acting more like an ecosystem in which all parts depend on each other.

The first focus of the design process was an understanding Melbourne's climate and weather patterns. In the roaring forties, pressure cells move from west to east causing Melbourne's famous north to south shift—the hot and dry desert wind from the north suddenly shifts to the southern cold wind. Despite high temperatures in summer—Melbourne's climate is normally cool-temperate.—Melbourne's climate is very variable: a problem that was turned into an opportunity enabling the building design to be developed around cold energy storage.

Features of benefit to staff include:

- a comfortable temperature range of 21 to 25 degrees Celsius (best practice range for office conditions)
- 100% fresh air supplied from the floor, with two air changes every hour
- no recycling of used air, or mixing of fresh and used air
- shared space on every floor, with access to outdoor balconies
- a rooftop garden
- increased bike storage and associated lockers, changing rooms and showers.

Heating and cooling

The air-conditioning system has a number of features, including:

- thermal storage batteries of Phase Change Materials (PCMs) used to capture cool conditions at night for use during the day for office cooling
- vaulted concrete ceilings that have a high thermal mass, thus modifying temperature and contribute to comfort conditions within the office space
- 'night flushing' of the building, allowing cooler night air to remove heat from the surface of the concrete ceilings via high-level windows that automatically open when the external air temperature falls below the temperature of the ceilings
- chilled ceilings, rather than providing chilled air directly to the office space
- evaporative cooling in the form of 'shower towers' on the building's facade
- recycled timber shutters that automatically close to protect the west elevation from the afternoon sun.

Energy

A gas-fired co-generation plant on the roof is used to generate electricity and heat, reducing reliance on the public electricity grid. The co-generation plant has much lower CO_2 emissions than coal-fired electrical generation and provides 60kVA of electricity—meeting about 30% of the building's needs.

Heat from the co-generation pant is used to assist the building's air-conditioning plant. This heat is used directly for heating or, via an absorption chiller, for cooling. It is estimated that the co-generation plant satisfies 80% of the building's fresh air heating/cooling requirements by using waste heat.

CH2 use about 26 m^2 of photovoltaic cells on the roof to generate about 3.5kW of electricity. Six wind-driven turbines, which assist with the exhaust air system (especially at night), are also used to generate power.

About 60% of the buildings hot water supply will be provided by 48 m^2 of solar hot water panels on the roof, which are supplemented by a gas boiler.

Air and light

On the building's north and south facades, the walls are alternate vertical bands of glass between thick concrete walls containing supply air ducts (on the south side) and exhaust air shafts (on the north side). To facilitate air supply and exhaust, the shafts widen towards the upper levels. Conversely, the windows are wider at street level and become narrower higher up the building. This achieves the dual goals of admitting more light to the lower

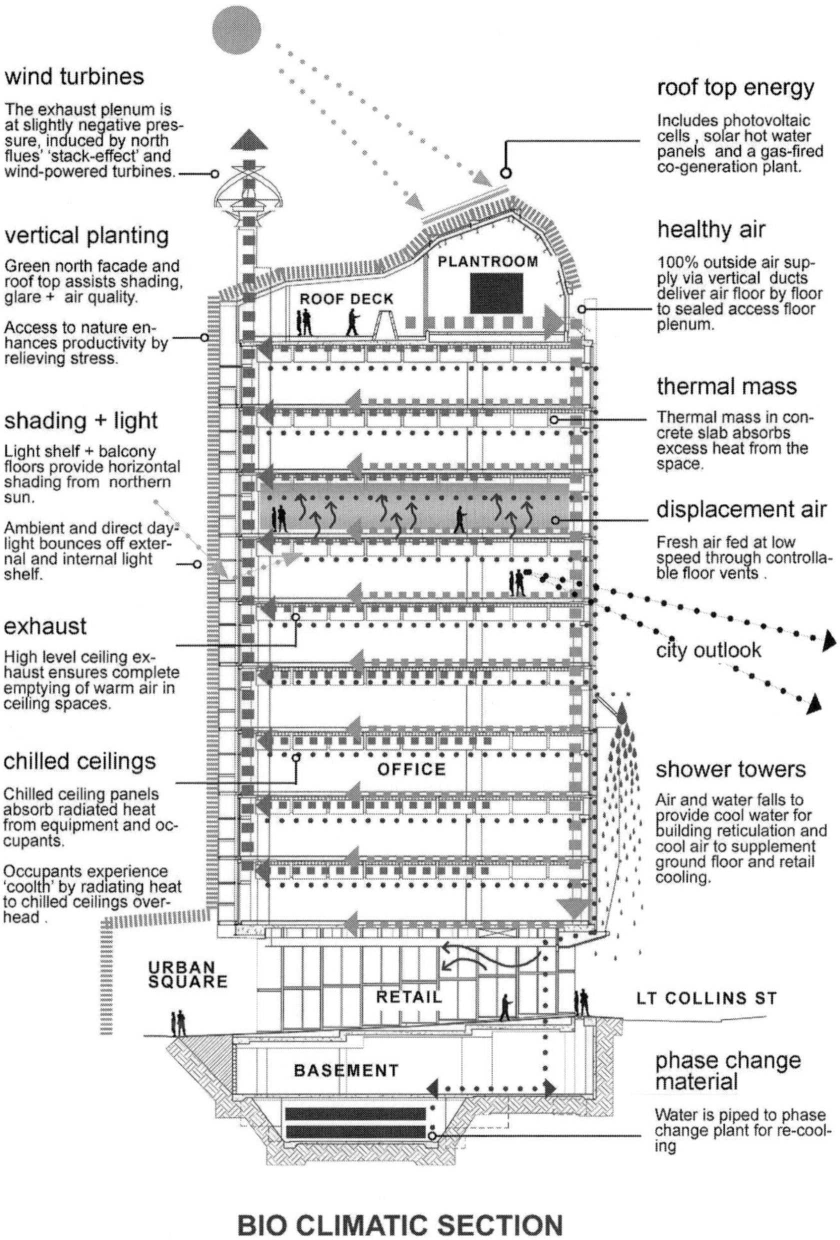

Figure 6: Green building features of CH2.

levels, where external light is restricted by surrounding buildings and where there is less air demand at the end on the air ducts.

Other features include:

- fresh air supplied from the floor, not the ceiling, and diffusers located on movable floor tiles allowing individual placement to suit workstations
- north facing windows shield direct sunlight and deflect light into the internal office space
- south facing windows admit reflected light, while at the same time reducing internal glare
- artificial lighting allows individual adjustment to suit tasks and ambient light is designed to support VDU usage with minimum glare.

Water

The most significant component of the water recycling initiatives in CH2 is the water mining project, which will deliver up to 100,000 litres per day of Class A water reclaimed from the from building waste and from the sewer in the adjoining Little Collins Street. This water will be used within the building for toilet flushing, plant watering and in the cooling towers and a considerable amount will be exported for use in park and street tree irrigation, fountains, street cleaning, and so on.

Outcomes

It has been estimated that commercial buildings in Australia generate more than 35 million tonnes of CO_2 a year (Property Council of Australia, 2001). The City of Melbourne's CH2 project not only establishes a benchmark in designing for reduction in greenhouse gas emissions and energy efficiency but provides a first-rate working environment for its occupants.

Conclusion

If we are to adopt a pathway to sustainable urban development, which reduces the pressure for more large-scale centralised generation and reduces our emissions of greenhouse gases, then we have to think differently about the supply–demand relationship for energy.

There is an alternative—distributed generation—which embodies a new model of supply and demand, where more choice and control is passed to consumers and local communities and the benefits of reduced emissions and lower-cost energy supply are enjoyed by all. It heralds a new era of business opportunity for utilities and responsible urban development.

References

ABB Inc. Corporate Communications (2007). Energy efficiency in the power grid. Report 1080. Norwalk, CT. <http://www.nema.org/prod/technologies/upload/TDEnergyEff.pdf>

Australian Business Council for Sustainable Energy (2006). Clean Energy Report, May 2006. Melbourne, Australia.

Blyth A (2006). Developments and challenges for energy distribution in Australia. Speech to New Zealand Electricity Networks Association (ENA) CEO Forum. Auckland, New Zealand. <http://www.ena.asn.au/udocs/ena_092106_104019.pdf>.

Boshier J (2007). Australia's electricity generators release crucial study on emissions. National Generators Forum (NGF). Media Statement. Canberra, Australia.

Ecogen (2007). *Ecogen magazine*, **Mar/Apr 2007**, 28–29

Hoch L, Chattopadhyay D, Fazio J, Fulford B and McCall K (2006). Assessing the value of demand response in the NEM. Final Report by CRA International. Melbourne, Australia. <http://www.demandresponseresources.com/>.

National Electricity Market Management Company Limited (NEMMCO) (2005). An Introduction to Australia's National Electricity Market. Melbourne, Australia. <http://www.nemmco.com.au/000-0187.pdf>

National Electricity Market Management Company Limited (NEMMCO) (2007). Annual Report 2007. Melbourne, Australia. <http://www.nemmco.com.au/000-0253>

Prime Ministerial Report of the Task Group on Emissions Trading (2007). Australian Government, Department of the Prime Minister and Cabinet. Canberra, Australia.

Sustainability Victoria (2007). Community Wind Park First in Victoria. Media Release. Melbourne, Australia. <http://www.sustainability.vic.gov.au/resources/documents/Media_Release_Daylesford_Community_wind_Park_0806.pdf>.

Weekend Australian (2007). How we use energy at home. *Weekend Australian*, **3 March**.

27

Energy-Efficient Housing and Sub-division Design

Michael Ambrose

The benefits of energy-efficient residential design and attempts to appropriate the associated principles and practices into new housing development are being increasingly recognised by both the house building industry and government. Recent inclusions in the *Building Code of Australia* have tried to address the historically poor performance of Australian housing in terms of energy efficiency, with mandatory energy performance standards now in place nationally. In addition, state governments have introduced their own requirements to complement those within the national Building Code. These have included tougher energy performance standards, mandatory solar hot-water services, rainwater tanks and energy-efficient lighting and appliance requirements.

Australia's record on tackling residential energy efficiency has been lacklustre. Dwellings tend to be poorly designed for both energy and water efficiency and our love of the detached house on the 'quarter acre' block has resulted in sprawling low-density cities with outer suburbs poorly serviced by the public transport network, resulting in a heavy dependence on private cars. Consequently, the energy consumption figures for Australia tend to be among the highest in the world. Latest estimates put our energy use at 265.8 GJ/person—significantly higher than the United Kingdom's average of 163.8 GJ/person, but still lower than the USA, which averages 341.1 GJ/person (Australian State of the Environment Committee 2006). Unfortunately, Australia's heavy reliance on fossil fuels, especially coal for electricity production, has resulted in Australia's greenhouse gas emissions being the highest in the world on a per capita basis. Data submitted to the United Nations Framework Convention on Climate Change, estimated Australia's greenhouse gas emissions to be 26.1 tonnes of CO_2 equivalents per person, compared with the USA total of 23.4 tonnes/person (United Nations Framework Convention on Climate Change 2004).

Improving the energy efficiency of our homes is a critical step in reducing Australia's overall energy use profile. Energy efficiency is one of the most cost-effective ways of tackling global warming issues and slowing our energy consumption rates. Existing technologies and energy-efficient design methods already exist and have been well tested and proven in Australia and internationally. Incorporating them more widely into local building practices, and having them accepted by the general public, are the principal challenges that lie ahead.

Key principles of energy-efficient design

Energy-efficient design is not a new idea. Its principles have been in existence for decades and have remained virtually unchanged. Indeed, many of the principles are taken from traditional building practices that date back centuries. Before the advent of mechanical climate control systems and artificial lighting, dwellings had to be designed to cope naturally with the environment in which they were located. For example, for centuries Eskimos in Alaska built semi-earth-covered homes that used the soil as a highly effective insulating barrier against the extreme cold of an Alaskan winter. Early European settlers reported that the inside of these these traditional homes remained between 15 and 20°C using only two or three seal-oil lamps and that the homes employed a natural ventilation system that ensured that fresh air was always circulating through the dwelling (Seifert 2007).

In Australia, the idea of building structures designed for the climate is also not new. Early 19th century settlers quickly saw the need for wide verandahs to provide shade. Elevated light-weight homes, such as the 'Queenslander', have been around for over a century in the warmer climates. In the cooler southern states, too, the use of thermal mass has long been well understood and homes using mud brick and *pise de terre* type construction were introduced into Australia in the early 1930s by Justus Jorgensen who founded the Montsalvat artists' colony at Eltham in Victoria and whose work would later influence designers such as Alistair Knox.

The basic principles of energy-efficient design involve several key aspects:

- **orientation and siting**—maximise solar gain during the winter and allow shading during the summer
- **layout and zoning**—internal layouts that place living areas to the north and provide for good natural lighting and ventilation
- **insulation**—good insulation of walls, ceilings and floors to slow the transfer of heat
- **windows and shading**—appropriately designed windows to allow natural lighting and passive solar heat gain; shading devices that shield windows from summer sun
- **ventilation and draught proofing**—design to capture natural cooling breezes during summer and ensure that air conditioned spaces do not 'leak' to the outside through gaps
- **thermal mass**—correct use of thermal mass to store heat
- **landscaping**—to provide appropriate shading and channel breezes.

For modern sub-division developments, the issue of lot orientation is one of the key areas that can influence the operating energy performance of housing. Orientating and siting a home so that it can maximise solar heat gain during the winter and be shaded during the summer is one of the core principles behind passive solar design. House designs that incorporate such principles will have lower operating energy costs than houses that are orientated poorly.

To site a home correctly needs not only an appropriate design, but also an appropriate parcel of land. Urban developers need to ensure that modern housing estates are designed in such a way that they provide an optimal lot layout. Effectively, this means having lots along the east–west axis, which are sufficiently wide to allow unimpeded access to the north.

The basic principles of energy-efficient subdivision design are:

- Lots should be orientated with their long axis close to east–west or north–south, and with an adequate north–south capacity in all cases.
- To facilitate the best lot orientation, streets should also be orientated as close as posible to east–west or north–south.
- Use of southern slopes should be minimised, and preferably used only for larger lots.
- Where two-storey developments occur, additional space to the south will be required to provide adequate solar access for the southern neighbours (Loder and Bayly Consulting Group *et al.* 1991).

Again, like house design, these concepts are not new and have been trialled by developers over the last couple of decades. The *Kingsley With The Sun* development in Wannaroo, Western Australia, was developed in 1986 and was the first sub-division in Australia that sought to protect solar access for all dwellings (Issac 1989). Others followed, including *Bila Village* in Forest Hill, Victoria which was a demonstration project undertaken between the state government and the Housing Industry Association—the principal residential building industry group. This estate was the first in Australia to integrate both advanced sub-division planning techniques with solar-efficient estate design, ensuring that all lots were able to provide passive solar design opportunities to future housing. It is interesting to note that this development also incorporated 14 display homes that all achieved a 5-star energy rating (the maximum rating available), yet most of the additional homes that were constructed in the development followed traditional 'standard plans' provided by builders and failed to take full advantage provided by the site layout (Issac 1989).

Unfortunately, the majority of today's greenfield land sub-divisions continue to rate solar orientation well down on their priority list, with many giving it no consideration at all.

Sub-division drivers

The sub-division of land for residential use involves many steps, including the provision of new roads, building lots and services such as electricity, water and sewerage. Depending on the scale of the sub-division, it may also involve the provision of public open space (parks), community facilities (schools, sporting grounds) and retail areas. At the largest scale, a sub-division can result in the creation of an entire new suburb; and, in Australia, large master-planned communities have been established by specialised urban development companies. Many of these developments contain in excess of 10,000 home sites and concern has been raised about how sustainability issues have been addressed by the developers—not only at the allotment scale, but also in terms of incorporating mixed-use development and provision for public transport. Mead and Wales (2004) report that the main drivers for land developers when configuring allotments within development sites are:

- **yield**—overwhelmingly identified as the most important driver when configuring allotments within development sites. Developments involve significant investment and financial outlay and the projected timeframe for return on investment are paramount
- **location**—the location of the development site in relation to market demand is a prime consideration. To make the development financially viable, the location of the site has to be perceived as attractive to the local market. Larger developers will provide amenities and facilities within master-planned communities, while those with less resources, or with smaller development sites, will use locations near existing 'draw cards', such as schools, child minding centres and shops
- **topography**—one of the physical elements of a site that most influences yield as it presents environmental and engineering constraints unique to each sub-division development
- **zoning**—preference is for land that is self or code assessable
- **allotment size/density** (related to zoning)—in many instances there are local authority requirements that the developer provide low-density development while also providing services that cannot be easily supported by a low-density population
- **competition**—competition between developers can play a significant role in how developments are sub-divided, although with the recent development boom, competition has not been as critical.
- **aesthetics**—the appeal of mature vegetation and landscaped developments complement allotment sales. Even when all existing vegetation had been completely removed, it is normal practice to re-introduce vegetation for entry, roadway and parkland appeal.
- **supply and demand**—traditionally, land developers supply a familiar product to a familiar market – that is the project home and land package.

It is no great surprise that sustainability issues do not feature in these drivers, as fewer development requirements translate to lower costs; and consumers, especially first homebuyers, tend to buy where property is most affordable.

Sustainability barriers

Despite the apparent lack of interest in sustainability issues by land developers, many are beginning to understand the important role they play in delivering the sustainable suburbs of the future. However, according to Miller and Ambrose (2005), land developers have identified significant barriers to the widespread adoption of sustainability principles as part of their design process, including:

- the current energy-efficiency standards for residential construction can be achieved without any reliance on appropriate sub-division orientation (an indication that the current 5-star benchmark has been set too low)
- other than the star rating provisions of the Building Code of Australia, there is no clear

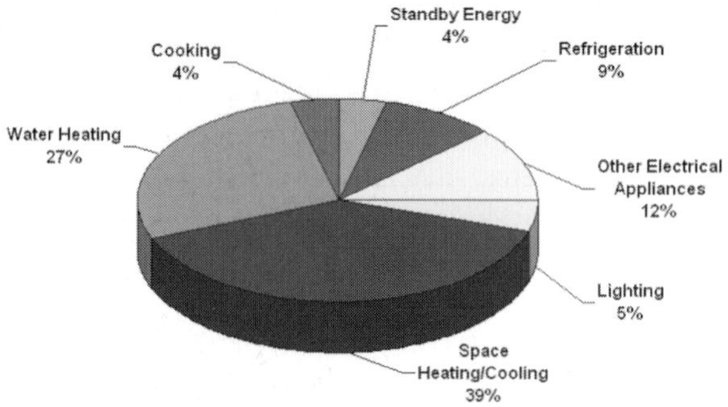

Figure 1: Average energy use in Australian dwellings (Source: Reardon 2001).

measure of how to achieve the necessary standard for planning sub-divisions, much less to strive for increased performance

- the absence of a benchmark for sub-divisions, so that developers can achieve a sustainable outcome without losing any competitive edge in the market
- an absence of tools that measure energy efficiency across an entire sub-division, and take into account factors such as construction materials, orientation, adjacent built forms, deciduous and evergreen vegetation, and how the home operates once occupied
- the lack of impact of sustainability/energy-efficiency information among consumers—industry will respond to consumer-driven demand
- the finance sector needs to recognise and value sustainable practices in order to overcome the financial barriers of up-front costs of implementing energy-efficiency measures

These barriers indicate that currently there is no clear understanding or path for developers, regulators and financiers to follow to enable energy-efficient and sustainable sub-divisions to occur as a function of contemporary practice. The current regulatory focus on buildings means that opportunities to improve energy efficiency at the urban planning stage are missed, and the relatively benign building energy provisions in the building regulations have meant that builders and designers have not needed to exert pressure on urban planners to provide more appropriately orientated blocks.

Urban developers have reported that financiers are a critical part in the development process and they tend to take a very conservative view when providing funding for property development. Any variation from the 'traditional' approach is likely to be subjected to very close scrutiny and developers may pay a premium for the additional 'risk' that financers may envisage.

Housing energy use

The use of energy (electricity and gas) for domestic purposes is the largest source of greenhouse gas emissions from Australian households and household energy use has been rising. In 2001, the average household's energy use was responsible for the production of about eight tonnes of greenhouse gases (carbon dioxide equivalent or CO_2-eq), per year (Reardon 2001). However, by 2006 this had risen to 9 tonnes—a 12% increase in just 5 years. Incorporating household transport in the calculations increases the levels even more—rising to 14 tonnes of CO_2-eq per household per year (Pears 2006).

Reducing energy consumption is by far the most practical and affordable way to reduce the greenhouse impact of residential development:

- Australia-wide, space heating and cooling, and water heating are the main energy consumers in residential dwellings (see Figure 1).
- Energy-efficient design that removes the need for heating and cooling systems and the use of energy-efficient lighting and appliances are solutions that are available immediately and often with little, if any, additional cost.

However, these national figures do not necessarily reflect energy consumption patterns in all climate zones. Traditionally, energy use in the warmer climate zones has been quite different from the pattern for the rest of Australia. Queensland's single biggest consumer of energy in the dwelling is not space heating, but hot water heating (Queensland Government 2004).

Figure 2 shows that space heating and cooling energy accounts for only 10% of the Queensland total, compared with 39% as an Australian average. This difference is due to the Queensland climate where homes have traditionally not had ducted heating or cooling systems like the southern states. However, this percentage is set to increase as Queenslanders install air-conditioning at an ever increasing rate. Indeed, data shows

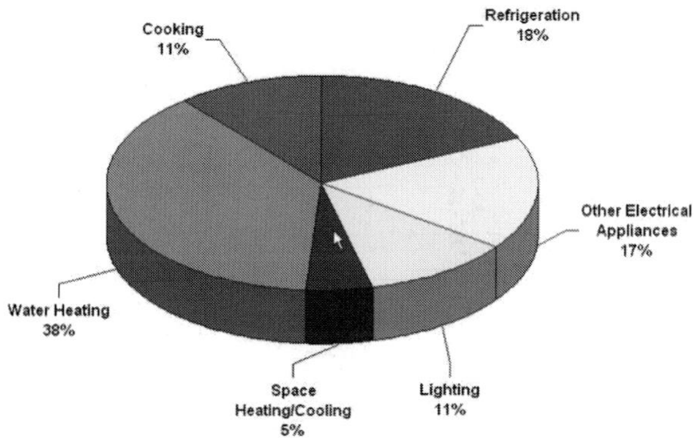

Figure 2: Queensland household energy use (Source: Queensland Government 2004).

that in 2001, around 28% of dwellings were air-conditioned. By 2004, this figure had increased to 36% and it is estimated that by 2007 more than 56% of Queensland homes were air conditioned (Mickel 2004).

Despite the introduction of energy-efficient requirements into building regulations, the actual amount of energy used by houses has increased over the last decade. Between 1983–84 and 2003–04, energy use in the residential sector grew by 52% or an average of 2.2% per year (Australian Bureau of Statistics 2006).

Much of this growth is linked to population growth, but it is also related to the growing size of the average home and the decreasing number of people per dwelling. Between 1984–85 and 2002–03, average house size increased from 162m^2 to 228m^2 (Figure 3) (Australian Bureau of Statistics 2005) while household size has been steadily decreasing and is currently just below 2.6 people per dwelling (Australian Bureau of Statistics 2007).

The long-term projections for energy use in the residential sector is that it will continue to

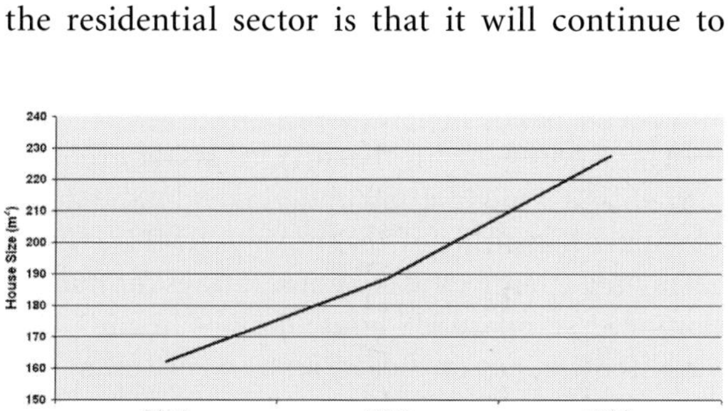

Figure 3: House size, Australia (Source: Australian Bureau of Statistics 2005).

increase at around 1.7% per year and that by 2030 the sector will be consuming 650 petajoules (1 petajoule equals 1 thousand million million joules) of energy or around 26.3GJ/person (Akmal and Riwoe 2005).

Currently, the majority of this energy comes from fossil fuel sources—both coal (via electricity generation) and natural gas. Together, electricity and natural gas account for 80% of a households energy consumption, with the remainder coming from wood, LPG and a very small proportion from solar energy (0.6%) (Australian Bureau of Statistics 2006). This high dependence on fossil fuels, and the trend of increasing dwelling size, but decreasing household size, presents enormous challenges for housing design into the future. The trend in energy-use patterns indicates that the majority of our energy requirements are used for space heating/cooling or water heating. Fortunately, both of these areas provide huge opportunities for energy reductions through existing technologies.

Traditionally, water heating has been provided by electric or natural-gas-powered hot water units. Although the trend has moved towards the more efficient natural-gas-fired units, these still use a fossil fuel and consequently produce greenhouse gas emissions. The obvious alternative is solar hot-water units, which use sunlight to provide the majority of the heating power and usually use an electric or gas booster system for the cooler periods of the year. Depending on the climate zone, solar hot-water units can provide in excess of 90% of a households hot water needs without the use of their booster systems. In states such as Queensland, a large scale adoption of such units could see a potential drop of 34% in fossil-fuel-based energy use.

The other significant energy consumer is space heating and cooling. Australia's relatively temperate climate has resulted in house building practices that have not focused adequately on thermally efficient buildings. Access to cheap energy has meant that heating or cooling homes has been easy to achieve through installed mechanical heating, ventilation and cooling (HVAC) systems. Data shows that even as recently 2005 only 61% of homes had some form of insulation (Australian Bureau of Statistics 2006).

New energy requirements for the Building Code of Australia (BCA) came into effect during May

Table 1: Comparison of Australian and USA insulation requirements for dwellings

Location	Building component	Australian R-value	USA R-value
Darwin	Ceilings	R 1.5	R 5.3
	Walls	R 0.9	R 2.3
	Floors	Nil	R 2.3
Brisbane	Ceilings	R 1.5	R 5.3
	Walls	R 0.5	R 2.3
	Floors	Nil	R 2.3
Adelaide	Ceilings	R 2.9	R 5.3
	Walls	R 1.0	R 2.3
	Floors	R 1.0	R 3.3
Perth	Ceilings	R 2.9	R 5.3
	Walls	R 1.4	R 2.3
	Floors	Nil	R 3.3
Sydney	Ceilings	R 2.5	R 5.3
	Walls	R 1.0	R 2.3
	Floors	R 1.0	R 3.3
Melbourne	Ceilings	R 3.4	R 5.3
	Walls	R 1.7	R 2.3
	Floors	R 0.3	R 3.3
Hobart	Ceilings	R 3.4	R 5.3
	Walls	R 1.4	R 2.3
	Floors	R 0.3	R 5.3
ACT	Ceilings	R 4.0	R 5.3
	Walls	R 1.9	R 2.3
	Floors	R 0.8	R 5.3

Source: ICANZ 2007

2006, which included the 5-Star energy rating system for homes. The BCA 2006 5-Star regulations rate the thermal efficiency of the building envelope (that is,its ability to resist the flow of heat) which is stated in terms of R-value. The higher the R-value number, the higher the envelope's thermal efficiency. The new regulations lift R values by only R0.5 in a typical home and when compared with European or US building standards, Australia is still well behind. A study undertaken by the Insulation Council of Australia and New Zealand (ICANZ) found that the new requirements lifted Australia from around 40% of the USA minimum standards to just over 50%, not even close to the lowest energy-efficient standards in the USA (ICANZ 2007). The study benchmarked Australian building regulations for the installation of thermal insulation in new homes against comparative US regulations for homes in equivalent climate zones. The city-to-city, climate-to-climate zone comparisons found that Australia has fallen sharply behind the USA in requirements for thermal insulation and performance of new homes.

The findings from the ICANZ study are supported by a similar study undertaken by the Royal Melbourne Institute of Technology (RMIT), which calculated energy efficiency star ratings for a selection of 51 standard US house designs in comparable climate zones to those found in Australia. The designs all met the energy standards required in their respective areas and were rated with AccuRate—the software required for BCA compliance.

Table 2: Star rating results for US house designs

Australian equivalent climate zone	Comparison location	Total number of plans rated	AccuRate stars: range	AccuRate stars: median	AccuRate stars: mean
Zone 1 Darwin	Florida	6	6–8.5	7.8	7.0
Zone 2 Brisbane	Texas	5	4.5–9	5.0	6.0
Zone 3 Longreach	North Carolina	5	4.5–6.5	5.5	5.4
Zone 4 Dubbo	Arizona	4	6.5–7.5	7.0	7.0
Zone 5 Perth	California (Bakersfield)	3	7–8	7.5	7.5
Zone 6 Melbourne	California (San Francisco Bay)	4	6–9	7.8	7.6
Zone 7 Hobart	UK/Canada	16	6.5–8.5	6.8	7.2
Zone 8 Thredbo	Pennsylvania: Massachusetts	8	4.5–9.5	6.5	6.8
ALL ZONES	–	51	4.5–9.5	7.5	6.8

Source: Horne *et al.* 2005.

The results found that overall the median rating was 7.5 stars and the overall mean rating was 6.8 stars—well above the 5-star requirements for Australian homes (Horne *et al.* 2005).

Sub-divisions and energy use

As mentioned earlier, one of the key principles for enhanced (passive) energy-efficient design of dwellings is the correct orientation of that building and allowing it access to sunlight and ventilation. The average size of a residential block has decreased over the period from 1993/94 to 2003/04—the average size in 2003/04 was nearly 8% smaller than the approximately 730m^2 average in 1993/94 of (Australian State of the Environment Committee 2006). By contrast, the average house size has increased over the same period. In addition, lot orientation is not seen as a critical driver for land developers and consequently is usually not consciously incorporated into sub-division design. These factors all combine to mitigate against the development of more energy-efficient housing.

A study by Miller and Ambrose (2005) assessed the impact of orientation on a dwelling's level of energy efficiency. A series of typical standard home designs were modelled at 45 degree increments throughout 360 degree rotation. Altering the orientation increased the annual total load (and decreased the energy efficiency) by between 10 and 32% compared with the optimum levels achieved for each of the dwellings.

The study also explored the combined impacts of increasing density, zero lot lines (where a dwelling wall is constructed along the boundary line), high solid fencing, close and dense foliage or high retaining walls on a home's energy efficiency. The study found that for dwellings separated by less than one metre, close proximity had the potential to greatly reduce the ability to capture breezes or sunlight for either of the dwellings along those shielded boundaries. Modelling of the impacts revealed that the annual total energy load increased by between 5 and 15%. This increase is relatively small and, in many instances, could be counteracted by using better design principles.

The impact of increasing housing density on energy efficiency is an area that has only just begun to be explored. It has generally been argued that increasing the density of dwellings will help lower the overall average energy consumption rates. Newton, Tucker and Ambrose (2000) report that apartments use 10% to 30% less energy than a detached house on a per occupant basis. However, this study did not cover operational energy outside heating and cooling requirements and Randolph and Troy (2007) argue that conclusive research to substantiate these claims remains elusive. Their review of studies suggests that per capita greenhouse gas emissions from high-rise flats are significantly higher than those for other forms of housing. Other reports also lend weight to these findings with Wright (2006) stating that the general trends suggest that although inner metropolitan medium-density housing consumes less operational energy (i.e. energy consumed within the home on an ongoing basis), than low-density urban fringe development; high-density, high-rise development consumes more operational energy than medium- or low-rise development. The suggested reason for this increased energy use is their high dependence on electricity (rather than the more energy-efficient natural gas) for the provision of their energy needs (Pears 2005). It is common for such developments to use reverse cycle air conditioners to provide an apartments heating and cooling needs, while hot water is provided by small inefficient electric units. In addition, the provision of communal services, such as lifts and swimming pools, adds considerably to the overall energy consumption of these apartments.

Socio economic factors are also cited as impacting on the expected energy consumption of dwellings, with homes in higher income low-density suburbs having significantly higher per capita consumption than those homes in low to medium income low-density suburbs (Randolph and Troy 2007). Reasons for this may well be related to the relative size of dwellings, with homes in higher income areas being larger, but still having the same occupancy levels as lower income areas.

Of course, these studies do not factor in transport energy and other studies have shown that the further people live from the city centre the less likely they are to use public transport and the more reliant they are on private cars (Newton 2001; Coalition of People's Transport 2004). High-rise developments tend to be close to city centres and/or good public transport access and, consequently, the likelihood is that travel distances and travel modes are significantly different for high-rise residents than outer urban residents, which will have an impact on their overall household energy consumption.

Tools for rating energy efficiency

Under the new Australian energy-efficiency building regulations, a star-based rating system is used to assess the relative energy efficiency of residential construction. The number of stars achieved is determined through a computer-based thermal modelling program that determines average annual energy usage required to maintain a house within a particular thermal comfort range, over a set time period and in a particular climate zone. The resulting energy metric, in MJ/m^2/annum, determines to which star band a house is assigned. AccuRate—an assessment tool developed by CSIRO—is the benchmark tool used for setting the standard rating system Australia-wide (Australian Building Codes Board, 2006).

Several states have developed additional systems to complement the national energy standard. BASIX, the Building Sustainability Index, has been developed by the NSW Government and is now a mandatory requirement for all new residential dwellings (including alterations and additions) throughout NSW. The BASIX tool assesses the performance of a proposed design against various sustainability indices, including water, thermal comfort and energy. BASIX is more flexible because, rather than insisting on specific design and fixture requirements, BASIX allows the building designer to select from a wide range of options. Once these selections have been made, they are listed on the BASIX Certificate, and must be shown on the project plans. A building inspector will check that the project is being built according to these specifications (NSW Department of Planning 2007) BASIX differs from AccuRate in that it is a prescriptive tool rather than a predictive tool and extends its influence beyond thermal efficiency to include a more holistic view of sustainable urban design.

The National Australian Built Environment Rating System (NABERS) is another rating system, which is similar in approach to that of BASIX. NABERS is a performance-based rating system for existing buildings and rates a building on the basis of its measured operational impacts on the environment. NABERS was initially developed for commercial buildings, but has now been extended to residential buildings. The system is intended as a voluntary tool that will provide information on the sustainability of existing buildings and provide the basis for promoting an improvement in the environmental performance of Australia's existing building stock (Department of the Environment and Water Resources 2007).

A rating system for sub-divisions is less well developed. Presently, there is no mandatory rating system employed anywhere in Australia for assessing the environmental performance of sub-divisions, but several voluntary systems have been developed. One of the first attempts was developed for Energy Victoria in the early 1990s to provide guidance to land developers to optimise the solar effectiveness of lots (Loder and Bayly Consulting Group *et al.* 1991). This system was developed further by Energy Victoria for the NSW based Sustainable Energy Development Authority (SEDA) and modified for NSW climate zones (Energy Victoria 1996).

The SEDA system used the following methodology to rate lots:

- the lots were rated on their ability to accommodate a dwelling with good solar access
- the rating scale was 1–5 stars, with 5 being the highest
- the tool applied to separate lots between 300 and 1000 m^2. For lots under 300 m^2, solar access was closely integrated with building design and siting, while lots over 1000 m^2 had a better chance of achieving good solar access without the need for rating
- the slope of the lot affected solar access; however, lots with a slope of over 20% automatically received a 1-star rating, regardless of slope orientation.
- the goal was to rate at 80% of the total sub-division as 5-star lots, with the remainder rating either 4 or 3 stars.

SEDA later modified this tool further into its current version, which has moved away from lot rating (Sustainable Energy Development Authority 2005). Rather than control lot size directly, the tool seeks to provide guidance on the placement of a dwelling on a lot to ensure solar access. The tool encourages dwelling shapes to be selected for certain lot orientations to allow good solar access to that dwelling, without compromising lot yields or an increase in lot size. It is based on common dwelling footprints and heights, yet has the flexibility to accommodate other variations. This tool specifies that each lot in a sub-division has both a Flexible Solar Access Zone (FSAZ) and a Minimum Solar Access Zone (MSAZ):

- the FSAZ is the reserved part of the lot that may not be built on, thereby allowing solar access to glazing and private open space.
- the MSAZ is the minimum area of the FSAZ that may not be built upon. The MSAZ can be moved to any place within the FSAZ at the development application stage to accommodate a range of house footprints. Once the MSAZ and the dwelling are located at development application stage, the FSAZ is no longer applicable and can be built upon.

All of these sub-division tools have focused on solar access and have not considered other sustainability issues, such as water management. However, the Queensland branch of the Urban Development Institute of Australia (UDIA) has recently released its *EnviroDevelopment* system which aims to provide a more holistic assessment process for residential sub-division development. *EnviroDevelopment* is performance based and applicable to a diverse range of development types and situations. It covers the broad spectrum of environment and sustainability issues from the initial conceptual stages of development, with elements devoted to ecosystems, water, energy, waste, materials and community (Urban Development Institute of Australia (Qld) 2007).

Renewable energy

The vast majority of Australia's electricity generation is sourced from large-scale fossil-fuel-fired power stations and delivered to consumers through an extensive national power grid. The latest energy data from the Australian Bureau of Agricultural and Resource Economics (ABARE 2007) reveals that only 7.4% of Australia's electricity comes from renewable sources, with the vast majority of this from large-scale hydro power stations located predominantly in Tasmania. Hydro power stations account for almost 86% of renewable electricity generation, with the remaining renewable energy sources accounting for only 1.0% of the nation's electricity generation (see Figure 4).

Despite the advancements made in renewable energy generation, it is very rarely used in urban environments where access to the national grid is available. However, communities and developers are now looking at ways that this technology can be incorporated into their built environments to help reduce the environmental impact that traditional electricity generation causes. Consumers wishing to source their electricity from renewable sources have several options available to them, the simplest being to select the 'green power' option from their current electricity provider; however, this option is

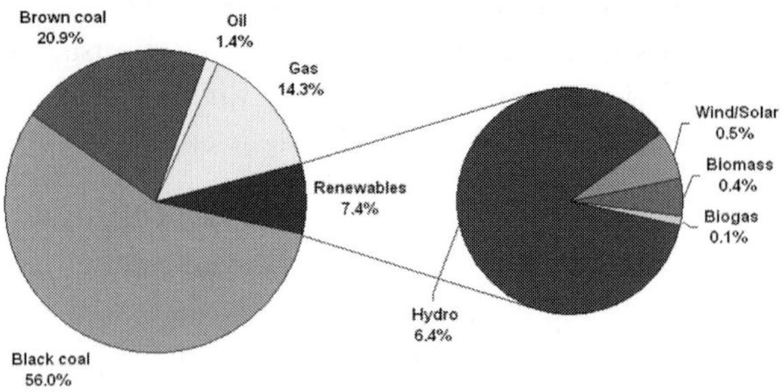

Figure 4: Electricity generation by fuel type (Source: ABARE 2007)

only available to individual consumers. For communities or urban developers wishing to incorporate renewable energy into a development project the only real option is to include on-site energy generation. Renewable on-site electricity generation use natural sources that are continuously replenishing, such as, sunlight, wind and water. A substantial selection of renewable energy generating technologies are now available including photovoltaic arrays, wind generators, geothermal systems and micro hydro units. The scale of these units also varies from small domestic systems designed to power a single home, to larger scale units that are capable of providing electricity for several houses or even an entire community.

Conclusion

The need to address the energy consumption patterns of the residential sector has never been greater than now. The impacts of global warming and climate change have become increasingly obvious and are driving governments and communities in all countries to take more urgent action. Residential energy use is a significant proportion of Australia's total energy consumption, but it also provides great opportunities for significant savings to be achieved. These opportunities are spread across all stages of the residential development process, including the selection of future development areas, how those areas are subdivided, what types of dwellings are built (considering both design and materials issues) and, finally, how those homes are used and operated by the occupants.

Currently, some of these stages are being tackled (although many people would argue that much more can be done), while others are being virtually ignored. The current energy-efficiency requirements for dwellings are well behind most other developed countries and this is reflected in the high domestic energy consumption values in Australia. Raising performance requirements will have flow on effects as to how sub-divisions are planned, by creating the need for developers to consider aspects such as solar access and natural ventilation opportunities. Broadening the overall scope of energy-efficiency requirements will also help incorporate more operational energy aspects into a performance assessment, such as lighting and appliances. It may also elevate the importance of factors that developers consider when selecting development opportunities, such as proximity to public transport rather than simply proximity to a freeway.

However, regulation alone is not the answer. The residential sector encompasses the entire population—we all live somewhere. A well-informed community constitutes a powerful advocacy group. As people's expectations increase, the market should respond and deliver better products and outcomes. The technologies for delivering more energy-efficient housing are already known—it is how these are moved into mainstream acceptance that is the challenge now. The critical question is 'how long will it take?'.

References

ABARE (2007). *Energy in Australia 2006.* Australian Government Department of Industry, Tourism and Resources, Canberra.

Akmal M and Riwoe D (2005). Australian *Energy, National and State Projections to 2029–30.* ABARE, Canberra.

Australian Building Codes Board (2006). *Building Code of Australia.* Australian Building Codes Board, Canberra.

Australian Bureau of Statistics (2007). *Year Book 2007.* Cat. no. 1301.0. Australian Bureau of Statistics, Canberra.

Australian Bureau of Statistics (2006). *Australian Social Trends, 2006.* Cat. no. 4102.0. Australian Bureau of Statistics, Canberra.

Australian Bureau of Statistics (2005). *Year Book 2005.* Cat. no. 1301.0. Australian Bureau of Statistics, Canberra.

Australian State of the Environment Committee (2006). *Australia State of the Environment 2006.* Department of Environment and Water Resources, Canberra.

Coalition of People's Transport (2004). *The Place to be on PT – A Vision for Greater Melbourne's Transport.* Coalition of People's Transport, Melbourne.

Department of the Environment and Water Resources (2007). *Building and Construction Website* <http://www.environment.gov.au/settlements/industry/construction/>, viewed 7 June 2007.

Energy Victoria (1996). *Solar Access for Lots.* Energy Victoria, Melbourne.

Horne RE, Hayles C, Hes D, Jensen C, Opray L, Wakefield R, and Wasiluk K. (2005) *International Comparison of Building Energy Performance Standards.* RMIT University, Melbourne.

Insulation Council of Australia and New Zealand (2007). *US Versus Australian Regulations.* Insulation Council of Australia and New Zealand, website <http://www.icanz.org.au/international/111/>, viewed 6 June 2007.

Issac T (1989). *Solar Efficient Residential Subdivision.* Energy Victoria, Melbourne.

Loder and Bayly Consulting Group, Sustainable Solutions Pty Ltd and Isaccs T (1991). *Energy Smart Lots – A Study of Solar Efficient Residential Subdivisions.* Energy Victoria, Melbourne.

Mead E and Wales N (2004). *Benchmarking the Practices, Perceptions and Design of Sustainable Subdivisions.* Cooperative Research Centre for Construction Innovation, for Icon.Net Pty Ltd, Brisbane.

Mickel J (2004). 'Queensland Minister calls for more economical cooling'. Media release. 27 October.

Miller A and Ambrose MD (2005). *Sustainable Subdivisions: Energy-efficient Design.* Cooperative Research Centre for Construction Innovation, for Icon.Net Pty Ltd, Brisbane.

Newton PW *et al.* (2001). *Australia State of the Environment Report 2001: Human Settlements* CSIRO Publishing, Melbourne.

Newton PW, Tucker SN and Ambrose MD (2000). Housing form, energy use and greenhouse gas emissions. In *Achieving Sustainable Urban Form.* (Eds K Williams, E Burton and M Jenks), pp. 74–83. E & FN Spon, London.

NSW Department of Planning (2007). *About the BASIX indices fact sheet, Sustainability Unit.* NSW Department of Planning, Sydney.

Pears A (2006). *Global Warming - Cool It.* Department of Environment and Heritage, Australian Greenhouse Office, Canberra.

Pears A (2005). Does Higher Density Really Reduce Household Energy Requirements? It Depends... *Urban Policy and Research* **23(3)**, 367–369.

Queensland Government (2004). 'Towards sustainable housing in Queensland – discussion paper'. Department of Local Government and Planning, Brisbane.

Randolph B and Troy P (2007). 'Energy consumption and the built environment: A social and behavioural analysis, City Futures Research Centre, research paper No. 7', University of New South Wales, Sydney.

Reardon C (2001). *Your Home: Technical Manual: Design for Lifestyle and the Future.* Australian Greenhouse Office, Canberra.

Seifert R (2007). A history of northern energy efficient housing. In *Alaska Residential Building Manual.* (Ed. Seifert R) pp. 1–20. HCM-00051, Alaska Housing Finance Corporation, Fairbanks, USA.

Sustainable Energy Development Authority (2005). *Solar Access for Lots – Guidelines for Residential Subdivision in NSW.* Sustainable Energy Development Authority, Sydney.

United Nations Framework Convention on Climate Change (2004). *Greenhouse gas emissions data for 1990–2003,* sourced via website <http://globalis.gvu.unu.edu/indicator.cfm?Country=AU&IndicatorID=199>, viewed 7 June 2007.

Urban Development Institute of Australia (Qld) (2007). *Enviro development web site.* (http://www.udiaqld.com.au/), viewed 7 June 2007.

Wright K (2006). The relationship between residential density and non-transport energy use. *Australian Planner* **43(4)**, 12–13.

28

Transitions in Transport

David Lamb

The motor car has been described as 'the machine that changed the world'. Historians may feel that other devices, such as the steam engine, extraction of gas from coal or understanding dynamic electricity, are more deserving, but it is indisputable that the car has substantially changed life for a large proportion of the world's population during the 20th century. As well as offering personal mobility as never before, world trade depends on the internal combustion engine—from the trucks that haul the produce from farms to market to the ships carrying goods across the seas.

Early developers of the motor car thought that the preferred fuel would be alcohol, but oil became so abundant in the United States that it soon dominated wherever cars were made. The advent of mass-production technologies, together with readily available and inexpensive oil has fuelled the seemingly unstoppable penetration of cars to every corner of the globe. Every country aspiring to an improved standard of living wanted a car industry and car makers were offered a range of incentives to build car factories.

Over the 20th century car production grew from a handful to more than 60 million vehicles per year. Steam cars and electric cars were soon replaced by petrol-powered cars because the fuel was so convenient, widely distributed and easily stored within the vehicle. Early safety concerns based on flammability of the fuel were soon overcome and petrol service stations began to appear at major intersections in every city of the more advanced countries. Motorisation became one of the indicators of a modern, developed nation.

Giant corporations grew out of the profits to be made from producing simple, but affordable, cars such as the Ford Model A through to the luxury limousines such as Packard and Rolls Royce. With each new model there seemed to be more mechanisation. Electric horns, starter motors, headlamps and windshield wipers became standard and made motoring easier. In the second half of the 20th century power steering, power-assisted brakes, air conditioning and computer-controlled fuel management were all added to make cars more reliable. Car makers competed by adding convenience features, such as radios, cigar lighters, electric seats, cassette tape players and even automatic canopies. An entire new industry emerged to provide accessories for cars.

In Australia, Ford began assembling cars in Geelong in 1925 from imported components. However, Australia had already built an industry to make replacement parts for cars. The long supply line from Europe and USA meant that local component manufacturers of seats, springs, tyres and other damage-prone components could easily compete with overseas competition.

In 1949, Holden began production of the first car that could be called 'Australian', although Ford could claim that the utility version of their passenger cars was an Australian invention.

The 1960s was a time of high protection from import competition and, with the incentive of high import duties and regulations setting minimum levels of local content, Ford began production of its American 'compact-sized' car—the Falcon. Other manufacturers, such as British Motor Corporation, Chrysler, Volkswagen, Toyota and Renault, set up local assembly plants to overcome the protective barriers, but Holden and Ford held the major share of the market and also dominated the development of the local components industry.

About half way through the 20th century, car construction methods departed from having a separate body mounted on a rigid chassis by adopting monocoque body design, whereas trucks continued to rely on the chassis for its basic rigidity. Advances in metal-forming technology opened the way for more stylish body designs and more automation of the body building process for cars.

As the number of cars grew and they offered more power and speed, the cost of vehicle crash trauma started to cause public concern. In the United States, Ralph Nader's 1965 book *Unsafe at Any Speed* examined the built-in dangers of American automobile design and inspired a demand for research into the causes of road accidents. Regulations requiring improved vehicle occupant protection followed. In Australia, road accidents reached a peak in 1970, with 3798 fatalities at a time when the total number of vehicles was 3.74 million. Over the next 36 years until 2006, the combined effects of improved vehicle design, better roads and regulations have reduced the road toll in Australia to 1605—a reduction of 58%. Over the same period the number of vehicles increased 260% to 13.49 million.

But, for all the innovations incorporated into cars, the means of providing the power remained the job of the internal combustion engine and, while the modern engine delivers a higher power-to-weight ratio, it is still readily recognisable as the engine developed a hundred years ago. Refinements to car engines seem to have been aimed mostly at extracting more energy from the fuel. Car makers promoted speed and acceleration as 'performance'. Marketers used gains in engine output to sell bigger cars to performance-hungry buyers.

A similar change was occurring in trucks, where increased power allowed an increased payload. With less inhibition on engine size, truck engines could get bigger and pull bigger loads. The diesel engine proved superior in engine efficiency in large trucks, mining machinery and industrial applications and today all long-haul trucks are powered by diesel engines. Businesses concerned with the cost per tonne of freight are more likely to pay a higher price for a fuel-saving technology than car buyers who tended to be more concerned with styling and image aspects.

In November 1997, Toyota published a supplement to *Time Magazine* announcing that it intended to begin production of a different kind of car. The supplement presented facts and figures about global climate change, the role of automotive emissions as a contributing factor, and the need to take corrective action. The world's first mass-production hybrid car, called the Prius, rolled off Toyota's production line on schedule at the end of 1997. Initially the car was sold only in Japan and production volume was modest—around one thousand per month—but after proving reliability in the Japanese market production volumes were steadily increased. Honda soon became the second car maker to produce a hybrid car when it announced its Insight model. Both marques were introduced to the American market by 1999 and in 2001 Honda introduced the Insight to the Australian market at the Melbourne Motor Show. Toyota followed suit later in the year.

The purpose of hybrid-electric vehicle technology is to capture energy that would normally be wasted when a car is slowing or braking by converting that energy to electrical energy and storing it ready to be re-used when the car accelerates. There are variants on the principle, such as switching off the engine when the car is stationary to save more energy saving or using electric power to propel the car when there is sufficient stored energy available.

Hybrids, although much reviewed for their technological difference from conventional power trains, were not in great demand in the market but the two Japanese car makers continued their development and refinement to improve the performance of the battery pack and to increase reliability. Some of the early technical reviews of both models had drawn attention to the high cost of the battery pack and some prospective buyers may have been concerned over the possibility of early failure. The car makers countered by offering as much as 8 years warranty on the battery pack. This was a very significant advance on any warranty that had previously been offered unconditionally. Some years ago, car warranties were limited to 12 months or 12,000 kilometres, with tyres and batteries often excluded. Only in the past decade have warranties increased to as much as 5 years on vehicles.

Climate change and peak oil

Concern about global climate change did not become a matter of widespread public interest until

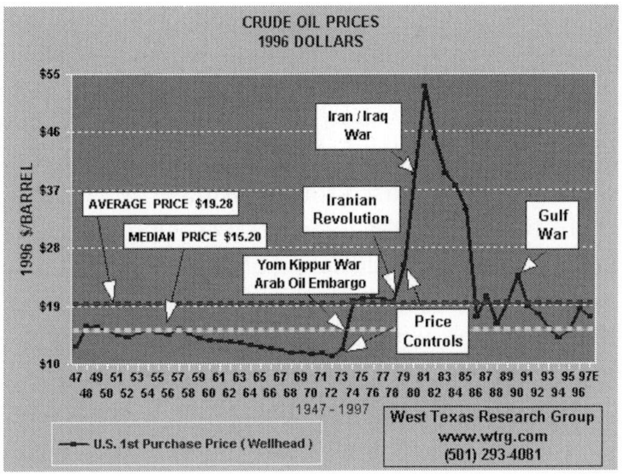

Figure 1: Oil prices 1947–1997

around the end of the 20th century. With oil prices remaining low, hybrids were popular only with environmental enthusiasts and movie stars. For the public at large, it was difficult to justify a 20% or more price premium to get better fuel efficiency. There had been oil price spikes before, such as in 1973 when the Organisation of Petroleum Exporters cut back production to force a price rise, and during the first Gulf War, as shown in Figure 1, but after each peak, the price settled back to around US$20 per barrel and remained at that level until the second Gulf War. However, from 2003 to 2006 the price of oil rose steadily and reached US$75, as shown in Figure 2. Although the price eased in late 2006, most commentators predict that the price will never again fall to $35 or below.

The price of oil had already reached US$60 per barrel when an event occurred that focused even more attention on our dependence on oil. In late August 2005, Hurricane Katrina devastated New Orleans and other towns in Louisiana, Mississippi and Alabama in the United States. As well as causing the deaths of 1836 people and damage costing more than US$100 billion, Katrina damaged or destroyed 30 oil platforms and caused the closure of nine oil refineries. Television footage of freeways out of New Orleans, jammed with cars that had run out of fuel brought home vividly the realisation of just how dependent we are on oil. Conditions were hot and steamy, but people kept their engines and air conditioners running without realising that they would soon be unable to go anywhere when the car ran out of fuel.

Climate scientists have for some years been predicting that the greenhouse effect will result in more severe storms. Although Hurricane Katrina could not be blamed directly on greenhouse gas emissions, there were enough newspaper reports about the rising sea temperature in the Gulf of Mexico and the effect on storm formation to cause many people to think that the increasing severity of storm activity was related in some complex way to human activity. The insurance industry was quick to draw the connection and the level of seminar activity on the topic accelerated.

The jolt of an interruption to the supply of oil in the US gave rise to a surge of articles about 'peak oil', which is based on the theory that the amount of oil under the ground is finite. The theory is named after American geophysicist Marion King Hubbert who created a method of modelling a peak rate of production given an assumed ultimate recovery volume. Hubbert's theory was initially greeted with scepticism by many in the oil industry after a 1956 paper he presented to the American Petroleum Institute, which predicted that production of oil from conventional sources would peak in the continental United States around 1965–1970 and that production would steadily decline from then. Others extended the prediction to the world oil supply and forecast dire economic consequences when demand overtakes supply. Oil production per capita actually peaked in 1979 because of world population growth (Duncan 2000). It could be argued that per capita oil production is more important than the absolute peak, and that its peaking wasn't noticed due to global socioeconomic inequality. Consumption has increased while new discoveries have declined, but as long as the oil price remained low, it seems everyone,

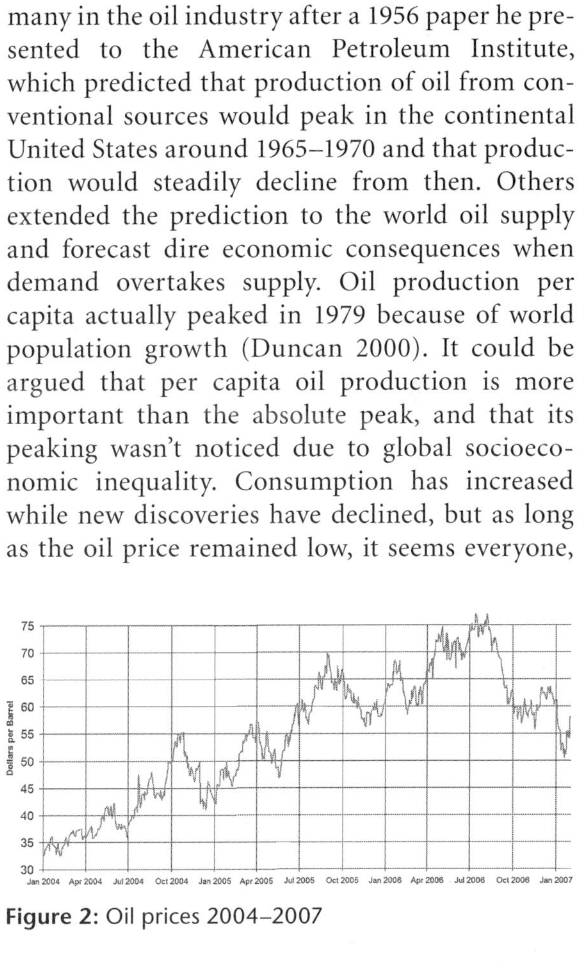

Figure 2: Oil prices 2004–2007

including governments, were content to let the laws of supply and demand prevail.

In its 2007 February *Highliqhts* report, the United States Government Accountability Office, says that most studies estimate that oil production will peak sometime between now and 2040. The wide range of estimates is because the timing of the peak depends on several unknown factors that will determine how quickly remaining reserves are used:

- how much of the remaining oil can be produced and at what cost
- environmental challenges
- the political and investment conditions in some countries with oil deposits
- the future demand for oil.

Demand for oil will, in turn, be influenced by future global economic growth and may be affected by government policies on the environment and climate change and consumer choices about conservation. 2040 may seem a long way off, but if oil consumption continues to rise, the peak will be much sooner and oil scarcity could become the dominant factor in world economics well within the lifetime of most people alive today.

International oil prices increased because of the interruption to supply in the US, but the market became much more sensitive to any event that could exacerbate security of fuel supplies. Recent events in the Middle East have done nothing to calm a nervous market. A breach to a pipeline in Alaska caused a jump of several dollars per barrel overnight. In Australia in 2004 and early 2005, there wasn't much of a reaction at first to the rising petrol price. Perhaps fluctuating price at the fuel pump was something the public had come to accept, or perhaps petrol was so relatively cheap, people were not overly concerned. Many commentators suspected collusion in the system somewhere, but the price watchdog, the Australian Competition and Consumer Commission had never been able to find evidence to support a case against oil companies. However, by mid 2005, the steady rise in petrol prices was making front-page news almost every day.

Australia was largely self-sufficient in oil during the last decades of the 20th century, but new discoveries have been in ever-more remote locations or deeper at sea. Technologies for extracting oil from difficult locations have improved, but they are increasingly expensive and the result has been a decline in Australian oil production. In 2005, Australia produced 78% of its oil requirements, but having to import the remaining 22% at a rising world price has caused some alarm.

In the 12 months to September 2006, Australia had a petroleum trade deficit of $8.25 billion. There was even an overall energy trade deficit of more than $5 billion, with the value of oil imports swamping the value of coal, gas and uranium exports.

Geoscience Australia, the national agency for geoscience research and geospatial information, forecasts that by around 2020 Australia's self-sufficiency in oil will continue to decline. The Australian Bureau of Agriculture and Resource Economics has published forecasts showing Australia's fuel consumption growing steadily year-by-year. The implications are that either more new oil will need to be discovered, or some low-cost alternative will be found or the Australian public will have to become more tolerant of increasing deficits in the balance of trade, with the consequent effect on their standard of living.

All the evidence points to an increasing scarcity of oil. The price of oil over the long term will rise until consumption is less than supply capacity and within the limits of exploitable reserves. As we shall see later in this chapter, there is no alternative fuel that can completely replace oil in the short or medium term. There are several alternatives that together could replace up to, but probably not more than, 20% of our oil needs, but without very substantial technological breakthroughs, there is no alternative that could replace a larger proportion of our oil use, and there is no likelihood of any new alternative ever being competitive with oil at today's prices.

Emissions

Since the 1960s there has been a trend towards regulating emissions from motor vehicles. The State of California took an early lead in research and legislation, driven by the high car density in Los Angeles and high level of congestion that combined to cause significant damage to the health of the residents breathing the polluted air.

Early legislation was aimed at improving air quality by reducing the contaminants of vehicle

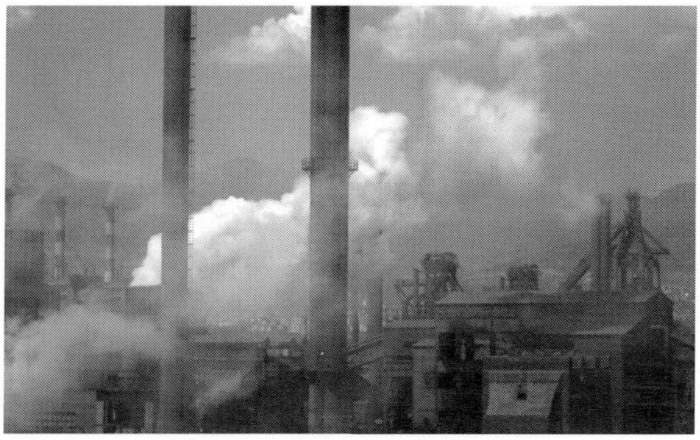

Figure 3: Air pollution in the mid-20th century.

exhaust known to be harmful to human health—mainly oxides of nitrogen and carbon particulate matter. The biggest single step forward was the introduction of the three-way catalytic converter that, using the heat of the exhaust gases and precious metals as a catalyst, breaks down the harmful oxides of nitrogen into nitrogen and oxygen. The gases emitted are nitrogen, oxygen and carbon dioxide, none of which is harmful to human health. At a time when air quality was high priority, the community solved the air-quality problem, but at the expense of increasing greenhouse gas emissions (additional carbon dioxide, plus some of the nitrogen and oxygen re-combine as nitrous oxide which is 310 times as potent as carbon dioxide in greenhouse gas terms).

Until very recently, American car buyers have shunned diesel-powered cars. This may be because diesel fuel pumps were only to be found at the less-attractive corner of a fuel station used mostly by truck drivers, or it may be an outcome of the fact that American carmakers did not offer diesel engines in cars. On the other side of the Atlantic, European car buyers have been buying as many diesel-powered cars as petrol ones. As a health issue, carbon particles are much more significant in diesel fuel than in petrol, so the popularity of diesel engines has prompted the European Community to take legislative action to reduce carbon particle emissions.

In one sense it is unfortunate that the term 'emissions' has come to be used both for contaminants that are harmful to human health and also used as a collective term for the gases that are known to contribute to global climate change. The issues would be less confusing if we expanded the descriptions to 'air-quality emissions' and 'greenhouse-gas emissions'.

Transport emissions are known to be an important contributor to overall levels of the gases that are causing global climate change. In the 1970s and 1980s our oil-driven transport was regulated to reduce air quality emissions, such as the oxides of nitrogen and particulate carbon. Table 1 shows the pattern of regulation to limit 'greenhouse' emissions and emissions that are harmful to air quality from cars and trucks in Europe. We can expect the rest of the world to follow suit—sooner in the case of cities in advanced economies and perhaps later in the less-developed nations.

Many technologies to achieve lower emissions are known. The only reason for not introducing them into production cars is that car makers believed that customers would not be prepared to pay the additional cost. Single technologies may only be able to contribute a few percentage points of improvement to fuel efficiency. Introducing them into production vehicles is a matter of investment and cost, but, as fuel prices rise, the penalty costs of emission-reduction technologies, which almost invariably offer better fuel economy, become more tolerable.

Toyota and Honda have now been making hybrid cars for 10 years, but other car makers have been slow to offer hybrid versions of their products. Part of the reason is that, at the time vehicle emissions gained wider interest in the community, American and European car makers were convinced that hydrogen offered a better solution to vehicle emissions. It is interesting to see that many of those car makers who presented spectacular concept vehicles preparing the public for vehicles powered by hydrogen fuel cells are now scrambling to regain some momentum with hybrid electric models. Just about every car maker has hybrid car development programs underway.

Hybrid cars use less fuel by re-using regenerative braking energy to assist the engine during acceleration. There is no disputing the fact that hybrids offer better fuel economy in congested traffic conditions. There are still some commentators who cite the superior fuel efficiency of diesel engines, but usually they fail to recognise the special conditions under which each technology

Table 1: European Union Emission Standards in place since 1996 (Source: CIA countries report).

Emissions standard	Particulate matter (PM)/ (mg/km)		Oxides of nitrogen (NOx) (mg/km)		Hydrocarbons (HC) (mg/km)	
	Diesel	Petrol	Diesel	Petrol	Diesel	Petrol
Euro 2 (1996)	80–100	–	–		–	-
Euro 3 (2000)	50	–	500	150	–	200
Euro 4 (2005)	25		250	80	–	100
Euro 5 (2009)	5	5	180	70	–	100
Euro 6 (2014)	5	5	80	70	–	100

performs best. In cities, where stop-start driving is common and many of the journeys are short, hybrid cars can save fuel and emissions. However, for long journeys on highways where speed is fairly constant, a hybrid uses just as much fuel as its conventional counterpart and more fuel than an equivalent non-hybrid diesel car. We can expect car makers to adopt the best of both technologies by using a small diesel engine to generate the electricity for a full hybrid car.

The degree of electrical assistance so far available in a hybrid car has been fairly modest, leading to the term 'mild hybrid'. The term 'full hybrid' implies that more electrical energy can be stored for re-use during acceleration. Idling wastes as much as 17% of a car's energy in urban operations, so turning off the engine whenever the car is not moving becomes an essential feature of a hybrid vehicle. A full hybrid provides the opportunity to operate the engine at its most efficient settings, using the stored electrical energy to accommodate the varying power requirements. A full hybrid only operates the engine when it is needed to supplement the battery energy; in other words it is an electric vehicle with auxiliary power supply.

Over the last couple of decades we have seen the increasing electrification of the motor car. Hybrid-electric technology relies on batteries that have increased energy storage capacity and, as batteries take over more of the load, the opportunity to re-charge the batteries from the mains power supply becomes more attractive. In this way, the trend towards full electrification is assured. Eventually, we can expect the electrical systems to take over the role of propelling the vehicle, while engines will become smaller and smaller until they are virtually an auxiliary power source. When plug-in hybrids and electric cars make up a significant proportion of the fleet, it will become possible to integrate cars with the electricity grid to share energy at times of peak load. This concept, known as V2G is gaining credibility rapidly in the United States.

As was demonstrated earlier in this chapter, road accidents were reduced by 58% over a period when the number of cars more than doubled. This dramatic improvement was achieved through research, regulation, enforcement and improvements to vehicle design. This should give us confidence that we can cope with the massive changes that will be needed to cut back on greenhouse gas emissions if we are to stabilise climate change.

The electric car and hydrogen fuel cells

Some years ago when California State was considering new legislation to compel car makers to offer vehicles with lower emissions, the notion of an all-electric car became popular and there was much debate about whether an electric car should be classed as a 'zero-emission vehicle'. Critics claimed that charging car batteries overnight from the mains grid would simply shift the emission burden from the tailpipe of the car to the power station, so more electricity would need to be generated. However, with improved power-management technology it is now more feasible to balance electrical load to avoid overstretching the grid, especially in urban environments. As more homes become partly self-sufficient with solar panels and perhaps a small wind turbine, much of the electricity needed for our cars could be self-generated. Once the challenge of on-board storage of electricity is overcome sharing energy becomes much more manageable.

One of the biggest benefits of a plug-in electric vehicle is that it can be refuelled almost anywhere. Parking stations could offer a recharge service; in fact a vehicle's batteries could be recharged anywhere the vehicle is stationary for an hour or so.

Electric vehicles can be far superior in energy-efficiency than equivalent oil powered vehicles. A very fuel-efficient petrol car, such as the Honda Civic—rated at 4.8 L/100 km—can travel 0.62 km per megajoule of energy[3]. An electric equivalent car, using electricity generated by an efficient natural gas fired electric generator of the type commonly used in Australia during periods of high demand, could travel around 2 km per megajoule, about three times more energy-efficient as its oil-based counterpart. (The energy content of petrol is 42 MJ/L and about 20% of the energy is lost in transport and refining by the time it reaches the fuel pump. The Honda Civic, using 4.8 L/100km travels 20.8 km per litre of petrol, so its energy efficiency is 20.8 km/L / (42 × 80%) or 0.62 km/MJ.)

For some years we have been told that hydrogen fuel cells will power our cars in the not-too-distant future. California's Governor Schwarzenegger, a strong supporter of the so-called hydrogen economy, has encouraged development of hydrogen-powered cars and has publicised the opening of a hydrogen refilling station in his state. However, the International Energy Agency predicts that it will take three decades or more for hydrogen to become a significant fuel because of the hurdles of producing hydrogen sustainably, the cost and reliability of fuel cells and the technologies for transporting and storing hydrogen under great pressures. The challenge therefore is to fill the gap in a time of shrinking oil supplies.

Ultimately, hydrogen may be the dominant fuel or energy carrier, but if sustainable hydrogen were available, it would seem to be easier to use static hydrogen fuel cells to generate electricity and store electricity in the car than to build a new distribution system for hydrogen and store hydrogen in cars to generate electricity on board the vehicle. Perhaps both alternatives will occur because long distance vehicles would be limited by their ability to store electricity, while urban vehicles would be limited by their ability to carry a store of hydrogen.

Alternative transport fuels

The slow transition to renewable hydrogen or renewable electricity means that transport globally will continue to rely heavily on oil for the next 30 years or more. There are other alternatives to oil; for example, LPG and natural gas, but they aren't renewable. Biofuels can be renewable, with ethanol and biodiesel the most promising candidates. There are other alternatives, but none that is readily available or likely to be available in the short or medium term. More research on these possible fuels of the future is needed.

CSIRO is developing a 'Fuels Ready Reckoner' that compares one fuel against another to show the relative benefits and costs across a range of criteria, such as production cost, environmental impact, sustainability, safety and public acceptance. Early results show that biofuels can only be expected to provide around 15% of our oil needs because of the limits of suitable land and water supply. Even this may be optimistic—the Fourth Report of the Intergovernmental Panel on Climate Change indicates that 10% is the likely limit around the world, and Australia is particularly constrained by limited rainfall for growing crops.

The picture may improve depending on research into genetic modification of crops to suit their conversion to fuel or using algae in the fuel production process. Even so, there will be physical limits of water supply or land use. It would be too optimistic to forecast that anything over 20% of Australia's liquid fuel needs could be supplied from crops, no matter how good the technology. To put this into perspective, CSIRO studies show that using all Australia's arable land, with today's crops and technologies, could only provide 67% of liquid fuel needs at today's rate of consumption. That proportion would decrease in drought years and as consumption increases.

Liquefied Petroleum Gas (LPG)

LPG is an attractive short-term alternative because Australia has the world's most extensive LPG distribution system and has reserves that can sustain likely demand for some years. Despite claims that a large proportion of Australia's petrol use could be replaced with LPG, this is unlikely to happen because about 20% of the LPG supply comes from petroleum refining and the other 80% comes from the excess propane found in natural gas basins. Australian reserves of LPG are one-tenth those of the natural gas reserves, and even by the most optimistic estimates, it is unlikely LPG could make up much more than 20% of Australia's demand for petrol.

Compressed Natural Gas (CNG)

CNG is an attractive alternative to petrol or diesel because Australia has large reserves of natural gas and internal combustion engines can easily be modified to use it. There are three obstacles to rapid take-up:

- Most of Australia's natural gas reserves are offshore on the north-west coast, while the greatest need is in the south-east corner of the continent.
- There is only a handful of outlets for vehicles to refuel.
- The amount of CNG sufficient for a range of 200 to 300 km requires a tank so large that it would take up most of the load carrying space of a car. CNG is suitable for buses and depot-based heavy vehicles such as garbage trucks because space for the fuel tank and the need for frequent refuelling are not an issue.

CNG could be an important transitional fuel because there are so many internal combustion engine vehicles in Australia's fleet and it will take 12 to 15 years to replace the fleet with lower emission alternatives. Using a home compressor, people who can live with a limited vehicle range could find CNG an attractive and cost-effective solution if oil becomes scarcer.

Liquefied Natural Gas (LNG)

Natural gas is liquefied and shipped overseas in purpose-built vessels with double hulls to contain the great pressure. Long haul trucks can be adapted to carry sufficient LNG to be refuelled at a depot with the special facilities for liquefaction at each end of their journey, but it is unlikely that this approach could be adopted for smaller trucks or cars because of the infrastructure cost that would be required.

Ethanol

Ethanol is attractive as an octane enhancer and also because it produces less particulate matter when burned in an internal combustion engine. As it has fewer effects on human health, increased use of ethanol has been strongly advocated by some health professionals. As will be explained later, there is a benefit to air quality from using ethanol and there is also a greenhouse-gas benefit when high levels are mixed with petrol. However, at lower levels, such as e10, the benefit is small.

Ethanol is sometimes claimed to be a suitable substitute for petrol in Australia on the basis that ethanol made from sugar cane has replaced 70% of Brazil's reliance on petrol. This is true, but it took 40 years to reach that level. The US government has set a target of replacing 25% of its petrol with ethanol over the next 25 years. However, the Australian land, climate and water supply are very different.

CSIRO has assessed the potential in Australia for growing crops suitable for making ethanol. By taking the average crop yield over a 5-year period, it is possible to calculate the amount of ethanol that could be produced. These calculations, shown in Table 2, indicate that if all the crops grown in Australia were used to produce ethanol, this would replace only 75% of Australian demand for petrol. Clearly, this would not be practicable, but if it were decided to make ethanol instead of exporting Australia's harvest, only 49% of petrol could be replaced by ethanol. There are other effects of using crops to make fuels. Some US states offer financial incentives to companies making ethanol from corn—the preferred feedstock in the US. Mexican farmers have found it more profitable to send their corn crop to the US so the price of tortillas—a Mexican staple food made from corn—has doubled. On 21 February 21 2007, the *Age* newspaper carried a photograph of a large crowd of citizens protesting over the price of tortillas.

It is clear that ethanol produced from Australia's harvest could replace only a small proportion of petrol, unless a significant technological breakthrough is discovered in making fuel from biomass (see Table 2). Much research is needed in this area. Any ethanol produced in Australia would probably be used in a low-percentage blend and sold in the region close to the point of mixing. There is little to be gained from standardising a particular mix percentage, so there will probably be no change to the current legal upper limit of 10%, (e10) except for specific exclusions.

Three factors are involved in the uptake of an alternative fuel such as ethanol:

- vehicles that use the fuel must be available for purchase
- the fuel must be available at sufficient outlets that a motorist can feel confident of being able to purchase it
- the consumer must want to purchase both the type of car and fuel, as opposed to purchasing any of the competitor fuels.

When ethanol first reached broad public awareness as an alternative fuel for cars, there was a great deal of confusion about whether car makers would void warranty cover of their products if motorists used non-standard petrol. This fear has largely dissipated now that car makers have published lists that show that at least 60% of the car fleet can safely use e10 without any fear of engine or other damage. Higher concentrations of ethanol require modification to the fuel supply system to avoid the risk of corrosion to fuel pipelines and damage to plastic components in the fuel system.

Ethanol has not become very widely available around Australia—production volumes have not been sufficient to replace even 1% of petrol's usage, so the issue has largely disappeared. As has been shown, it is unlikely for many years that Australia could support enough ethanol production for everyone to use e10, so it is more likely that pockets of e10 availability will be limited to the regions where the ethanol is being produced.

Biodiesel

Biodiesel has a special significance because long-haul road transport and farming and mining equipment are dependent on diesel fuel. In the event of oil scarcity, whatever oil is available would presumably be preserved for these applications. Diesel fuel constitutes more than 40% of Australia's oil use and is less readily substituted by biofuels than petrol.

Biodiesel can be made from any oil-rich crop, such as Australian canola, as well as from tallow. The supply of Australian tallow and waste-cooking oil (which can also be used to make biodiesel) is limited. To date, canola has been too expensive to use for making biodiesel so that the expansion of Australian biodiesel production is dependent on the supply and import of palm-oil from Malaysia or Indonesia. There is a great deal of controversy over the environmental credentials of Malaysian palm oil because some of it is sourced from recently cleared tropical rainforest and there is great concern over the loss of habitat for tropical species and other environmental impacts. There is also concern about rising food prices, with trouble having already occurred in India over the rising cost of cooking oil.

CSIRO has calculated the amount of biodiesel that could be produced from Australia's crops using typical land yields (see Table 3). The most optimistic calculations show that biodiesel could replace only 10% of Australia's diesel oil unless new technologies are found for sustainably converting biomass to liquid fuel.

Many farmers are attracted by the idea of producing their own biodiesel. Especially in remote regions, fuel can be comparatively very expensive and it makes good economic sense for farmers or small communities with a common interest to form cooperatives to grow crops that can be processed into fuel for their needs. As fuel prices rise in the future, more such production could be expected. However, national regulation is needed to ensure that fuel produced under such arrangements is either required to conform to national quality standards or not allowed to be mixed with fuel that is sold on the national market. Increasingly, engines of the future will demand very high consistency in fuel quality. Using sub-standard fuel could cause severe damage to the sophisticated engines of tomorrow. Makers of diesel engines will need to be aware of their customers' needs and respond with assistance instead of threats to void warranties.

Oil from coal

It is possible to make oil both from natural gas and from coal. It is widely known that during World War 2, Nazi Germany was deprived of oil from the Middle East and used coal to produce fuel oil using a process known as the Fischer-Tropsch process. South Africa, under embargo during the apartheid era, used the same process and is possibly now the world leader in coal-to-liquids technology. Much research is underway in the coal-rich countries to develop the process to the point where it is price-competitive with oil, but without exacerbating greenhouse gas emissions. Critics claim that 'clean coal' is an impossibility, and the future of coal depends on finding successful and reliable means

Table 2: Physical limits for producing ethanol from agriculture using current technologies.

Feedstock	Con (L/t)	Australian production (Mt)	Ethanol ALL feedstock (ML)	% 04–05 petrol equiv.	Export (Mt)	Ethanol EXPORT feedstock (ML)	% 04–05 petrol equiv.
Sugar	620[E]	5.0 (4.26.5)[A]	3,075 (2,590–3,386)	15	3.8 (3.1–4.2)[A]	2,375 (1,914–2,600)	12
Molasses C	270–290[D]	0.6–1.2[B]	280	1.4	0.5B	140	0.07
Wheat	360[D]	20.6 (10.1–26.1)[A]	7,419 (3,648–9,408)	37	14.8 (9.1–17.9)[A]	5,337 (3,278–6,432)	27
Coarse grain	360	11.3 (6.9–15.6)[A]	4,083 (2,493–5,637)	21	5.5 (3.8–7.2)[A]	1,978 (1,355–2,587)	10
TOTAL		14,857	75%	9,690	49%		

[A] ABARE Australian Commodity Statistics 2005, 2000–01 to 2004–05 data used.

[B] CSIRO/BTRE/ABARE 2003 350 ML report p 49 (Bullock 2001 60 ML 36 c/l but ABARE say that this is an overestimate).

[C] DEUS 2005.

[D] Erecon.

Source: CSIRO unpublished preliminary calculations

Table 3: Physical limits for producing biodiesel from agriculture using current technologies.

Feedstock	ConvL/t	Australian production (kt)	Biodiesel ALL feedstock (ML)	% 04–05 diesel equiv.	Export (kt)	Biodiesel EXPORT feedstock (ML)	% 04–05 diesel equiv.
Waste cooking oil – Australia	870A	90–105[A]	72–84 90–105[A]	0.6			
Tallow – Australia	894 A	500 Lo grade 260 Hi grade 240[A]	447	2	340[A]	304	2
Oilseed crops (including canola, cottonseed and others)	400	2,533 (1,532–3,094)[A]	1,013 (613–1,238)	7	1,498 (887–2,189)[A]	599 (355–876)	4
TOTAL for Australia	na	na	1,538 (1,132–1,769)	10%		903 (659–1,1 80)	6%

Source: CSIRO unpublished calculations

of capturing the emissions from coal-based electricity generation and sequestering the greenhouse gases in secure, long-term storage in expired mines or gas wells or under the sea.

Oil from natural gas

Research is underway to find suitable catalysts for the conversion of natural gas to oil and means of minimising greenhouse gas emissions.

Even experts cannot predict success in the short term and the technology cannot be regarded with a high degree of confidence for the medium term, but the opportunity for Australia to be more self-sufficient in transport fuel makes this a national imperative for research.

Butanol

Butanol is an industrial solvent and is also used as a base for perfumes. It is usually made from fossil fuels, but can also be made from fermentation of biomass by bacteria. It has been claimed that butanol could be synthesised from ethanol, and that this would provide a suitable substitute for aviation fuel. As most of the claims relate to production from corn, it appears that it is unlikely to have a high potential as a transport fuel for Australia. For some time, Virgin Airlines chairman, Sir Richard Branson, has been proposing butanol as a replacement for aviation fuel because it does not solidify at the cold temperatures experienced in high-flying jet aircraft. A company named Green Biologics has been set up to develop 'the new low-cost next generation' biofuel with funding from the British Department of Trade, shareholder and business investors. The company claims to have access to a breakthrough technology using genetic engineering in a novel fermentation process based on cornstarch.

Methanol

Methanol is a liquid fuel that can be used in its pure form, or mixed in any proportion with petrol, and combusted in a traditional internal combustion engine. Methanol is usually made from natural gas, but it can be made by biochemical means. Due to its lower vapour pressure, cold-weather starts can sometimes be difficult using pure methanol in a traditional engine. An M-85 blend (85% methanol, 15% petrol by volume) overcomes this problem, although engine re-design would easily obviate the need for any petrol component. The energy density of methanol is approximately 50% that of the same volume of petrol (whereas ethanol is approximately 65% by volume).

However, methanol has a number of other favourable combustion properties, such as a high octane number, a very high heat of vaporisation and a greater tolerance to lean fuel mixes than petrol. This means that dedicated methanol engines can be run at greater fuel efficiency, and with lower emissions, than conventional petrol engines. In part, this could compensate for the lower volumetric energy density. Methanol fuel cells are being manufactured for laptop computers and similar devices. It remains to be seen whether the technology can be scaled up to compete with hydrogen fuel cells.

Dimethyl ether

This fuel, which is very similar to diesel oil, can be made from many different feedstocks. It can be used as a fuel in diesel engines, petrol engines (30% DME / 70% LPG) and gas turbines. It works particularly well in diesel engines due to its high cetane number, which is greater than 55 compared with diesel which is 38–53. Currently the European Union is considering DME in its potential biofuel mix in 2030.

Fuel quality

As oil becomes more expensive or scarcer, the attraction for alternatives will increase and it has been shown above that many alternatives will be needed to replace even a small proportion of Australian fuel use. As more and more alternatives become available, the need for quality control will increase. This is especially so because future engine technology will demand ever-increasing fuel quality. This will be a challenge to the alternative fuel industry and to regulators.

Aviation

Around 6% of Australia's oil is used as aviation fuel. More people are flying today than at any time in history and the growth of air travel, encouraged by very low price flights, is a concern to environmentalists. Emissions from domestic air travel in Australia rose 76% between 1990 and 2005—faster than any other category. This is despite improvements in engine efficiency and better logistics management. It is common for airlines to sell older aircraft to less-developed nations, or to low-cost carriers or freight carriers, but this may change in future if fuel becomes much more expensive. The international nature of the aviation industry probably means a greater likelihood of international agreements aimed at reducing emissions such as those under consideration in Europe in 2007.

Rail

Rail transport accounts for less than 0.5% of Australia's oil consumption, but much of the energy used by rail networks is electricity. For many years, road transport has held a major share of the growth in freight carriage. New highways and efficient distribution networks have provided an efficient and low-cost flow of goods around the country. When oil prices rose in 2004 and 2005, many road transport companies who were operating on fixed price contracts tried to transfer cargo onto rail for long-distance carriage, only to find that the rail system, which had suffered from years of under-investment, could not meet the demand. Despite this, there has been little movement to invest in the rail networks to attract a higher share of traffic. It seems the price of fuel will need to rise considerably higher for the nation to make the major investments required.

Ocean shipping

Around 4% of the world's oil is used in ships. Ocean freight consumes the majority, but pleasure cruising has grown in recent years. Containerisation contributed to improved throughput at terminals and reducing overall logistics cost. While this has given us wider choices than ever before in the range of goods and food available, it has brought with it some behaviours that may be seen in a different light in the emissions-conscious world of the future. In a world facing the dire consequences of climate change, shipping bottled water around the world or merely sailing the seas to overcome boredom may be seen as avoidable or even irresponsible.

The car of the future

So much of our infrastructure has been built to accommodate the motor car. Even in the extreme case, if we made immediate decisions to use cars less, eliminated all avoidable car and truck journeys and only bought small cars, it would take 12 years or more to replace our car fleet. We have seen that hydrogen-powered cars are at least 30 years away and that we will be dependent on oil to a large extent for many years.

Oil scarcity and the need to reduce greenhouse gas emissions will accelerate the change to more efficient, hybrid or electric drive trains, with city cars leading the way. If fuel becomes extremely scarce, it would probably be conserved for trucks to transport goods. As we have seen, replacing diesel is more difficult than finding alternatives to petrol. So, if freight transport is given priority, urban car journeys will have to be drastically reduced, which implies a major shift to public transport. Fuel scarcity will not only increase transport costs, it will inevitably affect the value we place on transport.

In this chapter we have concentrated on road transport, particularly cars because they make up such a large proportion of the vehicle-kilometres travelled. However, all forms of transport will see big changes. Rail will become more competitive for long distance freight as the cost of road transport rises. Public transport will continue to increase its share of journeys in cities, but the biggest changes will be the way in which cars are made, the way they are refuelled and, most of all, the way they are used. As well as concerns about global warming, increasing recognition of the importance of transport fuel security and its impact on the Australian economy will force governments to develop a comprehensive plan to encourage appropriate investment and to discourage inappropriate choices.

Reference

Duncan RC (2000). Pardee Keynote Symposium, 13 November 2000, Reno, Nevada.

SECTION 6

Transition to Integrated Urban Water Systems

Australia requires a transition in its sources of urban water supply—given current forecasts of a 1260 GL deficit in the national urban water balance by 2030 (based on future growth in population, current patterns of consumption and a forecast reduction in rainfall and run-off for many catchments, according to Maheepala and Blackmore). In responding to this challenge, authors in this section argue that each urban region will need to develop their most eco-efficient water supply solution, taking a systems view that includes both water and energy (de-salination faces a major test here, unless powered by renewable energy).

Demand management by industrial and domestic sectors via water efficient appliances and behaviour change represents continuing conservation potential, but there is evidence of demand hardening (Inman). Deeper understanding is needed concerning how water is used and how that service is valued in the community.

In the future, *fit-for-purpose water use* should mean that potable water is not employed for such tasks as toilet flushing and gardens, which account for over 50% of all domestic water use. Transition to an *integrated urban water system* that incorporates *stormwater* (city as catchment) and *recycled wastewater* represents the basis for a sustainable water supply for Australia's cities and avoids the need for ongoing water restrictions as a permanent water conservation method (Maheepala and Blackmore; Diaper *et al.*; Speers).

Again, there are a range of barriers to be overcome, including: community acceptance of an entirely new water supply system, which may involve recycled water for potable use (Syme and Nancarrow); conflicts between lifestyle values that govern household consumption and conservation values that support maintenance of garden amenity in the suburbs; and the capacity of current institutional structures and administrative processes to accommodate a shift to more decentralised, closed-loop urban water systems.

PWN

29

Urban Water Futures

Andrew Speers

This chapter sets out some of the key issues that will arise in the future with regard to the management of urban water systems. Foremost among these is the need to place urban water systems on a more sustainable footing and, in doing so, to design systems that suit local conditions—whether they be economic, social or environmental—rather than providing a 'one size fits all' solution.

It is useful to begin by considering the context of current urban water systems and the drivers for change. The systems we currently employ are not substantially different from those established in the early part of the last century. The goals of those systems were to separate pure drinking water from wastewater by as great a distance as possible and to remove pathogenic waste from the urban environment as safely as possible. Our goals remain the same today, so it is therefore not surprising that our approaches have not changed greatly. We still seek to protect human health, to provide water for drinking and hygiene and for the production of goods and services.

What has changed substantially is our appreciation of three critical factors:

- the demands we place on urban water systems are not sustainable. We often require more water than can be reliably extracted from the environment without seriously degrading it.
- the disposal of wastewater can have acute and chronic impacts on the waters in the area of discharge
- there is significant uncertainty about the impacts of climate change on the environments we use for our water supply and on wastewater disposal, such that there is a significant risk of systemic failure should conditions deteriorate.

In seeking solutions to these problems, we need also to appreciate that funds required to make the transition to new urban water systems are not unlimited and that community preferences must be considered in their design.

However, in considering the future of urban water systems, the assumption should not be made that all water systems are unsustainable, or that the degree of unsustainability is the same in each system. Consider, for example, some of the water systems serving parts of the south-western coast of Victoria, such as Portland and Pt Fairy. These use groundwater supplies that are plentiful and which emerge under pressure. Water is reticulated using gravity and wastewater is collected and treated prior to discharging at high rates of dilution to an ocean environment that is well mixed and turbulent. In these systems, rates of water use are appropriate to the supply and neither acute nor chronic water pollution is evident. While more could to be done to reduce power consumption, use greener sources of energy and reduce materials usage, it would be hard to describe these urban water systems as unsustainable.

Even where unsustainable conditions do exist, there is no single solution. Recycling of water, use of rainwater tanks and similar measures are appropriate only in the right conditions. Careful analysis is required, taking into account community preferences, to design a system that is well used, cost effective and which mitigates negative impacts. Such systems will differ from location to location and would be affected by factors such as cost, geography, geology, climate, customer-base, demand, and water source and disposal options, among others.

A further note of caution is required with regard to the goal of sustainability. Considerable

Table 1: Key water issues

Issue	Importance
Pricing	Economically efficient means of rationalising demand; driver for the development of economically efficient alternatives and management options
Demand management	Economically efficient; often represents 'low hanging fruit' or 'no regrets' options
Advanced treatment options and Alternative supplies	Membrane technology improvements and reduction in cost opens the way for alternative sources to be exploited
System configuration	The development of improved treatment technologies opens up the opportunity to re-configure water systems, localising sources of supply and potentially revealing efficiencies
Systems approach	Integration of our understanding of systems (e.g. impact of water treatment and use on energy consumption) are important for development of sustainable outcomes

effort has been put into defining the term—suggestions of great subtlety and elegance have been made. Valuable work has also been done in developing indicators of sustainability so that progress toward the goal might be measured. However, our most strenuous efforts should be directed to moving *away* from *un*sustainability rather than working out what sustainability will look like when we get there. It is not, therefore, the intent of this chapter to suggest what a sustainable urban water system will look like, but an attempt is made to recognise some of the issues and alternatives that we might consider in improving upon current, unsustainable conditions.

We must at least understand that the concept of sustainability requires the integration of social, economic and environmental factors. Solutions need to take into account social impacts, both on current and future generations, and on groups within society. They must recognise that resources are not limitless and, consequently, we must ensure that solutions are economically efficient and that they protect or enhance the environment's quality and productive capacity.

The remainder of this chapter examines some of the key responses that might be developed to improve upon those urban water management systems and approaches that are demonstrably unsustainable. The selection of these factors is subjective, but a justification for their inclusion here is outlined in Table 1.

This chapter concludes with a brief discussion of decision-making processes for urban water sustainability and future research needs.

Pricing

During the early 1990s, the Council of Australian Governments (COAG) initiated a far reaching program of water restructuring, a key component of which was price reform (COAG 1994). Whereas domestic water prices had traditionally been based on property values—and were therefore a form of taxation—the COAG reforms encouraged the adoption of pay-for-use pricing. Industrial and commercial users also benefited from water becoming a controllable cost. Overall, the general intent of price reform was to have the prices charged reflect the actual cost of providing the service—including capital, operating and maintenance costs—and a market return on the investment made. The reforms sought to use price as a means to rationalise consumption and ensure that utilities were financially self-sustaining.

It is clear that consumption per capita has fallen, but the research has not been done to determine the contribution of price reform to this fall. Studies of the price elasticity of demand have been carried out—notably by Hunter Water Corporation. However, this has not been a consistent area of study and a conclusive understanding of the impact of price on the various market sectors—domestic, industrial and commercial—is not available. Our understanding has been confused by the introduction of a number of other educational and motivational campaigns at the same time, and by water restrictions linked to drought and climate fluctuation. Nevertheless, it would be most surprising if establishing water as a cost that can be reduced through frugality did not provide a significant incentive for individuals and industry to reduce consumption or for service providers and appliance manufacturers to respond by developing more water efficient appliances and services.

A critical component of water price reform cited by the COAG was the incorporation, where possible, of environmental 'externalities' in water prices. While the National Water Commission (2005) reported that progress has been made in line with expectations, there is still much to be done. The term 'externalities' refers to those costs and benefits that accrue to third parties for which they are not compensated or do not pay. An angler whose catch falls because of pollution from a sewage treatment plant is bearing a cost of the operation of that plant and its supporting systems. Similarly, the demolition of a redundant urban power plant and its replacement with parkland would confer a benefit on householders whose view was improved, but for which they do not pay. Typically, externalities associated with urban water systems are negative, involving as these systems do, abstraction of resources from the environment, the significant alteration of those environments and the disposal of wastewater

The COAG water reform agenda allows an extended period for utilities and governments to incorporate environmental externalities in price. This reflects the fact that externalities are difficult to quantify and, as a result, it is difficult to set a price accurately reflecting their value that would be accepted in the electorate. Nevertheless, a range of mechanisms exist to internalise externalities not all of which require direct increases in price. Young (2000) suggested that non-financial means of internalising externalities would include:

- *motivational schemes*—these might include, for example, programs like 'Tidy Towns' to encourage water conservation or wastewater minimisation/re-use. Clearly, the lower the demand placed on these systems, the smaller the value of the 'externality' created
- *informational campaigns*—these are not dissimilar to motivational campaigns, and would include public education campaigns highlighting the value of responsible use of water and wastewater systems
- *regulation*—the introduction of more targeted and stringent regulation can reduce or eliminate externalities. Returning to the earlier example of the angler whose catch is reduced due to water pollution, it is clear that the externality would no longer exist if regulation prevented these discharges from occurring.

While these are non-market mechanisms, it should not be assumed they have no impact on price. To be efficient, the cost of the motivational, information and regulatory campaigns should be passed onto consumers (in case these costs become 'externalities' themselves) and the cost of improving treatment processes or purchasing water efficient appliances of course represents the cost of internalising externalities.

Young (2000) also refers to what he calls 'right-market mechanisms' (also referred elsewhere as economic instruments, although Young feels the term is misleading). Right-market mechanisms change or introduce a restraint on market processes. Young suggests that in the urban water context they can be particularly useful in reducing externalities associated with existing urban water systems. Examples include:

- *Off-set mechanisms.* These require action to be taken to off-set the impact of an activity or development. For example, a contribution may need to be made to the restoration of a creek to off-set the installation of a hard surface that increases run-off elsewhere.
- *Load-based licensing.* These licences are structured so that there is a cap on concentration and total load of pollutants discharged to the environment. There may be an implied trade-off between the pollutants and loads, and the process establishes the basis for a trading scheme involving more than one source of discharge
- *Tradeable water rights.* In Australia, water trading schemes are now well established and used, for example, extensively throughout the Murray Darling basin. The extent to which they might be used in an urban area is debatable. However, it is clear that the potential exists for trading between urban and rural areas. The more extensive use of market mechanisms tends to increase water prices, reflecting the true value

and scarcity of water. Under such arrangements, stormwater, for example, becomes valuable and may be cleaned and re-used rather than being discharged in an unregulated way. Negative externalities can thereby be incorporated in price.

Demand management

A corollary to price reform and the internalisation of externalities as described above is demand management. While not suggesting that demand management is subordinate to price reform, it also reduces externalities and it is often a very cost effective way to extend water supplies.

Demand management has been widely practiced in Australia for a number of years and should continue to be a critical component of the management of urban water systems into the future. Demand-management techniques include education, regulation and restriction, installation of water efficient appliances and practices, and their subsidisation to stimulate market development and substitution (such as replacing flushing toilets with composting toilets in appropriate situations, or using means other than water for cooling industrial processes). The specific techniques of demand management are not, however, the subject of this discussion. Instead we must look at the relative cost effectiveness of demand-management options, which, in turn, justify a significant and ongoing effort with regard to the promotion of demand-management techniques.

Table 2 compares various demand-management options with some alternative supply options. This table is adapted from a paper by Turner *et al.* (2005) which contributed to the development of a long-term water supply strategy for Canberra. Thus, it should not be taken that the costs presented in this table will apply to all circumstances. Furthermore, the table does not include options that might be available in certain areas, but not in Canberra, and which may be more expensive than the options presented here, making these listed here relatively more attractive. This table presents the outcomes as the 'levelised cost' of water saved or supplied. Levelised cost is the present value of the stream of all costs over a set period divided by the present value of the stream of water demand saved or supplied. It takes into account 'whole of society costs', which include capital and operating costs of a scheme and, in the case of options such as demand

Table 2: Levelised cost of water management options for Canberra

Option	Levelised cost (AUD/kL)	Savings assumptions
Demand management		
Pricing and information awareness	0.08	Simplified assumption of 5% of 75% of single residential houses
AAA rated showerhead rebate	0.22	16.5 kL/household/pa
AAAA washing machine rebate	1.02	50% reduction compared with top loading machines
Minimum water efficiency performance standards (MWEPS)	0.03	Simplified assumption where MWEPS provide savings in new washing machines and maintain savings in other end uses
Source substitution		
Rainwater tank rebates (existing houses)	10.62	35 kL/household/pa for a 5kL tank in a single residential household
Rainwater tank rebates (new houses)	4.45	55 kL/household/pa for a 10kL tank in a single residential household
Greywater system rebates (existing houses)	5.13	50% of outdoor use in single residential household
Greywater rebates (new houses)	4.87	50% of outdoor use in single residential household

Source: adapted from Turner *et al.* (2005)

management, the costs of marketing project management and evaluation (Turner *et al.* 2005)

The table highlights the potentially significant differences in levelised unit cost between supply-side options and demand-management options. Recognising that this table says nothing about what the best options for Canberra might be—as this is only a partial list and the contributions to be made by demand management and other options to Canberra's long-term water supply needs are not quantified—there is nevertheless a stark difference in demand-management costs and supply-side costs. This would suggest that demand-management options should always be considered in planning a water management strategy.

With regard to deployment of demand-management options, it is important that the education and awareness-raising campaigns are not a 'one-off' effort. While replacement of inefficient household appliances with more efficient models may produce a permanent reduction in water demand, there is an ongoing need to provide incentives and educational materials so that awareness is maintained. The AAA appliance water rating scheme is an example of a valuable awareness raising program that could be maintained and expanded.

It is regrettable that many commentators and members of the community do not view demand management in the same light as 'built' solutions. While the latter are more tangible, the volume of water saved by demand management can be as significant as the volume produced, and often at lower cost and environmental impact. Demand-management schemes will continue to be a part of any future urban water system.

Alternative sources

Urban water systems tend to be linear in design—taking water from a remote location, delivering it to homes and businesses for single usage applications and returning it to the environment at a location unlikely to have a human health impact. Not all systems are as straightforward as this description, of course, with users often supplied from numerous points and cross-connection existing between systems. However, the basic configuration is as described.

Opportunities to obtain water from alternative sources are well known and increasingly discussed. They include:

- recycled water—either for non-potable or potable purposes
- greywater—re-used for second quality purposes (such as cooling water having a second life as wash-down water) or for garden watering in the domestic environment
- desalinated water—sourced from oceans or estuaries and used for potable or non-potable purposes
- groundwater—sourced from previously untapped supplies
- stormwater—gathered from urban environments, for non-potable or potentially potable purposes following treatment
- rainwater—gathered at the household or community level and used for non-potable purposes such as garden watering and toilet flushing
- dams—containing water from previously unregulated rivers.

There is not the space here for an analysis of the efficacy of each of these options. They may all have potential in certain circumstances and may be sustainable under those conditions. In considering future urban water systems, however, a number of issues are usefully highlighted

Transport costs

Underground infrastructure—pipes, pumps and associated facilities—needed to transport water to customers and to remove wastewater from their premises typically represents 50–75% of a utility's costs (both operating and capital). Thus, in considering alternative water sources, attention will need to be paid to the transport costs associated with accessing them. That is, the cost of water is not merely the cost of production (such as impoundment and treatment) but also the cost of getting water to and from consumers. In some instances, transport costs can militate against the use of certain water sources.

Consider, for example, two potential new water sources: desalinated sea water and recycled wastewater. Over the past two decades, the costs of desalination have fallen significantly as improvements in membrane technology have occurred and as the market for such technology has expanded. Zhou and Tol (2004) cite a cost for the production of desalinated water using sea water as a source as $US1/m^3 (and $US0.60 if brackish water is used). However, the cost of transport of desalinated water, which, by its nature is sourced from the lowest point in the environment, may range from a few cents per cubic metre to more than $US1. Consequently, desalinated water may be prohibitively expensive for interior areas and areas of high elevation. In comparison, water sourced from wastewater may frequently be available in an area where access to existing infrastructure might be available and/or at an elevation where gravity can aid flow.

Desalinated water is being considered for widespread use in jurisdictions throughout Australia. The above discussion is not intended to suggest that such schemes are not viable. Rather, it suggests that careful consideration be given to the cost of transporting water from any source, as these costs can be high in many circumstances.

Treatment costs

A further consideration is the cost—both in financial and environmental terms—of treatment of alternative water sources. Present debate in Australia appears to be favouring those sources considered most 'pure' by the community (or perhaps politicians interpreting their attitudes!) Thus, sea water, river water and groundwater are favoured as sources over stormwater and wastewater and other second quality sources. It is not the case, however, that these favoured sources are always the most 'pure' or the cheapest to treat.

In the Singapore NEWater™ system a 'multiple' barrier approach is employed comprising conventional wastewater treatment, followed by microfiltration, reverse osmosis and ultra-violet disinfection. The resulting high-quality water is then used for non-potable purposes with the 2010 goal of substituting 15% of Singapore's non-potable water demand with NEWater. Because it is already treated to a high level, using conventional treatment processes, the quality of the source water is already quite high. Due to advances in membrane technology, many sewage treatment facilities in Australia and elsewhere already employ a level of treatment beyond conventional systems, producing an even higher effluent quality. The marginal cost then of taking treated effluent to potable water quality – whether or not the product is used for potable water supply – then becomes relatively small and may be very competitive against other more favoured sources.

Once again a comment is required lest the above be taken as an assault on the appropriateness of using sea water, river water or groundwater as a source of future supply. Any of these sources may be suitable in the right circumstances, and community opposition to other alternatives may mean that these sources are seen as the only available options. However, as treatment technologies improve and our understanding of the risks of using particular supplies improves, the range of options available to us should be expanded. If we seek efficient options, the cost of delivering water from the full range of sources available at the quality demanded must be carefully considered.

Localisation

The development of more effective sewage treatment technologies also opens the way to decentralisation of facilities. Stormwater, treated wastewater or greywater, lower quality groundwater, and so on, all become potential sources of supply. Furthermore, as the costs of production fall, the opportunity exists to source waters locally as the economies of scale historically present in sewage treatment facilities are eroded. For example, in 1999 CSIRO analysis suggested that the optimum size for a sewage treatment facility incorporating recycling to consumers was 10,000 connections or less. This finding was also influenced by the reduction in transport costs that arose though localisation.

Taken to its ultimate extent, localisation may mean treatment of water on site (probably at industrial premises where large volumes are involved) and extensive re-use. It might be possible to have an enclosed system in the future, with only limited input of new water. In this regard, regulation is an important driver. Part of the regulatory push in Australia has been to improve the quality of effluent discharged by industrial and commercial premises. Trade-waste policies, and similar mechanisms, have employed economic instruments and more conventional regulatory mechanisms to encourage or require discharges to an improved quality level. Producers have responded by installing pre-treatment facilities and by reformulating industrial processes. In future water systems there will be significant opportunity to re-use industrial wastewater in industrial processes, either within the premises at which they were produced, or at neighbouring facilities. Once again, the avoidance of transport costs (possibly including avoided discharge fees) will make improved treatment of wastewater and its re-use that much more viable.

It short, the development of ever cheaper and more effective treatment processes opens up a myriad of water sources to potential exploitation, either for potable or non-potable purposes. The effects of these developments may be more extensive in new development areas where infrastructure has not yet been established—existing infrastructure represents a major investment that is not going to be lightly abandoned or reconfigured. Nevertheless, there are chances to exploit the new technologies extensively, such as 'sewer mining', involving abstraction of sewage from existing mains close to users and its treatment for re-use in industrial processes.

Systems thinking

Systems approaches are dealt with more fully in other chapters. These require consideration of the overall operation of water systems, what these systems represent in terms of the provision of services, and the relationship between water systems and other systems. Thus, for example, we should be forced to think about the relationship between water abstraction, treatment and use and energy consumption. In doing so, we might begin to consider the location of water that we use and the energy consumed in delivering it to us. In a systems approach, we might – if energy were our only consideration – choose to use stormwater more widely as the energy involved in transportation might be less that transportation of that water from a remote groundwater source.

Therefore the systems approach becomes important if we are to understand infrastructure not in isolation, but in the context of a complex urban environment where causes can have widespread effects. One way of thinking of systems is to develop the concept of urban metabolism, which draws its inspiration from a comparison with the biology of the human body (Gandy 2004). A biological entity requires inputs of energy, water and materials to function, but the configuration of these and the relationships between them define the form and performance of the entity and its fitness to survive. We can measure the metabolic rate of the entity by measuring inputs and understanding transformations. So, too, it is suggested, can we understand the metabolism of the city and the interactions between systems by measuring the city's inputs and outputs and monitoring transformations.

As can be seen from the simple example above, systems thinking encourages us to consider the range of influences on a particular system, not just that system in isolation. Thus, systems thinking would encourage us to understand that water systems in an urban environment such as tropical Cairns could differ significantly in form and function from one operating in Melbourne. For example, rainwater tanks may be more economically viable in Cairns because of the more plentiful rainfall (greater input to the system) than in Melbourne where tanks might have to be supplemented frequently by reticulated supplies.

Despite these benefits, however, any modern conception of urban water systems needs to consider the *social* conditions in which systems operate—and this is too often forgotten. The current debate in Australia about the use of recycled water as a source of drinking water highlights the importance of considering community views in designing systems.

Proponents of using recycled water for drinking water and those that claim such use will damage human health are equally strident. Politicians respond to this debate by either not participating in it or suggesting that the public will not accept recycled water for drinking. Consumer surveys detect a difference of opinion about recycled water use when it has been perceived to have been transformed by the environment such as when it is discharged to a river and abstracted later for use) and when it is proposed to be injected directly into the drinking water supply, even though it can be conclusively demonstrated that there is no difference in water quality and risk (or that the former approach might actually increase risks). Thus, designing a system that is efficient in terms of urban metabolism—that is, which minimises use of water, energy and materials, and which reduces or eliminates the production of waste products—may not produce a usable system because it may be rejected by consumers either at the concept phase (before construction) or—more damagingly from an economic and political perspective—when the system is commissioned. See Po *et al.* (2005) for an analysis of individual's responses to use of water from various sources and the key determinants of those attitudes.

A second limitation with a systems approach is that we must recognise that much of the infrastructure of our cities is already in place. In the case of water systems too, the infrastructure is of such a monumental and inflexible nature, that it cannot be easily re-designed. What systems thinking might now suggest we do if we could start from scratch might not be economically or physically possible. This is not to discard the concept, but to recognise that we need to engage not only in the conceptual but also the concrete—and that the most efficient outcomes of our systems thinking might often be generated at the margin, such as in new developments on the urban fringe.

Decision making

The discussion above highlights the complexity of urban water systems development and design. Because the nature of sustainability requires consideration of economic, social and environmental factors, new approaches are required that enable dissimilar factors to be compared using a common denominator. One such technique is multi-criteria analysis (MCA). MCA describes any structured approach that seeks to determine overall preferences among alternative options, where the options accomplish several objectives. In MCA, objectives are identified and corresponding attributes or indicators are developed. Different environmental and social indicators may be developed alongside economic costs and benefits to measure progress toward the identified objectives. MCA includes techniques (scoring, ranking and weighting) for comparing and ranking different outcomes.

Other techniques, both proprietary and public, are available with the goal of promoting effective decision making where multiple competing objectives exist. These techniques are being more widely used. While they can be criticised for their complexity, and the difficulty of getting all interested stakeholders involved, they should be an important and well-developed part of decision making, particularly when decisions of a fundamental nature—such as the development of a water supply strategy for a region—are being considered.

Research requirements

The preceding discussion highlights some of the opportunities available to develop more sustainable urban water systems. The market for more sustainable systems is being stimulated by consumer concern, regulation and price. These trends will continue and will be further stimulated by the development of better pricing structures that internalise externalities, and improvements in membrane technology that will enhance performance and reduce costs of treating wastewater. More innovative approaches to system design, which capitalises on these improvements, will enable a greater range of sources to be tapped, potentially localising supplies and reducing transport costs.

The full potential of these developments will, however, only be realised if a number of critical issues are tackled. Foremost among these as far as the community is concerned is the safety of drinking water supplies, in particular supplies sourced from recycled water. A number of high-profile direct potable re-use schemes, most recently in Toowoomba in Queensland, have been defeated due to community resistance. It is evident, however, that the debate over the safety of this source of supply was characterised by misinformation and misunderstanding. A significant and public research effort should be initiated in an attempt to prove or disprove the safety of direct potable re-use.

The costs and benefits associated with sustainable water supplies and wastewater disposal systems are not efficiently captured by those benefiting or being negatively affected. For example, in many jurisdictions, a company significantly reducing its dependence on the water and wastewater systems, may not capture all of the savings generated, as the water utility may still base charges on meter size or availability of the service, whether or not it was used. Such policies may be justified on the grounds that all must contribute to community assets, but they do work against the provision of incentives to the installation of more self-contained or sustainable systems. Research into the best means of stimulating the market through the efficient allocation of benefits and costs would be valuable.

More holistically, research into efficient means of internalising externalities would be valuable. Important research has already been done in this regard (Young 2000) but general principles have not yet been widely applied. Internalisation of externalities would rationalise water consumption by setting a price for water that more accurately reflects the long-run marginal costs of the provision of water services, including currently unaccounted for costs and benefits. As an adjunct to this research, it would be useful to determine what opportunities might be available for water trading within urban areas. While significant difficulties exist in this regard, the potential availability of new supply sources opens opportunities up to more extensive water trading.

Conclusion

Future urban water systems need to be more organic—that is, developed in response to the context in which they are situated. We strive for sustainable water systems, but in doing so we will need to avoid 'one-size-fits-all' solutions. As we develop such responses, we will be able to use new sources of water, including stormwater and treated wastewater. However, in selecting the most appropriate system, we should pay close attention to the costs of transport of the water and the environmental impacts of this transport. We will,however be in a position to reconsider the scale at which water systems operate—and decentralisation may provide cost and sustainability opportunities not previously available, with self-containment on a single site or among a number of co-located industrial premises becoming feasible.

More sustainable systems will, however, not be as simple in their design, nor will the 'preferred' approach necessarily be understood or supported by stakeholders. It will therefore be incumbent on practitioners to improve decision-making processes, to properly research alternatives and to search widely for a comprehensive understanding of the impacts and costs of alternative approaches.

References

COAG (Council of Australian Governments) (1994). Communiqué 25 February,1994. Attachment A – Water Resource Policy. <www.coag.gov.au/meetings/250294/attachment_a.htm>

Gandy M (2004). Rethinking urban metabolism: water, space and the modern city. *City* **8(3)**, 363–379.

National Water Commission (2005). *Water Reform Assessment Framework*. National Water Commission, Canberra.

Po M, Nancarrow B, Leviston Z, Porter N, Syme G and Kaercher J (2005). 'Predicting community behaviour in relation to wastewater reuse. What drives decisions to accept or reject?' CSIRO Water for a Healthy Country Flagship Report. CSIRO, Perth.

Turner A, White S and Bickford G (2005). The Canberra least cost planning case study. In *Proceedings of the 2nd International Conference on the Efficient Use and Management of Urban Water*. International Water Association Publishing, London

Young M (2000). *Managing Externalities, Opportunities to Improve Urban Water Use*. CSIRO Land and Water, Adelaide

Zhou Y and Tol RSJ (2004). *Evaluating the Costs of Desalination and Water Transport* <www.uni-hamburg.de/Wiss/FB/15/Sustainability/DesalinationFNU41_revised.pdf>

30

Integrated Urban Water Management

Shiroma Maheepala and Jane Blackmore

Water supply for Australia's cities is reaching crisis point. Current sources are inadequate to meet projected demand, and the situation is exacerbated by climate change. The compound effects of population growth and diminishing supply, coupled with the acknowledged needs of the environment, both within and outside our cities, mean that solutions based either on single components of the water cycle or solely on traditional surface and ground water sources are no longer feasible. To drought-proof our cities for future generations, and to maintain ecological health and diversity, it is essential that we adopt a holistic approach to urban water management. Such an approach requires a major change to the way we view and manage supply, use and disposal of water in our cities, and cooperation and collaboration at all levels of urban water management. We call such an approach Integrated Urban Water Management (IUWM). This chapter provides details of the IUWM approach, including its definition and key features, and discusses its relevancy to water management in Australian cities. Furthermore, we discuss factors to be considered and processes to be followed when adopting the IUWM approach at a city scale.

What is Integrated Urban Water Management (IUWM)?

IUWM is a process that encourages water utilities to plan and manage water supply, wastewater and stormwater systems in a coordinated manner to minimise their impact on the natural environment, to maximise their contribution to economic development and to engender overall community wellbeing and improvement. The IUWM approach considers the entire urban water cycle, including:

- all possible water supplies (such as surface water and groundwater, stormwater, reclaimed wastewater and desalinated water)
- human and ecological demands for water
- structural and non-structural measures to improve the efficiency of water consumption, distribution, treatment and disposal
- the local context
- stakeholder views
- access to safe, reliable and economical urban water services without degrading freshwater and terrestrial ecosystems.

A number of alternative terms to IUWM can be found in the literature. These terms include *Integrated Water Cycle* (Coombes *et al.* 2003; NSW Department of Utilities, Energy and Sustainability 2004) and *Integrated Urban Water Resource Management* (Global Water Partnership 2007; United Nations Environmental Program 2007). While the processes and systems to which these names apply have much in common with IUWM, the use of alternative terms indicates a need for a universally accepted definition of the IUWM approach.

Integrated Water Resource Management (IWRM) is a term that is commonly used in planning at river basin level (Jønch-Clausen 2001; Davis 2007). It is defined as a process that promotes the coordinated development and management of water, land and related resources, in order to maximise the resultant economic and social welfare in an equitable manner without compromising the sustainability of vital ecosystems (Global Water Partnership Technical Advisory Committee 2000).

The IWRM approach deals with the water allocation problem at regional and/or river basin level. A region might include a number of urban areas as well as rural, hydro-electricity and agricultural users. The IWRM approach helps in making decisions on regional water allocations by considering the needs of all the competing users. IUWM can be viewed as a subset of IWRM that is concerned with the management of water supply, wastewater and stormwater in urban areas within the boundary conditions set as part of the IWRM process (such as total annual urban water allocation). Close interaction and communication between IWRM and IUWM is critical if each is to be successfully implemented.

Features of IUWM

Conventional urban water systems generally consist of a water supply system, a wastewater system and a stormwater system. In general, a water supply system consists of headwork infrastructure to abstract water from surface and ground water sources, treatment infrastructure to produce water to drinking or potable water quality and distribution infrastructure for reticulation of potable water to end uses. Both wastewater and stormwater systems consist of pipes, pumps, temporary storages and treatment infrastructure for collection, treatment and disposal of sewage and stormwater. In many countries, stormwater is not treated before disposal. Some countries, particularly in the USA and Europe, use a combined sewer system, which is a single system for collection of both stormwater and wastewater. The conventional water systems are centrally managed systems with infrastructure to bring in water for human consumption and remove wastewater and stormwater generated in urban areas as quickly as possible to inland or marine receiving water bodies.

The *traditional paradigm* for urban water management takes the view that the urban water system consists of water supply, wastewater and stormwater system and the main objective of managing it is to improve the health and hygiene of urban areas from waterborne diseases and floods. Unfortunately, methods used to achieve this objective have caused many adverse environmental and economic impacts (Pinkham 1999; Marsalek *et al.* 2001; Vlachos and Braga 2001). They include irreversible modifications to natural ecosystems with negative consequences for native flora and fauna in and around waterways, river basins and coastal waters, and rising costs of managing assets that transport potable water, wastewater and stormwater over long distances.

The *IUWM approach* has emerged as an alternative to the traditional paradigm due mainly to its potential to provide solutions to multiple objectives and reduce such adverse impacts (Pinkham 1999). Features of the IUWM approach include:

1. **Providing water security through source diversification.** This feature implies use of a variety of supply sources, such as surface water, groundwater, recycled water, stormwater, roof water, grey water and desalinated water, to meet the urban water demand. Some sources, such as stormwater, grey water, recycled water and roof water, originate from urban areas, which implies considering urban areas as watersheds or catchments, rather than sinks that consume water. Some of these non-conventional sources, such as recycled water and grey water, have the potential to provide a reliable water supply even in times of prolonged drought, because of their non-dependence on rainfall. Increased community concern about the use of ongoing water restrictions as a permanent water conservation measure makes the concept of source diversification an attractive alternative to the traditional approach to provide urban water services (WSAA 2007).
2. **Reducing demand on potable water supplies through seeking to provide water that is fit for purpose.** About half of the urban water demand is currently used for purposes that do not require potable water.
3. **Enhancing urban amenity values** by improving urban landscape and providing habitat for native flora and fauna. Adverse impacts of stormwater discharge on urban waterways and rivers can be addressed using concepts such as *day-lighting* and *water sensitive urban design.* Day-lighting is the deliberate exposure to light of a previously covered river, stream or stormwater drain in a more natural form, where feasible, or by providing a new channel or waterway

(Pinkham 1999). Water sensitive urban design (WSUD) is design of urban landscape to provide measures, such as grass swales, bio-filtration, ponds, wetlands and lakes, to attenuate and reduce discharge of urban run-off and associated pollutants to receiving waters (Wong 2006). WSUD is also known as Sustainable Urban Drainage Systems or SUDS in the UK and Low Impact Developments or LID in the USA (Department of Environmental Resources 1999).

4. **Promoting the use of a broad range of components that can be mixed and matched to provide a water service that is appropriate for local conditions.** Components can be structural and non-structural. The structural components include the infrastructure required to provide water conservation and efficiency, recycling, stormwater harvesting, greywater use and rainwater tanks, as well as traditional tools for water supply, treatment and disposal. The non-structural components include education, pricing, incentives and regulations. Alternative packages of IUWM components can be assembled, evaluated and adjusted to achieve optimal outcomes for the multiple functions required of urban water services (Mitchell 2006). These outcomes include amenity values, ecosystem protection, energy use and greenhouse gas emissions, public health protection and flood protection. This contrasts to the traditional approach in which public health and flood protection are the main objectives.
5. **Seeking to provide localised and decentralised urban water services** that can reduce long-distance water, wastewater and stormwater transportation costs.
6. **Seeking cooperation between governments and utility sectors to make multi-objective decisions that are aligned with the principles of sustainability.** This requires coordination, collaboration and participation in the management of water supply, wastewater, stormwater and receiving waters in urban areas
7. **Avoiding short-term, localised and single-sector-based decisions**, which often result in undesirable outcomes at the system or the whole-urban-system level.
8. **Providing more sustainable water systems**, which are defined as systems that are designed and managed to contribute fully to the objectives of society, both now and in the future, while maintaining their ecological, environmental and hydrological integrity (ASCE and UNESCO 1998)

Why do Australian cities need IUWM?

The current population in Australia is about 20 million. About 64% of Australians live in urban areas, and this is expected to increase by 35%, or 4 million people by 2030 (see Table 1). Of Australia's urban population, 63% live in Sydney and Melbourne, and 27% live in Perth, Adelaide and Brisbane. The highest growth rates will be in

Table 1: Population in Australian cities in 2006 and the estimated population growth by 2030

City	2006 Population (000s)	Projected population in 2030 (000s)	Increase in population by 2030
Adelaide	1,133	1,260	11%
Brisbane	1864	3,071	65%
Canberra	330	447	35%
Darwin	115	200	74%
Hobart	206	254	23%
Melbourne	3,683	4,870	32%
Perth	1,512	2,299	52%
Sydney	4,308	5,387	25%
Total	12,774	17,287	35%

Source: data from WSAA (2007)

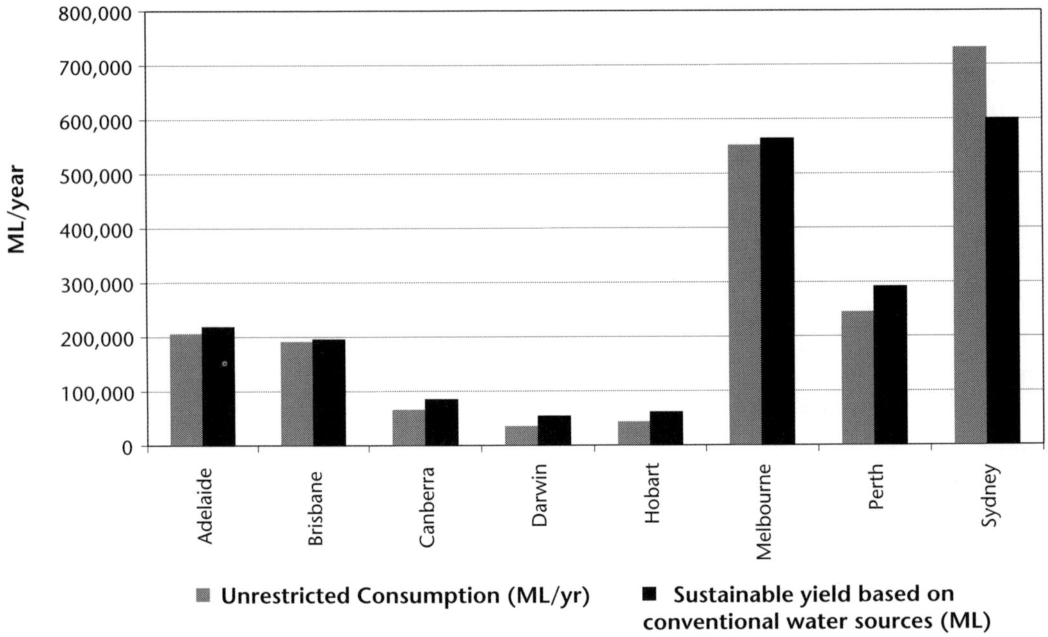

Figure 1: Comparison of 2001–02 financial year unrestricted consumption and sustainable yield (based on conventional surface and ground water sources) in major cities (see Table 2 for sources of data)

South-East Queensland (Brisbane and surrounding areas) and some 80% of the additional population will live in Sydney, Melbourne and Brisbane (see Table 1). Therefore, the key challenge faced by the urban water industry in Australia is to ensure that adequate water supplies are available to meet economic and social needs of the growing population in major cities without degrading hydrological integrity of water resources and the surrounding natural environments.

Do Australian major cities have adequate water supplies to meet needs of the expected population growth? Figure 1 and Table 2 attempt to provide an answer to this question by examining supply and demand balance at annual scale in major cities. Brisbane, Sydney and Melbourne, which currently account for 71% of the total urban water consumption for Australia, each have unrestricted water consumption close to their sustainable yield. Seventy-seven per cent of Australians live in Sydney, Brisbane and Melbourne and about 80% of the additional population is expected to live in these cities by 2030. With water supplies just sufficient to support current populations, additional sources and alternative ways of providing services must be found if these major cities are to support expected population and economic growth. The situation is similar for all other major cities in Australia (see Figure 1 and Table 2).

Population growth is not the only reason for the water shortage in Australian cities. Prolonged drought over the last 10 years has contributed enormously to the current state of water availability in major cities. In 2006–07, rainfall in some water supply catchments serving major cities was 80% below the long-term average (WSAA 2007). This meant significant reductions in inflows to storages that had already experienced nearly a decade of drought. The result was an unprecedented depletion of available water. As reported in WSAA (2007), in 2006–07, storage levels dropped from 48% to 35% in Melbourne; 51% to 41% in Canberra; 81% to 77% in Hobart; 100% to 95% in Darwin; 31% to 25% in Perth and from 31% to 17% in Brisbane. Sydney and Adelaide experienced some wet periods in 2006–07; consequently, storage levels increased from 42% to 57% in Sydney and 52% to 75% in Adelaide.

Climate change will exacerbate the supply shortages. The predicted change in mean annual rainfall and run-off in water supply catchments in Brisbane, Sydney, Melbourne and Perth due to

Table 2: Unrestricted consumption and sustainable yield in 2001–02 and average annual sewage collected in major Australian cities over the last five years (2000–01 to 2005–06)

City	Unrestricted[1] consumption (ML/yr)	Sustainable yield based on conventional water sources[2] (ML)	Average wastewater volume generated during 2000–01 to 2005–06 (ML/yr)	Average wastewater generated as a percentage of unrestricted consumption
Adelaide	205,137	218,000	96,830	47%
Brisbane	190,156	195,000	109,074	57%
Canberra	64,570	84,000	30,067	47%
Darwin	34,222	53,420	14,119	41%
Hobart	42,509	60,749	21,255[3]	50%
Melbourne	551,097	564,000	336,767	61%
Perth	244,081	291,000	106,248	44%
Sydney	729,953	600,000	477,998	65%
Total	2,061,725	2,066,169	1,192,358	58%

Data sources: [1] based on 2001–02 WSAA Facts; [2] based 2001–02 WSAA Facts and Australian Water Resource Assessment (www.water.gov.au/RegionalWaterResourcesAssessments; [3] estimated by authors

2030 climate change scenarios is shown in Table 3. All major cities are expected undergo reduction in run-off in water supply catchments. The highest reduction is expected to occur in Perth. Evidence suggests that Perth has already undergone a step-change in climate. Inflows to major storages have decreased from an average of some 338 GL per year in the 1911–1974 period, to some 177 GL in the 1975–1996 period, and to some 114 GL per year during the drought of 1997–2005 (see Figure 2).

Detailed studies were undertaken in some Australian cities, such as Perth (http://www.ioci.org.au), Melbourne (Howe *et al.* 2005) and Benalla—a major town in north-east Victoria (Maheepala and Perera 2003; Maheepala *et al.* 2003a)—to quantify the impact of climate change on run-off in water supply catchments, water consumption and sustainable yield. Such studies provide methods for quantifying the potential impact of climate change by combining the latest IPCC (International Panel on Climate Change) projections with hydrological and water supply system simulations.

The potential impact of both population growth and climate change on the aggregate supply and demand balance of all major Australian cities is given in Table 4. At present, the amount of water

Table 3: Predicted percentage change in mean annual rainfall and run-off by 2030 in water supply catchments in some major Australian cities

City	Brisbane	Sydney	Melbourne	Perth
Mean annual rainfall (mm)[1]	1189	1220	661	869
Mean annual minimum and maximum temperature[1]	9.5°C–29.2°C	6.7°C–26.3°C	5.8°C–25.7°C	8.1°C–31.5°C
Predicted change in mean annual rainfall from the long-term average rainfall[2]	–2% to +2%	–2% to +2%	–5% to –2%%	–10% to –5%
Predicted percentage reduction in mean annual run-off from the long-term average rainfall (corresponding to the change in mean annual rainfall)[3]	12%	9%	11%	24%

Sources: [1] based on 'Climate education: Australia – climate of our continent' (http://www.bom.gov.au/lam/climate/levelthree/ausclim/zones.htm); [2] based on http://www.climatechangeinaustralia.gov.au/futureclimate.php (2007) and [3] based on Chiew (2006).

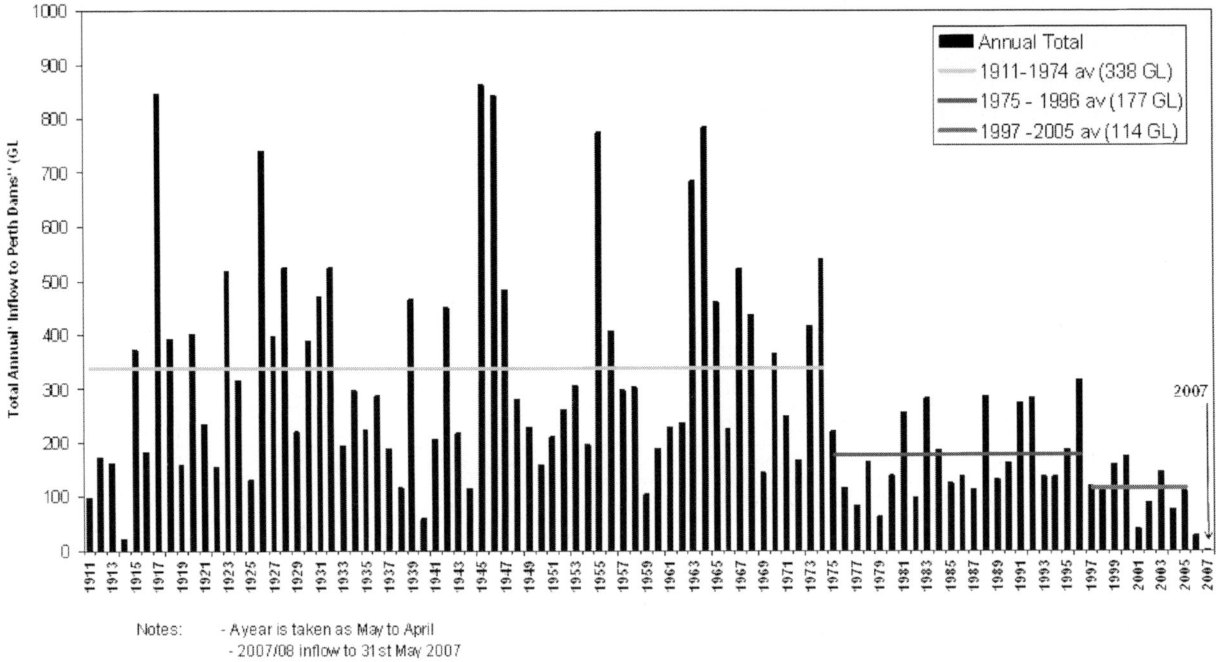

Figure 2: Annual inflows to surface water storages in Perth from 1911 to 2007 (Source: http://www.watercorporation.com.au).

Table 4: Urban water balance—Australia wide

Description	Available water (GL)	Consumption	Total
Current			
Population of Australian capital cities (million)			13.1
Sustainable yield	2066		
Unrestricted consumption		2062	
Difference between supply and demand (GL)			4
Future—2030 with business as usual			
Population of Australian capital cities (million)			17.6
Sustainable yield (25% reduction to account for potential climate change impacts)	1550		
Consumption based on 2004 per capita consumption rates		2811	
Difference between supply and demand (GL)			–1261
Future—2030 with measures identified in urban water resource strategies			
New sources of water (new/larger dams, inter-regional/inter-basin transfers, groundwater)	684		
Alternative sources of water (stormwater, recycled water, roof water, grey water)	195		
Stage 1 desalination (in Perth, Sydney, Melbourne and Brisbane)	210		
Water conservation measures		–326	
Total supply/demand	2639	2485	
Difference between supply and demand (GL)			154

available from conventional sources is just sufficient to meet urban water consumption. However, by 2030, the aggregate annual consumption will exceed the amount of fresh water available from conventional sources. The total expected annual water shortage in 2030 is 1261 GL (see Table 3). Finding solutions to such a vast water shortage in the face of drying and warming climatic condition represents a major national challenge.

Water supply shortfall is not the only issue of concern to Australian cities. The need to protect urban ecosystems from stormwater and wastewater discharges, rising costs of maintaining ageing water infrastructure, understanding community expectations and minimising greenhouse gas emissions from water projects are also critical challenges (WSAA 2007).

The traditional pipe-in–pipe-out management approach cannot provide solutions to these challenges. Fundamental changes to the way we think and a new, scientific approach to the way we manage water supply, wastewater and stormwater as one integrated system are needed. The IUWM approach is equipped with principles and tools to address current challenges confronting urban water management practice. Accordingly, all major Australian cities have begun to embrace various elements of the IUWM approach. *Fit-for-purpose, source diversification* and *WSUD* are the first set of IUWM concepts that have been considered for adoption by the Australian water industry.

Changing priories for sourcing urban water supplies in some major Australian cities are shown in Figure 3, which compares urban water supply sources identified in 2005 and 2007 urban water strategies of Sydney, Melbourne, Brisbane, Perth and Adelaide. It clearly shows embracing of the concept source diversification. The use of this concept in practice (that is, new/large dams, inter-regional/inter-basin transfers, groundwater, stormwater,

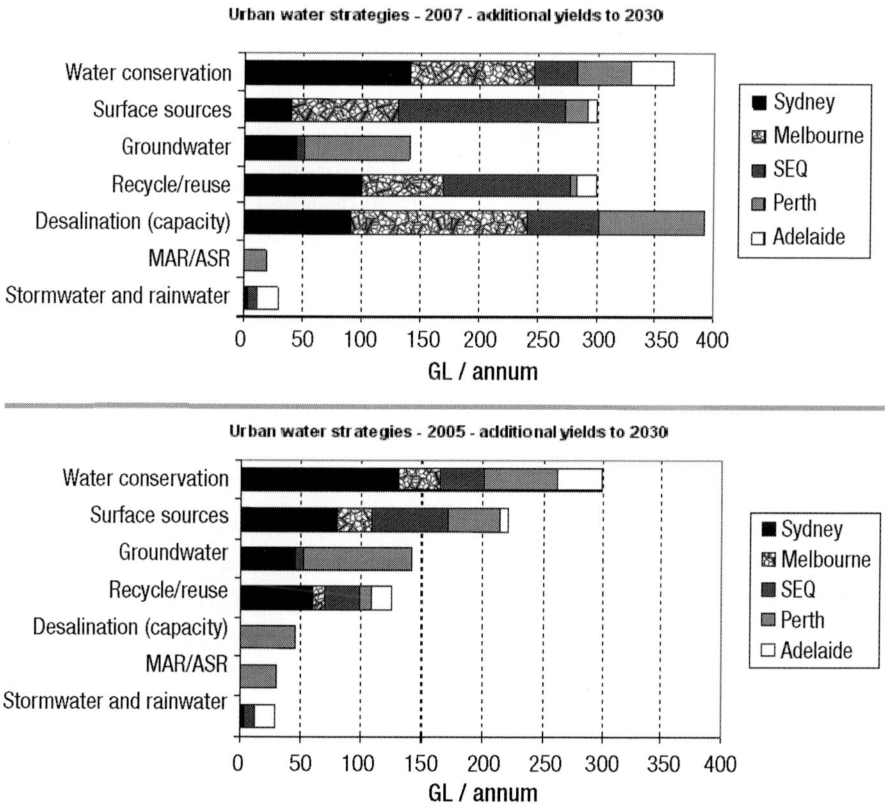

Figure 3: Changing priorities for sourcing urban water supplies up to 2030 in some Australian major cities, as identified in urban water management strategies in 2005 (bottom bar chart) and 2007 (top bar chart); (Data sourced from Urban Water Strategies of Sydney, Melbourne, Brisbane, Perth and Adelaide (references given in Table 5) and compiled by Alan Gregory, CSIRO).

recycled water, roof water, grey water and water conservation strategies shown in Table 4) in major Australian cities has the potential to generate 879 GL per year (that is, the sum of 684 and 195). Combination of new water (879 GL/year) and water conservation (326 GL/year) has the potential to reduce the predicted water shortages in 2030 from 1261 GL/year to 56 GL/year (that is, 1261 GL/year – 879 GL/year – 326 GL/year).

The reason for considering *fit-for-purpose, source diversification* and *WSUD* concepts is mainly because of the realisation that:

- the amount of wastewater generated from Australian cities represents about 50% of the amount of potable water that they use (see Table 2 and Figure 4 for the average volumes of wastewater generated by the major cities in Australia during 2000–01 to 2005–06);
- the amount of stormwater run-off from Australian cities is about equal to the amount of potable water they use (Commonwealth of Australia 2002)
- about 60% of the potable water is currently used for purposes that can be met with lower quality water than the potable water (that is, outdoor and toilet water uses; see Figure 5 for how water is used in Australian households).
- WSUD can improve stormwater quality, protect waterway health and downstream ecosystems and improve the urban landscape.

The use of desalinated water is also being considered as a supply option in major Australian cities (WSAA 2007). A desalination plant has just been commissioned in Perth that has the capacity to produce 45 GL/year. Planned desalinated plants in Sydney, Brisbane and Melbourne will produce 90 GL in 2008, 45 GL in 2008 and 75 GL in 2010, respectively, which will have the potential to convert 56 GL/year Australia-wide water-supply shortfall into a surplus of 154GL/year in 2030 (see Table 4).

While the use of IUWM concepts such as water *fit-for-purpose* and *source diversification* at the planning level clearly indicate a positive step towards the adoption of an IUWM approach, it does not mean that urban water management in Australian cities has changed from traditional to integrated management. Integrated management requires not only the use of these concepts, but also a coordinated and participatory approach to managing the water supply, wastewater and stormwater systems by considering system interactions, flow-on impacts and system risks. However, institutional structures and administrative processes currently in place have not been designed to support the coordinated and par-

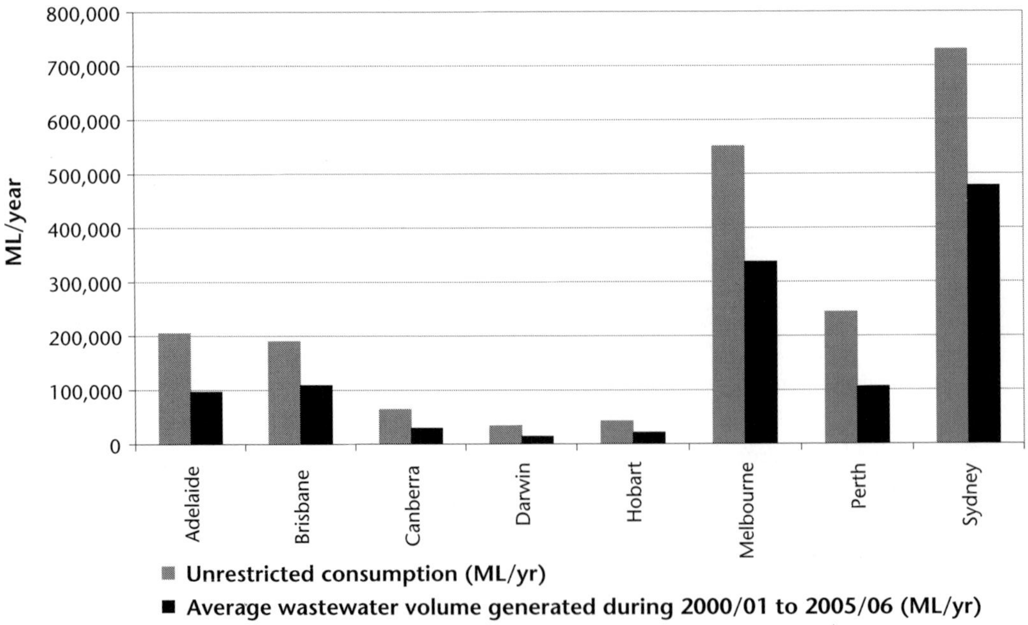

Figure 4: Average wastewater volumes generated in major Australian cities (see Table 2 for sources of data) over the last five years (2000–01 to 2005–06).

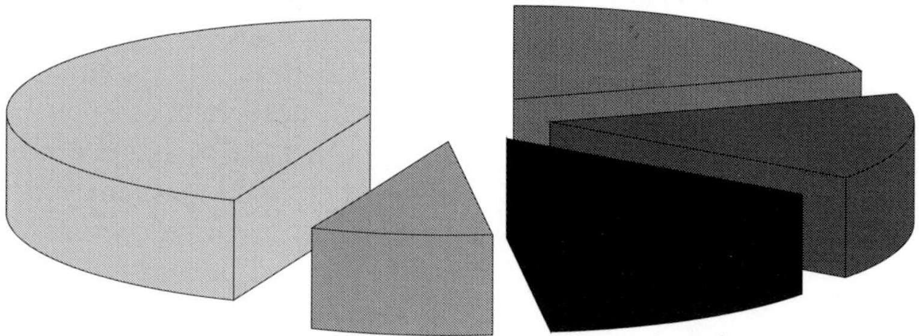

Figure 5: Nature of water use by the urban residential community in Australia (Source: based on WSAA 2007).

ticipatory approach required to adopt the IUWM approach (Brown 2005). Adoption of the IUWM approach is not trivial and it involves a long-term commitment from many stakeholders.

In our view, Australian cities have not yet adopted the IUWM approach, but have begun to embrace the IUWM approach, particularly at the strategic planning level. A list of recently released urban water planning documents by the utilities in major Australian cities is given in Table 5. They all recognised the need for an IUWM approach and attempted to adopt it for the development of urban water management strategies. While there are some differences in the processes being adopted by the various utilities, they all seems to follow a common framework, which involves setting objectives for the development of strategies in line with sustainability principles and evaluation of the feasibility of a wide range of alternative urban water management options against multiple objectives with the aim of recommending a set of preferred strategies. The methods used for evaluation are either qualitative or semi-quantitative. The recommended strategies include water conservation, diversification of supplies by considering wastewater recycling, rainwater harvesting, increased water transfers and the use of non-structural measures such as rebates, taxes and various pricing structures (see Figure 3).

Table 5: Recently released urban water management strategies by major cities

City	Strategic planning document
Adelaide	Water proofing Adelaide (2005)
Brisbane	Water for today and tomorrow (2004)
Canberra	Think water Act water (2004)
Darwin	The Darwin water story (2006)
Hobart	The Water Reform Package (2001)
Melbourne	The Water supply demand strategy for Melbourne 2006–2055
Perth	Securing our water future (2003)
Sydney	Metropolitan water plan (2006)

In addition to strategic planning, some Australian utilities have begun testing of IUWM concepts and tools at a local scale. As mentioned earlier, while the use of IUWM concepts and tools does not imply adoption of the IUWM, local or development scale implementations enable utilities to identify limitations and feasibility of different tools and concepts before implementing them at the city scale. Key features of, and lessons learned from, local- or development-scale implementation of IUWM concepts and tools, and the challenges associated with planning and implementation of the IUWM approach at the city scale, are discussed below.

Learning thus far with IUWM concepts and tools

A number of applications of the IUWM concepts exist at sub-division and large-neighbourhood scale (Mitchell 2006). These applications focus on technical aspects of total water cycle integration (that is, water supply, wastewater and stormwater service provision), rather than on a coordinated approach to urban water services through the engagement of all the relevant stakeholders.

Water quantity and quality benefits of adopting various IUWM tools at development scale include (Mitchell 2006):

- potable water usage savings of up to 20%
- reductions in stormwater volumes leaving the site, ranging from minor to near elimination
- significant reduction in wastewater flows leaving the site due to recycling and the use of water-efficiency measures.

It is important to note that only a very few sites are currently monitoring system performance in terms of flow and contaminant fluxes. Therefore, benefits reported above are estimates only.

The Pimpama–Coomera development (www.aila.org.au/QldMark/2006/Pimpama.pdf) is one example of attempting to adopt IUWM concept at regional development scale. It was designed to accommodate some 150,000 people—up from the current population of approximately 5000—and covered about 6000 hectares in a major growth area on the Gold Coast. The planning process was completed in 2004, and included a range of IUWM tools to reduce potable water usage, source control for stormwater flow and water quality management, and a coordinated approach for master planning that engaged all the relevant stakeholders. The following IUWM tools were recommended for use in the Pimpama–Coomera development:

- Installing rainwater tanks to provide locally captured water for selected household uses (such as in gardens) as well as slowing roof runoff to reduce stormwater flows
- Supplying recycled water into the homes via a third pipe, for toilet flushing and outdoor use, as well as fire fighting; potable water is then used only for drinking, cooking and bathing
- Redesign of the wastewater system, with reduced stormwater inflow and sewer spills
- WSUD techniques, such as swales, bio-retention, wetlands and basins, to enhance waterway health
- The compulsory use of water-efficient products along with enhanced education on the value of water and improved landscaping practices
- A reduction in water supply pressures to minimise water losses.

A key lesson so far is identification of the need for considerable change in the approach to urban water system modelling and analysis, which will require methods to account for the interactions between water system components. Currently there are few analytical tools and procedural frameworks to aid the planning, management and integrated analysis required for adopting the IUWM approach. Such tools and decision-making frameworks will be discussed in the next section.

Another key lesson is the operation and maintenance concerns associated with local scale application of IUWM concepts, such as decentralised recycling and grey water use schemes. The operation and maintenance concerns include odour problems with dry composting toilets and maintenance management problems. The concept of centralised management of decentralised and integrated water, wastewater and stormwater systems can be a solution to this concern because of the community trust associated with centralised management bodies (such as existing large water utilities). However, changes to current roles, responsibilities and organisational structures of existing water utilities are required if this concept is to be implemented—an aspect that appears to have low priority within the Australian water industry, which is dominated by supply-related problems created by the current drought conditions.

Planning and implementing IUWM in Australian cities

The importance of planning

To achieve IUWM, the urban water system must be considered as a whole—each component, whether structural or not, contributes to its overall performance. Furthermore, the aim of IUWM is to fulfil multiple objectives, some of which might potentially conflict with others or with the aims of the individual organisations involved. Goals need to be set and many stakeholders need to cooperate if the transition to this new form of management is to be effective. It is of paramount importance that the planning process is careful, thorough and, above all, supported by appropriate scientific analysis. The problem is too big for any piecemeal approach to succeed.

The solution to each city's problems will be different. Available water sources, infrastructure condition, population attitudes, housing densities, regulations and distribution of responsibilities are just a few of the parameters that will influence selection of the *best* alternative. The current situation dictates the starting point from which the city will make the transition to fully integrated management. Despite inter-city differences, a standardised approach to management, with well-articulated steps and supporting advice on who should participate, what data is needed and how to undertake the necessary analyses will contribute greatly to achieving the required outcome—that is, a flexible, adaptive process to provide sustainable water management.

Factors to be considered

Factors that need to be considered in the planning process can be broadly considered under four main headings: physical (including environmental), institutional, social and financial.

Physical factors include the condition and extent of existing infrastructure. This can hinder the introduction of new approaches and technologies, or indicate problems, such as overflows or flooding, that need to be overcome. Historic or environmentally sensitive sites should be given due consideration, and the needs of the environment ascertained. The topography, geology and hydrology of the region, together with availability and accessibility of new water sources, will also need to be considered. There can be technological advances on the horizon that will provide ideal solutions, and that can be factored in to the planning process.

Institutional factors include existing legislation at the federal, state and local levels. Current legislation might not support the use of alternative water supplies, and might need to be amended. The speed with which water restrictions can be introduced might need amendment, and legislation might need to be put in place to ensure compulsory maintenance of decentralised water-treatment facilities. Institutional needs will vary depending on the options selected, and action to change restrictive legislation or introduce new controls might need to be taken early in the development of an alternative system.

Successful adaptation to the IUWM approach requires cooperation between many different stakeholders, who might have conflicting aims. Buy-in from all relevant stakeholders is needed for successful implementation. To achieve this, the boundaries of the system should be identified and relevant stakeholders brought into the planning process right from the start. The transition to IUWM might, for example, affect downstream uses (for example, those dependent on waterways for their living, such as agriculturalist). Domestic and industrial users who will be asked to accept a new approach to their water service provision are key stakeholders, and their support is important. The attitude of the local community, and any special cultural or social issues should be carefully assessed and consultation encouraged.

Economic factors will dictate who bears the costs of the project. This will require detailed analysis, because the complex interactions of the system make the evaluation of direct costs and externalities no easy task. Pricing structures, and the use of incentives, need to be evaluated.

The planning process

The steps in the planning process for IUWM have evolved through experience derived from a series of development case studies led by different organisations, mostly at the suburb or development scale. Although there are some differences in the approaches used, the basic steps are similar and follow a conventional pathway to option selection. The steps described below are based on the experience of CSIRO, which has undertaken a number of case studies over the past 6 years (Mitchell *et al.* 2003; Grant *et al.* 2006a, 2006b; Maheepala *et al.* 2006a, 2006b; Sharma *et al.* 2007). The process is currently being formalised in an Integrated Urban Water Management Manual, which is being developed under a project co-funded by CSIRO and AwwaRF (www.awwarf.com/research/TopicsAndProjects/PSB.pdf).

The IUWM planning process has the following key tasks:

- **Initiating the project and formulating a key stakeholder group.** Although the project will generally be initiated within one organisation, IUWM will not be achieved without extensive collaboration. The formation of a key stakeholder group, which initiates appropriate activities, monitors progress, is informed by technical and expert analysis and advises the decision makers on critical issues, is essential to the success of the project

- **Agreeing on objectives, measures and criteria.** The objectives of IUWM are agreed at the start of the process. Objectives generally include provision of a sustainable water system, as well as the solution of site-specific problems. The metrics or measures by which achievement of each objective is to be assessed are agreed, together with criteria or performance limits (such as water quality standards, limits on capital cost or an acceptable number of flood events per hundred years).
- **Understanding the current system, and identifying constraints and opportunities.** A detailed understanding of all aspects of the existing system— including geology, topography, infrastructure condition (for example, failure rates), legislative and institutional arrangements and demographics—provides a base case against which alternatives can be assessed and an indication of possible solutions.
- **Generating, evaluating and selecting options.** This step is the main technical and scientific hub of the planning process. A number of potential options are selected, and their performance is analysed in terms of the agreed measures. The performance of different options is compared, using increasingly sophisticated optimisation processes, and a preferred option is selected for development. Performance assessment of IUWM options is described in detail below.
- **Implementing the chosen option and reviewing its performance.** Once a preferred option has been agreed upon, detailed design is undertaken and construction and implementation can begin. Staging of the introduction of different aspects of the preferred option will be carefully analysed, and performance should be monitored to ensure that the anticipated outcomes are being achieved.

Throughout the process there is ongoing consultation with the community and relevant technical groups. Data and information are collected and managed, and key decisions are made and communicated to all relevant parties. Many of the steps are run in parallel, feeding back into each other, and the key stakeholder group continues it activities throughout the life of the project.

Performance assessment

Understanding the performance of the urban water system under any management method is essential for effective and efficient planning and operation. This is true for IUWM, too, but the main difference between traditional management and IUWM is the need to account for interactions among components of the water cycle. This requires significant change to the way urban water systems are viewed, and the way data on urban water systems are collected and used. Many models are available to assist in the assessment of hydrological, contaminant, economic and other aspects of system performance. Where models are not available, use can be made of consultative processes using expert opinion or consensus.

Models that evaluate the complex interactions between different aspects of the system at the metropolitan scale are not yet readily available, but research is being undertaken into the formulation of such integrated models. One such model is Hydro Planner (Maheepala *et al.* 2005; Grant *et al.* 2006c; Maheepala *et al.* 2007). A comprehensive review of models that can be used for IUWM is given in Mitchell *et al.* (2007).

Outputs of system performance models feed into integrated decision-making frameworks, which allow decision makers to consider multiple objectives and determine preferences and priorities for alternative IUWM configurations. Multi-objective decision frameworks based on sustainability principles (Hellstrom *et al.* 2000; Maheepala *et al.* 2003b, 2006a, 2006b and 2003b; Ellis *et al.* 2004; NSW Department of Utilities, Energy and Sustainability 2004) are appropriate in this regard to evaluate relative strengths and limitations of alternative IUWM configurations.

Risk and resilience in IUWM

If robust urban water options are to be adopted, and potential system failures are to be anticipated and avoided, it is not sufficient to take a deterministic approach to the evaluation of system performance. The decision makers need to understand the risk of system failure, which includes understanding the resilience of the system. A less risky, more resilient system might be preferable to one that outperforms it, but that has a high risk of failing to function. Comprehensive risk assessment that is informed by system monitoring provides continual feedback on the *riskiness* of the system, and allows controls to be adjusted before breakdown occurs. This adaptive management process should be applied to the entire system—monitoring aspects such as technology take-up (for example, are households taking up greywater treatment at the anticipated rate?) and the success of financial incentives—as well as aspects of the physical functioning of the system.

Implementation of IUWM

The implementation of IUWM involves detailed engineering design of the options chosen as part of the planning process—undertaking appropriate measures to implement them (for example, construction for structural options such as pipes and treatment plants; and legislation and processes for non-structural options such as rebates, taxes and capacity building through education) and development of operation and maintenance procedures and administration processes to support the adoption of IUWM.

None of these tasks is trivial because institutional and legislative structures and infrastructure that are currently in place have been designed to provide a system that does not account for system interactions, flow-on impacts, re-use and feedback loops. For example, Brown (2005) examined institutional structures currently in place in Sydney for stormwater management and concluded that the inertia within the public administration of urban stormwater inherently privileges and supports traditional stormwater management, and it requires a considerable change to support integrated stormwater management. In our view, this is also true for urban water management in water supply and wastewater contexts.

Therefore, it is important to consider the implementation aspects at the planning stage. The key implementation aspects are scientifically robust operation and maintenance procedures and institutional structures to administer these procedures. Evidence for this claim can be seen clearly from the development level implementations in Australia, described earlier this chapter. Another example is the SEQ Water Grid which is a proposed water supply method to tackle the water supply shortfalls in the South-East Queensland region. Its objective is to facilitate water sharing across the region and is currently at the planning stage. One of the planning objectives is to establish operating principles that link major water storages and ensure that water can be moved to where it is needed and to develop an institutional structure called a Grid Manager to administer operation of the SEQ Grid (www.qwc.qld.gov.au).

Reviewing and updating IUWM Strategies

The main objective of the IUWM approach is to provide an urban water system that is more sustainable than the current urban water management practice. Because the state of sustainability is like a moving target due to changing nature of the surrounding milieu, achieving sustainability can be viewed as a journey. The journey towards sustainability needs periodic reviews of IUWM strategies to understand the performance of IUWM that has been already implemented. Amendments and new strategies to address challenges and needs that are of high priority at the time of undertaking the review can then be developed.

Triple-bottom-line (TBL) reporting (Kenway *et al.* 2007) can be used for understanding performance of the urban water system. TBL is a framework for measuring and reporting performance against economic, social, and environmental parameters. Trends in TBL indicators can be used to check whether targets, which are defined as part of the planning process, have been achieved.

Therefore, TBL reporting and IUWM planning form a continuous cycle that allows a utility to move towards a more sustainable state. This is shown in Figure 6. This continuous cycle includes design and implementation of IUWM options or strategies, monitoring system performance and gathering data, analysing data to report against TBL indicators, analysing trends in data and feeding that information to reviewing IUWM strategies, which leads to development of amendments and/or new strategies to help move towards a more sustainable state.

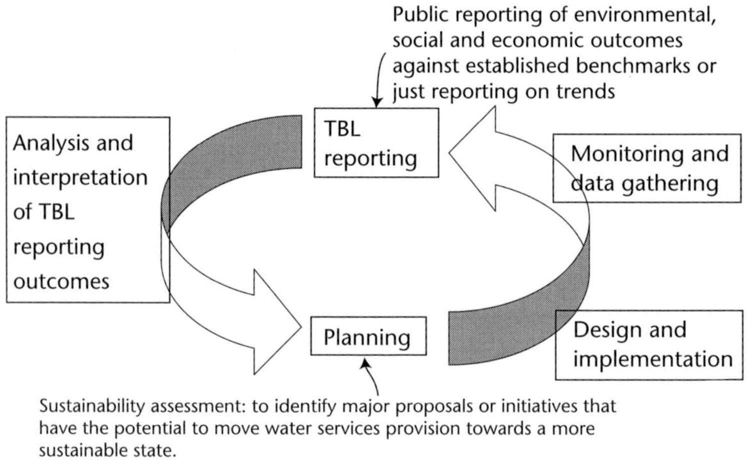

Figure 6: Illustration of continuous cycle of planning, implementation, monitoring, reporting and reviewing IUWM strategies

Conclusions

Integrated Urban Water Management is a complex process that involves a long-term commitment from many players. Rather than making short-term, isolated decisions on individual components of the water cycle, it demands complex scientific analysis of the functioning of the entire system, and an equally scientific approach to the planning and management processes. It requires politicians, utilities, scientists and financial and other organisations to work together towards the achievement of a higher goal: that of providing water services in a sustainable way. Although our understanding of what is needed to provide water services in a sustainable manner is currently limited, it is incumbent on us to use our current knowledge to the best advantage, and to set up planning and management processes that allow for the introduction of new ways of doing things as they become available. The IUWM approach is one such process, and its introduction opens up exciting possibilities for making the transition to a sustainable future.

References

ASCE and UNESCO (1998). 'Sustainability criteria for water resource systems'. Prepared by the Task Committee on Sustainability Criteria, Water Resources Planning and Management Division, ASCE, Washington DC.

Australian Capital Territory Government (2004). Think Water Act Water. ACT Government publishing Service. <www.thinkwater.act.gov.au>

Brown RB (2005). Impediments to integrated urban stormwater management: the need for institutional reform, *Journal of Environmental Management* **36(3)**, 455–468.

Chiew FHS (2006). Estimation of rainfall elasticity of streamflow in Australia. *Hydrological Science Journal* **51**, 613–625.

Commonwealth of Australia (2002). 'Introduction to urban stormwater management in Australia, Department of Environment and Heritage'. A report prepared under the urban stormwater initiative: living cities program. <http://www.environment.gov.au/coasts/publications/stormwater/pubs/stormwater.pdf>.

Coombes PJ, Boubli D and Argue J (2003). Integrated water cycle management at the Heritage Mews development in western Sydney. In *28th International Hydrology and Water resources Symposium*, 10–14 November 2003, Institution of Engineers, Wollongong, New SouthWales.

CSIRO and Melbourne Water Corporation (2005). *Melbourne Climate Change Study.* <http://www.melbwater.com/content/library/publications/reports/climate_change_study/Climate_Change_Study.pdf>.

Davis MD (2007). Integrated water resource management and water sharing. *Journal of Water*

Resources Planning and Management **133(5)**, 427–445.

Department of Environmental Resources (1999). *Low-impact Development: an Integrated Design Approach*. Department of Environmental Resources, Prince George's County, Maryland.

Ellis JB, Deutsch JC, Mouchel JM, Scholes L and Revitt MD (2004). Multicriteria decision approaches to support sustainable drainage options for treatment of highway and urban runoff. *Journal of Science of the Total Environment* **334–335**, 251–260.

Global Water Partnership (2007). Toolbox Integrated Water Resource Management. <http://www.gwptoolbox.org/>

Global Water Partnership Technical Advisory Committee (2000). 'Integrated water resource management'. TAC background paper no. 4. Global Water Partnership, Stockholm, Sweden.

Government of South Australia (2005). Water Proofing Adelaide : A Thirst for Change, Government of South Australia publishing Service. <www.waterproofingadelaide.sa.gov.au/WPA/Publications/>.

Grant A, Maheepala S, Leighton B, Mirza F, Rahilly M and Rahman J (2006a). Hydro Planner: providing an improved process for assessing urban water supply-demand balance. In *Proceedings of 30th Hydrology and Water Resources Symposium*, 4–7 December 2006, Launceston. The Institution of Engineers, Barton, Australian Capital Territory, Australia.

Grant A, Sharma A, Mitchell VG, Grant T and Pamminger F (2006b). Designing for sustainable water and nutrient outcomes in urban developments in Melbourne, Australian *Journal of Water Resources* **10(3)**, 251–260.

Grant T, Opray L, Grant A, Sharma A, Mitchell VG and Pamminger F (2006c). Life cycle aspects of alternative water and sewage servicing. In *Proceedings of Urban Drainage Modelling and Water Sensitive Urban Design Conference 2006*. International Water Association, Melbourne.

Hellstrom D, Jeppsson U and Karrman E (2000). A framework for systems analysis of sustainable urban water management. *Journal of Environment Impact Assessment Review* **20**, 311–321.

Jønch-Clausen TFJ (2001). Firming up the conceptual basis of integrated water resources management. *International Journal of Water Resources Development* **17(4)**, 501–510.

Kenway S, Howe C and Maheepala S (2007). *Triple Bottom Line Reporting of Sustainable Water Utility Performance*. CSIRO and Australia and American Water Works Association Research Foundation, Denver, CO.

Le Gal La Salle C, Vanderzalm J, Hutson J, Dillon P, Pavelic P and Martin R (2005). Isotope evolution and contribution to geochemical investigations in aquifer storage and recovery: a case study using reclaimed water at Bolivar, South Australia. *Hydrological Processes* **19(17)**, 3395–3411.

Maheepala S and Perera C (2003). Climate change and reliability of water supply. *Water Science and Technology* **47(7–8)**, 101–108.

Maheepala S, Snell DG and Perera C (2003a) Effect of climate change on security of Benalla water supply system. In *Proceedings of 28th International Hydrology and Water Resources Symposium*, 10–14 November 2003. Institution of Engineers, Wollongong, NSW.

Maheepala S, Speers A, Booker N and Mitchell VG (2003b). A framework for assessing sustainability of urban water systems. In *Proceeding of 28th International Hydrology and Water Resources Symposium* 10–14 November 2003. The Institution of Engineers, Wollongong, NSW.

Maheepala S, Leighton B, Mirza F, Rahilly M and Rahman J (2005). Hydro Planner – A linked modelling system for water quantity and quality simulation of total water cycle, In *MODSIM 2005 International Congress on Modelling and Simulation Society of Australia and New Zealand*, December 2005, (Eds A Zergerand RM Argent) pp. 683–689. <http://www.mssanz.org.au/modsim05/papers/maheepala.pdf >.

Maheepala S, Sharma A and Diaper C (2006a). A practical framework for planning and conceptual design of stormwater utilisation schemes. In *Proceedings of 7th International Conference on Urban drainage Modelling and the 4th International Conference on Water Sensitive Urban Design*. 2–7 April, 2006, Melbourne (Eds A Deletic and T Fletcher). International Water Association, Melbourne.

Maheepala S, Evans M, Sharma A, Gray S and Howe C (2006b). Assessing water service provision scenarios using the concept of sustainability. In *2nd IWA Leading-edge Sustainability in Water Limited Environment.* (Eds MB Beack and A Speers). Water and Environmental Management Series, International Water Association Publishing, London.

Maheepala S, Leighton B, Mirza F, Rahilly M and Rahman J (2007). Hydro Planner – for assessing water quantity and quality aspects of integrated and diversified urban water systems. *Proceedings of OzWater 07 Conference, March 2007.* Australian Water Services Association, Sydney.

Marsalek DJ, Rochfort MQ and Savic PD (2001). Urban water as a part of integrated catchment management. In *Frontiers in Urban Water Management, Deadlock or Hope.* (Eds C Maksimovic and JA Tejada-Guibert) pp. 37–83. IWA Publishing, London.

Mitchell VG, Duncan H, Inman M., Rahilly M, Stewart J, Vieritz A, Holt P, Grant A, Fletcher T, Coleman J, Maheepala S, Sharma A, Deletic A and Breen P (2007). In *Proceedings of the conference on Integrated Urban Water Modelling – Past, Present, and Future**, Rainwater & Urban Design 2007, Joint 13th International Rainwater Catchment Systems Conference and the 5th International Water Sensitive Urban Design Conference, Sydney, Australia, 21–23 August, 2007.

Mitchell VG (2006). Applying integrated urban water management concepts: a review of Australian experience. *Journal of Environmental Management* **37(5)**, 89–605.

Mitchell G, Gray S, Shipton B, Woolley R, Erbacher J, Egerton G and McKnoulty J (2003). Evaluating integrated urban water systems alternatives for Brisbane, *Water Science and Technology* **47**, 1–9.

New South Wales Department of Utilities, Energy and Sustainability (2004). Integrated water cycle guidelines for NSW local utilities. <http://www.deus.nsw.gov.au/publications/IWCM%20Guidelines%20-%20October%202004.pdf>

New South Wales Government (2006). Metropolitan Water Plan: February 2006 Progress Report, New South Wales Government Publishing Service, <www.waterforlife.nsw.gov.au/__data/assets/pdf_file/0016/1447/water-progressreport_2006.pdf>

Pinkham R (1999). '21st century water systems: Scenarios, visions and drivers'. 20 pp. <http://www.rmi.org/images/other/Water/W99-21 21CentWaterSys.pdf>

Queensland Government (2004). *Water for Today and Tomorrow. An Integrated Water Management Strategy for Brisbane.* Queensland Government Publishing Service, (www.brisbane.qld.gov.au).

Sharma A, Grant A, Tjandraatmadja G, Grant T, Opray L and Pamminger F (2007). Towards sustainable water and sewerage servicing options for a multi-story infill development. In *Water Management Challenges in Global Change - Computing and Control for the Water Industry,* CCWI2007 and Sustainable Urban Water Management SUWM2007 Conference, De Montfort University, Leicester, UK. A.A. Balkema Publishers and Taylor & Francis Group, The Netherlands.

Uddameri V (2007). A dynamic programming model for optimal planning of aquifer storage and recovery facility operations. *Journal of Environmental Geology* **51**, 953–962.

United Nations Environmental Program (2007). <http://www.unep.or.jp/ietc/Brochures/IUWM. pdf>

Victorian Government (2006). Water supply-demand Strategy for Melbourne 2006–2055, Victorian Government Publishing Service. www.watersmart.vic.gov.au/downloads/Strategy_WebVersion.pdf

Vlachos PE and Braga PB (2001). The challenge of urban water management. In *Frontiers in Urban Water Management, Deadlock or Hope.* (Eds C Maksimovic and JA Tejada-Guibert), pp. 1–34. IWA Publishing, London.

Water and Power Corporation (2006). The Darwin water story. Northern Territory Government Publishing Service, Darwin <www.nt.gov.au/powerwater/docs/campaigns/drw_water_story/the_darwin_water_story_2006.pdf>

Western Australian Government (2003). Securing Our Water Future: *A State Water Strategy for Western Australia.* Western Australia State Government Publishing, <http://portal.water.wa.gov.au/portal/page/portal/PlanningWater-

Future/StateWaterStrategy/Content/State_Water_Strategy_complete_001.pdf >.

Wong THF (Ed.) (2006). *Australian Runoff Quality.* Institution of Engineers, Sydney.

WSAA (Water Services Association of Australia) (2007). 'Report card for 2006/07: performance of the Australian water industry and projections for the future'. WSAA, Melbourne.

31

Decentralised Water and Wastewater Systems

Clare Diaper, Ashok Sharma and Grace Tjandraatmadja

Conventional urban water systems date back 3500 years (Walski 2001), with little change having taken place over the last 100 years. The drainage of rainwater and wastewater was evident in the earliest civilisations of the Middle East, the Mediterranean and the Orient (Delleur, 2003). The urban water system consists of the water supply, wastewater and stormwater systems, which have historically been managed, planned and designed as separate systems (Figure). These systems require large-scale water extraction, treatment and distribution and also large-scale collection, treatment and disposal of wastewater and stormwater.

The conventional urban water system was designed primarily to protect human health, but its long-term operation has been seen to have many adverse environmental, social and economic consequences. These include the impacts on native flora and fauna in and around waterways, the removal of urban streams and water amenities and the high economic costs because the provision of current services has to increasingly rely on ageing water infrastructure. In addition to these impacts, conventional systems have limited capacity to allow for increased demand for drinking water (in the context of projected declines in sustainable water yield)—and concomitant increase in the flows and contaminants from wastewater—caused by increasing urban growth and higher density residential development.

These factors, in combination with the increasing pressure for more sustainable urban water systems that protect and enhance the environment, have led to the development and implementation of a range of structural and non-structural techniques that provide an alternative to the existing conventional infrastructure. These techniques are fundamental in the transition of our cities water services to more resilient and sustainable systems.

In this chapter, a particular suite of water servicing techniques and technologies—called decentralised or distributed systems—are described. These techniques are structural interventions that help alleviate the pressure on existing water, stormwater and wastewater infrastructure. Non-structural techniques, such as education, pricing and legislation, are not dealt with in detail here, as they are generic approaches that could equally apply to transport or energy systems.

Decentralised systems are the structural component of the Integrated Urban Water Management (IUWM) approach (see Chapter 30), which is being considered as an alternative to the conventional centralised approach for providing urban water services (Figure 1). The IUWM approach:

- considers all parts of the water cycle
- considers all human and ecological water requirements
- aims to minimise impact on the natural environment and maximise economic benefits
- applies a broad range of structural and non-structural practices
- considers all water resources based on water fit-for-purpose usage both in quality and quantity.

In addition to water resource issues, IUWM should also consider energy use, even though the energy demand for providing current urban water services, including both water supply and wastewater treatment, is only a small percentage (1.5%) of the total household energy use in Australia (PMSEIC Report 2007). As alternatives may increase energy demand, assessment of energy use

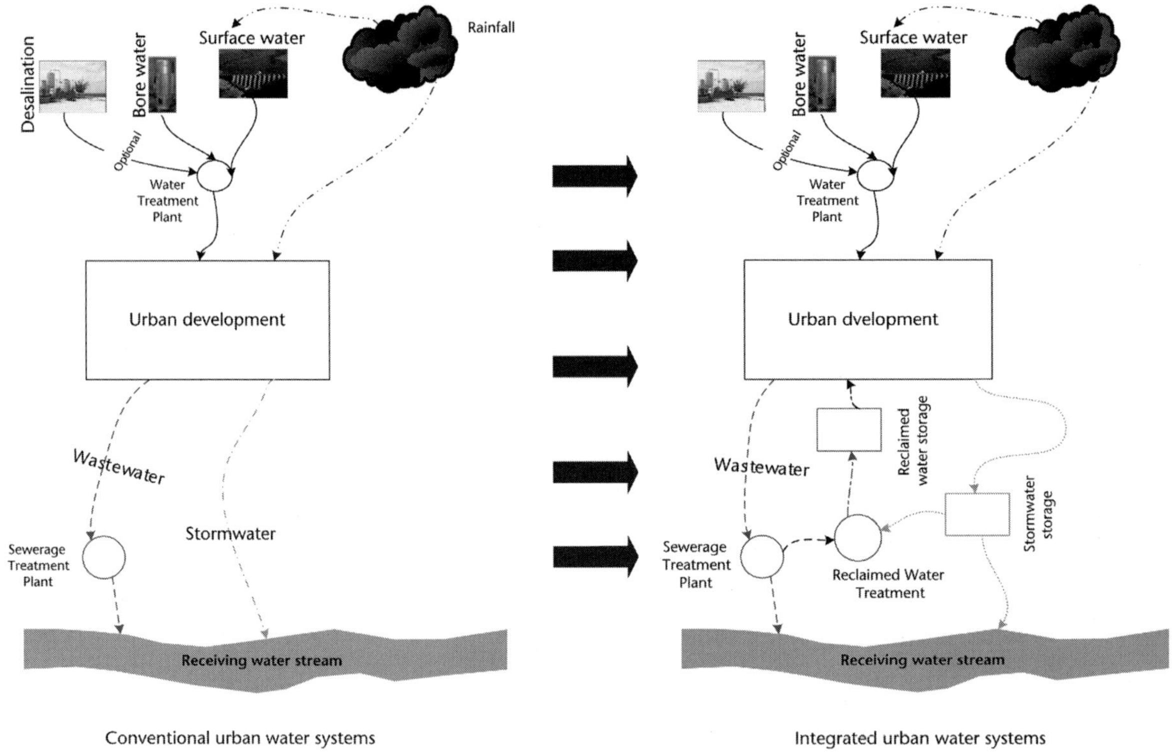

Figure 1: Examples of conventional and integrated urban water systems

should be undertaken to allow equitable comparison of different options.

Decentralised approaches are applicable to stormwater, wastewater and water systems in the urban water cycle and this chapter provides details of generic principles of decentralised systems that apply to all urban water flows. As such, the following terms will be used:

- '*urban water*' will be used when referring to all three water types
- *wastewater* describes the total flows of used water
- *blackwater* describes waste collected from the toilet alone
- *greywater* describes all other wastewater, not including toilet wastes
- *rainwater* is the water collected from roofs of buildings
- *stormwater* is the water that runs off from all other pervious and impervious surfaces, such as roads, paths, driveways, open space and gardens.

Decentralised systems can be defined as the collection, treatment and use of rainwater, stormwater, groundwater or wastewater at different spatial scales—from individual homes, clusters of homes, urban communities, industries, or built facilities, as well as from portions of existing communities either independent from, or as part of, a larger system (adapted from Crites and Tchobanoglous 1998). Decentralised systems operate and distribute water near their point of origin, although some residual components can be transported to a centralised point for further treatment and re-use.

Decentralised systems have historically been provided in semi-urban, rural and remote areas, where the provision of centralised systems is not technically, economically or environmentally feasible. However, rapid urbanisation, increased pressure on water resources and the need to minimise contaminant loads to receiving waters has meant conventional centralised systems are not always the most appropriate solution for urban development. As a consequence, innovative decentralised systems are being planned and implemented for new and future urban developments—either as separate facilities or in combination with a centralised system.

Decentralised systems have a number of positive impacts on the water cycle by reducing inputs to, and outputs from, the development area. The extraction of freshwater resources is reduced due to substitution of mains water with rainwater, stormwater or reclaimed wastewater. Where wastewater is treated and substituted for mains water uses, contaminant loads and volumes to wastewater treatment facilities are reduced, potentially postponing or negating the need for a capacity upgrade. Stormwater flows and pollutant discharge to the environment is reduced and potential flooding issues due to increased urban development can be avoided. In addition, decentralised water systems can improve urban amenity, reduce or prevent wastewater overflows, reduce mains water, wastewater and stormwater peak flows and potentially reduce costs and energy use.

With the advent of this new and innovative approach to water servicing, there is a need for detailed system assessment both pre- and post implementation, as there is currently limited information about their operation, maintenance, reliability and performance. Many assessment tools have been developed to examine water and contaminant flows in the urban environment (Mitchell *et al.* 2007) and life cycle costing has been adopted as an appropriate economic assessment. Life cycle assessment and multi-criteria analysis are tools that have been used to compare the performance of these new systems with the conventional approach. Within these assessment techniques, it is vital to ensure the human health and environmental aspects are taken into account and that the impacts of the decentralised system on the larger system and existing infrastructure are assessed.

This chapter discusses decentralised techniques at a range of scales and then provides details of specific prevention, collection, storage, treatment and distribution techniques for urban water. The advantages of these techniques, factors to be considered in their implementation and their applicability in making the transition from the existing urban water infrastructure to a more sustainable state are discussed.

The scale of decentralised systems

The decentralised systems concept is applicable to a single dwelling, to small and medium residential developments or to multi-storey buildings. At the single house scale there are many options for alternative water servicing (Figure 2). The diagram shows many potential options, but, it is unlikely that all options would be applied to one individual property.

Rainwater can be collected from the roof and used for garden irrigation, toilet flushing or in the laundry. Greywater can be collected from the laundry, bath or shower and either re-used directly for subsurface irrigation, or treated and then used for toilet flushing or garden irrigation. Overflows and unwanted materials produced by the greywater treatment system can be discharged to the sewer. Blackwater can be collected and combined with kitchen wastewaters for onsite treatment, which can then be used for subsurface irrigation. Stormwater run-off from the property can be reduced by replacing impervious paved areas with pervious paving systems. Outside the property boundary stormwater flows can be directed through swales, which remove some particulate material and allow infiltration.

Restrictions due to existing building infrastructure may limit the application of some of these options to new construction. For example, there may be difficulties in collecting greywater separately from blackwater in slab-built properties as modifications to pipe layout will be difficult. The direct diversion of individual greywater streams (that is, from the washing machine), is likely to be more readily achieved with detached properties.

The feasibility of application of wastewater or greywater to gardens for irrigation will be dependent on a number of factors, including; the soil type, local climatic conditions, application area available, treated water quality and the level of groundwater. Current legislation in Australia prohibits the installation of a wastewater treatment system on an existing property that has access to sewerage, but there are exemptions to this legislation—the 'Sustainable House' in Sydney, successfully treats all wastewater on site (Mobbs 1998).

Rainwater collection is appropriate for most housing types, but the efficiency of collection and the reliability of supply will be limited by the roof

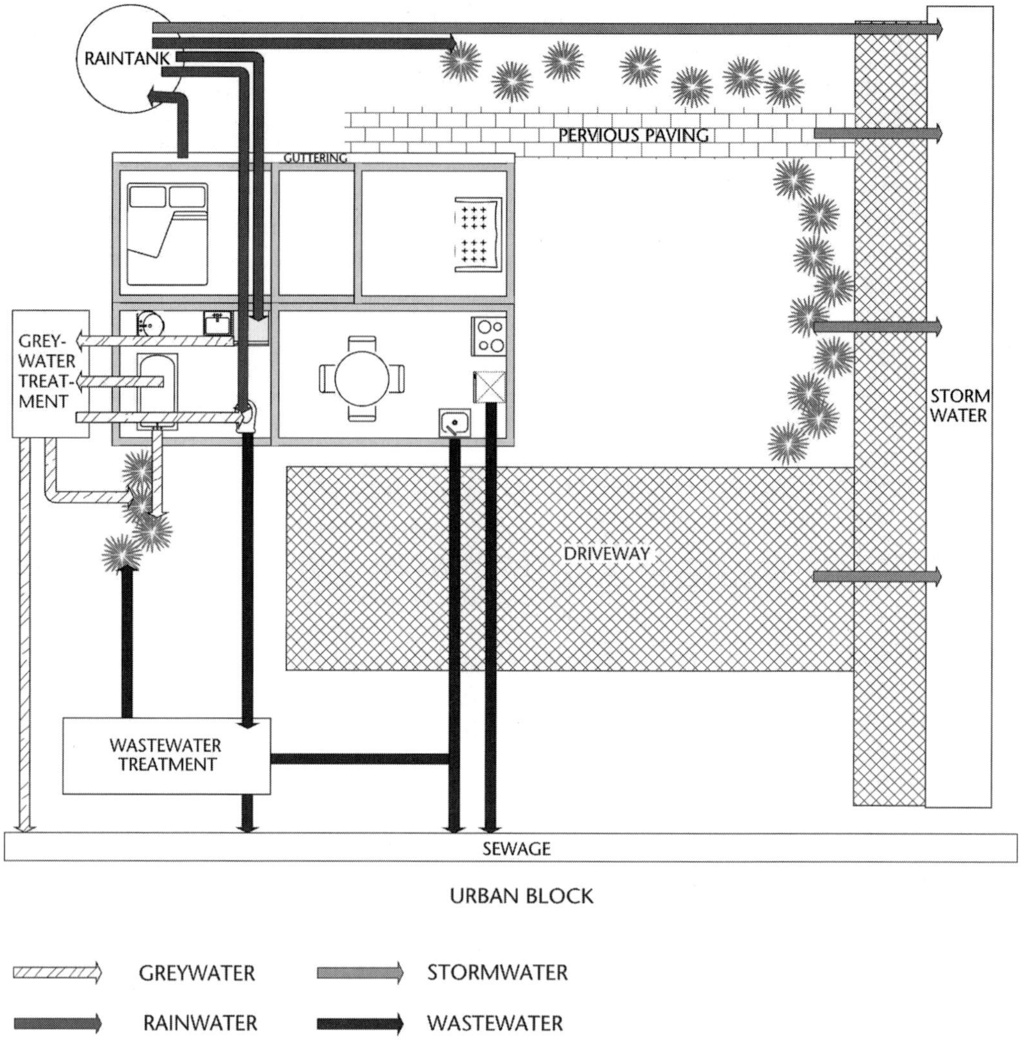

Figure 2: Options for decentralised water servicing at the household

area available for collection and the area available for a storage tank. Rainfall patterns in the diverse climate regions of Australia will also need to be considered when assessing the feasibility of rainwater collection. The replacement of impervious areas with pervious options is also appropriate for both new construction and retrofit options.

In addition to the issues associated with the different decentralised options at the household scale, the impacts on the larger urban water system will also need to be tackled. For example, the use of rainwater and greywater as substitute for indoor and outdoor mains water use, will reduce the flows in the mains water network. The impact of these reduced flows on mains water quality, particularly residual disinfectant concentrations, should also be assessed. Reduction in water flows or use of greywater or wastewater for irrigation will have impacts on wastewater flows and quality in the sewage network. The potential for blockages, changes to wastewater quality, increased anaerobic activity and impacts on centralised treatment facilities also need to be considered.

The range of options increase at a larger scale, as household options can be combined with neighbourhood or cluster-level systems. Figure 3 shows a number of potential options for water supply, wastewater and stormwater collection, storage, treatment and distribution. In option (A) wastewater is collected from a high-density building, treated and re-used on site (see, for example, 60L in the Case Study list). Greywater can also be collected at this larger scale for treatment and redistribution (see Inkerman Oasis in the Case Study list), but will be limited by the ability to source separate greywater from blackwater. Option (B) is a decentralised water supply for a small cluster of houses, where water is abstracted from groundwater or an aquifer, then treated and distributed locally. This type of decentralised water supply option is most often combined with an injection well for treated wastewater or stormwater, which replenishes the aquifer (see Bolivar Case Study). Another option is that of sewer mining in which wastewater is abstracted from the sewer main and treated at, or close to, the abstraction point for local use. These is shown in option (C) where a wastewater treatment plant is fed with wastewater from the trunk sewer and the treated water is distributed to non-potable uses and irrigation of public open space (for example, Kings Domain trial and CH2 in Melbourne).

Another alternative for wastewater is cluster scale collection—Option (D)—where wastewater is collected from a number of houses and then re-used locally for irrigation of sports fields or open space (note: there is no connection to the trunk sewer in this example). In addition to wastewater use for outdoor purposes, there are examples of this type of scheme where wastewater is piped back to homes for toilet flushing (for example, Pimpama Coomera). Option (E) demonstrates the potential for stormwater collection and use.

As with households, the feasibility of these options in existing development may be limited due to the constraints of existing infrastructure, road and housing layout and availability of suitable areas for storage or treatment plant location. For example, the availability of a suitable area and access for an on-site recycling system in an existing high rise or high density development may be limited or suitable land area required for storage of stormwater may not be available in an existing catchment. In addition to the restrictions of the existing local infrastructure, the impacts of these different techniques on the larger water system also need to be assessed. For example, the reduction in sewer flows needs to be considered, as does the collection of stormwater on downstream catchment flows.

How these components of decentralised systems fit into other non physical aspects of the urban water system should also be considered when assessing their feasibility. The institutional arrangements and roles and responsibilities of stakeholders in system implementation and operation— such as system operators and householders—need to be clearly defined. This aspect of decentralised water servicing is still evolving in Australia and, at present, there are no guidelines or standards for generating appropriate institutional relationships. Regulatory, economic and social issues also need to be incorporated in the development of decentralised water services. These are discussed in more detail in the following section.

Factors affecting selection of decentralised systems

There are a number of factors that will influence the selection of decentralised or integrated urban water management measures for a particular site, and these are dependent on both local physical conditions and broader system institutional, regulatory, social and economic considerations. Physical conditions that impact on the selection of systems include climate, topography, soil types, lot size and layout, occupancy and water use, and road layout and streetscape.

Climate

Climate influences the selection of approaches to water servicing, including storage needs and treatment options. Average annual rainfall can be used as an initial indicator of available water in new and existing sub-divisions but temporal distribution (daily, weekly, monthly and seasonal), also needs to be considered, as do the contaminants associated with the rainfall and run-off and the temperature and evaporation effects.

Topography and soil types

Building and infrastructure layout is dependent on the landscape topography. Steeper slopes impose constraints on style and form of construction, increasing surface flows and loads and potentially increasing sewage pumping and costs. However, steep slopes can be exploited to capture gravity feed in collection and supply of stormwater

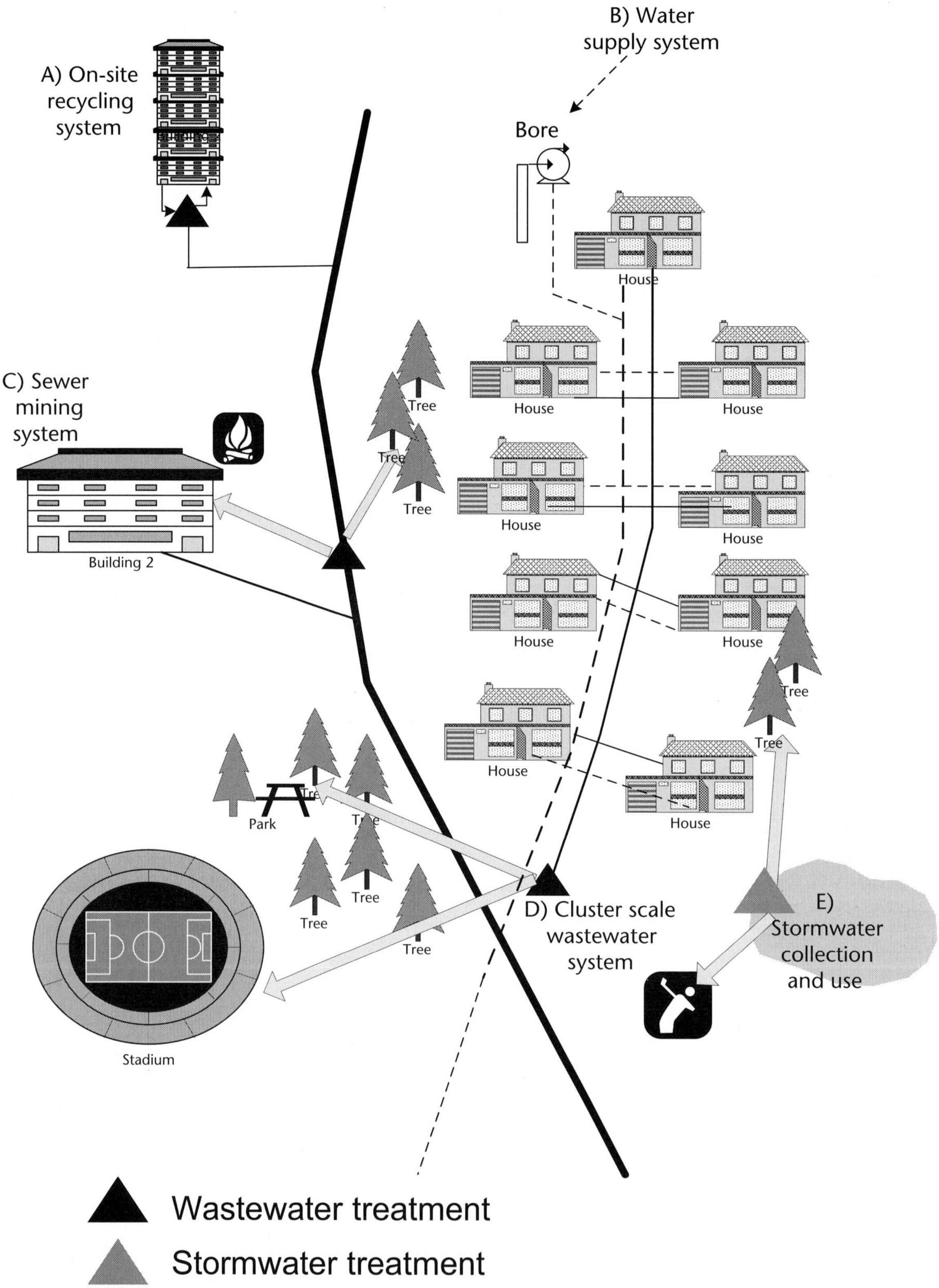

Figure 3: Options for decentralised systems at the neighbourhood scale

or wastewater flows. At a single lot scale, sloping terrain can improve the potential for storage systems underneath houses (see, for example, Payne Road in Case Study list). The soil type is important in assessing the infiltration of urban water flows and the potential for soil degradation, increased run-off and ease of installation of subsurface systems.

Lot size, layout and density

For most developments, lot size is strongly linked to financial viability for the developer, but changing demographics, range of dwelling types, consumer expectations and affordability are also influential. Lot size affects the selection of water-servicing strategies—potential catchment, storage and irrigation areas should be considered. For example, a large irrigation area is required to distribute and absorb treated wastewater for many on-site wastewater treatments and rainwater tanks need large surface areas to capture and store significant water volumes for reliable supply, depending on regional precipitation patterns. The layout of housing lots can also impact stormwater flows by reducing impervious areas to reduce run-off and incorporating natural drainage corridors.

Road layout and streetscape

The road layout and streetscape can affect the capacity for large-scale management of storm- and wastewater, which often requires large-scale storage capacity to provide supply in dry periods. For example, location of roads adjacent to open space allows temporary storage or treatment of stormwater at, or close to, the source, and integration of Water Sensitive Urban Design (WSUD) features with the landscape also reduces stormwater flows and contaminant discharge (Wong, 2006)

Water use and water quality

Household occupancy patterns influence the consumption of mains water and the quantity and quality of the wastewater produced. Key factors when selecting water servicing options are:

- Per capita drinking water consumption and wastewater generation generally decrease with increasing occupancy levels.
- When installing water-efficient appliances to reduce the volumes of mains water and wastewater, there is likely to be a concomitant increase in the concentration of contaminants in wastewater. Wastewater treatment systems need to be able to treat these higher concentrations.
- Seasonal and climate impacts on water demand needs to be investigated, as generally water consumption is higher in summer, and lower in winter.

Data and information regarding the quality of urban water collected for subsequent treatment and use is vital in assessing the type and required performance of a treatment system, the potential for source control and the potential end uses. Matching the quality of the urban water with appropriate end uses is known as the 'fit-for-purpose' approach, and can minimise or eliminate the requirements for treatment.

Water storage and availability

As water demand is dependent on a number of factors—such as occupancy, water use patterns, block and garden size—and is subject to seasonal and diurnal patterns, the cycles of water demand and water supply are often not naturally aligned. Water storage is important to help provide a balance between demand and supply.

At a sub-divisional level, storage of alternative water sources is required during wet months to ensure supply when demand increases during dry periods. Stormwater or treated effluent can be stored in lagoons, ponds or in underground aquifers.

Treated wastewater can provide a reliable source of water, but cannot be adopted as the sole indoor supply source and there are issues associated with the build-up of contaminants in the recycling loop and the disposal of sludges produced by treatment. Recycled water is often blended with other sources such as groundwater, stormwater and traditional mains water supply.

Legislation and regulation

The legislation surrounding domestic and commercial water use and wastewater re-use in Australia has been rapidly evolving as the demand for alternative water sources increases. The current legislation is highly complex and varies from state to state and, at times, from council to council. While great inroads are taking place in regards to water re-use, integration of waste services has been developing at a slower pace as water supply, stormwater and sanitation have traditionally been dealt as separate entities under the responsibility of different bodies (for example, stormwater was the responsibility of councils and water and wastewater rests with water authorities). There is a strong need for the development of national approaches and guidelines for decentralised and integrated water service systems and this is currently occurring through the development of Australian Guidelines for Water Recycling (Environment Protection and Heritage Council *et al.* 2006). A full review of the complex legislative processes is not attempted here, but national documents that provide some background to these issues are:

- The Environment Protection and Biodiversity Conservation Act 1999 (the EPBC Act) and the Trade Practices Act 1974 relating to wastewater re-use
- 'Introduction to Urban Stormwater Management in Australia' (Department of the Environment and Heritage 2002)
- Urban Stormwater—Best-Practice Environmental Management Guidelines (CSIRO 1999).

Current legislation on decentralised systems focuses mainly on the performance and licensing requirements for on-site treatment systems in regards to health and traditional environmental impacts (such as nitrogen, phosphorus and suspended solids) and not on the overall integration.

Implementation and management strategy

The implementation and management strategy for any development with an alternative water service needs to incorporate new approaches to water servicing—generally requiring a higher level of stakeholder involvement and the development of strong partnerships or alliances not necessarily present in traditional project management strategies. In addition, the investment required for infrastructure installation and maintenance in decentralised systems can be staged more readily than centralised systems. Preliminary infrastructure can be implemented and then later expanded when growth occurs, requiring less initial expenditure on capital works. The adoption of decentralised stormwater and wastewater management can also decrease the need for overcapacity in sewer and stormwater collection design and allow the installation of smaller size pipes at shallower depths.

While decentralised systems often transfer a greater responsibility for system operation to localised stakeholders, studies on the feasibility of urban decentralised system implementation indicate that these systems can be effectively operated under a centralised management structure. Organisations such as the body corporate, water provider or a service contractor can be made responsible for management and maintenance of the system, thus providing the required safeguard for public health and reliability (Bradley *et al.* 2002, Hedgeland 2004). An example of such strategy is the management of residential on-site wastewater systems with telemetry and remote monitoring in Mobile, Alabama, USA (White 2005).

Financial and economic considerations

Traditionally, the economic viability of urban water infrastructure has been assessed on capital and operating costs only. In some circumstances, this method will demonstrate that decentralised systems are economically viable because features, such as staged costs can have a positive effect on the net present value of the investment for the lifetime of the infrastructure. However, capital and operating cost assessment alone does not provide a full picture of the true benefits of a project—there are added benefits, such as local environmental services, social and other community impacts, or larger system savings, as achieved by deferring the need for dam construction or an upgrade of wastewater treatment plant capacity. The need to incorporate these other factors in the fair and equitable assessment of alternative options has led to the development of other performance assessment methods such as total resource costing or life cycle costing (CRC Water Quality and Treatment 2007).

Social and public perceptions

There has been an increasing need to incorporate social aspects of alternative urban water servicing in the assessment of the feasibility of such systems. The community requires service quality and reliability and that environmental quality is maintained or improved (Syme and Robinson 1988). Thus, the implementation of new approaches is in part reliant on community acceptance and understanding of the system. There have been a number of studies assessing community acceptance of recycled stormwater and wastewater for a variety of end uses (ARCWIS 1999; Po *et al.* 2003). The general trend is greater acceptance of less personal contact end uses, such as garden watering, and less acceptance of more personal end uses, such as showering or drinking. The degree of acceptance is also dependent on the source water quality. These perceptions may affect the success of decentralised servicing options and community education and involvement in the decision making through transparent and inclusive approaches is recommended (see, for example Toowoomba in Case Study list).

Urban water systems

There are many different options for decentralised systems in the urban context, including rainwater, stormwater, greywater and wastewater techniques, and a variety of different end uses. There are also many different types of collection, treatment, storage and distribution systems—and various demand-management and source-control options can be used in conjunction with these. The following section describes some of these options and provides detail of relevant case studies.

System components

Figure 4 is a generic flow diagram showing the different steps of service provision for all water, wastewater and stormwater options. Systems can be categorised into:

- prevention measures, where reduction of contaminants or pollutants occurs at the source
- collection systems, which transport water to storage or treatment
- storage systems, either before or after treatment, to balance flows and provide a more reliable supply
- natural and engineered treatment systems—both for the removal of contaminants
- distribution systems, which supply the water to appropriate end uses.

Treatment systems can be further sub-divided into primary, secondary and tertiary treatment methods, with methods being combined—from primary to tertiary—to provide the required water quality. The combination of treatments that delivers the best overall performance is often referred to as a 'treatment train'. Each of these levels of treatment can be based on physical, chemical and/or biological processes.

There is some overlap in these categories. For example, rainwater collection systems can be designed to also provide some source control and stormwater storage systems may also provide some level of treatment due to sedimentation or UV exposure. However, this broad categorisation serves to describe system components in a simple and structured way.

Prevention—demand reduction and source control

Demand reduction and source control are complementary strategies aimed at minimising the

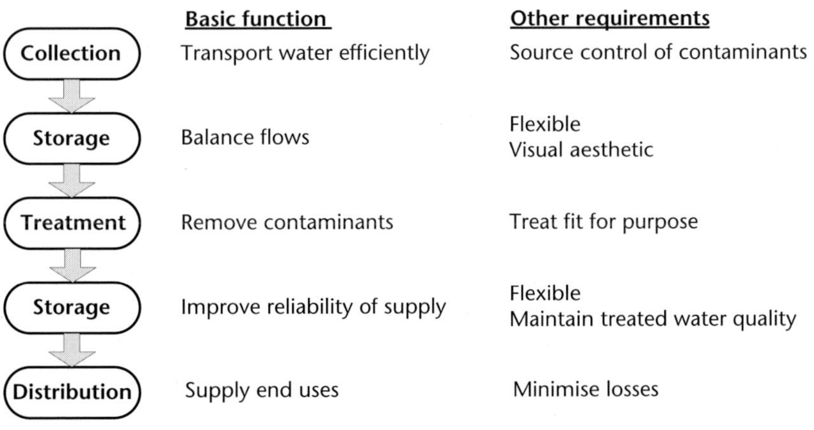

Figure 4: Generic stages of service provision for water, wastewater and stormwater

consumption or generation of water and the input of contaminants into urban wastewater streams. Demand reduction is generally implemented due to water scarcity and also has an impact on wastewater flows (such as water restrictions or the mandatory adoption of industry standards for appliances and fittings that use less water).

Source control aims to reduce concentration, load or flow of contaminants in urban water by implementing strategies that minimise or eliminate contaminant input at, or close to, their point of origin. In general, urban water streams closer to the source have higher contaminant concentrations and less cross-contamination, thus reducing the energy and chemical requirements for treatment and recovery. In addition, the requirements for downstream treatment and processing can be reduced. Examples of source-control strategies include:

- physical systems, such as urine separating toilets for differentiated treatment to reduce the nitrogen load at the treatment plant (Otterpohl *et al.* 1999) and the adoption of sediment traps in dental practices for amalgam and mercury separation in Europe (Icon 2001)
- legislative systems, such as regulations prohibiting the disposal of paint or toxic compounds to wastewater
- voluntary processes, such as national eco-labelling schemes to promote more environmentally friendly household products (Lauridsen and Røpke 2005).

Source-control systems also improve the potential for resource recovery, whereby the contaminant collected has potential value as a resource. An example of this is nutrient recovery from urine (van Betuw *et al.* 2007), energy from blackwater (Hanæus *et al.* 1997; Otterpohl *et al.* 1999) and fat and grease collection from restaurants and industrial kitchen traps for biofuel production (Thailand) (Stoll and Gupta 1997).

Collection systems

In the past, the only choice for the collection of domestic wastewater was through a gravity sewer system, either with pumping stations within the system for transporting the wastewater from houses to centralised treatment plants or direct feed to a septic tank and effluent disposal trench. Conventional sewers were generally used in cities and large towns—with septic tanks in rural areas—and these systems were seen as disposal options rather than as providing a resource for use.

Historically, stormwater collection systems in urban areas have consisted of pipes and channels—often concrete lined—with the primary aim of transporting stormwater flows away from developed areas to reduce the potential for flooding. The new model of WSUD, which integrates the management of the urban water cycle into the design of the built form in an urban environment, incorporates the collection, infiltration and treatment of stormwater within the development. The concept also aims to maximise visual and recreational amenity.

These large-scale conventional wastewater and stormwater systems provide reliable and feasible solutions, but can have the following limitations:

- in low density development, long lengths of pipe work are required, resulting in high cost
- satellite development may be costly to connect to a centralised supply, treatment or disposal facility
- high cost due to deep excavation in rock strata
- in areas with high groundwater tables, there may be high infiltration
- inadequate treatment to deal with contaminants or flows from unspecified sources
- not suited to undulating terrain or certain soil types
- affected by climate variability.

For wastewater, the cost of installing gravity sewers may be too expensive, particularly for small unsewered communities and alternative wastewater collection systems can be economically feasible in these areas. Alternative wastewater collection systems are described in various textbooks and literature (for example, USEPA 1981; DPIE 1988; Crites and Tchobanoglous 1998). The alternative collection systems are summarised in Table 1.

Collection systems for stormwater often incorporate some element of treatment or flow reduction through infiltration. For example, swales and buffer strips are vegetated channels that direct stormwater flows to downstream water ways or treatment systems. They are also designed to promote infiltration, attenuate flows and reduce and remove pollutants.

Rainwater collection systems in the form of gutters and downpipes can also be designed to

Table 1: Characteristics of alternative sewer systems

Sewer type or combination	Ideal topography	Construction cost in rocky, high groundwater sites	Sulphide potential	Minimum slope or velocity required
Conventional gravity	Downhill	High	Moderate	Yes
Septic Tank Effluent Pumping (STEP)	Uphill, undulating	Low	High	No
Septic Tank Effluent Gravity (STEG)	Downhill	Moderate	High	No
Grinder pump	Uphill	Low	Moderate–High	Yes
Vacuum	Flat	Low	Low	Yes
STEG-STEP	Undulating	Low–Moderate	High	No
Conventional-GP	Undulating	Moderate–High	Moderate	Yes
Conventional-Vacuum	Undulating	Moderate–High	Low–Moderate	Yes

Source: adopted from Crites and Tchobanoglous (1998)

incorporate treatment and removal of contaminants. Meshes can be placed over gutters or downpipe breaks to prevent leaf litter entering the system and first flush devices can be used to divert the first volume of a rainfall event. The efficiency of these devices in removing contaminants or in collecting water is not fully understood, although existing first-flush devices have been found to reduce catch efficiency (Gardner *et al.* 2004).

Storage systems

Storage of stormwater and treated wastewater is often required to provide reliable supply, to match supply–demand balance, to balance flows, for emergency needs and during the implementation of new treatment measures when diversion of existing flows is required. The size of storage is directly proportional to the amount of urban water collected and the requirements to meet peak demands. Emergency storages are provided to meet demand during failure of supply from source, pump failures or natural disasters. The size of storage is generally based on the assessment of risk, desired degree of reliability and cost considerations. Storages are also provided to manage the seasonal variations in supply and demand. The operational capacity of a storage is determined analytically or graphically using mass balance as the basis of analysis (Peavy 1985).

Rain storages are provided to collect rainwater from the roof for various indoor or outdoor usages. Rainfall is variable and the demand changes with the seasons and with various end uses. The size of a rainwater tank will depend upon rainfall, area of roof available to capture rain, end usage, number of occupants and available space for raintank installation. The optimal size of a rainwater tank is often based on volumetric reliability, which is defined as the fraction of total demand sourced from the rainwater tank. The optimal size of rainwater tank is the volume beyond which any significant increase in size will have marginal increase in volumetric reliability.

As with rainwater tanks, the provisions of stormwater collection ponds and storages are based on the climate, demand, losses and catchment area. The optimal size of stormwater and wastewater storages can be estimated on the basis of volumetric reliability considerations. However, due to the marginal variation in daily wastewater flows, reclaimed water storages are significantly smaller than stormwater storages for similar end uses.

Small storages are also provided for greywater use at the single household scale to increase availability of greywater. However, in most states in Australia, untreated greywater cannot be stored for more than 24 hours due to water-quality concerns. Health and environmental issues should always be considered when planning for storage of untreated and treated alternative water resources.

Treatment systems

Decentralised treatment systems vary in complexity, operational and maintenance requirements, cost and energy usage depending on the source water quality and the required treated water quality or end use. Treatment systems consist of three basic processes—biological, chemical and physical— and all treatment systems for rainwater, stormwater, greywater, blackwater and total wastewater can be defined as combinations of these processes. Water, wastewater and greywater systems tend to be highly controlled systems technologically—involving tanks, pipes and valves. However, stormwater treatment is often through engineered natural processes, such as wetlands, swales and bioretention systems.

Physical treatment

Physical treatment processes are those that rely on physical forces to provide treatment, such as sedimentation, screening and filtration, flotation and mixing. The two main processes used in decentralised systems are sedimentation and filtration. Sedimentation is used in water, stormwater and wastewater treatment processes to remove settleable solids and can be combined with chemical treatment in order to target the precipitation and settling of particular components of water and wastewater streams.

Chemical processes

There are a wide range of chemical processes used in the treatment of water and wastewater, but they are rarely applied to stormwater treatment. The most commonly used techniques are chemical addition for flocculation, coagulation or disinfection. In addition, adsorption processes—in which contaminants adhere to the surface of a media such as activated carbon—are used in the treatment of drinking and are now gaining popularity in the treatment of wastewater. There are disadvantages of the chemical treatment techniques, such as the transport and use of chemicals, the disposal of flocculated material and the energy requirements for treatment techniques.

Biological processes

There is a huge variety of biological process used primarily for treating wastewater and stormwater. Biological treatment can be broadly categorised into aerobic and anaerobic processes and the aim of the process is to reduce the organic content and the levels of nitrogen and phosphorus. Both aerobic and anaerobic processes can be further categorised into suspended-growth and attached-growth systems—each having specific benefits and application to different urban water streams. The majority of stormwater treatment processes are aerobic attached-growth processes, where the microorganisms are attached to a medium such as gravel. Anaerobic processes are generally only applied to wastewater and are used for the treatment of sludge and wastes high in organic matter. Anaerobic wastewater processes have the advantage that they also produce methane, which can then be used as an energy source. A disadvantage of biological treatment processes is that they can be sensitive to changes to inlet flows and quality.

Combined processes

While the chemical, physical and biological description of treatment processes is useful in describing the various types of systems, in reality, combinations of these processes tend to occur. For example, in a stormwater biofiltration system, both physical removal of suspended material and degradation by biological processes will occur. Membrane bioreactors, which are often used for treatment of wastewater in decentralised systems, combine biological treatment of organic material and a physical separation by a micro or ultrafiltration process. Different treatment processes are also combined to form treatment trains to provide water of a required quality, with primary through to tertiary and advanced treatment measures. This combination of processes gives a more robust result, with multiple barriers to ensure required removal of contaminants. For all treatment processes, their physical size is an important factor in assessing their feasibility as a decentralised option in the urban environment. For wastewater the membrane bioreactor is often selected for this reason, as the technology is relatively compact and provides a high product water quality.

A summary of wastewater treatment processes used for wastewater re-use and their contaminant removal is given in Metcalf and Eddy (1991) and Asano *et al.* (2007), while stormwater use systems are discussed by Wong (2006) and Hatt *et al.* (2004).

Distribution systems

Distribution systems for indoor use of treated urban water will be similar in design to traditional mains water systems. The dual reticulation system, in which two pipe systems are installed—one for mains water and one for treated water—are the most common method of supplying two sources of water with differing qualities to one building. Installation of dual pipe systems in existing buildings may be restricted by the building envelope and construction, and are most commonly used in new developments (such as Pimpama Coomera and Rouse Hill). The avoidance and pre-emptive management of cross connections in a dual supply system is a necessary component of risk mitigation. The quality of the treated urban water also needs to be considered in the distribution system and the potential impacts on scaling, corrosion and biofilm growth.

Distribution systems supplying vital end uses, such as toilet flushing, will require a back-up supply system to allow for outages in recycled water flows. Landscape and irrigation end uses can generally withstand outages of supply, so back-up systems may not be necessary for these applications.

End uses

Substitution of mains water use with treated wastewater or stormwater has the double benefit of reducing mains water demand and reducing the discharge of wastewater or stormwater flows to the environment.

Treated urban water streams can be used for environmental purposes, such as irrigation and maintenance of green space or maintenance of environmental flows. Water quality requirements are generally lower than for other end uses, although there are often requirements for sub-surface irrigation or holding times before use in irrigated areas, in order to reduce potential human exposure to the treated water. Treated urban water of a suitable quality has also been used for cooling purposes in air conditioners and cooling towers, fire protection, toilet and urinal flushing, commercial car washing, commercial laundries, water features, dust suppression during construction, snow melting, water features and street cleaning (Asano *et al.* 2007). Important factors when assessing the use of urban water for these applications are the infrastructure requirements, water quality and required supply reliability and public acceptance.

When used in industrial applications, the impact of treated urban water on corrosion, scaling, biofilm formation and potential issues associated with industry product quality will need to be considered—depending on the specific industry use.

For all end uses of urban water from a decentralised system, the social, environmental and economic comparison to conventional processes should be made and an assessment of the larger system impacts should also be undertaken.

Conclusions

There are a plethora of options for decentralised urban water systems, which deal with different aspects of the transition of urban water services to more sustainable systems. Many options are suitable for both new buildings and retrofit situations, such as rainwater tanks, sewer mining and some aspects of WSUD, such as pervious paving. However, many of the more advanced wastewater and greywater collection, storage and treatment systems may only be appropriate for installation in new building developments, as existing plumbing and space requirements may not allow their implementation. Changes to traditional plumbing and drainage design are required to provide housing stock that is more amenable to alternative systems.

Decentralised systems are not suitable for every single development. However, when appropriately explored and designed for context specific conditions, they can maximise the investment and improve sustainability of urban development. One of the greatest challenges faced in the adoption of decentralised systems is the shift of mind set required to *integrate* centralised and decentralised technologies. This shift needs to tackle the design and maintenance of integrated systems and include the development of appropriate structures for their management and operation, instead of systems that consider one or the other in isolation.

References

ARCWIS (1999). *The Social Basis for Urban Water Provision in the 21st Century.* CSIRO Urban Water Program, CSIRO Land and Water, Canberra.

Asano T, Burton FL, Leverenz HL, Tsuchihashi R and Tchobanoglous G (2007) *Water Reuse: Issues, Technologies and Applications*. McGraw-Hill, NY.

van Betuw W, Mels A and Braadbaart (2007). Technology selection and comparative performance of source separating wastewater management system in Sweden and The Netherlands. In *Proceedings of Advanced Sanitation*. ISSN 0342-6068. Aachen, Germany.

Bradley BR, Grainger G, Rubin R and Tchobanoglous G (2002). Evaluation of onsite wastewater treatment technologies using sustainable criteria. *Clean Technologies and Environmental Policy* **4**, 87–99.

CRC Water Quality and Treatment (2007). 'Costing for sustainable outcomes in urban water systems: a guidebook'. Research Report. University of Technology, Sydney.

Crites R, and Tchobanoglous G. (1998). *Small and Decentralized Wastewater Management Systems*, McGraw-Hill, NY.

Delleur JW (2003). The evolution of urban hydrology: past, present, and future. *Journal of Hydraulic Engineering-Asce* **129(8)**, 563–573.

Diaper C, Tjandraatmadja G and Kenway S (2007). 'Sustainable Subdivisions – Review of technologies for integrated water services'. Research Report. CRC for Construction Innovation, Brisbane.

EPA (US) (1991). *Alternative Wastewater Collection Systems*. EPA 625/1-91/024, Centre for Environmental Research Information, US Environmental Protection Agency, Cincinnati, Ohio.

Gardner T, Baisden J and Miller G (2004). Rainwater first flush devices: are they effective? In *Proceedings of Sustainable Water in the Urban Environment*, August, Sunshine Coast.

Hanæus J, Hellström D and Johansson E (1997). A study of urine separation system in an ecological village in northern Sweden. *Water Science and Technology* **35(9)**, 153–160.

Hatt B, Deletic A and Fletcher T (2004). 'Integrated stormwater treatment and reuse systems, inventory of Australian practice'. CRC Catchment Hydrology Industry Report, Canberra.

Hedgeland RM (2004). Benefits and versatility of decentralised sewerage. In *Proceeding of the First International Conference on Onsite Wastewater Treatment and Recycling and 6th Specialist Conf. on Small Water Systems and Wastewater Systems*, Fremantle, Western Australia, 11–13 February.

Icon (2001). 'Pollutants in Urban Waste and Sewage Sludge'. Final Report. Imperial College Consultants Ltd, London.

Lauridsen PV and Røpke I (2005). *The Journal of Transdisciplinary Environmental Studies* **4(3)**, 2–18.

Mitchell VG, Duncan H, Inman M, Rahilly M, Stewart J, Vieritz A, Holt P, Grant A, Fletcher T, Coleman J, Maheepala S, Sharma A, Deletic A and Breen P (2007). Integrated Urban Water Modelling – Past, Present and Future. In *Proceedings of Rainwater and Urban Design 2007, Joint 13th International Rainwater Catchment Systems and 5th International Water Sensitive Urban Design Conference*, Sydney.

Metcalf and Eddy (1991). *Wastewater Engineering: Treatment, Disposal and Reuse*. McGraw-Hill, NY.

Mobbs M (1998) *Sustainable House*. Choice Books, ACA,Sydney

NSW Department of Environment and Climate Change (2006). Website. Nabers homepage. <http://www.nabers.com.au/, last accessed 18/08/2007>.

Otterpohl R, Albold A and Oldenburg M (1999). Source control in urban sanitation and waste management: ten systems with reuse of resources. *Water Science and Technology* **39(5)**, 153–160.

PMSEIC (Prime Minister's Science, Engineering and Innovation Council) (2007). *Water for Our Cities: Building Resilience in a Climate of Uncertainty*. <http://www.dest.gov.au/sectors/science_innovation/publications_resources/profiles/water_for_our_cities.htm>.

Peavy HS, Rowe DR, and Tchobanoglous G (1985). *Environmental Engineering*. McGraw-Hill, NY.

Po M, Kaercher JD and Nancarrow BE (2003). 'Literature review of factors influencing public perceptions of water reuse'. Technical report 2003/54. CSIRO Land and Water, Canberra.

Stoll U and Gupta H (1997). Management strategies for oil and grease residues. *Waste Management and Research* **15**, 23–32.

Syme GJ and Robinson S (1988). 'The evaluation of social impact in Australian water resources management'. Australian Water Resources

Council Water Management Series No. 12. AGPS, Canberra.

Tchobanoglous G (1981). *Wastewater Engineering: Collection and Pumping of Wastewater.* McGraw-Hill, New York.

Walski TM, Donald VC, and Savic DA (2001). *Water Distribution Modeling* 1st edn. Haestad Press, Waterbury, Connecticut.

White KD (2005). Decentralized wastewater cluster management: operation and maintenance: experience and costs. In *Proceedings of WEFTEC 2005,* pp.1682–1692. Water Environment Research Foundation, Alexandria VA, USA.

Wong THF (2006). *Australian Rainfall Quality – A Guide to Water Sensitive Urban Design.* Engineers Media,Engineers Australia, Canberra

Case studies

60L. <http://www.60lgreenbuilding.com/> Accessed 06/09/07

Inkerman Oasis. <http://www.portphillip.vic.gov.au/design_case_studies.html#L2> Accessed 06/09/07

Bolivar. <http://www.clw.csiro.au/research/urban/reuse/projects/asr_water_quality_ watereuse.html> Accessed 06/09/07

Kings Domain. Melbourne Water, *The Source* **19 (May 2002)**

CH2. <http://www.melbourne.vic.gov.au/info.cfm?top=171&pg=1933> Accessed 06/09/07

Pimpama Coomera. <http://www.goldcoast.qld.gov.au/t_gcw.asp?PID=5885> Accessed 06/09/07

Payne Road. <http://www.payneroad.com.au/mainpages/home.htm Accessed 06/09/07>

Rouse Hill. <http://www.sydneywater.com.au/SavingWater/RecyclingandReuse/RecyclingAndReuseInAction/RouseHill.cfm> Accessed 03/10/07

Toowoomba. <http://www.toowoombawater.com.au/> Accessed 03/10/07

32

The Water-efficient City: Technological and Institutional Drivers

Matthew Inman

Australians have long recognised the need to conserve water given the regular occurrence of periods of drought (Keating 1992). While the strength of this conviction has varied over the last few decades, water efficiency, as a component of an adaptive management approach to water planning, is now common place in Australian cities.

In looking to the future of these cities, and how the transition to more sustainable urban centres will be influenced by resource efficiency, there are some key questions that emerge from a water perspective:

- **knowledge of where water is used**—strategies for water efficiency require a detailed understanding of where, how and why water is used. What approaches will be required to deliver this knowledge?
- **potential to deliver efficiency**—what is the water conservation potential for different end uses and how should that shape planning for a water-efficient city?
- **demand during drought**—how is the role of emergency response measures (such as water restrictions) affected by longer term water efficiency reform?

This chapter investigates water efficiency and the role demand management will play in the future of Australian cities. Specific focus is given to the residential sector and the culture of water efficiency in the community. While recognising that industrial and commercial sectors are significant water consumers, it is the residential sector that is the largest (and growing) water user in Australian cities. And when considering transitions, it will be the households and communities of Australian cities where the change and innovation defining this transition will be most widely, and keenly, felt.

Understanding demand: end-use analysis

There has been a significant shift in the approach to residential water metering as the demand/ supply balance has come under pressure in metropolitan areas. Residential water use is the major water-use sector in Australian cities, representing some 59% (Rathjen *et al.* 2003) of water supplied (or close to 70% of metered consumption in some of the largest cities; Sydney Water 2006). Planning for the future requirements of major metropolitan areas necessitates a framework to understand the complex drivers of residential water use. In analysing these drivers, there has been a steady realisation of the need to understand and measure water use at the level of each appliance or fixture. End-use analysis is central to the development of projections for water demand.

The annual meter read cycle for residential properties was replaced in many Australian cities by quarterly meter reads almost two decades ago. The premise at that time was not so much to smooth water-demand projections, but rather to smooth cash-flow projections. Since that time, there have not been many broad-scale advances in the frequency of customer meter readings. An exception is Wide Bay Water, which installed metering that enables weekly collection of water use data. However, even in this case, the resolution of the metering technology is not sufficient to provide the level of detail required to classify water consumption by end use (Sands 2007).

Table 1: Methods for developing an end-use breakdown of residential water demand

Type	Description (with reference)	Pros	Cons
Third-party survey (phone or face-to-face interview)	Australian Bureau of Statistics Household Water Use Survey (ABS, 2002)	Comparatively low cost and rapid on a per household basis	Appliances can be wrongly characterised by householders; data on usage and behaviour not quantitative
Custom survey with household audit of appliances/ fixtures	Custom-designed survey delivered door to door by trained assessors (Roberts, 2004)	Comparatively low cost and rapid on a per household basis; accurate assessment of the appliance stock	Usage and behaviour data not quantitative
Smart metering (high resolution flow meters with signature / pattern recognition of water use events by end use)	Installation of high resolution meters and data loggers to record water use by end use; household audit of appliances / fixtures (Loh and Coghlan, 2003)	Usage and behaviour data available by end use for major end uses.	High cost; current technologies require significant time and effort to complete signature/ pattern recognition process; minor end uses (such as taps) more difficult to determine from flow data signatures.

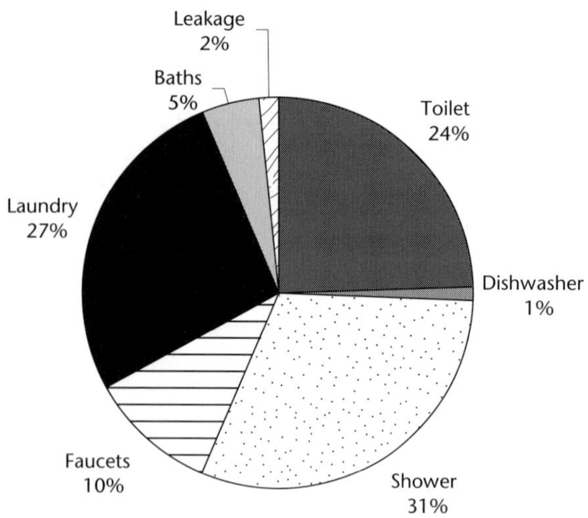

Figure 1: Profile of end uses for water, Sydney households (2002).

Collecting data to develop an understanding of water use at the level of individual appliances and fixtures remains a complex challenge, which requires integration of a range of data sources. While smart-metering technologies have been used in some Australian cities (Loh and Coghlan 2003), the sample size has not been sufficient to provide a statistically representative analysis and breakdown of demand by end use for the entire city. Some of the methods and approaches that have typically been adopted by water authorities in Australia to build a picture of demand by end use are listed in Table 1.

The output from the development of an end-use assessment of water demand is used to help shape priority areas for demand management initiatives. This issue is developed further in later sections of this Chapter. A comparative chart containing data from Sydney (reproduced from White *et al.* 2004) is presented in Figure 1.

Metering technologies to capture end-use data

Metering technology is continually advancing: wireless sensors and other devices, as well as improvements to analytical techniques, provide opportunities to more accurately and rapidly capture data on residential water consumption by end use. A key issue that defines the approach offered by new metering technologies relates to whether the data is to be collected close to the point of use or whether the data will be collected from close to the billing meter at the property boundary. These two alternatives offer very different opportunities for collecting end-use data—and neither of them can yet be considered as straightforward or ideally suited to the task.

- **Point of use data collection.** This approach involves installation of small (ideally wireless) flow meters close to the point of use so that each flow meter only records consumption associated with a single end use (a cold water tap at a basin or a dishwasher, for example). In this way, there is no chance of multiple, overlapping events reducing the accuracy of data collected
- **Property boundary data collection.** This approach involves installation of a high resolution flow meter in conjunction with, or on the household side of, the billing meter. The high resolution data (for example, 10 second sample interval for Trace Wizard©; Loh and Coghlan, 2003) is matched against flow data signatures / patterns to capture key water use events and end uses.

With an increasing need for end-use data, and a decreasing cost of implementation for much of this technology, it is clear that smart metering will play an increasingly important role in the management of urban water systems, both from a utility planning perspective and the way people use water.

There are three key factors that are considered to have the potential to drive a transformation in the collection and utilisation of end-use data.

- **Ease of installation** (particularly for the existing residential dwelling base). Metering solutions that monitor water consumption at the point of use will not see widespread adoption if they do not conform to the aesthetics of bathroom or kitchen design as perceived by the householder. Wireless technologies installed close to the point of use could have a role for some appliances / fixtures (such as dishwashers or washing machines), but installation issues will remain when considering end uses such as showers for existing dwellings. Conversely, a smart meter installed close to the billing meter at the property boundary is more readily used for existing dwellings.
- **Accuracy of data collected** (particularly for the existing residential dwelling base). Metering solutions that provide point-of-use flow metering inherently provide a high level of accuracy in distinguishing water-use events. The issue of overlapping water-use events is much more significant when relying on signature/pattern analysis of flow data collected from a meter close to the property boundary. For example, two related events such as toilet flushing followed by hand washing can be difficult to separate on the basis of flow signatures and can result in an over-estimation of water use for a given event (Koeller and Gauley 2004). Another issue is associated with differentiating between similar events taking place in different parts of the house, such as kitchen, laundry and bathroom tap usage.
- **Turnaround time for the data and presentation of the interpreted results.** With new approaches to pattern recognition and software analysis, there is the potential to dramatically shorten the time between the actual water use event and when the data on this event is made available to inform decisions. Systems that can analyse, interpret, transmit and present data on end use of water at intervals during the day would redefine water metering for an urban centre.

With broader adoption of metering technologies that enable water use to be measured over much shorter time periods (hours as opposed to months), many of the ideas and concepts that have been raised to transform the urban water market, such as cap and trade (Young *et al.* 2007), peak pricing (Sands 2007) as well as different pricing for indoor and outdoor water use (currently achieved through dual meters in some cities in America), could be implemented. Equally, with increased accuracy of end-use metering techniques that are more readily applicable to existing dwellings, there will be opportunities to drive broad adoption of these technologies. In this way, it is clear that advances in metering and the reforms afforded by increased water use data resolution will be one of the key transitions that drive water efficiency within our cities.

The role of appliance performance

One of the foundations of water efficiency in the broader context of resource use is the recognition that it is linked to the provision of services (cleaning clothes or dishes), rather than merely a product (water) (White *et al.* 2004). In this way, the focus for efficiency is on providing the same, or greater, level of service with less water, rather than assuming that with water saving comes a reduction in the level of service. This allows a much broader approach to be taken in recognising new technologies that may play a role in driving efficiency in our homes and businesses in the future.

A well-defined demand-management program incorporates options that tackle components of the water use profile in order of priority. In the case of the residential sector, these priorities are defined through analysis of the end uses of water as described in Figure 1. Appliances and fixtures such as washing machines or showers stand out as high water users from this level of analysis. There is, however, a second level of assessment required to appropriately allocate priorities for these residential end uses within a broad, city-wide demand management program.

The choices around an implementation model, and the likely performance of a demand management option targeting a specific residential water using appliance or fixture, is affected by three key issues:

- **Technical performance specifications for the efficient version of the appliance/fixture in comparison to other versions in the marketplace.** The technical performance of common household water using appliances and fixtures has improved significantly over recent years. In Australia, the Water Efficiency Labelling Scheme (WELS) has provided a framework—in conjunction with the relevant standards—for rating of appliances and fixtures against water use criteria (WELS 2007; WELS Determination 2007). In some instances, the WELS water use criteria significantly exceed the performance of the most efficient appliances and fixtures currently on the market, suggesting a pathway for further advances in efficiency. Showerheads are a good example where current best-practice products measured against WELS criteria receive a 3-star rating. The extent to which a higher rating can be achieved for this fixture type will depend on the set of defined performance criteria. Meanwhile, there are some examples of 6-star-rated appliances/fixtures already available on the market for the commercial sector—for example, urinals or commercial tap sets.
- **Household usage and behavioural patterns for the appliance/fixture.** A key aspect affecting the likely performance of demand management options relates to the frequency of use and general behaviour patterns associated with the appliance/fixture. While there are some clear links to issues associated with the use of appliances such as running washing machines without a full load, there are also some examples of where behaviour drives efficiency beyond that which would be expected from an analysis of the technical specifications. Shower usage is a good example of this issue, where inefficient showers are often not used at maximum (or capacity) flow rates. The term 'throttle back' has been used to describe adjusting the taps to reduce the flow rate from capacity to provide a more comfortable shower. Evidence of this can be seen in studies of shower flow rates for non-efficient showers in particular (Roberts 2004).
- **Market penetration (installed base and sales) for the various products in the appliance or fixture category.** With sales of appliances / fixtures occurring in the background of more formalised demand-management programs, the performance of these structural initiatives can vary dramatically – both between options and over the lifetime of an option. Performance in this context is the cost-effectiveness of the investment for the volume of water saved. An example of the changing impact of proportionally reduced sales motivated by the rebate (due to broadening market acceptance and penetration) on the performance of the option is presented in Figure 2.

Following on from the example provided in Figure 2, the market penetration (in terms of installed base) for front loading (efficient) washing machines has historically been low and relatively stable—growing slowly between 1987 and 2002 (ABS 1987; ABS 2002 respectively). In this context, options that could be implemented across a city or jurisdiction to improve the efficiency of this residential end use (such as rebates to influence purchasing decisions) perform well from the perspective

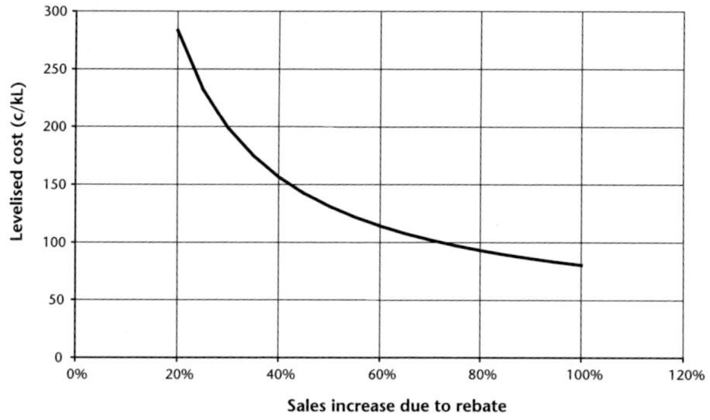

Figure 2: Change in cost-effectiveness of rebate option with changing level of sales due to rebate (levelised cost is a unit cost measure used when comparing options where the annual operating costs are not constant; Howe and White 1999).

of cost-efficiency. The relative strength of these options will change though as the marketplace changes—different implementation approaches will perform better (from a unit cost perspective) depending on the market penetration and sales rates. The most recent data from the Australian Bureau of Statistics (ABS 2006) indicates that the market penetration for front loading washing machines has grown significantly—up to 15.4 % in Sydney (2006). The differences in performance of rebate programs over a product lifecycle are described in Table 2.

In looking at future opportunities to drive increased water efficiency associated with residential water using appliances and fixtures, the product lifecycle outlined in Table 2 will continue to define the extent to which these opportunities will affect the overall efficiency of a city. In 2004, Caroma released a dual flush 4.5 / 3 litre toilet pan and cistern onto the Australian market. The evolution of products from Caroma, and other companies, in this market has seen average flush volumes reduce from 11 litres to 3.8 litres for a 6/3 litre dual flush toilet (Roberts 2004). There are companies currently developing products that push efficiency further for this product type (for example, Propelair 1.5 litre flush—see www.propelair.com and Environment Agency 2006).

Understanding the extent to which technologies for each end use can be improved and enhanced in terms of water efficiency enables a snapshot of the conservation potential to be developed (Snelling *et al.* 2007). In this context, 'snapshot' is the most appropriate way to consider conservation potential

Table 2: Matching approaches to delivery of demand management options with market dynamics for water using appliances and fixtures.

Product lifecycle	Installed base	Sales	Comments on option implementation method
Newly released on market	Low	Low	For a newly released product, a rebate can be very effective at encouraging purchases. The performance of the option can decline with increasing sales rates where there is a strong value proposition for the product irrespective of the presence of a 'point-of-sale' rebate.
Market recognition builds	Growing	Growing	The term 'free-riders' is often used to describe those people who had already decided to purchase the appliance. During this phase of the product lifecycle, the cost-effectiveness of a 'point-of-sale' rebate model for the option declines in proportion to the growing rate of underlying sales of the product (to 'free-riders'). At this point, a shift to product-exchange options (where the rebate is only provided on receipt of an inefficient product that is being replaced) provides a more targeted approach to driving adoption of efficient appliances.
Established product with high performance	Large	Dominant market share	In this market, there is very little value in targeted interventions to drive further adoption of an already high-selling, efficient appliance/ fixture. Development of performance standards is likely to deliver the most benefits as this would ensure all products met the performance benchmark in a manufacturing context where this benchmark was well established.

Table 3: Example of approach to determining conservation potential for a key end use (toilet flushing). Note that the dataset has been simplified for ease of presentation. Average flush data and frequency for dual flush toilets from Roberts (2004).

	Market penetration					
Scenario	**11 litre single flush (11 litres per flush average)**	**9/4.5 litre Dual Flush (6.5 litres per flush average)**	**6/3 litre dual flush (3.8 litres per flush average)**	**4.5/3 litre dual flush (3.4 litres per flush average)**	**1.5 litre single flush (1.5 litre per flush average) assumed as next generation technology**	**Conservation Potential assuming full adoption of next generation technology (litres per fixture per flush)**
Current	15 %	10 %	70 %	5 %	0 %	6.2
Widespread adoption of current best practice	5 %	5 %	30 %	60 %	0 %	4.9
Widespread adoption of next-generation technology	5 %	5 %	5 %	15 %	70 %	2.8

as there are many uncertainties in projecting likely efficiency gains that could be secured through technologies and behaviours not yet established in the marketplace. Table 3 presents these concepts with a hypothetical example focusing on a change in performance of the technology, as opposed to changing water use behaviours.

The example presented in Table 3 highlights the conservation potential for this key component of water demand. At a point in time, the conservation potential is specified as full adoption of a particular technology—here assumed to be a 1.5-litre single-flush toilet (modelled on the Propelair example referred to earlier). Therefore, in the current scenario, the conservation potential is significant due to the predominance of the 6/3 litre dual flush toilet. With widespread adoption of next-generation technology, the conservation potential is reduced to 2.8 litres per fixture per flush in this hypothetical example. In developing this approach further, we would need to consider potential changes in usage behaviour that might reinforce or reduce the impact of the improved technology.

Of course, while a 1.5-litre flush system might be regarded as evolutionary technology for cistern-based systems, there are many products available on the market that are waterless (see Your Home 2007). In exploring the conservation potential for an end use, scenarios where water demand can be reduced entirely may not be too distant a reality. While the examples listed in Table 3 deal with technologies where water serves to transport waste, a true transformation might change that model for the city as well and dramatically increase the conservation potential for that end use.

At the centre of this 'conservation potential' approach remains an understanding that a service is ultimately sought from water—rather than just a specific volume. Therefore, conservation potential will only be limited when opportunities to improve the efficiency of the delivery of that service have been exhausted.

Demand during drought

In the classic model of rain-fed water supply systems, water restrictions are an operational reality. The calculation of yield from storages contains an explicit assumption regarding the extent to which voluntary and mandatory water restrictions will drive water savings. Such measures have long been considered as a short-term intervention or an emergency response— the exception rather than the norm, with no more than 10 years in 100 on

Table 4: Periods of water restrictions for Sydney (various public sources).

Date	Stage of restrictions
July 1994 to November 1994	Voluntary
November 1994 to January 1995	Stage 1
January 1995 to March 1995	Stage 2
March 1995 to October 1996	Stage 1
November 2002 to October 2003	Voluntary
Oct 2003 to June 2004	Stage 1
From June 2004 to June 2005	Stage 2
From June 2005 to present (June 2007)	Stage 3

average requiring restrictions, for example (Varley, 2003).

This model is being challenged in Australia due to prolonged drought. Table 4 indicates the periods of water restrictions for Sydney since the completion of Warragamba Dam in 1960.

The challenges to this model of water restrictions are two-fold: maintaining the savings during the current extended period of drought; and the longer term impact of demand hardening on performance of restrictions during future drought periods.

Firstly there is the short-term effort to maintain the water savings during extended periods of restriction and across several stages of increasing severity of these restrictions. While the threat of fines, regular patrols by water restrictions officers and ongoing print, radio and television education campaigns all contribute to the maintenance of water savings, several factors are testing the effectiveness of these measures.

- **Value of outdoor spaces (amenity and investment).** Across Australia, there has been substantial investment in residential property renovations and rebuilds in recent years as the property market has experienced sustained growth. Investments in landscaping of the order of many tens of thousands of dollars and beyond are commonplace in a growing number of suburbs. This investment provides amenity and correspondingly contributes to the market value for the property. In this way, maintaining this investment is no longer seen as discretionary.
- **Differences in rainfall between the demand and supply catchments.** For many Australian cities, such as Sydney, the demand catchment and the supply catchment are geographically separate. Heavy rainfall in the demand catchment is quite often not matched by similar falls in the supply catchment. As such, there can be the perception that the drought has broken or there is no longer a need for restrictions. Advertising campaigns and prominent placement of data on storage levels (nightly television news for example) are used to reinforce the message to a community that may be experiencing average rainfall conditions.
- **Uncertainty around the model for restrictions beyond discretionary uses.** If long-held assumptions about the performance of severe water restrictions are no longer valid then the yield estimation methodology necessitates more conservative savings, which drives a more conservative yield calculation result for surface water storages.

The implications of these issues for our understanding of the role of water restrictions in urban water management strategies are significant. Coupled with findings that show how a small reduction in rainfall results in a much larger reduction in inflows to storages (see Spurling *et al.* 2007), there is considerable uncertainty surrounding long-term projections for the yield from surface water storages.

Demand hardening

The issue raised above of the reality of the performance versus the theoretical water restrictions regimes has broader implications, because discretionary use is reduced through water efficiency improvements across the residential water use spectrum. The traditional approach has been that during periods of unrestricted demand, there was a level of inefficiency built into this use. In this way, during periods of drought, limitations could be imposed on water usage that would not dramatically affect the homeowner (by targeting discretionary usage) resulting in short-term water savings that would enable a city to manage water supplies through the drought period.

The implementation of broad demand-management programs that drive improved efficiency have fundamentally altered the water use profile for cities. There is a realisation that driving significant improvements in the efficiency of use during periods of unrestricted demand will dramatically affect the ability of consumers to respond during the periods of drought. This issue is referred to as demand hardening.

A theoretical question therefore arises as to the extent to which some slack—or inefficiency—should be left in the system in terms of discretionary water use, which can then be 'saved' during drought periods versus striving to optimise efficiency for periods of unrestricted demand. This issue has been brought into sharp focus recently with the moves in several Australian cities to introduce permanent water savings measures. These measures have typically targeted outdoor water use as presented in Table 5 and capture many of the rules that have been commonly used for Stage 1 restrictions.

When water-demand savings that were once comparatively straightforward to achieve with Stage 1 or 2 water restrictions become 'locked in', assumptions regarding the yield for a rain-fed water supply system need to be revisited. Specifically, a reduction in the performance of water restrictions will reduce the theoretical yield for the water supply system. Equally, water restrictions regimes would increasingly be required to target non-discretionary uses in much earlier phases of a drought period.

The response to this series of issues regarding yield from storages (demand hardening; decrease in run-off) has been a deliberate shift to diversify water sources and reduce reliance on rain-fed systems (in the supply catchments). Large-scale recycling and desalination are two key measures that have seen significant investment in Australia in recent years.

Equally, at the household scale, there has been a diversification of water supply sources to provide greater self-sufficiency. Good rainfall in the demand catchments of some east coast Australian cities has seen rainwater tank rebates promoted during much of the current drought. There has also been an increase in the number of households using a greywater diversion device (sometimes no more sophisticated than a bucket in the shower). In both of these cases, the primary impact is seen in a reduction in the use of potable water for outdoor uses—again contributing to the 'lock-in' of savings that might be expected to be delivered from water restrictions regimes in the future.

In this respect, the concept of water restrictions as a component to drought response will disappear as discretionary use is more tightly managed and non rain-fed sources are brought on-line. While not the first time such a statement has been made (Keating 1992), this current drought period has driven structural changes to demand and supply that reduces the significance of restrictions at the fundamental level of water supply operations.

The water-efficient city

The potential for sustained and short-term demand reductions through water efficiency has long been recognised. However, the exact size of the savings, or conservation potential, remains, and will remain, a product of the time. The extent to which water use associated with a given appliance or fixture can be reduced and made more efficient is very closely tied to the technologies available. The conservation potential identified today will itself appear conservative in 10 years as new technologies develop.

Co-benefits

Increasingly there is a recognition of the wider benefits delivered through water efficiency—beyond a reduction in potable water required from bulk water sources and the household water bill. For example, the NGAC (New South Wales Greenhouse Gas Abatement Certification) scheme has recognised the energy savings that are derived from a reduction in hot water usage through monetary credits. Efficient use of water also improves the performance of local alternative water supply sources. For example, where a rainwater tank is supplying some indoor and outdoor uses, an efficient toilet or washing machine will ensure a greater volume of water available from the rainwater tank for outdoor uses.

Potential impacts on water and wastewater systems

Provision of a service, as opposed to a volume of water, is a key principle in water efficiency. Once

Table 5: Comparison of water restrictions policies and permanent water savings measures for Melbourne and Canberra.

City	Stage 1 restrictions (residential focus)	Stage 2 restrictions (residential focus)	Permanent water saving measures
Melbourne (source of information – water utility media releases)	• Hand watering is allowed any time, • Manual sprinklers from 5am to 8am and from 8pm to 11pm • Automatic sprinklers from 11pm to 6am. • Topping up of pools and spas is allowed but permits are required to fill or refill. • Cars can be washed only by using a watering can or bucket. Hoses can be used for rinsing only.	• Private lawns cannot be watered at any time. • Vehicles can only be cleaned with a bucket, watering can, or high-pressure water-cleaning units. No hoses can be used unless connected to a high-pressure water-cleaning unit. • Apply to fill a new pool	• Use manual watering systems only between 8pm and 10am • Use automatic watering systems only between 10pm and 10am • Fit your hose with a trigger nozzle • No hosing paved areas • Apply to fill a new pool
Canberra (source of information – water utility media releases)	• Sprinkler and other irrigation systems from 6pm to 9am on any day. • Trigger hoses held by hand and buckets anytime. • A bucket and mop or high pressure, low volume cleaner can be used to wash paved areas at any time. • Otherwise, water must not be used to clean paved area unless cleaning required as a result of accident, fire, health hazard or other emergency.	During the week· • Ban on sprinkler and irrigation systems other than drippers· • Hand-watering and drippers between 7am-10am and 7pm-10pm on alternate days as per the 'odds and evens' system General Exemption apply for weekends • Water systems such as sprinklers between 7pm and 10pm on Saturdays for even numbered houses and Sundays for odd numbered houses • Pools can't be filled or emptied without an exemption • Vehicles can only be washed once per month unless washed at commercial car wash that recycles water or holds an exemption to use potable water.	• At all times gardens, lawns and plants may only be watered without causing pooling or runoff. • (1 September - 31 May) Sprinkler and other irrigation systems can be used to water lawns and plants only between 6pm and 9am on any day. • (1 June - 31 August) No limitation on watering systems during this period. • Hand watering - a hand-held hose fitted with a trigger nozzle, a bucket or a watering can may be used at any time. • If not washed at a commercial car wash, any vehicle should be washed on a lawn or other porous surface wherever practicable and then may only be washed by using either a bucket or watering can; a high pressure low volume cleaner or a hand held hose fitted with a trigger nozzle

the water used in homes and businesses enters the wastewater network, it continues to provide a service in the form of waste transport. There will be an increasing focus on ensuring that existing and new systems and networks can be operated and maintained effectively in the context of a water-efficient city. Other issues (such as greywater recycling and sewer mining) will be coupled with the potential for reductions in sewer flows resulting from more efficient use of water—all contributing to an increase in the complexity of management and operation of wastewater systems in particular.

Metering

Technologies for water metering and sensing, data analysis and transmission have seen some significant improvements in recent years. The current challenges, centring on turnaround time for information on water use and the automated identification of end uses, will be overcome. This will drive a proliferation of data-gathering exercises and a dramatic increase in data availability. This, in turn, will affect the planning that takes place in water utilities, where the current regime has very poor data—a legacy of quarterly meter reads. Overcoming these hurdles in data capture and analysis will offer up new approaches to water pricing and models for water trading within a demand catchment.

Adaptive management will remain a central component to the implementation of demand reduction programs for water-efficient cities of the future. In striving for cost-effective approaches to tackle residential end uses, there will be a clear understanding of how the performance of an option will change over time and the associated conservation potential. This understanding is built from three key tenets of residential water efficiency—the performance of the technology, water-use behaviour and market penetration.

As for water restrictions—structural changes to demand and supply will mark the end of their significance as a component of urban water management and system operation. In this way, the water-efficient city of the future will be fundamentally different: more complex in terms of diversity of water sources; less structured with new approaches to water pricing and urban trading; and with a conservation potential that will continue to evolve in time—re-cast as technology and water use behaviour responses to a myriad of familiar, and unfamiliar, drivers.

References

Australian Bureau of Statistics (1987). *Domestic Water Use in NSW.* Cat. no. 8719.1. ABS, Canberra

Australian Bureau of Statistics (2002). *Domestic Water Use, New South Wales, Oct 2002.* Cat. no. 4616.1. ABS, Canberra.

Australian Bureau of Statistics (2006). *Environmental Issues: People's views and practices.* Cat. no. 4602.0. ABS, Canberra.

Environment Agency (2006). WC Innovation at WRC. *Water Demand Management Bulletin* **7**, 6.

Howe C and White S (1999) Integrated resource planning for water and waste water: Sydney case studies. *Water International* **24(4)**, 356–362

Keating J (1992). *The Drought Walked Through: A History of Water Shortage in Victoria.* Department of Water Resources, Melbourne.

Koeller J and Gauley W (2004). 'Effectiveness of data logging residential water meters to identify and quantify toilet flush volumes: a pilot study'. Report prepared for the Metropolitan Water District of Southern California <www.cuwcc.org> viewed July 2007.

Loh MTY and Coghlan P (2003). *Domestic Water Use Study In Perth, Western Australia 1998-2001.* Water Corporation, Perth.

Rathjen D, Cullen P, Ashbolt N, Cunliffe D, Langford J, Listowski A, McKay J, Priestley A and Radcliffe J (2003). 'Recycling water for our cities'. Report to Prime Minister's Science, Engineering and Innovation Council, 28 November 2003, Canberra, Australia. <http://www.dest.gov.au/sectors/science_innovation/publications_resources/profiles/recycling_water_for_our_cities.htm#publication> viewed July 2007.

Roberts P (2004). *2003 Appliance Stock and Usage Patterns Survey.* Yarra Valley Water, Melbourne.

Roberts P (2005). *2004 Residential End Use Measurement Study.* Yarra Valley Water, Melbourne.

Sands O (2007). The swing to smart solutions. *Water21: Magazine of the International Water Association* **April**, 26–28

Snelling C, Turner A, Riedy C, White S and Cummings S (2007). Where does the water go? Best practice stock modeling for water using appliances. *Proceedings of 4th IWA Specialist Conference on Efficient Use and Management of Urban Water Supply*, 21–23 May 2007, Jeju Island, Korea, International Water Association and the Korean Society of Water and Wastewater, Korea

Spurling T, Shrinivasan M, Coombes P, Cox S, Dillon P, Langford J, Leslie G, Marsden J, Priestly T, Roseth N, Slatyer T, Wong T and Young R (2007). 'Water for our cities: building resilience in a climate of uncertainty'. Report to Prime Minister's Science, Engineering and

Innovation Council, June 2007, Canberra, Australia. <http://www.dest.gov.au/sectors/science_ innovation/publications_resources/profiles/water_for_our_cities.htm> viewed July 2007.

Sydney Water (2006). 'Water conservation and recycling implementation report 2005–2006'. Sydney Water.

Varley I (2003). 'Water Supply System Reliability'. Discussion Paper. Prepared for the Independent Expert Panel on Environmental Flows for the Hawkesbury Nepean, Shoalhaven and Woronora Catchments.

WELS (Water Efficiency Labelling and Standards) Determination (2007). Legislative Instrument - F2007L00330.

WELS (2007). Website. <www.waterrating.gov.au> viewed July 2007.

White S, Milne G and Riedy C (2004). End use analysis: issues and lessons. *Water Science and Technology: Water Supply* **4(3)**, 57–65

Young M, McColl J and Fisher T (2007). Urban water pricing: How might an urban water trading scheme work? *Droplet newsletter* **5 (4 February)**.

Your Home (2007). Website. <www.greenhouse.gov.au/yourhome> viewed July 2007.

33

Changing Attitudes to Urban Water Use and Consumption

Geoffrey Syme and Blair Nancarrow

The recent enduring drought has highlighted the long-term issues facing Australian urban water users in the future. While droughts come and go, the impacts of climate change and population growth in urban centres in Australia will result in an increasing environmental and social footprint, especially if per-capita demand returns to pre-drought patterns. The difficult environmental, economic and social trade-offs associated with providing new dams or groundwater supplies, or simply increasing the movement of water from regional to metropolitan centres, will become evident. Innovative solutions will be required. The adoption of these solutions is likely to require attitudinal, decision-making and behavioural change by all parties—domestic consumers, water utilities and regulators, town planners, appliance manufacturers and architects. The issues confronting the implementation of the National Water initiative, which to this point has largely been focussed on rural issues, will become increasingly relevant to the cities.

The obvious starting point to alleviating urban water problems is that of encouraging improved efficiency in water use among domestic consumers, commerce and industry. While in the interests of a coordinated response, demand management needs to be spread across all sectors, over 60% of urban water is used by domestic households in Australia (see Engineers Australia 2006 for details of urban consumption patterns). This disparate group of domestic consumers requires more detailed and complex assessment and understanding of demand management in terms of potential policy instruments than, for example, major industry where water quality can vary, which opens up more options for recycling and re-use. For this reason, this chapter focuses largely on domestic use.

Interest in attitudes, however, should not limit itself to attitude change as it pertains to current water use behaviours themselves. We need to understand more about how families view longer term issues such as setting the levels of service required and attitudes towards the potential to augment supplies or adopt alternative supply systems. We also need to know if current attitudes and values have the capacity to change on major issues such as suburban landscape preference and finally how, at an institutional level, we can formulate and implement decision making processes that allow beneficial long-term change to occur.

This chapter is therefore divided into four parts. Firstly, we will discuss the current attitudinal and lifestyle determinants of water use and their implication for household conservation and the formulation of demand-management programs in Australia. Secondly, we will illustrate and assess the potential for more detailed public input into the setting of levels of service for the various facets of water supply. Thirdly, we will examine the potential for in-depth attitudinal analysis to assess whether innovative modes for delivering water supply will be acceptable and how they may be introduced in a participative manner. Finally, it will be apparent from the above discussions that a major driver of acceptance of change is that users must trust in institutions and the decision-making processes. The chapter concludes with a discussion of the significance of the need to meld both professional and lay attitudes and knowledge resiliently into participative adaptive learning processes if water reform and change is to become an integral part of our future urban water delivery system.

The drivers of domestic consumption and the potential for demand management in the Australian urban community

Perceptions of water as a resource

Marketing and other studies conducted by social researchers have regularly used multivariate statistical techniques to segment their consumers. For instance, there is a standard fourteen item 'Ways People Think about Water' inventory available (Nancarrow *et al.* 1996) which can provide a quick and systematic method for observing changes in patterns of thinking over time and between cities. These items cover varying contexts from the 'rights to water', its environmental and aesthetic significance, to aspects of water as a domestic commodity. From the initial development of this inventory with respondents from three Australian cities, four basic clusters of people were identified:

- *Self interested* (33%) [20%]—people with a general lack of strong thought about water in any context other than 'their rights'.
- *Earthy* (29%) [23%]—individuals who generally thought about water in its aesthetics, conservation and utility contexts, but not in terms of additives or water rights contexts.
- *Environmentalists* (18%) [32%]—respondents who were primarily concerned about water in terms of aesthetics and conservation, rather than utility or water rights contexts.
- *Service oriented* (20%) [25%]—people who generally thought about water in all its contexts, but more so in terms of rights, additives and utility, and least likely in the conservation context.

The approximate percentages in each cluster in the early 1990s (in round brackets) show that the largest segment (approximately one-third) were most concerned about their rights in regard to water services, although there was not a huge interest by this group in water issues. Nevertheless, about half of the urban community was thinking about the wider roles of water in aesthetic and environmental terms. The percentages in square brackets were derived from the same questions from the second Perth Domestic Water Use Study (Loh and Coghlan 2003) from which very similar clusters emerged. It can be seen that at that stage (measures taken over the 1999/2000 period, several years after the initial study), the urban community was starting to think more about water conservation and more interested in water issues than in the earlier study.

Predicting household water consumption

The two Perth Domestic Water Use studies conducted in the early 1980s and at the turn of the century (Water Authority of Western Australia 1983; Loh and Coghlan 2003) provide very detailed bases on which to predict water use. It is likely that many of the findings, especially for internal (indoor) water, use, are applicable to other Australian cities.

In the first study, detailed diaries of water useage patterns were collected. In the second, smart metering was available to track water use. In addition to details of appliance use, demographic and income details were collected as well as block size, garden area, and so on. Pertinent to this chapter, detailed data on attitudes was gathered during family interviews in the first use study and for a family member in the second (Syme *et al.* 1990; ARCWIS 2002). These attitudes could be related to actual overall household consumption in both studies. Furthermore, accurate estimates of external water use could be obtained from the second study.

In the first study, total household consumption could be predicted from people's attitudes towards the recreational value of their garden, the perceived contribution of the garden to the re-sale value of the house and their concern about the price of water. Thus the outdoor lifestyle, price and investment were prime motivators of use. At this stage, attitudes towards conservation did not correlate significantly with household consumption. The first Domestic Water Use Study enabled improved measures to be used in the second, by which time one quarter of the variance in water consumption in Perth could be explained by attitudes. Again 'green lifestyle' was a significant predictor of household water use, as was attitude to price. A new indicator of water use also emerged—that of attachment towards home as an environment. That is people who spent a lot of time at home were greater users of water than those who spent more time away from home and who attached more significance to the wider environment.

The second Domestic Water Use Study also provided the opportunity to develop a predictive model of external water use. This is important, as external water use is more elastic than internal water use and is highly affected by the restrictions that have been introduced by drought. As expected, the model (Syme *et al.* 2004) showed that block size, income, garden condition and irrigation technology were all predictors of external water use. But lifestyle, garden interest, outdoor recreation and conservation attitudes also predicted external water use, largely independently. Because about half the variation in external water use was accounted for by attitudinal and lifestyle variables, this indicates that water resource managers need to carefully assess the social costs of restrictions. The model developed provides a means of doing so.

Predicting domestic water conservation

There have been a number of studies that have assessed the determinants of water conservation in urban households. As for the ways people think about water, Moore *et al.* (1994) have shown that there have been increasingly positive attitudes towards conservation over time. In an application of a reasoned action model for water saving (Kantola *et al.* 1982), both social norms and individual attitudes (such as attitudes towards price) affected intentions to conserve. When it came to persuasion, Kantola *et al.* (1984) showed that the most effective form of message was one of a 'citizen's duty'.

Aitken *et al.* (1994) demonstrated that attitudes towards water conservation are important—when people were made aware that their attitudes and their water behaviours were not consistent, householders cut back somewhat on their water consumption. Finally, Gregory and di Leo (2003) found that water use (particularly indoor use) can be viewed as a habitual behaviour and is thus not susceptible to easy change. This is why the best means for encouraging indoor water conservation are efficiency measures (technology) rather than persuasion (Syme *et al.* 2000).

Implications for demand management

In explaining both household consumption and intentions to conserve from a social psychological perspective, it is evident that it is likely that internal use will be harder to modify than external use because of its habitual nature. Therefore, appliance and technology innovations are most likely to change indoor useage. On the other hand, while drip irrigation and other technologies will help outdoor water use, it would seem to be more elastic in that regimes of irrigation, plant types and the ratio of lawn can all be changed. Outdoor water use is perceived as less essential and therefore should have a higher degree of elasticity. This is, of course, why outdoor restrictions on irrigation are supported by utilities and, generally, by the public (e.g. Nancarrow *et al.* 2002) as being the most acceptable policy in the short term. Such an observation is supported by the findings of Thomas and Syme (1988) in their assessment of price elasticity of water for varying applications in Perth. Nevertheless, there are likely to be limits to tolerance of ongoing restrictions given the value of the garden for lifestyle and recreation and the ongoing financial investment in maintaining the garden, not only for aesthetics and amenity, but also in terms of its contribution to the value of the home as an economic asset (Thomas *et al.* 1987).

There is thus some general understanding of the components of demand management, but integrated approaches to planning will be needed, combined with supply strategy and the application of an orderly evaluation program (Baumann *et al.* 1998; Troy and Holloway 2004). It is important to be specific about the objectives of the demand-management program, particularly in the long term (see, for example, Opitz and Dziegielewski 1998). These objectives will need to be more comprehensive than simply balancing supply and demand by developing new water supplies when current standards of reliability cannot be met. Wider consideration of the community's acceptance of the environmental and energy footprints of any new supplies will need to be balanced against the community's willingness to conserve and accept modified levels of service (Kenway *et al.* 2007).

Learning theory provides perhaps the best integrating concept—at least to integrate the elements of demand management and to promote enduring efficient water use at a household level—that of applied behaviour modification. A basic model is shown in Figure 1. In their review of individual motivators for conservation, Syme *et al.* (2005) found that 27 variables had been identified in the

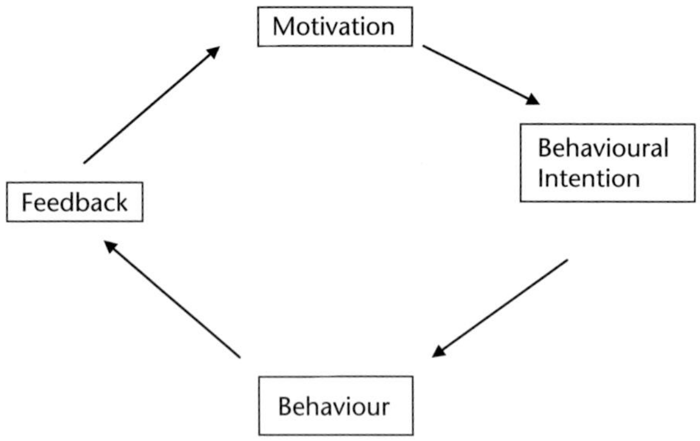

Figure 1: A basic learning model for achieving efficient water use. In short, people require motivation to intend to save water and change their behaviour. If this is to be maintained over time, they also require feedback to show that they have been successful.

environmental psychology literature. These included: individual variables, such as intrinsic ethically based motivators; social variables, such as perceived social norm motivators; and wider institutional considerations, such as trust in local and state government. An individual's intentions to perform conservation behaviours were largely mediated by the perceived difficulty or 'cost' of performing a behaviour and the resources, such as income. Whatever the behaviour required (including decisions to purchase water-saving technologies), all four elements in Figure 1 operate to either maintain or instigate changed behaviour at a household level. It is very rare that all, (or, sometimes, any) of these elements are systematically investigated to provide a sound behavioural basis for demand-management programs in different locations and in particular environments.

One clearly evident problem is the lack of timely feedback for the behaviour, either in terms of the quantity of water used or saved or the cost incurred. Basic learning theory indicates that the closer the feedback, either positive or negative, to the performance of the behaviour, the more likely the behaviour will be learned (Geller *et al.* 1982). In the case of water, this rarely occurs because meters, if they are in place, are usually situated outside the home and therefore not in a conspicuous position. Water bills can arrive on a three monthly or even annual basis, so feedback on the results of the behaviour is far from immediate and therefore unlikely to assist in an adaptive learning role to promote conservation. This may also be part of the reason why studies have universally shown that the price elasticity of water is low. The pricing signal arrives well after consumption. One means of solving this issue is to locate meters in a conspicuous place within the dwelling to provide rapid feedback both on consumption and on cost. At the very minimum, the price elasticity of informed consumers under differing tariff structures could then be determined.

Better understanding of the issues that motivate voluntary conservation in a variety of situations can guide information and intervention initiatives, especially in relation to the degree of change that can be expected at the household level. None of the environmental/psychological determinants of behaviour are hard to measure, and understanding these variables will greatly enhance our ability to set realistic and sustainable levels of consumption. Nevertheless, as discussed in the next section, there may be water-service issues that represent non-negotiable levels of service of supply, which may vary for differing com-

Table 1: Scoping levels of service

	DIMENSION 1 **What comes in the household's door**	**DIMENSION 2** **Costs of what comes in the household's door**	**DIMENSION 3** **Symbolic, aesthetic, moral or ethical Issues**
Quality of service delivery	time taken to answer phones response time to complaints design of bill	loss of environment risk of dam failure water treatment recreation	equity who pays definition of community service obligations regional differences in levels of service
Quality of product	pricing drinking water quality water pressure reliability of supply	waterways quality risks in implementation confidence in institutional arrangements	appropriate degree of recycling support of urban amenity

munities, but underpin responsiveness to demand management.

Setting levels of service

Provision of water services is the most complex challenge facing any utility. As in the provision of energy, a continuous supply is needed. But there are other dimensions to supplying water:

- health risks
- aesthetic considerations, such as taste and odour
- the potential of additives, such as fluoride to enhance health
- varying levels of flow and pressure associated with delivery
- the ability to selectively restrict use.

Currently, the establishment of the levels of service required by urban consumers is set through a combination of three components. These are:

- traditional engineering practice and regulator response to health and environmental considerations
- economic analyses and regulator perspectives on the benefits versus costs of the service supplied (including pricing)
- complaints data.

Community data, where available, usually comprises generalised satisfaction ratings or willingness to pay for changes in service. Presumably, if people are willing to pay, the levels of service being offered are appropriate. However, it must be acknowledged that much decision making is influenced by the utility's own perception of acceptable risk in construction, reliability of supply and experience of what the community traditionally has been prepared to accept or endorse. In general, a well-performing utility is seen as one in which political visibility is low.

As the effect of climate change becomes more apparent, and the environmental footprint of developing new water sources or information on over allocation of existing ones becomes more evident to the community, community involvement in the wider trade-offs involving levels of service will become more evident. In this case, there will be a need for a more accountable process for deciding on appropriate levels of service. Increasingly these standards will need to be set within an overarching sustainability framework and will include more detailed social analysis of what are seen to be appropriate trade-offs, not only in terms of levels of service delivered to the customer, but also in terms of the social, economic and environmental trade-offs associated with new sources.

Fortunately, we now have a wide variety of social assessment techniques that can take us much further than complaints data and willingness to pay (Syme *et al.* 1992; ARCWIS 1999; ARCWIS 2000; Syme *et al.* 2004). The ARCWIS review suggested that there was a need for a simultaneous holistic view of the overall levels of service supplied by the utility. Most available studies tend to cover one or two dimensions of levels of service at a time (such as satisfaction with water quality or reliability). The framework suggested is shown in Table 1. It should be noted that within the framework for the overall levels of service model, the first dimension—that of household delivery— maintains the greatest emphasis. However, it is becoming evident that the wider externalities associated with new sources of supply, such as wider regional effects on catchments and estuaries, requiring analysis for public accountability (Jenerette and Larsen, 2006) are becoming of concern, as are all the variables related to the wider costs associated with urban water supply. Symbolic, aesthetic, moral or ethical issues, which are difficult to measure, are also becoming of concern, with the competition between regions becoming elevated as we search for overarching principles for our urban water outcomes (Sustainability Panel 2007).

Assessing innovative water supply systems

As current water supply systems increasingly come under stress, water utilities are being faced with the need to introduce innovative water delivery systems. It is also becoming evident that the community is showing increased interest—both positively and negatively—in new systems. While, there are increasingly positive attitudes towards conservation, all innovation will not necessarily be accepted. The need to understand and incorporate community attitudes is vividly illustrated in the area of

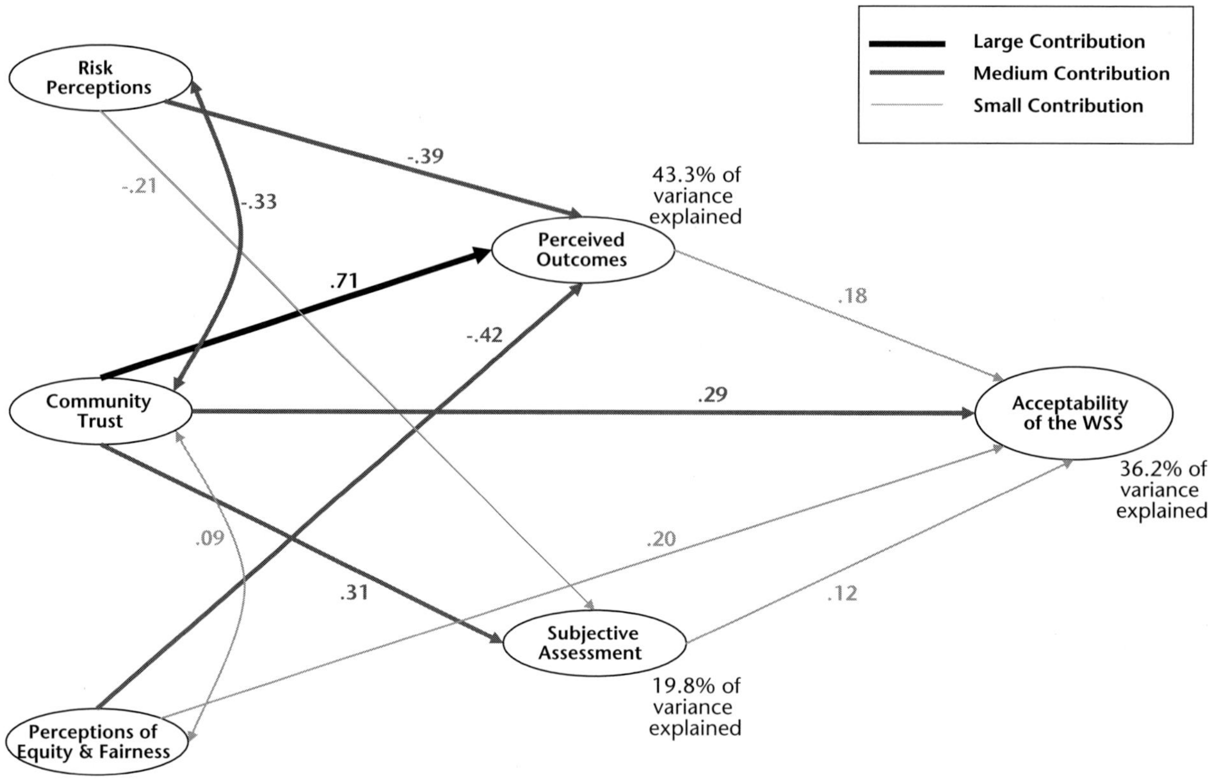

Figure 2: A structural equation model of attitudinal determinants of Melbourne's current water supply system (WSS)

recycling treated wastewater for potable use. This has been demonstrated by the widespread debate generated in cases such as Toowoomba and Goulburn where such innovation has been considered. This is, perhaps, not surprising given the attitudinal data presented by Syme and Nancarrow (2006). From these data (Table 2), there is clearly a long way to go before the community as a whole will feel comfortable with drinking recycled wastewater and the reason for such concern can be seen. Therefore, if innovation in water delivery is to be achieved, we need to move beyond simply measuring appropriate levels of service into a technological design phase—where the ingredients required by the community are assessed in detail—before projects are presented to the public as a *fait accompli*.

While in the past it has been appreciated that the success of new water supply systems is dependent on incorporating consumer or household preferences and improved community engagement, there has been very little comprehensive research that has attempted to examine acceptance of an entirely new water supply system. We will now outline an approach capable of being used to examine the social psychological basis of people's reaction to alternative urban water systems. The model is shown in Figure 2 (Leviston *et al.* 2006).

It can be seen that risk, trust and perceptions of fairness of particular systems are important in judging overall acceptability of a water system. This shows that innovative water supply systems will be judged in terms of the wider institutional and community setting, as well as in regard to their specific characteristics. The 'perceived outcome' relates to beliefs about the characteristics of the system itself. Is it seen to be reliable and sustainable in the long term? 'Subjective Assessment' refers to the specific judgment as to whether the benefits of the proposed system outweigh the risks. It can be seen that this judgment has a quite discernible effect from that of risk when considered alone. The strength of the relationship between the variables will differ according to the particular technology or water supply system being evaluated. But it has been

Table 2: Community acceptability of a series of technological approaches to re-use water (Approaches shown in descending order of acceptability).

Technological approach	Percentage acceptable
Re-use stormwater that has been treated to approved health standards on golf courses and ovals	98
Re-use stormwater that has been treated to approved health standards for fire fighting	97
Re-use stormwater that has been treated to approved health standards on your home garden	96
Re-use wastewater that had been treated to approved health standards for fire fighting	95
Re-use wastewater that has been treated to approved health standards on golf courses and ovals	95
Re-use stormwater that has been treated to approved health standards for toilet flushing	95
Re-use wastewater that has been treated to approved health standards for toilet flushing	92
Install a small, enclosed and quiet wastewater treatment plant in your neighbourhood to allow for re-use of wastewater in local parks and gardens	89
Re-use wastewater that has been treated to approved health standards on your home garden	88
Store stormwater that has been treated to approved health standards in wetlands for re-use at a later time	81
Install and regularly maintain a domestic wastewater treatment unit underground at your property to allow for re-use of the wastewater in your garden	72
Install, but have maintained by approved authorities, a domestic wastewater treatment unit underground at your property to allow for re-use of the wastewater in your garden	71
Store wastewater that has been treated to approved health standards in wetlands for re-use at a later time	70
Re-use stormwater that has been treated to approved health standards in the laundry for washing clothes etc.	68
Re-use wastewater that has been treated to approved health standards in the laundry for washing clothes etc.	51
Re-use stormwater that has been treated to approved health standards in the bathroom for personal washing etc.	50
Install a composting toilet where approved authorities maintain it and dispose of the compost	35
Install a composting toilet and use the compost on your property	33
Re-use wastewater that has been treated to approved health standards in the bathroom for personal washing etc.	31
Re-use stormwater that has been treated to approved health standards for drinking	29
Re-use wastewater that has been treated to approved health standards for drinking	16
Install a domestic urinal	13
Implement rostered water use, where you would be provided with specific times for such activities as washing or garden watering	12
Buy bottled drinking water which would allow lesser quality water to be provided through the water supply system	11

Source: Syme and Nancarrow 2006, p. 190

shown in a number of case studies—varying from neighbourhood stormwater schemes in Brisbane, through managed aquifer recharge in Perth, to recycling schemes in high-rise apartment blocks in Sydney and in a potential sewer-mining scheme in Melbourne—that the basic building blocks of the model remain constant (Leviston *et al.* 2006). The model therefore applies at all scales and to all systems and can therefore provide a basic evaluation tool for innovative systems before they are built.

Thus the advantage of having this model is that possible schemes can be evaluated beforehand.

Not only can the likely degree of acceptance be measured, but the specific social psychological determinants of that acceptability can be ascertained. This enables the system designer to modify the scheme or focus precisely on the risk, trust or equity issues to improve the design from both the community and professional viewpoint. Thus adoption issues become in-built and, as a result, innovation is more likely.

In general terms, the model is also of great interest because of the nature of the variables that were not found to be significant. The researchers first hypothesised that such systems would be assessed mainly in terms of the personal benefits the householder would receive. It was found that measures of water-related lifestyle and the desired level of service for households that had been successfully measured in relation to consumption studies did not relate to overall system evaluation. Instead it was found that people tend to look at long-term change and innovation in terms of societal, rather than personal, benefit and this clearly incorporates the notion of trust in decision makers and sustainable outcomes from the water supply system.

Changing attitudes through procedural justice and a sustainability perspective

We have seen that there is an increasing tendency for the community to be actively concerned with the wider context of urban water use and the environmental effects of new sources and the downstream effects of urban water use. Fewer residents of our settlements are taking water supply for granted and considering it from a purely utilitarian viewpoint. This has been acknowledged by the urban water industry, both nationally and internationally, and is reflected in the rise in interest in applying the concept of sustainability to broaden the breadth of planning and service delivery (Ashley *et al.* 2004).

Initially, sustainability was often considered in the context of evaluating plans and practice through the triple bottom line of environmental, economic and social analyses. Increasingly, there is an acknowledgement that a key element of sustainability is institutional (see, for example, Hedelin 2007). That is, it is recognised that quality of the decision making process is a highly significant precursor to implementing sustainable practice. Institutions are not simply construed as formal government organisations, but more as the interfaces of various sectors of the community and government to obtain outcomes that reflect the long-term aspirations of the community for the benefits it wishes to attain from its water resources (Syme *et al.* 2007). One of the key elements of quality decision making in this context is the achievement of procedural justice in the involvement the community has with the water utility.

Procedural justice has a number of elements (Lawrence *et al.* 1997) and these support acceptance of the decision making process and its outcomes (Tyler and Blader 2000). While most of this research has been conducted in traditional social service areas, it is also relevant to urban water management and other natural resource management issues. For example, in a long-term planning exercise for creating strategic plans for the use of wastewater with the West Australian Water Corporation, perceptions that the public involvement process was procedurally just led to greater commitment for involvement in future decision making activities (Syme and Nancarrow 2002).

Because there is public voice, this enables information exchange and can include local knowledge that can enhance the possibilities for adaptive learning. While the concept is appealing, it has not really occurred in many places but it is vital that urban utilities grasp the opportunity. The recent debate on whether to pipe water to Perth from the large South West Yarragadee aquifer some 300 kilometres from the metropolitan area provides a good example of the widening context within which urban utilities will be evaluated. The early social research on this issue indicated concern by both South West Yarragadee and Perth residents that there was too little knowledge concerning: the aquifer, fairness issues relating to the concept of 'reasonable regional needs' (that is, was it fair to take the water from that community?) and finally the doubt that the Water Corporation would reverse its decision if data showed that it was not sustainable for the aquifer. These issues continued to drive the community debate as the hydrological modeling became more detailed and the social and economic analysis increased. Finally, it became evident that the limits to usefulness of modeling had been reached and that only monitoring the effects of the abstraction of water would validate the decision. Therefore, in their acceptance of the proposal, the Independent Sustainability Panel established by the state government recommended a series of public accountability and involvement processes designed to ensure that adaptive learning through procedurally just processes could occur. The issue for the community then became whether they could trust the utility or the regulators to effectively implement the recommendations. Comparisons were made between the recommended adaptive learning approach and the widely publicised deterioration in the Gnangara mound aquifer in Perth where breaches in extraction rules had become publicised. Thus the issue of trust was a paramount one for the Water Corporation and for the community when assessing new developments.

Given the growing discussion of reallocation of water from rural to urban communities, similar cases are likely to occur. Thus the water industry has been moved by circumstance and the community to become much more visible player in the overall movement towards sustainability. It is no longer sufficient for urban utilities to think on a project-by-project basis and to deliver levels of service that satisfy only the utilitarian aspects of water supply. Increasingly, issues of urban and environmental amenity and the integrity and quality of decision making are becoming central concerns for the community.

Conclusion

Household water consumption is an important component of overall water use in Australia's cities. Until now, the traditional water utility, with a centralised supply and wastewater stream, has served the community well. It is clearly evident though that the development of new sources needs to be considered much more carefully than in the past. Non-contested surface water dams are becoming scarce, groundwater management for water supply in cities such as Perth is being carefully scrutinised and even technological 'fixes' such as desalination plants are being debated in terms of energy costs and environmental impacts.

While these challenges are very evident to engineers, they are also of concern to household water consumers. As shown in the systems evaluation model, notions of the long-term sustainability of water supply alternatives are important determinants of the acceptability of systems. In this context, community attitudes are moving away from the purely utilitarian view of tap-water to one in which the wider footprint of water supply is being more carefully considered.

There is an increasing tension between the *lifestyle values* that govern household consumption and *conservation values* that support maintenance of amenity in suburban and regional landscapes. This highlights the need for increased effort in three areas. Firstly, increased systematic use of social science is needed in the planning and implementation of demand-management strategies, establishing levels of service and in designing new systems. Secondly, there is a need to embed adaptive learning principles into the changing water utilities role at the individual household level and at the wider regional water planning level. Finally, while the community is prepared to acknowledge the 'expert' professional role of the utility, key water decisions involving trade-offs and significant values or ethics must involve auditable and procedurally just processes. The public will wish to ensure that the criteria (not necessarily the analysis) established by the decisions reflect their view of long-term sustainability (Syme and Nancarrow 2005). Public trust for utilities will become more central to implementing change, and possibly harder to earn.

References

ARCWIS (1999). *The Social Basis for Urban Water Provision in the 21st Century.* CSIRO Urban Water Program, CSIRO Land and Water, Perth.

Aitken CK, McMahon TA, Wearing AJ and Finlayson BL (1994). Residential water use: predicting and reducing consumption. *Journal of Applied Social Psychology* **24**, 136–158.

ARCWIS (2000). *Drinking Water Aesthetics: A Summary of an Integrated Methodology to Determine Community Preferences and a Policy Direction.* CSIRO Urban Water Program, CSIRO Land and Water, Perth.

ARCWIS (2002). *Perth Domestic Water-Use Study: Household Appliance Ownership and Community*

Attitudinal Analysis 1999–2000. CSIRO Land and Water, CSIRO Urban Water Program, Melbourne.

Ashley RM, Blackwoord DJ, Butler D and Jowitt P (2004). *Sustainable Water Services: A Procedural Guide*, IWA Publishing, London.

Baumann DD, Boland JJ and Haneman WM (Eds) (1998) *Urban Water Demand Management*, McGraw Hill, New York.

Engineers Australia (2006). *Water and Australian Cities*, Institution of Engineers, Canberra.

Geller ES, Winett R and Everett PB (1982). *Preserving the Environment: New Strategies for Behaviour Change*, Pergamon, New York.

Gregory GD and di Leo M (2003). Repeated behaviour and environmental psychology: The role of personal involvement and habit formation in explaining water consumption. *Journal of Applied Social Psychology* **33**, 1261–1296.

Hedelin B (2007). Criteria for the assessment of sustainable water management. *Environmental Management* **39**, 151–163.

Jenerette GD and Larsen L (2006). A global perspective on changing sustainable urban water supplies. *Global and Planetary Change* **50**, 202–211.

Kantola SJ, Syme GJ and Campbell NA (1982). The role of individual differences and external variables in a test of the sufficiency of Fishbein's model to explain behavioural intentions to conserve water. *Journal of Applied Social Psychology* **12(10)**, 70–83.

Kantola SJ, Syme GJ and Nesdale AR (1984). An informational analysis approach incorporating Fishbein's model in a test of the effects of appraised severity and efficacy in promoting water conservation. *Journal of Applied Social Psychology* **13**, 164–182.

Kenway S, Howe C and Maheepala S (2007). 'Triple Bottom Line Reporting of Sustainable Water Utility Performance'. Awwa Research Foundation, in press.

Lawrence RL, Daniels SE and Stankey GH (1997). Procedural justice and public involvement in natural resource decision making. *Society and Natural Resources* **10**, 577–589.

Leviston ZL, Porter NB and Nancarrow BE (2006). *Interpreting Householder Preferences to Evaluate Water Supply Systems. Stage 3.* CSIRO Water for a Healthy Country National Research Flagship, Land and Water, Perth.

Loh M and Coghlan P (2003). *Domestic Water Use Study: In Perth, Western Australia, 1998–2001.* Water Corporation, Perth.

Moore S, Murphy M and Watson R (1994). A longitudinal study of domestic water conservation behaviour. *Population and Environment: A Journal of Interdisciplinary Studies* **16(2)**, 125–139.

Nancarrow BE, Kaercher J and Po M (2002). 'Community attitudes to water restrictions policies and alternative sources. A longitudinal analysis, 1988–2002'. Report to CSIRO Land and Water Consultancy, Perth.

Nancarrow BE, Smith LM and Syme GJ (1996). The ways people think about water. *Journal of Environmental Systems* **25(1)**, 15–27.

Opitz EM and Dziegielewski B (1998). Demand management planning methods. In: *Urban Water Demand Management.* (Eds DD Baumann, JJ Boland, JJ and WM Haneman). McGraw Hill, New York.

Sustainability Panel (2007). *Sustainability Assessment of the South West Yarragadee Water Supply Development.* Department of Premier and Cabinet, Government of Western Australia, Perth.

Syme GJ and Nancarrow BE (2002). Evaluation of public involvement programs: Measuring justice and process criteria. *Water* **29**, 18–24.

Syme GJ and Nancarrow BE (2005). Sharing the pain of water reallocation: creating consent by taking fairness and justice seriously. In: *Consent and Consensus, Politics, Media and Governance in Twentieth Century Australia* (Eds D Cryle and J Hillier) pp. 371–387. API Network, Curtin University of Technology, Perth.

Syme GJ and Nancarrow BE (2006). Social psychological considerations in the acceptance of reclaimed water for horticultural irrigation. In *Growing Crops with Reclaimed Water* (Ed. D. Stevens) pp. 189–196. CSIRO Publishing, Melbourne.

Syme GJ, Seligman C and Thomas JF (1990). Predicting water consumption from homeowners' attitudes. *Journal of Environmental Systems* **20(2)**, 157–168.

Syme GJ, Nancarrow BE, Bishop, BJ and Vanderwal P (1992). 'Community analysis of household water pressure satisfaction'. Report No. 40. Urban Water Research Association of Australia Research, Melbourne.

Syme GJ, Nancarrow BE and Seligman C (2000). The evaluation of information campaigns to promote voluntary household water conservation. *Evaluation Review* **24(6)**, 539–578.

Syme GJ, Shao Q, Po M and Campbell E (2004). Predicting and understanding home garden water use. *Landscape and Urban Planning* **68**, 121–128.

Syme GJ, Jorgesen BS and Po M. (2005). 'Community and behavioural aspects of state of environment reporting: development of a WA community state of environment inventory'. Report to WA department of Environment, CSIRO Land and Water, Perth.

Syme GJ, Porter NB, Goeft U and Kington EA (2007). Integrating social wellbeing into assessments of water policy: Meeting the challenge for decision makers. *Water Policy*, in press.

Thomas JF and Syme GJ (1988). Estimating residential price elasticity of demand for water: a contingent valuation approach. *Water Resources Research* **24(11)**, 1847–1857.

Thomas JF, Syme GJ and Salerian SN (1987). Factors influencing bore ownership of domestic households in Perth. In *Domestic Water Use in Perth, Western Australia*. Water Authority of Western Australia, Perth.

Troy P and Holloway D (2004). The use of residential water consumption as an urban planning tool: a pilot study in Adelaide. *Journal of Environmental Planning and Management* **47(1)**, 97–114.

Tyler TR and Blader SL (2000) *Cooperation in Groups: Procedural Justice, Social Identity and Behavioral Engagement.* Psychology Press, Philadelphia.

Water Authority of Western Australia (1983) *Domestic Water Use in Perth, Western Australia.* WAWA, Perth.

SECTION 7

Transitioning Waste to Resources

Australians generate approximately 32 million tonnes of solid waste annually (1.6t/person/pa)—a rate of increase of 7.5% annually over the past decade (ABS 2007, Australia's Environment: Issues and Trends, Cat. No. 4613.0,). More than half of this solid waste is disposed to landfill. A key 21st century transition is for governments, industries and communities to view waste streams as resource streams as a means of increasing resource productivity and reducing resource consumption.

As in the previous sections that deal with material flows of energy and water, there remains significant scope for consumer and community response to reducing consumption, as well as reducing waste through re-use and recycling (the 3Rs). There are two additional directions for innovation that enables a transition from the traditional industrial era process of: extract–manufacture–use–dispose. One involves *cradle-to-cradle* initiatives for individual manufacturers that require a complete reconfiguration of industrial processes around end-of-life strategies for products that 'close the loop' on the product life cycle and thus significantly reduce their ecological footprint (Kaebernick *et al.*). The other involves development of *eco-industrial clusters*, whereby new industries and products are created by identifying synergistic uses for multiple waste and energy streams in a region. This represents a Horizon 3 challenge, given the need for multiple stakeholders to become involved: mapping and characterising waste streams, developing technologies to create useable materials around which new products can be manufactured, as well as planning and constructing new industrial parks for this new class of green industries—in job poor outer metropolitan and peri-urban regions. Batten *et al.* provide some examples of successful implementation of industrial ecology principles.

PWN

34

Cradle-to-cradle Manufacturing

Hartmut Kaebernick, Scott Ibbotson and Sami Kara

Over the past 200 years of industrialisation there has been a significant increase in prosperity, manufacturing capital, standards of living and real wages. As a result, four critical factors have been identified that society currently faces (Westkamper 2000):

1. a rising consumption of natural resources
2. the dramatic increase in the world population
3. environmental impacts
4. an unstoppable worldwide globalisation.

New methods and strategies are being developed to deal with these problems and challenges—transforming the present cradle-to-grave economy into a cycle or cradle- to-cradle economy. One area that can assist in this transformation is that of product manufacturing. Current manufacturing systems are considered to be open loop in style: where manufacturers' main interests are focused in the areas of design, development, sales and distribution as can be seen in Figure 1 (Kara *et al.* 2004). Through research conducted in sustainable manufacturing, this open-loop approach is seen as obsolete if a manufacturer wishes to adopt a sustainable approach to product development.

Coupled with this, there is a greater drive by today's society to urge manufacturers to move away from this system of manufacturing and adopt an approach that considers the whole product life cycle (Suzuki *et al.* 2000). In general, a product's life cycle in general, consists of material extraction, manufacturing, assembly, use and end-of-life (EOL) (Rose *et al.* 2002). As shown in Figure 2, a shift to a closed-loop manufacturing system, is proposed to enable manufacturers to move away from open-loop manufacturing and take into account the whole product life cycle (Suzuki *et al.* 2000). This closed-loop system encourages the notion of product take-back, where the responsibility of disposing of a product shifts from the consumer to the manufacturer (Kara *et al.* 2004). The producer then becomes accountable for material recovery and recycling or product disposal, rather than the consumer who currently usually disposes the product to landfill at its EOL. As a result, manufacturers adopting a closed loop system can reduce the impact a product can have on the environment. This is achieved because there is less dependence on the use of raw materials, as well as a reduction in the amount of waste to landfill. Subsequently,

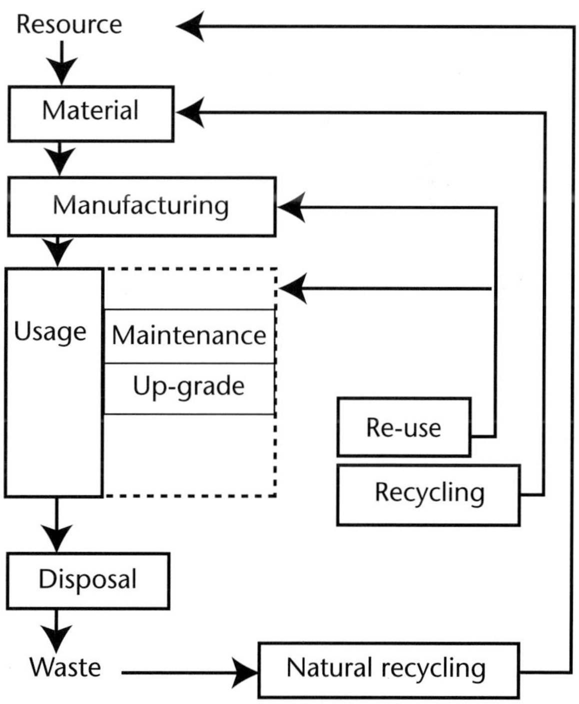

Figure 1: Open-loop manufacturing (Source: Kara *et al.* 2004).

this system has the potential to increase profits and/or improve a manufacturer's competitiveness through appealing to environmentally conscious consumers (Mangun and Thurston 2000).

Closed-loop manufacturing

Drivers for implementation

Government legislation and policies

To encourage manufacturers to take on a closed-loop product life cycle, government legislation and policies have been introduced in many European countries, as well as in Japan and in Australia (Scheidt *et al.* 1994; Klausner *et al.* 1999; Boks *et al.* 2000; Rose 2000). This was done not only to push the notion of product take-back by manufacturers, but also to help in the reduction in the amount of landfill and minimisation of wasted resources (Yamagiwa *et al.* 2001).

In Europe examples of legislation and policies include: Waste of Electrical and Electronic Equipment (WEEE), the Restriction of use of certain Hazardous Substances (RoHS) and End-of-Life Vehicle (ELV), which are all directives of the European Union (EU) (Feldmann *et al.* 2002; O'Neill 2003; Feldmann and Melzer 2004; Przekop and Kerr 2004; Shangguan 2004). While WEEE and RoHS are aimed at the electrical and electronics equipment producers, ELV is aimed primarily at the automotive industry. This directive bans the use of heavy metals—such as lead, mercury, cadmium and hexavalent chromium—in components sold in Europe. Only under exceptional circumstances are manufacturers allowed to use these metals. This directive also specifies goals for overall vehicle recyclability and recycled content, as well as requiring vehicle manufacturers to develop dismantling manuals for their components (Przekop and Kerr 2004).

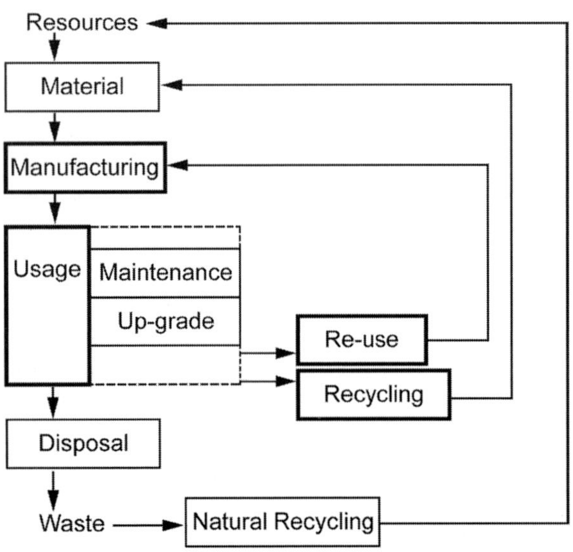

Figure 2: Closed-loop manufacturing (Source: Suzuki *et al.* 2000).

In Japan there is a concept of 3R (reduce, re-use, recycle) (Daimon *et al.* 2004), which is the basic law for establishing a recycling based society (Kato and Kimura 2004). There is also a law on recycling of End-of-Life Electronic Home Appliances (EL-EHA) (Nakamura and Kondo 2004), which was introduced in April 2001. It was the first law introduced to compel manufacturers of televisions, washing machines, air conditioners, refrigerators and freezers to be responsible for the EOL treatment of their products (Nakamura and Kondo 2001).

Compared with other countries, environmental legislation in terms of product take-back is in its infancy stage in Australia. The current focus in Australia is more on waste reduction than prevention—the majority of initiatives are focused on recycling. Preventative programs centred on waste avoidance are almost non-existent and at best have only minor efficiency gains rather than wholesale reconfiguration. There is little evidence to suggest widespread re-use, refurbishment or re-manufacturing activity that can be classed as anything other than cottage-based or boutique in its orientation (Gertsakis and Lewis 2003). Legislation is in place throughout the various states and territories. For example, in New South Wales (NSW) there is the 'Extended Producer Responsibility Priority Statement 2004' which is part of the *Waste Avoidance and Resource Recovery Act 2001* (EPA NSW 2004a). However, this legislation, and others, is not mandatory and therefore only a few companies have taken on the proposed strategies.

Some examples of current policies around the world include Integrated Product Policy (IPP) (Jeswiet 2002) and Eco-efficiency (Feldmann *et al.* 2002). These policies deal with the whole life cycle of the product, including the EOL decision for that product. The Extended Producer Responsibility (EPR) (Scheidat and Shuqiang 1994) also holds manufacturers responsible for the entire life cycle of the product, including disposal—forcing manufacturers to take product recovery into account during product development.

Consumer awareness

Legislation and policies are not the only forces driving manufacturers to take-back and determine the best EOL strategy for their products. Consumers are more aware of the problems that can arise from neglecting the environment (Kara *et al.* 2004). As a result, it has become more enticing for manufacturers to design and market environmentally friendly or 'green' products to gain a marketing edge against competitors. Furthermore, consumers around the world have a strong motivation for environmental issues and are willing to pay a higher price for products that are more environmentally conscious (Stevels 2000). Consumers' purchasing decisions are increasingly influenced by the environment so there are opportunities for companies to use a 'green' marketing strategy for their products (Ottman 2004).

Environmentally Conscious Manufacturing and Product Recovery (ECMPRO)

Increased environmental awareness has resulted in the development of what is known as Environmentally Conscious Manufacturing and Product Recovery (ECMPRO) (Gungor and Gupta 1999). ECMPRO involves incorporating environmental thinking into the development of new products, which adds a new dimension to this process.

The automotive industry is one of the leaders in this area, with researchers at Ford, Chrysler and General Motors (GM) developing easier ways to disassemble their cars and take an 'ease-of-destruction' approach into consideration, along with taking 'ease of construction' (Gungor and Gupta 1999). For example, in Australia the components in the Ford Territory SUV are 75% recyclable, lead free e-coating and water-based paints are used and 60% of all waste related to manufacturing of the vehicle is recycled (Wheels Automotive Design Awards 2006).

Consumer domestic and electronic equipment industries are also involved in the development of cleaner environmental practices for their products. Companies such as Canon, Proctor & Gamble and Digital are inventing ways of making their products more environmentally friendly (Gungor and Gupta 1999). Digital has a 6R approach—which stands for recycle, reclaim, refurbish, re-manufacture, re-sell and re-use, to deal with their products at the end of the first life cycle (Dorgelo 1996). In Australia, Dell Computers has initiated a recycling program for the environmentally safe disposal of computers. This is dependent on the age and configuration of the used computer. A cash rebate is offered for the old computer during the purchase of a Dell computer (EPA NSW 2004b).

Ecodesign / Design for Environment

Ecodesign is seen as one of the tools manufacturers can use to assist in the adoption of a closed-loop manufacturing system. The concept behind Ecodesign is that it considers the environmental aspects at all stages of the product development process, with the ultimate aim of developing products that cause the lowest possible environmental impact throughout the product life cycle (The United Nations Environmental Program 1997). In order for Ecodesign to be effective, it needs to be integrated into the early design stage of product development. A core component of Ecodesign is that of 'Design for Environment (DfE)'. DfE is the systematic consideration, during new product and process development, of design issues associated with environmental safety and health over the whole product life cycle (Kuo *et al.* 2001).

There are three main goals of DfE (Kuo *et al.* 2001):

- minimise the use of non-renewable resources
- effectively manage renewable resources
- minimise toxic release to the environment

The origin of DfE stems from the traditional engineering concept of Design for 'X', which is an integrated approach to designing products and processes for cost effective, high-quality life cycle management (Rose 2000). As a result, there are a number of DfX tools that focus on the environmental impact of a product. These include not only DfE, but also Design for Recycling (DfR), Design for Disassembly (DfD), Design for End-of-Life (DfEOL) and Design for Reuse (DfRe).

Design for Recycling (DfR)

The goal of DfR is to encourage pre-production planning for safe and efficient recycling, through

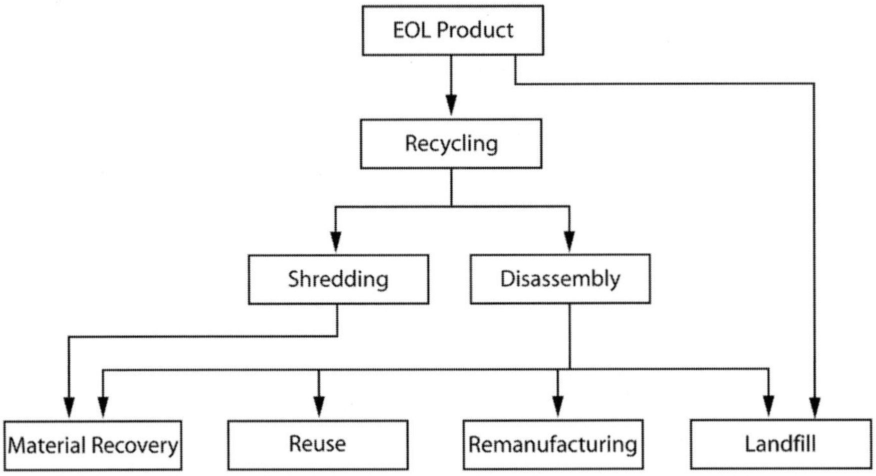

Figure 3: The role of disassembly in EOL decision making (Source: Zussman *et al.* 1994).

eliminating as much hazardous and non recyclable materials as possible from the manufacturing process (Weiner 2001). Products designed for recycling can be classified into two categories—recycled products and recyclable products. Recycled products are those that are manufactured using recycled materials or from components no longer in use (The Catalan Centre for Recycling 2006). Alternatively, recyclable products are ones that are designed to be recycled at the end of their useful life. To achieve this, a number of factors are considered, including the use of mono-materials (such as a single type of plastic for the whole product), eliminating toxic and hazardous substances and a code system is used to identify materials that are difficult to recycle (The Catalan Centre for Recycling 2006).

To assist designers to use DfR, a number of guidelines have been established (Gungor and Gupta 1999). Further guidelines are available on DfR for plastic components (Masanet *et al.* 2002) as well as items made of metal (Henstock 1988). There are an increasing number of companies around the world adopting these principles to reduce the environmental impact of their operations. For instance, Mercedes Benz has adopted a DfE and DfR approach to its car manufacturing. This is highlighted with its use of mono-material systems in its under-body protection and bumpers (Mercedes Benz 2006), which allows ease of disassembly and high-quality material recycling. Another example of DfR is provided by BMW, in which nearly 90kg of the 162kg of plastics in a BMW-3 series can be economically recycled so as to adhere with its own recycling standards (Veshagh 2006). However, there has been an increased trend in the last decade in using reinforced plastics in industries such as aircraft, car and electronic industries. This has been directly translated into increased number of discarded products with reinforce plastics. Recycling of reinforced plastics poses a challenge—because thermosetting plastics are highly resistant to mechanical, thermal and environmental effects. Although, the introduction of green composites is promising, more work is needed in the optimisation of the properties of recycled composites, in the development of economical processing and in educating the market to accept the reused products (Baillie 2004).

Design for Disassembly (DfD)

The concept of DfR can be used in conjunction with DfD, as the principles of DfD are mainly concerned with disassembly for recycling. However, it can also be used for disassembly, re-manufacture and reuse (Santos and Ferrao 2005). Disassembly plays a vital role in decisions regarding EOL options as shown in Figure 3.

Disassembly can be defined as a systematic method for separating a product into its constituent parts, components, subassemblies or other groupings (Moore *et al.* 1998). Disassembly can involve the removal of some components from the product (partial disassembly) or complete disassembly. A product designed for disassembly can be quickly and easily broken down so that the components can be recycled (Siddique and Rosen 1997). There are two main methods for disassembly—reverse disassembly and destructive disassembly (Kuo *et al.* 2001).

An example of reverse disassembly is a fastener that can be screwed in and then screwed out again. Additionally, if two parts are snap-fitted together, then they can be snapped apart. Destructive disassembly on the other hand, involves either deforming or cutting the part. DfD has many advantages, including reduction of work needed in part and material recovery, uniformity and predictability in product configuration, simple and quick disassembly operations, ease of handling of disassembled parts, ease of separating and treating removed materials and residuals, and reduction in product variability (Pornprasitpol 2006).

From a product designer's perspective, while it may be desirable to integrate DfD principles into the design of a product, it can be quite challenging. The reason is that designers generally do not have the experience in disassembly and recycling to determine the impact of various design aspects on the difficulty and effort required at the disassembly stage (Hesselbach and Kuhn 1996). A number of tools have been developed to assist designers to integrate DfD into their design, such as DfE software (Boothroyd Dewhurst Inc 1996). The software simulates the EOL disassembly of a product and illustrates to designers quantitatively the economic and environmental affects of their decisions. The effectiveness of such tools has been demonstrated with a number of cases, such as washing machines, refrigerators and other household appliances.

Another tool used to aid in DfD is known as 'selective disassembly'. The aim is to disassemble components identified as having reuse potential, and then determining a disassembly sequence for the selected components with minimal removal of other components. Currently very little research has been conducted in this area, which can be attributed to the time-consuming and costly nature of the analyses, as well as the high number of probable disassembly sequences (Santochi *et al.* 2002). However, this approach is known to dramatically reduce the traditional problems associated with complete disassembly.

Design for End-of-Life (DfEOL)

DfEOL looks at determining a suitable EOL strategy for a product at the early design stage. One such tool to aid designers in DfEOL was developed by Stanford University's graduate level Design for Manufacturability course. The software is known as the End-of-Life Design Advisor (ELDA). ELDA was developed for Product/Industrial designers to forecast an EOL strategy for a particular product during the conceptual design phase. This phase is seen as the area where major decisions and changes can be made to the product with minimal impact on development costs. ELDA has an 89% success rate in predicting EOL strategies with industry best practice and uses statistical analysis to map product characteristics to EOL strategies (Rose *et al.* 2000). The EOL treatment options ELDA recommends are: re-use, service, re-manufacture, recycle (disassembly first) and recycle (shred first).

Design for Re-use (DfRe)

The principle behind DfRe is to design a product or component for the sole purpose of re-using it at the end of its first life cycle. At present there has been little research that examines designing a product or component from a re-use perspective and very little work has been done regarding tools that could be used to design a product for re-use purposes.

Two existing tools are a tolerance based model, which is used to assess a component's suitability for reuse (Suzuki *et al.* 2000). The other is a 'closed rapid product life cycle' concept, in which product users are motivated to return and upgrade their products to the latest model. In turn, the product parts are quickly taken back and re-used to avoid them becoming obsolete or sustaining serious damage (Kimura *et al.* 1998).

However, neither of the two methods provides a way of indicating to a designer which parameters are crucial to extend the life of a product so that it can be re-used. Such a methodology would greatly aid in the area of DfRe.

One recent solution to tackle this problem was through a methodology to determine the extended design life of a component based on design parameters and failure mechanisms (Ibbotson *et al.* 2007). To achieve this, firstly the components with similar failure mechanisms were clustered into groups and design parameters linked with those components

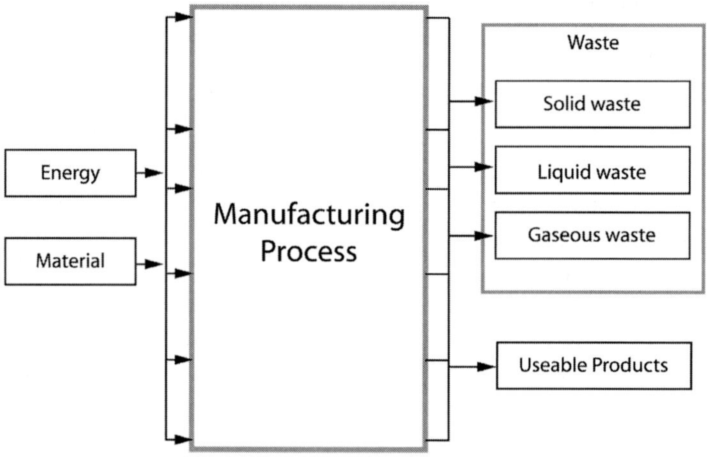

Figure 4: Input and output of manufacturing processes (Source: Bartel 2003).

were identified. Mathematical models were then developed for each cluster group to establish the relationships between failure mechanisms, design life and the associated design parameters. Finally, these relations were represented in graphical format to help designers to calculate a particular design life for the associated design parameters. The applicability of the methodology was demonstrated on electric motors. The results showed that the methodology can aid designers in estimating the design life and the associated design parameters with great accuracy (Ibbotson *et al.* 2007).

Life Cycle Assessment (LCA)

LCA is one of the major tools used in implementing Design for Environment and Cleaner Production (Sun 2004). The methodology behind LCA is to assess all the environmental impacts of a product (or service), from 'cradle-to-grave'—that is, from the initial extraction and processing of raw materials to final disposal (Ayres 1996). The International Organisation for Standardisation (ISO) defines LCA to be (International Organisation for Standardization 1996):

A holistic environmental accounting procedure which quantifies and evaluates all wastes discharged to the environment and energy and raw materials consumed throughout the entire life-cycle, beginning with sourcing raw materials from the earth through manufacturing and distribution to consumer use and disposal.

There are four basic stages of LCA, which include goal definition and scoping, inventory analysis, impact assessment and interpretation (Wenzel *et al.* 1997). LCA has been used extensively to determine the environmental impact of various products such as nickel-cadmium batteries (Rydh and Karlstrom 2000), produce packaging made from polystyrene (Zabaniotou and Kassidi 2003), photocopying machines (Fuji Xerox 2007), refrigerators, washing machines and televisions (Doowdell *et al.* 2000). The results achieved from an LCA are deemed to be very precise and acceptable (Chen and Chih-Wei 2001). However, LCA does have its limitations. A full analysis can be quite time consuming, costly and the data required needs to be very specific, which is normally not available in the early stages of product development (Hoffman 1997). This has lead to much research in the area of Simplified LCA (SLCA) (Kaebernic *et al.* 2003). However, these methods are not a substitute for a full LCA analysis, but they are tools that can be used by designers to assist in environmental decision making during the conceptual design stage (Kara *et al.* 2007).

Cleaner production technologies

In order to manufacture products, inputs—such as raw material, energy, manpower, capital, time and fluids—are converted into outputs, which are products, capital revenue, experience, knowledge, energy and industrial waste. This conversion process may refer to a whole factory or a simple process as shown in Figure 4.

The environmental impacts of industrial activities occur at the input stage, with the extraction of energy and material, and at the output stage, with the generation of waste, which includes solid, liquid and gaseous waste. Cleaner production technology is aimed at manufacturing processes that reduce the quantity of polluting residues and make the most rational use of raw materials and energy—all at a reasonable cost. Thus, cleaner technology means the integration of the environmental constraints into the production process in order to produce better while polluting less. This requires modifications and improvements of the manufacturing processes with the aim to minimise the use of auxiliary materials and energy. Process modifications and improvements can be realised by using alternative manufacturing technologies, which can help generate benefits of process optimisation, quality control, energy conservation and preventive management (Bartel 2003).

Reducing the number of steps in production—using the lowest possible number of production techniques— and optimising the usage of manufacturing consumables results in further improvements. Optimising manufacturing processes to reduce the 'non-product output' of waste and emissions per production unit, increases the efficiency of material use and decreases the amount of materials sent to landfill. To achieve this objective, designers and engineers should select shapes that eliminate waste-generating processes, such as sawing, turning, milling, and punching, and look for opportunities to recycle manufacturing residue in-house. Optimising manufacturing processes also means optimising energy consumption by using cleaner energy sources, such as natural gas, wind, hydro or solar energy.

Collection strategies for EOL products

Another key component in closed-loop manufacturing entails the collection of products once they have reached the end of their first life cycle. It is no longer viable for manufacturers to only consider the forward logistics distribution chain. They must now also take into account the complete environmental effects of all the products and processes until the product is returned at the EOL (Kara *et al.* 2007). This is known as 'reverse logistics'. The driving forces behind reverse logistics are economic as well as environmental. The economic reasons include the recovery of the value still incorporated in the used product, along with key savings in materials and components (Gonzales-Torre *et al.* 2004). Environmental pressures arise from concern about solid waste pollution, rising levels of products going to landfill or lack of raw materials (Ginter and Starling 1978). Reverse logistics systems, however, are inherently more complex than forward logistic systems because of the unknown quantity and quality of the returned products (Gungor and Gupta 1999). Previous research has suggested that it is advantageous to develop models that integrate both forward and reverse flows of new and used products, through designing reverse logistics networks that have independent forward and reverse distributions (Ginter and Starling 1978).

There are a number of solutions that have been developed to tackle some of the problems linked with reverse logistics networks. One such solution is a simulation-based approach for white goods collection in the Sydney Metropolitan Area (SMA). The model consists of a reverse logistics network made up of transfer stations, drop-off points and a disassembly plant (Kara *et al.* 2007). The most significant attribute of the proposed solution is the ability to easily comprehend how the system works through changing various parameters, so the most important factors can be further analysed.

Another issue concerning reverse logistics is that of managing the return of products. There are two types of product returns depending on the condition of the product—EOL and warranty returns. EOL products are those that have reached the end of their useful life and are disposed by their users. On the other hand, warranty returns are products that have been returned by the user within the specified warranty period. The underlying problem facing both types is the difficulty in quantifying the number of returned products. Knowing the rate of product return is critical for planning the whole collection system, from transportation through to resource allocation. A further hurdle regarding reverse logistics is that different parts of the world have their own collection strategies, which are affected by the country's intrinsic factors, such as income, education, compliance with WEEE or its environmental practices. For an optimum collection strategy to be successful, it is

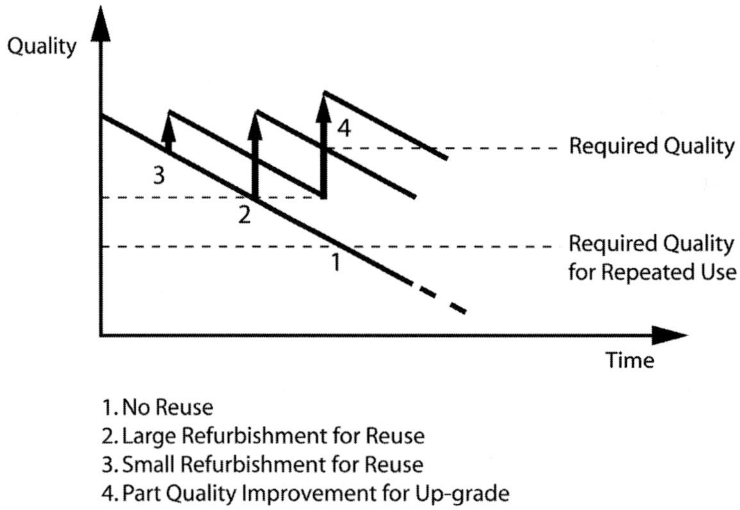

Figure 5: Quality of Reused Parts (Source: Kimura *et al.* 1998).

imperative that it is based on the characteristics of a particular country, state, suburb or city (Hanafi *et al.* 2007).

The previously mentioned 'closed rapid product life cycle' concept can also be used in deciding suitable collection strategies. In this strategy, users are motivated to return and upgrade their products (Kimura *et al.* 1998). An example of this is demonstrated on a single-use camera as shown in Figure 5.

The quality of a product can deteriorate quickly over time due to wear, but, in this case, it is mainly due to rapid technology development. If quality falls below the required level for repeated use, the product cannot be reused. Depending on the time of product return, small or large refurbishment or even improvement for upgrade can increase the quality above the required level, so that reuse is possible. This strategy encourages users to return the product at an appropriate time suitable for reuse.

EOL strategies (3R)

As mentioned earlier, product responsibility in terms of product take-back and EOL decisions has shifted from the consumer to the manufacturer. EOL is the point in time when a product no longer satisfies the initial purchaser (Rose *et al.* 2000) or when the product no longer performs the intended functions due to failure or wear-out. Depending on the product type, there are various EOL scenarios to select from, including re-use of recovered parts and subassemblies, repair, re-manufacturing, primary and secondary recycling, incineration and disposal (Mazhar *et al.* 2005).

Refurbish/re-manufacture

Re-manufacturing is the process of rebuilding a product at the end of its life cycle. Re-manufacturing is also known as refurbishing, reconditioning and overhauling. During this process the product is disassembled, cleaned and inspected. Defective components are replaced and the product is reassembled. It is then tested and inspected again, to ensure it meets, or exceeds, the standards of a newly manufactured product (Sundin *et al.* 2000). A flow chart of a typical re-manufacturing system is shown in Figure 6. Ordinarily, it is difficult to use established conventional production systems for re-manufacturing (Gungor and Gupta 1999) because of the need for higher flexibility to allow for uncertain qualities and quantities.

Re-manufacturing is different to repairing because it creates a full next life for a product, whereas repair only extends the useful life (Steinhilper 2001). Additionally, a repaired product will be returned to the original owner, while a re-manufactured product is anonymous and ownerless—just like a new one. Re-manufacturing is seen as being both profitable and less harmful to the environment than conventional manufacturing due to the reduction in landfill, the saving of virgin material and a decrease in energy consumption. The

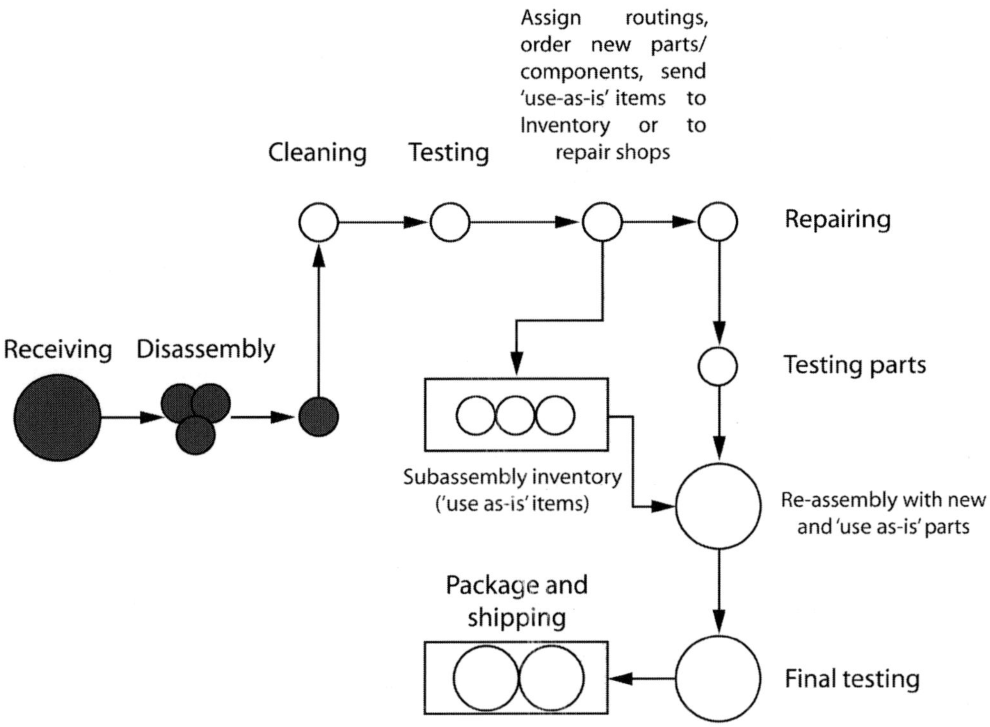

Figure 6: A re-manufacturing flow diagram for a product or part (Source: Gungor and Gupta 1999).

concept of re-manufacturing is preferable to recycling because it adds value to waste products by returning them to working order, while recycling only reduces the used product to its raw material value (Gehin *et al.* 2007).

Re-manufacturing is used in many industries, including the automotive, electronics and office furniture industries. The automotive industry has a long association with re-manufacturing and accounts for almost two-thirds of all re-manufacturing (Ijomah *et al.* 2007). Components typically re-manufactured include brake shoes, clutches, drive shafts, differentials, engines, fuel injectors, oil pumps and turbochargers. In the electric and electronics industry, a number of products have been identified as having good re-manufacturing potential, including photocopiers, vending machines, ATMs, computers and mobile phones. Office furniture suitable for re-manufacturing includes desk chairs, filing cabinets, desks and meeting tables. Re-manufactured office furniture looks identical to brand new furniture and performs in exactly the same way (Steinhilper 2001).

Re-use

The concept behind re-use is based on the fact that some components/parts and subassemblies have a design life that exceeds the life of the product itself (Kara *et al.* 2004). For instance, there is empirical evidence indicating that the lifetime of a motor often exceeds the lifetime of the product using it (Klausner *et al.* 1998). Therefore the re-use of electric motors in consumer appliances taken back by the manufacturer is technologically feasible. It can be associated with significant manufacturing cost savings potential and it does not compromise product quality. Economic advantages derive from a reduction of the unit manufacturing cost as a result of the decrease in material cost. Furthermore, there is a reduction in the inventory holding cost due to a higher inventory turnover (Reyes *et al.* 1995). A number of products and components have been identified as having good re-use potential, such as electric motors, gearboxes, LCD monitors, mobile phones and washing machines (Mazhar *et al.* 2005).

Companies which have implemented a re-use strategy for their products include Fuji Xerox for their photocopiers (Reyes *et al.* 1995), Kodak for their single-use cameras (Kara *et al.* 2004), and Weinmann—a German medical company—for an oxygen-supplying machine to aid people with severe lung disease (Griese *et al.* 2004). Siemens—in particular, its Medical Solutions division—is another company that is focusing on the re-use of parts, either as spare parts or standard parts for their products. At present approximately 50% of the parts found in an X-ray tube are reused by the company (Griese *et al.* 2004).

Despite the benefits associated with the reuse strategy there are several limitations that need to be tackled before re-use can be successfully used as an EOL strategy (Klausner *et al.* 1998):

- The time required for testing a product may be too long leading to potentially higher labour costs.
- A product may need to be disassembled to access and remove a component before it can be tested.
- The majority of information is based on indirect measures of degradation in order to assess the re-use potential due to insufficient availability of direct measurements.
- It is often difficult to identify parameters that reliably indicate the degree of degradation at the time of product return.
- Usually small sample sizes are used to relate the tested parameter(s) to the reuse potential.
- Due to rapid technology development, old parts become obsolete quickly.
- It is difficult to implement quality assurance for old parts.
- In many cases, the cost of collecting and replacing old parts is currently higher than the cost of fabricating new parts.

In addition, companies are reluctant to adopt reuse as it can reduce the market demand for new products (Linton 1999). Innovation is also said to suffer, as it is claimed that competition among companies inhibits the reuse of recovered components (Scheidt and Shuqiang 1994). However, this could be challenged with the example of the Wienmann company and the benefits that were achieved by applying the reuse strategy. Customer perception, as well as the fact that some products simply do not survive another life cycle, is another factor that hinders the adoption of a reuse strategy as an EOL option (Reyes *et al.* 1995).

Recycling

Recycling can be defined as the process of using component materials and processing them to produce the same materials or useful degraded materials (Linton 1999). Primary recycling is known as processing of materials into another 'high-value' application, whereas secondary recycling is processing of materials into a 'low-value' application (Ijomah *et al.* 1999). The goals and overall benefit of recycling include:

- reduction in greenhouse gas emissions
- consumption of resources and energy products
- conservation of landfill
- minimisation of the generation of toxins and pollutants
- conservation of non-renewable virgin materials (Rose 2000).

With these advantages of recycling in mind, this EOL option does have its limitations when compared with re-manufacturing and reuse (Bras and Mcintosh 1999):

- Recycling does not reduce material waste and the amount of landfill as effectively as re-manufacturing or reuse.
- Recycling only recovers materials and does not save the manufacture of new components (as is the case for re-manufacturing and re-use).

The most common types of recyclables are household recyclables (such as PET bottles, glass containers and aluminium cans) and electronic products. The two differ in that household recyclables are generally composed of a single material, while electronic waste is a composite of many materials, some of which are considered to be hazardous. Additionally, household recyclables are made of low-value materials, resulting in a limitation on the amount of post-processing steps needed to economically separate the material. In contrast, electronic waste and EOL goods can have a combination of precious and non-precious metals, often in small quantities. The typical metal breakdown in electronics is shown in Figure 7.

RICOH uses a comprehensive recycling strategy—a 'comet circle' as shown in Figure 8. This is a concept to demonstrate how a recycling-based society should operate, with the aim of producing

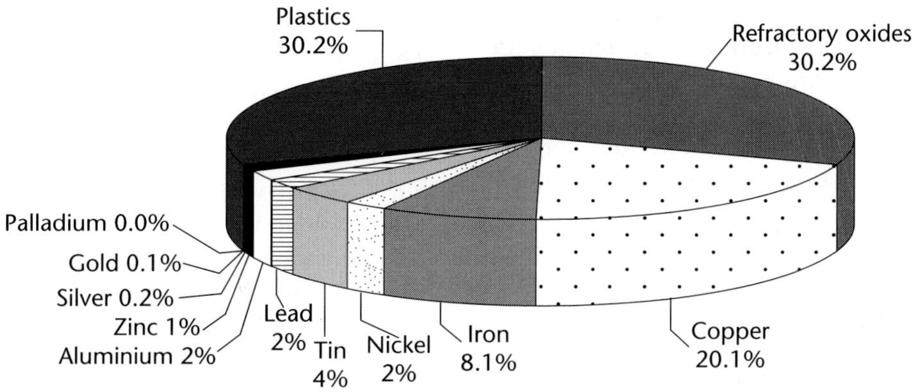

Figure 7: Typical material composition of electronic scrap (Source: Sodhi and Ramer 2001).

maximum effect with minimal resource use. The stages of this process array in a comet shape forming recycling loops (or orbits) around the customer, with items leaving these orbits destined for final disposal (Rose 2000). RICOH employs this system of recycling in all five of its operational regions (Japan, The Americas, Europe, China and the South Pacific). Through the use of this system RICOH achieved a reuse and recycle rate of 96.4% for its copiers in the Japanese financial year 2000.

Philips Consumer Electronics participated in a study looking at the effectiveness of television recycling. The investigation found that television recycling can be a profitable EOL option if certain criteria were met (Stevels *et al.* 1999): the value of materials disassembled per minute exceeds the labour and processing cost to recover those materials; and DFE guidelines are applied, using easily recyclable materials and having substantial economies of scale (>300,000 television sets/year).

Implications for the Australian manufacturing industry

The implications of environmentally sustainable production strategies on the Australian manufacturing industry depend heavily on future government policies and consumer behaviour. Because

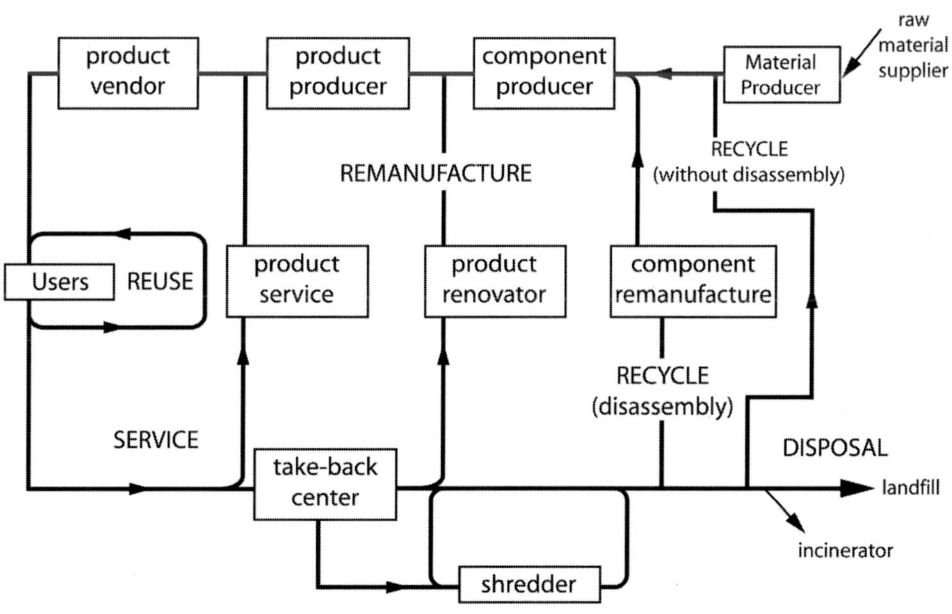

Figure 8: RICOH comet circle (Source: Rose 2000).

the current environmental legislation in regard to industrial production in Australia is still in its very early stages, and on many aspects is expressed more like a recommendation rather than a mandatory law, the current effects on manufacturing companies are relatively minor compared with the situations in other countries around the world. Similarly, consumer behaviour in purchasing and using eco-friendly products is largely at the stage of awareness, rather than active participation. Consumers are still very reluctant to spend more money for an eco-friendly product or service. Consequently, the overall impacts of environmental strategies on the Australian manufacturing industry are still very low, and focussed on higher impact point solutions in only some areas.

However, the development trajectories for Australian industry will have to follow the worldwide trends, which are clearly set to drastically increase the impacts of environmental requirements on the manufacturing industry in the near future. Therefore, the Australian manufacturing industry should prepare now for the implementation of more comprehensive *cradle-to-cradle* strategies in order to avoid falling further behind current and future development trends.

Another point of interest is the two-sided nature of industrial production and usage in Australia. On the one hand, there is a manufacturing industry producing industrial products for the local market. This industry must focus its efforts on the design and manufacture of products with low environmental impacts—a real cradle-to-cradle approach. On the other hand, Australia is a major import-oriented economy as far as industrial products are concerned. This means that Australian consumers use a large number of products from overseas manufacturers. In this case, we cannot change the design of the product or its usage performance, but the end-of-life strategies for these products become a major problem. This problem is even greater than for an export-oriented economy, because any end-of-life strategies have to cope with a wide range of foreign products, the design of which cannot be influenced. Therefore the development of end-of-life techniques and strategies should become a major focus for the Australian industry, which could potentially lead to the formation of a new branch of manufacturing companies.

Conclusion

The cradle-to-cradle manufacturing strategy is one of the approaches for industry that comes closest to enabling a sustainable industrialised world. It may seem like an idealistic vision for some industries, but it has been proven to work successfully in several manufacturing organisations. Many suitable techniques are already available, but driving legislation and economic implementation remain as challenges to this important transition.

References

Ayres RU (1996). Life cycle analysis: a critique. *Resources, Conservation and Recycling* **14**, 199–223.

Baillie C (Ed.) (2004). *Green Composites: Polymer Composites and Environment.* Woodhead Publishing, Cambridge UK.

Bartel A (2003). Analysis and classification of design and manufacturing technologies for cleaner production. Diploma Thesis, Technical University of Hamburg, Germany.

Boks C and Stevels A (2000). Lessons learned from 10 years take-back and recycling. In *Proceedings of the 7th CIRP International Seminar on Life Cycle Engineering*: *IEEE.* pp. 30–37. The University of Tokyo, Tokyo, Japan.

Bothroyd Dewhurst Inc. (1996). *DFE User Guide.* Boothroyd Dewhurst Inc., Wakefield Rhode Island.

Bras B amd McIntosh MW (1999). Product, process, and organizational design for remanufacture – an overview of research. *Robotics and Computer-Integrated Manufacturing* **15**, 167–178.

Chen JL and Chih-Wei L (2001). A simple life cycle assessment method for green product conceptual design. In *Proceedings Second International Symposium on Environmentally Conscious Design and Inverse Manufacturing.* pp. 775–780. Union of Eco Designers, Tokyo, Japan.

Daimon T, Kandoh S and Umeda Y (2004). Proposal of decision In *Proceedings of the 11th International CIRP Life Cycle Engineering Seminar on Product Life Cycle – Quality Management.* University of Belgrade, Belgrade, Serbia.

Dorgelo R (1996). 6R-principle of digital. In *Proceedings of the First International Working Seminar on Reuse.* pp. 85–106. Technical University of Eindhoven, Eindhoven, The Netherlands.

Dowdell DC, Adda S, Noel R, Laurent D, Glazebrook B, Kirkpatrick N, Richardson L, Doyle E, Smith D and Thurley A (2000). An integrated life cycle assessment and cost analysis of the implications of implementing the proposed waste from electrical and electronic equipment (WEEE) directive. In *Proceedings of the 2000 IEEE International Symposium on Electronics and the Environment.* pp. 1–10. Institute of Electric and Electronic Engineers (IEEE), San Francisco, CA.

EPA NSW (2004). *Extended Producer Responsibility Priority Statement 2004.* Department of Environment and Conservation (NSW), Sydney.

EPA NSW (2004a). *Frequently Asked Questions About EPR.* NSW Government, Sydney. < http://www.environment.nsw.gov.au/waste/faqsepr.htm>.

Feldmann K and Melzer K (2004). Implementation of e-waste legislation in EU members states, North America and China. Presented at *Proceedings of the 11th International CIRP Life Cycle Engineering Seminar on Product Life Cycle – Quality Management.* University of Belgrade, Belgrade, Serbia.

Feldmann K, Melzer K and Trautner S (2002). Integrated product policy – challenge in the scope of technology, economy and environment policy. In *Proceedings of the 9th CIRP International Seminar on Life Cycle Engineering.* pp. 1–11, FAPS, Erlangen, Germany.

Fuji Xerox (2007). *Suzuka Fuji Xerox Starts Operations at Plastic Recycling Factory in Yokkaichi.* Fuji Xerox, Japan <http://www.fujixerox.co.jp/eng/headline/2006/0418_prcy.html>.

Gehin A, Zwolinski P and Brissaud D (2007). A tool to implement sustainable end-of-life strategies in the product development phase. *Journal of Cleaner Production*, in press.

Gertsakis J and Lewis H (2003). Sustainability and the waste management hierarchy. In *EcoRecycle Victoria.* pp. 1–16. RMIT, Melbourne, Australia.

Ginter PM and Starling JM (1978). Reverse distribution channels for recycling. *California Management Review* **20**, 72–82.

Gonzalez-Torre PL, Adenso-Diaz B and Artiba H (2004), Environmental and reverse logistics policies in European bottling and packaging firms. *International Journal of Production Economics* **88**, 95–104.

Griese H, Poetter H, Schischke K, Ness O and Reichl H (2004). Reuse and Lifetime Extension Strategies in the Context of Technology Innovations, Global Markets, and Environmental *Legislation.* In *Proceedings of the 2004 IEEE International Symposium on Electronics and the Environment.* pp. 173–178. Institute of Electric and Electronics Engineers, Washington DC.

Gungor A and Gupta SM (1999). Issues in environmentally conscious manufacturing and product recovery: a survey. *Computers Industrial Engineering* **36**, 811–853.

Hanafi J, Kara S and Kaebernick H (2007). Generating fuzzy coloured Petri net forecasting model to predict the return of products. In *Proceedings of the 2007 IEEE International Symposium on Electronics & the Environment.* pp. 245–250. Institute of Electric and Electronics Engineers (IEEE), Orlando, Florida.

Henstock ME (1988). *Design for Recyclability.* Institute of Metals, Maney Publicatiions London. UK.

Hesselbach J and Kuhn M (1996) Disassembly evaluation of electronic and electrical products. Presented at *Proceedings of the 1998 IEEE International Symposium on Electronics and the Environment.* Institute of Electric and Electronics Engineers, Oak Brook, Illinois.

Hoffman III W (1997). Recent advances in design for environment at Motorola. *Journal of Industrial Ecology* **1**, 131–140.

Ibbotson S, Kara S and Kaebernick H (2007). A methodology to analyse the critical design parameters for "reuse". *International Journal of Environmentally Conscious Design and Manufacturing,* in press.

Ijomah W, Bennett JP and Pearce J (1999). Remanufacturing evidence of environmental conscious business practice in UK. In *EcoDesign '99: First International Symposium on Environmentally Conscious Design and Inverse Manufacturing.* Tokyo. pp. 192–196. Tokyo, Japan. IEEE Computer Society Piscataway New Jersey.

Ijomah WL, McMahon CA, Hammond GP and Newman T (2007). Development of design for remanufacturing guidelines to support sustainable manufacturing. *Robotics and Computer Integrated Manufacturing*, in press.

International Organization for Standardization (1996) *Environmental Management - Life Cycle Assessment*. ISO/DIS 14040 series. Sydney.

Ishii K, Eubanks CF and Di Marco P (1994). Design for product retirement and material lifecycle. *Materials & Design* **15(4)**, 225–233.

Jeswiet J (2002). Tools for product and supply chain focused policies and tools for sustainable development: part II, IPP, EPR and eco-efficiency. In *Proceedings of the 9th CIRP International Seminar on Life Cycle Engineering*. pp. 33–47. FAPS, University of Erlangen, Erlangen, Germany.

Kaebernick H, Sun M and Kara S (2003). Simplified lifecycle assessment for the early design stages of industrial products. *Annals of the CIRP* **52**, 25–28.

Kara S, Manmek S and Kaebernick H (2007) An integrated methodology to estimate the external environmental costs of products. *Annals of the CIRP* **56(1)**, 9–12.

Kara S, Mazhar I and Kaebernick H (2004). Lifetime prediction of components for reuse: an overview. *International Journal of Environmental Technology and Management* **4**, 323–348.

Kara S, Rugrungruang F and Kaebernick H (2007). Simulation modelling of reverse logistics networks. *International Journal of Production Economics* **106**, 61–69.

Kato S and Kimura F (2004). The product life cycle design method using a strategic analysis. In *Proceedings of the 11th International CIRP Life Cycle Engineering Seminar on Product Life Cycle – Quality Management*. pp. 45–48. University of Belgrade, Belgrade, Serbia.

Kimura F, Hata T and Suzuki H (1998). Product quality evaluation based on behaviour simulation of used products. *Annals of the CIRP* **47(1)**, 119–123.

Klausner M, Grimm W and Hendrickson C (1998). Reuse of electric motors in consumer products: design and analysis of an electronic data log. *Journal of Industry Ecology* **2(2)**, 89–102.

Klausner M, Grimm W and Horvath (1999). An integrating product takeback and technical service. In *Proceedings of the IEEE International Symposium on Electronics and the Environment*, pp. 48–53. Institute of Electric and Electronics Engineers, Danvers, Massachusetts.

Kuo T-C, Huang SH and Zhang H-C (2001). Design for manufacture and design for 'X': concepts, applications, and perspectives. *Computers and Industrial Engineering* **41**, 241–260.

Linton J (1999). Electronic products at their end-of-life options and obstacles. *Journal of Electronics Manufacturing* **9**, 29–40.

Mangun D and Thurston DL (2000). Product portfolio design for component reuse. In *Proceedings of the 2000 IEEE International Symposium on Electronics and the Environment*. pp. 86–92. Institute of Electric and Electronics Engineers (IEEE), San Francisco.

Masanet E, Auer R, Tsuda D, Barillot T and Baynes A (2002). An assessment and prioritization of "design for recycling" guidelines for plastic components. Presented at *2002 IEEE International Symposium on Electronics and the Environment*. Institute of Electric and Electronics Engineers, San Francisco, CA.

Mazhar MI, Kara S and Kaebernick H (2005). Reusability assessment of components in consumer products – a statistical and condition monitoring data analysis strategy. In *Proceedings of the 4th Australian LCA Conference*. pp. 1–8. ALCAS – Australian Life Cycle Assessment Society, Sydney.

Mercedes Benz (2006) *End-of-Life Vehicles at Mercedes Benz Design for Recycling and the Environment*. Mercedes Benz, U.K. <http://www2.mercedes-benz.co.uk/content/unitedkingdom/mpc/mpc_unitedkingdom_website/en/home_mpc/vans/home/aftersales/customer_support/end_of_life_vehicles/design_recycling.html>

Moore KE, Gungor A and Gupta SM (1998). Disassembly process planning using Petri nets. Presented at *Proceedings of the 1998 IEEE International Symposium on Electronics and the Environment*. Institute of Electric and Electronics Engineers, Oak Brook, Illinois.

Nakamura S and Kondo Y (2001). Waste Input-Output Analysis of Disposal, Recycling, and Extended Life of Electric Home Appliances. In *Proceedings of the 2nd International Symposium on Environmentally Conscious Design and Inverse Manufacturing (EcoDesign'01)*: pp. 814. Union of Eco Designers, Japan.

Nakamura S and Kondo Y (2004). Recycling and extended life of electric home appliances: hybrid analysis based on the waste input-output model. In *Proceedings of the 2004 IEEE International Symposium on Electronics and the Environment*. pp. 104–105. Institute of Electric and Electronics Engineers (IEEE), Washington DC.

O'Neill M (2003) WEEE and RoHS and trends in the environmental legislation in European Union. In *Proceedings of EcoDesign2003: Third International Symposium on Environmentally Conscious Design and Inverse Manufacturing*. pp. 5–8. Union of Eco Designers, Tokyo, Japan.

Ottman J (2004). *Green Marketing: Opportunity for Innovation* (2nd edn). BookSurge Publishing: Chicago, Illinois.

Pornprasitpol P (2006). Selective disassembly for re-use of industrial products. ME thesis, University of New South Wales, Australia.

Przekop LA and Kerr S (2004). Life cycle tools for future product sustainability. In *Proceedings of the 2004 IEEE International Symposium on Electronics and the Environment*. pp. 23–26. Institute of Electric and Electronics Engineers, Washington DC.

Reyes W, Moore M, Bartholomew CRM, Currence R and Siegel R (1995). Reliability assessment of used parts: an enabler for asset recovery. In *Proceedings of the IEEE International Symposium on Electronics and the Environment*. pp. 89–94. Institute of Electric and Electronics Engineers, Orlando, Florida.

Rose C (2000). Applying Environmental Value Chain Analysis to Product Take-Back Systems. In *Proceedings of the 7th CIRP International Seminar on Life Cycle Engineering*. pp. 38–45. Institute of Electric and Electronics Engineers, San Francisco, Florida.

Rose C (2000). Design for environment: a method for formulating product end-of-life Strategies. PhD thesis, Stanford University, USA.

Rose CM, Ishii K and Stevels A (2002). Influencing design to improve product end-of-life stage. *Research in Engineering Design* **13**, 83–93.

Rose CM, Stevels A and Ishii K (2000). A new approach to end-of-life design advisor (ELDA). In *Proceedings of the 2000 IEEE International Symposium on Electronics and the Environment*. pp. 99–104. Institute of Electric and Electronics Engineers, San Francisco, CA.

Rydh CJ and Karlström M (2000). LCA for Recycling Portable Nickel-Cadmium Batteries. In *Proceedings of the 7th CIRP International Seminar on Life Cycle Engineering*. University of Tokyo, Japan.

Santochi M, Dini G and Failli F (2002). Computer aided disassembly planning: state-of-the-art and perspectives. *Annals of the CIRP* **51**, 1–23.

Santos E and Ferrao P (2005) DfR and DfD applied to electrical and electronic equipments resulting environmental life cycle performance - a case study for Portugal. In *Proceedings of the Fourth International Symposium on Environmentally Conscious Design and Inverse Manufacturing*. Union of Eco Designers, Tokyo, Japan.

Scheidt L and Shuqiang Z (1994). An approach to achieve reusability of electronic modules. In *Proceedings of the IEEE International Symposium on Electronics and the Environment*. pp. 331–336. Institute of Electric and Electronics Engineers, San Francisco.

Shangguan D (2004). Environmental leadership in electronics manufacturing: lead-free and beyond. In *Proceedings of the 2004 IEEE International Symposium on Electronics and the Environment*. pp. 33–39. Institute of Electric and Electronics Engineers. Washington DC.

Siddique Z and Rosen DW (1997). A virtual prototyping approach to product disassembly reasoning. *Computer-Aided Design* **29**, 847–860.

Sodhi MS and Reimer B (2001). Models for recycling electronics end-of-life products. *OR Spectrum* **23**, 97–115.

Steinhilper R (2001). Recent trends and benefits of remanufacturing: from closed loop businesses to synergetic networks. In *Proceedings Second International Symposium on Environmentally Conscious Design and Inverse Manufacturing*. pp. 481–488. Union of Eco Designers, Tokyo, Japan.

Stevels A (2000). Green marketing of consumer electronics. In *Proceedings of the International Congress on Electronics Goes Green*. pp. 1–5. Fraunhofer IZM, Berlin, Germany.

Stevels ALN, Ram AAP and Deckers E (1999). Take-back of discarded consumer electronic products from the perspective of the producer – conditions for success. *Journal of Cleaner Production* **7**, 383–389.

Sun M (2004). A simplified assessment approach for environmentally sound product systems design. PhD thesis, University of New South Wales, Sydney, Australia.

Sundin E, Bjorkman M and Jacobsson N (2000). Analysis of service selling and design for remanufacturing. In *2000 IEEE International Symposium on Electronics and the Environment*. pp. 272–277. Institute of Electric and Electronics Engineers, San Francisco

Suzuki H, Matsuda M, Kimura F (2000). Prediction of used parts deterioration based on a tolerance model. In *Proceedings of the 33rd CIRP International Seminar on Manufacturing Systems*. pp. 149–154. Stockholm, Sweden.

The Catalan Centre for Recycling (2006). Design for recycling. pp. 1–6. Waste Agency of the Department of the Environment, Government of Catalonia, Spain.

The United Nations Environmental Program (UNEP) (1997). *Ecodesign: A Promising Approach to Sustainable Production and Consumption*. United Nations Environmental Program (UNEP), Paris, France.

Veshagh A (2006). Survey of eco design and manufacturing in automotive SMEs. In *13th CIRP International Conference on Life Cycle Engineering*. pp. 307–310. KU Lueven, Leuven, Belgium.

Wenzel H, Hauschild M and Alting L (1997) *Environmental Assessment of Products Volume 1 Methodology, Tools, and Case Studies in Product Development*. Chapman and Hall, UK.

Westkämper E (2000). Life cycle management: approaches and visions towards sustainable manufacturing. In *Proceedings of the 7th CIRP International Seminar on Life Cycle Engineering*. pp. 1–16. University of Tokyo, Japan.

Wheels Automotive Design Awards (2006). *Ford Territory Turbo*. Standards Australia, Sydney. <http://www.wheels.australiandesign.org.au/entries.asp?entryID=30>.

Wiener RK (2001). Needed: Automotive design for recycling. *Iron Age New Steel* **17**, 32.

Yamagiwa Y, Numata J and Namiki Y (2001). Knowledge management in ecological design environment friendly product design strategies and design enabling both easy assembly and disassembly. In *Proceedings of International Conference on Management of Engineering and Technology (PICMET '01)*. Volume 1, p. 497. Portland International Centre for Management of Engineering and Technology, Portland, USA.

Zabaniotou A and Kassidi E (2003). LCA applied to egg packaging made from polystyrene and recycled paper. *Journal of Cleaner Production* **11**, 549–559.

Zussman E, Krivet A and Seliger G (1994). Disassembly oriented assessment methodology to support design for recycling. *Annals of the CIRP* **34**, 9–14.

35

Eco-industrial Development

David Batten, Dick van Beers, Glen Corder and Steve Cook

Traditionally, business, the community and the environment have been viewed as separate systems, which operate independently of—or even in opposition to—one another. For example, industrial systems tend to emphasise the independence and competitiveness of firms, whereas ecological systems emphasise interaction and interdependence. However, awareness of the interdependencies between these systems is increasing—highlighting the need for a business framework that protects the natural environment while improving business performance. A symbiotic approach, which is based on industrial ecology, is evolving to unite the needs of industrial and natural systems. By working together, a community of businesses can enjoy a greater, collective economic and environmental benefit than the sum of the benefits each business would realise if it optimised its own performance alone.

An ambitious future goal for a sustainable Australian society—*zero-waste* for our urban industries and regions—is consistent with the ambitions of industrial ecology. To achieve this goal, new technologies, new firms and new communities need to emerge that convert various low-value by-products into products of higher value. Some challenges of re-use are technological: the task being to find new materials and new products that re-use solid by-products in clever, practical ways. The challenges grow more complex if the by-products are fluids or gases. For successful re-use, waste streams rely on exchange processes, mutual reciprocity and broader networks of cooperation between private firms, public organisations and local communities.

A systems view suggests that recycling and re-use efficiencies may be improved markedly by strengthening local, inter-firm networks and adopting principles of industrial ecology. In this evolutionary perspective, the potential synergies of local eco-industrial complexes contribute towards the creation of a cleaner, less wasteful and more sustainable society. The industrial fringes and suburbs of Australia's mega-metropolitan regions are potentially fruitful areas for eco-industrial experiments of this kind (Newton 2002). Such experiments yield a better understanding of the economics of re-use in systems terms. We recognise that certain industrial by-products—such as sand, ash and waste heat (steam)—can be used more efficiently as intermediate inputs to other economic processes nearby.

The aim of this chapter is to build a case for the wider application of principles of industrial ecology to urban, suburban and community planning in Australia – including the development of incubators to promote eco-industrial principles to Australian industry and government. The chapter begins with a short primer on industrial ecology, followed by a discussion of approaches that can help to capture regional resource synergies. Local examples of regional symbiosis are based on case studies of current eco-industrial developments in three Australian states. Two are underway in established areas of resource-intensive activity: Kwinana in Western Australia and Gladstone in Queensland. The third is a new eco-industrial park being planned for Dandenong—a major manufacturing centre near Melbourne.

Theory and principles of industrial ecology

A short primer on industrial ecology

Motivated primarily by the ecosystem metaphor, the field of industrial ecology has been building

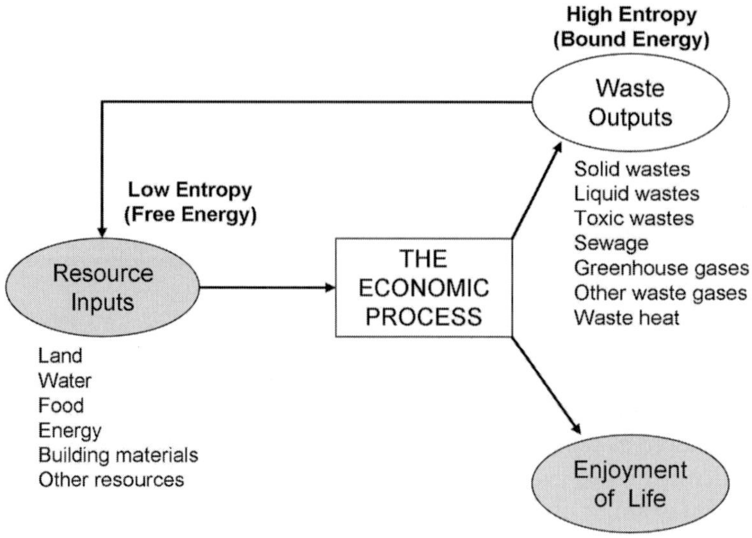

Figure 1: Closing the loop (Source: Batten 2005).

momentum since the late 1980s. Industrial ecology adopts a local and holistic view of a firm's resource streams. The broad goal is to maximise the re-use of by-products that are mostly dumped as wastes, thereby having a positive effect on environmental, human and social capital (Frosch and Gallopoulos 1989). Firms that exchange resource streams profitably because of their close proximity and output-to-input affinity are called *symbionts*. The term *symbiosis* has been borrowed from ecology, where it means 'living together'. In the Danish township of Kalundborg, for example, four symbiotic companies—a power station, an oil refinery, a pharmaceutical plant and a Gyproc plant—managed to work closely together over a period of 20 years to produce a mutually profitable role model for industrial symbiosis.

Eco-industrial partnering adds value to businesses and communities by optimising the use of energy, water, materials, human and other community resources. For business, value is added as by-products, water and energy are cycled back into the production stream to be used as raw materials for another product or firm. This 'closing of the loop' (see Figure 1) results in the conservation of natural resources as well as lower disposal and production costs. Eco-industrial development offers the potential to achieve greater efficiency by way of 'economies of systems integration' (Ayres 1995). Business partnerships meet common service, transportation and infrastructure needs. If organised well, partnerships become a built-in incentive to: (1) maximise the re-use of materials, energy, water and labour, and (2) grow the industrial symbiosis further.

Note how closely Figure 1 resembles the extended metabolism model of the city (see, for example, Wolman 1965; Newman 1999). Urban metabolism is a biological systems approach for examining resource inputs and outputs of urban settlements and their economies. The physical and biological processes associated with converting resources into useful products, by-products and wastes are like the human body's metabolic processes or those of a complex ecosystem. The metabolism model is one of the more pragmatic ways of addressing sustainability.

Eco-industrial development projects are learning systems that seek to gain multiple benefits from each economic activity (Batten 2005). For example, an integrated resource recovery system increases resource efficiency, lowers pollution, creates opportunities for entrepreneurs and workers and strengthens the local and national economy. Eco-industrial development also suggests that, in order to maximise efficiency of resource use, we should assess and optimise community assets that are often overlooked. Some of these are particular human resources, natural habitats, cultural and aesthetic resources and existing institutions.

Resources may include elements of a community that are typically considered to be detriments—such as brownfields or abandoned buildings—which can be revitalised, renewed or mined for materials. In the context of community development, lower costs for businesses translate into more opportunities to re-invest in new jobs, training and good environmental management practices.

Given their past tendency to view industrial by-products as wastes—thus delivering solids to landfill, liquids into our rivers and oceans, and gases into the atmosphere—Australian businesses are strong candidates for industrial symbiosis. Significant progress has been made in the case of kerbside recycling of household wastes, such as paper, cardboard, plastics and glass. The setting of targets for waste to landfill arose in the early 1990s, and states competed to see who could set the highest target. The original 50% target was ambitious for all cities except Canberra, but has been achieved through the success of kerbside and construction waste recycling. But the potential for profitable exchange of a much wider variety of by-products produced by Australian businesses has not yet been realised. Harvesting these valuable by-products could divert 80% or more 'waste' from landfill. This approach seems to be working in Canberra, New Zealand and Canada.

In Canada and the USA, several current eco-industrial projects aim to revitalise and improve economically distressed communities, with a particular focus on redevelopment of brownfields and decommissioned government properties (such as military bases). The potential benefits for communities are improved environmental health, enhanced resource efficiency, more jobs and sustainable businesses. Eco-efficiency corresponds to a 'win–win' situation for a community, its businesses and the environment. It is a pragmatic route to sustainable development.

Wider adoption of an eco-industrial approach in Australia has many applications. At a very basic level, it can be used to develop a strategy where each partner organisation seeks higher intra-firm performance. However, the most exciting, and potentially most profitable, eco-industrial activity involves moving to a new level of performance by increasing the interconnections and interdependencies *between* companies. Much of the success of an eco-industrial approach depends on how closely businesses are willing to work together for their mutual advantage. Promoting this broader interaction between companies can be a value-adding service provided by local government, the community or a developer to attract new companies and improve the performance of existing firms.

Eco-industrial development seeks to achieve:

- co-location of industries that benefit economically and environmentally from the exchange of their by-products and waste
- resource efficiency in materials, energy, water and transportation, with the cost savings gained through co-location and sharing of facilities
- cleaner production through good housekeeping, reduction/substitution of toxic materials, strict control of emissions, separation of by-products or residuals.
- use of renewable energy and recyclable materials to replace fossil fuel sources and finite material supplies.
- rehabilitation of existing buildings to higher energy and environmental standards and use of green architecture and engineering in new facility and infrastructure design
- enhancement of quality of life and economic development in neighbouring communities through cooperative projects between industry, government and community-based organisations
- ecological site planning and utilisation based upon clear understanding of the carrying capacity of air, water, and land systems and the nature of remaining native ecological systems.
- establishment of environmental management systems with objectives and indicators informed by eco-industrial development, not just compliance with regulations.

In Australian cities, many opportunities exist to re-use waste streams as resource inputs to other economic activities. This can reduce waste storage, extend the useful life of such inputs and create the potential for new industries and new communities. In some cases, however, the additional amount of free energy used is greater than the decrease in the entropy of what is re-used. A sound business case will require economically viable re-use and recycling opportunities for some of Australia's major waste streams. For example, there is scope for developing some areas adjacent to our power stations or wastewater treatment plants—that service some of the mega-metropolitan regions along our coastline—into anchor tenants for eco-industrial complexes (see Batten 2003, 2005).

Incubators can promote industrial ecology

An eco-industrial planning body (EIP) can enhance the likely success rate of new or existing businesses by setting up an *incubator.* This catalytic resource plays a vital role in linking suppliers to anchor tenants, developing a theme cluster (such as resource recovery or renewable energy), or helping to fill niches in the by-product exchange network. Many EIP projects currently underway in Europe, North America and South-East Asia are exploring the role incubators could play. In some countries, it means creating programs to support business cluster development that could potentially lead incubator development.

An incubator can take several different forms: a partnership between government and industry; a venture capital company; or an anchor tenant. The literature on recycling networks, industrial metabolism and industrial ecology stresses the need to identify one or more key actors or organisations, around which a sustainable recycling network of industrial actors can emerge. The key actor can assume various guises—such as a catalyst, a coordinator, a symbiotic factor or a support system. A typical anchor tenant facilitates the use of by-products as resource inputs, of residual waste energy as valuable outputs, and helps small- and medium-sized businesses to explore potential linkages with other, potentially symbiotic, businesses in the partnership.

One possible anchor tenant that could serve as an incubator is a regional power plant. An example of this kind of incubator is the energy supply system powering the Finnish city of Jyväskylä (see Korhonen 2001). This co-generation plant captures waste energy from electricity production to produce heat for productive purposes, instead of dumping it into the air or into the marine ecosystem. It uses a heterogeneous mix of fuel inputs, such as biomass or waste fuels. In the case of the Jyväskylä industrial ecosystem, cascading of energy in the power plant and the re-use of local by-products as fuels in the plant lead to savings of approximately 40% in the consumption of coal and oil compared with the consumption of these fuels without these features. SO_2 emissions are more than 50% lower, and CO_2 emissions more than 30% lower, than they would be without these features.

In an ecosystem, the flow of material and energy from producers (plants) to recyclers (micro-organisms, bacteria, fungi) is very large, whereas the corresponding flow in a typical industrial system is practically non-existent. In the latter, the flow from consumers to recyclers is also small. Therefore, an important goal must be to increase the flow from production and consumption to the recyclers and from recyclers back to the producers, thereby closing the loop more efficiently (see Figure 1). In order to minimise the use of energy for transportation, and maximise use of local labour, these interlinked activities—production, consumption and recycling—should be located close to one another (as in the Kwinana and Gladstone industrial areas).

Another important characteristic of organisms, and of an ecosystem as a whole, is the ability to adapt to conditions in their surrounding environment. In this respect, the sustainability of any system can be understood as successful, local adaptation. For example, a new eco-industrial park being planned in accordance with principles of industrial ecology via local adaptation is Dandenong LOGIS in Victoria.

Experiences in Europe and other parts of the world suggest that there is generally a unique, evolutionary path behind any successful eco-industrial development—although typically nested within a generic set of industrial ecology enablers. Path-dependence is a key feature of most dynamic systems, but other factors are also important. For example, success stories emphasise the importance of agglomeration economies (that is, economies of scope and co-location) and industrial symbiosis. In the next section, we shall review some tools and techniques that have helped identify key incubators and other symbiotic industrial partners.

Capturing regional resource synergies

Types of regional synergies

Although only one of several theories and concepts within industrial ecology, industrial symbiosis is perhaps its best-known and most widely applied principle. Sometimes called *inter-enterprise cooperation*, it is mainly concerned with the cooperative management of resource flows through networks of businesses—as a means of approaching ecologically sustainable industrial activity (Chertow *et al.* 2004). This symbiosis is achieved in practice by identifying and implementing *regional resource synergies*. By recognising and strengthening various links among its firms, an industrial area may be transformed into an *industrial ecosystem* through the process of *industrial symbiosis*. Chertow provides the following definition:

Industrial symbiosis engages traditionally separate industries in a collective approach to competitive advantage involving physical exchange of materials, energy, water, and/or by-products. The keys to regional synergies are collaboration and the synergistic possibilities offered by geographic proximity. (Chertow 2000)

Regional synergies have been categorised as follows (van Berkel 2003):

- **Supply chain synergies:** involving a local manufacturer and a dedicated supplier of principal reagents for core process industries (such as production of ammonia and chlorine for industrial use).
- **By-product synergies:** involving the use of previously disposed by-products (a solid, liquid or gas) from one facility by another facility to produce a valuable product. By-product synergies can be elaborate with regard to industrial gases and inorganic process residues (such as recovery and on-selling of carbon dioxide and hydrogen).
- **Utility synergies:** involving the shared use of utility infrastructure, mainly water and energy (e.g. water recovery and co-generation).

Wherever a business realises a benefit by co-locating near its main customers, it accrues agglomeration economies (Desrochers 2004). Although important for reducing a firm's costs, such supply-chain synergies do not meet the criterion of '*by-product exchange between traditionally separate industries*', which is the distinctive feature of industrial symbiosis (Chertow 2000).

Developing by-product synergies in areas with a concentration of large industries in close geographic proximity can provide a range of sustainability benefits. An example is improved utilisation of by-products that would otherwise be discharged to landfill or into the atmosphere or marine environments. Such an approach to resource utilisation fits with the overall aim of sustainable resource processing: namely to progressively and systematically eliminate wastes and emissions, while at the same time enhancing business performance and meeting community expectations (Herbertson and Sutton 2002). Both the producer and consumer of these by-products can benefit from such a relationship.

The much-quoted iconic example of regional synergy development or industrial symbiosis is the Danish town of Kalundborg. This town is viewed as demonstrating best-practice because it shows how sustainability can be integrated into environmental and economic systems (see Ehrenfeld and Gertler 1997; Roberts and Wadley 2002). There are four main industries – Asnaes Power Station (a 1500MW coal-fired plant), a large oil refinery operated by Statoil, Novo Nordisk (a leading maker of pharmaceuticals and enzymes) and Gyproc (a plasterboard manufacturer)—as well as several other farms and firms within the municipality. They trade, use each other's waste streams and energy resources, and turn by-products into raw materials (see Figure 2).

Symbiosis in an industrial context signifies the development of a cooperative network of exchange processes among a group of industries that benefit mutually through trade in by-products. Industrial symbiosis at Kalundborg evolved gradually over 30 years or more. There was no master plan or grand design. The manager of Asnaes Power Station believes that the existing economic incentives were generally sufficient to generate most of Kalundborg's symbiosis. A Novo Nordisk vice president claimed that:

At the time we were just doing what was profitable and what made sense...We do what pairs of us think is a good idea.

It seems to have paid off, because the total investment by the network partners amounted to US$60 million in 1994, generating partnership returns of approximately US$120 million.

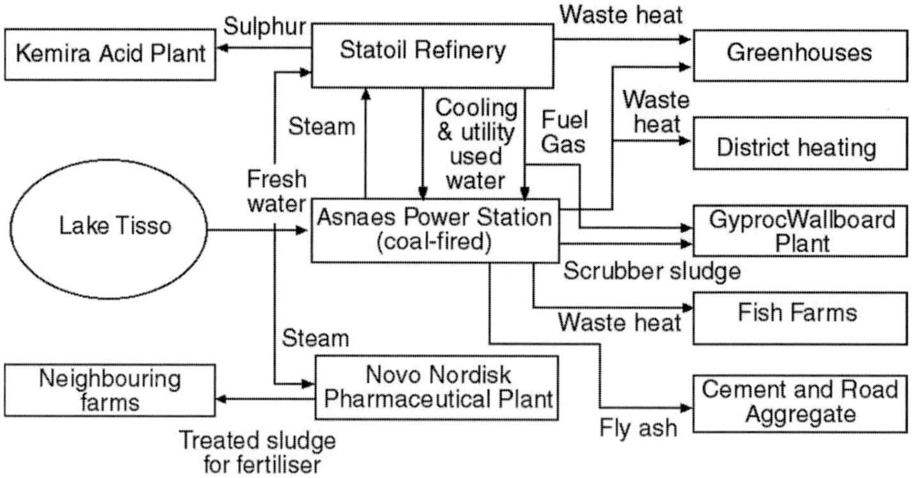

Figure 2: Industrial symbiosis at Kalundborg (Source: Ehrenfeld and Gertler 1997).

The power station was the first industrial tenant in the town, and can be viewed as the incubator of the whole exchange system. By exporting part of its formerly wasted energy, Asnaes has reduced the amount of available energy discarded by 80%. It delivers process steam to its neighbours: Statoil refinery receives 40% of its steam requirements and Novo Nordisk all of its steam requirements. The power station' scrubber provides a gypsum-containing feedstock to Gyproc, thereby satisfying two-thirds of their gypsum needs. Fly ash and clinker (from coal burning at Asnaes) are sold for road building and cement production. Since 1972, Statoil has piped gas to Gyproc to fire wallboard-drying ovens, all but eliminating the common practice of flaring waste gases. Some of Statoil's gas is clean enough to be burned by Asnaes as well. Statoil has also been piping cooling water to Asnaes (for use as boiler feed-water) and making wastewater available to Asnaes for cleaning purposes. The town is a unique example of an industrial ecosystem that was a 'grass roots' development, which is driven by market forces rather than top-down planning or premeditated design. There is a message here for would-be imitators. The evolutionary pattern at Kalundborg may not be easily replicated elsewhere. Cultural differences are a key factor when it comes to the successful application of industrial ecology principles.

Regional by-product exchanges have developed in other industrial areas, such as Rotterdam (Baas 1998), Styria (Schwarz and Steininger, 1997), Landskrona (Starlander 2003), Jyväskylä (Korhonen, 2001), Montreal (Dias and Yates 2001), the Forth Valley (Harris and Pritchard 2004) and Tampico (Young 1999), among others. Kwinana and Gladstone are two major, minerals-processing-intensive regions in Australia with many successfully operating by-product synergies. In addition, both areas possess other potential synergy opportunities which, if implemented, could benefit industry, the environment and the community.

By way of contrast, Dandenong LOGIS is a new, large-scale commercial project on a former Dandenong treatment plant site, which is 2 kilometres south of Dandenong's town centre and 30 kilometres south-east of Melbourne. It is expected to become a leading example of sustainable, eco-industrial development in Victoria. Melbourne Water is responsible for site rehabilitation and VicUrban will develop the eco-industrial estate that will employ up to 3000 workers.

Eco-industrial tools and approaches

Some of the earlier work in industrial ecology identified four distinct approaches (van Berkel et al, 1997). One is *materials-specific*. It analyses the way a material or substance flows through an economy in order to identify, evaluate and implement improvement opportunities—for example, through dematerialisation and resource cascading. A second approach is *product-specific*. It analyses the ways in which different component material flows of a selected product may be modified to improve product functionality and environmental interactions. A third approach is *regionally focused*. It aims at optimising the exchange of materials and energy between industries at the local level, through the recovery and re-use of valuable components from waste streams. The fourth approach is *actor-specific*. It investigates opportunities and constraints facing different firms in order to change material and product flows in an environmentally compatible direction. When applied to producers, this latter approach has much in common with cleaner production, pollution prevention and waste minimisation.

In the case studies already discussed, the primary emphasis is on the third approach: regionally focused exchanges of materials and energy, by way of local recovery and re-use of valuable by-products from waste streams that would otherwise pollute our atmosphere, soils, rivers or oceans. A number of tools have been developed over the past decade to assist with the identification and quantification of regional synergies. Some of these are listed below (examples a, b, and c are taken from Bossilkov *et al.* 2005):

(a) Industrial Materials Exchange (IME) tool

The IME tool developed by Bechtel Corporation appears to be the most widely cited tool for industrial ecology design. Intended for use in the identification and analysis of by-product synergies, as well as for planning new eco-industrial projects, it consists of a technology database of over 300 primary industrial processes as well as recycling and reprocessing operations, and a set of scenario development and optimisation tools for evaluating alternative configurations. The tool can be used to design new eco-industrial parks, or to identify new synergetic interactions between existing industrial facilities. It can also be used to search a previously assembled database for a given area to identify possible material exchange links within a region. It has been used in Brownsville, Texas and in a demonstration project in Tampico, Mexico (Young 1999). In the latter case, the number of synergies was more than doubled using brainstorming workshops, compared with those generated by the IME tool.

Further collaboration between Bechtel and the US Department of Energy has resulted in a 'next generation' version of IME, known as the Dynamic Industrial Materials Exchange (DIME). DIME incorporates dynamic simulation of material flows to aid the design and analysis of by-product synergies that are affected by fluctuation in material availability or process requirements.

(b) The Matchmaker! system for creating virtual eco-industrial parks

Another spin-off of Bechtel's initial work was the MatchMaker! system for determining input–output matches among companies and organisations (Brown *et al.* 1997). This computer-based tool is designed as a material exchange system to find outlets for re-using materials based on generic descriptions of candidate firms. These were dubbed 'virtual' eco-industrial parks because, by concentrating on exchanges of materials (rather than water or energy), firms within a much larger geographical area (within 50–100 km) could participate in the industrial symbiosis.

(c) The US EPA's Toolkit – FaST, DIET and REaLiTy

The US Environmental Protection Agency has commissioned a more elaborate set of computer models for input–output matches, including the Facility Synergy Tool (FaST), the Designing Industrial Ecosystems Tool (DIET) and the Regulatory, Economic, and Logistics Tool (REaLiTy). These tools allow a community to investigate whether the addition of specific types of industries might enhance industrial symbiosis (Industrial Economics 1998). A similar matching tool to FaST is currently under development for identifying potential synergies between candidate firms for the Dandenong LOGIS site.

(d) The CSRP regional synergies toolkit

Bossilkov *et al.* (2005) highlighted a need for structural development of synergy opportunities, with particular relevance to mineral-processing regions. As a result, the Centre for Sustainable Resource Processing (CSRP) has been developing a practical toolkit using a rigorous, systemic approach to regional synergy identification. The toolkit focuses on two key components:

- Targeted and customised identification, prioritisation and evaluation of potential synergy opportunities

- Assessment of technology needs and opportunities for regional synergies involving water, energy and materials recovery and processing.

Results from the trial of the prototype toolkit on CSRP case-study projects in Kwinana and Gladstone (discussed in the next Section) demonstrated its ability to streamline and enhance the identification and evaluation of synergy opportunities (Bossilkov 2006). The toolkit is described in more detail in van Beers *et al.* (2007), and efforts are well underway to apply it elsewhere in Australia and internationally.

Australian case studies

The Kwinana Industrial Area (KIA)

Heavy process industry in Western Australia is concentrated in a few industrial areas, of which the Kwinana Industrial Area (KIA) is the largest and most diverse. Established in the 1950s, the area is located 40 km south of Perth on the shores of the Cockburn Sound—a sensitive marine environment. It has a deep-water port, and is therefore strategically placed for export markets in Asia. About 4800 people work in the area's core industries, and many more in related sectors and service jobs. The total economic output of the area exceeds A$15 billion annually (SKM 2007). Thus the KIA plays a very important role in the economy of Western Australia, and even more so in the local community.

The KIA features a diverse range of industries—ranging from fabrication and construction through to high technology chemical and biotechnology plants and large resource processing industries. These are complemented by a variety of utility operations, including two power stations, two co-generation plants, two air-separation plants, port facilities, and water and wastewater treatment plants. Historically, there is considerable integration between these industries in the area. Figure 3 provides an overview of the location of companies in Kwinana.

With no formal industry association for the KIA, some core firms established Kwinana Industries Council (www.kic.org.au) in 1991. The KIC's original purpose was to organise air and water monitoring collectively for firms in the area. It now considers a broader range of issues common to Kwinana's major industries, and fosters positive interactions between member companies, government and the broader community.

Based on the desire of local industries and the state government to understand and document the full economic and social contributions of the area, the KIC initiated regional economic impact studies, including analysis of the principal material and energy flows within the area and the level of industrial integration. The most recent study (SKM 2007) illustrates the growing complexity of inter-industrial relationships over a 15-year period. Upon completion of the second economic impact study (SKM 2002), the KIC initiated the Kwinana Industries Synergies Project (Taylor 2002). This later merged with the activities of the Centre for Sustainable Resource Processing.

Existing synergies

The recently updated inventory of existing regional synergies (van Beers *et al.* 2005) showed that these synergies are diverse. There are 47 synergy projects in place, including 32 by-product synergies (Figure 4) and 15 utility synergies (Figure 5). Each synergy can involve multiple material, water and/or energy flows, such as when a by-product is being returned (perhaps in a processed form) to the producer, or forwarded to another company for further processing.

Some illustrative by-product and utility synergies in Kwinana are:

- **By-product synergy.** The CSBP chemical works supplies gypsum for residue area amelioration at Alcoa's alumina refinery. CSBP produced gypsum (calcium sulphate dehydrate) as a by-product of the manufacture of phosphoric acid. Even though this practice ended in the 1980s, there remains a stockpile of some 1.3 million tonnes of gypsum. After extensively reviewing options for re-using this material, CSBP found that Alcoa could use it to assist in plant growth and soil stability in its residue areas. Alcoa takes approximately 10,000 tonnes each year, on an ongoing basis.
- **By-product synergy.** Since 1990, Air Liquide has purified and compressed process CO_2 received from CSBP's ammonia plant, and other industrial facilities in Kwinana, to a standard that can be used as food grade CO_2

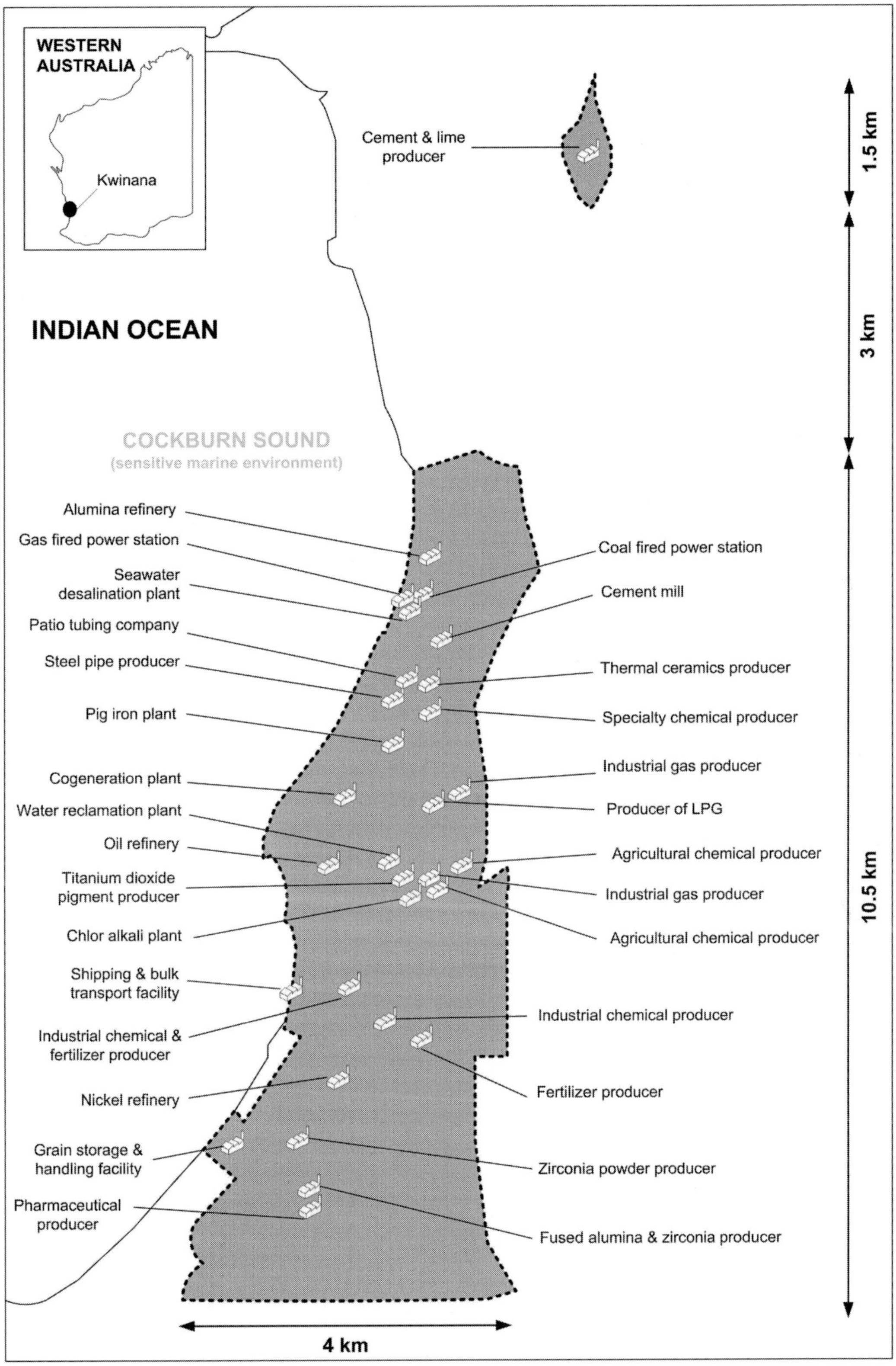

Figure 3: Location of companies in the Kwinana Industrial Area (Source: van Beers 2007).

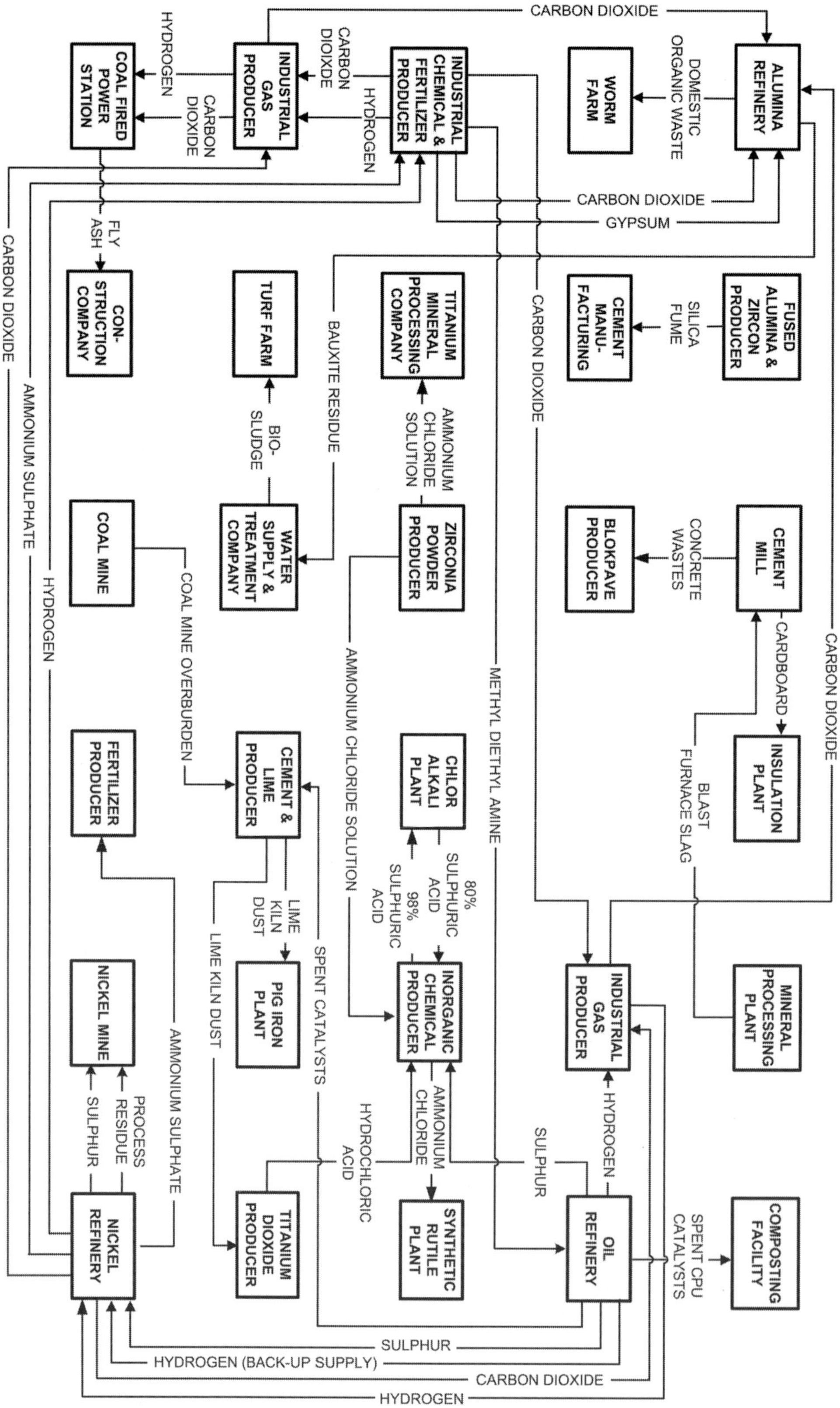

Figure 4: Existing by-product synergies in the Kwinana Industrial Area (Source: van Beers *et al.* 2007).

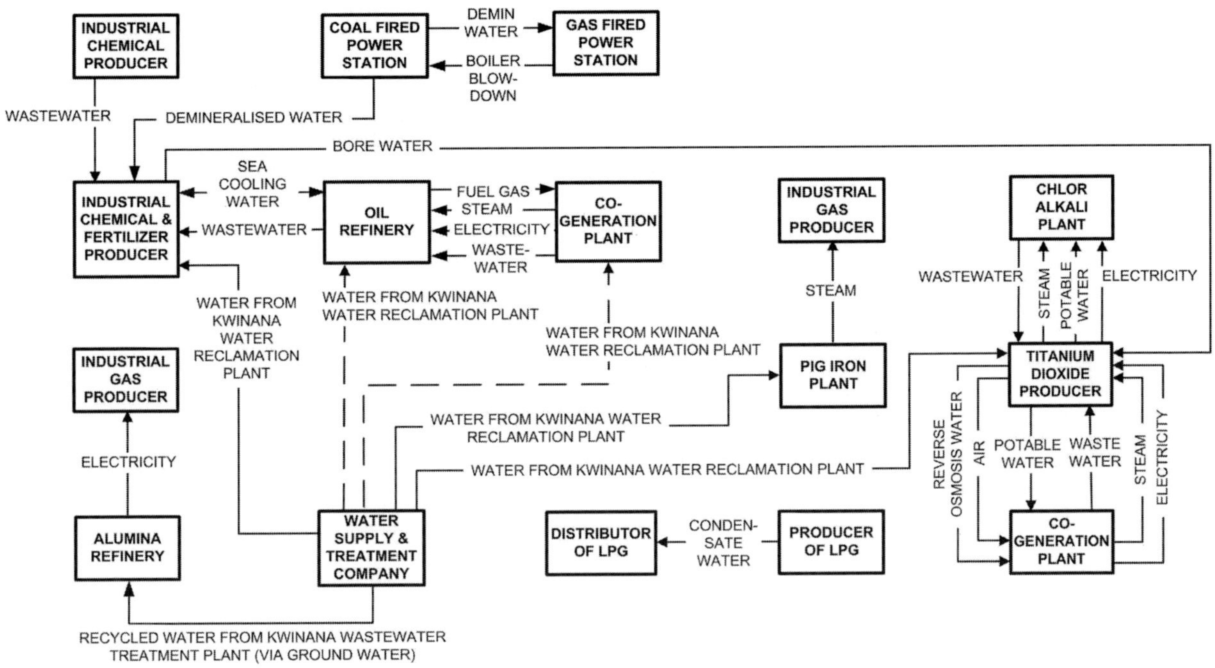

Figure 5: Existing utility synergies in the Kwinana Industrial Area (Source: van Beers *et al.* 2007).

for soft–drinks and beer. CO_2 is also used in dry ice and water treatment. This initiative reduces the emissions of carbon dioxide to the atmosphere, while also avoiding energy use that would otherwise be required to produce CO_2 from air.

- **Utility synergy**. The Kwinana Cogeneration Plant (116 MW capacity) produces all the process steam for the refinery, and generates electricity for BP and the grid. The cogeneration plant is fired with natural gas, supplemented with excess refinery gas, replacing BP's steam boilers (which were in need of replacement at the time). The total benefit has been estimated as a reduction of about 170,000 tonnes of CO_2 emissions per year. It is estimated that this synergy saved the refinery about A$15 million in capital expenditure, while ensuring a reliable, cost-competitive source of steam and electricity for their refinery. Moreover, the refinery has achieved greater process efficiencies as a result of this synergy. The co-generation plant discharges its wastewater to BP's wastewater treatment facility.
- **Utility synergy.** The Kwinana Water Reclamation Plant (KWRP) takes secondary treated effluent from the nearby Woodman Point wastewater treatment facility, to produce a low TDS (Total Dissolved Solids) water supply. This water will be used by CSBP, Tiwest, Kwinana Co-generation Plant, BP and HIsmelt to replace potable water (6 GL/year = about 2–3% of the total potable water use in the drought-affected Perth metropolitan area). The low TDS will enable the process plants to cut their use of chemicals in cooling towers and other process applications, thereby reducing metal loads in their effluents. In exchange for taking water from the KWRP, the industries will be able to discharge their treated effluents into the deep ocean outfall through the Water Corporation pipeline, thereby eliminating the current discharges of treated process water into the sensitive Cockburn Sound (Water Corporation 2003).

The diversity and maturity of existing synergies is quite remarkable, both in absolute terms, and compared with leading international examples such as Kalundborg (Denmark), the Forth Valley (UK) and Rotterdam (Netherlands) (Bossilkov *et al.* 2005). A number of factors unique to Kwinana have doubtless contributed to regional synergy development. They include:

- the diverse mix of major processing and manufacturing industries that primarily produce for international markets (having limited local competition)
- the relative isolation from other major industrial centres in Australia
- the vicinity to Perth as a major metropolitan centre
- the gradual urban encroachment around the industrial area
- the growing recognition of the recreational significance of Cockburn Sound
- a member-based organisation (KIC) that provides a platform for industry interaction and stakeholder engagement.

As well as the Kwinana Industries Council, key industries in the area have made it a strategic priority to further develop resource synergies in Kwinana.

Sustainability benefits of synergies

A synergy is unlikely to be implemented unless all involved parties perceive some benefits (direct or indirect). For all synergy examples discussed in this chapter, there are tangible economic benefits and less-tangible, environmental or community benefits (such as reputation).

Table 1 presents a summary of the commercial, environmental and community benefits for some existing synergies in Kwinana. It reveals that types of benefits can vary greatly—often going well beyond the conventional business case benefits. Security of water and energy supply, increased energy efficiency, lower operational costs for energy use and reduced storage costs for the inorganic by-products are key benefits from these synergies. In addition, all of these synergies had environmental and community benefits. These case studies exemplify that the benefits from regional synergies are not just commercial, but also strategic, leading to reduced exposure to risk and improved reputation. The critical ingredient in initiating a regional synergy is for all the involved parties to appreciate the range of benefits, both direct and indirect, that will result from its implementation (Corder *et al.* 2006).

Progressing new synergies

Despite the growing number of regional synergies, it appears that only direct re-use opportunities

Table 1: Illustrative benefits of existing synergies in Kwinana (adapted from Corder *et al.* 2005)

Synergy	Commercial benefits	Environmental and community benefits
CSBP gypsum re-use at Alcoa residue area	Reduced costs to manage gypsum stockpile (long term) Lower cost gypsum source for alumina refinery	Reduction of stockpiled gypsum onsite at chemical plant Increased soil stability and plant growth at Alcoa residue area
Air Liquide using by-product CO_2 from Kwinana industries	Costs savings for industrial gas company to produce food grade CO_2 from otherwise emitted CO_2 emissions	Proportion of CO_2 emissions from industries are no longer emitted to the atmosphere Avoidance of energy use that would otherwise be required to produce CO_2 from air
Two Cogeneration facilities (BP, Tiwest)	Increased energy efficiency Reliable source of electricity and superheated steam Sales of electricity and steam from cogeneration plant	Reduced greenhouse gas emissions Increased energy efficiency Employment
Kwinana Water Reclamation Plant (KWRP)*	Water security for industry users Availability of high-grade water for industry	Potable water conservation Redirection of discharges of treated industrial effluent from coastal zone into deep ocean Water quality improvement in sensitive coastal zone

* A more comprehensive assessment of the triple-bottom-line impacts of these regional synergies can be found in Kurup *et al.* (2005).

available through relatively well-established technologies (such as heat exchange, water re-use and co-generation) have been realised. This implies that even though Kwinana is a good example of international best practice in regional synergies, technology and engineering challenges have prevented many other potential opportunities from being realised.

To enhance the further development of new regional synergies, the Centre for Sustainable Resource Processing has undertaken research to facilitate the process of identifying and evaluating potential synergy opportunities and assisting the industries with feasibility studies and implementation of selected synergy projects. Over 90 synergy opportunities have been identified through 4 activities: a resource and process flow database, focused opportunity identification workshops, on-site company visits and literature reviews. A preliminary screening exercise was conducted to eliminate synergies that could readily be identified as infeasible or without significant sustainability benefits (for example, only very small quantities of materials are involved). As a result, the synergy opportunities were consolidated in a list of 25 priority projects for consideration by the KIC.

Based on the interest and perceived importance by the KIC and local industries, current development efforts focus on 15 short-listed priorities, with regard to water (such as re-use of treated industrial effluents, demineralised water and treatment of oily wastewater), energy (such as energy recovery from flue gases) and by-products (such as recovery and re-use of catalysts, organic waste and ammonium sulphate) (van Beers 2007). The feasibility assessments of these synergies comprise evaluations of the by-product streams, review of potential uses, evaluation of processing and source treatment needs, concept designs, and assessments of the economic, technical, environmental,and social aspects. The KIC, the industries and the CSRP are committed to implement further synergies and thereby enhance the sustainability of the KIA. In addition to supporting synergy opportunities, the research contributes to the development of collective strategies for the Kwinana region on the recovery and re-use of inorganic by-products, water and energy. These efforts have a more long-term and strategic focus and are conducted in close cooperation with the KIC.

The Gladstone Area

Queensland is second only to Western Australia in terms of geographical size and mineral resource reserves. Gladstone, which is located 550 km north of Brisbane, is one of the most industrially intense areas in Australia, generating 30% of Queensland's exports and 5% of Australia's total exports value. Although major industrial operations have been part of this region since Queensland Alumina Limited commenced operation in 1967, a 21,000 hectare industrial land bank – the Gladstone State Development Area – was established in 1993. This secured a large area of suitable land for major industrial development over a 30–50 year timeframe (GEIDB 2007). Gladstone has ready access to a deep-water port as well as world-class road and rail transport infrastructure.

Figure 6 shows the location of the main process industries in Gladstone. Several major industries operate in the region and they have formed an association called the Gladstone Area Industry Network (GAIN). Major industries in GAIN include Queensland Alumina Limited (alumina refinery), Rio Tinto Alcan (alumina refinery), Gladstone Power Station (coal-fired), Cement Australia (cement and lime plant), Boyne Smelters Limited (aluminium smelter), Orica Chemicals (ammonium nitrate, sodium cyanide and chlorine plant), plus the largest Queensland multi-cargo port (Central Queensland Ports Authority) and Awoonga Dam (Gladstone Area Water Board). Also, there is a proposed shale oil mine and processing plant (Queensland Energy Resources) and a waste management company (Transpacific Industries). Apart from the utility suppliers, the major value-chain integration is the supply of alumina from Queensland Alumina Limited to Boyne Aluminium Smelters. Several smaller companies support the major industries.

From 2004 to 2007, the CSRP supported a research project investigating opportunities for industrial synergies in the Gladstone region (with GAIN). The main aim of the project was to assist operations to achieve greater efficiencies in energy, water and materials consumption, and reductions in waste and emission generation. The

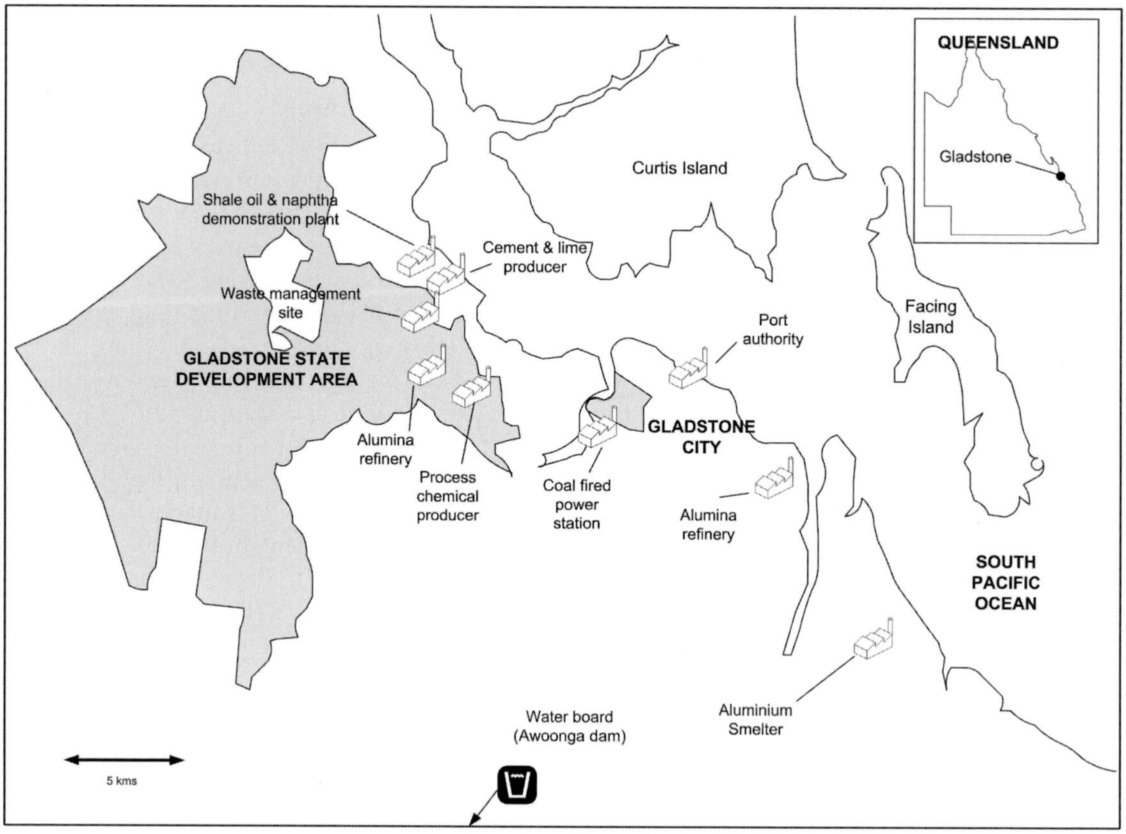

Figure 6: Location of companies in the Gladstone Industrial Area (Source: Corder and Moran 2006).

project was managed through the University of Queensland's Sustainable Minerals Institute (www.smi.uq.edu.au). The aim in the first year was to identify potential synergy opportunities based on collected data, interviews with GAIN members and independent research. A list of short-term (relatively easily implemented) and long-term (needing more research, but having large benefits) synergies were identified. The key short-term opportunities were the consolidation of wastes for use as alternative fuel source (reducing the use of fossil fuels, such as coal) and re-use of water. Long-term opportunities related to the re-use of large volume waste streams (bauxite residue and fly ash) and gaseous emissions (carbon dioxide and sulphur dioxide). Although no new synergies were implemented by the end of the project, it raised the awareness of regional synergy opportunities with the local industries and left them well positioned to act on any potential future initiatives.

Existing synergies

Currently, there are five main synergies in Gladstone (van Beers *et al.* 2007):

1. **Alternative fuels for cement clinker production.** Cement Australia, which is voluntarily committed to use alternative fuels, has a program to re-use wastes/by-products with high calorific value as a partial substitute for coal used in their clinker kilns. Operating at high temperatures (1500 to 1800°C) and long residence times (4 to 6 seconds) allows cement kilns to destroy wastes and by-products while extracting the calorific heating value. The Gladstone operation uses:
   - domestic tyres, which have typically been disposed to landfill
   - solvent-based fuels, prepared from hazardous and combustible wastes
   - spent cell linings from Boyne Aluminium Smelters—a waste generated in alumina

reduction to aluminium, containing silica, alumina and fluoride (a benefit for the manufacturing of cement).

2. **Water re-use.** In 2002, Gladstone suffered a severe drought, which initially brought about 10% water restrictions, then 25% restrictions 6 months later. Queensland Alumina Limited (QAL) coped with the initial 10% water restrictions, but risked production losses if forced to operate with 25% less water. QAL decided to fund a project to build an 8.5 km pipeline so that secondary treated effluent from the Calliope River Sewage Treatment Plant could be used for its mud-washing process. This scheme not only resulted in a water savings for QAL of about 6.5 ML/day, but also ended the discharge of treated effluent from the treatment plant to the local waterway, thus eliminating any future needs to upgrade the plant to tertiary treatment (Australian Aluminium Council 2004).
3. **Waste separation and re-use.** Transpacific Industries established a waste sorting and transfer facility at QAL in June 2003 to divert waste from landfill. Most of QAL's waste streams—apart from old asbestos waste, bauxite residue and fly ash—are now sorted for re-use, recycling or reselling. By early 2004, QAL reported the total elimination of waste going to the company's landfill, recycle rates of 85% (mostly from metal, cardboard and wood), and that the facility was cost neutral (QAL 2004a). Recovery costs are mainly recouped from the reselling of scrap metal. After 12 months of operation, the recycle rate had passed 90% (QAL 2004b).
4. **Fly ash re-use.** Pozzolanic Enterprises, a subsidiary of Cement Australia, collects fly ash from the Gladstone Power Station for use as cement additive in Cement Australia operations. The properties of fly ash—in particular its shape and fine size—are beneficial in concrete production. Pozzolanic selects fly ash that meets the required specifications, with the remaining fly ash being discharged to local ash dams. The power station benefits from reduced fly ash disposal and cement operations benefit from improved product quality as well as reduced use of conventional raw materials.

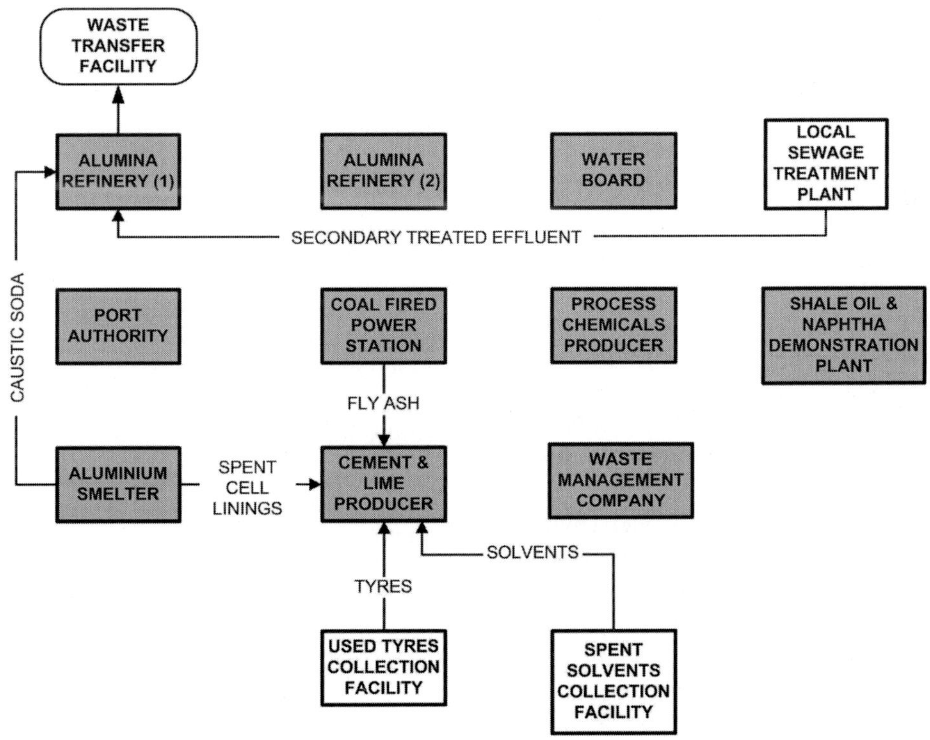

Figure 7: Existing by-product synergies in Gladstone (Source: van Beers *et al.* 2007).

Table 2: Illustrative benefits of existing synergies in Gladstone (adapted from Corder *et al.* 2006)

Synergy	Commercial benefits	Environmental and community benefits
Queensland Alumina Limited effluent re-use	During drought alumina refinery was able to continue to operate at full production No need for council to install tertiary treatment if or when required	No effluent discharge to local waterways Greater availability of water for region
Alternative fuels at Cement Australia	Savings in fuel costs for cement plant Savings in waste disposal costs	Improved emissions compared with traditional fuel, coal Improved re-use of wastes, historically bound for landfill Increased life span of local landfill site
Gladstone power station fly ash re-use	Revenue for power station Improvement of cement product quality	Reduced disposal of fly ash to local bunds Reduced use of raw materials by cement plant

* A more comprehensive assessment of the triple-bottom-line impacts of these regional synergies is available from Kurup *et al.* (2005).

5. **Caustic recovery**. At Boyne Aluminium Smelters, by-product caustic soda (sodium hydroxide) is produced from the treatment of spent cell linings to remove leachable fluorides and cyanides. The recovered caustic soda is used as a small supplementary source of caustic soda at the nearby QAL refinery.

A schematic overview of the current synergies in Gladstone is presented in Figure 7.

As mentioned earlier, all involved parties must benefit in some way for a by-product synergy to be successfully implemented. Table 2 presents a summary of the commercial, environmental, and community benefits from some of the existing synergies in Gladstone.

Dandenong LOGIS

In terms of the domestic market, Melbourne remains Australia's manufacturing heartland. Greater Dandenong is a key region in this heartland, ranking third behind Melbourne and Monash in terms of total jobs. Its major industry employers are manufacturing, retail and wholesale trade. Compared to the Melbourne Statistical Division, Greater Dandenong has a significantly higher proportion of its jobs in manufacturing, trade, transport and storage. Particularly strong manufacturing sectors include: wood and paper products; petroleum, coal, chemicals and associated products (such as plastics); metals; minerals; machinery and equipment; and miscellaneous manufacturing. Motor vehicle services, road transport, storage and logistics are also above the MSD average. The same sectors have contributed to recent jobs growth in Greater Dandenong (SGS Economics and Planning 2006).

Dandenong LOGIS is a new, large-scale commercial project being planned for the former Dandenong treatment plant site, which is located 2 kilometres south of Dandenong's town centre and 30 kilometres to the south-east of Melbourne. Comprising 180 ha, it is expected to become a benchmark example of sustainable, eco-industrial development in Victoria. Melbourne Water is responsible for site rehabilitation and VicUrban will develop the eco-industrial estate, which will employ up to 3,000 workers, housed in buildings that demonstrate leading edge sustainability initiatives within a commercial context.

A map of the site, showing the planned, coarse-grained distribution of uses, is shown in Figure 8. LOGIS will target advanced manufacturing (39%), logistics users (60%) and provide a small service centre for passing trade and employees (1%). Currently, 34% of all employees in Dandenong work in manufacturing, so it is critical to Dandenong and Victoria's long-term success that land be provided for these and other new businesses. Dandenong LOGIS will help to satisfy this need in a sustainable way. The sustainability goals of LOGIS are to deliver an eco-industrial park that:

- reduces the ecological footprint of its occupiers
- facilitates cooperation between its business users to create competitive advantages

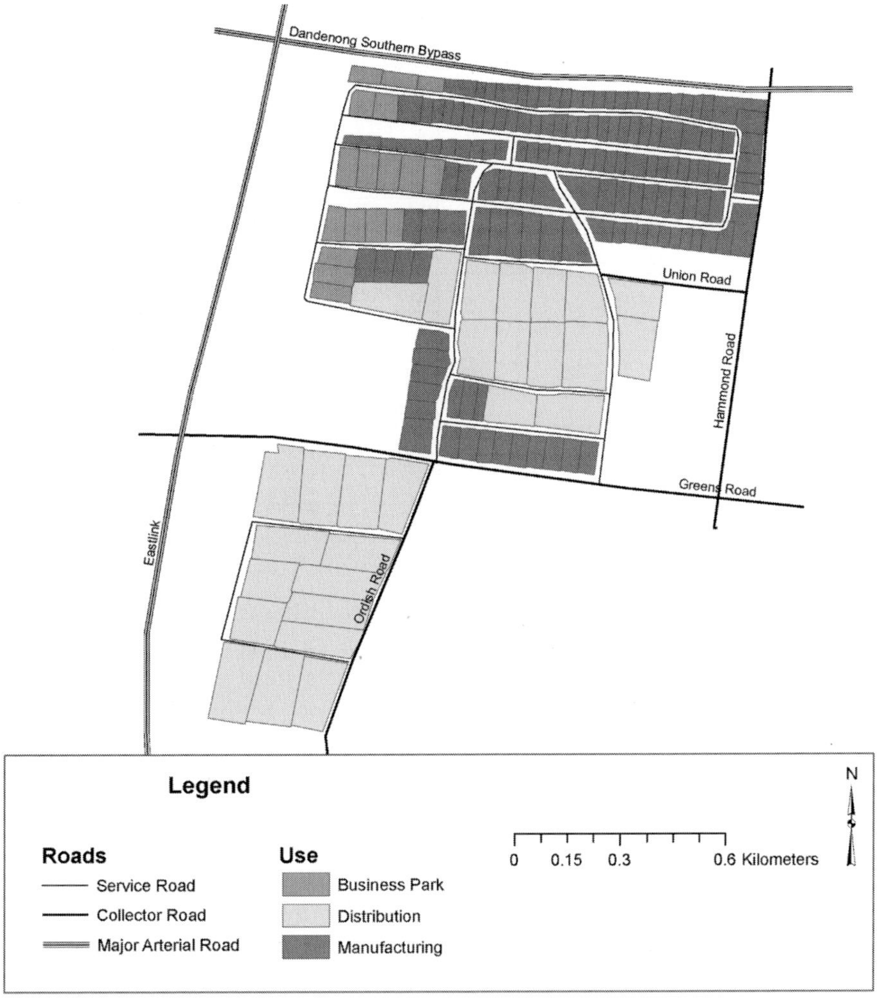

Figure 8: Proposed allocation of uses for Dandenong LOGIS (Source: Cook *et al.* 2007).

- strengthens linkages to the local and metropolitan business community.

Industrial clustering

In the project's early planning stages, the concept of *clustering* was adopted. Industrial ecology is linked intrinsically to clusters (Wallner 1999; Roberts 2004). There is a large amount of literature on the importance of clusters in supporting economic development. Firms that are arranged in clusters are able to achieve synergies and leverage economic advantage from shared access to information and knowledge networks, supplier and distribution chains, markets and marketing intelligence, special competencies, resources and support institutions available in a specific locality. Industrial ecology offers unique opportunities to add value to manufacturing firms located in a cluster, because the clustering of firms with similar waste and by-product streams creates opportunities to concentrate and minimise the re-use costs of the by-products and the collection costs of the wastes in one location.

With the help of an earlier report proposing a cluster strategy for the South East Region (SGS 2000), VicUrban concluded that the clusters that stand the greatest possibility of succeeding in Dandenong are those shown in Table 3:

Industrial Symbiosis Research for LOGIS

Late in 2006, the Temaplan Group and CSIRO Land and Water were commissioned by VicUrban to: (1) explore the potential for industrial symbiosis and (2) assess infrastructure systems that could facilitate the creation of an eco-industrial

Table 3: Preliminary Clustering Strategy for Dandenong LOGIS (Adapted from VicUrban documents)

INDUSTRIAL CLUSTER	ANZSIC SECTORS
1. Automotive and transport equipment	Transport equipment
	Basic metal products
2. Manufacturing systems and components	Industry machinery and equipment exporters
3. Domestic appliances and household goods	Plastics
	Chemicals
4. Food technology and systems	Textiles, clothing and footwear
	Glass Products
5. Building and construction	Electronic, electrical and equipment exporters
6. Logistics	Scientific equipment
Food warehouses and distribution operations	Food ingredients providers
Retail goods warehouse and distribution operations	
Logistic company container interchanges facilities	

park on the LOGIS site. The goal of (1) is to identify 10–12 key businesses that offer the greatest opportunity for industrial symbiosis within the identified boundaries and constraints of the LOGIS site.

In the first stage of the research, a sample survey was designed and distributed to 24 firms located near the site. This survey obtained data about: (a) annual inputs of materials, energy and water and (b) annual outputs of products and by-products. Based on the responses and interviews with key firms of interest, facility profiles were prepared to characterise the annual input–output flows of various facilities in and around the LOGIS site. Figure 9 shows a facility profile for a manufacturer of glass products used primarily in automotive applications. Each facility profile includes the main mix of materials used by the firm, plus their annual water and energy requirements. These are shown by arrows pointing into the box. Arrows pointing out of the box depict the firms' products and by-products. Profiles like this are useful for identifying potentially symbiotic linkages (that is, output-to-input matches) among the facilities near the site, and also for identifying potential recruits that may be targeted for the EIP.

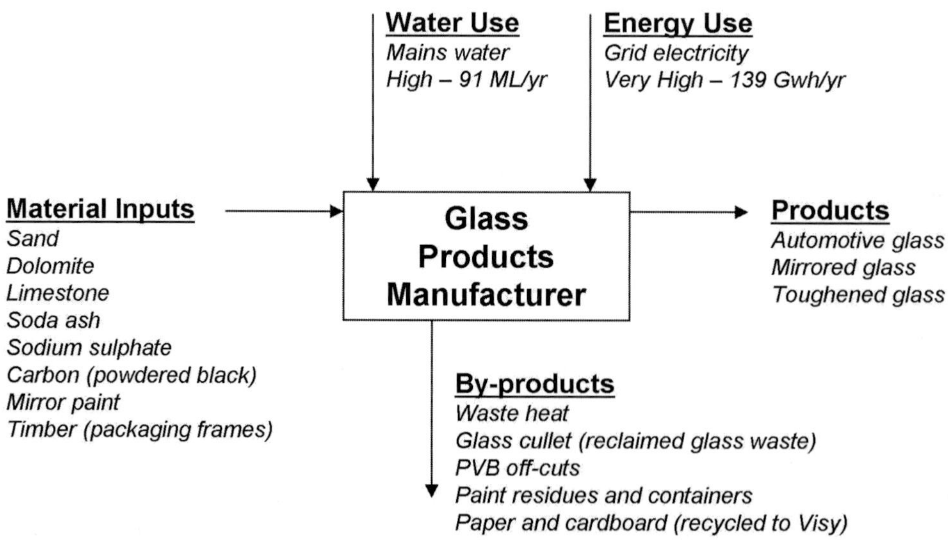

Figure 9: Facility profile for a glass products manufacturer (Source: Batten and Cook 2007).

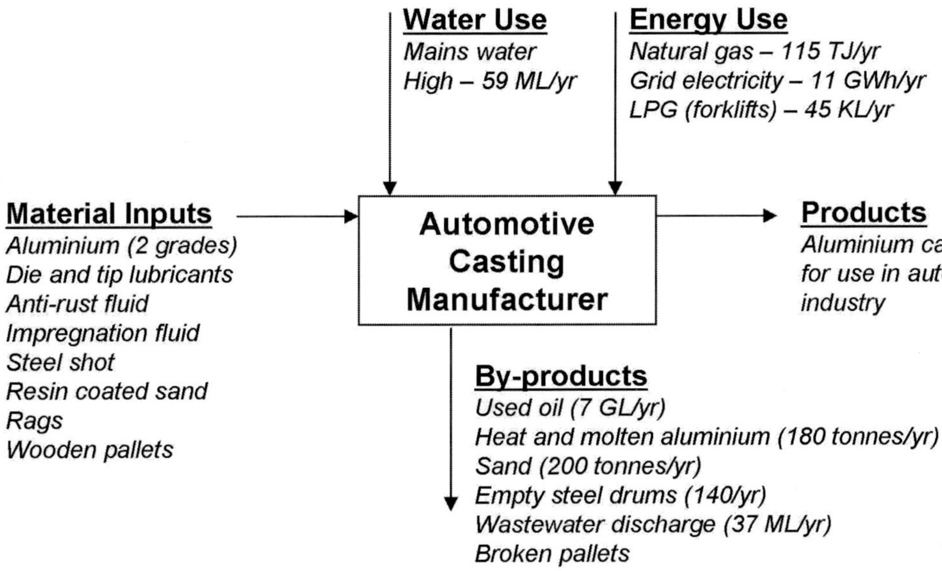

Figure 10: Facility profile for an automotive casting manufacturer (Source: Batten and Cook 2007).

Note that some of this manufacturer's by-products – paper and cardboard – are currently recycled to a large paper and board recycler located nearby. In several businesses close to the LOGIS site, there is an understanding that the most desirable destination for by-products is another firm nearby. Such a co-location ambition is a basic principle of industrial symbiosis.

In this way, a facility profiles database can be constructed. The database being created for the LOGIS project is similar to the EPA's database tool (*FaST*) discussed earlier. The goal is to find matches between by-products and the resource requirements (materials, energy and water) of various industrial processes. This allows the user to quickly identify potential by-product synergies between facilities or the types of industrial partners that can be recruited as beneficial recipients for by-product streams from existing facilities.

Figure 10 depicts a facility profile for a representative manufacturer of metal castings used in making automobiles. Note that the energy requirements of this firm are quite diverse (natural gas, electricity and LPG), yet a significant amount of heat is wasted when it could be recovered and used as a source of power, steam or cooling.

Figure 11 illustrates some potential synergies based on the facility profiles of several large manufacturers located near the LOGIS site. Note that some synergies are already in place (solid arrows), whereas others are only potential linkages at this stage (arrows with broken lines). These potential linkages warrant further investigation in the next stage of the project, to determine if any are mutually feasible and profitable.

Other potential synergies are listed in Table 4. The existing and potential supply, by-product and utility synergies are already diverse. Examples of each may be observed in Figure 11. Three firms produce products that are key inputs to the automotive industry and, to a lesser extent, other transport industries. Moreover, eight of the fifteen firms that responded to the sample survey are engaged in manufacturing activities involving vehicles or other transport equipment. This affinity reinforces VicUrban's idea that an *Automotive and Transport Equipment Cluster* could be one worthwhile anchor cluster of firms on and near the LOGIS site.

LOGIS Scenarios

Three potential scenarios for the LOGIS project have emerged. Although it is too early to be sure that these three scenarios offer the greatest potential for symbiosis within LOGIS, they are attractive because each, in its own way, could build on industrial clusters that are well-established close to the LOGIS site or elsewhere in the Greater Dandenong region. Building on these strengths is analogous to adaptive ecological design principles.

The three suggested scenarios are:

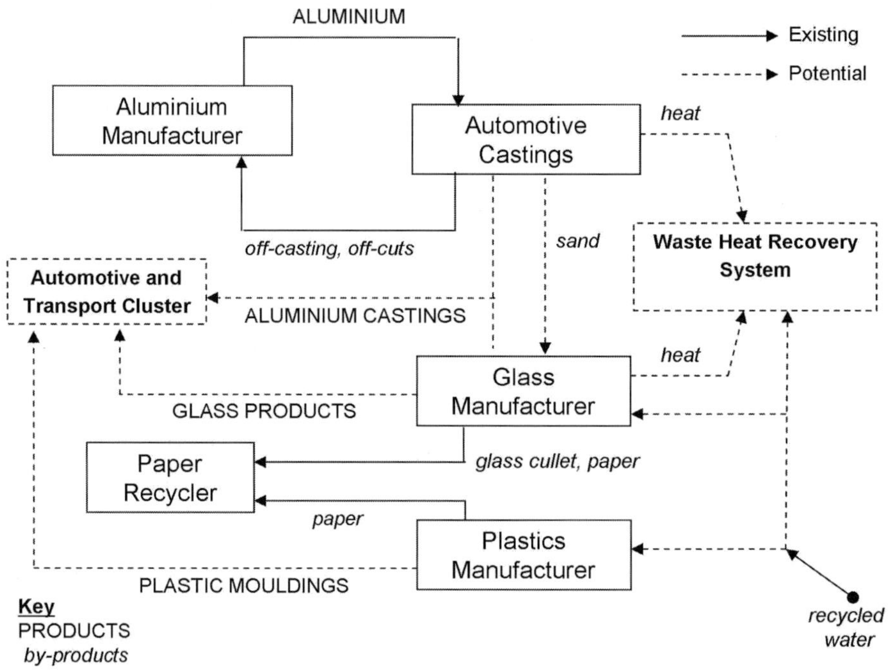

Figure 11: Some existing and potential synergies near the LOGIS site (Source: Batten and Cook 2007).

1. **A Smart Transport and Energy scenario.** Anchor tenants for this scenario would be the Automotive and Transport Equipment Cluster and a Waste Heat Recovery System, as depicted in Figure 12. In Dandenong South, there is a larger-than-expected set of firms associated with automotive and/or transport parts and equipment—possibly the largest of all industries in the Greater Dandenong region. Thus there is an established foundation to build on for the sharing of inputs and exchange of products and by-products. Regardless of the final

Table 4: Some Potential By-product Synergies near LOGIS.

Output from	By-product	Input to	Application(s)
Nissan Casting	Waste heat	Heat recovery system	Power, steam or cooling
Pilkington Glass	Waste heat	Heat Recovery System	Power, steam or cooling
Steele and Lincoln Foundry	Waste heat	Heat recovery system	Power, steam or cooling
Nissan Casting	Sand	Boax Constructions or Zaganite	Asphalt or road-marking materials
Steele and Lincoln Foundry	Sand	Boax Constructions or Zaganite	Asphalt or road-marking materials
Pilkington Glass	Glass cullet and off-cuts	Visy or Alex Fraser	Recycling Road-base mix
Jayco Caravans	Aluminium	Nissan Casting	Aluminium casting
Eastern Treatment Plant	Treated wastewater	Pilkington Glass	Cooling
Eastern Treatment Plant	Treated wastewater	Poly Pacific Plastics	Cooling
Castricum Brothers Abbatoir	Tallow, intestines	Fuel producer Soap producer Tennis racquet maker	Biodiesel, glycerine, Soap, candles, racquet strings

Source: Batten and Cook 2007

tenants, heat is produced in significant amounts nearby, and may be sourced from several large companies already (such as Pilkington Glass, Nissan Casting, Steel and Lincoln Foundry). As energy recovery from waste heat is on the horizon for renewable and distributed energy systems, there are potential synergies to be gained between several firms involved in the automotive and transport cluster (producing heat but currently wasting it into the atmosphere) and the heat recovery process.

2. **A Smart Food and Pharmaceutical Systems scenario.** Key anchor tenants would be the Food Technology and Systems Cluster identified in VicUrban's cluster analysis and a Chemicals and Pharmaceuticals Cluster. These two clusters are of interest because they are gradually merging. Leading agrochemical companies, such as Monsanto and Novartis, use knowledge of living organisms to produce seeds and agrochemicals for plant production, veterinary products, and diagnostic and therapeutic products for human health care. This rapidly growing, global industry is capitalising on the trend towards eating healthier food and taking drugs and supplements to live longer. Neutraceuticals are a recent example – food products that function as, and substitute for, pharmaceuticals. Re-use of by-products in this complex area is a significant challenge.
3. **A Smart Building and Recycling Technology scenario.** Key anchor tenants would be a Building and Construction Cluster and a Materials Recycling Cluster. These clusters would showcase recycled metals and mineral products, reconstituted plastics, glass products, industry machinery and equipment and electronic, electrical and scientific equipment exporters. We would add recycled products that are reconstituted from used products for further use as building materials. For example, automobile tyres may be shredded and crumbed, then used as brake pads, for insulation and in road repairs. However, they can also be used in innovative building applications. For example, Ecoflex (www.ecoflex.com.au) converts recycled tyres into building products such as road pavements, retaining walls, drainage channels, scour protection mats, culverts and the 'E-Pod'—a method for constructing strong concrete slabs that uses less concrete.

Conclusion

In this chapter, we have built an argument for the wider application of principles of industrial ecology to the industrial planning of Australian cities and their surroundings – including the development of key incubators to promote eco-industrial principles to Australian industry and government. Examples of regional symbiosis were based on case studies of current eco-industrial developments in three Australian states.

In each case, the key to industrial symbiosis, or the development of any regional synergy, is that each of the participating parties derives a benefit. The types of benefits can vary greatly and often stretch well beyond the conventional economic ones. Security of water/energy supply, increased water/energy efficiency, lower operational costs, and reduced storage costs for by-products are some key benefits from the synergies. Aside from mutual profitability, synergies provide environmental and community benefits. The benefits can also be strategic—leading to reduced exposure to risk and a better reputation with key stakeholders.

A recent review of international case-studies on enabling mechanisms for regional synergies (van Berkel 2006), came to the conclusion that the realisation of successful synergies depends on three key aspects (see Figure 12):

1. **Proven technology:** appropriate processes and equipment are available to ensure that the resource synergy eventuates, so that the resource is being transferred between the companies involved and converted as and when required.
2. **A convincing business case:** giving an opportunity to reduce costs, generate new revenue and/or secure access to vital resources that can be materialised by one or more businesses.
3. **License to operate:** at the minimum, all government approvals should be in place, but also endorsement from affected communities.

Nature is clever. We can make great progress by following nature's footprints to a less wasteful and more sustainable future. These resourceful footprints are everywhere. Nature is complex, yet all

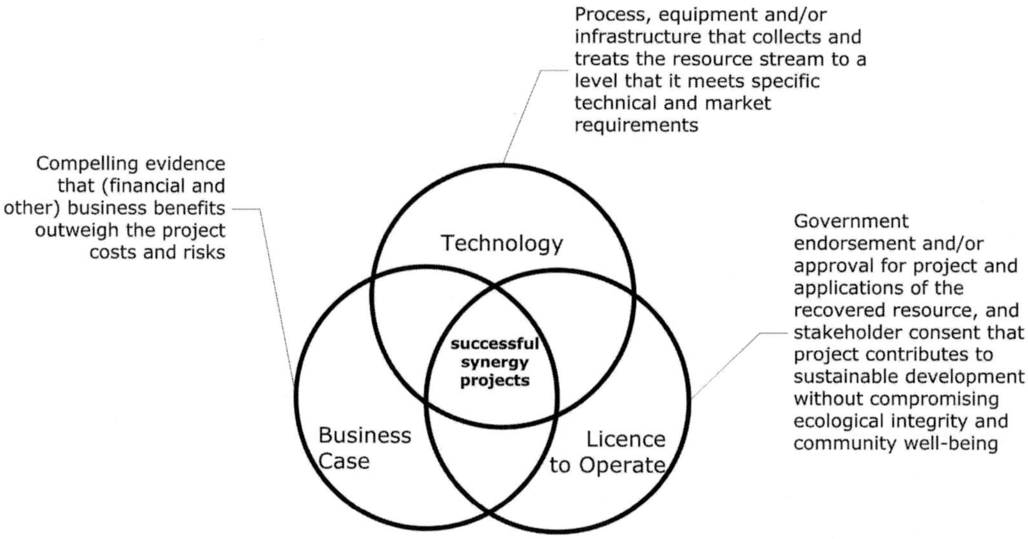

Figure 12: Success Factors for Regional Resource Synergies (Source: van Berkel 2006).

waste is being used somewhere. Just as everything is used at one point or another in a food chain, there is little doubt that in a number of industrial areas—such as Kwinana, Gladstone, and Dandenong—eco-industrial complexes are evolving in which the 'wastes' from manufacturing and processing plants become the feedstock for other industries. To what extent these initiatives succeed depends on the energy and ingenuity of industrial leaders and politicians who hold regional destinies in their own formative hands, perhaps with a little help from the scientific community.

References

Altham J and van Berkel R (2004). 'Industrial symbiosis for regional sustainability: an update on Australian initiatives'. Paper presented at the International Sustainable Development Research Conference, University of Manchester, UK.

Australian Aluminium Council (2005) *Sustainability Report 2004*. Australian Aluminium Council, Canberra.

Ayres R (1995). Industrial metabolism: theory and policy. In *The Greening of Industrial Ecosystems* (Eds D Richards, B Allenby and R Frosch), National Academy of Science, Washington DC.

Baas L (1998). Cleaner production and industrial ecosystems, a Dutch experience, *Journal of Cleaner Production* **6(3–4)**, 189–197.

Batten D (2003). Converting waste into wealth: the scope for eco-industrial complexes in Australian regions, In *Proceedings of the World Congress of the International Solid Waste Association (ISWA)*, Melbourne.

Batten D (2005). 'Applying eco-industrial principles to urban and regional development: elements of the business case'. Report for the Urban Water Infrastructure Team, CSIRO Manufacturing and Infrastructure Technology, Highett, Vic.

Batten D and Cook S (2007). 'Industrial symbiosis for Dandenong LOGIS: sample survey results, facility profiles and preliminary scenarios', Confidential interim draft report, March.Temaplan/CSIRO/VicUrban, Melbourne.

Bossilkov A, van Berkel R and Corder G (2005). *Regional Synergies for Sustainable Resource Processing: a Status Report*. Centre for Sustainable Resource Processing (CSRP), Perth, WA.

Chertow M (2000). Industrial symbiosis: literature and taxonomy. *Annual Review of Energy and Environment* **25**, 313–337.

Chertow M, Ashton W and Kuppalli R (2004). 'Advancing the study of industry and environment', Industrial symbiosis research symposium. Report No. 3, Yale School of Forestry and Environmental Studies, New Haven.

Cook S, Sharma A, Burn S and Batten D (2007). 'Infrastructure initiatives assessment for Dandenong LOGIS'. Land and Water Science Report, June. CSIRO, Melbourne.

Corder GD and Moran CJ (2006). 'The importance of water in the Gladstone industrial area'. Paper presented at the Water in Mining Conference, Brisbane, Qld, 14–16 November.

Corder GD, Van Beers D, Lay J and Van Berkel R (2006). *Benefits and Success Factors of Regional Resource Synergies in Gladstone and Kwinana.* Paper presented at the Green Processing Conference, Newcastle, NSW, 5–6 June.

Côté RP and Cohen-Rosenthal E (1998). Designing eco-industrial parks: a synthesis of some experiences, *Journal of Cleaner Production* **6**, 181–188.

Desrochers P (2004). Industrial symbiosis: the case for market coordination. *Journal of Cleaner Production* **12(8–10)**, 1099–1110.

Dias S and Yates R (2001). *Advancing Cooperative By-Product Synergy Programs Between Canada, Mexico and USA*. Hatch, Canada.

Ehrenfeld J and Gertler N (1997). Industrial ecology in practice: the evolution of interdependence at Kalundborg. *Journal of Industrial Ecology* **1**, 67–79.

Frosch R and Gallopoulos N (1989). Strategies for manufacturing. *Scientific American*, **Special Issue, September,** 97–108.

GEIDB (2007). *A World Class Investment Location: Gladstone Economic and Industry Development Board*. See http://www.gladstoneindustry.org.au/gladstone-state-development-area/

Harris S and Pritchard C (2004). Industrial ecology as a learning process in business strategy. *Progress in Industrial Ecology* **1 (1/2/3)**, 89–111.

Herbertson J and Sutton P (2002). 'Foundations of sustainable resource processing'. Paper presented at the Green Processing Conference, Cairns, Qld, 29–31 May.

Jacobsen NB (2003). Industrial symbiosis in Kalundborg, Denmark. In *Eco-industrial Strategies*, (Ed. E Cohen-Rosenthal) Greenleaf Publishing, Stevenson, Washington.

KIC (2005). *Kwinana Industries Council.* See <http://www.kic.org.au>

Korhonen J (2001). Co-production of heat and power: an anchor tenant of a regional industrial ecosystem. *Journal of Cleaner Production* **9**, 509–517.

Kurup B Altham W and van Berkel R (2005). 'Triple Bottom Line Accounting Applied for Industrial Symbiosis'. Paper presented at the 4th Australian Life Cycle assessment Conference, Sydney, Australia, 23–25 February.

Lowe E, Warren J and Moran S (1997). *Discovering Industrial Ecology: An Executive Briefing and Sourcebook*, Battelle Press, Columbus, Ohio.

Newman P (1999). Sustainability and cities: extending the metabolism model. *Landscape and Urban Planning* **44**, 219–226.

Newton P (2002) Urban Australia 2001. *Review and Prospect* **39,** 37–45

QAL (2004a). Media Release, 13 January 2004. Retrieved from <www.qal.com.au>.

QAL (2004b). Media Release, 30 June 2004. Retrieved from <www.qal.com.au>.

Roberts B (2004). The application of industrial ecology principles and planning guidelines for the development of eco-industrial parks: an Australian case study, *Journal of Cleaner Production* **12**, 997–1010.

Roberts B and Wadley D (2002). Planning for sustainable industry development, Royal Australian Planning Institute, Brisbane.

Schwarz E and Steininger K (1997). Implementing nature's lesson: the industrial recycling network enhancing regional development. *Journal of Cleaner Production* **5**, 47–56.

SGS Economics and Planning (2000). 'Cluster identification strategy: South East Region,' Draft Report for South East Development' March. SGS, Melbourne.

SGS Economics and Planning (2006). 'LOGIS Site: Economic Analysis,' Report prepared on behalf of the Hornery Institute, April. SGS, Melbourne.

SKM (2002). *Kwinana Industrial Area Economic Impact Study: an example of industrial interaction.* Sinclair Knight and Merz for Kwinana Industries Council, Perth, WA.

Starlander J-E (2003). Industrial symbiosis: a closer look on organisational factors. Unpublished Masters Thesis, The International Institute for Industrial Environmental Economics, Lund University, Sweden.

Taylor M (2002). 'Waste and emission reduction through regional synergies in the Kwinana Industrial Area'. Paper presented at the Green Processing Conference, Cairns, Qld, 29–31 May.

Tibbs H (1992). Industrial ecology: an environmental agenda for industry. *Whole Earth Review* **4**, 19–20.

van Beers D (2007). 'Status report on regional synergies in the Kwinana Industrial Area, Perth, Western Australia'. Centre for Sustainable Resource Processing (CSRP), Perth, WA.

van Beers D, Bossilkov A and van Berkel R (2005). 'Status report on regional synergies in the Kwinana Industrial Area, Perth, Western Australia'. Centre for Sustainable Resource Processing (CSRP). Perth, WA.

van Beers D, Corder G, Bossilkov A and van Berkel R (2007). Industrial symbiosis in the Australian minerals industry: the cases of Kwinana and Gladstone. *Journal of Industrial Ecology* **11(1)**, 55–72.

van Berkel R, Willems R and LaFleur M (1997). The relationship between cleaner production and industrial ecology, *Journal of Industrial Ecology* **1(1)**, 51–66.

van Berkel R (2003). 'The Kwinana Industrial Area: an evolving example of an eco-industrial Park'. Paper presented at the Industrial Ecology for a Sustainable Future, International Society for Industrial Ecology, 29 June–2 July. University of Michigan, Ann Arbor.

van Berkel R (2006). *Regional Resource Synergies for Sustainable Development in Heavy Industrial Areas: An Overview of Opportunities and Experiences.* Curtin University of Technology, Perth, WA.

Wallner H (1999). Towards sustainable development of industry: networking, complexity and eco-clusters, *Journal of Cleaner Production* **7**, 51–66.

Water Corporation (2003). 'Kwinana Water Reclamation Plant (KWRP): expanding Perth's water options, Public Environmental Review'. Water Corporation, Perth, WA.

Wolman A (1965). The metabolism of the city. *Scientific American* **213**, 179–180.

Young, R. (1999). *By-Product Synergy: A Demonstration Project Tampico, Mexico.* Business Council for Sustainable Development, Gulf of Mexico.

SECTION 8

Infrastructure and Urban Planning Transitions

Australia faces the twin challenges of transitioning to sustainable and competitive 21st century urban infrastructures as well as adopting a mode of urban planning that is reflective of 21st century needs and challenges compared to those of the previous century.

Part of the infrastructure challenge involves redressing the national *shortfall in infrastructure expenditure*—estimated at between $25 and $38 billion (Hardwicke; Regan)—which has produced an indifferent Infrastructure Report Card from Engineers Australia and is reflected in the congestion and system failures regularly catalogued in the metropolitan media. A related part of the challenge flows from a general reluctance on the part of national and state governments to initiate *major long-term nation-building investments in critical Horizon 2 and Horizon 3 future infrastructure platforms* (catalogued in Sections 5, 6 and 7 of this book)—which are fundamental to any sustainability transition of our cities.

The challenges for urban planning are many, as cities are where an increasing proportion of the world's—and Australia's—populations are concentrated. A major persisting challenge involves linking places of work and places of residence with modes of transport other than that of the now-ubiquitous motor car, around which many decades of mid-to-late 20th century urban and infrastructure (road) planning was based and where the initial challenges to urban sustainability came to be centred (Newman). More recent challenges to 20th century urban planning have focused on the decline in population health reflected in a growth of chronic non-communicable diseases (termed 'lifestyle diseases', such as obesity, diabetes, cancer and cardiovascular disease; McMichael; Timmins *et al.*)—most of which have strong links to our high consumption, sedentary, drive-everywhere urban lifestyles. An alternative is '*smart growth*' (Falconer and Giles-Corti), which refers to a new urban planning paradigm that seeks to *re-cycle the city* by an increased focus on urban renewal, (mixed land use, mixed mode—especially active and public transport— and mixed housing); and *design liveable walkable neighbourhoods* for existing, as well as greenfield, suburbs. The jury is still out concerning the extent to which alternative models of urban form (Scheurer) can best accommodate these principles and generate positive outcomes related to reduced ecological footprint and improved human health (among others). The paradigm of low density car-based urban sprawl, however, needs to be consigned indefinitely to the 20th century.

PWN

36

Urban Form, Sustainability and Lifestyles in 21st Century Cities

Jan Scheurer

Early 21st century cities offer an ever-increasing patchwork of physical urbanisation models, as well as social and lifestyle opportunities, which presents a major challenge in determining what directions urban development is taking. The need for policy makers and concerned community members to envision a viable path to achieving sustainability adds further complexity to the issue.

This chapter examines a variety of approaches and models to link the present urban experience with the requirements of a more resource-efficient and liveable urban future. It is emphasised that for a process of sustainable transformation to be approached successfully, it is paramount that contemporary physical and socioeconomic trends in cities be viewed from a user perspective. The spectrum of lifestyles manifest in a city—and the forces that shape them and distribute them spatially and socially—hold important clues to understanding the functional realities of contemporary cities and their further evolution.

During the transition from an industrial to an information-based society in Western cities in the past 30 years—and exacerbated by the transition to a market economy in the formerly communist countries—the paradigmss guiding urban expansion and development for much of the 20th century are now severely out of tune with the contemporary forces that shape cities. It has become obvious that unlimited growth, which is supposed to benefit society as a whole, meets formidable environmental, social and economic constraints in an age when humanity's aspirations, activities and problems are increasingly global in character. Near-universal employment has given way to a highly fragmented labour market with inherent uncertainties and disparities, while cultural change, female empowerment and economic necessity have made it less and less acceptable to uphold the traditional division of labour between sexes in the nuclear family. Neighbourhoods—once built for the convenience of women and children at home—have gradually become dormitory suburbs largely devoid of daytime activity. Simultaneously, small units of one of two members have emerged as the dominant household form in cities, caused by falling birth rates and a growing proportion of aged people. This is associated with a stronger presence of conspicuously non-family oriented lifestyle groups—alternative, hedonistic and pragmatic—which have special requirements of consumption, cultural diversity and practical convenience to be met in their residential and working environment. This new plurality has established local social milieux, which determine the contemporary character of neighbourhoods rather than earlier distinctions largely based on income, life cycle stage and ethnicity. These milieux have also become valuable incubators for the urban economy, particularly in innovative knowledge-oriented sectors that have taken the lead from industrial mass-production to top the value-adding pyramid, generating a more hybrid and heterogeneous picture of urban space (compared with the 20th century model of functional segregation), increasingly blurring the line between private life (which is carried back into the urban arena to some degree) and work (which is making inroads into the home following trends to more telecommuting, contract work, self-employment and the life-long need for further education).

The implications of these trends on the future form of cities, however, are contested. This is particularly so with regard to the sustainability agenda, which mandates cities to reduce their flow of material resources while simultaneously increasing their liveability. In this chapter, we will discuss a number of models of sustainable urban development that are relevant for cities in the developed world under the set of societal conditions described above, and explore their potential and limitations. Experiences from concrete examples of neighbourhoods in selected cities where residents have made conscious lifestyle choices in tune with the sustainability agenda, and created a more liveable urban environment in the process, will conclude with the optimistic note that, despite analytical confusion on the part of urban research, positive changes on the ground remain feasible.

Sustainability and the urban future—four approaches

The Decentralised City

The Decentralised City can be characterised as a continuation, and intentional exacerbation, of the trends that led to suburbanisation and the spatial dispersal of settlements in many Western cities, largely after 1945. The trends included aspirations for home ownership, relative spatial independence and self-determination, and growing demand for residential space in both quantity and quality. It seeks to counter claims regarding the inefficiency and high resource use inherent in the dominant urban configuration of the late 20th century (Newman and Kenworthy 1999; Hall 2002) through a reduction of central control to strengthen local and sub-regional autonomy. This is also connected to a vision of relative social equality where home ownership and tenure over land are accessible to all population groups. The model can historically be traced back to Frank Lloyd Wright's 1940s concept of the 'Broadacre City' (Fishman 1977; Hall 2002) and was promoted within the sustainability discourse of the 1990s by groups such as the Californian researchers around Gordon and Richardson (1989), and by Troy (1996) in Australia.

The position can be related to sustainability mainly in two ways. Firstly, there is the expectation that, in the absence of development constraints, land values and housing costs will be kept moderate and spatially balanced, which translates into high standards of living at greater affordability. In Australia, this stance is promoted in the current housing affordability debate by, among others, the Institute of Public Affairs (Moran and Staley 2006). Secondly, in providing individuals with sufficient space and control (ownership), they are encouraged to pursue sustainability goals within their own backyards, without relying on collective efforts within a local community and its possible vagaries (Troy 1992, 2003). It is furthermore maintained that dispersed and decentralised settlement patterns reflect a market preference, and any attempt at tampering with this force would be an authoritarian intervention that stifles economic efficiency and individual freedom (Gordon and Richardson 1997).

Thus, the Decentralised City assumes a 'weak' stance on sustainability (see Zethoven 1990; Jacobs 1999). It accepts a relatively high level of resource use built into the idealised urban form in exchange for the equity and economic gains associated with the model, and points to the empowerment of the individual as consumer/reproducer to adapt to more resource-conscious lifestyles and (eventually) choose technologies that have a lower environmental impact (such as home-based renewable energy systems, backyard organic food production and zero-emission vehicles). While car dependence is inevitable, confidence is expressed that ongoing decentralisation will disperse the present concentrations of destinations (jobs, retail, services and recreation) and thus substantially shorten the average distance of journeys.

The model, however, appears to come at a macroeconomic cost that is commonly disregarded by its proponents. Low levels of accessibility in dispersed outer suburbs—compared with their more centrally located inner urban counterparts—put pressure on households and businesses to allocate a high and increasing proportion of their resources on transport. Simultaneously, the theoretical trend for employment to follow the dispersal of housing tends to be riddled with mismatches in skills and demographics, as well as sheer numbers (Maher 1994; Dodson 2003). There is now talk of increasing locational inequality in Australian cities (Dodson and Sipe 2005; Gleeson 2006b). It remains doubtful to what extent markets would continue to favour peripheral low-density living if the costs of such inequalities were more explicit, or as the costs of transport go up along with fuel prices. In addition, it is often unclear whether households' decisions to relocate outward are sufficiently informed to take into account all implications and potential inconveniences of suburban living, or whether they merely express a 'least-worst' solution—reflecting a lack of attractive and affordable housing options nearer to the city centre (Hesse and Schmitz 1998). The same inability to find optimal locations under contemporary market conditions may hold true for land-intensive industries that are commonly driven to suburbia by land value pressures (Kahnert 1998). Market demand for peripheral urban growth may therefore be more fragile than it seems, if and when policies are put in place to internalise societal, macroeconomic and environmental costs, and to overcome market distortions, or when the cost of transport becomes more significant. It is fair to say that this less-than-confident outlook into a future of increasing resource constraints and socioeconomic pressures has begun to discredit traditional suburban development—even in its regional strongholds across North America and Australia—providing momentum for developers, planners and the community to take the exploration of alternative urbanisation models seriously.

The Smart-Growing City

The Smart-Growing City, as the terms suggests, is essentially a model applicable under growth conditions, and can be clearly distinguished from the more 'European' models of the Compact City and the Dispersal-Pragmatic City (see below), which take into account stable (or even declining) urban populations. Much of its rationale is derived from the sustainability shortcomings experienced in traditional suburban development (Gillman 2002). Hence, quality public spaces under local stewardship and a variety of housing forms are included to combat social isolation. A mix of uses and flexibility for functional change are a basis for increased economic opportunities. Walkable neighbourhoods and the availability of user-friendly public transit are expected to break the dominance of the car. Respect for, and gentle integration of, environmental features within the urbanised area—on a par with their preservation beyond it—are intended to maintain a better balance between the natural and the built environment (Hall 2003). These goals are being pursued by allowing stakeholders to have a greater say in the process of urbanisation, which is traditionally a highly corporatised and top-down undertaking—at least in the suburbs of New World cities—without denying the formidable differences in culture and urban tradition between North America and Australia (Gleeson 2006a). Smart-Growth principles were first elaborated by planning and design practitioners under the banner of New Urbanism—for example, Calthorpe (1993) in the US and Morris and Kaufman (1996) in Australia— and have recently made substantial inroads into government programmes and development codes throughout the New World. For example, this occurred in Western Australia where the planning system was comprehensively reformed along New Urbanism principles (WAPC 2004).

It can be argued that the Smart-Growing City occupies an 'intermediate' stance on sustainability (see Zethoven 1990)—it embraces the concept of growth as paramount for the wellbeing of the city, but regards it as imperative that this growth changes fundamentally in character. There is a clear expectation that these proposed changes will have synergistic and self-reinforcing effects on the welfare of communities and improve liveability in all areas. Similarly, it is anticipated that Smart Growth will, over time, relieve society from the external costs of suburban development (see above) and 'pay its own way' when its impact on the urban economy and, more specifically, government budgets are considered (Benfield *et al.* 1999).

The same claim, however, cannot necessarily be made for the aspect of resource use. While Smart Growth poses no insurmountable barriers to housing development that encompasses urban ecology approaches aimed at resource savings or mobility management, outright encouragement of such solutions or behaviour is not an integral component of the concept. Rather, the focus is on the incorporation of passive measures needed to enable more sustainable lifestyles—such as energy-efficient building design—without guaranteeing them, while retaining every opportunity for the user to maintain energy-intensive behaviour patterns; or increasing the availability of non-car transport modes while continuing to generously supply parking and roads. More bluntly, Smart Growth attempts to hand out the 'carrots' of the transition towards sustainability, while being confident it can hold back the 'sticks' and still achieve tangible outcomes. Only ongoing experience can tell whether this model will work.

The Compact City

The Compact City describes an attempt to bring the qualities of historic European and Middle Eastern cities—that evolved during many centuries of low resource use and relative social stability—back into contemporary city building. Key factors in this process are the existence of a high-quality public facilities and spaces which users can appropriate for productive and recreational, formal and informal activities alike (Jacobs 1961; Gehl 2001; Engwicht 1999), increases in residential and employment density in established areas accompanied by constraints to outward urban growth, and substantial shifts from cars to other modes of transport. This vision is supported by evidence from existing cities of relative compactness that show significantly lower energy use and more balanced travel choices than their more dispersed counterparts (Newman and Kenworthy 1989, 1999; Apel *et al.* 1997). To varying degrees, the Compact City philosophy came to characterise urban development practice in many European cities during the 1980s and 1990s, and is also applied at the regional level in the model of 'Decentralised Concentration' (Holz-Rau 1997).

The Compact City's main claim towards sustainability consists in its objective to preserve and rehabilitate open space through concentration of urban development at high densities, and the substantial savings in travel demand its proponents anticipate will result from such practice. This is linked to the belief that the Compact City—like the Smart-Growing City—can unleash immense synergistic potential with regard to social and economic interaction in these multifunctional, walkable environments and thus make a direct contribution against socially undesirable processes of alienation, segregation, lack of opportunities and urban crime.

The Compact City's stance on sustainability can be characterised as 'deep' or 'strong' (Zethoven 1990; Jacobs 1999). Its stated aim is to fundamentally reconfigure present settlement and transport patterns without necessarily tying this process to the coincidence of economic or population growth, or, even more generally, to a responsiveness to the market forces that shape current urban development trends. Ultimately, the Compact City would overcome the phenomenon of suburban dispersal through land use diversification and increased density of some areas, and the ruralisation, or gradual retreat, of urban settlements from others (Register 2006). It would also break the dominance of the car—the use and ownership of which would face increasing physical, legal and financial restrictions and be reduced to a supplementary transport mode in a system based on walking, cycling and (mostly rail-based) public transit. As the argument goes, the result would be both enhanced resource efficiency—from transport effects as well as from 'economies of scale' when implementing energy efficiency technologies into compact development—and an enhanced quality of urban life, facilitated by a more versatile and adaptable public realm (White 1988; Gehl 2001).

The main dilemma facing Compact City models, however, is their applicability under present urban development pressures and conditions. Restricting urbanisation to high-density nodes and corridors is liable to price out land-intensive uses, such as some industries, big-box retail facilities or single-family housing. Some of these uses would invariably seek to subvert Compact City planning frameworks by locating in the least protected 'undevelopable' (read cheap) areas—unless, of course (and as expected by some Compact City proponents), they eventually disappear altogether. There is an implicit presumption here that the market demand for land uses incompatible with the Compact City vision is as volatile as it is undesirable. From a contemporary perspective, and depending on the political weight afforded to it, this attitude may appear either unrealistic and powerless—or prescriptive and authoritarian (Jessen 1997; Sieverts 1997). This extends to the field of transport as well, where the question is yet to be answered whether the travel savings observed in existing compact cities really materialise when translated into a strategic program to transform dispersed cites (Hesse 1999). It is possible that the Compact City—in a reversal of the Smart Growing City's attitude (see above)—might be determined to apply the 'sticks' for enforcing more sustainable urban structures, while largely neglecting to focus on the 'carrots' that would make them palatable. For this reason, the Compact City often struggles to convince planning practitioners—and the public at large—that it provides a compelling and tangible image of the future city, although its relevance in theory is indisputable.

The Dispersal-Pragmatic City

The model of the Dispersal-Pragmatic City, at a glance, does not fundamentally differ from the sustainability goals pursued in the Smart-Growing City and the Compact City. Again, it is intended to enhance the complexity of the urban fabric—and the synergistic opportunities it provides—through allowing for a richer mix of land uses and urban activities within localities, while alleviating socio-economic problems in the process. A higher degree of local self-sufficiency in precincts or suburbs thus transformed would make proximity-based interaction more relevant, resulting in travel savings and potentially the empowerment of communities. In contrast to both the Smart-Growing and the Compact City, however, the Dispersal-Pragmatic City focuses on the existing middle and outer suburbs as the primary laboratory of change, and is reluctant to accept the relevance of historically motivated re-urbanisation models for the majority of contemporary planning tasks. Its early proponents include a network of German researchers around Sieverts (1997, 1998) who coined the term of the 'Zwischenstadt', but there is now also discernible influence on planning policies in the Netherlands, for example (Priemus 1999; Bertolini 2005). Australian theory in this vein is present in the work of Gleeson (2006a, 2006b).

The Dispersal-Pragmatic City's stance on sustainability can aptly be classified as 'intermediate' (Zethoven 1990), because it aims to achieve sustainability goals using current market trends, economic pressures and lifestyle choices in the process of urban development. In fact, it derives much of its justification from the perceived failure of Compact City ideals to account for these fundamental conditions, and attempts to liberate the pathway towards sustainability from what is seen as 'ideological baggage', resulting from an excessive focus on the virtues of the pre-industrial city. In line with its emphasis on peripheries, where the boundaries of urban and rural dissolve, it is conversely propagated to break down the distinction between the built and the natural environment and work towards fine-grained integration of the two. A closer connection of productive activities in residential areas (such as energy generation and food cultivation) is likely to put a limit to the kind of densities seen as desirable in the Compact City.

There is, however, no clear picture of future urban form in the Dispersal-Pragmatic City. Rather, the model is open to include elements from any of the other models discussed here, which would then coalesce throughout a metropolitan region in a collage-like fashion. The importance of constantly improving and rebuilding existing urban fabric and—not least in the face of declining public funds for infrastructure investment—the need to rely on networks already in place are stressed.

The lack of imagery may work to the detriment of the Dispersal-Pragmatic City as a perceivable urban development model, while simultaneously offering sufficient scope to pursue and implement sustainability policies at a local level, at an appropriate pace and in collaboration between the set of stakeholders relevant in each case, without there having to be a metropolitan-wide 'grand plan' in place as a prerequisite for successful transformation. Dispersal-Pragmatic policies can therefore incorporate both 'carrots' and 'sticks' where appropriate, and produce a broad diversity of outcomes and approaches on the ground. Whether this actually happens, however, will largely depend on political decisions at all levels—particularly in (continental) Europe and Australia—where government control of the planning agenda remains relatively firm and where the extent of self-regulation in urban development that can be observed in most US metropolitan regions is unlikely to occur.

Making changes—urban sustainability in practice

Resource efficiency from a user perspective

Innovations in residential development with the aim of reducing resource use have taken two contrasting approaches to meet their objectives

Table 1: Overview of urban futures approaches

Urban futures approach	The decentralised city	The smart-growing city	The compact city	The dispersal-pragmatic city
Origin	North America, Australia	North America, Australia	Europe, Australia	Europe, Australia
Sustainability attitude	Weak	Intermediate	Strong	Intermediate
Settlement pattern	Dispersed Low densities Decentralised Coarse-grained Patchwork of monofunctional uses	Dispersed and compact Hierarchical densities Polycentric Fine-grained mix of land uses	Compact High densities Concentric and polycentric Maximum diversification of land uses	Dispersed and compact Varying densities Decentralised and polycentric Mix of land uses both coarse- and fine-grained
Rationale	Assumed market preference Greater affordability of land through reduced restrictions on supply Greater sub-regional autonomy and job-housing balance	Better local communities Mitigation of social problems Revival of local economy Enhanced travel choices Greater job–housing balance	Opportunities for synergy between uses Land savings Assumed travel savings Building on strengths of 2000 years of urban evolution	Assumed market pressure Urban reconstruction most needed in middle and outer suburbs Greater functional balance across region Enhanced local opportunities
Transport model	Car with technical improvements Flexible public transit supplementary	Fixed and flexible public transit Walking and cycling Taming of car, but no restriction	Fixed public transit Walking and cycling Car restrictions essential	Fixed and flexible public transit Walking and cycling Car restrictions where feasible
Policy Recommen-dations	Reduce planning interventions Empower individuals Regulate limited assets through pricing	Strengthen and create local Centres Apply growth boundaries Establish strategic public transit system Improve housing choices	Revitalise and increase density of urban and regional centres Apply growth boundaries and dismantle some sprawl Improve strategic public transit system Restrict car use and access	Take development pressure from centres Functionally enrich periphery Reconcile urbanised land and interspersed nature Build mainly on existing infrastructure

(Gestring *et al.* 1997). The technological approach attempts to equip buildings with appliances and construction techniques that minimise their demand for energy or water, and are designed to function independently from user perception or understanding of these technologies. The social–behavioural approach attempts to influence residents' awareness and behaviour to save resources through adjustments of their lifestyles, and to instigate technological innovations according to their own needs and capabilities. It has been argued that while the technological approach has appeal from an engineering point of view, and is more susceptible to government policies with their legal and financial tools, it is questionable whether the goals of a sustainability agenda taken seriously can be achieved entirely without a contribution of the users by changing some aspects of their everyday behaviour (Gestring *et al.* 1997).

This debate corresponds to the urban futures models described above insofar as the different approaches make varying reference to technological and behavioural aspects of sustainable housing. In the Decentralised City, there is broad confidence that resource efficiency will be achieved through behavioural changes and consumer preferences for less energy-intensive products and services, without providing explicit support for comprehensive technological solutions through public or private bodies with sufficient capitalisation and expertise to do so. Conversely, in the Smart-Growing City resource savings are induced by technology and urban form, but leave adjustments in behaviour almost entirely up to the users. The technological approach is also prevalent in the Compact City, but here substantial behavioural shifts are equally significant and anticipated through strong institutional, financial and physical discouragement of resource-intensive lifestyles. In the Dispersal-Pragmatic City, greater freedom exists towards applicable pathways for more resource-efficient housing because small, local steps can sometimes have a significant impact.

Communities and liveability

There is a long-standing tradition in community-oriented housing in European cities, which can be traced back to the first housing co-operatives and share-owned associations at the beginning of the 20th century. Many of these early innovations enabled residents to gain access to services and facilities, which would be out of reach financially or geographically unless supplied on a shared basis at the neighbourhood level. There is also strong potential for increased casual interaction and social activities in neighbourhoods designed around such shared spaces and facilities. This is in marked contrast to the post-war planning tradition, when growing household affluence and autonomy in tandem with a sense of mistrust in the benefits of community life—particularly in countries where this had been abused by totalitarian systems in the past—led to a high degree of individualism and alienation in the relations between the (private) home and the (public) city.

Urban life has always been characterised by the dichotomy of the competing needs to relieve the individual from routine household labour and to create a living environment that reflects and supports an individual's or a group's basic values and aspirations. As such, a residential neighbourhood and the surrounding city are both a 'machine'—offering services to its inhabitants' convenience and facilitation of everyday activities—and a 'home'—functioning as a target for self-expression, invigoration and creativity (Häußermann and Siebel 1987; Gestring *et al.* 1997). In the Decentralised City, the strong functional and spatial segregation between the 'home' concept, which is largely confined to a household's private domain, and the 'machine' concept, which starkly dominates the public arena, is upheld in the tradition of the modernist city. Talen (2005) maintains that this tradition generates a rationalistically determined public realm with little or no room for place-making or symbolic expressions of culture. In the Smart-Growing and the Compact City, 'home' and 'machine' functions are more interwoven and overlapping, allowing residents to move domestic labour and creative activities into the public realm and thus influence its form and character (if within certain confines of functional and legal rigidity). In the Dispersal-Pragmatic City, again, a range of possibilities is imaginable and offers important scope for experimentation.

Towards mobility-conscious lifestyles

While recent policies to improve energy-efficiency in buildings, and to appeal to their users' behaviour, have made it clear that resource use at home can be reduced dramatically, a similar effect is still far from being achieved in the field of personal mobility which, after all, is tightly connected to the location and accessibility of the home itself. In ecologically oriented housing developments without travel-demand management, energy use in transport commonly substantially exceeds energy use in the home (Scheurer 1998a) and becomes the most formidable challenge when promoting sustainable urban living. In addition, the accommodation of the car in residential environments poses significant costs in the provision of parking—particularly in higher-density situations—and often leads to unsatisfactory design solutions involving the space between buildings (Scheurer 1998b). Curbing excessive car use in favour of walking, cycling and the use of public transit—and fostering less distance-intensive activity patterns—does, however, require a fundamentally different policy approach than saving resources within the buildings. While pedestrian-friendly urban design and the existence of useful services and facilities within walking distance from homes can contribute to more balanced transport behaviour patterns, a strategy of mobility management that is to be successful on a neighbourhood level will need to approach the issue from the perspective of the users and their mobility needs.

In the Decentralised City and the Smart-Growing City, there is no serious intent to reduce the current extent of car ownership and use, but to merely mitigate its most damaging effects by technological and urban design measures—a stance that appears questionable and insufficient from a sustainability perspective (Apel *et al.* 1997). In the Compact City, substantial reductions in both car ownership and use are part of the model, but their implementation via democratic means—and in the absence of a trend-breaking urban mobility crisis (see earlier chapters by Dodson and Sipe and Newman)—remains uncertain. However, mobility management schemes in housing areas are beginning to provide possible solutions to this dilemma and could be a valuable contribution to tackle the problem in the Dispersal-Pragmatic City.

Conclusion

A more sustainable city is unlikely to materialise in the absence of a regime of powerful incentives and disincentives. However, attempts to implement a more resource-efficient urban form against current market forces and user needs are liable to jeopardise the process, as they meet formidable political, commercial and community resistance. Perspectives of future urban development will need to take into account both the need for regulation and the need for stakeholders' freedom to pursue—in a multilateral process of mediation—the most desirable solutions for all users of the urban environment.

While each model contains valuable elements to inspire the transition of cities towards greater sustainability, the Decentralised City, the Smart-Growing City and the Compact City include different barriers that constrain the forces of society from making the full transition. Provocatively worded, the Decentralised City negates the productive role of communities and coerces people into increasing household labour in the interest of urban ecology. The Smart-Growing City suggests to users that their behaviour is irrelevant for the sustainability of the city, or distrusts their ability and determination to adjust it. The Compact City ignores the economic, social and cultural trends that have led to the de-intensification of cities during the past 200 years.

At the same time, the Decentralised City illustrates the opportunities that exist for individuals to lead a more sustainable life. The Smart-Growing City demonstrates the synergistic effects that can come into place where urban form is reconfigured to reflect the needs of local communities, a balanced transport system and a vibrant local economy. The Compact City provides an inspirational and archetypical vision of what could be achievable if both elements—resource-saving urban form and resource-saving user behaviour—ideally coincided.

If the Dispersal-Pragmatic City is to emerge as an overarching model of urban future at the beginning of the new century, its value with regard to sustainability will be determined by the extent to which it can learn from these important messages of the preceding models, and to which it can incorporate successful elements into its collage-like vision of the city-region. The Dispersal-Pragmatic City's strength lies in its openness to pursue different paths simultaneously, as closely to the users' needs as possible, as long as—and only as long as—these continue to contribute to the wellbeing of the city-region as a whole, and do not fall victim to the temptation of promoting exclusivity and fragmentation. Its weakness lies in its current lack of imagery—for the time being, the physical and social visions from the Decentralised City, the Smart-Growing City and the Compact City will need to continue to fill this void.

References

Apel D, Lehmbrock M, Pharoah T and Thiemann-Linden J (1997). Kompakt, mobil, urban: Stadtentwicklungskonzepte zur Verkehrsvermeidung im internationalen Vergleich. Deutsches Institut für Urbanistik (difu), Berlin.

Benfield FK, Raimi MD and Chen DDT (1999). *Once There Were Greenfields. How Sprawl is Undermining America's Environment, Economy and Social Fabric.* National Resources Defence Council/Surface Transportation Policy Project, New York /Washington DC.

Bertolini L (2005). Cities and transport: exploring the need for new planning approaches. In *The Network Society. A New Context for Planning.* (Eds L Albrechts and SJ Mandelbaum) pp. 67–80. Routledge, Oxford, UK.

Briggs G (2005). The intelligent city: ubiquitous network or humane nnvironment? In *Future Forms and Design for Sustainable Cities* (Eds M Jenks and N Dempsey) pp. 31–53. Architectural Press, Oxford, UK.

Calthorpe P (1993). *The Next American Metropolis. Ecology, Community and the American Dream.* Princeton Architectural Press, Princeton, New Jersey.

Dodson J (2003). *Is there a Spatial Mismatch between Housing Affordability and Employment Opportunity in Melbourne?* State of Australian Cities Conference. University of Western Sydney, Parramatta, New South Wales.

Dodson J and Sipe N (2005). 'Oil vulnerability in the Australian city'. Research Paper 6. Griffith University, Urban Research Program, Brisbane.

Engwicht D (1999). *Street Reclaiming. Creating Liveable Streets and Vibrant Communities.* New Society Publishers, Annandale, New South Wales.

Fishman R (1977). *Urban Utopias in the 20th Century: Ebenezer Howard, Frank Lloyd Wright and Le Corbusier.* Basic Books, New York.

Florida R (2002). *The Rise of the Creative Class.* Pluto Press, North Melbourne, Victoria.

Gehl J (2001). *Life Between Buildings.* Using Public Space. Danish Architectural Press, Copenhagen.

Gestring N, Heine H, Mautz R, Mayer HN and Siebel W (1997). *Ökologie und urbane Lebensweise. Untersuchungen zu einem anscheinend unauflöslichen Widerspruch.* Vieweg, Braunschweig/Wiesbaden.

Gillman, O (2002). *The Limitless City: A Primer on the Urban Sprawl Debate.* Island Press, Washington DC.

Gleeson B (2006a). Towards a new Australian suburbanism. *Australian Planner* **43(1)**, 10–13.

Gleeson B (2006b). *Australian Heartlands. Making Space for Hope in the Suburbs.* Allen and Unwin, Sydney.

Gordon P and Richardson H W (1989). Gasoline consumption and cities. A reply. *Journal of the American Planning Association* **55(3)**, 342–346.

Gordon P and Richardson H W (1997). Are compact cities a desirable planning goal? *Journal of the American Planning Association* **63(1)**, 95–106

Hall P (2002). *Cities of Tomorrow.* 3rd Edn. Blackwell, Oxford, UK.

Hall P (2003) Smart growth on two continents. In *Urban Villages and the Making of Communities* (Ed. P Neal) pp. 33–47. Spon Press, London.

Häußermann H, Siebel W (1987). *Neue Urbanität.* Suhrkamp, Frankfurt (M), Germany.

Hesse M and Schmitz S (1998). Stadtentwicklung im Zeichen von 'Auflösung' und Nachhaltigkeit. In *Stadt – Landschaft.* (Ed. Bundesamt für Bauwesen und Raumordnung) pp. 435–453. Informationen zur Raumentwicklung, Bonn, Germany.

Hesse M (1999). Die Logik der kurzen Wege: Räumliche Mobilität und Verkehr als Gegenstand der Stadtforschung. *Erdkunde* **53**, 317–329.

Holz-Rau C (1997). *Siedlungsstrukturen und Verkehr.* Bundesforschungsanstalt für Landeskunde und Raumordnung (BFLR), Materialien zur Raumentwicklung Heft 84, Bonn, Germany.

Jacobs J (1961). *The Death and Life of Great American Cities.* Penguin, New York.

Jacobs M (1999). Sustainable development as a contested concept. In *Fairness and Futurity. Essays on Environmental Sustainability and Social Justice.* (Ed. A Dobson). pp. 21–45. Oxford University Press, Oxford, UK.

Jessen J (1997). Führt das städtebauliche Leitbild der kompakten und durchmischten Stadt zur Stadt der kurzen Wege? In *Die unaufhaltsame Auflösung der Stadt in die Region?* (Ed. M Bose) pp. 77–100. Harburger Berichte zur Stadtplanung Band 9, Hamburg, Germany.

Kahnert R (1998). Wirtschaftsentwicklung, Sub- und Desurbanisierung. In *Stadt – Landschaft.* (Ed. *Bundesamt für Bauwesen und Raumordnung*) pp. 509–520. Informationen zur Raumentwicklung, Bonn, Germany.

Landry C (2006). *The Art of City Making.* Earthscan, London.

Maher C (1994). Residential mobility, locational disadvantage and spatial inequality in Australian cities. *Urban Policy and Research* **12(3)**, 185–191.

Moran A and Staley L (2006). *Fixing the Crisis: A Fair Deal for Homebuyers in WA*. Institute for Public Affairs: Melbourne (Vic).

Morris W and Kaufman C (1996). *Mixed Use Developments. New Designs for New Livelihoods.* Department of Tourism, Small Business and Industry, Brisbane.

Newman P and Kenworthy J (1989). *Cities and Automobile Dependence. An International Sourcebook.* Gower, Aldershot, UK.

Newman P and Kenworthy J (1999). *Sustainability and Cities. Overcoming Automobile Dependence.* Island Press, Washington DC.

Priemus H (1999). Dutch experience with compact city policy and ABC location policy. In *Stadt der kurzen Wege. Zukunftsfähiges Leitbild oder planerische Utopie?* (Eds J Brunsing and M Frehn). Institut für Raumplanung (IRPUD), Dortmund, Germany.

Register R (2006). *Ecocities. Rebuilding Cities in Balance with Nature. Revised Edition.* New Society Publishers, Gabriola Island, British Columbia.

Scheurer J (1998a). *Evaluation of Danish Ecological Housing and Planning.* ISTP, Murdoch University, Perth and Forskningscentret for Skov, Landskab og Byplanlægning, Hørsholm, Denmark.

Scheurer J (1998b). Carfree housing in Europe. A new approach to sustainable residential development. *World Transport Policy and Practice* **4(3)**, 16–20.

Sieverts T (1997). *Zwischenstadt. Zwischen Ort und Welt, Raum und Zeit, Stadt und Land.* Vieweg, Braunschweig/Wiesbaden, Germany.

Sieverts T (1998). Die Stadt in der Zweiten Moderne, eine europäische Perspektive. In *Stadt – Landschaft.* (Ed. *Bundesamt für Bauwesen und Raumordnung*) pp. 455–473. Informationen zur Raumentwicklung, Bonn, Germany.

Talen E (2005) *New Urbanism and American Planning: The Conflict of Cultures.* Routledge, New York.

Troy P (1992). The new feudalism. *Urban Futures* **2(2)**, 36–44.

Troy P (1996). *The Perils of Urban Consolidation. A Discussion of Australian Housing and Urban Development Policies.* Federation Press, Sydney.

Troy P (2003). The virtues of suburbia. *Australian Financial Review*, 5 December.

WAPC (Western Australian Planning Commission) (2004). *Liveable Neighbourhoods. A Western Australian Government Sustainable Cities Initiative.* 3rd Edn, WAPC, Perth.

White WH (1988). *City. Rediscovering the Centre.* Doubleday, New York.

Zethoven I (1990). Sustainable development – a critique of perspectives. MEnvSt thesis, University of South Australia, Adelaide.

37

The Urban Environment and Health

Anthony McMichael

And so we find that this grand and wonderful defining artefact of civilisation is at best a glorious muddle in which each generation attempts to function efficiently with the product of ideas that were up to date a century before, but could have become obsolete or downright obstructive in the meantime. And, as if that were not enough, there are always some parts of the city's past that are deemed worthy of preserving for the future.

(Reader 2004, p. 305)

During the centuries before the modern Scientific Age in western society, the promotion of health and prevention of disease was not a central consideration in the development and administration of cities. In his famous work *The Republic*, Plato made his utopian proposal for an 'ideal' city, or, failing that, a 'just' city. Aside from discrimination against women, the non-propertied and slaves, these classical Greek aspirations for cities and their communities were largely framed in terms of politics, ethics and social relations. A little over a half-millennium later, imperial Rome had an increasingly gargantuan struggle to provide local shipping and storage facilities for feeding its swelling population, which necessitated importing 1000 tonnes of grain daily. Rome also needed massive sewage disposal facilities—achieved via their *cloacae*, which emptied indelicately into the River Tiber. In general, in those classical times, little attention was paid to issues and policies concerning urban environmental influences on health and disease.

Over the following centuries, food shortages, famines and epidemic disease outbreaks were regarded as being in the lap of the gods—indeed, often as the price exacted by Divine Retribution. Even so, and boosted by dramatic experiences such as the Black Death (bubonic plague) in 14th-century Europe, the idea of quarantine became widely adopted, albeit with little understanding of how it worked. However, this was part of civic administration, which became especially important in times of crisis and had little connection with urban planning.

Throughout most of history, cities have generally arisen via the growth of former rural, river-port or coastal towns, usually without much planning or restriction on land-use. This process continues today in much of the developing world, where urban growth and form is predominantly driven by land markets, speculation, automatic adoption of 'western' forms (roads and cars) and innovative entrepreneurs. The local accumulation of domestic and workplace wastes, and the pollution of air and waterways, are considered accidental side-effects of this unplanned growth.

The fledgling practice of epidemiological research into the causes of disease, and the associated emergence of the formal social agencies and processes of public health, had their origins early in the 19th century. The practice didn't mature to become mainstream until the rampant industrialisation and accelerating urbanisation in the mid-19th century. The crowding, squalor, deprivation and life-shortening characteristics of urban-industrial life could no longer be ignored—especially when epidemic diseases (such as cholera, 'consumption' (TB), measles and smallpox) threatened both rich and poor. The combination of malnutrition and recurrent infections weakened the poorly paid workers, exacerbated poverty and threatened the success of industrial capitalism.

Despite the risks of contagion in crowded settings, villages, towns and cities have been growing and evolving for the past ten millennia as populations have slowly become more urbanised. Many people are drawn to living in cities for diverse reasons—both voluntary and involuntary. Yet this is a relatively new environment for our human species. Within the long evolutionary timeframe, cities are cultural artefacts that have arisen very late in the evolution of our species. Not surprisingly, these artificial environments confer both benefits and risks to human wellbeing, health and survival.

Because cities are primarily manifestations of culture, this human inclination towards settled higher density communal living has a different basis from that of other social creatures, such as ants, bees and termites. The processes of natural selection have endowed those organisms with an in-built instinct for building their types of 'cities', which are perfectly suited to those organisms' biology, healthy functioning and survival. (For example, each of the giant termite mounds in northern Australia is the product of instinctive construction by millions of toiling insects. Elliptical in cross-section, the mounds are oriented side-on to minimise exposure to the midday sun, while the vertical ribbing maximises the cooling effect of passing breezes. By building these structures, the termites are merely doing the bidding of their genes.)

Humans, although clearly inclined towards social living, carry no genes specifically for city building. It is only through inspiration, chance, trial-and-error and technological innovations that we have found ways to live in settled communities, build houses and large public buildings, transport ourselves around our cities, import and distribute food and material needs, and dispose of our wastes.

The motivations for urban living are complex. They include security, commercial opportunity, wage-labour subsistence, cultural and aesthetic fulfilment, access to education and health care, and others. An important question arises: has a concern for better health played any part in the creation, planning and management of cities? Or have the health benefits and detriments to the population been incidental to urbanisation?

This question is now becoming urgent, as concern and discussion about *sustainability* gathers momentum. We must understand that population health is much more than an incidental consideration in the sustainability debate. A mature understanding of sustainability will recognise that its ultimate purpose is to improve and support human experience—especially health and survival. Wikipedia wisely defines sustainability as depending on a system of 'parallel care and respect for the ecosystem and the people within'.

Sustainable cities attuned to the fundamental need to sustain human wellbeing and health, should therefore embody urban environments that:

1. support healthy living, now and into the future. This would include ready and equitable access to good foods, encouragement of physical activity, social cohesion, a clean environment, minimised microbial transmission and aesthetic and cultural fulfilment
2. minimise the ecological footprint of cities, such that the world's natural environment remains health-supporting for present and future generations.

To better understand and foster this ecological perspective and thus attune urban policy and planning processes to human biological, psychological and social needs, we can learn from recent history. Firstly, though, it is important to clarify what is meant by 'the ecological perspective'.

Cities and health: thinking ecologically

Within popular culture, we usually think of health and disease at the personalised, individual level. However, the relationship between urban environments, city living and health needs to be viewed more broadly—it needs to be understood within a 'human ecology' framework. The urban environment exerts various systemic influences that affect the *rates* of disease in the urban population at large. While factors at the individual level (such as behaviours, genes and chance) influence which particular individuals get sick, the population's overall rate of disease reflects the properties of the shared physical, social and cultural environment: that is, the community's overall way of living.

Consider the contemporary and widespread problem of the rise of obesity, which is evident in Australia (Figure 1) as in most other countries around the world. The combined processes of industrialisation, urbanisation, modernisation and

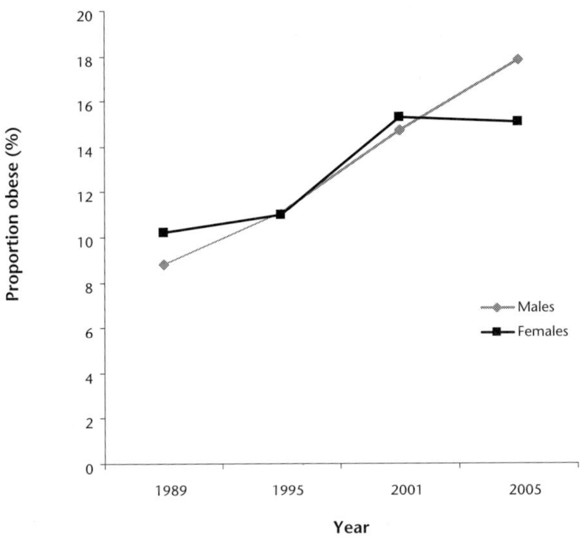

Figure 1: Time-trend in prevalence of obesity in Australian adults, by sex 1989–2005. Data (self-reported height and weight) are from the periodic National Health Survey (based on Dixon and Broom, 2007, p. 154). (Note: self-report data typically provide estimates approximately one-third lower than the proportion estimated by objective anthropometry.)

the rise of consumer culture have influenced both discretionary components of the energy balance equation: food energy intake and physical activity. While being overweight is not confined to urban populations, cities have been the engine room of social and technological change leading to an 'obesogenic' modern environment. The problem of obesity is primarily one of a systemic change in our way of living and the associated shifts in cultural values and behavioural priorities (Dixon and Broom 2007), rather than a consequence of defective individual behaviour (although the latter is the preferred neoliberal explanation).

A systems-based approach also enables us to view the various aspects and impacts of the urban transport system within a more integrated framework. This approach highlights the great diversity of direct and indirect health impacts from our transport-related behaviours (Figure 2). For urban planners and managers, it also makes clear just how pervasive are the influences—direct and indirect—of social decisions about transport systems and usage choices on human wellbeing and health.

This approach raises questions about how other systemic influences on health—the main 'urban health penalties'—have changed over recent time.

The changing profile of urban health penalties

Two hundred years ago during the early stages of industrialisation, the crowding, squalor, poverty and industrial environmental blight led to two great public health 'urban penalties'. Various infectious

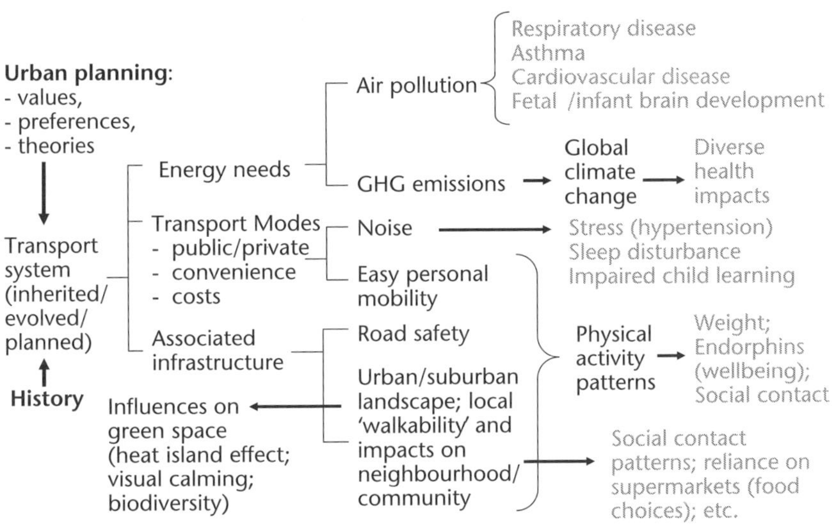

Figure 2: Urban transport—an integrated view of the determinants and the social and health impacts of transport modes in cities. The diagram suggests an ecological perspective, viewing transport and related behaviours as a 'system'. Outcomes on wellbeing and health are shown in grey (modified from McMichael 2007).

diseases became rife and were major killers. Meanwhile, in factory towns and cities, the air became black with smoke. By the mid-20th century, those two great health hazards of early urban-industrial life had largely been controlled in developed countries.

There is an analogous but unresolved situation today. Two of the greatest health-endangering aspects of urban environments and living are obesity (discussed above) and the increasing contribution of cities to greenhouse gas emissions and the attendant risks to safety, health and survival.

An important new dimension of these two modern health problems is that both extend well beyond the bounds of cities, and are affecting the population at large. For example, energy use in cities and the resultant greenhouse gas (GHG) emissions have serious consequences—via climate change—for humans everywhere. The resultant health risks include:

- heatwaves, especially in cities
- exacerbation of local air pollution
- mudslides endangering shanty towns
- intensified extreme weather events
- heightened transmission of temperature-sensitive infections.

The contribution of the world's cities to GHG emissions and the resultant climate change is now becoming a dominant component of the huge and ever-growing 'ecological footprint' of urban populations. This means that large cities have both an urgent responsibility and, because of density and economies of scale, a relatively cost-efficient opportunity to cut back on their massive GHG emissions. In particular, the cut-back of emissions is made easier by virtue of the high energy-density of modern cities, which facilitates the use of new energy systems and renewable energy. London's mayor, Ken Livingstone, has forged an energy strategy that will achieve a 20% reduction in emissions by 2010 compared with 1990 levels. In the USA, Seattle claims to have already achieved a 60% reduction. There are some relatively easy solutions for cities, one of which is methane capture from massive urban land-fills—as demonstrated by Toronto over the past decade.

Before considering in more detail how these systemic urban-environmental public health problems have arisen—and have been responded to—we should also note the apparent rise of two further contemporary 'urban penalties'.

First, infectious diseases pose an unexpectedly large, resurgent, health threat in modern cities. While the 'classical' water- and food-borne infections due to poor hygiene have receded, respiratory infections have the potential for rapid spread in population-dense settings. This was demonstrated by the urban-based outbreak and spread of SARS (Severe Acute Respiratory Syndrome) in 2003. Other aspects of urban culture, including sexual networking and illicit drug use, potentiate the spread of various infections, including HIV/AIDS and hepatitis C.

Second, there appears to be an increase (albeit still inadequately researched) in the prevalence of mental health disorders in urban populations, most notably depression. The connection is not straightforward. But cities are the centre of the consumer culture and its associated 'emptiness' of spirit—a culture reinforced in the urban setting via sophisticated and pervasive advertising and marketing and where shopping (on credit and by car) is easy. There is evidence that this urban consumer culture fosters dissatisfaction, alienation and mental health problems (Eckersley 2006).

The long-term trends in these major 'urban health penalties' are summarised semi-quantitatively in Figure 3. The main infrastructural changes and social policies that ameliorated these urban environmental risks to health are shown at the bottom of the figure.

Historical review: England and Australia over the past two centuries

To what extent have considerations of health risks have guided urban planning and management in modern times?

Infectious diseases: miasmas, germs and people

The need for domestic hygiene and public sanitation was increasingly recognised by European governments from the mid-19th century. Sanitary reforms and new infrastructure yielded health gains. Motives were mixed: they included personal self-protection, enhanced economic productivity and environmental amenity.

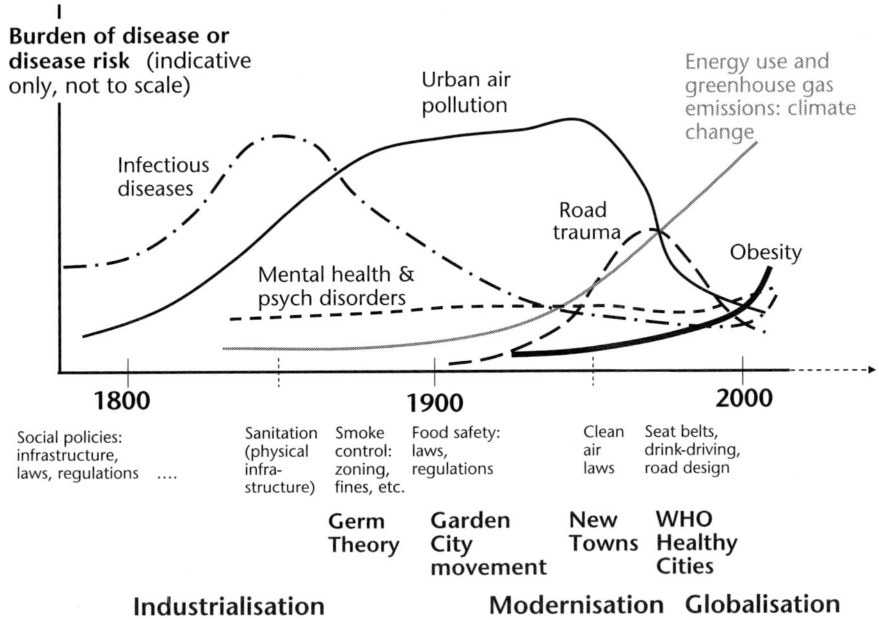

Figure 3: The rise and (for some) fall of major 'urban health penalties' (in today's developed countries) during the past two centuries. The main remedial responses and the influential emergent theories of urban design, wellbeing and health are shown at the bottom (modified from McMichael 2007).

The record is well documented in England. The miasma theory, which could seemingly explain why cholera and other diseases were epidemic in places where the water was undrained and foul-smelling, motivated the mid-19th century English sanitary reformers. In the 1850s, William Farr, the assistant commissioner for the 1851 London census, implicated miasmas in the spread of cholera in London. His extensive statistical analyses for London's major suburban regions showed a close inverse correlation between the cholera death rate and height above sea level—reflecting, he believed, the menace of the deadly miasmas that arose from the fetid Thames River. More generally, the miasma theory stimulated improvements in municipal sanitation, along with other important public health reforms and the beginnings of environmental hygiene. The ensuing reduction in cholera in the later decades of the century in England appeared to corroborate the theory.

Edwin Chadwick's celebrated 1842 *Report on the Sanitary Conditions of the Labouring Population of Great Britain* (Finer 1952) suggested that local atmospheric miasmas—arising from putrefaction, filth and excreta—were the cause of the 'endemic and contagious diseases' that were rife within the poorer crowded sections of the urban population of London. The Public Health Act of 1848 flowed from Chadwick's report, which resulted in the establishment of the General Board of Health, and gave local boards of health the power to install and improve local sewage and sanitation. In most towns, the supply of water was inadequate—often limited to a public stand-pipe that flowed intermittently. Open sewers, which were originally land drains, usually dried out in summer; while closed sewers were poorly designed and often collapsed or became blocked. Chadwick's initiatives were the main driving force for the formal legislation-based sanitation—a movement that was reinforced when the Public Health Act was strengthened in 1875.

While this legislative initiative may seem to have viewed good health as an important goal for society, Chadwick's stronger motivation was utilitarian. (Indeed, the young Chadwick had previously been secretary to Jeremy Bentham, who was the pioneering proponent of utilitarianism. Under this philosophy, it was desirable to find ways of reducing poverty and disadvantage, and thus maximise social harmony and happiness. Thomas Malthus, in contrast, had famously argued several decades earlier that poverty was inevitable and that it was a major source of 'checks'—via disease and early death—on the growth of the human population.) In this age of rapid industrialisation, a healthy workforce was needed. The exacerbation of chronic poverty by illness and incarceration in dismal crowded work-houses (ironically a product of Chadwick's 1830s administration of the Poor Law) was an impediment to factory output.

In the final quarter of the 19th century, the germ theory was propounded, which was based on the pioneering work of Louis Pasteur and Robert Koch. 'Contagious' diseases were caused by 'germs', via a process of person-to-person 'infection'—they did not arise *sui generis* from miasmatic emanations. Though guided by the now superseded (miasmatic) theory, Chadwick had—in strict biomedical terms—done the right thing for largely the wrong reason (although there is no doubt that unhygienic squalid and damp conditions are breeding grounds for microbes). In terms of the main question in this chapter, Chadwick had initiated major change in urban infrastructure explicitly on behalf of the health of the urban population.

The 19th-century Australian experience

In Australia in the late 19th century, urban epidemics of diarrhoeal and respiratory diseases persisted. The stench of Melbourne's inner suburbs was dire. In 1876, after two bad years of epidemics of measles and assorted streptococcal infections, Melbourne's Board of Health reported:

The contagium which causes epidemics is the offspring of insanitary habits, and is nurtured and spread by the impure air of unventilated houses. ... In a great many parts, the houses are not only damp, but impregnated with poisonous gas from pent-up sewage. (McCalman 2005)

Miasma theory was evidently central to public health thinking in Melbourne at that time and helped to propel the introduction of municipal sewerage. This almost certainly contributed to the marked fall in infant mortality that began around 1890—and had halved rates of death in Victoria overall by around 1920 (Cumpston 1989).

Meanwhile, in Europe the ideas of Pasteur and Koch were replacing miasmatic thinking with the specific concepts of the germ theory. This nurtured a significant shift in public health thinking, bringing a new focus on specific disease agents and the possibilities for prevention at individual and family levels.

Cities as 'dark satanic mills': coal combustion, particulates and health

During those infection-blighted decades of early sanitary reform in Europe, another great environmental and public health problem was the escalating concentration of black smoke and respirable particulates in urban-industrial air.

At lower levels, coal smoke had been a longstanding environmental problem, as evident in the historical records. The burning of coal in England dates from at least one thousand years ago. During the High Middle Ages and Renaissance centuries, coal smoke in London was increasingly regarded as a nuisance, and royal edicts were issued in late 16th century to curb the burning of coal. Queen Elizabeth I complained that coal smoke caused her grievous annoyance. In 1661, the famous diarist, John Evelyn, published *Fumifugium, or the Inconvenience of Aer and the Smoake of London Dissipated* (Evelyn MDCLXI). This colourfully worded tract highlighted the link with adverse health outcomes. Of Londoners, Evelyn wrote:

... that her inhabitants breathe nothing but an impure and thick Mist, accompanied with a fuliginous and filthy vapour ... corrupting the Lungs, and disordering the entire habit of their Bodies; so that Catharrs, Phthisicks, Coughs and Consumption rage more in this one City than in the whole Earth besides.

Regrettably, little notice was taken of Evelyn's critical words. A hundred years later, a group of concerned physicians and scientists re-published his text, but, again, there was little impact on public concern and policy. Epidemiological research did not yet exist, and ideas about the causes of disease were very different from modern understanding. For example, upper respiratory tract infections were often cured by dangling a 'frog in the throat'.

In the mid-19th century, rapidly-industrialising Britain encountered the pall of William Blake's 'dark satanic mills' in the industrial midlands cities. This dense smoke pollution in many cities posed a different sort of challenge. While sewage and sanitation was a problem of infrastructure, which government had to tackle, smoke was politically more difficult to deal with because of its association with desired economic expansion (Kessel 2005). Today, a stroll in many of China's big cities underscores this point.

Official awareness of the health risks from air pollution emerged in the 1840s. In 1850, London's Medical Officer of Health, John Simon, petitioned the Commissioners of Sewers for the abatement of London's smoke. The Smoke Acts of the 1850s were targeted specifically at London—empowering the police to enforce provisions against factories, furnaces, public baths, and steam-boats on the Thames. Enforcement was to be by the 'best practicable means'—rather elastic wording that has remained the Achilles heel of much modern environmental legislation.

It was to take another hundred years before serious attention was paid to controlling rampant urban air pollution and its often dramatic health impacts. That turnabout in public thinking and official policy finally occurred in the wake of the notorious London Smog of 1952, which caused an excess of 4000 deaths immediately and an estimated 8000 additional deaths in the ensuing three months. The long-overdue *Clean Air Act* was passed in the late 1950s. Similar legislation was enacted during the 1950s, '60s and '70s in most other industrialised countries.

Note, however, that the building of sewers, the supply of reticulated disinfected drinking water, and the attempted curtailment of smoke did not in themselves have much affect on the general planning and layout of cities and suburbs. The miasma theory, however, implanted a resilient belief in the health-giving properties of open space and clean air. This, as Newman and Kenworthy (1989) have argued, turned out to be mixed blessing: a legacy for town-planners that nurtured suburban spread and its eventual dependence on private vehicle transport.

The story continues

The story of urban planning and reform in relation to considerations of health continued to evolve in Britain. In 1875, Benjamin Ward Richardson—an English physician-sanitarian, polymath and social reformer—proposed his vision of 'Hygiea'. His proposal was for medium-density cities of 100,000 persons, green spaces, good transport system, and clean air and water.

In the 1890s, Ebenezer Howard proposed the building of health-supporting 'garden cities', for populations of approximately 32,000. Howard prescribed a quite specific layout for these carefully planned cities, with concentric layering of commercial, green space, market-garden and residential areas. Garden cities were to be self-contained communities, surrounded by green belts, and with efficient public railway transport around the periphery, including interconnections to nearby garden cities.

In 1898 Ebenezer Howard published *To-Morrow: a Peaceful Path to Real Reform* (this was reissued in 1902 as *Garden Cities of To-Morrow*) (Howard 1898). In the following year, he organised the Garden City Association, and went on to found two garden cities in England: Letchworth in 1903, and Welwyn Garden City in 1920. Both designs—though not full realisations of Howard's ideals—have proved to be durable successes, leading to stable and cohesive communities today. Howard's garden city idea was influential in the United States (for example, Radburn, NJ) and in Germany, where it influenced the worker housing built in the Weimar years.

In Australia, the design of the new national capital, Canberra, was heavily influenced by Howard's garden city movement. The American architect, Walter Burley Griffin, drew on Howard's ideas in his innovative planning during 1910–11. The city of Canberra incorporates significant areas of natural vegetation that have led to it being called the 'bush capital'.

The garden city concept was revived again in England after World War II, when the New Towns Act (1946) triggered the development of many new communities based on Howard's egalitarian and health-promoting vision. Also crucial to this New Towns thinking was the Abercrombie Plan for London (1944), which envisaged moving 1.5 million people from London to new and expanded towns. These were to be satellite cities outside the large metropolis, echoing of some of Howard's garden city ideas. During the 1940s there was a strong consensus in England in favour of halting the uncontrolled sprawl of London and other large cities. The assumption was that *if we can build better, we can live better.* On the other side of the world, South Australia responded in the 1950s and built Elizabeth—a satellite city to the north of Adelaide— which was opened by the young Queen Elizabeth II in 1954.

But have garden cities been successful? It is evident that this could not easily be said of Elizabeth? Canberra, too, has long had its doubters and detractors. The 'build better to live better' rhetoric can nurture an uncritical assumption that the right physical layout will, of itself, promote health. Reality is more complex: social environments, population mix, economics, history, good planning and inspired local leadership are all important.

The linkages between urban planning, urban living and health were also being paid increasing attention in North America. In early twentieth century a 'healthy cities' movement evolved in North America (somewhat inspired by the Hygiea idea). In the 1920s, Toronto chlorinated its drinking water, improved sanitation, promoted family hygiene and pasteurised the city's milk supply.

Subsequently, in the 1970s to 80s, a new emphasis on health-related characteristics of the world's burgeoning cities emerged. This resonated with the rise of community-oriented values in health promotion, as were enshrined in WHO's Ottawa Charter (1985): the need for greater engagement, consultation and empowerment. Recognition of the health implications of rapid urban population growth, associated changes in family and social relations, the environmental impacts of road-based transport, and other features, stimulated WHO to launch the international 'Healthy Cities' program. Drawing inspiration from WHO's 'Ottawa Charter' (Health Promotion), this program sought to empower urban communities and engage them in shaping their social and physical environments, in the interests of health. This program has provided frameworks and practical guides for governments and city planners to promote the development of healthy urban environments (Hancock 1996). More generally, this program has stimulated development of theories about multiple and interacting determinants of health-theories that are more accessible to policy-makers, planners and implementers.

Today there are still concerns about the burgeoning size of major cities, the health impacts of outer-suburban isolation, the escalating costs of extended infrastructure support, the future of urban transport and the magnitude and reach of a typical city's environmental impacts (via its energy use, waste generation and greenhouse gas emissions). Certainly, cities can achieve efficiencies of both density and scale – but that may be little more than an offset against the increasing level of consumption stimulated by modern urbanism.

While these features of our modern built environment all entail risks to human wellbeing and health (both now and into the future), so far the public discourse about sustainable cities pays relatively little attention to the criterion of health. The recent rise in rates of chronic diseases (such as cardiovascular disease, diabetes, and cancer) and of (now widely remarked) obesity—along with growing recognition of health risks from global climate change—may soon correct that way of thinking.

'Sustainable cities'

Australia has recently had a federal parliamentary enquiry into *Sustainable Cities 2025* (House of Representatives Standing Committee on Environment and Heritage 2005). The real index of sustainability should be the quality of human experience and is durability across generations. The widely invoked 'triple bottom-line accounting' (comprising indices of economic activity, environment and social conditions) actually refers to *intermediate* markers—markers of the conditions that determine human experience. Therefore, population health must now be recognised as a key criterion of sustainability, and be brought into the mainstream of the planning and management of our cities.

Our current pre-occupation with 'sustaining' the economic engine, creating wealth, maintaining population growth, enhancing tourism, extending road networks for a growing car fleet, and so on, is missing the real point of sustainability. Urban planning, sanitation and public health were closely entwined in the early days of the industrial revolution, when infectious disease epidemics depleted and weakened the workforce. That link between planning and health has weakened in recent times, particularly against a neoliberal backdrop of expec-

Urban air pollution and health – the need for continuing, classical, epidemiological vigilance

Alongside the need for a wider application of an ecological perspective in how we think about and study the influence of urban environments on patterns of health and disease, important research questions remain about specific relationships between urban environmental exposures and health. Those questions remain the focus of mainstream reductionist epidemiological research.

This continuing need for focused, specific, epidemiological research is well illustrated by the widening range of recognised health risks from urban air pollution from transport and industrial sources. Let's consider two examples. First, beyond the serious non-communicable (mostly cardio-respiratory) disease consequences of air pollution, which cause an estimated 800,000 premature deaths annually in cities around the world (Prüss-Üstün and Corvalán 2006), there is now a better understanding of the role of various air pollutants in triggering asthma attacks (especially childhood asthma). A major pan-European study has reported that rates of asthma attacks in children and in adults increase by around 3% for each 25 micrograms/m3 increase of respirable PM10 particulates (Atkinson et al. 2001). Second, a recent study in New York City (Perera et al. 2006) has shown that children whose mothers were exposed to the upper quartile of polycyclic aromatic hydrocarbons (PAH) concentration during pregnancy were at almost three-fold greater risk of cognitive developmental delay at age 3 years than were children whose maternal exposure was in the lowest quartile.

tation that individuals, not communities-at-large, are responsible for the pursuit of good health.

The real measure of sustainability is the quality of human experience and the capacity to maintain it at a high level. Therefore, we should be wary of conventional 'triple bottom-line accounting', since indices of economic activity, environment and social conditions are only intermediate markers that determine consequent levels of human opportunity, security, wellbeing and health. They are means, not ends. We should therefore be thinking creatively about how to mainstream population health into the planning, modification and management of this century's overwhelmingly dominant human habitat, and develop correspondingly appropriate guidelines and indicators for monitoring the sustainability of cities.

There are some encouraging signs. Early in 2006, several hundred US city mayors formulated the US Mayors Climate Protection Agreement. Representing around one-fifth of the national population, these mayors committed their cities to meet, or beat, the US emissions reduction target in the Kyoto Protocol—despite their national government's refusal to ratify that treaty. Many of the participating cities are those with explicit awareness and concern about the physical risks posed to their communities via cyclones, rising seas and heatwaves. The generally shared strategies included more efficient generation and use of electricity, greater reliance on renewable energy sources, increased motor vehicle efficiency, improved public transportation systems, curbing suburban sprawl and encouraging walking and cycling. This is 'cities-as-habitat' and 'cities-with-responsibilities-to-future-generations' thinking—in short, such thinking can truly tackle 'sustainability'.

The role of researchers

If health has not been accorded as much priority in urban planning as public health proponents would wish, then perhaps there have been deficiencies in how urban environmental health research has been conceptualised, carried out and communicated.

There is a long tradition in epidemiology—reaching back to mid-19th century (and in England especially)—of comparing indices of health in urban and rural populations, and between cities, and between socioeconomic sub-populations. During the early industrial period, health and life expectancy were lower in cities than in the countryside. In England, there were great differences between Manchester, Liverpool and London. Around 1840, there was around a two-fold difference in life expectancy between labourers and gentry. Edwin Chadwick (discussed earlier) showed that average life expectancy in agricultural Rutland was 52 years for a gentleman and 38 years for working man. In the towns, the corresponding figures were 38 years and (shockingly) 17 years, respectively (Finer 1952). Furthermore, time-series data showed that death rates were again rising after a century-long decline up to the early 1830s. Chadwick concluded that as urban populations had increased, their sanitary infrastructure could no longer cope.

In 1845, Friedrich Engels described Manchester's factory-working masses as 'pale, lank, narrow-chested, hollow-eyed ghosts' afflicted with rickets and scrofula, and subject to early death from consumption (tuberculosis), typhus and scarlet fever. His intuitive, basic, epidemiological research revealed that, within the working-class quarter of Manchester, death rates were twice as high in the worst houses in the worst streets as in the best houses in the best streets (Engels 1845/1987). Nevertheless, most of the above-mentioned descriptive comparisons failed to seek explanatory factors—and, even less, to seek understanding of the underlying social, cultural and economic contextual influences on these factors

Over the past half-century, a rather superficial, often itemised, approach has prevailed. This reflects—in the words of Sandro Galea and colleagues (Galea *et al.* 2005)—'both the number and complexity of determinants and of the absence of a unified framework that integrates the multiple factors that influence the health of urban populations.' Those authors point out that multiple factors encompass not only local community initiatives (such as neighbourhood watch schemes), but municipal facilities and policies, national trends and governmental policies, and the increasingly pervasive consequences of globalisation. They describe as example, how food affects the health of urban populations:

Food availability depends on international distribution systems, tariffs, and trade routes. Global trends in liberalisation or restriction of trade then affect movement of food between countries and markets. Within cities, efficacy of food inspection mechanisms depends to a large extent on municipal policies and resources made available for regular food inspections. In turn, food intake has been shown to differ between neighbourhoods within cities. Therefore, factors at multiple levels all ultimately affect food choices in cities and contribute to health or disease. Researchers studying the issue of how food affects health in urban areas can consider global and national levels, municipal levels, and inter-neighbourhood variability, as they are related to the risk for heart disease, cancer, obesity, or chronic hunger within cities. Findings can guide national or state food policies, municipal-level interventions, or community campaigns.

Conclusion

Humans are social animals: we seek comfort, security, variety and opportunity. We are attracted to settling in villages, towns and cities. Worldwide, as cities proliferate, we have now become a predominantly urban species. We must now learn to shape and manage the urban environment to accord with the needs of human biology and of the ecosphere (McMichael 2001).

In early 19th century, health problems were conceptualised and tackled at the population/community level. The germ theory redirected attention towards specific individual-level factors. Much of modern epidemiological research has continued in that individualist vein—focusing on personalised behaviours and circumstances that account for why some people have heart attacks or cancer and others don't. Today, we are again recognising the fundamental role of environmental systems and processes as *ecological* determinants of population health.

Our growing awareness that health risks to whole populations arise from changes in ways of living, in cultural priorities, and from the ever-widening impacts of humans on environmental assets and systems (including the climate system) underscores the urgent need to understand that population health is the central criterion of 'sustainability'. Our prime, anthropocentric, reason for seeking social stability—a congenial and safe urban environment—and the maintenance of nature's life-support systems is to ensure the protection and improvement of human wellbeing, health and survival.

References

Atkinson RW, Ross Anderson H, Sunyer J, Ayres JON, Baccini M, Vonk JM, Boumghar A, Forastiere F, Forsberg B, Touloumi G, Schwartz J and Katsouyanni K (2001). Acute effects of particulate air pollution on respiratory admissions. Results from aphea 2 project. *American Journal of Respiratory and Critical Care Medicine* **164(10)**, 1860–1866.

Cumpston J (Ed.) (1989). *Health and disease in Australia*. AGPS, Canberra.

Dixon J and Broom DH (2007). *The 7 Deadly Sins of Obesity. How the Modern World is Making us Fat.* University of New South Wales Press, Sydney.

Eckersley R. (2006). Is modern western culture a health hazard? *International Journal of Epidemiology* **35**, 252–58.

Engels F (1987). *The Conditions of the Working Class in England [1845]*. Penguin, London.

Evelyn J (1661). *Fumifugium, or the Inconvenience of Aer and the Smoake of London Dissipated.* W Godbid for Gabriel Bedel and Thomas Collins, London.

Finer SE (1952). *The life and times of Sir Edwin Chadwick*. Methuen and Co., London.

Galea S, Freudenberg N and Vlahov D (2005). Cities and population health. *Social Science and Medicine* **60**, 1017–1033.

Hancock T (1996). Planning and creating healthy and sustainable cities: The challenge for the 21st century. In *Our cities, our future*. (Eds C Price and A Tsouros). WHO Healthy Cities Project Office, Copenhagen.

House of Representatives Standing Committee on Environment and Heritage (2005). *Sustainable cities 2025*. Government of Australia, Canberra.

Howard E (1898). *To-morrow: A peaceful path to real reform*. Reissued 2003. Routledge, London.

Kessel A (2005). *Air and public health*. Cambridge University Press, Cambridge.

McCalman J (2005). 'All just melted with heat': mothers, babies and 'hot winds' in colonial Melbourne. In *A Change in the Weather*. (Eds T Sherratt, T Griffiths and L Robin) pp. 104–115 National Museum Australia Press, Canberra.

McMichael AJ (2001). *Human frontiers, environments and disease. Past patterns, uncertain futures.* Cambridge University Press, Cambridge.

McMichael AJ (2007). Will consideration of environmental sustainability revitalise the policy links between urban environment and health? *NSW Public Health Bulletin* **18(4)**, 41–45.

Newman PWG and Kenworthy JR (1989). *Cities and automobile dependence*. Gower, Aldershot, UK.

Perera FP, Rauh V, Whyatt RM, Tsai WY, Tang D, Diaz D, Hoepner L, Barr D, Hsuan Tu YH, Camann D and Kinney P (2006). Effect of prenatal exposure to airborne polycyclic aromatic hydrocarbons on neurodevelopment in the first three years of life among inner-city children. *Environmental Health Perspectives* **114(8)**, 1287–1292.

Prüss-Üstün A and Corvalán C (2006). *Preventing disease through healthy environments. Towards an estimate of the environmental burden of disease.* World Health Organisation, Geneva.

Reader A (2004). *Cities*. Heinemann, London.

38

Smart Development: Designing the Built Environment for Improved Access and Health Outcomes

Ryan Falconer and Billie Giles-Corti

Deficiencies in conventional land use and transport planning

Land use and transport policy in Australia's major cities have been modelled on the US, and are characterised by low-density housing development and separation of land uses. Until the end of Second World War, there was little car ownership, urban areas were relatively concentrated and public transport was the principal mode of transportation (Laird and Newman 2001). In Melbourne, for example, only one in every four households owned a car at the end of the war'. Moreover, streets tended to be multi-functional: they were play spaces for children and places for public transport, walking and cycling, as well as for cars (Davison 2004).

In the period immediately following the Second World War, a number of significant events signalled a change in city planning. First, there was a rapid growth in suburbia when the War Service Homes Scheme financed land purchases and detached dwellings for returning servicemen (Gleeson 2006). Second, the US model for urban growth, which involved separating land uses, building to low densities and investing in roads was replicated in Australia (Newman and Kenworthy 1999; Gillham 2002; Frumkin *et al.* 2004). Third, innovation and increasing competition in the motor vehicle sector, combined with mass marketing, helped to sell the virtues and reduce the price of private motor cars, making them more accessible to the less affluent. The result has been a long-term

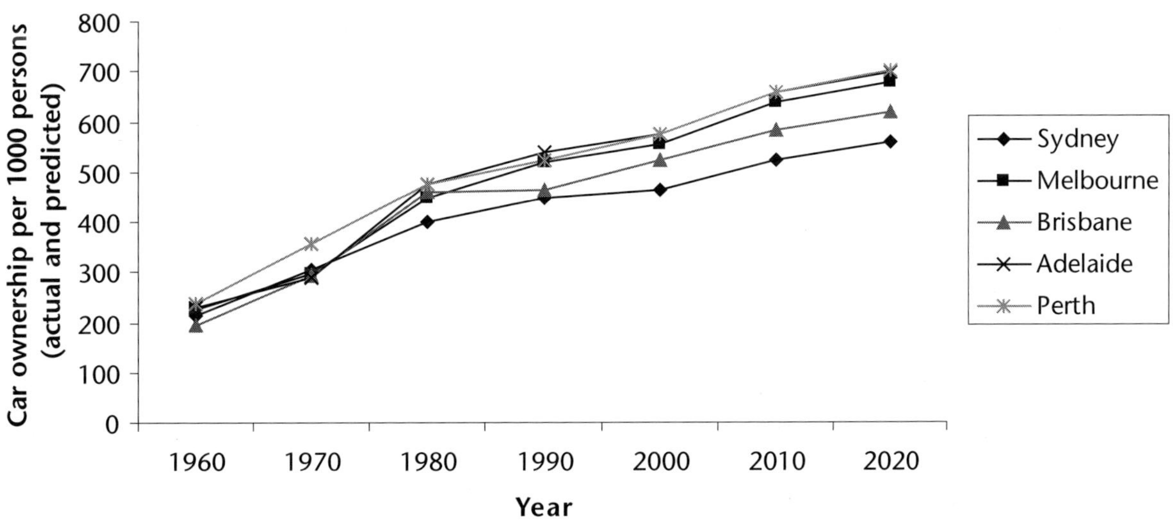

Figure 1: Historic and projected vehicle ownership in major Australian cities (Source: Kenworthy and Laube 1999; Australian Bureau of Statistics 2001; Cameron 2004. Data for 1961, 1971, 1981 and 1991 have been used as proxy for 1960, 1970, 1980 and 1990 respectively, to ensure continuity.)

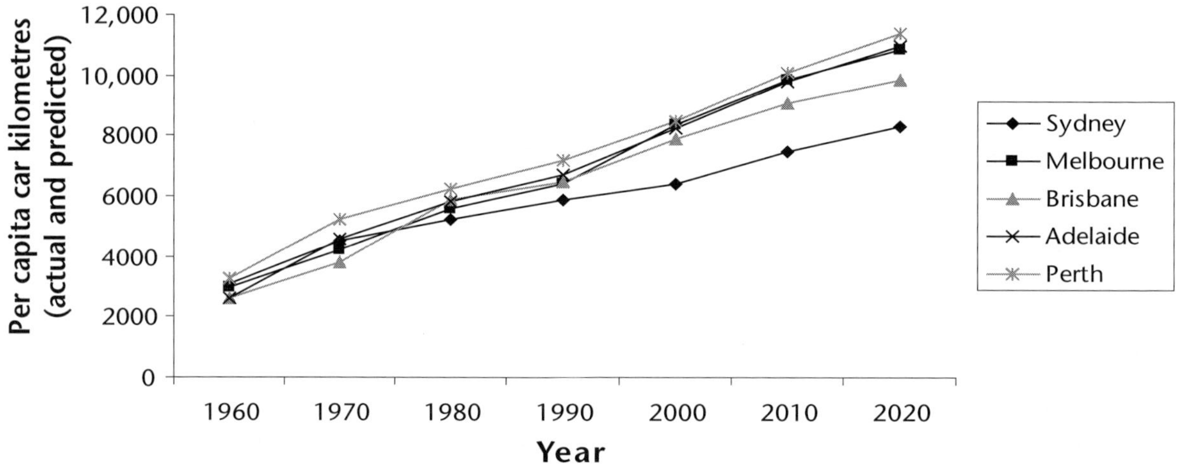

Figure 2: Upward trend in per capita car kilometres (per annum) for major Australian cities. (Source: Kenworthy and Laube 1999; Australian Bureau of Statistics 2001; Cameron 2004.)

upward growth in vehicle ownership. In Perth in 1961 there were 239 passenger cars per 1000 people, rising to 357 per 1000 in 1971, 475 per 1000 in 1981 and 523 per 1000 in 1991 (Kenworthy and Laube 1999). Nowadays, estimates of ownership range from 630 cars per 1,000 (Cameron 2004) to 679 per 1000 persons (Ashton-Graham *et al.* 2005). A similar trend is evident for other Australian cities (Newman and Kenworthy 1989; Newman and Kenworthy 1999).

While at some point the market may become saturated with vehicles, it has been argued that this 'saturation point' is nowhere in sight, with vehicle ownership projected to continue rising (Cameron 2004). Combining data from Kenworthy and Laube (Kenworthy and Laube 1999), the Australian Bureau of Statistics (Australian Bureau of Statistics 2001) and Cameron (Cameron 2004), Figures 1 and 2 shows projections for vehicle ownership in five Australian cities and the concurrent rise in annual per capita car kilometres, respectively.

As private motor vehicles became more affordable and more widely used, it also became necessary to facilitate motor vehicle mobility through land-use planning. These planning policies were linked to an increase in road infrastructure to support motor vehicle travel, including drive-in facilities, franchised petrol stations and extensive car parking (Davison 2004).

The rapid uptake of private motor vehicles offered people unprecedented mobility enabling them to live further away from their places of work and from services. Residing within a short distance of essential facilities and places of work became unnecessary, as did proximity to public transport. The form of residential development that resulted increased dependence on cars for travel (Newman and Kenworthy 1989).

Car dependence has also been perpetuated by government expenditure, which has favoured development of the road network at the expense of public transportation infrastructure—once again showing parallels with the US experience (Davison 2004). In 1969, for example, the State Government of Victoria conceived Melbourne's Metropolitan Transportation Plan, which allocated A$1.675 billion of a A$2.6 billion transport funding package for a 494-kilometre urban freeway system. The key aim of the Plan was to facilitate commutes between decentralised residential areas and decentralised places of employment. By comparison, only A$0.355 billion was allocated to public transport (Davison 2004). Similarly, from the 1970s to the 1990s, the Australian Federal Government took on national responsibility for developing roads and interstate highways, with the aim of providing more seamless motor vehicle travel. Road networks were considered to be the modern transport framework (Newman 2005).

The preoccupation with providing infrastructure to make commutes fast and more convenient may partly explain community preferences for private motor vehicle travel. The convenience and quality of infrastructure also encourages additional (and often unnecessary) car trips (Bachels and Newman 2001). Motor vehicle affordability, combined with increasing affluence and the provision of supportive infrastructure, has increased the opportunities (and the expectation) to drive (Cameron *et al.* 2004).

Recent evidence provides insights into the extent to which the motor car has become the preferred mode for urban travel. In 2005, the Planning and Transport Research Centre (PATREC) estimated that around 4.8 million trips were made daily in the Perth metropolitan region, of which 83% were made by car, as either driver or passenger (PATREC 2005). The strong preference for private motor vehicle travel in Perth is highlighted in the 2001 data, which suggested that, on average, each Perth resident made about 803 trips by car per annum, compared with only 159 trips by foot, 65 by public transport and 32 by bicycle (James *et al.* 2001).

However, approximately 250,000 of the daily car trips are 1 kilometre or less (Premier's Physical Activity Taskforce 2006). Based on these estimates, around 5% of the 4.8 million daily car trips made in Perth could potentially be substituted by an active mode, such as walking or cycling. While some short trips would involve the movement of heavy goods, such as groceries, or goods that would be awkward for pedestrians or cyclists to carry, a significant proportion could undoubtedly be substituted for an active mode. Giles-Corti (2006) estimated that even if one quarter of these daily short trips were converted to a 10 minute walk, it would increase daily walking in Perth by some 625,000 minutes of walking, which could have a potentially significant health benefit at a population level.

But why choose to drive over using active modes? In a society that has favoured planning for the motor vehicle at the expense of alternative modes, motor vehicle use has become so habitual and convenient that driving is as much a part of the Australian culture as it is in the US (Bachels and Newman 2001; Rees 2003). This has become so entrenched that driving may be now seen as 'necessary' even when there are alternatives available (Handy *et al.* 2005b). There is evidence that people drive more than even *they* think they *need* to (Handy 2003; Handy *et al.* 2005a) and changing these habits may prove challenging (Frank *et al.* 2004).

For decades, the dependence on cars for daily travel needs and the preference to drive, have generated concerns about the sustainability of Australia's form of urban development. However, more recently there have been growing concerns about how decisions about land use, transportation systems and urban design affect the population's health. The section that follows considers the public health implications of sprawl and car dependence.

Sprawl and car dependence: implications for public health

It is now recognised that urban sprawl and car dependence harms human health by decreasing physical activity—in particular walking—and by increasing levels of obesity and respiratory problems (Giles-Corti *et al.* unpublished). It has been found that the time spent driving is related to levels of obesity by increasing sedentary behaviour and decreasing time available to be active (Frank *et al.* 2004). A growing body of evidence suggests that aspects of the built environment, such as residential density, street connectivity, mixed use planning and neighbourhood design, influence levels of walking, cycling and public transport use (Transportation Research Board 2005).

So are these trends important? Physical inactivity is second only to tobacco smoking as a behavioural risk factor in terms of its contribution to the burden of disease in Australia (Begg *et al.* 2003). Physical inactivity is an independent risk factor for major chronic diseases, including cardiovascular disease, diabetes, poor mental health and some cancers (US Department of Health and Human Services 1996) A recent Access Economics report estimated that the direct and indirect cost of obesity in Australia is around $21 billion annually (Access Economics 2006), while the direct cost of physical inactivity is $377 million (Stephenson *et al.* 2000).

Just under half of adult Australians undertake insufficient recreational and trip-related physical activity to benefit their health (Armstrong *et al.* 2000)—that is, as little as 30 minutes of moderate physical activity (even walking) on 5 or more days per week. Alarmingly, between 1997 and 1999, the proportion of Australians who reported being completely inactive increased from 13% to 15%, while the proportion of those sufficiently active (that is, 30 minutes on 5 or more days/week) declined from 62% to 57% (Armstrong *et al.* 2000).

Increasing levels of inactivity and obesity are of concern globally, For more than a decade, governments around the world have become increasing concerned about an unrelenting increase in the proportion of adults and children who are either overweight or obese (World Health Organization 2000). In Australia, two-thirds of men and half of women are overweight or obese (National Health and Medical Research Council 1997) as are 19%–23% of the Australian children and adolescents (Booth *et al.* 2001). In Western Australia, the proportion of overweight or obese boys jumped from 9.3% to 21.7% between 1985 and 2003; and in girls from 10.6% to 27.8% (Hands *et al.* 2004). Increasing physical activity is a key strategy to curb the global epidemic of obesity and overweight (NHMRC 1997; WHO 2000; Access Economics 2006).

How do these trends relate to the built environment? While numerous explanations have explored the rapid change in weight status, one plausible explanation is that the obesity epidemic is, at least in part, a 'physiological response to a toxic environment'—an 'obesogenic' environment that discourages physical activity (Hill and Melanson 1999; Hill and Peters 1998; Swinburn *et al.* 1999), encourages sedentary behaviour and over consumption of food. In 2002, the International Task Force on Obesity developed a framework that attempts to articulate the complex web of societal forces contributing to obesity levels (Kumanyika *et al.* 2002). As can be seen from Figure 3, land use and transportation systems are implicated.

The built environment either facilitates or discourages participation in physical activity. For example, in the last decade or so, there has been a rapid decline in the number of children walking or cycling to school (Harten and Olds 2004). Western Australian data from 2002 suggest that only 27.8% of adults report walking for transport and only 2.6% cycling for transport (McCormack *et al.* 2003). There is now strong evidence that the built environment—land use systems, transportation systems and urban design—affects the transport-mode choices of both adults (Buxton 2000; Kenworthy and Laube 1999; Naess 1993; Newman and Kenworthy 1989; Newman and Kenworthy 2000) and children (Centres for Disease Control and Prevention 1992; Lynch 1977).

Pedestrian-friendly neighbourhoods are an example of urban design that encourages physical activity. These pedestrian-friendly neighbourhoods have:

- destinations required for daily needs close to the housing
- connected street networks that provide alternative routes to local destinations
- aesthetic streetscapes
- footpaths
- streets and parks that are designed to encourage surveillance.

These neighbourhood attributes not only encourage local walking, but are also said to help create social capital and feelings of safety among local residents.

Social capital is the feature of social life—networks, norms and social trust—that enables communities/populations to effectively pursue joint objectives (Putnam 2000) and to cooperate for mutual benefit. Leyden found that social capital was higher in more walkable neighbourhoods (Leyden 2003), while Lund (Lund 2002) showed that regular walkers experienced a higher sense of community in the pedestrian-friendly environments. However, real and perceived traffic and crime-related safety issues are also associated with walking in local neighbourhoods for adults (Giles-Corti and Donovan 2003) and children (Timperio *et al.* 2006). From a crime perspective, there is growing interest in the relationship between sustainable urban development and crime prevention through environmental design (Cozens 2002). It is likely that residents living in neighbourhoods they perceive to be safe are less likely to constrain their behaviour and are more likely to walk than those who are fearful (personal communication (Skogan and Maxfield 1981; Ferraro 1995).

Tackling deficiencies: smart-development strategies

The links identified between land use, transport and public health suggest the need for a revised planning agenda that not only aims to improve urban sustainability, but also public health. In this section, we discuss some of the dimensions of 'smart growth development' and consider examples of where smart growth development may redress deficiencies associated with conventional planning practice and produce enhanced sustainability and health outcomes.

A range of design alternatives have emerged to counter urban sprawl and encourage travel by active modes (Giles-Corti *et al.* unpublished). These alternatives have been labelled Transit-Oriented Development, New Urbanism and Smart Growth (Krizek 2003). Residents of pedestrian-friendly neighbourhoods with higher residential density, mixed land-use, and accessible shops and transit are more likely to use non-motorised forms of transport than those living in conventional suburbs poorly served with these characteristics (Ewing *et al.* 2003; Frank *et al.* 2003; Krizek 2003). Based on this premise, new developments are being trialled with the hope of countering the negative influences of urban sprawl.

Although there is a paucity of longitudinal evidence on the impact of the built form on travel patterns, land use configuration has been identified by some US government agencies as one mechanism to improve transport sustainability (Handy *et al.* 2005a). Cross sectional evidence suggests that improved access, and hence opportunities for more sustainable travel behaviour (particularly walking and cycling), can be achieved by mixing land uses, increasing development densities and improving street network permeability (Newman and Kenworthy 1989; Cervero and Radisch 1996; Newman and Kenworthy 1999; Egger *et al.* 2001; Frumkin *et al.* 2004; Naess 2005). Where there is greater activity intensity (that is, a greater mix of land uses and higher development densities) residents have access to a variety of opportunities in close proximity (see Llewelyn-Davies 2000; McIndoe *et al.* 2005; Naess 2005), which theoretically reduces the need to drive. Importantly, the creation of a pedestrian-friendly development is a sustainable and passive public health intervention— increasing incidental daily physical activity in the course of people undertaking 'their daily business. Aside from improving transport sustainability, pedestrian-friendly development can also improve urban vibrancy, with associated benefits for community life.

In the US, New Urbanism and Smart Growth are two well-known urban reform approaches. New Urbanism has become increasingly popular during the last 20 years. The movement advocates design qualities reflective of small US cities *circa* 1900 to 1920 (Beatley 2004; Frank 2004). In particular, New Urbanist developments are intended to increase social interaction because they encourage more walking—thereby generating more social capital than may be expected in conventional neighbourhoods (Urban Land Institute 2005).

The defining characteristics of these developments are argued to be increased walkability, high permeability, increased land use mix, local character and compactness—all of which combine to provide residents with a highly liveable community (Charter for New Urbanism undated). These features are supportive of active transport and public transport (Bressi 1994). Moreover, such developments are often built to a walkable scale and anchored by a multi-use neighbourhood centre.

In contrast to the many advocates of New Urbanism, Marshall (2001) is somewhat more sceptical. He argues that New Urbanist developments are all-too-often located within an outdated transport system. In his view, there is insufficient attention given to sustainable transport in New Urbanist planning. That is, insufficient opportunities tend to be provided to allow people to walk or cycle for shorter trips and use public transport for longer trips. This is a significant point and one that needs to be explored in the context of any smart growth development strategy.

Smart Growth is said to be somewhat different to New Urbanism, in that it is focused as much on long-term as on short-term outcomes (Smart Growth Network undated). Gillham (2002) defines Smart Growth as:

Managed growth that attempts to fulfil the need to provide for growth (both economic and in population) while at the same time limiting the undesirable impacts of that growth.

The Smart Growth network was formed in 1996 in the US by the Environmental Protection Agency. It now includes more than 30 different organisations with an interest in sustainable growth (Frumkin *et al.* 2004). Frumkin and others espouse higher level strategic planning as an integral part of smart growth. This is to guard against piecemeal development and to ensure a high level of connectedness between new and existing neighbourhoods. Moreover, smart growth can best be realised by a focus on infill, rather than greenfields, development. A major advantage of smart growth development is that it takes advantage of existing infrastructure rather than greenfields development on the urban fringe, which usually requires starting from scratch with the creation of transportation, utilities, social and community infrastructure. Thus, van Vliet and Gade (2001) argue that '*urban renewal is a good example of recycling*'. By implication, cities reflecting smart growth patterns will be more compact.

In Australia, smart-growth strategies have been contextualised to produce vibrant flagship developments. Two examples are considered here: Subiaco, Perth and Brisbane Urban Renewal, in inner north-east Brisbane. Subiaco is situated 3 kilometres from the centre of Perth. Development of 'Subiaco Centro' has occurred since the mid-1990s following the re-location of the Perth-Fremantle rail line underground, which previously divided the suburb. Subiaco Centro includes more than 1000 dwellings, with provision for 350 more (Australian Council for New Urbanism 2006). Importantly, it is characterised by a mix of housing opportunities, including apartment-style living and town houses, a mixture of uses, with provision of street-level commerce and retail, and relatively high densities by Perth standards, with lots generally varying between 223m^2 and 315m^2 and a maximum plot ratio of 1:1 (Subiaco Redevelopment Authority 2005). The dwellings are in close proximity to a vibrant well-established 'main street', which offer ready access to retail and entertainment and shops and services required for daily living.

The vibrancy of Subiaco Centro likely relies as much on street-level design as on density and mixing of uses. Local roads tend to be narrow and much consideration has gone into balancing the needs of pedestrians and cyclists with motorists. Public transport users are well catered for with the rail line and ancillary bus services. Additionally, thoughtful planting and other aesthetics, including sympathetic street lighting and architecture, make Subiaco Centro an appealing community. While Subiaco Centro exhibits many of the characteristics that seem likely to facilitate sustainable transport, there has as yet been little research into the travel patterns of members of the community.

The Brisbane Urban Renewal project was initiated by a City Council taskforce, established in 1991. The localities of Fortitude Valley, New Farm, Teneriffe, Newstead and Bowen Hills—a combined area of around 730 hectares—was originally earmarked for renewal, but the target area has since grown to exceed 1000 hectares (Brisbane City Council undated). The Australian Council for New Urbanism (2006) reports that:

Over the past 12 years the area has been transformed from a declining, outdated and unattractive precinct to a vibrant, diverse and increasingly high amenity inner urban area.

This area shares many characteristics in common with Subiaco Centro, which together improve local vibrancy. A new railway station in Fortitude Valley provides a strong anchor point, and increased urban density and mixing of uses throughout the renewal area are positive. Street-level design is well managed, with infrastructure to support active mode users. Particular areas of renewal, including Fortitude Valley's Brunswick Street West, Brunswick Street Mall and Chinatown Mall have been reinvigorated with lighting, landscaping and footpath improvements (Brisbane City Council undated).

Their positive qualities aside, smart-growth developments could be criticised for giving insufficient attention to socially equity. While developments such as Subiaco Centro and inner north-east Brisbane have superior community environments, this can translate to exclusionary housing costs when the market sets the prices. US research, for example, has found that land value near to rail stations is generally at a premium (Cervero *et al.* 2004). It is therefore important that governments engage in partnerships with land developers to ensure that a minimum amount of affordable housing is provided, to ensure that lower socioeconomic groups are not excluded. Clearly, consideration of social equity issues will be a priority within smart-growth developments, and there needs to be ongoing evaluation to ensure there are housing opportunities for all members of the community. For example, the website for the Brisbane City Council (Brisbane City Council undated) states that there are:

... low-cost housing units, boarding house rooms and detached houses [in the renewal area], whereby developers contribute to the Brisbane Housing Company for low-cost dwellings.

Subiaco Centro and inner north-east Brisbane are both showcase examples of smart growth developments. The *Network City* planning strategy (Department for Planning and Infrastructure and Western Australian Planning Commission 2004) is a step towards sustainable and integrated land use and transport planning at a wider scale across the greater Perth metropolitan region. Significantly, it recognises that a nodes-and-corridors approach to urban growth and change is necessary and that existing areas must be better used. It targets 60% of new growth in established areas, leaving 40% to be in areas of new growth (DPI and WAPC 2004). The strategy includes objectives designed to reduce car dependence, enhance public transport and not '*inequitably limit accessibility based on location or access to a private car*' (DPI and WAPC 2004: p66). The strategy recognises that to realise these objectives, improvements to city-wide alternative transport mode networks are necessary.

Furthermore, *Liveable Neighbourhoods* (LN) has been developed to tackle the need for a 'smart' statutory planning *mechanism*. LN remains an alternative to conventional standards—standards that have facilitated sprawl and car dependence. Of particular interest is its potential to facilitate more travel by alternative modes. The Western Australian Greenhouse Task Force (2004) reports that LN is:

... a voluntary planning design code that promotes the development of sustainable communities with mixed land use and a balanced transport system. It encourages reduced car usage, better use of public transport, more walking and cycling, improved access to services and more efficient land use.

LN has links to New Urbanism, Transit-Oriented Development and Smart Growth. A trial of the code began in February 1998 and, after a number of reviews over the last decade, the code will be mandated in late 2007. However, until now developers have had the option of voluntarily accepting the LN code over the conventional code, as an *optional set of standards*. The voluntary nature of the code has been a significant challenge to it being a useful part of the sustainability agenda.

The design code was developed by the Department for Planning and Infrastructure through a process of retooling the *Australian Model Code for Residential Development 1995* to facilitate a better fit for the West Australian context. State planning authorities recognised the need to deal with issues associated with conventional planning practice, especially rapid fringe development. Such issues include the high cost of providing services and facilities, including utilities, to outlying developments, a lack of local employment opportunities, relatively poor public transport provision, car dependence, and questionable social, economic and environmental sustainability.

The LN aims to be a performance-based vehicle to meet the objectives of the State Planning Strategy. It is intended to facilitate the development of more sustainable communities, as part of the vision for Western Australia 2029. Some of the key outcomes envisaged for LNs are increased support for active travel modes and public transport, higher development densities and increased lot diversity. Development is intended to be focused round activity centres and public transport nodes (Western Australian Planning Commission 2004). The design code is to be applied to development proposals on *greenfields* sites encompassing two or more lots and larger infill sites (Western Australian Planning Commission 2004). Moreover, the code is a regulatory tool—not simply an advisory document and it can be applied at a variety of scales, with the intention being that development is coordinated.

There is a *prima facie* case to suggest that the smart-growth developments and policies presented are more sustainable than conventional developments and policies and will have a positive impact on health. Nevertheless, as yet there is little supporting evidence. Assessment of sustainability and health impact requires a concise understanding of what needs to be achieved and how success can be measured (Renne 2007). It would seem that key indicators of sustainability would include lower vehicle kilometres travelled and a modal split more in favour of alternative modes—particularly active modes—for the health benefits that might result. However, as yet, there has been little research to show how well the policy is being implemented and whether it is facilitating more sustainable transport behaviour. For example, are land uses and transport being coordinated in new neighbourhoods? Is the design code being consistently applied? Importantly, what is the impact on local residents?

Changes to the built environment, such as development of new neighbourhoods affect transport systems and vice versa. Many researchers acknowledge this association, but remain undecided about the exact influence of the built environment on travel patterns (Newman and Kenworthy 1999; Crane 2000; Frank 2000; Ewing and Cervero 2001; Saelens *et al.* 2003; Sallis *et al.* 2004). One key reason is because most research to date has been cross-sectional rather than longitudinal, which limits the conclusions that are capable of being drawn on causal links between built environment, transport, mode-choice or behavioural outcomes.

The extent to which a new government policy can influence active transport behaviour is the subject of a project funded by the Western Australian Health Promotion Foundation (Healthway) and the Australian Research Council. The Residential Environment Project (known as RESIDE) (Giles-Corti *et al.* 2007) commenced in 2003 (http://population-health.uwa.edu.au/reside). RESIDE's principal aim is to study the impact of the state government's Liveable Neighbourhood (LN) Community Design Codes on the walking, cycling and public transport use behaviour of local residents.

RESIDE was designed to redress the need for causal evidence on relationship between the built form and physical activity (Ewing *et al.* 2003; Handy *et al.* 2005a). Its longitudinal design will enable examination of a critical unknown: do residents choose neighbourhoods that support their preferred travel behaviour or does the neighbourhood change their behaviour?

RESIDE is being conducted over 5 years. Study participants are people building homes in 74 new housing developments (total number of houses =1813): 18 of which have been designed according to the Liveable Neighbourhood Guidelines (LND), 14 of which are hybrid neighbourhood developments (HND) and the remaining are conventional neighbourhood developments (CND). Study participants are first surveyed before they move into their new homes and then again 12 and 36 later.

RESIDE is an ecological study (von Bertalanffy 1968; McLeroy *et al.* 1993; Stokols 1996; Sallis *et al.* 2006) studying the multiple-levels of influence on behaviour: individual, social environmental and physical environmental. Each time they are surveyed, RESIDE study participants complete a comprehensive questionnaire that measures individual-level (such as demographic and health-related characteristics, attitudes and perceptions); social environmental (such as social support for physical activity); and physical environmental (such as perceptions of the local neighbourhood) factors that that might influence participation in active modes of transportation. Study participants provide comprehensive information about the frequency and duration of walking and cycling undertaken within, and outside, their neighbourhood (Giles-Corti *et al.* 2006), as well as information about their where they work, how they travel to work and how much time it takes. A sub-sample of study participants has also participated in a Transport Sustainability and Health Study, and has completed a comprehensive 2 or 7-day travel diary.

In addition to collection of self-reported data, a key objective of RESIDE is to develop objective measures of the built environment surrounding the study participants' homes (both at baseline addresses and in new neighbourhoods), using Geographic Information System data. A walkability index based on that developed by Frank (Frank *et al.* 2004; Leslie *et al.* 2005) has been developed (Learnihan 2007). The index combines measures of residential density, mixed use planning and street connectivity. The index will be used to assess the extent to which local walking is associated with the walkability index of the local neighbourhood built environment.

Longitudinal studies like RESIDE are important to inform the development of future policy and practice with the aim of optimising neighbourhood designs to produce health, planning and sustainability outcomes and to avoid unintended negative consequences. However, a key feature of RESIDE is that it collects information on the multiple levels of influence of behaviour in order to better understand the independent impact of the built form, over and above individual and social-environmental determinants of behaviour. This information will assist in decision-making about future educational and policy interventions.

Cities at the crossroads: making smart development into common practice

Australian cities are at a crossroads. With sustainability now firmly entrenched in policy rhetoric, it is time to consider the future development of our cities and how this will affect the economy, environment, society and the health of Australians. While there have been various deficiencies in the way in that Australian cities have been, and are being, planned, strategies are emerging to redress these problems, including flagship 'smart growth' developments. The case studies presented here are testimony to their potential and to the vibrant landscapes that can be created when the rhetoric is put into practice.

Nevertheless, to have a real impact on the community, smart growth development needs to become more widespread. The examples presented, while significant, do not reflect common development practice. Moreover, strategic coordination is essential, with a link between regional and neighbourhood planning to facilitate switches to active modes (Williams 2001; Shore 2006). Accordingly, smart-growth development needs to be city- or regional-wide rather than governed by individual local government authorities. A regional approach offers greater potential to develop strong transport links between sub-centres of cities, enabling greater emphasis on public transport. Individual neighbourhoods may themselves be strong anchor-points for public transport, but for the system to work, they have to be linked to other equally well-designed neighbourhoods.

When smart growth development principles begin to be considered at the metropolitan level, challenges facing planning authorities can be recognised. Increasingly, governments will need to consider whether there is need to retrofit areas of the city that are poorly served by public transport and lack the urban characteristics that make them pedestrian- and transit-friendly.

It remains important, too, that smart-growth developments and the strategies that underlie them are regularly re-evaluated. RESIDE demonstrates that it is possible to establish a study to evaluate the impacts of government urban planning policy with a view to informing policy into the future. Further research and evaluation in this area is essential. Evaluation will enable intended and unintended consequences to be monitored to ensure that any changes observed are consistent with changing sustainability and health priorities. Increasing urban populations and related demand for housing, for example, are likely to create new challenges for land use and transport planners in the future.

What will be the intended—and unintended—environmental and health consequences of building on the urban fringe or even retrofitting existing neighbourhoods? Will higher density developments generate fewer motor vehicle trips and increase walking and cycling? And if not, why not? Will obesity levels in these neighbourhoods decline because people are more active? If successful, will it be possible to produce similar outcomes in disadvantaged neighbourhoods? On the other hand, will there be unintended negative consequences of higher density contemporary living? What are they, and what can we do about them? For example, will we have the wisdom of the 19th century planners to ensure that there is sufficient public open space that provides a high-quality restorative nature experience that gives people a sense of novelty and surprise, and reduces stress and aggression? Planning has the potential to have a major positive impact on the health and wellbeing of people in the 21st century.

References

Access Economics (2006). *The Economic Cost of Obesity.* Diabetes Australia, Melbourne

Armstrong T, Bauman A and Davies J (2000). *Physical Activity Patterns of Australian Adults. Results of the 1999 National PA Survey.* AIHW, Canberra.

Ashton-Graham C, John G, Radford T and Rampellini P (2005). 'TravelSmart + TOD = Sustainability and Synergy'. Paper presented at the Transit Oriented Development (TOD) conference 'Making It Happen', 6–8 July, Fremantle, Western Australia. <http://www.dpi.wa.gov.au/mediaFiles/ts_tod.pdf>

Australian Bureau of Statistics (2001). *Motor Vehicle Census*, ABS publication 9309.0, ABS, Canberra.

Australian Council for New Urbanism (2006). *Australian New Urbanism: A Guide to Projects* 2nd edn. Ecologically Sustainable Pty Ltd, Fitzroy, Victoria.

Bachels M and Newman P (2001). Cities back on track: using a systems approach for sustainable city policy. In *Back on Track: rethinking transport policy in Australia and New Zealand* (Eds P Laird, P Newman, M Bachels and J Kenworthy) pp. 133–155. University of New South Wales Press, Sydney.

Beatley T (2004). *Native to Nowhere: Sustaining Home and Community in a Global Age*. Island Press, Washington DC.

Begg S, Vos T, Barker B, Stevenson C, Stanley L and Lopez A (2003). *The Burden of Disease and Injury in Australia 2003*. AIHW, Canberra.

Booth M, Armstrong T, Chey T, Hesketh K and Mathur S (2001). The epidemiology of overweight and obesity among Australian children and adolescents 1995–97. *Australian and New Zealand Journal of Public Health* **25(2)**, 162–169.

Bressi T (1994). Planning the American dream. In *The New Urbanism: Toward an Architecture of Community* (Ed. P Katz) pp. 25–42. McGraw Hill, New York,.

Brisbane City Council (2007) *About Urban Renewal Brisbane*. <http://www.brisbane.qld.gov.au/BCC:STANDARD:990592241:pc=PC_1727>.

Buxton M (2000). Energy, transport and urban form in Australia. In *Achieving Sustainable Urban Form* (Eds K Williams, E Burton and M Jenks) pp. 54–63. E and FN Spon, London.

Cameron I. (2004). Understanding, modelling and predicting transport mobility in urban environments. PhD thesis, Murdoch University, Perth.

Cameron I, Lyons T and Kenworthy J (2004). Trends in vehicle kilometres of travel in world cities, 1960–1990: underlying drivers and policy responses. *Transport Policy* **11**, 287–298.

Centres for Disease Control and Prevention (1992). Barriers to children walking and biking to school-United States, 1999. *Morbidity & Mortality Weekly Report* **51(32)**, 701.

Cervero R. Arrington G, Smith-Heimer J, Dunphy R, Murphy S, Ferrell C, Goguts N, Tsai Y.-H, Boroski J, Golem R., Peninger P, Nakajima E, Chui E, Meyers M, McKay S and Witenstein N (2004) *Transit Oriented Development in America: Experiences, Challenges, and Prospects*, TCRP Report 102, National Academy Press, Washington DC.

Cervero R and Radisch C (1996). Travel choices in pedestrian versus automobile orientated neighbourhoods. *Transport Policy* **3**, 127–141.

Charter for New Urbanism (2007). <http://www.cnu.org>.

Cozens P (2002). Sustainable urban development and crime prevention through environmental design for the British City. Towards an effective urban environmentalism for the 21st century. *Cities* **19(2)**, 129–137.

Crane R (2000). The influence of urban form on travel: an interpretive review. *Journal of Planning Literature* 15(1), 3–23.

Davison G (2004). *Car Wars: How the Car Won Our Hearts and Conquered Our Cities*. Allen & Unwin, Crows Nest, NSW.

Department for Planning and Infrastructure and Western Australian Planning Commission (2004). *Network City: Community Planning Strategy for Perth and Peel*. WAPC, Perth.

Egger G, Donovan R, Giles-Corti B, Bull F and Swinburn B (2001). Developing national physical activity guidelines for Australians. *Australian and New Zealand Journal of Public Health* **25(6)**, 561–563

Ewing R and Cervero R (2001). The influence of land use on travel behaviour. *Transportation Research, Policy and Practice* **33**, 823–845.

Ewing R, Schmid T, Killingsworth R, Zlot A and Raudenbush S (2003). Relationship between urban sprawl and physical activity, obesity and morbidity. *American Journal of Health Promotion* **18(1)**, 47–57.

Ferraro KF (1995). *Fear of Crime: Interpreting Victimization Risk*. State University of New York Press, New York.

Frank L (2000). *How Land Use and Transportation Systems Impact Public Health: A Literature Review of the Relationship Between Physical Activity and Built Form*. Centers for Disease Control and Prevention, Atlanta, Georgia.

Frank L, Andresen M and Schmid T (2004). Obesity relationships with community design, physical activity and time spent in cars. *American Journal of Preventive Medicine* **27(2)**, 87–96.

Frank L, Engelke P and Frank L (2003). *Health and Community Design. The Impact of the Built Environment on Physical Activity.* Island Press, Washington, D.C.

Frumkin H, Frank L and Jackson R (2004). *Urban Sprawl and Public Health: Designing, Planning and Building for Healthy Communities.* Island Press, Washington DC.

Giles-Corti B (2006). People or places: what should be the target? *Journal of Sports Science and Medicine* **9(5)**, 357–366.

Giles-Corti B and Donovan R (2003). Increasing walking: the relative influence of individual, social environmental and physical environmental factors. *American Journal of Public Health* **93(9)**, 1583–1589.

Giles-Corti B, Knuiman M, Timperio A, Van Niel K, Pikora TJ, Bull FCL, Shilton T and Bulsara M (In press). Evaluation of the implementation of a government community design policy aimed at increasing local walking: Design issues and baseline results from RESIDE *Preventive Medicine.*

Giles-Corti B, Timperio A, Cutt H, Pikora T, Bull F, Knuiman M, Bulsara M, Van Niel K and Shilton T (2006). Development of a reliable measure of walking within and outside the local neighborhood: RESIDE's Neighborhood Physical Activity Questionnaire. *Preventive Medicine* **42(6)**, 455–459.

Gillham O (2002). *The Limitless City: A Primer on the Urban Sprawl Debate.* Island Press, Washington DC.

Gleeson B (2006). *Australian Heartlands: Making Space for Hope in the Suburbs.* Allen & Unwin, Crows Nest, NSW.

Handy S (2003). Driving less. *Access* **23**, 20–25.

Handy S, Cao X and Mokhtarian P (2005a). Correlation or causality between the built environment and travel behaviour? Evidence from Northern California. *Transportation Research Part D* **10**, 427–444.

Handy S, Weston L and Mokhtarian P (2005b). Driving by choice or necessity? *Transportation Research Part A* **39**, 183–203.

Harten N and Olds T (2004). Patterns of active transport in 11–12 year old Australian children. *Australian and New Zealand Journal of Public Health* **28**, 167–172.

Hill J and Melanson E (1999). Overview of the determinants of overweight and obesity: current evidence and research issues. *Medicine & Science in Sports & Exercise* **31(11 Supp.)**, S515–521.

Hill J and Peters J (1998). Environmental contributions to the obesity epidemic. *Science* **280(5368)**, 1371–1374.

James B, John G and McKaskill J (2001). 'Potential for increasing walking trips'. Paper presented at the International Walking conference 'Walking in the 21st Century', 20–22 February, Perth. <http://www.dpi.wa.gov.au/mediaFiles/walking_21centconf01bpaper_james.pdf>.

Kenworthy J and Laube F (1999). *An International Sourcebook of Automobile Dependence in Cities 1960–1990.* University Press of Colorado, Boulder.

Krizek K (2003). Residential relocation and changes in urban travel - Does neighborhood-scale urban form matter? *Journal of the American Planning Association* **69(3)**, 265–281

Kumanyika S, Jeffery R, Morabia A, Ritenbaugh C and Antipatis V (2002). Obesity prevention: the case for action. *International Journal of Obesity* **26(3)**, 425–436.

Laird P and Newman P (2001). How we got here: the role of transport in the development of Australia and New Zealand. In *Back on Track: Rethinking Transport Policy in Australia and New Zealand* (Eds P Laird, P Newman, M Bachels and J Kenworthy) pp. 1–21. University of New South Wales Press, Sydney.

Learnihan V (2007). The physical environment as an influence of walking in the neighbourhood: objective measurement and validation. Master of Public Health thesis, University of Western Australia, Perth.

Leslie E, Saelens B, Frank L, Owen N, Bauman A, Coffee N and Hugo G (2005). Residents' perceptions of walkability attributes in objectively different neighbourhoods: a pilot study. *Health & Place* **11**, 227–236.

Leyden K (2003). Social capital and the built environment: the importance of walkable neighbourhoods. *American Journal of Public Health* **93(9)**, 1546–1551.

Llewelyn-Davies Consultants (2000). *Urban Design Compendium*. English Partnerships, London.

Lund H (2002). Pedestrian environments and sense of community. *Journal of Planning Education and Research* **21(3)**, 301–312.

Lynch K (Ed.) (1977). *Growing up in Cities: Studies of the Spatial Environment of Adolescense in Cracow, Melbourne, Mexico City, Salta, Toluca and Warszawa*. MIT Press, Cambridge, Massachusetts (UNESCO, Paris).

Marshall A. (2001). *How Cities Work: Suburbs, Sprawl and the Roads Not Taken*. University of Texas Press, Austin.

McCormack G, Milligan R, Giles-Corti B and Clarkson J (2003). *Physical Activity Levels of Western Australian Adults 2002: Results from the Adult Physical Activity Survey and Pedometer Study*. Government of Western Australia, Perth.

McIndoe G, Chapman R, McDonald C, Holden G, Howden-Chapman P and Sharpin A (2005). *The Value of Urban Design: the Economic, Environmental and Social Benefits of Urban Design*. Ministry for the Environment, Wellington, New Zealand. <http://www.mfe.govt.nz/publications/urban/value-urban-design-full-report-jun05/>.

McLeroy K, Steckler A, Simons-Morton B, Goodman R, Gottlieb N and Burdine J (1993). Social science theory in health education: time for a new model? *Health Education Research* **8**, 305–312.

Naess P (1993). *Energy Use for Transport in 22 Nordic Towns*. Norwegian Institute for Urban and Regional Research, Oslo.

Naess P (2005). Residential location affects travel behaviour – but how and why? The case of Copenhagen metropolitan area. *Progress in Planning* **63**, 167–257.

Newman P (2005). 'Transit Oriented Development: an Australian overview'. Paper presented at the Transit Oriented Development (TOD) conference 'Making It Happen', 6–8 July, Fremantle, Western Australia. <http://www.patrec.org/conferences/TODJuly2005/papers/Newman%20paper%20REV.pdf>.

Newman P and Kenworthy J (1989) *Cities and Automobile Dependence: An International Source-book*. Gower, Aldershot, UK.

Newman P and Kenworthy J (1999). *Sustainability and Cities: Overcoming Automobile Dependence*. Island Press, Washington DC.

Newman P and Kenworthy J (2000). Sustainable urban form: the big picture. In *Achieving Sustainable Urban Form* (Eds K Williams, E Burton and M Jenks) pp. 109–120. E and FN Spon, London.

National Health and Medical Research Council (1997). *Acting on Australia's Weight: A Strategic Plan for the Prevention of Overweight and Obesity*. Commonwealth of Australia, Canberra.

Planning and Transport Research Centre (2005). *Newslines*, Number 2, June, PATREC, Perth <http://www.patrec.org/news/newsletters/Newsletter02.pdf>.

Premier's Physical Activity Taskforce (2006). *Western Australia State Walking Strategy, 2005–2020*. Government of Western Australia, Perth.

Rees W (2003). Ecological footprints and urban transportation. In: *Sustainable Transport: Planning for Walking and Cycling in Urban Environments* (Ed. R Tolley) pp. 3–19. Woodhead Publishing, Cambridge, UK.

Renne J (2007). 'Evaluating Transit-Oriented Development Using a Sustainability Framework: lessons from Perth's network city'. Paper presented at 13th Annual International Sustainable Development Research Conference, 10–12 June, Västerås, Sweden.

Saelens B, Sallis J and Frank L (2003). Environmental correlates of walking and cycling: findings from the transportation, urban design and planning literatures. *Annals of Behavioural Medicine* **25(2)**, 80–91.

Sallis J, Cervero R, Ascher W, Henderson K, Kraft M and Kerr J (2006). An ecological approach to creating active living communities. *Annual Review of Public Health* **27**, 297–322.

Sallis J, Frank L, Saelens B and Kraft M (2004). Active transportation and physical activity: opportunities for collaboration on transportation and public health research. *Transportation Research Part A* **38**, 249–268.

Shore W (2006). Land-use, transportation and sustainability. *Technology in Society* **28**, 27–43

Smart Growth Network (undated) *Smart Growth on-line* <http://www.smartgrowth.org>.

Skogan WG and Maxfield MG (1981). *Coping with Crime: Individual and Neighborhood Reactions.* Sage Publications, Beverley Hills.

Stephenson J, Bauman A, Armstrong T, Smith B and Bellew B (2000). *The Cost of Illness Attributable to Physical Inactivity in Australia: A Preliminary Study.* Commonwealth Department of Health and Aged Care, and the Australian Sports Commission, Canberra.

Stokols D (1996). Translating social ecological theory into guidelines for community health promotion. *American Journal of Health Promotion* **10(4)**, 282–298.

Subiaco Redevelopment Authority (2005). *Centro Village Site Design Guidelines, Subiaco,* SRA: Perth. <http://www.sra.wa.gov.au/NR/rdonlyres/269DE04E-8A4D-47E5-941E-2F27FA1B7A86/0/051215CentroVillageDGfinal.pdf>.

Swinburn B, Egger G and Raza F. (1999). Dissecting obesogenic environments: the development and application of a framework for identifying and prioritising environmental interventions for obesity. *Preventive Medicine* **29(6 (Part 1)**), 563–70.

Timperio A, Ball K, Salmon J, Roberts R, Giles-Corti B, Simmons D, Baur L and Crawford D (2006) Personal, family, social, and environmental correlates of active commuting to school. *American Journal of Preventive Medicine* **30(1)**, 45–51.

Transportation Research Board (2005). *Does the Built Environment Influence Physical Activity? Examining the Evidence.* TRB, Washington DC.

Urban Land Institute (2005). *Higher-Density Development: Myth and Fact* <www.uli.org/Content/ContentGroups/PolicyPapers/MFHigher010.pdf>.

US Department of Health and Human Services (1996). 'Physical activity and health. A report of the Surgeon General'. Department of Health and Human Services, Centres for Disease Control and Prevention, National Center for Chronic Disease Prevention and Health Promotion, Atlanta, Georgia.

van Vliet D and Gade T (2001). Sustainable urban renewal: Kolding, Denmark. In *Achieving Sustainable Urban Form* (Eds K Williams, E Burton and M Jenks) pp. 310–318. E and FN Spon, London.

von Bertalanffy L (1968). *General SystemTtheory. Foundations, Development, Applications.* George Braziller, New York.

Western Australian Greenhouse Task Force (2004). *Western Australian Greenhouse Strategy.* Government of Western Australia, Perth.

Western Australian Planning Commission (2004). *Liveable Neighbourhoods.* 3rd edn. WAPC, Perth. <http://www.wapc.wa.gov.au/cgi-bin/index.cgi?page=/udmp/index.html>.

World Health Organization (2000). *Obesity. Preventing and Managing the Global Epidemic.* WHO Technical Report Series 894. WHO, Geneva.

Williams K (2001). Does intensifying cities make them more sustainable? In *Achieving Sustainable Urban Form* (Eds K Williams, E Burton and M Jenks) pp. 30–45. E and FN Spon, London.

39

Transitions to Smart, Sustainable Infrastructure

Leanne Hardwicke

Is Australia's infrastructure smart or sustainable? Apparently not, according to reports by bodies such as Engineers Australia (2001, 2005), the Business Council of Australia (2006) and the Committee for Economic Development of Australia (2005).

World-class sustainable infrastructure is vital to the Australian economy. It underpins the delivery of essential services, drives economic growth, supports social needs and is closely linked to the high quality of life that Australians enjoy today.

Sustainable development of infrastructure is essential. Infrastructure is defined as sustainable if it meets the needs of the present without compromising the needs of future generations. In operational terms, this means ensuring that our infrastructure is environmentally, socially and economically sustainable in the 21st century.

Infrastructure has generally been described as the structural elements of an economy that facilitate the flow of goods and services between buyers and sellers, producers and consumers. In this sense, infrastructure includes such assets as roads, railways, airports, ports, energy systems, telecommunications, water, wastewater and sewerage systems —and is now commonly termed economic infrastructure.

More recently, infrastructure has been described more broadly to include the basic services of a country that make economic and social activities possible. This description includes housing, health and education services, together with other institutions that facilitate economic and social interaction—and is termed social infrastructure.

This chapter focuses on economic infrastructure.

For most of the 20th century, economic infrastructure in Australia was used to promote national and regional development. As well, its production and maintenance provided employment opportunities and was often a major focus of government investment during recessions as a means of 'pump-priming' the economy (Productivity Commission 1999).

However, Australia has a relatively low rate of investment in and a low stock of public infrastructure. While there was a major expansion in Australia's infrastructure during the 1950s and 1960s, the proportion of public investment in infrastructure fell in later years. The average age of Australian infrastructure has been increasing since the 1970s (Commonwealth Department of Treasury 2007) and in the 1990s concern was expressed that the infrastructure built in this earlier period was reaching the end of its economic life, with Australia being faced with a huge infrastructure replacement cycle.

Recently, there has been an upward shift in the amount of public spending on much neglected infrastructure, but overall, Australia's investment in public infrastructure has been declining as a proportion of GDP. According to the Commonwealth Treasury, Australia's expenditure on infrastructure investment as a proportion of GDP is at the low end of OECD countries (Port Jackson Partners 2005).

There has also been a shift in the way in which infrastructure is delivered in Australia in the last 15 years, from mostly public provision to provision by public and private suppliers (termed public private partnerships or PPPs).

The quality of Australia's infrastructure is an indicator of the nation's current and potential economic viability. The benefits and costs associated with infrastructure performance are continuous and, over time, assume major economic significance. It is therefore vital that Australia's infrastructure meets today's needs and through careful planning, maintenance and construction, tomorrow's needs as well.

It is now widely acknowledged that there is a positive relationship between investment in infrastructure and productivity. Cross-country differences in productivity growth can be explained by differences in levels of infrastructure investment. Infrastructure is a key determinant of comparative advantage between countries and, in turn, international trade.

A review of Australian studies has found the following results:

- There is a statistically significant positive relationship between investment in road and other infrastructure and private sector output (Allen Consulting Group 2003).
- A 1% increase investment in road infrastructure increases private sector total factor productivity (output) by 0.27% (Allen Consulting Group 2004).
- An additional $1 billion investment in roads would yield annual increases in GDP. These might range from $810 million for urban arterial roads to $270 million for local roads. In addition, some 19,000 jobs could be created for arterial roads and some 2,400 jobs could be created for local roads (Allen Consulting Group 2004).
- Further infrastructure reform in the energy sector will benefit GDP by 0.41% (ACIL 2002).
- Rail reform will benefit GDP by 0.15% (Port Jackson Partners 2005).
- Urban water reform will benefit GDP by 0.35% (Productivity Commission 2004).
- Urban transport reform will benefit GDP by 0.13% (Productivity Commission 2004).

Infrastructure has significant implications for the achievement of sustainable development objectives, because it underpins many aspects of economic and social activity.

Challenges of the 21st century

The geography of Australia sets it apart from many other nations. It is a vast continent with a small, low-density population, requiring a large infrastructure network.

The 21st century brings with it many challenges that will have an impact on Australia's infrastructure in terms of what will be required to service current and future needs, and how that might be funded.

Some infrastructure sectors are presently under strain. For instance:

- many communities are facing serious water shortages
- electricity capacity is low in several states in peak periods such as summer
- the cost of urban traffic congestion is rising in major cities
- ships are queuing off some of our major ports.

Australia's infrastructure must be brought up to a standard that meets current needs as a matter of priority before facing further challenges into the century.

The Intergenerational Report (Commonwealth Department of Treasury 2007) states that demographics will represent a major challenge for Australia in this century. Australia's population is currently around 21 million and is estimated to grow to 28.5 million by 2047. Population growth is a strong factor that influences the level of required infrastructure. The development of new suburbs creates the need for roads, reticulated water, electricity generation and distribution, telecommunications, waste- and stormwater infrastructure and other public facilities.

There is also a changing pattern in population settlement, with pressures now increasing in coastal regions and the largest cities.

As well, the Australian population will continue to age because of declining mortality rates and below-replacement fertility rates. This will place pressure on government spending in areas such as health and aged care. Australian government spending on health is projected to increase as a proportion of GDP from 3.8% in 2006–07, to 7.3% in 2047. Demand for government spending on services and infrastructure for the ageing population will compete with future economic infrastructure demands. The ageing population and low fertility rates also create particular challenges with regard to the skills available to provide infrastructure for Australia, as is currently the case.

Public attitudes—particularly for a cleaner environment—are placing, and will continue to place, pressure on infrastructure, as will climate change and drought.

Droughts not only results in water shortages—and the need for new infrastructure and technologies to supply the population with water for all its activities—they also affect energy infrastructure.

Hydroelectricity provides 17% of Australian electricity generation capacity—mostly from the Snowy Mountains or Tasmanian hydroelectricity systems. Large quantities of water are also essential in coal-fired electricity generation.

Recently, NEMMCO (2007) prepared a report for the Ministerial Council on Energy Standing Committee of Officials on the potential of drought to restrict electricity generation. The report provided the results of an analysis of two rainfall scenarios on electricity generation in the immediate future, which provides an indication of the long-term impact if we experience increased drought conditions in Australia into this century.

For instance, under a low rainfall scenario there will be restrictions on generating capacity averaging 1070 MW, or 2.6%, of national electricity market (NEM) capacity from the second quarter in 2007 through to the second quarter in 2008. From the third quarter 2008 to the first quarter 2009, these restrictions rise to 2130 MW, or 5.1%. All NEM regions are affected—some more than others. For example, NSW could lose 14.4% of its generating capacity. Under the average rainfall scenario, capacity is reduced by an average 4.9% from the second quarter 2007 to the second quarter 2008 and the situation improves to an average loss in capacity of 0.9% until the first quarter 2009. In this scenario, the impacts are confined to Victoria and Queensland.

The demand for energy will increase throughout the century. Projections over the next 25 years show strong growth in the use of all forms of energy. Electricity demand is expected to increase by 62%, demand for petroleum fuels by 41% and natural gas by 87%. A large part of expanded demand is due to Australia's failure to deal with energy conservation. Accommodating these changes will require large infrastructure investments, and will require massive financial and technical responses unless efforts are made to moderate this growth through demand management and other measures.

Climate change will be the biggest challenge to Australia's future infrastructure. According to the Intergovernmental Panel on Climate Change (IPCC 2007), global atmospheric concentrations of carbon dioxide, methane and nitrous oxide have increased markedly as a result of human activities since 1750 and now far exceed pre-industrial values determined from ice cores spanning many thousands of years. The IPCC concluded that increases in carbon dioxide concentrations are primarily due to fossil fuel use and land use changes, and increases in methane and nitrous dioxide concentrations are primarily due to agriculture.

The IPCC believes with 90% probability that these changes have led to global warming. Global warming is expected to bring about:

- more frequent very hot days
- more frequent and longer droughts
- more frequent and more intense heavy rain
- more intense tropical cyclones
- more intense storms
- higher peak wind speeds, and
- higher storm surges
- an increase in mean sea level

When planning infrastructure that is designed to last for the next 20 to 100 years, these factors must be taken into account.

Research sponsored by the Association of British Insurers (2007) has shown that the damage caused by severe storms could increase by two-thirds by the end of this century, even with quite small increases in intensity. The most extreme storms have the capacity to create losses so large that insurance markets become very volatile as the demand for capital by the insurance industry, and consequently its costs, rises. The same research shows that both climate change adaptation and mitigation have the potential to significantly reduce damage and the associated costs.

In March 2007 the Victorian Government accepted delivery of a report it had commissioned on an Infrastructure and Climate Change Risk Assessment (CSIRO *et al.* 2007). The report found that climate change posed particular challenges for infrastructure owners, managers and decision makers. The modelling undertaken for the report makes it clear a number of climate change risks would be seen by the courts as reasonably foreseeable, posing negligence litigation issues for all involved in ownership, development and management of affected infrastructure should there be a failure to deal with climate change risks properly.

Adaptation responses discussed in the report include:

- changes in materials selection to cope with alterations in temperature, solar radiation, humidity, soil moisture levels, inundation and the frequency and intensity of extreme events
- design standards will need to be reviewed to build the capacity of infrastructure to withstand a wider range of extreme and changing conditions
- maintenance regimes will need to adapt to acceleration in the degradation of materials and structures
- technology may need to be upgraded, for example to maintain water quality after severe bush fires in catchments
- planning considerations will be vital, for example, essential services planning may need to adapt to manage changed resource demand pressures
- cultural change and engagement mechanisms will also be important within different professions and between professions and the community at large.

Specific challenges for particular infrastructure sectors are set out below and in Table 1. Engineers Australia's assessment of the state of infrastructure in 2001 and 2005.

Transport infrastructure

Sustainable transport systems should include use of non-polluting, non-depleting energy sources and the least carbon and pollutant intensive infrastructure investment, transport technologies and modal options. The following sections set out the major challenges facing roads, rail, ports and airports in achieving sustainability for the future.

Roads

In 2001, the Engineers Australia ratings for roads ranged from C to D. The ratings in 2005 were: national roads C+, State roads C, and local roads C–, which have all improved marginally since 2001.

Overall, national roads are only adequate, despite upgrade work on the eastern seaboard. State roads vary greatly in quality and increased traffic is reducing local amenity.

In Australia's major cities, the physical growth and social attitudes have developed around the rel-

Table 1: Engineers Australia Infrastructure Report Card results

Infrastructure category	Sub-category	Aus 2005	Aus 2001	NSW 2003	Qld 2003	Vic 2005	SA 2005	NT 2005	WA 2005	Tas 2005	ACT 2005
Roads	National	C+	C	C+	C+	C	C	B–	B–	B+	
	State	C	C–	C+	C	C–	C–	C–	B–	C	
	Local	C–	D	C–	C	C–	D	C–	C+	D+	
	Overall	C	C–	C	C	C–	C–	C	B–	C	B
Rail		C–	D–	D	C+	C–	C	A	C+		
Electricity		C+	B–	B	D+	C	B–	B–	B–	B–	B
Gas		C+	C		C	C	B+	A	B+		A–
Ports		C+	B		B–	C		B+	B–	B	
Water	Wastewater	C+	C–	C–	C+	B	C	C	B–	D+	B
	Potable water	B–	C	C	B	B	C+	B–	B–	D+	C
	Stormwater	C–	D	D	C	C–	D	C+	C+	C–	C
	Irrigation	C–	D–		C+	D		B	C+		
	Overall	C	D+	C–	C+	B–	C–	C+	C+	D	C+
Airports		B	B		B			B+			B
Overall		C+	C	C–	C+	C	C	B	B–	C–	B–

ative convenience of the private vehicle, placing demands on the road system that are becoming unsustainable. There are both current and future issues with congestion. The total cost of congestion is estimated to be $3 billion per year and is expected to triple in the next 20 years if remedial measures are not taken. Initiatives are needed to reduce private vehicle usage and congestion in metropolitan areas of Australia.

Freight traffic and freight volumes in Australia are expected to increase, with the freight task expected to double in tonnage terms by 2020. Therefore, action must be taken to review freight routes and improve inter-modal transfers, particularly at ports.

The highly decentralised nature of Australia's population means that the transport industry is pivotal to the effective functioning of the economy. Future traffic growth will affect the entire network, not just those areas surrounding population centres.

Some state roads, particularly those in rural areas, require attention to raise them to an acceptable level of service. A lack of renewal expenditure on the local road network has led to the generally poor quality of local roads.

Improvements to substandard sections of state roads and bridges are progressing at a slower rate than desirable and local government infrastructure standards are falling behind (mainly in the rural areas).

A national framework for planning and allocating priorities for road funding has been lacking.

Rail

In 2001, Engineers Australia rated rail between A and F with an overall rating of D–. In 2005, rail improved its rating to C–. There have been some notable improvements, but widespread delays remain and there are uncertainties with new investments.

There is emerging congestion in many metropolitan rail networks. As well, an integrated transport network is needed to connect regional activity centres, and to encourage the transfer of commuters from road to rail.

Improved infrastructure is required to support the transfer of inter-state freight from road to rail and rail access to ports should have an appropriate level of investment. The Melbourne to Sydney to Brisbane corridor is sub-standard and is seriously affecting rail freight viability. There is an urgent need for funding.

Solutions include urgent upgrading of existing infrastructure, implementation of proposed projects, better national coordination and a less complex regulatory environment. A very high speed train between Melbourne to Sydney to Brisbane is one of the options that needs serious consideration.

Ports

In 2001, Ports were rated B by Engineers Australia. The rating declined in 2005 to C+. The main concerns with ports are coordination with land and air transport systems and urban encroachment, limiting the ability to expand.

Channel deepening is required for the Port of Melbourne to meet future growth in the size of ships, but taking into account environmental concerns. The impending deepening of Melbourne Port will necessitate deepening of Fremantle Port to maintain competitiveness.

The current and continuing boom in the minerals sector in Western Australia, particularly those industries at the Burrup Peninsula, will necessitate the continued upgrading and provision of facilities to ensure that state's ports can compete domestically and internationally.

The current system, whereby common user facilities are built only with a guaranteed return on investment (that is, a commitment from industry is required before more infrastructure is built), could create a bottleneck, with facilities not being provided in a timely manner. There are localised bottlenecks, with some ports facing immediate capacity restraints, as identified by the Prime Minister's Exports and Infrastructure Taskforce (2005). For some ports, the time and cost issues with obtaining permits for strategic developments, capital works and maintenance works, including dredging, require resolution.

Airports

Airport infrastructure is generally in good shape and was rated B in 2001 and achieved the same rating in 2005. Certified and registered airports are required to undertake annual technical and safety inspections. Airports are heavily regulated for safety and security, which tends to mean that assets are in a good condition.

However, future expansion will be needed to support increased passenger movements. Expansion will be accompanied by noise impacts on nearby residential areas, and this will remain a problem. Solutions to date include quieter aircraft, noise insulation and flight path sharing. These measures assist, but a total solution is a long way off.

Water infrastructure

Water infrastructure includes potable water, wastewater and stormwater systems. Sustainability in this infrastructure sector would require initiatives involving alternative water supplies, such as re-use, stormwater harvesting and promotion of sustainable suburbs through water-sensitive urban design.

Potable water

Metropolitan Australia is spreading and much of the water infrastructure in metropolitan Australia is old. Catchment areas for water harvesting are limited. The barriers to providing new water solutions are generally not technical barriers, but rather political, environmental and social.

Potable water was graded C in 2001 by Engineers Australia. In 2005, potable water improved to B–. The latest rating recognised increased investment in renewing pipe networks, improved treatment and reduced water losses from system leakage.

However, spending on renewals is not keeping up with the rate of asset deterioration. Problems still exist with excess water use and encroachment on catchments by urban areas. Both can be solved through appropriate policy and planning of development, such as reconsideration of pricing principles to reflect the true cost of water and establishment of urban growth boundaries.

Record droughts in recent times have highlighted the need for new sources of supply. The magnitude of anticipated water efficiencies through demand management is ambitious and will require permanent community behavioural change. Water restrictions and other conservation measures will not solve all of the problems.

Planning and development of water sources and infrastructure to keep up with increasing population growth, demand for water and declining rainfall and run-off into reservoirs has established that major capital expenditure is necessary for many years to come. There is a need to develop water-treatment technologies to a higher level of efficiency and to lower costs.

Wastewater

The rating given by Engineers Australia for wastewater in 2001 was C–, which marginally improved to C+ in 2005. This result has been achieved largely through increased investment since 2000. Rehabilitation of existing infrastructure and improved treatment have resulted in reduced discharge of pollutants into waterways. But problems still exist.

The level of re-use has been disappointing. However, there are increasing instances of local government operating wastewater re-use schemes. This is a positive development within the water industry, which protects the nation's natural waterways and also leads to greater sustainability of Australia's water resources.

Many collection systems are old and suffering from advanced deterioration. As many sewer mains are operating at their capacity limits—and many of those systems are nearing the end of their useful service life—there is a need for detailed strategic planning of the upgrading and replacement of those assets.

Stormwater

In 2001, stormwater was rated D by Engineers Australia. In 2005, it was rated C–, which was a marginal improvement. The quality of stormwater infrastructure varies widely. Much stormwater infrastructure is old and funds for maintenance, repairs and renewals are lacking.

Damage from cyclones and floods damage is expensive, with flood damage averaging over $300 million per year. Stormwater pollution also contributes to waterway degradation, which is a major issue in urban areas such as Sydney. Fragmentation of responsibilities for management, regulation and control is also a problem. Solutions include coordinated catchment management, improved land-use policies and implementation of systems that reduce run-off and replicate natural water cycles.

New development in the Coomera area between Brisbane and the Gold Coast is a good example of integrating potable and recycled water with stormwater management (see Chapter by Diaper *et al.*).

Knowledge on the condition of many stormwater assets is limited, and increasing demand and the ageing of assets has created a need for improved strategic planning in their management.

Energy infrastructure

Energy infrastructure includes electricity and gas. Energy sustainability is about mitigating Australian and global environmental impacts from energy use, notably the greenhouse impacts of energy production, transformation, supply and use.

Electricity

The quality of the infrastructure in the electricity sector was rated B– in 2001. In 2005, the rating was C+, which reflects distribution problems, higher peak loads and difficulties related to future investment.

Growing summer demand for electricity presents challenges to the adequacy of the existing infrastructure. Given Australia's projected growth in energy use and the current dependence on high-emission fossil fuels, the required increase in electricity production presents a significant challenge. For instance, Victoria is faced with particular challenges associated with its reliance on brown coal for generation. Forecasts suggest that Australia will require between 74 and 102% more electricity to be produced by 2020.

For the transmission network, the next decade is likely to present significant new challenges as the current configuration adjusts to changes in the pattern of supply and demand and, in particular, as the future of brown-coal generation becomes clearer.

For the distribution networks, maintaining network performance and responding to the dynamic changes flowing from the introduction of interval metering will provide new challenges. There are problems related to the lack of interconnections between the various state grids. This lack of interconnectivity inhibits the full benefit of the National Electricity Market being realised and contributes to lower supply reliability than is desirable.

In the future, the electricity industry must reduce greenhouse gas emissions, accelerate the development and use of sustainable energy resources, and actively educate consumers to reduce consumption.

There will continue to be an increased uptake of renewable sources as utilities work towards renewable energy targets and public sentiment encourages more sustainable activity. However, the distribution networks are not easily able to accept power from renewable sources such as wind or hydro in some instances.

Gas

In 2001, gas was rated C. In 2005, the rating was C+. However, while the condition of assets is good, many pipeline systems do not have adequate capacity to meet future demand.

Demand for natural gas is projected to increase due to the growing availability and competitiveness of natural gas and increased incentives for, and competitiveness of, low-emission energy sources. Much of the new demand is predicted to come from a growing share of gas-fired electricity generation

At projected levels of gas consumption for eastern Australia, gas reserves in the Otway and Gippsland Basins will be depleted between 2015 and 2030, unless significant new discoveries are made. Over the longer term—in the absence of significant discoveries—eastern Australia will rely on gas shipped from other markets.

There is a need for continued investment as gas supply and delivery infrastructure assets age. This, coupled with the natural process of resource depletion (requiring development of higher cost reserves and/or more remote greenfield reserves), will make maintaining competitive prices more challenging over time.

Solution and vision

While noting the generally sound state of the nation's infrastructure, significant parts are ageing and nearing the end of their economically useful lives. To have sustainable infrastructure for the 21st century, infrastructure providers must reduce greenhouse emissions, and lower pollutant levels in stormwater and effluent discharges into rivers and oceans. Resources (particularly water and energy) are limited and need to be managed through conservation, re-use and renewable strategies.

Improving social sustainability outcomes includes reducing commuter times, increasing road safety, improving air quality and providing access to broadband communication to all citizens.

Economic sustainability means ensuring that we have taxation and regulatory systems that promote new private sector investment in all types of infrastructure capable of generating adequate returns on investment. Lack of funding for infrastructure is a fundamental issue. Budgetary commitments to critical infrastructure elements are often only short term. Long-term planning is critical. As well, governments have been reluctant to use public debt funding.

There are numerous competing priorities for limited funds and there is little provision for funding to tackle changing community expectations and levels of service, such as for effluent re-use or improved public transport.

The provision of infrastructure grants often only cover capital works, with no allowance for ongoing maintenance. Infrastructure renewal studies undertaken within various infrastructure sectors throughout Australia have generally found that the level of investment in infrastructure renewal and maintenance is not sufficient to maintain service level standards or achieve the best life-cycle cost outcomes. There is an immediate need for increased funding for maintenance and renewals.

Econtech (2005) has estimated that significant infrastructure investment backlogs had developed in Australia. In particular:

- electricity has an under-investment of $1.15 billion
- gas has an under-investment of $2.6 billion
- road has an under-investment of $10 billion
- rail has an under-investment of $8.06 billion
- water has a potential under-investment of $3 billion
- the total estimated under-investment is $24.81 billion.

Recent public debate has emphasised the need for much better funding mechanisms to provide for current infrastructure needs and, just as importantly, to provide infrastructure for future generations.

There must be a significant increase in the amount of investment for both new infrastructure and ongoing maintenance of existing infrastructure. Funding must come from both the private sector and the public sector if Australia is to have the infrastructure it needs to be socially, environmentally and economically healthy.

There is an increasing recognition that the declining state of existing infrastructure will not be tackled without government leadership and coordination. There is recognition that greater public investment in infrastructure is both justified and effective and there is a realisation of the need for a better balance between public and private investment, and a more realistic assessment of the potential of PPPs.

The increased need for infrastructure investment will not go away. Billions of dollars will be needed to tackle the backlog of work, as well as meeting the changing needs caused by the ageing and growing population, and its move to new housing estates and to the coastal fringe.

More than 100 years after Federation, the development and maintenance of our infrastructure assets suffer from a lack of integration and coordination. This is directly related to our form of government, with the three tiers each having separate, and sometimes overlapping, roles and responsibilities. This issue is magnified by the general lack of long-term planning. Strategic planning needs to incorporate updates to accommodate changes in strategies and include long-term (at least 20-year) schedules of works and budgets to be effective.

There are some encouraging examples emerging of both longer term planning and infrastructure wide plans in the states and territories:

- Victoria—Melbourne 2030 (a 30-year strategic plan); and the Victorian Government White Paper on Water
- South Australia—South Australian Strategic Plan; the Strategic Infrastructure Plan for South Australia; the Regional Overview and the Draft Planning Strategy; and the Urban Stormwater Management Policy for South Australia
- Queensland—Integrated Transport Planning Framework; the Queensland State Infrastructure Plan; the South East Queensland Infrastructure Plan 2005; and the appointment of an Infrastructure Coordinator General
- Northern Territory—long-term strategic planning over a 20-year time horizon

- Western Australia—A Department for Planning and Infrastructure; a State Government Strategic Asset Management Framework; and an Investment Advisory Group
- NSW—State Infrastructure Strategic Plan 2002; NSW Infrastructure Council; and the recent announcement of the establishment of a new major projects unit.

With these initiatives occurring at a state and territory level, it is difficult to understand why there is such difficulty in taking a practical national approach to infrastructure.

The Council of Australian Governments released a communiqué on 3 June 2005 indicating agreement in principle to:

- hasten the long-term planning being undertaken under AusLink
- extend AusLink planning and coordination to ports and associated shipping channels
- each jurisdiction providing a report to COAG every five years on infrastructure
- the Commonwealth facilitating the establishment of groups to coordinate logistics chains of national importance
- re-invigorate the agenda for harmonising road and rail regulations
- establish 'one-stop shops' in each jurisdiction for project facilitation and approvals.

Engineers Australia believes this must be taken further, and has recommended the establishment of a National Infrastructure Council to ensure proper integration and coordination. The new Labor Government has appointed a Minister of Infrastructure, and has made a commitment to form a new statutory authority—Infrastructure Australia—to undertake a planning role for infrastructure projects of national significance. The new government has committed to converting the future fund into a building Australia fund.

All infrastructure sectors require significant enhancement before they meet Australia's current and future needs. The only way that this will occur is if infrastructure planning, maintenance and development become a priority for federal, state and local governments.

Adequate infrastructure underpins the Australian economy and the standard of living of all. As the current generation greatly benefits from the vision and commitment of those from the past, it is incumbent on them now to provide at least an equal legacy for future generations.

References

ACIL (2002). *Towards a Truly National and Efficient Energy Market*. COAG Energy Review Final Report. Commonwealth Government, Canberra.

Allen Consulting Group (2005). 'Investing in Australia's economic infrastructure'. Report to the Government of Victoria, Melbourne. <www.allenconsult.com.au/publications/view.php?id=304>

The Allen Consulting Group (August 2003). 'Funding urban public infrastructure'. Report for the Property Council of Australia, Sydney. <www.allenconsult.com.au/publications/view.php?id=258>

Association of British Insurers (2007). *Adapting to our Changing Climate: A Manifesto for Business, Government and the Public*. Association of British Insurers, London.

Business Council of Australia (2006). *Benchmarking the Progress of Infrastructure Reform: Challenges, Milestones and Outcomes*. Business Council of Australia, Melbourne.

Centre for International Economics (2006). 'Building wealth through infrastructure: setting priorities and valuing gains in New South Wales'. May 2006. Report prepared for Property Council of Australia, Sydney.

Committee for Economic Development of Australia (2005). *Infrastructure: Getting on with the Job*. Committee for Economic Development of Australia, Melbourne.

Commonwealth Department of Treasury (2007). *Economic Roundup Summer 2007*. Commonwealth Department of Treasury, Canberra.

Commonwealth of Australia (2007). *Intergenerational Report 2007*. Australian Government, Canberra.

Econtech (2005). Infrastructure in Australia's economic growth. In *Infrastructure: Getting on with the job*. Committee for Economic Development Australia, Melbourne.

Engineers Australia (2001). *2001 Australian Infrastructure Report Card*. Engineers Australia, Canberra.

Engineers Australia (2005). *2005 Australian Infrastructure Report Card*. Engineers Australia, Barton.

Commonwealth Government Exports and Infrastructure Taskforce (May 2005). 'Australia's Export Infrastructure – Report to the Prime Minister'. Australian Government, Canberra.

Port Jackson Partners (2005). 'The future for freight: economic analysis of the cost of moving freight on the inter capital city corridors'. Report prepared for the Australian Railway Association, Kingston, ACT.

Productivity Commission (1999) Microeconomic Reforms and Australian Productivity: Exploring the Links. Research Paper, AusInfo, Canberra.

Productivity Commission (2005) Modelling impacts of infrastructure industry change over the 1990s. Supplement to *Review of National Competition Policy Reforms*. Productivity Commission Inquiry Report, No. 33, Canberra.

40

Critical Foundations: Providing Australia's 21st Century Infrastructure

Michael Regan

Infrastructure is undoubtedly the least understood of the major asset classes in Australia. A tradition of public ownership and operation, its status as a public good and a lack of information about its investment characteristics in both public and private hands has contributed to limited recognition of its role in national and regional economies. However, this situation is changing. A coincidence of political, economic and financial events in the lead up to the worldwide economic recession of the late 1980s and Australia's microeconomic reforms of the 1990s bought into sharper focus the central role that infrastructure plays in both developed and developing economies. Evidence indicates that infrastructure contributes to an economy's performance, productivity and rate of growth. This suggests that establishing an optimal level of investment, improving coordination and planning by governments and greater private sector provision in supply and operation of infrastructure has important implications for Australia's future economic development and standard of living.

Dimensions

It is difficult to accurately measure Australia's infrastructure capital stocks and investment as a distinct group of assets. Infrastructure is not defined for international accounting or statistical purposes and values must be estimated from existing industry classifications. For statistical purposes, infrastructure includes the industry groups of gas, water and electricity utilities, transport and storage (economic infrastructure), which accounts for around 9.5% of gross domestic product (GDP), and health and education (social infrastructure), which accounts for a further 10.2% of GDP (ABS 2006a). Infrastructure generally refers to the structural framework of a country that facilitates economic and social activity. In economic terms, it means investment in fixed and durable assets, such as plant, equipment, buildings and civil works, in sectors such as road and rail transport, water resources, energy, urban transport, ports and airports and telecommunications. In social terms, infrastructure also describes services to human capital in the form of investment in fixed assets such as hospitals, public buildings and educational facilities.

Infrastructure accounts for around 23% of gross fixed capital formation—which reflects its capital-intensive characteristics—21% of net capital stock, 8.1% of employee remuneration and 12.2% of business gross operating surplus. Social infrastructure accounts for a further 25.6% of national employee remuneration, which reflects the labour-intensive nature of these industries (ABS 2006a, 2006b).

Infrastructure also has significant indirect impacts on the economy in a number of ways. First, the empirical evidence suggests that new capital spending on public infrastructure is linked to productive capacity, output, growth and productivity. It also lowers private sector costs, affects incomes and employment and it plays an important role in the efficiency and cost structures in most industries. However, infrastructure is not a group of homogenous assets. There are fundamental differences in industry structures, investment economics, regulation and supply chain networks. Nevertheless, infrastructure assets share several characteristics including:

- the requirement for large amounts of capital
- long investment horizons
- high levels of vertical integration in unbundled form
- low demand price elasticity
- low variable costs
- limited competition
- direct regulation
- a tradition of public sector provision and management.

Second, infrastructure also accounts for a significant proportion of the nation's capital stock and improvements in investment and operational efficiencies have implications for Australia's economic performance and international competitiveness. Third, infrastructure output and pricing are an input to most other sectors of the economy, with consequential impacts on inflation and the cost structure of nearly all industries. Fourth, evidence is emerging about the high multiplier value and positive externalities generated by infrastructure investment and improvements to the effectiveness of existing assets. This connection is shared with other developed economies although Australia's physical and economic geography, size and low population density suggests that infrastructure is especially important in this country.

Around 72% of economic infrastructure in Australia is owned by the public sector and operated directly by government (40%) or indirectly through the agency of government business enterprises (GBEs) (32%). The majority of state assets are in water resources, ports, energy utilities, railways and, urban transport (Regan 2005).

The public sector generally, and the GBE sector in particular, underwent significant reform during the 1990s. The strategy pursued by the Australian Government and, subsequently, the Council of Australian Governments (CoAG) was directed at improving microeconomic efficiency using a number of policy instruments including greater domestic competition, the corporatisation of GBEs, deregulation and the unbundling of vertically-integrated industries such as energy. During this period, Australia and other developed economies were embracing liberalisation as a solution to low levels of economic growth, low productivity, high unemployment and inflation that prevailed in the late 1980s.

The political economy of liberalisation favours a reduced role for the government in the economy and a greater role for markets. Further policy initiatives included the privatisation of GBEs and assets such as the capital city airports, the introduction of 'user pay' principles for government services and encouragement of private sector infrastructure provision.

The level of investment

Infrastructure is one of Australia's largest asset classes with a value of around $555 billion in 2006. Investment in infrastructure in Australia since the 1950s has declined in GDP terms (see Table 1). Aggregate public and private investment was at its highest level in the 1960s and early 1970s at around 6% of GDP. However, from the mid-1970s, aggregate investment declined to around 3% of GDP rising to 3.8% in recent years—well below the levels of the post-war years.

The decline in public infrastructure investment in Australia also occurred in other developed economies (Affuso *et al.* 2002; Coombs and Roberts 2007). A number of studies attribute lower infrastructure spending in developed economies to a group of factors, including:

1. The reform of public finances that occurred after the global recession of 1990 designed to correct poor rates of growth, large deficits and high levels of public debt. Governments are now reluctant to fund infrastructure through large deficits or taxation (Mehrotra and Valila 2006)
2. Microeconomic reform and changes in political economy designed to reduce the role of the

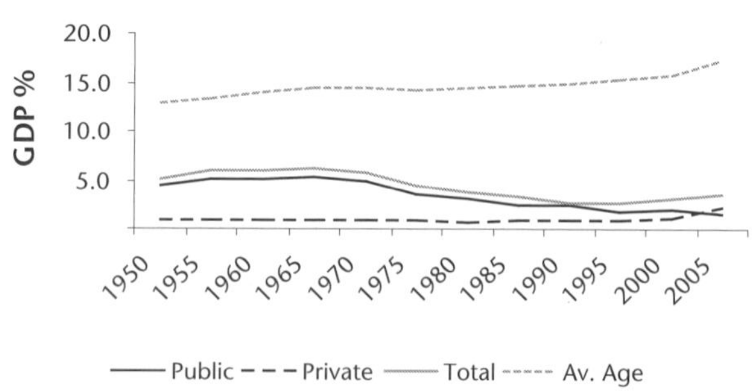

Figure 1: Infrastructure investment, Australia, 1950–2005 as a percentage of GDP. (Source: ABS 2006c; RBA Australian Economic Statistics 1950–1996, AGPS, Canberra).

Table 1: Gross Fixed Capital Formation, Australia, 1998–2006

$ million	**1998**	**2000**	**2002**	**2004**	**2006**
Electricity, gas and water	5,243	6,902	8,071	9,629	11,868
Transport and storage	10,629	12,030	13,007	19,078	22,722
Communications	4,128	6,016	5,397	5,426	6,858
Education	2,749	3,268	4,247	4,849	5,393
Health	3,157	4,052	4,722	5,342	6,689
General government	1,832	3,690	3,596	5,145	4,935
Total	27,738	35,958	39,040	49,469	58,465
GFCF	156,791	177,376	176,022	217,903	250,522
Infra GFCF%	17.69	20.27	22.18	22.70	23.34
GDP	703,258	769,045	813,542	873,197	921,747
GFCF:GDP %	22.29	23.06	21.64	24.95	27.18

Source: ABS 5204.0 2005–06 Chain Volume Measures

state in the economy and increase the role of markets through privatisation, competition policy, public private partnerships and the corporatisation of state-owned business enterprises (Bell 1997)

3. Improvements in technology and efficiency place greater emphasis on the selection of priorities for infrastructure spending and the efficiency with which infrastructure assets and services are managed (EPAC 1995a; Hulten 1996; Aschauer 1998)
4. Changes in the demand for infrastructure services as a result of changes in population growth, distribution and ageing in OECD countries (EPAC 1995a, p. 53)
5. Greater awareness of the negative effects, in economic and political terms, of new investment in assets such as motorways, airports and coal-fired power stations (EPAC 1995a)
6. Conversion of many infrastructure services to quasi-public goods and the negative effects of market pricing on demand (EPAC 1995a).

There are two further factors that have affected the level of investment in Australia. The first is Australia's unique geography, which includes a wide range of climatic variations, a large land mass, a small population and a low average density, although over 70% of the country's population is located in urban areas (O'Connor *et al.* 2001; Coombs and Roberts 2007). The second is a number of structural changes in the transition from manufacturing and heavy industry to the 'new' technology-based and services economy (Nieuwenhuysen *et. al.* 2001; Stillwell 2002). The drivers of structural change also affect the demand profile for infrastructure services and it is argued that the investment levels of 40 years ago do not provide an accurate guide to planners attempting to profile future demand for infrastructure services (Brunker 1992; Albala-Bertrand 1999).

In developed economies, the effects of new investment are less than in developing economies—returns are not much greater than for investment generally and the positive impacts in terms of output, growth and productivity are generally short lived (Barro 1996, p. 113; Demetriades and Manuneas 2000). A consequence of declining investment is an increase in the average age of infrastructure assets. In Australia, the average age of national capital stock increased from 13 to 17.4 years between 1950 and 2004 (ABS 2006a). The increasing age of Australia's infrastructure stock is reflected in high levels of capital consumption, with 56% of Australia's investment in the electricity, gas and water industries in 2006 accounted for by depreciation.

The condition of Australia's infrastructure

In GDP terms, infrastructure investment increased from 22.3% in 1998 to 27.2% in 2006 although it declined as a share of total capital stock. This

Table 2: Net Capital Stock, Australia, 1998–2006

$ million	1998	2000	2002	2004	2006
Electricity, gas and water	59,515	61,023	61,235	64,106	65,456
Transport and storage	163,789	169,139	173,983	189,175	208,178
Communications	51,107	57,143	62,122	65,240	70,052
Education	62,115	64,265	67,681	71,714	75,553
Health	49,754	53,558	58,304	63,469	69,812
General government	59,515	61,023	61,235	64,106	65,456
Total Infra NCS	445,795	466,151	484,560	517,810	554,507
NCS	1,918,195	2,050,648	2,155,626	2,318,284	2,649,236
Infra NCS %	23.2	22.7	22.5	22.3	20.9
Population	18,711	19,153	19,641	20,092	20,609
Infra NCS per capita	23.83	24.34	24.67	25.77	26.91

Source ABS 5204.0 2005–06 Chain Volume Measures

suggests that new investment is mainly being absorbed by capital consumption', which is demonstrated by an increase in the average age of infrastructure from an average 12.7 to 17.4 years in the 8 years to 2006 (ABS 2006a).

The repair and maintenance backlog for publicly owned infrastructure is a major concern in most developed economies. The backlog was created by cyclical cutbacks in maintenance expenditure programs and low short-term risks associated with reductions in routine repairs and maintenance compared with new capital investment (HM Treasury 1995). Repairs and maintenance are a significant component of public capital outlays—accounting for 46% of United Kingdom construction output in 2003 (NAO 2005a). Over the 20 year life span of a typical commercial building, ownership and operation (including opportunity rent costs and finance), repairs and maintenance may be five times the original cost of the building (NAO 2005a, p. 27). Accordingly, ownership and operation cost structures are an important component of infrastructure investment economics.

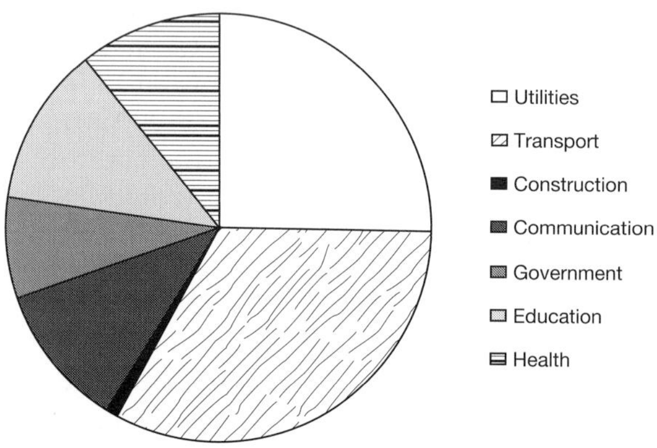

Figure 2: Infrastructure Sector, Industry Composition, Australia, 2006

The backlog is greatest for public buildings, water resources, ports, roads and railways. In the United Kingdom, the National Audit Office found that shortcomings in maintenance and repair were contributing to patron dissatisfaction and holding back urban and intercity patronage growth (NAO 2005b). PPPs and other alternative procurement mechanisms are tackling this problem by providing certainty for whole-life asset renewal, repairs and maintenance (Fitzgerald 2004).

A further consideration is the need for remediation work on Australia's existing infrastructure assets badly affected by under-investment and poor maintenance, resource management and land use practices. For example, the direct cost of landscape degradation in Australia is estimated at A$1.5 billion per annum. The reversal of existing rural degradation, including recovery from salinity problems, is estimated to cost A$8 billion and the loss of potential agribusiness production around A$1.7 billion annually. The capital estimates to restore Australia's 'core' economic infrastructure is around A$37.6 billion, of which roads will require A$16.8 billion over 22 years and water supply, drainage and irrigation around A$19.8 billion (IEA 2001).

There is currently much debate about public underspending on the nation's infrastructure (IEA 2002; Layton and Morton 2004; CEDA 2005; CIE 2005; CIE 2006; IPA 2007). Recent reports point to the poor condition of Australia's infrastructure assets (IEA 2001, 2002, 2005). The Institution of Engineers Australia issued its national report card in 2005 and identified improved ratings in some sectors (national highways, potable water, waste-water, railways, stormwater, irrigation and, gas production and distribution) and lower ratings for electricity and ports. However, the overall ratings are poor with no sector performing above the C level, which indicates that major changes are required to enable infrastructure to be fit for its current and anticipated purpose.

The Report Card identifies distribution problems, higher peak loads and impediments to future investment as problems for the electricity sector (IEA 2005). The other difficulties for this sector include its strong reliance on carbon fuels, low levels of investment in new generation and transmission assets, weaknesses in the operation of the national electricity market and the foreshadowed introduction of caps on greenhouse carbon emissions and a national carbon trading scheme.

In the port sector, the lower rating was attributed to increasing urban encroachment, which lowers the opportunity for ports to expand and coordinate with land and air transport systems (IEA 2005, p. 6). Ports face additional problems including strong growth in freight traffic and the reliance placed on supporting supply-chain networks—mainly road and rail services. International estimates of future demand anticipate a doubling of road and rail freight in OECD economies over the next 12–15 years (Cambridge Systematics 2004). This future demand cannot be met with existing or planned land transport infrastructures in Australia (Regan 2007). Ports are constrained by congestion on urban roads and capacity constraints on urban and regional rail networks. The much-publicised bottlenecks at the Port of Newcastle and Dalrymple Bay suggest that Australia's ports are not well equipped to deal with short-term changes in trade composition or demand. A further difficulty for the ports sector is regulatory failure and uncertainty governing future regulatory policy following the Commonwealth's announcement of its intention to transfer jurisdiction from state regulatory agencies to the Australian Consumer and Competition Commission.

Infrastructure and the economy

What role does infrastructure play in the broader economy and why are under-investment, poor coordination and planning a problem? The objective of most national economic policy frameworks is to encourage economic growth—or the increase in real GDP per capita. Growth theory and its application to developed economies has been the focus of extensive research for over 45 years. The interest in drivers of growth is understandable—increases in real output per capita mean better standards of living and a more prosperous economy. It follows that policies designed to foster growth and economic development are an important priority in most countries.

Contemporary growth theory had its origins in early work by Ramsey (1928) and Keynes whose work came to dominate macroeconomic management in the years following the Great Depression. In the 1950s, work by Solow (1956) and Swan (1965) laid the foundations for neoclassical growth theory, which offers a number of growth predictions including conditional convergence—the lower the starting level of per capita GDP, relative to the long-run or steady-state position, the faster the growth rate. The second prediction is that, in the absence of continuing improvements in technology, per capita growth will cease. Both predictions assume a diminishing return to capital. A problem for neoclassical growth theory was that some economies, such as the United States and Australia, continued to grow over long periods of time without experiencing diminishing returns to capital. This suggests wide differences between economies largely driven by political economy, institutional development, state interventions (such as policies to stimulate investment, competition, research and development) and the initial stock of human capital.

In the 1980s, Romer (1986), Lucas (1988) and Rebelo (1991) established endogenous growth theory. Central to this approach is the proposition that growth may continue indefinitely because the returns to investment in a broad class of capital goods (including human capital) do not diminish as economies develop (Barro and Sala-i-Martin 2001). This is offset by spillovers of knowledge across the economy and externalities generated by advancing human capital (especially health and education). Romer (1987, 1990) and others incorporated research and development and competition into the endogenous growth framework. They argued that technological advance results from purposive research and development activity that creates a form of ex post monopoly power. As long as ideas and inventions flourish, the long-run growth rate can remain positive. The neoclassical growth model maintains that the long-term growth rate depends on government-sponsored institutions, policy frameworks and interventions such as:

- an effective and investment neutral taxation system
- the rule of law
- the provision of essential infrastructure – ports, railways, roads, telecommunications, energy and water
- preservation of property rights; regulation of foreign trade; the level of development of capital markets (Barro and Martin 2001).

Recent research based on an empirical analysis of the determinants of growth for a sample of 87 developed and developing countries identified the following growth drivers:

- initial per capita GDP
- education (male upper level schooling)
- life expectancy
- fertility rate
- government consumption ratio
- the rule of law
- democracy (electoral rights)
- international openness
- change in terms of trade
- public and private investment
- monetary management (inflation) (Barro and Sala-i-Martin 2001).

It is against this background that interest developed in the role that infrastructure played in relation to productive capacity, output and growth.

Investment and growth

A recent subset of mainstream growth research in recent years has been the role of investment. The literature has focussed on three types of investment—investment in fixed assets, investment in human capital and investment in technology. The literature doesn't use the term 'investment' when it refers to the roles that human capital and technology play in economic development. However, the simple fact is that both require the allocation of significant resources—in the case of human capital development, it is significant state investment in health and education infrastructure and services is required. Technology also requires investment, public and private, as well as policy frameworks to encourage specialisation and provide strategic direction (such as the Queensland Smart State initiative). State investment is also necessary for supporting infrastructures, urban and regional planning and the creation of industrial clusters and transport nodes.

Research since the 1980s supports the view that investment in fixed assets, including information technology, improves productivity in developed economies and therefore makes a significant contribution to output, growth, private sector profitability and returns (De Long and Summers 1991).

Infrastructure and growth

In the late 1980s, researchers began to take a greater interest in the specific relationship between public infrastructure investment and economic growth. The literature on the subject is now extensive and probably accounts for most of the recent research around the drivers of economic growth. Allowing for significant differences between economies, there is now a significant body of international evidence that identifies the important relationship and interdependence between public capital in the form of infrastructure and regional economic performance. The overall conclusion that can be drawn from this evidence is that there is a relationship between investment in public infrastructure, output and growth. Infrastructure investment also affects the location of economic activity and can promote higher returns from industry specialisation and the formation of industry clusters, with both economic and social benefits. Infrastructure is also correlated with a reduction in private sector costs, private investment, productivity, employment and regional development (Regan 2005).

Although the evidence establishes a positive correlation in developed and transition economies, causation is believed to run in both directions—the benefits may be short-term in developed economies and the returns are diminishing over time. The other significant finding is that infrastructure investment per se is not necessarily effective in stimulating output or growth. It does require the concurrent operation of a number of other institutional and policy factors. There are also important differences between industries, with investment in land transport and telecommunications generating the greatest medium-term benefits (Fox and Smith 1990).

The evidence also suggests that infrastructure affects economic performance both directly and indirectly through the agency of capital deepening and other drivers of productivity (Louca 2003; Zugasti, Garcia and Moldonado 2001). What is also becoming clearer is that this is a complex relationship: causation runs in both directions and the results can vary between regional economies and over time. The evidence also suggests that the effectiveness of public infrastructure spending is influenced by a number of factors including economic and physical geography, industry structure and levels of urbanisation and industrialisation. The relationship between these variables is also dynamic, and the results of recent studies confirm that public infrastructure investment has differential impacts between and within regional economies over time.

The evidence is not unequivocal and considerable debate continues over measurement methods. Early research into the role of infrastructure was based on simple production function using time series macroeconomic data and a focus on output growth and productivity (Aschauer 1989a). The research that followed established a link between public infrastructure and these variables although estimates of the effect were excessive and causation remained a problem. In recent years, the research has examined causation issues—there has been wider use of disaggregated data and both value and physical measures of infrastructure investment. Single nation case studies and a growing body of evidence for regional economies are providing fresh insights. In particular, the role of endogenous and institutional growth theory, the effectiveness with which infrastructure is used, industry differences, the role of development policy and in particular, the role of private capital investment are now being explored.

A review of the empirical evidence to date suggests that, as a general rule, economic and social infrastructure contributes to an economy's productive capacity and is positively associated with the following indicators:

- output and growth (Aschauer 1989; Erenburg 1993; Flores de Frutos *et al.* 1998; Sanchez-Robles 1998; Kamps 2004; Pereira and Andraz 2005)
- productivity (Aschauer 1989; Wylie 1996; Everaert 2001; Pereira and Roca-Segales 2001)
- embedded technology (McCarthy 2001; Pakko 2002)
- private firm operating costs, returns and profits (Albala-Bertrand and Mamatzakis 2004; Ezcurra *et al.* 2005)
- employment and incomes (Peri and Cunat 2000; Seitz 2000; Chu 2001; Pereira and Roca-Segales 2001)
- private sector investment (Aschauer 1989; Erenburg 1993; Bosca *et al.* 2002)
- differences in regional development (Nadiri and Mamuneas 1996; La Ferrara and Marcellino 2000)
- the spatial development of industry and communities (Haughwout 2001).

Studies point to a positive and causal association between public investment in economic infrastructure and all of the above indicators (Gramlich 1994 Regan 2005). The empirical evidence suggests that there are several additional conclusions that can be drawn from international and single-country studies:

- There are major differences in the returns offered by different infrastructure industries—land transport and telecommunications generally offer greater productivity and growth returns than other industries (Canning 1989; Canning and Pedroni 1999; Affuso *et al.* 2000; Canning and Bennathan 2000; Datta and Agarwal 2004)

- A significant component of state-owned infrastructure services are not priced on the basis of production cost or opportunity cost (PC 2005)
- The positive association between infrastructure spending and growth does not occur in isolation, but requires the concurrent operation of other complementary factors, such as a competitive economy, effective regulatory frameworks, capital deepening, favourable development policies and private investment (Guild 2000; Rapin *et al.* 2000; Bassanini and Scarpetta 2001; Henisz 2002)
- Infrastructure generates higher returns in urban than regional areas (Aschauer 2000)
- Countries and regions pay a growth penalty for inefficient use of infrastructure (Bjorvatn 2000; Rioja 2003).

Optimal investment

Early studies by Aschauer (1989), Munnell (1990) and Morrison and Schwartz (1996) put the case that investment in the United States between 1970 and 1989 was sub-optimal and was responsible for a slow-down in the national GNP growth and productivity performance (Aschauer 1992). If the connection is important, how does a nation such as Australia establish an optimal level of infrastructure investment? Most international comparisons employ GDP per capita comparisons over time. However, the GDP measure fails to take into account differences between countries and regional economies in economic, physical and human geography, including land form, topography, population density and degree of urbanisation, the level of industrialisation, climate and the structure of the economy. These studies relied on a comparison of historical investment levels and used aggregate production data and, although these studies established a connection between the variables, the results were overstated.

More recent studies employing disaggregated regional data point to the importance of efficiency in infrastructure investment and management (Alexander and Estache 1997; Albala-Bertrand 1999; Rioja 2003). Hulten (1996) studied the effects of infrastructure inefficiency and identified a penalty of around 40% of the potential growth benefits of investment. Under-investment and poor reliability may also impose penalties on growth, output and private sector costs (Priyantha, Wijayatunga and Jayalath 2004). Hulten argues that how well an economy uses its infrastructure may be more important than how much it has.

Several recent studies have attempted to identify optimal levels of infrastructure using a number of alternative indicators of economic performance, including employment, productivity and social return (Barro 1997; Conrad 1997; Aschauer 2000; Albala-Bertrand 2001). Invariably, these models run into the difficulty of differences in the composition and structure of regional economies. Alternative suggestions favour centralised long-range planning although this gives rise to two particular sets of problems. The first is the role of the Commonwealth, state and local governments within a federal constitutional framework and the potential for different and, at times, competing objectives. From the Commonwealth's perspective, infrastructure affects Australia's economic performance, its international competitiveness, productivity, output and growth. It is also central to the Commonwealth's ongoing agenda for microeconomic reform. From the state's perspective, the priorities are regional – servicing population growth and distribution, urbanisation, regional output, employment and private sector costs. For local governments, the priorities are those of the state at sub-regional level together with economic development. The difficult task for state governments is striking a balance between investment to meet the demands of a growing population for energy, roads, schools, health services, urban transport and water and investment that promotes structural adjustment and economic development, such as telecommunications, services to international trade, long-haul freight rail, ports, airports, interstate highways and urban motorways.

Infrastructure investment must also be aligned with cyclical influences illustrated by present supply-chain bottlenecks experienced by coal exporters in New South Wales and Queensland, which demonstrates the difficulty in reconciling the long-term planning of infrastructure supply and short-term demand volatility. Recent increases in the price of coal and iron ore has led to increased production by Australia's coal and iron ore producers and exerted pressure on existing rail and port bulk-loading facilities. Commodity prices are volatile and producers can increase output at relatively short notice. For example, the price of coking coal averaged around US$50 a tonne between 1998 and mid-2004, increased to US$60 in mid-2003 and US$125 in mid-2005. Coking coal and iron ore doubled in price in the 12 months to mid-2004 (RBA 2007). The lead time to install new railway track and rolling stock is around 3 years, the expansion of storage facilities at ports is 18 months and expansion of port bulk-loading facilities takes around 3 years (ACPI 2006). It is simply not possible to put in place an adequate supply chain response in less than 3 years and that assumes that all of the actors in the supply chain were in agreement and willing to invest immediately.

New investment in capital-intensive rail and port facilities also involves other factors—stakeholder coordination and planning, speedy regulatory approvals and agreement on access and tariffs. The economics of new investment may require the re-writing of long-term contracts to mitigate the uncertainty of a drop in commodity prices and throughput. These uncertainties suggest that infrastructure supply involving commodity exports, international trade and the resources sector will rarely be in equilibrium with demand.

The Australian federal constitution gives responsibility for the provision of most infrastructures to the states. However, the division between Commonwealth and state powers is less clear today than it was 50 years ago and the High Court of Australia has shown a willingness to extend the Commonwealth's reach over activities that cross state borders. The state and territory governments carry responsibility for coordinating and planning infrastructure provision, frequently in conjunction with local governments. The interstate networks that support nearly all regional infrastructures need long-term planning and regulation well beyond state borders. Presently, state regulatory frameworks offer significant differences in their policy templates, criteria and timing for price determinations. The development of a national energy market, the Commonwealth's national water initiative, national highways and uniform regulation of telecommunications, aviation and ports suggests that the Commonwealth and Council of Australian Governments will assume greater responsibility for the coordination, planning and funding of networked infrastructure in Australia than has been the case in the past.

Supply will also be difficult to predict where climatic influences play a role. South East Queensland has the highest ratio of water storage per capita in Australia, but is facing severe shortages as a result of a prolonged drought, extreme variability in annual rainfall and high population growth. It is highly improbable that these conditions could have been anticipated by planners in the early 1990s. In industries such as urban transport, aviation and energy, where demand is less influenced by exogenous factors, supply can be planned over long lead times. However, even in these industries, changes to the regulatory environment to tackle greenhouse gas emissions will affect investment economics and require high levels of investment in fuel substitutes and cleaner fuel technologies. Energy prices will also have an impact.

Policy makers face additional problems. The underlying principle of fiscal federalism is that the primary role of the central government is income redistribution and the role of the states is economic development (Peterson 1995). The Australian model is characterised by vertical fiscal imbalance—the Commonwealth collects the majority of revenue and the states undertake most of the spending. In Australia, the state and territory governments are responsible for the provision of infrastructure and the Commonwealth provides the funds through a combination of avenues—reimbursement of GST revenue and untied and tied grants. This provides the Commonwealth with a degree of leverage in relation to state capital spending choices and priorities.

Governments must also source the finance for higher levels of infrastructure spending. This is influenced by a number of factors. The first of these is the capacity and willingness of governments at all levels to make long-term investments in capital-intensive assets involving long pay-back periods. Governments face short-term electoral cycles of 3–4 years, which puts capital spending on assets where the benefits are long term and difficult to measure at a disadvantage. Government finances infrastructure from debt or taxation or from consolidated revenue in competition with other programs. Fiscal management and taxes are sensitive issues at election time and undoubtedly contributed to the downturn in public capital spending in Australia over the past 15 years.

A further task for policy makers is finding an appropriate role for private investment. Private investors have a strong appetite for public infrastructure and offer a number of advantages to government—capital, management efficiency, improved service delivery and better alignment between incentive and performance than that offered by state-owned enterprises. However, the privatisation of existing assets and public private partnerships are not popular with sections of the community. Private investment is also not suited to public goods and, unlike government, requires a return commensurate with risk. Private investment is framed more by agreements with the state than market forces and, accordingly, special arrangements must be made for regulatory certainty and insulation from many state interventions that arise outside the scope of a project. Private investment must also be 'ring fenced' from existing networks, which is an impediment to long-range planning. For example, a tollway in a metropolitan area may involve closure of existing roads or it may limit the state's options for building connected or parallel non-tolled roads in the future.

The resolution of what is an optimal level of infrastructure development is not easily resolved. The evidence suggests that Australia should be considering a variety of methods to assess future infrastructure needs:

1. The efficiency of resource allocation and infrastructure management
2. The privatisation of GBEs in key infrastructure service areas where there is the potential for competition to improve performance and productivity at the enterprise level
3. Improved coordination and planning of Australia's infrastructure industries
4. The uniform regulation of infrastructures servicing interstate networks
5. Investment in research that examines enterprise returns, employment and productivity at the regional and firm levels.

Public investment

Australian governments have provided most public infrastructure since colonial settlement and around 71% of economic and social infrastructure in Australia is currently financed and operated by governments or government business enterprises (GBEs) (ABS 2006b). Traditional state ownership and operation of GBEs in the infrastructure services area has attracted extensive recent research. Empirical evidence from a number of countries suggests that privatisation improves microeconomic efficiency and where firms are trading as vertically-integrated monopolies, improvements at the firm level flow to the broader economy (Megginson 2005). However, this improvement may come at the cost of higher regional unemployment. At the enterprise level, privately owned firms outperform GBEs across all standard performance measures—revenue, employee incomes, gross operating profit, productivity, return on equity and investment (Megginson and Netter 1998).

Why is traditional state procurement and provision of infrastructure so inefficient? Several reasons are put forward for this—GBEs are saddled with non-commercial objectives, unfunded community services obligations, output pricing is not linked to costs of production, boards of management are selected for political allegiance over competence and pork-barrelling practices govern asset allocation decisions. Recent studies point to a range of other factors. In the United Kingdom, Mott McDonald (2002) reviewed the outcome of traditional public procurement projects in the United Kingdom over a 20-year period. The study used a sample of 50 projects exceeding £40 million in value (2001 prices). The report identified cost over-runs in 73% of traditional construction projects. The results supported similar reviews by the National Audit Office in 2001 and 2003 (NAO 2001, 2003). The primary cause of cost over-runs was specification changes during construction by state agencies (Allen 2003) (Table 3). Late delivery occurs in 70% of traditionally procured projects.

Table 3: Performance of traditional and PFI (PPP) procurement 2002–04

	Traditional		PFI-PPP	
	a	*b*	*a*	*b*
Cost overruns	24–51%	73%	2%	22%
Late delivery	4–39%	70%	–16%	24%

Source: (a) Mott McDonald 2004; (b) NAO 2003.

The Mott McDonald report also examined PPP projects and found that 22% experienced cost overruns and 24% were delivered late. The cost overruns on PPPs were mainly the result of changes to specification or variations instigated by a public agency or third parties. Cost overruns on PPP projects were found to be significantly less than for traditionally procured projects.

The Mott McDonald Report also identified high levels of optimism bias with traditional procurement leading to underestimation of project costs, completion periods and operating expenditures and underestimation of project cost. This supports similar evidence by Flyvbjerg *et al.* (2003). The incidence of optimism bias was much lower with PPP projects and this is attributed to the rigor applied to the procurement process, the preparation of the public sector comparator (PSC), the identification and costing of risk and risk transfer.

State-owned enterprises and agencies do not necessarily set prices to align with production costs. Prices are determined by government or state regulators and there is wide use of cross-subsidisation to support uniform pricing for remote and disadvantaged areas and the provision of unfunded community service obligations. Where pricing is not referenced to actual costs of production, investments returns may be sub-optimal.

In Australia, GBEs in 2005 achieved a return on assets greater than the bond rate in only two sectors—energy (7.8%) and ports (7.2%). In the same period, average returns on the ASX for listed infrastructure averaged 19% within a range 14.7 to 57% (Regan 2007). However, this is changing with the private sector making a greater contribution to infrastructure provision in recent years than the public sector (ABS 2006a). The change is driven by a number of factors including the privatisation of GBEs in the 1990s, wider use of franchising and public private partnerships (PPPs) to procure large and costly infrastructure projects and the strong private sector appetite for infrastructure investments (Watters 2007).

Private investment

The first wave of microeconomic reforms in the 1990s included the corporatisation of GBEs and the subsequent divestment of those trading in competitive markets such as banking and finance, shipping, airlines and airports, energy generation and distribution, insurance, telecommunications and gaming. The privatisation program raised $144 billion for the Commonwealth and State Governments in the 12 years to 2002. At the same time Australia was seeking greater private participation in infrastructure investment, increases in the superannuation guarantee levy were accelerating domestic savings and superannuation fund managers were developing an appetite for long-dated investments offering stable, indexed revenue streams. In 1995, the Commonwealth commissioned the Economic Planning and Advisory Committee (EPAC) to establish a task force to investigate and report on private sector involvement in the funding, management and control of infrastructure in Australia. The report found that the private sector has an important role to play in helping to meet Australia's future infrastructure needs (EPAC 1995). Fund managers invested most of their capital in listed equities and the Australian Stock Exchange (ASX) market capitalisation increased from $288 billion in 1994 to $682 billion in 2000.

The growth in private investment in infrastructure also occurred with private provision of infrastructure assets using franchises such as build-own–operate-transfer (BOOT) and public private partnerships (PPPs). In the late-1990s, several portfolio investment vehicles were established to acquire assets in energy, airports, ports, public transport and motorways (Infrastructure Trust of Australia 1996; Australian Infrastructure Fund 1997; Infratil 1998). A second wave of new listings from 1999 featured single-asset investments (Hills Motorway, Transurban, Macquarie Airports, ConnectEast and RiverCity Motorway). Many assets were acquired by unlisted vehicles including Brisbane, Melbourne and Perth airports, several of the Victorian-based energy generators, the Sydney Harbour Tunnel, Melbourne's public transport operators and Adelaide water. Many infrastructure services in the accommodation, utilities, and health and education sectors were provided by unlisted private corporations.

The privatisation of GBEs and assets in the 1990s led to rapid growth in capital markets and strong investor interest in listed infrastructure investments. In 1995, the new Infrastructure Index of the

Australian Stock Exchange had a market capitalisation of $17.2 billion. The Index was the strongest performing sector on the ASX over this period, highlighting strong institutional investor appetite for revenue-stable, long-term investments and the emergence of increasingly sophisticated hybrid investment structures designed specifically for infrastructure assets. In 2002, a proxy for the Index, which was discontinued in 2001, had a market capitalisation of $25.6 billion and in 2007, this stood at $1.04 trillion (excluding the telecommunications sector) (AFR 2007). An estimated $550 billion in assets is held in unlisted private ownership.

Alternative procurement methods

In Europe, early models of private infrastructure were franchises involving medium-term outsourcing arrangements that employed build, own, operate and transfer (BOOT) principles. These transactions were employed by state and local governments as a substitute for public capital and were based on an input specification for the asset. Following the Private Finance Initiative that was introduced in the United Kingdom in 1992, Victoria commenced using public private partnerships methods for outsourcing procurement of large and complex projects in the late 1990s. In 2000, Victoria developed a comprehensive policy framework, which has evolved into the most advanced template for this form of procurement in Australia and has provided the template for similar programs overseas.

PPPs developed from the asset financing model and represent a more rigorous approach to state procurement. PPPs are a form of outsourcing for large and complex projects such as tollways, public hospitals, water supplies, corrective services, public buildings, convention and conference centres and transport facilities. PPPs are not, in fact, partnerships nor are they the privatisation of state assets or businesses. They are contractual arrangements for the provision of goods and or services to, or on behalf of, government. PPPs feature a franchise, lease or licence for the financing and management of service delivery for a term of years. In the case of services such as policing, public hospitals and schools, the services delivered by the franchisee are non-core and relate solely to buildings, plant and support services. PPPs employ an output specification which creates the incentive for private firms in a competitive bid process to use new technology and innovation in their approach to service delivery.

A PPP is commissioned when the private bid offers a lower procurement cost than traditional public procurement using the PSC. The comparison takes into account risk transfer and is based on discounted cash flow measurement. The Australian PPP process observes the following protocols:

- the public sector prepares an output specification
- the public provision of the services is costed using value-for-money criteria, which require improved financial and public benefit outcomes over a risk-weighted model of traditional procurement
- the specification is put to public tender
- government decides whether to proceed with a PPP measured against value-for-money criteria (Dept of Treasury and Finance 2001, 2003).

PPPs are most commonly employed in the procurement of complex social and economic infrastructure facilities and services. These are the most difficult projects to deliver on time and within budget, and many public projects fail to achieve revenue and expense expectations (Mott McDonald 2002; Flyvbjerg *et al.* 2003). PPPs are typically long-term investment arrangements with tenure periods from 2 to 5 years in information technology and communications, 20 years with public hospitals and 25 to 35 years for tollways and car parks.

The history of PPPs is relatively brief and *ex post* determination of financial performance may prove difficult to measure once the project has been commissioned. From the public sector's perspective, a PPP is successful if it delivers the specified service outcomes over the life of the contract without full or partial service failure. Measures of service compliance include performance audits, consumer complaints and periodic user surveys (KPMG 2006; Mathias and Reddington 2006; Partnerships UK 2006). Service failure attracts financial penalties or payment abatement. From the private sector's perspective, a PPP is successful if the investors achieve the expected rate of return with minimal penalties or abatement of revenue. Arguably, the appropriate test for a satisfactory PPP is the service delivery record measured against the output specification and indicators of operational performance.

In the United Kingdom, the PFI program plays a limited, but important, part in public sector capital investment—accounting for around 12% of state capital spending in 1997–2003. However, it is expected to decline beyond 2007 (HM Treasury 2003a). In Australia, the Victorian government's 2005 budget papers indicate that PPPs account for around 10% of state capital procurement, 5% in New South Wales and less for the Commonwealth and other states.

Recent research suggests that PPPs are achieving value for money and other benefits in the delivery of public services compared with traditional procurement (ACPI 2006). This is mainly being achieved with risk transfer and the benefits include:

1. procurement cost savings in the range 9.0–19.7% (Allen 2003; Fitzgerald 2004; H.M. Treasury 2004)
2. asset delivery on time and within budget (Mott McDonald 2002; NAO 2003)
3. lifecycle asset costing (Fitzgerald 2004)
4. innovative design and service delivery methods and the introduction of new technologies (Fitzgerald 2004)
5. high user satisfaction levels (KPMG 2006; Serco 2006)
6. the sharing of windfall gains from revaluation
7. a low default rate compared with traditional outsourcing (Erturk and Gillis 2004).

The PPP model requires the responsible agency to prepare a risk-weighted model of traditional procurement (the PSC). The PSC is then compared with private bids within a competitive tender framework. The PPP procurement process includes a comprehensive business case and a series of 'gateways' or review stages. This process used for PPPs is now being employed in gateway programs employed in Britain and Victoria, with significant improvement to traditional procurement performance (NAO 2005a). PPPs are widely acknowledged as an improvement to the traditional method of procurement.

Conclusion

Infrastructure is one of Australia's largest and most important asset groups, which plays a central role in the nation's output, growth, productivity and microeconomic performance. It comprises a diverse group of hybrid assets whose performance is influenced more by industry structure, regulation and contracts than by market forces. There is considerable variation in the structure, economics, operational and investment characteristics of the various industries that make up the sector, suggesting that asset characteristics possess both generic and industry-specific attributes.

Australia faces many challenges in providing the infrastructure needed to meet future economic and social development. A guide to the management of future infrastructures is provided by the lessons of the past 50 years. This suggests that Australian governments have been under-investing in economic infrastructure since the early 1970s—our infrastructure is ageing and in poor condition and too much of current investment is being absorbed by consumption of fixed capital. The investment environment is affected by Commonwealth-state fiscal relations and vertical fiscal imbalance—governments are reluctant to use public debt to finance new investment even when the rate of return exceeds the cost of capital. Additionally, most infrastructure remains in the hands of GBEs, many of which are under-performing against most capital market benchmarks.

Australia possesses a unique human and physical geography and the primary challenge is to improve the efficiency of the nation's industries, productivity and international competitiveness. The empirical evidence suggests that institutional and policy reforms are central to these objectives, including programs to extend competition in service delivery, the uniform regulation of infrastructure businesses engaged in interstate trade, centralised long-range planning and coordination of key transport services and a greater emphasis on the allocation of resources and the efficiency with which infrastructure is managed. The success of PPP programs within Australia and overseas indicates a permanent role for this method of procurement. Infrastructure service delivery in the 21st century will require greater levels of public and private investment with the emphasis on the economics of investment and the composition of the economy as it continues the process of adapting to our increased global trade links, especially to the Asia Pacific region.

References

ABS (Australian Bureau of Statistics) (2006a). *Australian System of National Accounts, 2005–06.* Catalogue no. 5204.0, ABS, Canberra

ABS (2006b). *Australian National Accounts, National Income, Expenditure and Product.* Catalogue no. 5206.0. ABS, Canberra

ABS (2006c). *RBA Australian Economic Statistics 1950–1996.* Catalogue no. 5205. AGPS, Canberra

Affuso L, Masson J and Newbery DM (2000). Comparing investment on new transport infrastructure: roads vs. railways? *Fiscal Studies* **24(3)**, 275–315.

AFR (Australian Financial Review) 2007, Smart Investor, June.

Albala-Bertrand JM (1999). 'Infrastructure shortage: a gap approach', Working Paper. Department of Economics, Queen Mary, University of London.

Albala-Bertrand, JM and Mamatzakis, EC (2004) The impact of public infrastructure on the productivity of the Chilean economy, *Review of Development Economics* **8(2)**, 266–278.

Alexander I and Estache A (1997). *Infrastructure Restructuring and Regulation: Building a Base for Sustainable Growth.* World Bank, Washington DC.

Allen G (2003). 'The Private Finance Initiative (PFI)'. Library of the House of Commons Research Paper 03/79, Economic Policy and Statistics Section, London. <www.parliament.uk>, viewed on 25 February 2005).

Aschauer DA (1989). Is public expenditure productive? *Journal of Monetary Economics* **23**, 177–200.

Aschauer DA (1992). Infrastructure, productivity and economic growth: fair dinkum? In *Proceedings from the Bureau of Industry Economics Infrastructure Forum.* Canberra 1–2 June. Occasional Paper 7, BIE, Canberra.

Aschauer DA (1998). 'Public capital and economic growth: issues of quantity, finance and efficiency', Working Paper no. 233. The Jerome Levy Economics Institute, Annandale-on-Hudson, New York.

Aschauer DA (2000). Do states optimise? Public capital and economic growth. *Annals of Regional Science* **34(3)**, 343–363.

Australian Centre for Public Infrastructure (2006). 'The economics of infrastructure networks and supply chains'. Technical Report 02/06, School of Enterprise, University of Melbourne.

Barro RJ (1996). *Getting It Right, Markets and Choices in a Free Society.* MIT Press, Cambridge MA.

Barro RJ (1997). *Determinants of Economic Growth, A Cross-Country Empirical Study.* MIT Press, Cambridge, MA.

Barro RJ and Sala-i-Martin X (2001). *Economic Growth.* 2nd edn. McGraw-Hill, New York.

Bassanini A and Scarpetta S (2001). *The Driving Forces of Economic Growth: Panel Data Evidence for the OECD countries.* Economic Studies no. 33, OECD, Paris.

Bell S (1997). *Ungoverning the Economy, the Political Economy of Australian Economic Policy.* Oxford University Press, Melbourne.

Bjorvatn K (2000). Urban infrastructure and industrialisation. *Journal of Urban Economics* **48(2)**, 205–218.

Bosca JE, Escriba FJ and Murgui MJ (2002). The effect of public infrastructure on the private productive sector of Spanish regions. *Journal of Regional Science* **42(2)**, 301–326.

Brunker D (1992). Commentary on Winston CM and Bosworth B, Public Infrastructure Investment in private provision of economic infrastructure. In Occasional Paper 7, Bureau of Industry Economics, Canberra.

Cambridge Systematics Inc., Economic Development Research Group Inc., Boston Logistics Group Inc. (2006). 'Guide to quantifying the economic impacts of federal investments in large-scale freight transportation projects'. Report prepared for the Office of the Secretary of Transportation. U.S. Department of Transportation, Washington DC.

Canning D (1989). 'The contribution of infrastructure to aggregate output'. Paper produced as part of a World Bank research project: Infrastructure and Growth, A Multi-Country Panel Study (RPO 680-89), Department of Economics, Queen's University, Belfast.

Canning D and Bennathan E (2000). 'The social rate of return on infrastructure investments'. Research Paper 1145, Public Economics Development Research Group and Infrastructure

Division, International Development Division, World Bank, Washington DC.

Canning D and Pedroni P (1999). 'Infrastructure and long run economic growth'. Paper presented at a conference of the Econometric Society, Midwest Macro Meetings at the University of Pittsburgh and the Indiana University macro workshop. World Bank, Washington DC.

Centre for International Economics (2006). *Building wealth through infrastructure, Setting priorities and valuing gains in New South Wales.* Centre for International Economics, Canberra.

Committee for Economic Development of Australia (2005). *Infrastructure, Getting on with the Job.* Growth 54, CEDA, Sydney.

Conrad K (1997). Traffic, transportation, infrastructure and externalities, a theoretical framework for a CGE analysis. *The Annals of Regional Science* **31**, 369–389.

Coombs G and Roberts C (2007). *Trends in Infrastructure, Economic Roundup, Summer.* Department of Treasury, Canberra.

Datta A and Agarwal S (2004). Telecommunications and economic growth: a panel data approach, *Applied Economics* **36(15)**, 1649.

Demetriades PO and Mamuneas TP (2000). Intertemporal output and employment effects of public infrastructure capital: evidence from 12 OECD Economies. *The Economic Journal* **110**, 687–712.

De Long JB and Summers LH (1991). Equipment investment and economic growth. *Quarterly Journal of Economics* **2**, 445–502.

Department of Treasury and Finance (2001). *Public Sector Comparator, Technical Note, Partnerships Victoria, Guidance Material.* DFT, Melbourne.

Department of Treasury and Finance (2003). *Public Sector Comparator, Supplementary Technical Note, Partnerships Victoria, Guidance Material.* DFT, Melbourne.

Economic Planning Advisory Commission (1995a). *Private Infrastructure Task Force. Interim Report.* AGPS, Canberra.

Economic Planning Advisory Commission (1995b). *Private Infrastructure Task Force, Commissioned Studies.* AGPS, Canberra.

Erenburg S (1993). The real effects of public investment on private investment. *Applied Economics* **25(6)**, 831–838.

Erturk E and Gillis TG (2004). Default behaviour of global structured finance securities. *The Journal of Structured Finance* **Fall**, 48-55.

Everaert G (2001). Public capital and productivity growth: evidence for Belgium 1953–1996. *Economic Modeling* **18(1)**, 97–116.

Ezcurra R, Carlos G, Pascual P and Rapun M (2005). Public capital, regional productivity and spatial spillovers. *Annals of Regional Science* **39(3)**, 471–494.

Fitzgerald P (2004). 'Review of partnerships Victoria provided infrastructure'. Final report to the Treasurer'. Growth Solutions Group, Melbourne.

Flores de Frutos R, Garcia-Diez M and Perez-Amaral T (1998). Public capital stock and economic growth: an analysis of the Spanish economy. *Applied Economics* **30(8)**, 985–995.

Flyvbjerg B, Bruzelius N and Rothengatter W (2003). *Megaprojects and Risk, an Anatomy of Ambition.* Cambridge University Press, Cambridge, UK.

Fox FWF and Smith TR (1990). *Public Infrastructure Policy and Economic Development, Economic Review, March/April.* Federal Reserve Bank of Kansas City.

Gramlich, E.F. 1994, Infrastructure investment: a review essay. *Journal of Economic Literature* **32**, 1176–1196.

Guild R (2000). Infrastructure investment and interregional development: theory, evidence and implications for planning. *Public Works Management Policy* **4**, 274–285.

Henisz WJ (2002). 'The institutional environment for infrastructure investment'. *Industrial and Corporate Change* **11(2)**, 355–389.

Haughwout A (2001). Infrastructure and social welfare in metropolitan America, Economic Policy Review, Federal Reserve Bank of New York, December. *Journal of Public Economics* **83(3)**, 405–428.

HM Treasury (1995). *Private Opportunity, Public Benefit, Progressing the Private Finance Initiative.* Private Finance Panel. Stationery Office, London.

HM Treasury (2003a). *PFI: Meeting the Investment Challenge.* Stationery Office, London. <www.hm-treasury.gov.uk> Viewed on 1 March 2005.

HM Treasury (2003b). *The Green Book, Appraisal and Evaluation in Central Government. Treasury Guidance.* Stationery Office, London.

HM Treasury (2004). *Standardisation of PFI Contracts, Version 3.* Stationery Office, London.

Hulten CR (1996). 'Infrastructure capital and economic growth: how well you use it may be more important than how much you have'. National Bureau of Economic Research Paper No. 5847. Cambridge MA.

Hulten CR (1996). 'Infrastructure capital and economic growth'. Technical Paper 5847. National Bureau of Economic Research. Cambridge MA.

Institution of Engineers, Australia (2001). *Infrastructure Report Card.* IEA, Canberra.

Institution of Engineers, Australia (2002). 'Public investment in infrastructure: justified and effective'. Discussion Paper, March. Institution of Engineers, Sydney.

Institution of Engineers, Australia (2005). *Infrastructure Report Card.* IEA, Canberra.

IPA (Infrastructure Partnerships Australia) (2006). *To Build A Nation.* IPA, Sydney.

Kamps C (2004). *The Dynamic Effects of Public Capital: VAR Evidence for 22 OECD Countries.* OECD, Paris.

KPMG (2005). *Effectiveness of Operational Contracts in PFI.* KPMG, London.

Layton A and Morton A (2004). 'The need for increased infrastructure investment in Queensland'. Report for the Civil Engineering Construction Alliance, Brisbane.

Louca J (2003). Multifactor productivity and innovation in Australia and its states. In *Productivity and Regional Economic Performance in Australia* (Eds C Williams, M Draca and C Smith). Office of Economic and Statistical Research, Queensland Treasury, Brisbane.

Lucas RE (1988). On the mechanics of economic development. *Journal of Monetary Economics* **22(1)**, 3–42.

Mathias M and Reddington E (2006). *Good people, Good Systems. What Public Managers Say.* The Serco Institute, January, London.

Megginson WL (2005). *The Financial Economics of Privatisation.* Oxford University Press, Oxford.

Mehrotra A and Valila T (2006). Public investment in Europe: evolution and determinants in perspective. *Fiscal Studies* **27(4)**, 443–471.

Morrison CJ and Schwartz AE (1996). State infrastructure and productive performance. *American Economic Review* **86(5)**, 1095–1111.

Mott MacDonald (2002). 'Review of large procurement in the UK'. Report for H.M. Treasury, Croydon.

Munnell AH (1990a). Why has productivity growth declined? Productivity and public investment. *New England Economic Review* **Jan/Feb**, 3–22.

Munnell AH (1990b). How does public infrastructure affect regional economic performance? *New England Economic Review* **Sept/Oct**, 11–32.

Munnell AH (1990c). Is there a shortfall in public capital investment? An overview. In *Proceedings of a Conference held in June 1990.* Conference Series no. 34. Federal Reserve Bank of Boston MA.

Munnell AH (1992). Policy Watch: Infrastructure investment and Economic Growth. *Journal of Economic Perspectives* **6(4)**, 189–198.

Nadiri I and Mamuneas TP (1996). Contribution of highway capital to output and productivity growth in the U.S. economy and industries. *Journal of Public Economics* **63(1)**, 57–81.

National Audit Office (2001). 'Modernising Construction'. Report by the Comptroller and Auditor General, HC87, Session 2000-01, 11 January. House of Commons, London.

National Audit Office (2003). 'PFI: construction performance'. Report by the Comptroller and Auditor General. HC 371. Session 2002-03, 5 February. House of Commons, London.

National Audit Office (2005a). 'Improving public services through better construction'. Report by the Comptroller and Auditor General. Case Studies, HC 354-1, Session 2004-05, 15 March. House of Commons, London.

National Audit Office (2005b). 'Maintaining and improving Britain's railway stations'. Report by the Comptroller and Auditor General, HC 132, Session 2005–06, 20 July. House of Commons, London.

Nieuwenhuysen J, Lloyd P and Mead M (Eds.) (2001). *Reshaping Australia's Economy.* Cambridge University Press, Oakleigh, Victoria.

O'Connor K, Stimson R and Daly M (2001). *Australia's Changing Economic Geography: A Society Dividing.* Oxford University Press, Melbourne.

Pacific Road Corporate Finance (2005). 'Exports and infrastructure'. Submission to the Infrastructure Task Force. Centre for International Economics, Canberra.

Pakko MR (2002). The high-tech investment boom and economic growth in the 1990s: accounting for quality. *The Federal Reserve Bank of St. Louis Review* **March/April**, 3.

Pereira AM and Roca-Segales O (2001). Infrastructures and private sector performance in Spain *Journal of Policy Modeling* **23(4)**, 371–384.

Peterson PE (1995). *The Price of Federalism*. The Brookings Institution, New York.

Ramsey F (1928). A mathematical theory of savings. *Economic Journal* **38**, 543–559.

Rapun M, Gil C and Pascual P (2000). 'Infrastructure and growth in the European regions'. A paper presented at the 40th European Congress, European Monetary Union and Regulatory Policy in Barcelona. European Regional Science Association, Brussels.

Rebelo S (1991). Long-run policy analysis and long-run growth. *Journal of Political Economy* **99(3)**, 500–521.

Regan ME (2005). *Infrastructure: A New Asset Class in Australia, Research Project 03/05*. Australian Centre for Public Infrastructure, University of Melbourne.

Regan ME (2006). 'Public private partnerships: adding value to public infrastructure procurement?' Research Paper 06/05. Australian Centre for Public Infrastructure, University of Melbourne.

Regan ME (2007). 'The Economics of the South East Queensland Infrastructure Plan 2007–2026. A Report prepared for the Department of Infrastructure, Queensland. Brisbane.

Reserve Bank of Australia (2007). Bulletin, February. RBA, Canberra.

Rioja FK (2003). The penalties of inefficient infrastructure 2003. *Review of Development Economics* **7(1)**, 127–137.

Romer PM (1986). Increasing returns and long-run growth. *Journal of Political Economy* **94(5)**, 1002–1037.

Romer PM (1990). Endogenous technological change. *Journal of Political Economy* **98(5)**, S71–S102.

Sanchez-Robles B (1998). Infrastructure, Investment and Growth: Some Empirical Evidence. *Contemporary Economic Policy* **16(1)**, 98–109.

Seitz H (2000). Infrastructure, industrial development, and employment in cities: theoretical aspects and empirical evidence. *International Regional Science Review* **23(3)**, 259–280.

Solow RM (1956). A contribution to the theory of economic growth. *Quarterly Journal of Economics* **70(1)**, 65–94.

Stilwell F (2002). *Political Economy*. Oxford University Press, Melbourne.

Swan TW (1965). Economic Growth and Capital Accumulation. *Economic Record* **32 (November)**, 334–361.

Watters JP (2007). 'Private sector financing and infrastructure funding models for sustainable economies'. A paper prepared for Committee for Economic Development of Australia, Brisbane.

Wylie P (1996). Infrastructure and Canadian economic growth 1946–1991. *Canadian Journal of Economics* **29 Special Issue**, S350–355.

Zugasti CAA, Garcia RG and Maldonadon JS (2001). The effects of public infrastructure on the cost structure of Spanish industries. *Spanish Economic Review* **3**, 131–150.

SECTION 9

Transitioning Attitudes to Sustainability

A majority of the contributions to the preceding sections of this book represent a predominantly supply-side perspective on pathways towards more sustainable cities. Many of the supply-side contributions have a strong technological focus, although each identifies a range of barriers to innovation diffusion that span the marketplace (pricing, incentives), government (procurement, regulation) and community (values, lifestyles, awareness, affordability). Foremost among these perhaps are the attitudes and values of those in government, industry and community towards the sustainability of their current practices and behaviours.

The chapters in this section explore progress in the extent to which sustainability principles and practices are registering on the radar screens of boardrooms, government departments and households. There is evidence of some progress, including growth in the number of companies now involved in corporate sustainability reporting (Nugent and Hughes); the emergence of the Australian Sustainability Index, which has been created to benchmark the triple bottom line performance of the top 200 locally listed corporations (Grey); and the role that community engagement can play by using both transmissive (centralised, top-down, broadcast) and transformative (bottom-up, community-based social marketing) models to change population behaviour in relation to consumption (Langworthy and Brunt). In all three domains it is clear that we are only at the beginning of the attitude transformation process.

However, when attitude and behaviour change is coupled with technology and design innovation, it then becomes possible to consider that the prospects for a sustainable planet Earth are real and not illusory.

PWN

41

Government and Sustainability Reporting

Marcia Nugent and Phil Hughes

The connection between sustainable urban development and government urban planning policy and regulation is an important one. Government is therefore a key player in the process of transitioning towards sustainable urban development in Australia. This chapter explores sustainability reporting as a method employed by government to integrate sustainability concepts and performance into their policy and operations.

Since the late 20th century, Australian governments have transitioned towards new models of governance and administration. (The term 'Australian governments' broadly refers to federal and state departments, agencies, government business enterprises, local governments/councils and government boards and committees.) Department/agency-based management or 'managerialism' (Nethercote 2004), business-orientated government, 'joined-up' government (Bogdanor 2005; Hood 2005), and service provision opportunities provided by technology (John 2002) have all been considered and implemented in different ways.

Increasing public awareness of environmental sustainability, perhaps reinforced through the current media and political focus on climate change and drought, has resulted in increasing pressure on the public sector to undertake another transition—this time towards sustainable development, which is broadly defined as an integrated approach to accounting for relevant economic, environmental and social dimensions. The call for transition to transparent and accountable government is also increasing. A number of governments have responded to the call for transparency and accountability through setting targets in published state strategies (for example, see South Australia's Strategic Plan 2006 and New South Wales State Plan 2006).

The government sector is large, with significant direct and indirect sustainability impacts. As at March 2007, over 1.6 million people in Australia were employed in the public sector—some 228,300 employees in the Commonwealth Government, 1,249,700 in state government and 168,400 in local government (Australian Bureau of Statistics 2007). Because it employs 16% of Australia's workforce, the scale of sustainability impacts of the public sector is significant. In addition, the public sector is responsible for providing advice and implementing government policy that directly affect the community, economy and environment. Furthermore, its capital spending programs can be a powerful vehicle for leadership in environmentally sustainable development.

How government facilitates and demonstrates accountability, transparency and leadership attitudes towards sustainability is a critical question in a transition to sustainability. A general trend in a number of large multi-national organisations in response to similar demands has been to develop strategic responses and internal sustainability programs, with the development of a sustainability report a key communication tool to report on performance change to stakeholders.

This chapter outlines progress in sustainability reporting and discusses the benefits, challenges and ways to develop sustainability reporting in the public sector as part of a broader response to sustainability.

Sustainability reporting—an overview

An abridged history of sustainability and sustainability reporting

At its most fundamental level, sustainability reporting is the reporting of organisational governance and environmental, social and economic risks, opportunities and impacts. The term 'sustainability reporting' in this chapter encompasses a range of alternative reporting terms, such as 'triple-bottom-line' reporting, 'corporate-responsibility' reporting, 'total-performance' reporting, and 'non-financial' reporting. It is not restricted to environmental sustainability, but encompasses all aspects of sustainability.

Compared with financial reporting, which has a held a legislated and mandatory standing in Australian organisations for well over 100 years, sustainability reporting is a relatively new field. Indeed, the modern concept of sustainability itself is relatively recent (see Kidd 1992 for information on the six roots of sustainability). The first widely acknowledged definition of 'sustainable development' was in Brundtland's 1987 report to the United Nations:

development that meets the needs of the present without compromising the ability of future generations to meet their own needs. (Brundtland 1987)

The Brundtland report articulated and extended the concept of responsible environmental management/development and subsequently led to the development of the concept of 'corporate sustainability' or 'corporate social responsibility' (Mays 2003; Porter and Kramer 2006). In 1999, the United Nations Global Compact outlined 10 principles to guide corporate behaviour relating to human rights, labour, the environment and anti-corruption—legitimising the international status of 'corporate social responsibility' (Sparkes 2006). The 2007 G8 summit officially recognised corporate social responsibility for the first time, producing a declaration regarding 'Growth and Responsibility in the World Economy' in which they invite:

companies listed on our Stock Markets to assess, in their annual reports, the way they comply with CSR standards and principles. (G8 Summit Heiligendamm 2007)

The concept of the 'triple bottom line' was also becoming increasingly popular in the late 1990s as a method of expanding the traditional financial bottom line focus to include social and environmental concerns (Elkington 1997; Norman *et al.* 2004).

More recently, the worldwide focus on climate change has further heightened public awareness of the concept of sustainability and the stewardship responsibility of organisations and individuals towards the environment. This rising general awareness and understanding of sustainability issues has increased stakeholder demand for transparent organisational sustainability reporting.

Sustainability in an organisation is a complex concept, but it is integral to all areas, activities and aspects of the organisation. Sustainability is 'all pervasive, a ball of wool that can't and shouldn't be unravelled' (Hunting and Tilbury 2006). Every division of an organisation should be involved in a move towards a greater organisational sustainability focus. Sustainability is a concept that typically emanates from the heart of an organisation (through culture and values statements) and out to the business operations (through budgetary processes, priority setting, policy development, business processes, and so on). Suggett and Goodsir (2002) identify the core characteristics present when an organisation embraces triple-bottom-line reporting as:

- accepting accountability
- being transparent
- integrating planning and operations
- committing to stakeholder engagement
- undertaking multi-dimensional reporting and measurement.

As such, sustainability reporting—the public communication of the sustainability values, plans and activities of the organisation—is more complex than simple quantitative impact reporting of such things as water use or Occupational Health and Safety (OHS) statistics.

Sustainability reporting has evolved to embrace the concepts of sustainability, corporate social responsibility and the triple bottom line. Environmental reporting represented the first transition (White 2007). Up until 1999, all non-financial reporting of the top 100 Companies in 16 countries (N100) and the top 250 of the Global

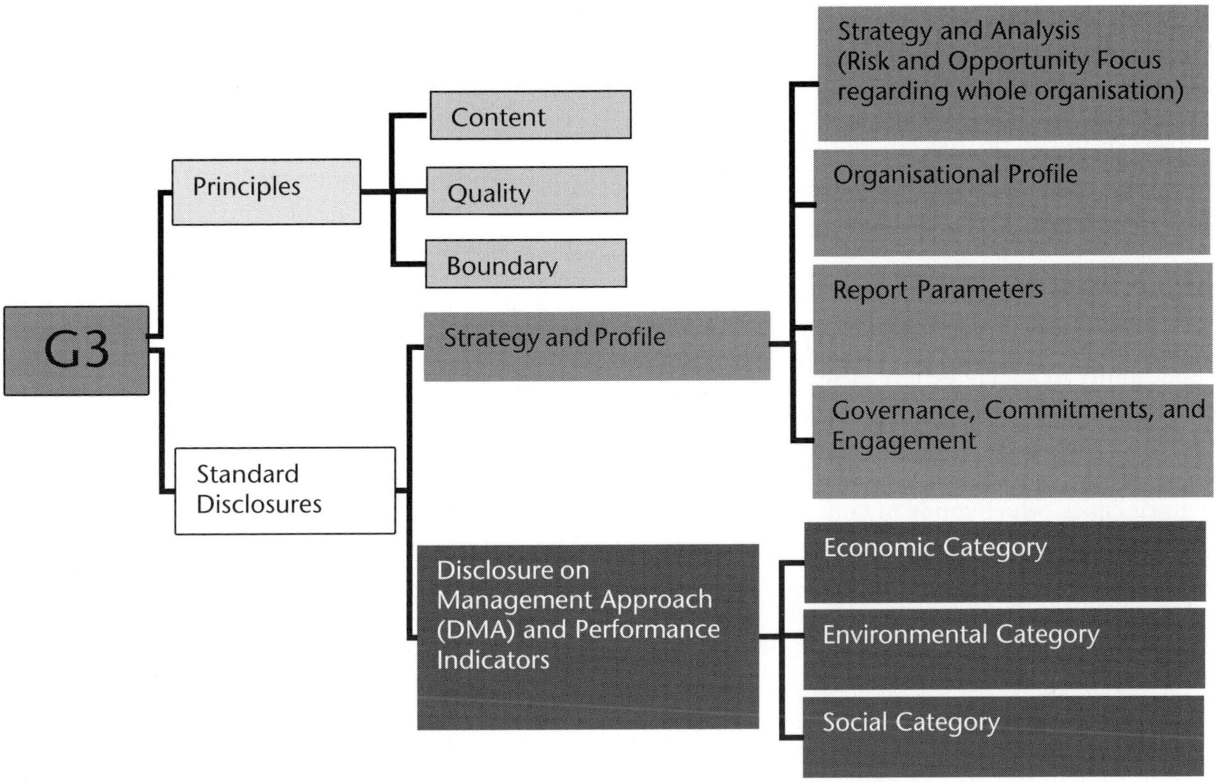

Figure 1: Outline of the Global Reporting Initiative (GRI) G3 Sustainability Reporting Guidelines (Source: adapted from: GRI 2006).

500 (G250) was environmentally focused. By 2005, 68% of non-financial reporting of the G250 and 41% of the N100 was badged as sustainability reporting (KPMG International 2005). Of the top 500 Australian companies, only six undertook sustainability reporting in 1995, but 65 produced a sustainability report in 2000 and 112 in 2005 (KPMG Australia 2005). Despite this considerable increase, Australia still has a relatively low level of sustainability reporting compared with international trends. Only 24% of the S&P/ASX 100 companies in Australia reported in 2005, compared with 81% in Japan, 71% in the UK, 43% in Finland and 37% in the USA (KPMG Australia 2005). However, sustainability reporting still remains more common in the Australian private sector than the Australian public sector, although Australian and New Zealand public agencies have shown a greater uptake of sustainability reporting than their international contemporaries (Centre for Public Agency Sustainability Reporting 2007a).

The Global Reporting Initiative (GRI)

The Global Reporting Initiative (GRI) was formed in 1997—founded by Ceres and the Tellus Institute (White 2007). Becoming an independent organisation in affiliation with the United Nations in 2002, GRI has developed a sustainability reporting framework that includes a suite of guidelines, protocols and supplements. Since 1997, over 1000 organisations have produced sustainability reports that reference the GRI Guidelines. The latest version of the GRI Guidelines (G3) was released in October 2006 and facilitates sustainability reporting based on the application of a set of sustainability reporting principles and standard disclosures (GRI 2006).

The G3 Guidelines consist of four report-content principles (materiality, stakeholder inclusiveness, sustainability context and completeness) and six report-quality principles (balance, comparability, accuracy, timeliness, clarity and reliability). Each of these principles has self-tests for organisations to determine how well the principle has been applied to their organisation's report. The GRI disclosure requirements are grouped into profile, management approach and performance indicators. The profile disclosures focus on the description, governance and stakeholder engagement of the organisation, while the management approach and performance indicator disclosures focus on organisational policy and procedure and impacts on the environment, economy and society (GRI 2006).

A major strength of the GRI framework is its applicability to a wide variety of sectors and types of organisations, thus promoting a standardised approach over sector and country borders. This is made possible particularly through the GRI reporting principles, which guide organisations through the interpretation and application of the guidelines, so that the report and reporting process is fit-for-purpose. Sector supplements, which provide focused instruction and indicators to discrete sectors—such as financial services, mining and metals, public agency and transport—further assist organisations to interpret the guidelines.

A review of the Sector Supplement for Public Agencies (GRI 2005) will be undertaken in 2007/08 to further develop the pilot version of the supplement to assist public agencies preparing sustainability reports at local, regional and national level. The pilot Sector Supplement for Public Agencies is based on the 2002 version of the GRI Guidelines and provides some additional guidance about the unique reporting components faced by public agencies as opposed to private enterprise. Rather than organisational impacts alone, a public agency must consider three levels of reporting—organisational, policy impact and contextual (GRI 2005). Contextual reporting may include jurisdictional reports like a State of the Environment report or community report and, although not every agency will undertake contextual reporting themselves, it is likely that each agency will somehow fit into a broader contextual reporting framework.

The uptake of sustainability reporting in the public sector

The uptake of sustainability reporting by the public sector has been considerably slower than the private sector (Dickinson *et al.* 2005; Calder and Culverwell 2005; Ball 2006). The role of government in the corporate responsibility and sustainability agendas seems to often focus on their regulatory and governance roles, rather than internal analyses of sustainability practices within their organisation and policies (Dickinson *et al.* 2005). The debate regarding the case for mandatory/voluntary reporting is an important consideration for the future of non-financial reporting, with public agencies participating primarily in their regulatory capacity commenting on the status of report-

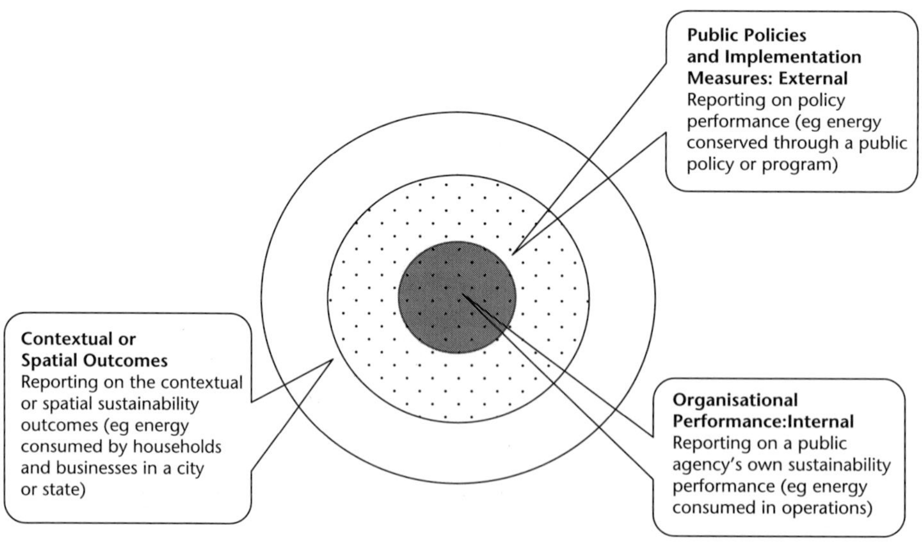

Figure 2: The three spheres of public agency reporting (Source: adapted from GRI 2005, p. 10).

ing requirements for the private sector (see KPMG and UNEP 2006).

The Centre for Public Agency Sustainability Reporting undertook a survey exploring the international take-up of public agency sustainability reporting in 2005. The report was described as a 'snap shot' of the experiences of public agencies undertaking sustainability reporting. It identified that, of the 60 targeted public agencies that provided data used in the quantitative analysis, 67% self-identified as having undertaken a sustainability report (Dickinson *et al.* 2005, p. 27). Although it used a very small sample, the study established that the greatest uptake year leading up to publication date was 2002, which mirrors the timing of the highest uptake in the private sector and the release of the 2002 GRI Guidelines (Dickinson *et al.* 2005, p. 28). It is noted that some of the self-identified sustainability reports were actually environmental reports or annual reports with a small number of environmental indicators. This affects the validity of the percentage uptake, but is also indicative of the level of general understanding of sustainability reporting in the public sector at that time.

Since the preparation of the survey in 2005, the Centre for Public Agency Sustainability Reporting has continued to engage with the public sector to encourage the uptake of sustainability reporting. As a general observation, the Centre has noted an emerging international groundswell of interest in 2007, perhaps influenced by the increasing profile of climate change and the resultant increasing demand from stakeholders and citizens for governments at all levels to report on their response. There is some risk in the coming years in Australia that the full breadth of sustainability reporting will be subsumed by the increasing interest in quantitative greenhouse-gas-emission reporting, which is only a small fraction of the reporting required to track and manage the social, economic and environmental impact of sustainability issues such as climate change.

Sparkes (2006) notes that a perception exists—particularly in the NGO sector—that a number of organisations are using non-financial reporting as a cynical public relations exercise. There is a real risk that sustainability reporting may be used as 'greenwash' by organisations looking to positively influence their reputation without substantially changing organisational behaviour. The GRI Guidelines are an important mechanism to reduce the risk of greenwash and enable those organisations embracing and reporting sustainability with the best of intentions to gain credibility. GRI G3 Guidelines set a best-practice benchmark for organisations and report users alike—providing a yardstick against which to measure the credibility of a report.

The 'public sector case' for sustainability reporting

Public agencies are formally accountable to legislative bodies and governments. They also have a civic responsibility to properly manage public goods, resources, and/or facilities in a way that supports sustainable development objectives and promotes the public interest. Public agencies are expected to be open and transparent in their management of public funds and assets. As significant employers, providers of services, and consumers of resources, public agencies also have a major impact on national and global progress towards sustainable development. Given their size and influence, public agencies are expected to lead by example in reporting publicly and transparently on their activities to promote sustainability. (GRI 2005:7–8).

The drivers and value of sustainability reporting are similar between public and private sector organisations in many ways. The business case for sustainability reporting appears to directly relate to an organisation's culture and may evolve over time in private sector organisations (UNEP *et al.* 2006). As described in Figure 3, it is suggested that the 'generation' in which an organisation's culture may sit determines the key concerns covered by a sustainability report and the value drivers underlying the report.

Drivers and benefits are often closely linked. Some fundamental benefits often cited about the preparation and production of a sustainability report include:

- allows holistic view of management implementation and integration of sustainable development
- potentially increases efficiency by reducing operating and compliance costs and improving operational management

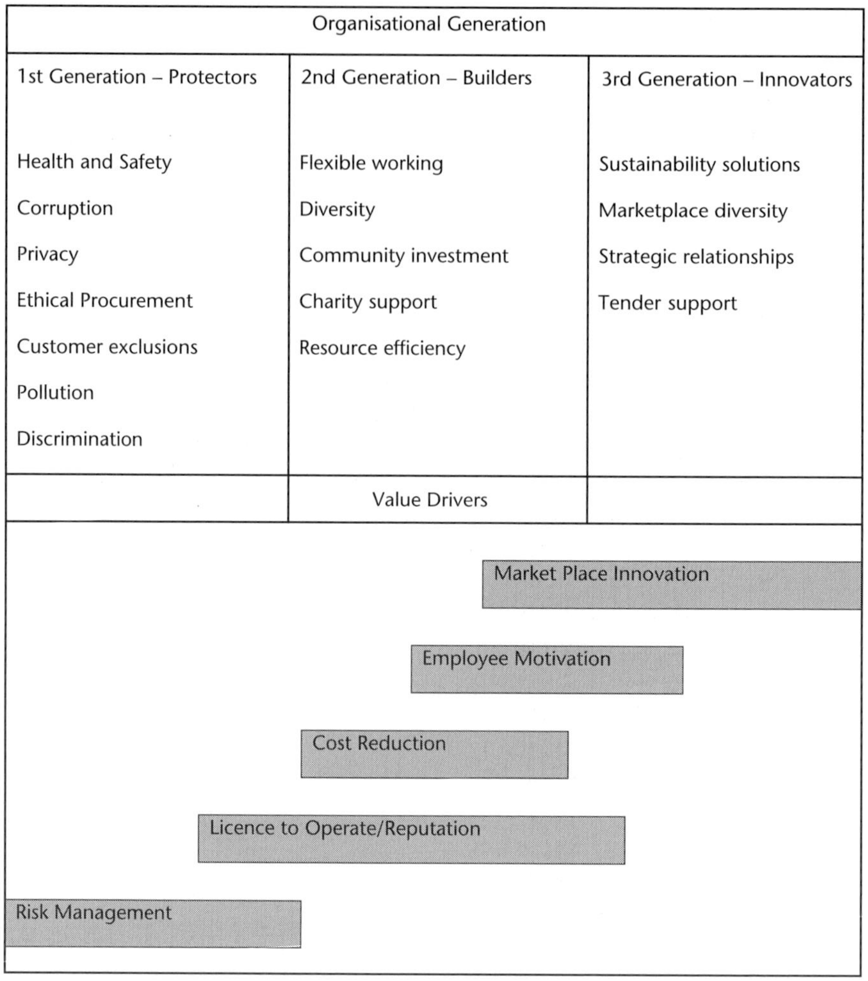

Figure 3: Business Case Evolution (Source: adapted from UNEP *et al.* 2006, Figure 4).

- attracts and retains high calibre employees as an employer of choice
- provides a business management tool to enhance knowledge and measurement of economic, environmental and social impacts and performance
- encourages organisational creativity and innovation.

The public sector has some particular opportunities to reap value from sustainability reporting. Public sector benefits may include:

- demonstrating performance and enhancing the business case for ongoing public funding and support
- allowing a sustainability filter to view operational risk and evaluating performance (internal and external policies)
- facilitating mutually beneficial stakeholder engagement and strengthening stakeholder relations and dialogue
- providing the ability to set benchmarks, both within and across sectors (with particular benefits for information flows between private and public organisations and inter-jurisdictional information flow)
- satisfying compliance and public policy commitment
- improving organisational accountability and transparency—promoting open and transparent governance
- demonstrating innovation and leadership by the public sector, and driving behaviour change (Centre for Public Agency Sustainability Reporting 2007b).

The 2006 Australian Commonwealth Parliamentary Joint Committee on Corporations and Financial Services report on Corporate Responsibility strongly supported the uptake of GRI and commended public agencies that were adopting sustainability reporting (Parliamentary Joint Commission 2006). The Corporation and Markets Advisory Committee also prepared a report in 2006 on corporate responsibility in relation to the Australian Commonwealth *Corporations Act 2001* (CAMAC 2006). The committee recommends that Government lead by example and promote responsible practices. This includes the:

application of performance and disclosure standards to department and other agencies and assessing, and reporting on, the workability and effectiveness of those standards. (CAMAC 2006, p. 169)

It seems, therefore, that the pressure felt by large private sector organisations to report is increasingly being felt in a range of ways by the public sector.

Beyond consideration of the tangible and immediate benefits gained through reporting, a number of intangible drivers that go to the heart of the role of government need to be considered. Beyond the scope of the political term of office and the life of public servant service, there is an ethical responsibility to ensure a government is not harming future prospects (social, economic and environmental) of the country, state or region it governs. It is the fulfilment of this ethical responsibility to future generations that is an intangible and fundamental benefit of managing organisational sustainability and undertaking sustainability reporting by a public agency.

Sustainability reporting challenges

Public agencies experience similar organisational challenges as private sector organisations when embarking upon sustainability reporting, such as cost and resource constraints, internal buy-in, availability of data, capacity and preparedness to report openly and transparently and costs of external verification (KPMG 2005). In addition, the public sector experiences a number of particular challenges, such as institutional arrangements, the politics of reporting and the embedding of sustainability in the machinery of government—all of which will be discussed in more detail below.

The Centre for Public Agency Sustainability Reporting works with a number of Australian and New Zealand public agencies (including federal, state and local governments and government-owned enterprises, such as water utilities) and its interaction with these agencies provides the basis of discussion in this section.

Organisational institutional arrangements, including processes, governance and structure, are important to smooth the uptake of sustainability reporting in any organisation. This is partially due to the interactive nature of sustainability reporting with the existing planning, budgeting and reporting mechanisms of the organisation, as well as the importance of good information flows and communication in the organisation to ensure quality reporting is achieved.

The political aspect of the budgeting and planning framework of a public agency adds a level of complexity to public agency sustainability reporting. Institutional arrangements—particularly in a state or federal model of central administrative control—may therefore present a particular challenge to public agencies. Linking sustainability reporting frameworks into an organisation's internal planning and budgeting framework is integral to achieving embedded sustainability reporting that will continue beyond the first flush of reporting endeavour and continue to add ongoing value to the public agency. Most importantly, sustainability reporting must make use of, be linked into and not duplicate existing corporate reporting systems and processes, which are typically complex in a public agency environment.

The institutional design of a public agency (within a portfolio or departmental structure) has also influenced the success of sustainability reporting for some public agencies. Inferred departmental responsibility appears to be affecting which state and federal departments/agencies are choosing to take up sustainability reporting. This is particularly visible in those agencies with environmental mandates. Shifting responsibilities, departmental restructures and perceived overlap of responsibilities has also negatively affected some agencies' sense of mandated ownership and leadership of sustainability reporting.

The perception that certain types of state and federal departments/agencies have greater obligation and use for sustainability reporting than others is contrary to the broad notion of sustainability reporting as an accountability document of organisational performance. Rather than being the responsibility of one department/agency within a government, the concept of sustainability requires that every area of government consider all aspects of sustainability in its decisions and operation, regardless of whether it is the core function of the department/agency. A material and balanced sustainability report for each department/agency of the government would be the ideal way to reflect a government that has transitioned to a sustainable development model. Any perceived obligation of the whole of a government to report cannot be fulfilled through one agency's report on their own impacts alone. Logically, a report of this type is not able to accurately represent the impact of an entire portfolio or government if focused on only one agency.

A sustainability report at a jurisdictional level may also have its place as unifying whole-of-government reports and would be an innovative reporting development. This type of report would also require input from all agencies, however, and a meaningful report would not be possible without integration of sustainability principles into the machinery of government and a joined-up government approach to sustainability reporting.

Local governments that are responding to sustainability reporting appear to be high performers already responsive to governance, community and environmental issues. Some ownership therefore already exists in the institution about issues central to sustainability reporting. Typically, a mature local government/council with good systems and processes is more likely to take-up sustainability reporting than one with basic institutional structure and systems—although there are some noted exceptions where local governments have recognised sustainability reporting as a transformational medium for organisational development. Particular challenges for local governments appear therefore to be institutional predisposition, capacity and funding. Local government is facing some particular financial sustainability issues in several states and a number are undergoing amalgamation and restructure, which may further affect individual council uptake of broader sustainability issues and reporting.

A fundamental challenge for local, state and federal agencies is managing the political context within which a sustainability report must exist. While a number of agencies promote transparency and accountability as a primary goal, the reality is that many political leaders appear less than comfortable with the notion of greater transparency, engagement and accountability. The political implication of reporting more than the good news—for the agency executive as well as for the responsible Minister or Councillor—needs to be assessed and understood as part of the sustainability reporting process. Ultimately, a sustainability report with obvious gaps and exclusions is likely to have greater negative reputation and credibility impact than an honest disclosure of any mistakes or less-than-perfect results and work in progress to tackle them.

Defining the composition of sustainability reporting is also a challenge for public agencies. In the public and private sectors, experienced reporters are trending towards full integration of sustainability and annual financial reporting (for examples of public sector integrated annual/sustainability reports, see Landcom (NSW); Hydro Tasmania; and Melbourne Water, Victoria). It seems, however, that a number of public agencies perceive the annual report to be a parliamentary document alone, which is restricted to parliamentary readership. There also seems to be a developing trend among state governments to reduce the amount of money and effort spent on turning the required annual parliamentary report into a readable report for a broader stakeholder base. The perception of the annual report as a communication vehicle of limited value may be affecting the perceived value of sustainability reports. Some fundamental behaviour shifts are therefore required to transition the annual report from the role of a wearying obligation into an opportunity to transform the sustainability of the agency.

A more insidious challenge is public sector bureaucratic cynicism regarding change. A 'wait-it-out' approach by some public servants, groups, directorates or agencies may not openly attack any attempt to transition systems and behaviours, but hinders uptake and may ultimately self-fulfil the prophesised failure. Along with a number of reform agendas underway in governments around Australia, the act of producing successful, material, credible, balanced and accessible sustainability reports may go some way towards restoring any lost internal employee confidence in the sector.

How to make progress

With so many challenges for public agencies undertaking sustainability reporting and embedded sustainability practice, how can progress be made and what foundational activities can be undertaken in the public sector to prepare the ground?

A renewed focus on good governance at the corporate level is fundamental for any organisation preparing for sustainability reporting. This is especially true for public agencies, where the political cycle and associated ongoing organisational change is a relatively constant challenge for governance arrangements. As noted above, good institutional arrangements promoting effective internal reporting and communication mechanisms are fundamental to the success of a sustainability report. So, too, are good internal governance arrangements. To actively seek sustainability in a public agency (therefore enabling a best practice sustainability report), a high level of commitment is required throughout the organisation, but particularly at the executive level. A governance framework that reflects sustainability at its foundations is a best-practice model to which public agencies may aspire, but some considerable effort and commitment is required from both the agency and its political leaders to move the organisation in this direction.

Good planning and reporting systems are also essential for effective sustainability integration and best practice reporting. An effective sustainability report reflects the planning framework of the organisation. It is therefore important that sustainability principles and outcomes are part of the organisation's highest-level planning documents. Linking sustainability down through the levels of organisational planning and budgeting into operational plans is an effective way of embedding sustainability thinking into a public agency. Although easier said than done, awareness of the breadth of the sustainability concept throughout the organisation is an important factor for the success of any new sustainability reporting endeavours.

In this vein, a review of reporting and reporting systems is an important early step towards sustainability reporting. A review of current systems to make use of data that is already gathered (such as OHS and human resources data) is one way to combat any sense of reporting fatigue within the

Table 1: Case Study: City of Melbourne

Public agency	City of Melbourne
Type of agency	City Council
Size	Revenue: $264.1 M (2006) Employees: 1100 (2006)
Number of sustainability reports	First sustainability report (2006)
Format	Stand-alone sustainability report
Length	116 pages
Availability	CD-Rom and web-based sustainability report
Number of GRI indicators	2 economic indicators, 13 environmental indicators, 15 social indicators, 7 public agency disclosures, and 30 standard disclosures; 12 indicators/ disclosures refer to the City of Melbourne Annual Report
Strengths	Use of Global Reporting Initiative and Sector Supplement for Public Agencies Explaining reasoning behind sustainability reporting and providing sustainability context of the organisation Structuring reporting around organisation's strategic priorities.
Challenges	Increasing stakeholder engagement around the report and gaining assurance Improving data collection and availability Expanding accessibility of report.

Table 2: Case Study: Melbourne Water

Public agency	Melbourne Water
Type of agency	Government-owned water management utility
Size	Revenue: $500 M (2006) Employees: 614
Number of sustainability reports	First integrated sustainability and Annual Report (has prepared social and environment reviews since 2001)
Format	Integrated sustainability and Annual Report
Length	124 pages
Availability	Hardcopy, online PDF and HTML
Number of GRI indicators	33 disclosures: 3 economic, 11 environmental, 8 labour practices, 2 product responsibility indicators sourced from GRI Draft G3 Guidelines
Strengths	Transparent reporting of 'key achievements', 'disappointments' and 'challenges' Consolidating all reporting into a Sustainability Report Use of GRI indicators linked with targets Public policy disclosures Assurance process
Challenges	Providing discussion on Melbourne Water's reasons for sustainability reporting Linking to the corporate plan Improving stakeholder engagement

agency and to reveal any reporting efficiencies. A review may identify reporting duplications in the agency or any historical processes/data that are now defunct.

Effective design and evaluation of public policy based on sustainability principles is a core consideration to propel a greater percentage of the public sector towards sustainability awareness and sense of internal responsibility, as well as sustainability reporting. While it may be argued that good public policy in itself achieves the policy-related goals of sustainability (consideration of long-term and system-wide impacts of the policy), effective policy evaluation is necessary to accurately assess the long-term sustainability impacts of a policy.

To facilitate public policy evaluation for sustainability reporting, sustainability and evaluation should be considered at policy inception and long-term evaluation plans (and the resources to undertake them) included in policy implementation plans from the outset. Evaluation of public policy is a core disclosure component of the GRI Sector Supplement for Public Agencies, however. Reporting outcomes (rather than outputs) is a key concept of a successful public agency sustainability report.

Tables 1 and 2 provide case studies from two public agencies that have undertaken sustainability reporting and illustrate some of the strengths and challenges of their approaches.

Conclusions

The power of enhanced performance reporting as a process for transitioning public agencies towards more sustainable organisational behaviour has been outlined throughout this chapter. Though uptake of this process appears to be greater in the private sector, the public sector is also experiencing some pressure from within, as well as externally, to undertake sustainability reporting. Despite some significant challenges, public agencies have a number of methods open to them to transition towards embedded sustainability and best-practice sustainability reporting. These include a focus on governance, internal planning, budgeting and reporting systems and effective public-policy evaluation.

Sustainability reporting provides a number of opportunities for public agencies. Through changing attitudes and underlying approaches of public sector governance frameworks—planning, budgeting and reporting—a public sector organisation can shift its focus towards long-term environmental, social and economic sustainability. In turn, the public policies, values and ultimate impact of public agencies will also be transitioned towards sustainability. The act of reporting itself promotes greater accountability and transparency of government and has the potential to re-energise engagement with staff, stakeholders and the community. Driving transformational change through greater performance and sustainability innovation in public agencies—while possibly not the original intent of an organisation undertaking sustainability reporting—is the ultimate result of attaining best-practice sustainability reporting.

References

Australian Bureau of Statistics (2007). *Wage and Salary Earners, Public Sector, Australia.* >Cat. no. 6248.0.55.001. ABS, Canberra.

Ball A (2006). Champion needed for public sector. *Chartered Accountants Journal* **85**, 18–20

Bogdanor V (Ed.) (2005). *Joined-up Government.* Oxford University Press, Oxford.

Brundtland G (Ed.) (1987). *Report of the World Commission on Environment and Development – Our Common Future.* Oxford University Press, Oxford.

Calder F and Culverwell M (2005). *Following up the World Summit on Sustainable Development Commitments on Corporate Social Responsibility – Options for Action by Governments.* Chatham House, London.

CAMAC (2006). 'The Social Responsibility of Corporations: Report December 2006'. Corporation and Markets Advisory Committee, Sydney.

Centre for Public Agency *Sustainability* Reporting (2007a). *Snapshot of Public Agency Sustainability Reporting 2006.* (not published)

Centre for Public Agency Sustainability Reporting (2007b). 'Fact Sheet 3 – Australia New Zealand Reporting Alliance Paper 2007' (not published)

Dickinson D, Leeson R, Ivers J and Karic J (2005). 'Sustainability reporting by public agencies: International uptake, forms and practice'. Research report by The Centre for Public Agency Sustainability Reporting™ (http://www.publicagencyreporting.org)

Elkington J (1997). *Cannibals with Forks: The Triple Bottom Line of 21st Century Business.* Capstone, Oxford, UK.

Forum for the Future (2006). *Sustainability – A Reporting Framework for the Public Services.* CIPFA, London, UK.

G8 Summit Heiligendamm (2007). *Growth and Responsibility in the World Economy – Summit Declaration* (7 June 2007) <http://www.g-8.de/Webs/G8/EN/G8Summit/g8-summit.html>

Global Reporting Initiative (2005). *Sector Supplement for Public Agencies – Pilot Version 1.0.* Global Reporting Initiative, Amsterdam.

Global Reporting Initiative (2006). *Sustainability Reporting Guidelines Version 3.0.* Global Reporting Initiative, Amsterdam.

Hood, C (2005) The idea of joined-up government: a historical perspective. In *Joined-Up Government* (Ed. V Bogdanor) pp. 19–42. Oxford University Press, Oxford.

Hunting SA and Tilbury D (2006). 'Shifting towards sustainability: Six insights into successful organisational change for sustainability'. Australian Research Institute in Education for Sustainability (ARIES) for the Australian Government Department of the Environment and Heritage, Sydney.

John P (2002). Who's in the Joined-Up Government? Australian Elected Representatives Adaptive Responses to Network Technology. In *Knowledge, networks and joined-up government : conference proceedings from the International Political Science Association Committee, Structure and Organisation of Government Research.* 3–5 June 2002. University of Melbourne (Ed. M Considine) pp. 109-131. Centre for Public Policy, Melbourne University, Melbourne.

Kidd C (1992). The evolution of sustainability. *Journal of Agricultural and Environmental Ethics* **5(1)**, 1–26.

KPMG Australia (2005). *The State of Sustainability Reporting in Australia.* Commonwealth of Australia, Canberra.

KPMG International (2005). *KPMG International Survey of Corporate Responsibility Reporting.* KPMG Global Sustainability Services, Amsterdam.

KPMG and UNEP (2006). *Carrots and Sticks for Starters – Current Trends and Approaches in Voluntary and Mandatory Standards for Sustainability Reporting.* KMPG's Global Sustaina-

bility Services and United Nations Environment Programme (UNEP), South Africa and France.
Mays S (2003). *Corporate Sustainability – an Investor Perspective: The Mays Report.* Commonwealth of Australia, Canberra.
Nethercote JR (2003). Australian public administration in perspective. In *The Australian Experience of Public Sector Reform* (Ed. Australia Public Service Commission Publications) pp. 11–20. Commonwealth of Australia 2003, Canberra.
Norman W and MacDonald C (2004). Getting to the bottom of "triple bottom line". *Business Ethics Quarterly* **14(2)**, 243–262.
Parliamentary Joint Committee on Corporations and Financial Services (2006). *Corporate Responsibility: Managing Risks and Creating Value.* Commonwealth of Australia, Canberra.
Porter ME and Kramer MR (2006). Strategy society – the link between competitive advantage and corporate social responsibility. *Harvard Business Review* **December**, 78–92.
Sparkes R (2006). From mortmain to corporate social responsibility: the historical background. In *Corporate Social Responsibility - Volume 1: Concepts, Accountability and Reporting* (Ed. J Allouche) pp. 38–77. Palgrave Macmillan: Hampshire, UK.
Suggett D and Goodsir B (2002). *Triple Bottom Line Measurement and Reporting in Australia.* Allen Consulting Group, Melbourne
UNEP, Standard & Poor's and SustainAbility (2006). *Tomorrow's Value: The Global Reporters 2006 Survey of Corporate Sustainability Reporting.* SustainAbility Ltd., London.
UN Global Compact and Global Reporting Initiative (2007). *Making the Connection: Using the GRI's G3 Reporting Guidelines for the UN Global Compact's Communication on Progress.* United Nations Global Compact, New York.
White A (2007). *The Quiet Revolution in Business Reporting.* Ceres, Boston.

42

Sustainable Corporations

Francis Grey

Westpac Bank recently advertised on national television that they were number one in social responsibility in banking. Meanwhile their billboards urged us to think of future generations as well as our own. It invoked the question in viewers' minds as to what had changed that this was so important. Should I really go to work today and think of future generations? If so, this constituted a significant moment in the transition to sustainability, or its converse, 'the war on dysfunctional organisations'—the source of much corporate unsustainability.

On organisations

Human achievement tends to be celebrated in terms of individuals, but much of it has been built by the 'organisation'. It is human organisation that has delivered great outcomes and great disasters and the continuum in between. When most people think of corporate environmental disasters or plant closures and the like, they are inclined to think of malevolent corporations that destroy in order to profit. What is less obvious is the systematic web of organisations and individuals that sustain that 'malevolent' corporation in their act of destruction. Furthermore, what is even less clear to the public are the 'small acts of unsustainability' in which we are all complicit each day—and how these are often forced on us by organisations that are more or less unconscious of, or desensitised to, their impact.

It is this state of 'unconsciousness' that is regarded by one writer as indicative of a dysfunctional state in an organisation. 'We know the good', Euripides wrote, 'but we do not practise it'. The true characteristic of consciousness is therefore not simply knowledge, but a balanced use of our qualities so that what we know and say is related to what we do (Saul 1997 p.186). In the dysfunctional organisation: 'The selfishness of the individual ego becomes more stubborn, more intractable, more determined as individuals come together in power structures. The whole becomes more resistant to the claims of social justice and more inherently recalcitrant than the sum of its constituent members, like some monster given life by Dr Frankenstein' (Young 2006 p.7). In theory at least, as individuals we can make our own choices, but as individuals grouped together in organisations the decision-making process can seemingly slide out of our grasp, along with any sense of responsibility. The paradox is that these decisions are still made by individuals, but once an organisation is involved, these individuals can be hard to locate.

If the public are confused about the supporting web of individuals or organisations that host the malevolent corporation, then they are also unclear about the role of organisations— as opposed to corporations—in the present unsustainable trajectory of societal development. The corporation has attracted the most criticism, with even a film, *The Corporation*, dedicated to this apparently systematic destroyer of human values. The premise of the film is: 'what is wrong with the corporation?' Equally, and more powerfully, one could have argued 'what is wrong with the organisation?', because corporations are just another form of organisation. Is it true to suggest that corporations are the source of unsustainable outcomes, while other organisations are not?

In the same way that there is a wide variation in the character of individuals, there is also a wide variation in the character of corporations. Indeed, there are corporations that could be diagnosed as acting in a psychopathic manner, but there are many more who have healthy characters—of varying qualities. Again, the corporation is only one sub-set among a wide range of organisations that drive human society on its unsustainable trajectory. Small businesses—or 'non-corporatised' business entities—and government departments are probably the most common. In addition, there are many other types of organisation that service the needs of humanity, all of which deserve scrutiny in terms of the degree to which they contribute to the unsustainable trajectory of human society.

Economics and the market

Economic *theory*—as encapsulated in the neo-classical economic framework, practised worldwide by governments and used by politicians and the mainstream public policy agencies such as the World Bank—provides a framework that prescribes the appropriate role for organisations. Organisations of all kinds ultimately serve the public good, and only exist because of that outcome. Public policy acts to ensure that the welfare of society is maximised and in that context public and private organisations serve their societal role. Public organisations by definition should not create negative externalities (negative impacts on society) hence they should not pose a theoretical problem for public policy. Private organisations act within the market. Economic theory holds up the 'efficient' market as a place where private organisations—acting in their own self interest and that of their shareholders—also deliver on the public interest. In an efficient market there are no negative externalities from private organisations. It should be noted that an efficient market is not always a 'free' market and a 'free' market is not always an efficient market.

However, in reality, private and public corporations both create negative externalities. These diminish the welfare of society and should be eliminated. In public policy it is generally held that governments regulate or develop market-based mechanisms to reduce negative impacts, but that if the cost of acting exceeds the benefits of action then the 'intervention' is not economically wise. This sets the scene for thousands of public policy disputes as people debate each negative externality, including among others, urban amenity, labour relations, child care policy and greenhouse emissions. In the meantime, the negative effects on public welfare continue to mount. If all organisations internalised their negative externalities and increased intra-generational welfare, there is still the problem of inter-generational welfare. Let's examine the oil market as an example. The internalisation of intra-generational externalities would probably slow oil consumption (by raising the price) but it still leaves unresolved the fundamental problem of intergenerational inequity. In theory if all consumers from all generations could bid for oil in today's market we could set prices for oil across time to reflect the inter-generational demand. Obviously the future cannot be present, except in the hearts of the present generation, and hence the present price for oil may not have any bearing on its future demand. However, at least inter-generational inequity—the toughest negative externality of all—can be remedied by organisations that take some responsibility to eliminate their own negative inter-generational impacts in a manner that enhances the organisation and its goals. Such an approach is consistent with the requirements of a neo-classical economic framework. If negative externalities can be eliminated (that is, achieve economic, social and environmental sustainability) in a profitable manner then they should be!

The challenge of environmental sustainability

In this book we have seen much discussion about the technical improvements and transitions required to achieve more sustainable outcomes. From a public policy perspective there is a clear argument in favour of eliminating unsustainable behaviour. Indeed, public policy would expect that all organisations seek to eliminate their negative externalities. None of this is possible unless our human organisations recognise the significance of sustainability, commit to delivering the outcome and, most importantly, improve their performance. For example, let's look at a sustainability challenge such as delivering sustainable stationary energy supplies. The demand for more energy is growing dramatically. Coal-fired power stations are being built at an extraordinary rate to meet this new demand. Meanwhile the requirement to reduce greenhouse emissions is growing and will become very severe in the foreseeable future. Any rational analysis of the global energy industry will conclude that we need major efficiency gains in the existing technology and installations, combined with radical improvements in efficiency plus a shift to renewables in the new plants being built, or we will face the unpalatable prospect of either ongoing excessive greenhouse emissions generation or shutting down valuable investments, and rationing electricity supply. Today, there is growing recognition of the need for the global energy industry to become more sustainable.

Global warming represents the most dramatic example of an exogenous shock on a collision course with our human drive to global economic prosperity. As the Stern (2006) report has made clear, only world wars and global economic depression have provided comparable challenges to the economic expansion set in motion by the industrial revolution. Like a relentless steam piston, the industrial revolution—based on fossil fuels—has set us on the path of a collision with the constraints of the global climate system. The window within which we can bring about a smooth transition is closing rapidly. The reckoning is likely to be an abrupt and brutal adjustment—either of our choosing or forced upon us as the global climate reacts. Our mission, when we choose to acknowledge it, is to bring about radical improvements in energy efficiency, plus a switch to renewables, within the next thirty years. For the moment the consensus view is that we need at least 60% reductions in greenhouse emissions by 2050, although given the history of economic forecasting (not global warming predictions), we can expect that this is a serious underestimate because of the unprecedented and accelerating pace and the absolute scale of Chinese and Indian economic expansion, not to mention other significant parts of the developing world. The consensus view has implications for each organisation, each household and each individual: on average each will be required to reduce their emissions of greenhouse gases by 60%. Anyone who doesn't comply will face at least higher costs, if not more stringent sanctions. A carbon-trading market will most probably distribute the pain to those with the least ability to respond, and provide rewards to those with the greatest ability to respond. In this new world, the ability to respond effectively is the key.

The pressures from developing countries

This ability to respond effectively is mirrored in other areas. With vast industrial potential blossoming on the China–India axis, companies from around the world will soon be responding to an innovation deluge as Chinese and Indian companies increasingly become innovators in their sectors, rather than just follower producers. There are already anecdotal signs that Chinese companies are becoming less cost competitive—the first sign that they are moving up the pathway to becoming innovators, in the same manner that Japanese companies made the transition from replicators to high quality innovators. When Japanese companies first achieved innovator status they brought with them major competitive hurdles and challenges for leading companies in western nations. Now Chinese companies in particular can be expected to upset the comfortable incumbents in many industry sectors. The incumbents will face waves of competing new companies seeking advantage in their sectors. The incumbents previously beat cost competition by outsourcing to China; now they will face product competition from China, and this will require greater organisational competence than they have previously demonstrated.

Ironically, the Chinese companies may well be better positioned to deal with the sustainability challenges of the 21st century. China's economic growth has driven it closer to a sustainability crisis than any other nation so far. The pressure on every natural resource, including fresh water and space, is intense. If China is to realise its ambition of high-level economic prosperity it must solve a series of environmental sustainability issues. According to the Chinese Vice Minster for the State Environmental Protection Agency, cited in the China Daily recently: 'traditional ways of development have caused the near breakdown of China's resources and environment, and people's lives are in great danger' (WME 2007). China's water availability is only just above northern Africa on a per capita basis. Apparently, however, 25% of this water supply is so contaminated it is not fit for any purpose (CLSA 2006).

Burgeoning industrial production—as presently constituted—requires high volumes of clean water. China needs to have high-efficiency technology in order to realise its dream. Chinese companies will have to innovate in this area simply to stay in business. All forms of environmental impact per unit of output will need drastic reductions or China will not achieve its economic ambitions. If the Chinese are successful in achieving these outcomes, Western companies will then face not only innovative Chinese competitors in the mould of Japan, but highly efficient producers with much lower carbon and ecological footprints, and hence—all other things being equal—a lower cost structure.

Meanwhile in developed countries

The writing is on the wall. Chinese companies and Indian companies, among others, are more likely to see that writing first, as they begin to live the nightmare of increasingly tight environmental constraints. Forward-thinking Western companies have begun to appreciate the coming environmental crunch as billions of people pursue their economic dreams. These developed world corporations are better positioned than other institutions and organisations, including governments, to respond to the challenges and opportunities that the situation represents. These corporations have resources, experience, knowledge, distribution and every other advantage—including incumbency—at their disposal if they seek to tackle this environmental crunch. More importantly, they have a singularity of focus to create profits by providing services, which means they can move with considerable effectiveness and speed. The crucial element is the ability to respond—the organisational capacity to recognise change and move to benefit from it rather than be damaged. The evolution of the industrial era has been seen as tumultuous, yet in many respects those changes have been glacial in pace as we moved from one technology to the next. Of late we (humans) have been impressed by the speed with which some new technologies, particularly IT, have been developed and deployed.

The car industry

By contrast, more traditional (older) industry has normally experienced a more leisurely pace. For example, the car industry develops new models and concepts that take many years to reach production—each one being more or less a small incremental improvement over the previous version. As a result, when oil prices suddenly surge some car makers are left with unsold stock as consumers rapidly move away from higher to lower energy-use options. Looking forward from 2007, it is clear to see that—with interruptions by the odd recession or two—the world is heading inevitably into a serious oil demand–supply crisis and the imposition of a global carbon cap. Yet, given the long lead times for car design, the vehicles of 2017 are already being worked on. Will these cars be suitable for the world in which they will be rolled out? How would car makers respond to government bans on all non-hybrid vehicles? Here is a fundamental test of car manufacturing organisations' robustness. How do they respond to such a massive level of uncertainty? The likely response is to try to pick winners. The smart response is to be flexible enough to move as required and the leadership position is to set the pace by actually designing the problem out of existence, through, for example incorporating very high energy efficiency in the designs for rollout in 2017, while reducing the time-to-manufacture schedule. With a shorter roll out schedule—or 'cycle time'—it is possible for companies to respond more effectively than their competitors to changing circumstances, and hence deliver vehicles more closely matched to market requirements, whether that is energy efficiency, roll protection or coffee cup holders.

Regulation

Similarly, when we look at a public sector responsibility in areas, such as building regulation, we see a glacial rate of change. Building regulation has been touted as the prime means of improving energy efficiency, at least in new and renovated buildings. Yet, despite the long-lived nature of these investments, the pressure to deliver higher performance has been relatively minimal given the required task of reducing energy use and greenhouse gas emissions. Until recently, the global warming debate has been so conflicted that it was reasonable to assume that this was the blockage to better outcomes. That blockage has now largely disappeared. It could be argued that the prime responsibility here rests with the politicians and industry lobbyists who constrain the building regulators. This is most definitely the prime blockage. A lack of leadership, however, is always the biggest impediment to change. Yet it is quite clear that building regulators have enormous capacity to analyse and prepare different regulatory scenarios in anticipation of coming changes, and to review existing approaches. The economic case for improved building efficiency is well-known and should provide a motivator in its own right to a robust and committed regulator. Unfortunately, at least publicly, most of this work has not been done. Regulators have not been motivated to actively seek to manage their domains. As organisations they are not necessarily driven to deliver better regulatory outcomes; instead they are seen as, and see themselves, as the passive tool of parliaments. In China this is having dire consequences as environmental damage is being done faster than the government can bring it under control.

Let's use Al Gore's Second World War analogy in reference to the dangers of global warming. In preparation for war, the British Government began to prepare technology for the country's defence. At first haltingly, and then with greater precision and speed, public agencies delivered answers to the problems of the day. These agencies seemed to work in partnership, as appropriate, with private organisations. The result was a very 'close run thing'. It looks like global warming will equally be a 'close run thing'. It begs the question of what it takes to get organisations to be ready for the big challenges of the 21st century.

In both examples—the car company and the building regulator—I have attributed poor performance to low organisational competence. This will be seen as controversial. Yet these organisations don't have much vision or energy to understand or engage with their domains. As a result, it is hard to expect them to then show much organisational drive in pursuit of outcomes that they don't envisage. This is very much a chicken and egg situation. Organisations are established in order to achieve goals. A significant part of the process of achieving goals is to be aware of your own domain, and the pressures and opportunities that act to constrain or support your goals. By logic of their establishment, organisations should be aware of, and seek to leverage, or remove, elements from their domain that respectively assist or impede their goals. For decades, Australian companies have debated industrial relations and yet at the same time they have failed to leverage the opportunities to improve corporate performance, by, for example, reducing their energy use. In another example, many Australian companies have now adopted a policy of zero deaths and injuries at work—a policy, which, in general, enhances corporate productivity. This begs the question of what the previous policy had been. From direct experience, companies find it difficult to answer this question—but by observation the answer is that there was an instinctual view that a non-zero deaths and injury policy was going to be costly. This is organisational dysfunction made manifest.

Organisational transition

The failure of organisations occurs in many ways. In essence the organisation was established to achieve a goal. In order to achieve that goal in the most effective manner, the organisation needs to continuously evaluate its actions, relative to that goal. The daily failure to achieve that goal happens in many ways. Most organisations would compare themselves with military hierarchies. Indeed, there have been anecdotal claims that our organisations have been modelled on military systems. Unfortunately, the cutting edge examples of modern military performance are the special forces teams that are routinely in the news. These teams are small, focused, highly trained, independently minded and distinctly non-hierarchical. They perform because the people within them are committed and trained to deliver high performance and are given the latitude to achieve it. The entire group have leadership qualities, each with their own specialisations. The group is given the objectives and the tools to undertake their mission, as well as the freedom to deliver the outcome as the situation requires. In many senses, the approach taken mimics that used on naval vessels. Ships crews can only be trained so far—eventually they must go to sea and undertake their role according to their own judgement. These approaches are the anti-thesis of the bureaucracy that characterises most contemporary organisations.

Stephen Young (2006), author of *Moral Capitalism* (Young 2003) and executive director of the Caux

Roundtable, suggests that power, fear and greed foster organisational entropy. Organisational structures that are bureaucratic enhance dysfunction by facilitating the pursuit of power and greed and the growth of fear. These practices undermine the mission, 'attract dysfunctional personalities as flowers do bees' and facilitate, and are facilitated by, the failure of leadership and governance. Young (2006) goes on to note that organisational dysfunction 'drives motivations and actions outside the organisation's value hierarchy, sabotages and destabilises the organisational purpose and occurs at every level of the organisation'. Dysfunction will manifest itself in many different ways. Organisational failure will also combine with the failure(s) and failings of individuals, which can enhance the level of dysfunction.

The organisation and its directors have the capacity to power up or disempower, as well as create very productive or unproductive organisations. If the 'power level' drops sufficiently, the organisation can be unsustainable and can behave in unsustainable ways. Such an organisation can become 'toxic' to its inhabitants; for example, leading to individuals becoming emotionally or even physically damaged through ill health. At moderate 'power levels' the organisation may be able to sustain itself, though given the on-going need to improve sustainability performance may see it left behind by more empowered competitors.

A sustainable organisation must be sustainable itself, as well as producing products that are sustainable. These two elements cannot be separated. The fuel cell company that damages the health of its employees through poor emission control or process design may be no more sustainable than the oil company it seeks to displace. A sustainable organisation would not fall into this trap because it would maintain a clear focus on its mission, remain 'conscious' of its reality (that is, chemicals used in fuel cells can be dangerous) and engage in Socratic questioning (is this the way we should do our business?) and accept the pervasive uncertainty that they won't have a safe product until they have made it safe! Sustainability embodies the cliché: it is a journey, not a destination. Stephen Young (2006) continues that leadership should, among other things:

- 'show no personal fear but acknowledge fear in others
- bring fears out in the open
- process fears through discussion
- ask for alternatives
- have a strategy rationally related to reality
- worry about the organisation's resource base, power and influence
- remember—all crises pass.'

All organisations have the opportunity to benefit by 'powering up' their entire web of connections across the economic, social and environmental dimensions. What matters here is that each organisation takes stock of its sustainability performance (defined as being the gap between where it is and where it would be if it could significantly reduce its negative impacts) and then implements changes to achieve a better outcome. The steps comprise:

1. Understanding the organisation's purpose
2. Identifying the challenges, stakeholders and uncertainties in relation to this purpose
3. Articulating the organisation's values
4. Identifying the organisation's contributions to its stakeholders and what it gets in return
5. Identifying gaps in the expectations of stakeholders and the outcomes
6. Tracking performance in closing the gaps
7. Reporting, critiquing, improving and publicly displaying the organisation's sustainability journey in all its detail: good and bad
8. Comparing with other similar organisations and benchmarking against industry performance.

Because I have a financial interest in this latter element, I should declare an apparent conflict of interest! However, this would be my view even if I didn't have the conflict of interest.

Measuring and monitoring performance of sustainability-oriented organisations

So far I have focused on organisations as a generic group because there has been too much of a tendency to let governments off the hook, ignore the non-corporate business community and demonise all corporations for irresponsible behaviour, when it is more a significant and shifting minority who are deeply problematic. As a group, all our organisations are contributing enormously to an unsustainable lifestyle by producing unsustainable goods and services. I would go as far to argue that the overwhelming bulk of consumers would buy sustainable products if they were available and offered at the same price as 'unsustainable' products. By reference to 'products' I am also including government departments, NGO's and all businesses. Consumer sovereignty alone won't drive product sustainability because producers have the strongest vantage point from which too understand what is possible in terms of delivering sustainable products. The only barrier to a sustainable human economy is the willingness of producers to get to work producing sustainable products (with the same price as unsustainable products—this latter part is a pre-condition for achieving sustainability with no loss in prosperity).

Creating sustainable products is a complex process that we will take many years to perfect, if ever. It will, however, require a 'whole-of-organisation' approach. Solutions advanced by the engineering department can be easily squashed by the legal or marketing department, though it tends to be the finance unit! Sustained delivery of sustainable outcomes will require a 'sustainable' organisation. In this way, the technological options offered in this book will fit within an effective organisational structure that can hone and deliver the product. Some organisations are aware of sustainability pressures and some are not. The quality of the organisation matters more than the awareness, although good organisation and awareness tend to go hand in hand.

The fitness of the world's organisations to receive the imminent deluge of sustainability focused innovation is minimal at best. In the private sector the Dow Jones Sustainability Index (DJSI) and the Australian SAM Sustainability Index (AuSSI) have been created to benchmark performance of the top 2500 listed corporations globally and top 200 listed corporations locally.

I have had some experience of this process as I was part of the team that founded the Australian SAM Sustainability Index (AuSSI—www.aussi.net.au) in Australia, and worked with the team behind the Dow Jones Sustainability Index (DJSI—www.sustainability-indexes.com). I should also reveal that I have a commercial interest in this work! However, in the same context one could also consider other indices (such as http://www.ftse.com/Indices/FTSE4Good_Index_Series/index.jsp) and financial services products as well.

Both the DJSI and AuSSI indices measure the financial value of the group of companies deemed, by benchmarking, to be the sustainability leaders in their industry sectors. In the DJSI, the top 10% of companies, broadly speaking, are chosen to participate in the index, from 58 industry sectors and 2500 of the largest listed companies in the world. I have called this process the 'corporate Olympics'. The sustainability assessment process seeks to identify sustainability leaders in their sectors. These companies, while often exhibiting 'feet of clay', are considered to be leading their sector in moving towards a sustainable future. The leaders are identified by testing them in response to a series of questions designed to separate leaders from laggards. The questions are centred around areas of significant economic, environmental and social challenge to the corporation and its particular sector. These are areas where corporations are struggling to avoid or reduce significant negative impacts and create profitable outcomes. The questions test the extent to which corporations have materially (that is, with respect to creating business value) tackled these challenges, allowing the identification of leaders and laggards. A recent study (Lee 2006) has analysed the DJSI and concluded that 'the DJSI is argued to be one of the world's best corporate sustainability indexes and is found to be consistent with modern portfolio theory (MPT) and the maximisation of shareholder wealth. Corporate sustainability is synonymous with socially responsible investing (SRI), whereby both strategies formally integrate economic, social and environmental concerns into the final investment decision.'

The natural reaction is to compare the value of the sustainability leaders with their peer group. The contribution of sustainability performance to organisational performance may show up in this comparison. The results would have implications for all organisations, not just corporations. Each index website contains data and graphs that show each group of leaders tracking their peer groups with considerable consistency. There is no clear graphical sign of significant under or out performance in this data.

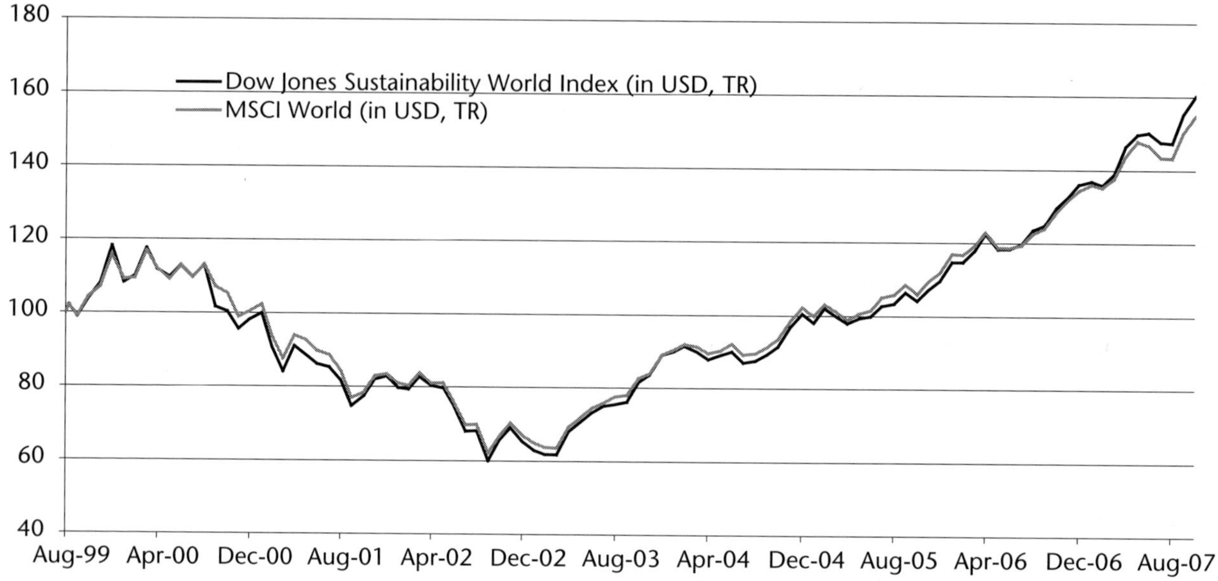

Figure 1: The Dow Jones Sustainability Index (DJSI) representing the value of the global sustainability leaders versus the market as a whole (represented here by the Morgan Stanley Capital Index —MSCI). Source: DJSI www.sustainability-index.com (about 300 companies).

In the absence of a clear graphical sign of sustainability leaders out/under performing the benchmark indices, it requires formal analysis to indicate whether or not value is being created or identified in this process. Lee (2006) has studied this question with respect to the DJSI. His analysis reveals that for 'a global BOS (best of sector) sustainability portfolio —namely the DJSI—when compared with a comparable unrestricted global portfolio, indicated that the DJSI does not underperform the market portfolio, suffer from higher levels of risk or result in increased diversification costs. Results … indicate SRI imposes no significant financial benefit or cost for the investors'. This

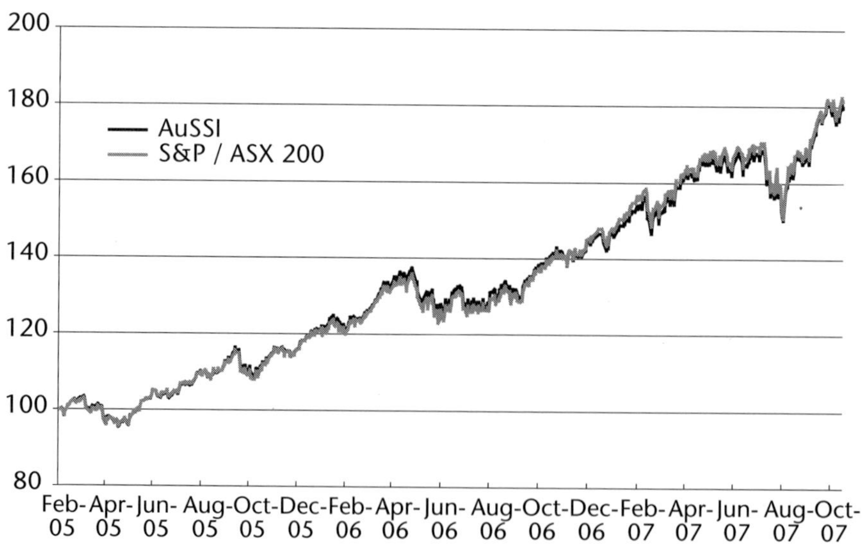

Figure 2: The AuSSI, representing the value of the sustainability leaders versus the value of the market as a whole (represented here by the ASX top 200 index) since its launch date. (Source: www.aussi.net.au (70 companies).

directly contradicts what certain, I suspect inaccurate, proponents of economic theory have claimed. Lee's work indicates that, at the *portfolio* level, it is not more *risky* to invest this way. The crucial element is that risk and return are the opposite sides of the same coin in an investment portfolio. The investment cannot be reviewed unless it is considered in a *return and risk* framework. Higher risk implies greater returns and vice versa. Thus, it is significant that the DJSI is found to be a concentrated portfolio, but with the same risk as the 'market' portfolio representing the underlying benchmark index.

Lee continues his analysis at the level of the firm. He prepared matching pairs of leading and lagging corporations as well as pairs of industry-based portfolios with leading and lagging companies. He also applied accounting tests (to see if performance differences showed through in accounting variables) and market tests (to see if performance differences showed through in investment market characteristics). 'Accounting test results find no difference with regard to the performances of leading and lagging CSP [corporate sustainability performance] firms, while market tests provide strong cross-sectional evidence of economically and statistically significant differences in performance. Leading CSP firms are found to provide a large returns discount relative to lagging CSP firms across all periods, portfolio types, and market tests. However, neither the accounting nor market tests support a systematic difference with regard to leading/lagging industry sorted portfolios' (Lee 2006, p. 168). That is, leading firms, at the individual level, showed that they had lower returns in the market test, which indicates the market has them priced as lower risk investment opportunities, though this was not captured when they were grouped into portfolios. It also means that the cost of capital for these leading firms is lower, increasing their capacity to grow compared with their more risky and lagging peers. 'There is strong evidence to suggest that leading CSP firms provide lower returns, as is consistent with hypothesis one, thereby suggesting a lower cost of equity capital' (Lee 2006, p. 169).

This counter-intuitive result by Lee (2006) is a significant, if early, insight that requires replication. Firstly, it indicates that we should consider financial markets assessment of 'value creation' with care. Secondly, and more importantly, the view that sustainability-leading companies are regarded by the market, with little or no knowledge of sustainability, as being less risky (and hence 'paying' lower returns to investors) is exciting. It suggests, should future research confirm Lee's results, that there is a real pay-off for being a sustainability leader. Any company that can access cheaper capital than its peers is in a good position to grow its business faster than its peers. In the words of Vicsuper—the largest client of the SAM Group (producers of the DJSI and the AuSSI)—if leading sustainability-oriented organisations do not deliver better outcomes than laggards, then humanity is lost (Welsh 2005, *pers. comm.*)!

Conclusion

In the transition to a sustainable economy, it needs to be remembered that improved sustainability performance must deliver a better economy in all other dimensions, or it is unlikely that the activity has a sustainable future. The work of Lee and others may indicate that—on a virtual blind test—the financial market is already rewarding the transition towards sustainability by individual firms. All other things being equal, this reward will get stronger as markets themselves become aware of sustainability and its significance for economic prosperity. Vicsuper is, a large Victorian based super fund (with whom I have a commercial relationship—that's another disclosure for transparency purposes), and the first super fund in the world to produce a sustainability report. The CEO stated recently 'our commitment to sustainability means that we're constantly improving our business operations, which in turn benefits all our members and employers. It's not about being seen to do the right thing; sustainability is now a business imperative' (Welsh 2007, p. 5). It is also an organisational imperative—'we must change'—and a key to continuous performance improvement—'we want to change'—if we are to create sustainable outcomes at all levels, from individual to global. The beauty of sustainable outcomes is that we need to achieve them, we want to achieve them, we will better off having achieved them—and it is likely that a sustainable world will be a very agreeable world as well.

References

Big Picture Media Corporation (2005). *The Corporation.* A film directed by Mark Achbar. <www.thecorporation.com>.

Cannon M (1977). *The Land Boomers.* Melbourne University Press, Melbourne.

Lee D (2006). An analysis of the sustainability investment strategy employing the Dow Jones World Sustainability Index. PhD Thesis. Department of Accounting and Finance, Monash University, Melbourne.

McDonough W and Braungart M (2002). *Remaking the Way We Make Things: Cradle to Cradle.* North Point Press, New York.

Rothman A, Montague J and Sandison D (CLSA) (2006). 'CLSA Asia Pacific markets, Asia in the World: Conviction ideas for global investors'. Comments on water availability on page 1, Shanghai. <www.clsa.com>.

Saul JR (1997). *The Unconscious Civilisation.* Penguin Books, Melbourne.

Stern (2006), *The Economics of Climate Change: The Stern Review.* Cabinet Office – HM Treasury. Cambridge University Press, Cambridge, UK.

Welsh B (2007). *Sustainability in Super. Your Agenda for Success.* Australian Institute of Management Victoria and Tasmania, July 2007.

WME (2007). Air pollution's deadly toll. News report. *WME Environment and Business magazine* **July**, 14. <www.wme.com.au>

Young SB (2003). *Moral Capitalism – Reconciling Private Interest with the Public Good*, Berrett-Koehler Publishers Inc. San Francisco.

Young SB (2005). 'Building sustainable organisations'. Presentation to the Australian Catholic University, 20 June 2005. Sydney.

Young SB (2006). *Moral Capitalism and the Unending Quest for Human Perfection.* 15 June 2006 address at iberAmericana University. Caux Roundtable, St Paul, Minnesota. <www.caux-roundtable.org>

43

Community Engagement and Behaviour Change

Anne Langworthy and Charlotte Brunt

Growing concern about society's lack of progress in relation to sustainable development has led to governments increasingly exploring 'demand side' solutions as well as their more traditional 'supply side' approaches to community growth and development. Governments have also focused on evidence-based decision making as a basis for targeting policy and programs. Given the high levels of unsustainable resource consumption by Australia's population over the past decades (Lawn 2003; Newton 2006), the Australian Government, like many internationally, has engaged in successive programs of indicator development (such as Measures of Australia's Progress and State of the Environment Reporting) as a basis for informing policy and strategy at national and state levels. In contrast, local-level indicators have emerged as a tool for empowering local communities to tackle issues that impact their population, economy and environment. The rhetoric relating to the impact and influence of indicators as an agent for change is compelling, but the evidence of the indicator information becoming public knowledge and triggering action in the absence of targeted programs is scant.

Indicators of community sustainability have been developed in Outer Eastern Melbourne and it is clear that developing and publishing the indicators alone does not necessarily engender behavioural change. This has led to the development of a model for indicator action that is focused specifically on action to tackle the issues of energy use and climate change.

In this chapter, we propose the integration of indicator development and community engagement as a means of initiating change in attitudes and behaviour of individuals and households in relation to resource consumption.

Drivers of change

Concern with issues of sustainability gathered international momentum in the last decades of the 20th Century. The World Commission on Environment and Development, chaired by Norwegian Prime Minister Gro Harlem Bruntland, and its report, *Our Common Future* (1987), popularised the notion of sustainable development and the principle of ensuring that our actions today should not limit the range of economic, social and environmental options open to future generations.

The work of the United Nations and the World Business Council for Sustainable Development, among many others, indicates that the 1990s were a time of international soul-searching in relation to the environment. In seeking to understand and strategically manage sustainability issues, organisations and communities have begun to focus on indicators. In response to the need for new measures of performance, John Elkington (1999) coined the term 'triple bottom line' (TBL), which provides a context for measurement of sustainability. Over the last decade, the conceptual understanding of economic, environmental and social (TBL) dimensions and sustainable performance has been continuously improving. TBL reporting is becoming accepted business practice, and a range of benchmarks and indicators are used by organisations that seek to rate business performance against the triple bottom line (Elkington 2001). Contemporary debate has refined TBL concepts, with consideration of a 'quadruple bottom line' or 'the five pillars of sustainability' that include an emphasis on culture and governance as added dimensions for measurement.

On a global scale, OECD and UN indicators measure a country's performance in relation to international conventions or agreed standards, often relating to sustainable development, human rights, safety and basic access to food, water and health care. National indicator suites often set targets for development and track progress towards the desired outcomes. Often these suites are assembled under the title of measuring progress—as in the case of Australia's *Measures of Australia's Progress* (Trewin 2006).

Influential in Australia's sustainability reporting, the Genuine Progress Indicator (GPI) attempts to deal with the inadequacy of the GDP as a measure of economic welfare by balancing the benefits of economic activity against the costs of economic growth (including the harmful effects of resource depletion, pollution, ozone depletion, loss of farmland, loss of wetlands, family breakdown and crime). GPI can be used to assess a nation's sustainable development. Calculations of the GPI for Australia for the period from 1966–1967 to 1994–1995 suggest that Australia's is failing to make the transition towards sustainable development (Lawn 2003).

In the United Kingdom the national indicator suite is driven by the government's sustainability agenda:

We have failed to see how our economy, our environment and our society are all one. And that delivering the best quality of life for us all means more than concentrating on economic growth. (Blair 1999, cited in Department of the Environment, Transport and the Regions 2001)

Canada goes further, providing a results-based perspective in their report *'Canada's Performance'*:

A healthy democracy such as Canada's requires the active engagement of its citizens in understanding the economic and social issues we face as a nation...we assess our progress in the following areas of Federal involvement: the economy; society, culture and democracy; health; and the environment. (Alcock, cited in Treasury Board of Canada Secretariat 2004, p. iii)

A non-government coalition of researchers, state agencies and Statistics Canada are now working on the Canadian Index of Well-being, for which an overall framework has been developed (Atkinson Charitable Foundation 2007). The first report is expected towards the end of 2007.

Indicator development

In June 2007, over 1000 people gathered in Istanbul, invited by the Organisation for Economic and Cultural Development (OECD) to attend the *Istanbul World Forum – Measuring and Fostering the Progress of Societies.* This second world forum followed a similar gathering in November 2004, where 550 invited experts from over 40 countries, including senior government bureaucrats, OECD officials and heads of national statistics institutes, met at Palermo Italy for the *Statistics, Knowledge and Policy, OECD World Forum on Key Indicators.* This forum marked the culmination of a decade of concern about indicators of sustainability and demonstrated the growing importance of comprehensive and reliable information to inform policy development and to track progress.

While the first world forum highlighted the development of indicator suites by the OECD, the United Nations (UN), the European Union (EU) and many countries including Australia, it also demonstrated an increasing trend towards local measurement of economic social and environmental sustainability in cities and regions around the world.

The Istanbul World Forum revealed that the momentum in indicator development has now grown to the point where there are indicator suites in many countries and on every continent. Where once this process tended to be centred in resource-rich nations, interest has spread to the developing world, where the UN or the EU are supporting the development of indicators. Indeed the UN Millennium Development Goals are an example of using indicators to both measure and facilitate progress globally. The eight goals—which range from halving extreme poverty, halting the spread of HIV/AIDS, providing universal primary education to ensuring environmental sustainability—and their 2015 targets, have the endorsement by all countries represented at the UN.

The OECD has committed to facilitating ongoing development of indicators because:

A culture of evidence-based decision making has to be promoted at all levels, to increase the welfare of societies. And in the "information age", welfare depends in part on transparent and accountable public policy making. The availability of statistical indicators of economic, social, and environmental outcomes and their dissemination to citizens can contribute to promoting good governance and the improvement of democratic processes. It can strengthen citizens' capacity to influence the goals of the societies they live in through debate and consensus building, and increase the accountability of public policies. (Istanbul Declaration, 30 June 2007)

Indicators as a tool for engaging communities

The real power of indicators is not merely in measurement, but in becoming a tool for engaging communities and fostering change. A major contention of the OECD Forum was that indicator development needs to turn statistics into knowledge and then into public knowledge.

Indicators of community sustainability are not an end in themselves—rather, they are a tool to measure progress and need to be strongly linked to the preparation of a vision for sustainable development. With a strong linkage to a community's sense of direction and existing institutional goals, indicators become meaningful and dynamic tools that help to guide progress, policy and direct resource allocation. One of the earliest articulations of this perspective on indicators is:

The indicators a society chooses to report to itself about are surprisingly powerful. They reflect collective values and inform collective decisions. A nation that keeps a watchful eye on its salmon run, or the safety of its streets makes different choices than does a nation that is only paying attention to its gross national product. The idea of citizens choosing their own indicators is something new under the sun – something intensely democratic. (Meadows cited in Sustainable Seattle 1998, p. 4)

While the sustainability agenda may have had its initial impetus over concern for the environment, the importance of the social and governance dimensions of urban development is increasingly discussed. Woolcock (2000) demonstrates the importance of local civic associations in terms of information dissemination and building social trust, which are underpinning conditions for effective governance and economic development. Social capital—which includes institutional capacity, social networks and community participation—has an important role to play in social and economic outcomes. Thus, there has been considerable research into the conceptualisation and measurement of social capital and related notions such as community strength and community building on the part of Australian governments:

Strong communities are constituted by people who understand the social, economic and environmental assets of the community and are working towards sustainability. Strong communities also understand and work with their most disadvantaged population to ensure minimum standards for all. To do these things, members of a strong community need to be engaged, involved, feel capable of working through issues and be supported through external partnerships. (Department for Victorian Communities, 2004, p. 4)

Local government and indicator action

In 1969, the Friends of the Earth, which has now grown into an influential international network of grassroots environmental groups in 70 countries, was founded on the slogan 'think globally, act locally'. We recognise climate change as a global issue, but we must act locally to change our behaviour in an attempt to mitigate the full force of climate change and develop adaptation strategies.

As the level of government closest to community, councils have an opportunity to foster local sustainability within their communities. Local government has long been concerned with engaging the community in order to ensure good governance and effect progress. Experience has shown that engagement results in the community, and other stakeholders, taking ownership of a direction, course of action or decision, and its implementation. One challenge is how to engage the hard-to-reach members of the community who tend not to be involved. Hard-to-reach community members are often disadvantaged people but they are also, increasingly, the time poor.

Community engagement theory has been evolving over recent decades. Arnstein (1969) developed the 'ladder of citizen participation' as a measure of the quality of community participation. Rocha's ladder of empowerment (1997) looks again at community participation in the context of political empowerment. The International Association of Public Participation (IAP_2) has developed the Public Participation Spectrum (2003), which is used internationally as a tool for increasing the level of public impact. In this spectrum, participation moves from the one-way process of informing and consulting to involvement, collaboration and, ultimately, empowerment.

Many contemporary government programs now recognise the power of local engagement. Neighbourhood Renewal in Victoria, for example is a state government program that targets disadvantaged communities and brings together the resources and ideas of residents, governments, businesses and community groups to tackle disadvantage in areas with concentrations of public housing. Often the small local initiatives to lift neighborhood appearance have much wider and longer term effects. They can increase pride and community cohesiveness. This capacity to effect change through local action is seen as very empowering (through the establishment of new enterprises; community job placements; public housing properties upgraded or built; energy improvements; reduction in crime and so on).

Burton (2007) argues that good leadership in the public sector is imperative for making the connection between global thinking and local action. However, successful engagement may mean relinquishing local government leadership and control. Local government becomes a stakeholder and partner, but leaves space for the emergence of new community leaders and distributed action.

Collaborative Regional Initiatives (CRIs) in California were founded to improve regional performance through the mobilisation of stakeholders. These organisations are deeply embedded in the business community, but focus on civic entrepreneurship to bring about economic vitality, social equity and protection of the natural environment:

Ultimately CRIs do what public agencies and legislative practices do not: help build knowledge, commitment and motivation among a substantial set of regional leaders who in turn can form the basis of a robust society. (Innes and Rongerude 2006, p. 15)

An evaluation of four CRIs found that while these organisations lacked formal power, they were able to bring about significant change—workable solutions to regional challenges. The most effective CRI leaders do not control the agenda but 'inspire, encourage and convene and support' (Innes and Rongerude 2006, p. 39).

Many indicator projects are run by organisations like the CRIs. In Australia, there are many indicator suites and initiatives which, in contrast, are often run by local government. Some of the earliest examples focussing on sustainability, can be found in capital cities, such as Brisbane, Perth and Melbourne, focus on sustainability. Local projects in Newcastle, Onkaparinga, Port Phillip and Moreland, for example, demonstrate the interest of local government in community sustainability or wellbeing indicators. Most recently the launch of Community Indicators Victoria (CIV 2007) provides a populated set of indicators for all Victorian local governments.

Have local indicator projects made a difference?

In the United States alone there are over 1000 indicator projects, many of which are represented in a Community Indicators Consortium, which is developing an active network and community of practice. The most effective indicator projects are dynamic, continually evolving and linked to a range of projects that tackle problems identified by the indicators. This is often, but not always, a contributing factor to the longevity of a project. For example, the Boston Indicators (10 years), the Santa Cruz Community Assessment Project (13 years), Sustainable Seattle (17 years) and the Jacksonville Indicators for Progress (22 years) are award-winning community indicator projects that have shown great longevity.

Much attention has been devoted to the developmental aspects of the Jacksonville project. However, little has been published in relation to the impact of the indicators on the community. While there appears to be some evidence of contribution to organisational decision making, especially for the Chamber of Commerce, there is some evidence of a disconnection with the wider Jacksonville community. The latest survey (2005) shows that of the 1196 community members contacted to complete the quality-of-life survey, 754 refused to participate (Jacksonville Community Council Inc. 2005).

In contrast, Sustainable Seattle, which is used as an exemplar worldwide, demonstrates strong evidence of continued community engagement. Like the initial process in Jacksonville, many people contributed to the development of an initial set of indicators. Over 200 people, including environmental and social activists, business leaders, educators, students, city and county government representatives, jointly contributed more than 2500 hours of volunteer time to produce and publish the first edition of *Indicators of Sustainable Community for Seattle* in 1993 and, subsequently, in 1995 and 1998—the last published report.

Since 2003, Sustainable Seattle has focussed on the collection of data and development of indicators at the neighbourhood level—establishing the Sustainable Urban Neighbourhoods Initiative (SUNI). A strong feature of SUNI is the street-level and neighbourhood surveys, which are delivered door–to-door, by telephone and online using open source software. Outcomes from the initiative include a plethora of neighbourhood projects, an identified set of core neighbourhood concerns—and a more responsive city (Sustainable Seattle 2006).

The Santa Cruz Community Assessment Project (CAP), which is run by the United Way (an independent non-profit organisation), began in 1994. Applied Survey Research (ASR)—a non profit social research firm that has completed a number of Community Assessment Projects—has researched and written a Santa Cruz report annually (Brutschy and Zachary 2007).

Characteristic of the CAP approach is use of the Results Based Accountability methodology created by Mark Friedman (cited in Brutschy and Zachary 2007): identifying shared program or community outcomes and establishing indicators to measure current and future outcomes. ASR has an action-oriented approach using the mantra, 'what gets measured, gets managed and what gets measured, gets done'.

The CAP is researched based: data is collected for 125 indicators in six domains (the economy, health, education, public safety, the social environment and the natural environment) before community goals are set for each of the themes. There are multiple champions for each goal and there is evidence that institutions, business and organisations are adopting individual CAP goals to bring about lasting change.

When an issue is identified as a priority, the data is monitored consistently so that groups receive feedback about progress. For example, a CAP priority action was an effort to tackle youth substance abuse (the 1995 report showed that 64% of young people surveyed had used alcohol, 55% had smoked cigarettes and 40% had smoked marijuana in the last 30 days). A coalition of over 110 agencies was mobilised to seek solutions to deal with youth substance abuse. Seven strategies were agreed, and the results can be seen in the significant decline in substance over time as shown by the survey (alcohol down to 47%, smoking to 44% and marijuana to 29% by 2004; Brutschy and Zachary 2007).

In contrast to neighbourhood-level initiatives, the impact of the Boston Indicators project is most evident at regional institution level, specifically the Metropolitan Area Planning Council and the City of Boston. The Boston College Citizen Seminars and the Leadership Forum bring together leaders from academia, business, unions and non-profit organisations to debate some of the pressing issues facing the Boston region.

The communication techniques used by the Boston Indicators project are crucial to the success of the project. Each report has a theme, which is a call to action. The titles of each of the reports are signposts in the journey taken over the last 10 years. The current report, *A Time Like No Other: Charting the Course of the Next Revolution* (2007), places Boston at the heart of what are essentially global challenges, but also celebrates the region's contribution to tackling these challenges.

An examination of these projects demonstrates that the themes, categorisation, and often the indicators themselves, used in the community indicators projects are often remarkably similar, although these common themes will be linked closely to the local character, community aspirations and priorities when action is mobilised.

If they are to make a difference, these projects will need strong partnerships and stakeholder involvement. While regional and city approaches have their place and affect institutional planning, often the most effective behavioural change happens at neighbourhood level.

The process of developing indicators of community sustainability in Outer Eastern Melbourne

From the beginning of this project, it was acknowledged that the development of the sustainability indicator suite has the capacity to fulfil a community-building role through the process of engagement. Historic evidence indicates that significant community development takes place only when local community members are committed to investing time and resources in the effort. This observation explains why successful communities are never built from the top down, or from the outside in. Effective community-development efforts are based upon an understanding of the community's assets, capacities and abilities, not deficiencies or needs (Kretzman and McKnight, 1993); that is, celebrating assets, connecting skills to the economy and inviting people to participate, allowing them to develop a vision for the future and harnessing their collective energy.

A vision for the regional future of Outer Eastern Melbourne aligned the community visions developed by each of the three constituent local governments (Knox, Maroondah and Yarra Ranges) in consultation with their communities. Residents and organisations that provide services to the community were stakeholders in the indicator-development process. The cooperation of stakeholder organisations was essential in the development of the indicator suite to highlight locally relevant issues, provide technical expertise, and for data collection and monitoring.

The project in the Outer East has now run for 4 years, led by local government champions who represent environmental, economic and social areas of responsibility within their local government in partnership, with a team from the Swinburne University Centre for Regional Development. It is only now that the indicators are being used in local government strategic planning. The engagement of councillors has been uneven and the impact of the indicators on councillor decision making has yet to be tested.

A regional report was produced and published online and individual reports were provided to each local government for its own use. The indicators revealed that the communities of the Outer East have a comparatively good quality of life, but there were some significant warnings for the community:

- the region has, on average high retail spending, but low levels of local employment
- school retention is a concern along with higher and further education and training participation rates
- households in the region have steadily increasing electrical energy consumption (8.2 tonnes green house gas emitted per household on average)
- waste was considered an issue
- while most residents felt that they could get help when they need it, they also felt disenfranchised and not able to have a real say about issues of importance
- despite lower-than-state-average and declining levels of crime, a large proportion of residents did not feel safe alone on their streets after dark. (Langworthy and Brunt 2005.)

The project demonstrated a sound process for the development of key indicators related to community sustainability, but not, as yet, how to use these indicators in galvanising action.

Changing community attitudes and behaviour to energy use

A model for moving from indicators to action is illustrated in Figure 1. Community participation is integral to each step in the process. The energy indicator is being used to test the model because climate-change mitigation and adaptation was identified as a priority theme by all local governments involved, given their increasing per capita rates of energy consumption.

Stages in the indicator—community action model

Stage 1. Indicators and reporting

Indicators and their reporting represent both the beginning and end of the cyclical process. The indicators provide the base data and focus for action and they provide the method of measuring outcomes, and thus progress.

A focus on climate-change mitigation provides the Outer Eastern Indicators project with an opportunity to test models of engaging communities with indicators to promote behavioural change. Energy-end-use efficiencies have become central issues for all levels of government in Australia, as governments develop both policy and program initiatives to lead by example in responding to greenhouse gas (GHG) emissions and climate change. Escalating energy use was identified as a red flag area in the indicators report and all three municipalities are tackling this important issue.

Energy use is a major environmental issue for Australia—continuing reliance on fossil fuels at a domestic level and the negative impacts of Australia's transport system have been largely responsible for an increase in fossil fuel usage by more than 25% in the decade 1991–2001 (Newton 2006). A 50% increase in energy demand is forecast by 2030. Energy is particularly relevant in outer eastern Melbourne, not only because regional consumption patterns are above the Victorian average, but also because peak oil will impact significantly on the region due to poor public transport services and existing travel patterns.

In addition, there have been legal precedents that pave the way for litigation against local governments in relation to the impacts of climate change arising from rising sea levels, flooding and extreme weather conditions and the increased risk of bushfire and vector-borne diseases. Action is required to manage these risks.

Stage 2. Leadership and stakeholder action groups

The next step in the engagement process is a regional forum, where the energy indicator and further information about energy use and climate change are discussed.

Those Invited to the stakeholder forum include people from:

- local government (officers and elected representatives)
- the education sector
- the energy sector (the energy network provider is already a partner in providing energy use data while the energy retail companies are yet to be engaged)
- wider business sectors
- interested community members (at least one township group is active in working to reduce local energy use).

The forum is designed to act as a catalyst for forming stakeholder action groups who will act as civic leaders, build their knowledge, identify

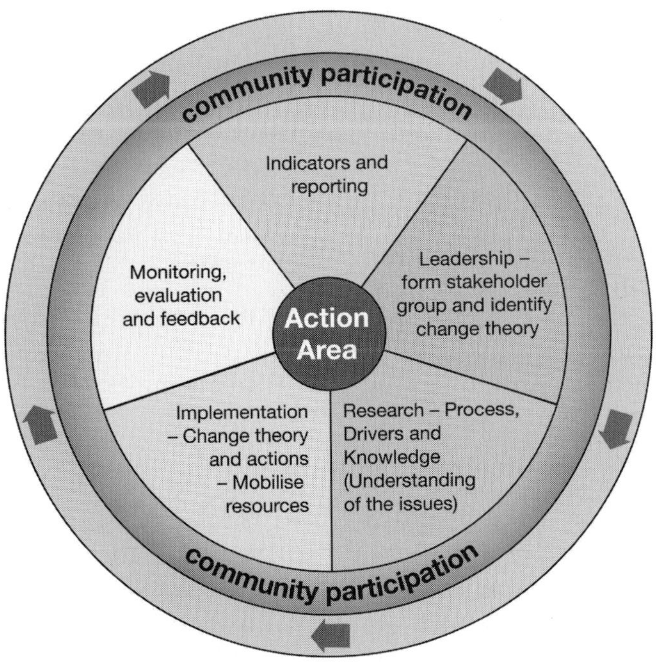

Figure1: Outer Eastern Melbourne indicator action model

strategies and access resources needed to bring about change at community level.

Stage 3. Research and knowledge

The forum initiates the process of building stakeholder knowledge. Disseminating scientific information alone, however, is not enough to ensure sustainability. The challenge is to change behaviour and community norms and, in this respect, the role of the social sciences has become increasingly important (Saunders *et al.* 2005). Thus, the information provided as a basis for discussion falls into two categories:

- information about the issue, consequences and potential action
- information change theory and drivers of behavioural change.

Many psychological processes simultaneously shape behaviour and influence choices. These include motivations, goals, attitudes and social norms. Three goal frames are important for understanding environmental behaviour:

- hedonic goals—to feel better right now
- gain goals—to guard and improve one's resources
- normative goals—to comply with social norms (Lindenberg and Steg, 2007).

Current awareness in the Outer East

Additional data have been collected about Outer Eastern attitudes and behaviours relating to energy use, which can be used to inform action. A preliminary survey of energy use and sustainable practices in Outer Eastern Melbourne has been undertaken by Swinburne University involving a sample of 80 people. The results show that most people were aware of the impact of climate change and felt that it was important to reduce their energy consumption. This is consistent with Australia-wide concern over climate change, with the Climate Institute reporting that it was 'more important to people than a wide range of issues including housing affordability and national security' (Climate Institute 2007, p. 3).

All those in the sample were making an effort to reduce energy use in one or more ways; for example, by replacing incandescent compact fluorescent globes, washing in cold water and/or limiting the use of air conditioning and tumble clothes dryer in summer. Over half the respondents claimed to have insulated their home specifically in response to climate change and almost 20% had switched to green energy.

When participants were asked what practices they were likely to undertake in the coming year to reduce energy consumption, the most common response was turning appliances off at the power point (38%). Twenty-nine per cent of respondents felt that they would invest in solar heating and 46% were intending to switch to green energy. (The Climate Institute (2007) also revealed strong public support for renewable energy sources.)

Interest in learning to abate the impact of climate change at a personal level was moderate, with 35% of respondents indicating that it was likely that they would improve their knowledge of energy-saving practices. The top three reasons for reducing energy consumption were to reduce power bills (28%), to abate diminishing water supplies (26%) and to reduce greenhouse gas emissions (12%). The least important reason cited was in response to government regulations (1%). The survey also identified cost as the biggest barriers to retrofitting.

Models of behaviour change

There are a number of models of behaviour change. Many derive from hierarchies of human need, such as those developed by Maslow (1970) and Max-Neef *et al.* (1991). These models usually identify enablers and barriers and provide insight into the steps or stages in the process of individual or community transformation.

Examples of models describing community change are:

- the hierarchical model for assessing and describing sustainability support in small town communities developed by MacGregor (2000)
- the stepwise model of behavioural change toward a pro environmental habit (Dahlstrand and Biel 1997, cited in Voronoff 2005)
- the model describing the conditions surrounding voluntary behavioural change (Ampt 2003) relating to individual transformation
- the Seven Doors Model for designing and evaluating behaviour change programs (Robinson 2007), shown in Figure 2, maps individual steps

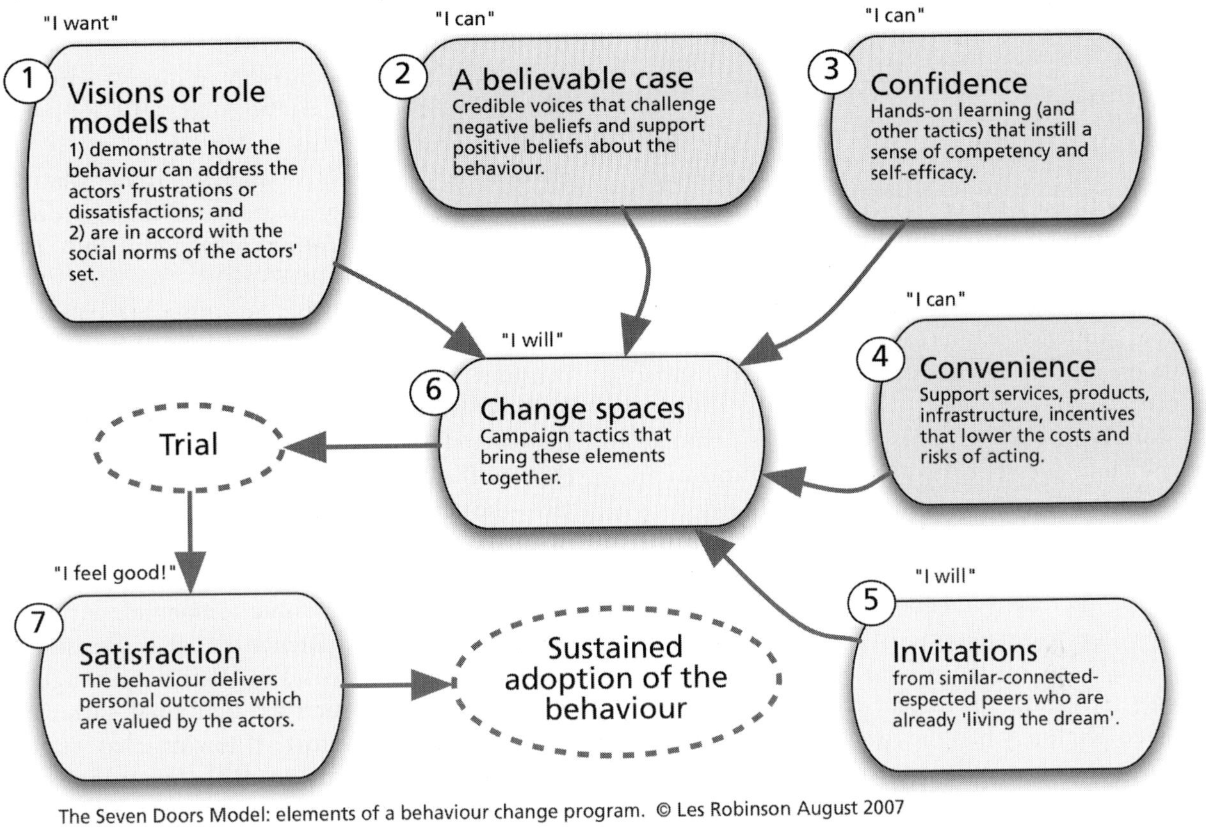

Figure 2: The Seven Doors model for designing and evaluating behaviour change programs

towards sustainable behavioural change, but also demonstrates the change the community as a whole must embrace.

Other change theory that underpins concepts such as community engagement—transformative education, community development, tipping points and social marketing—also assists in identifying strategies and approaches for stakeholder groups.

Stage 4. Implementation

It is important that the models of change theory and their actions are adopted by the stakeholder group, but it is clear that a concerted educative campaign is required. It is here that social marketing strategies may assist in guiding stakeholder group action.

Social marketing

Social marketing and community-based social marketing (social marketing initiatives delivered at the community level as described by McKenzie-Mohr (2004)) are strategies aimed at uncovering the barriers that hinder the adoption of more sustainable behaviours and provides an enabling framework for change. Social marketing is similar to commercial marketing, but the object is to achieve behavioural change rather than to influence product purchasing decisions (Ampt 2003). Ultimately, social marketing benefits not only the individual, but society as a whole and can lead to large-scale social changes (Dann 2005).

Social marketing has brought about change in community behaviour in a range of areas, such as food safety, food illness prevention, smoking, alcohol related problems, HIV, air quality and sexual abuse. Many social marketing campaigns in Australia relate to public health (for example, 'Slip, Slop, Slap' and 'Smoking Kills') and road safety ('Speed Kills', 'If you drink and drive, you're a bloody idiot' and 'Hector, the Road Safety Cat') where campaigns act in concert with legislation and punitive measures to change behaviour. Social marketing can also be used to tackle major social issues such as domestic violence ('Australia Says No to Violence').

Twenty years ago, the 'Grim Reaper' campaign spearheaded the Australian Government's attempts to deal with the issue of HIV AIDS. This campaign proved to be so successful that its resurrection is currently being considered now that infection rates are again rising. A key characteristic of the Grim Reaper campaign was its capacity to shock. Twenty years on would a similar campaign have the same effect?

Tom Lowe (2006), in his paper, 'Is this climate porn? How does climate change communication affect our perceptions and behaviour?', suggests that the shock/horror factor no longer has the same impact. While we might speculate about 20 years of increased media access, and violence depicted on television and in film, as reasons for desensitisation, his study found that extreme representations about the impact of climate change in the media actually had the reverse effect on members of the public. While the reporting promotes concern, and even anxiety, it creates an impression that the solution to the problem is beyond the reach of the ordinary citizen. Thus the public may have heightened awareness, and even deep concerns, about the problem, but will be unaware or unwilling to make sacrifices or change behaviour to tackle the issue.

It is important that individuals feel confident and optimistic that the changes they have made will have an impact. To be optimistic, a person must feel that they have the capacity to take action and that their action will make a difference (Voronoff 2005). When people feel disempowered, they are less likely to make changes, especially if the change is a difficult one. If threatening messages are used, they must be offset with mitigating actions that individuals can take to abate the threat (Robinson 2007). Information only campaigns, which do not cultivate personal control through participation, may create a sense of hopelessness (Voronoff 2005).

Environmentally damaging behaviours may be more difficult to change than health behaviours that have been the focus of many of the social marketing campaigns already mentioned. Pro-environmental decisions often require knowledge or information about the consequences of an action and associated behaviours to minimise environmentally harmful effects. Without sufficient information, people are incapable of acting in an environmentally sensitive way, even if they want to. In this case, social norms are ignored and hedonic goal or gain goal frameworks come into play (Lindenberg and Steg 2007).

The impacts of global warming are collective and spatially and temporally removed from many consumers and, at the present time, appear not to be directly affecting the daily lives of individuals—they therefore don't seem real to many people (Voronoff 2005). Essentially, this distances people—the perception being that these things happen in distant places, to other people, at some time far into the future.

While general pro-environmental attitudes have an important influence on behaviour, simply being pro-environmental is often not enough to overcome barriers such as additional inconvenience, expense or effort. Knowing that certain actions and behaviours harm the environment is generally not enough to effect behavioural change, especially if the activity is ingrained or if the competing 'more-sustainable' activity requires more effort. Our present lifestyle and economy is based on apparently insatiable patterns of consumption that are fuelled largely by a supply of non-renewable resources—middle Australia aspires to larger houses and more vehicles. Electricity is delivered unseen and is therefore a relatively intangible product and its cost is also separated temporally from its use, as a bill arrives in the post at quarterly intervals. Brown-coal electricity generation means that power in Victoria is particularly greenhouse-gas intensive, yet consumers incur a cost penalty for switching to green energy'.

Another factor that makes fostering pro-environmental behaviour even more challenging is dissent among professionals, scientists or politicians. Even a small amount of dissension can stymie activity and justify inaction.

However, successful environmental social marketing campaigns do exist. For example, TravelSMART, through its school, workplace and household programs has reduced reliance on the car through advice tailored to the target audience.

Similarly, the Save Water campaign in Victoria has reduced today's water consumption from 1990s levels by 170 ML (average of 1108 ML in 1990s to 938 ML for winter use in 2007), or approximately 15%, despite population and economic growth over this time.

This type of approach could be effective in tackling energy use in the Outer East, but will need resources accessible to the stakeholder action group (such as programs, incentives, examples of success stories and 'how to' guides).

Stage 5. Monitoring feedback and evaluation
The episodic nature of indicator reporting may mean that changes are not observed for some time. Indicator action groups will need to closely monitor progress, identify additional data to be collected and set up feedback mechanisms Additional data may include:

- the results of more frequent monitoring of regional electricity use
- measures of community awareness and attitude (especially related to any social marketing or educative campaigns undertaken)
- an audit of green energy adoption
- identifying changes to local planning processes
- street audits of energy-saving household retrofitting.

Conclusion

Indicators are an important tool in sustainable development and can be used to inform decision making and empower stakeholders, but they are just one component of a community-engagement process. The indicator is both the beginning and the end of the cycle—it provides information that establishes the problem and it assists in monitoring the results of subsequent action. For the community, the indicator becomes an important initiator and marker of progress.

References

Ampt R (2003). Understanding voluntary travel behaviour change. Paper presented at the 26th Australasian Transport Research Forum. Wellington, New Zealand.

Arnstein S (1969). Ladder of Citizen Participation. *Journal of the American Institute of Planners* **35(4)**, 216–224.

Atkinson Charitable Foundation (ACF) (2007). *Canadian Index of Well-Being: Measuring what Matters.* <http://www.atkinsonfoundation.ca/ciw>

Boston Foundation and Greater Boston's Civic Community (2007). *A Time Like No Other: Charting the Course of the Next Revolution. A Summary of the Boston Indicators Report 2004-2006.* A project co-sponsored by The City of Boston and The Metropolitan Area Planning Council.

Bruntland G (Ed.) (1987). *Our Common Future: The World Commission on Environment and Development.* Oxford University Press, Oxford.

Brutschy S and Zachary D (2007). Achieving outcomes from indicators: community assessment projects in the United States, A focus on Santa Cruz County. Paper presented at the OECD 2nd World Forum on 'Statistics, Knowledge and Policy', Istanbul, Turkey.

Burton D (2007) Evaluating Climate Change Mitigation Strategies in South East Queensland. Urban Research Program, Research Paper 11. Griffith University, Brisbane, Queensland. <http://www.griffith.edu.au/centre/urp/urp_publications/research_papers/URP_RP11_Burton_ClimateLocGovt_final.pdf>

Climate Institute (2007). *Climate of the Nation: Australians Attitudes to Climate Change and its Solutions.* The Climate Institute, Sydney, NSW. <http://www.climateinstitute.org.au/images/reports/climatenation.pdf>.

Community Indicators Consortium (2007). *CIC Web Page: Helping Communities Develop and Implement Community Indicators.* Community Indicators Consortium, Lynn, MA, USA. <www.communityindicators.net>

Community Indicators Victoria (2007). *CIV Web Page.* CIV, The McCaughey Centre, School of Population Health, University of Melbourne, Melbourne. <www.communityindicators.net.au/>

Dann SD (2005). Social change marketing in the age of direct benefit marketing - where to from here? Paper presented at the Social Change in the 21st Century Conference,. Queensland University of Technology, Brisbane.

Department for Victorian Communities Strategic Policy and Research Division (2004). *Indicators of Community Strength in Victoria.* State Government of Victoria, Melbourne.

Department of Human Services (2005). *Evaluating Neighbourhood Renewal: Passing the First Test.* State Government of Victoria, Melbourne.

Department of the Environment, Transport and the Regions (2001). *Achieving a Better Quality of Life: Review of progress towards sustainable development – Government Annual Report 2000.* UK Government, London. <http://www.sustainable-development.gov.uk/publications/pdf/susdevel.pdf>.

Elkington J (1999). *Cannibals with Forks – The Triple Bottom Line of the 21st Century Business.* Capstone Publishing, Oxford.

Elkington J (2001). *The Chrysalis Economy – How CEOs and Corporations Can Fuse Values and Value Creation.* Capstone Publishing, Oxford.

Innes J and Rongerude J (2006). *Collaborative Regional Initiatives: Civic Entrepreneurs Work to Fill the Governance Gap.* Institute of Urban and Regional Development (IURD), University of California, Berkeley.

International Association for Public Participation (2003) IAP Public Participation Spectrum. IAP, Westminster, USA. <http://www.iap2.org/associations/4748/files/spectrum.pdf>.

Istanbul Declaration (2007). <http://www.oecd.org/dataoecd/14/46/38883774.pdf>

Jacksonville Community Council Inc. (2005). *Quality of Life '05 Survey Final Results.* American Public Dialogue Inc., Jacksonville, Florida. <http://www.jcci.org/statistics/documents/2006_survey.pdf>.

Kretzmann J and McKnight J (1993). *Building Communities from the Inside Out: A Path Towards Finding and Mobilizing a Community's Assets.* The Asset-Based Community Development Institute, Evanston.

Langworthy A and Brunt C (2005). *Community Sustainability Summary Report: Outer Eastern Melbourne.* Centre for Regional Development, Swinburne University of Technology, Lilydale, Victoria.

Lawn PA (2003). Growth, technological progress and sustainable development: preliminary evidence of Australia's sustainable development performance. *International Journal of Environment and Sustainable Development* **2(2)**, 139–161.

Lindenberg S and Steg L (2007). Normative, gain and hedonic goal frames guiding environmental behaviour. *Journal of Social Issues* **63(1)**, 117–137.

Lowe TD (2006) *Is this Climate Porn? How Does Climate Change Communication Affect our Perceptions and Behaviour?* Tyndall Centre for Climate Change Research, University of East Anglia, Norwich, UK.

Macgregor C (2000). Applying an analytical model for assessing community sustainability: some preliminary results from northern Australian remote towns. I Paper presented at the First National Conference on the Future of Australia's Country Towns, Bendigo, Victoria.

Maslow AH (1970). *Motivation and Personality* (2nd edn). Harper and Row, New York.

Max-Neef MA, Elizalde A and Hopenhayn M (1991). *Human Scale Development: Conception, Application and Further Reflections.* The Apex Press, New York.

McKenzie-Mohr D (2004). *Quick Reference: Community-Based Social Marketing.* McKenzie-Mohr & Associates, Fredericton, Canada. <http://www.p2.org/summit2004/documents/Presentations/Fostering%20Sustainable%20Workshop%20Handout.pdf>.

Newton PW (2006). *Human Settlements: Theme Commentary Prepared for the 2006 Australian State of the Environment Committee at CSIRO.* Department of the Environment and Water Resources, Australian Government, Canberra. <http://www.deh.gov.au/soe/2006/commentaries/settlements/index.html>.

Robinson L (2007). *Introduction to the Seven Doors Model.* Enabling Change website. <www.enabling-change.com.au/7_doors_page.html>.

Rocha E (1997). Ladder of Empowerment. *Journal of Planning and Educational Research* **17(31)**, 31–35.

Saunders C, Brook AT and Myers Jr OE (2005). Using psychology to save biodiversity and human wellbeing. *Conservation Biology* **20(3)**, 702–705.

Sustainable Seattle (1993, 1995, 1998). *Sustainable Seattle and Indicators.* Seattle, USA, <http://www.sustainableseattle.org/Programs/RegionalIndicators>.

Sustainable Seattle (2006). *Sustainable Urban Neighborhoods Inititative: Year Three Report.* Seattle, USA. <http://www.sustainableseattle.org/Programs/SUNI/yearthreefinalreport>.

Treasury Board of Canada Secretariat (2004). *Canada's Performance; Annual Report to Parliament 2004.* Government of Canada, Ottawa. <http://www.tbs-sct.gc.ca/report/govrev/04/cp-rc_e.asp>.

Trewin D (2006). *Measures of Australia's Progress.* Australian Bureau of Statistics, Canberra.

Voronoff D. (2005). *Community Sustainability. A Review of What Works and How it is Practiced in Victoria.* Environment Victoria, Melbourne. <http://www.envict.org.au/file/EV_Community_Sustainability_Report_2005.pdf>

Woolcock M (2000). The Place of Social Capital in Understanding Social and Economic Outcomes. Paper presented at the International Symposium 'The Contribution of Human and Social Capital to Sustained Economic Growth and Well-Being'. March 2001. OECD, Quebec, Canada.

SECTION 10

Two Scenarios for the Future

Perspectives on the future always carry with them a measure of uncertainty, but are valuable to the extent that they can highlight future conditions that need to be avoided, as well as desirable future states that we should be actively working towards. There are *different options* for the future, and *different choices* that can be made that will radically shape that future. Previous sections of this book have outlined different options and pathways related to key urban domains such as water, energy, waste, infrastructure and urban planning.

In this final section, both Trainer and Ellyard identify the most significant challenge for 21st century Australians, their governments and their industries as being *a fundamental shift in their value systems*. For Trainer it involves *transitioning to a simpler way of life*, which our settlement systems will need to support and which stands in contrast to the current paradigm of over-consumption that currently appears to be hard-wired into our population and cities. For Ellyard, it involves *transitioning to sustainable prosperity* by adopting a new set of design principles and technologies to guide our local activities and urban development as well as a *global perspective* which is needed to avoid a 21st century repeat of the tragedy of the commons.

PWN

44

Transitioning to The Simpler Way

Ted Trainer

The almost universally accepted assumption is that the alarming problems consumer-capitalist society is running into can be solved in, and by, consumer-capitalist society. That is, it is taken for granted by politicians, bureaucrats, corporations and the population in general that solutions can be found and implemented that will enable the continuation of a society committed to affluent 'living standards' and constant and limitless increases in economic output. The purpose of the following discussion is firstly to explain why these assumptions are seriously mistaken and then to sketch the implications for radical change.

The situation

The range of major global problems facing us is widely recognised, but their magnitude and significance is not. There are two themes to be dealt with here. The first is to do with sustainability and the second concerns global economic justice

Sustainability

The way of life we take for granted in rich countries such as Australia is grossly unsustainable. The two clearest and most forceful instances are to do with the ecological footprint concept and the greenhouse problem.

It takes about 7 hectares of productive land to provide water, energy, the built environment and food for one person living in a rich country such as Australia. If the world's forecast nine billion people (in 2050) were to live as we do in Australia now, they would need about 63 billion ha of productive land, but that is about nine times all the available productive land on the planet.

The most disturbing argument relates to the greenhouse situation. It is generally accepted that global temperature increase must be kept below 2 degrees Celsius, and this would require the atmospheric concentration of carbon dioxide to be kept below 450 ppm, and possibly below 400 ppm. This, in turn, would require annual emissions of carbon to be reduced from the present level of about 8 billion tonnes to around 1.5 GT by 2050, and eventually to perhaps 0.5 GT (Baer and Mastrandrea 2006). If 1.5 billion tonnes of carbon were shared equally among the 9 billion people we are likely to have on earth after 2050, then the per capita average amount we could consume would be 0.17 tonnes, which corresponds to around 5% of the amount of fossil fuel Australians use now.

These are but two of the main limits-to-growth arguments that show that there is no possibility of all people rising to the living standards we take for granted today in rich countries such as Australia. The crucial point here is the magnitude of the overshoot. Most people have no idea of how far beyond sustainable levels of consumption we are—and how big the reductions will have to be. We seem to be 5–10 times over some crucial limits. Again, it is clear that consumer-capitalist society is grossly unsustainable.

The impossible implications of economic growth

The foregoing argument has only been that the *present* levels of production and consumption are quite unsustainable. Yet we are determined to increase living standards and levels of output and consumption as much as possible and without any end in sight. In other words, our supreme national goal is economic growth. Few people seem to recognise the consequences of pursuing the current trajectory of economic growth. If we have a 3% per annum increase in output, by 2080 we will be producing eight times as much every year. If by then all the expected 9 billion people had risen to the living standards we would have then, the total world economic output would be *more than 60 times* as great as it is today, even though the present level is unsustainable

For about 50 years, the limits-to-growth analysis has been pointing to these kinds of numbers and insisting that there is no plausible alternative to accepting the need not only to abandon the obsession with growth, but also moving to rates of per capita resource consumption that are perhaps one tenth of present amounts.

'But can't technical advance solve the problems?'

Most people assume that technological advance will defeat these limits-to-growth arguments and enable everyone to go on enjoying affluent lifestyles and pursuing limitless economic growth, by reducing the energy and resource inputs needed to produce things and the associated environmental impacts. However, the sheer magnitude of our over-consumption makes this impossible.

Perhaps the best known 'technical fix' optimist, Amory Lovins, claims that we could at least double global output while halving the resource and environmental impacts; that is, a 'Factor Four' reduction (Weiszacher and Lovins 1997). This is a plausible claim, but it is easily shown that such an achievement would be nowhere near enough to solve the forecast future problems.

Let us assume that present level of global resource use and ecological impact must be halved (although from above it is clear that this would not be sufficient). As has been explained, if we in rich countries average 3% growth, and 9 billion rose to the living standards we would then have by 2070, total world output would be 60 times as great as it is today. Now do you think technical advance will make it possible to multiply total world economic output by 60 while halving impacts—that is, not a factor 4 reduction but a factor 120 reduction? (For a critique of Lovins, see Trainer 2007a.)

Similarly, consider again the multiples evident in the above discussion of the greenhouse problem. The Australian Bureau of Agricultural and Resource Economics (2006) estimates that we are on a path to a total global carbon emission rate per year in 2050 that is an alarming 2.6 times as high as it is now, and that conservation effort is only likely to reduce the resulting 15 GT figure to 11.6 GT, when it was explained above that the target should be in the vicinity of 1.5 GT. For 9 billion people this averages 0.17 tonne per person, which is a factor 40 reduction on present rich-world emissions.

The main assumption underlying the 'technical fix' optimism is that we can change to renewable energy sources and thus avoid use of fossil fuels. There is a strong case that this unquestioned faith is invalid (see Trainer 2007b). Just consider the liquid fuel problem. We will probably be able to produce 7 GJ of ethanol per tonne of biomass (Fulton 2005), and to grow biomass at no more than 7 t/ha, given that the scale would have to be very large. (Note that some researchers do not think that the production of ethanol from woody inputs will be viable; see Augenstein and Benemann 2007.) To provide the 128 GJ p.a. of oil plus gas that Australians now use on average, we would need 2.6 ha of biomass plantation per person. To provide this energy to 9+ billion we would need some 23 billion ha of plantations … on a planet with only 13 billion ha of land (… and at the present rate of growth Australian energy demand will be about 2.5–3 times as great by 2050; ABARE 2006).

Clearly there is no possibility that biomass could provide all people with anything like the quantity of liquid fuel that we now take for granted in Australia.

'But won't we run everything on hydrogen?' It is not likely that we will be able to rely heavily on hydrogen in view of the difficulties and energy losses associated with producing and distributing the very light and diffuse hydrogen atom. For instance, Bossell *et al.* (2003) explain that a standard petrol tanker capable of delivering 20 tonnes of petrol could only deliver 320 kg of compressed hydrogen. To pump hydrogen to Europe from solar thermal plants in the Sahara might take most of the energy pumped. To run cars on hydrogen produced by windmills would deliver to the wheels about one-quarter of the energy the wind produced (Trainer 2007).

'Well then we'll have electric vehicles.' Bossel *et al.* (2003) explain that this would halve the losses associated with hydrogen-powered vehicles. Australian transport takes almost twice as much energy as electricity demand so to run the present transport system on electricity would require generation of about five times as much electricity as is generated now. Trainer (2007) argues that renewables will not be able to meet present electricity demand

Again, the problem is not whether renewable energy sources can meet present demand; rather it is whether they can fuel a society that is committed to growth. The International Energy Association (2006) projections indicate that Australian energy consumption is heading towards 2.5 times as great as it is now by 2050. If 9 billion people were to have that level of consumption, world demand would be about 20 times as great as it is now. Yet, by then, the supply of petroleum will probably be half as great as it is now.

This is not an argument against renewable energy sources. We must move to full dependence on them as fast as we can, as they could fuel a sustainable and just society. But the general conclusion from the foregoing discussion has been that they cannot sustain consumer society into the future.

Clearly, a huge and radical system change is needed. The problems cannot be solved solely by technical advances and more conservation effort on the part of individuals and firms *within* consumer-capitalist society. The overshoot is far too big for that. The problems are being caused by some of the fundamental structures of consumer-capitalist society. Consequently, most of what is currently said under the heading of 'sustainability' by industry and government is nonsense and most of the current efforts being made to 'save the planet' are completely inadequate. Most irritating are the 'what you can do in your own home' campaigners. 'Buy biodegradable wash up liquid, use a low-flow shower head, recycle your bottles, buy a smaller car, and so on.' Such efforts can make no more than a negligible difference to household impact, when we need something like a 90% reduction in national consumption. Nothing remotely like this is possible within a consumer-capitalist society committed to affluent lifestyles and limitless economic growth. It is only possible through dramatically reducing the volume of production and consumption and, therefore, by changing to a society that is about frugal, but adequate, 'living standards' and a stable economy.

The injustice of the global economy

Even if our rich-world way of life was not grossly unsustainable in terms of resource and environmental impact, a glance at the way the global economy delivers that way of life shows that it is not morally justifiable and that it inevitably generates an enormous problem of equity and security. The core problem can be put in terms of the contradiction between the market mechanism and the just distribution of resources.

In a satisfactory and sustainable society there might be a considerable role for markets, but only if they are carefully controlled. It is easily shown that the market system is responsible for most of the deprivation and suffering in the world. The basic mechanisms are most clearly seen when we consider what is happening in the developing world.

The enormous amount of poverty and suffering in the Third World is not due to lack of resources. There is, for instance, sufficient food and land to provide for all. The problem is that these resources are not distributed at all satisfactorily. Why not? The answer is that this is the way the market economy inevitably works.

The global economy is a market system and in a market scarce things always go mostly to the rich (that is, to those who can pay most for them). That's why rich countries get most of the oil produced. It is also why more than 500 million tonnes of grain are fed to animals in rich countries every year—over one-third of total world grain production—while 850 million people are hungry.

Even more important is the fact that the market system inevitably brings about inappropriate development in the Developing World (that is, development of the wrong industries). It will lead to the development of the most profitable industries, as distinct from those that are most necessary or appropriate. As a result there has been much development of plantations and factories in the Third World that will produce goods for local rich people or for export to rich countries. But there is little or no development of the industries that are most needed by the poorest 80% of their people. The Developing World's productive capacity—its land and labour—have been drawn into producing for the benefit of others, especially rich-world corporations and consumers. This is most disturbing and results in most of the best land being devoted to export crops.

Conventional development is driven by market forces on the promise that freeing the richest to maximise their incomes will generate wealth that will 'trickle down' to enrich all. For at least one billion people, the opposite happens. Consider the situation of the people in Bangladesh who are paid 15c an hour to produce shirts for export. Obviously it would be far better for them if they could put all their work time into small local farms and firms that used local land, labour and skill to produce for themselves the basic things they need. But where there is capitalist development, this is deliberately prevented. Third World ruling classes and rich-world governments will only support development that is determined by whatever will maximise the profits for investors. The conditions of the Structural Adjustment Packages imposed by the World Bank on indebted countries prohibit any other kind of development. Indeed they force poor countries to open their economies more to market forces and corporate investment and to deregulate (that is, to reduce state control over what is developed). The poorest 1 to 2 billion people live in countries where corporations can't make any profit, so there is almost no 'development' in them, when those countries could be solving their basic problems via appropriate development—quickly and without much capital or dependence on the global economy (for more on appropriate development, see Trainer 2005).

These are the inevitable consequences of an economic system in which what is done is that which is most profitable to the few who own capital, as distinct from what is most needed by people or their ecosystems (see Trainer 2007c). The Developing World problem will never be solved as long as these economic principles determine development and deliver most of the world's wealth to the rich. For these reasons, conventional Developing World programmes can be seen as a form of legitimised plunder. It gears their resources and productive capacity to the enrichment of the rich—not to meeting the needs of the poor majority. There are three major beneficiaries:

- the transnational corporations, who get maximum access to Developing World resources and markets
- the Developing World's ruling classes who participate in the deals
- the people who shop in the supermarkets of the rich countries which are stocked with goods flowing from the Developing World.

The main point of geopolitics and international relations has always been to get access to and control over the resources and markets of others. In our era, this is mostly achieved by allowing market forces to determine who gets what, because the richest always get what they want in a market situation. However, from time to time it becomes necessary to use force to keep in place the arrangements that suit the powerful. Thus the rich countries support dictatorial and brutal regimes willing to rule in our interests—the regimes enable, and actually engage in, terrorism and they invade and attack and kill thousands of innocent people, in order to ensure that regions keep to the economic and development policies that suit the rich countries (for more extensive documentation, see Trainer 2007d). The high 'living standards' we enjoy in rich countries could not be provided without these unjust and brutal processes. We could not get our cheap goods and most of the resources produced unless force is used to maintain the unjust systems delivering to us far more than our fair share of the world's wealth.

There is no possibility of satisfactory Third World development, or of eliminating the intense deprivation experienced by at least 3 billion people, until the rich countries stop hogging far more than their fair share of the world's resources, until development and distribution begin to be determined by need and not by market forces and profit, and until the rich countries stop taking far more than their fair share of global resources. These conditions cannot be achieved unless we develop a very different global economic system. Again, this must mean huge and radical structural change—moving to simpler ways of living and to an economy that focuses on meeting need rather than maximising profit.

Conclusions on the market system

The dominant ideology assumes that if we leave things to market forces—that is, to competition between individuals, corporations and nations all trying to maximise their self-interest—then we will end up with a satisfactory society. On the contrary, a free market will inevitably result in the strongest and richest winning, by taking even more and becoming even richer. The environment and social cohesion cannot be protected if the rules permit individuals to take as much as possible for themselves. Billions of people are unable to produce and sell the small quantities that would yield satisfactory incomes because, in this economy, a few giant corporations are allowed to sell things more cheaply and thereby take the markets and livelihoods.

Perhaps even more important are the damaging effects that the triumph of neo-liberalism is having on social cohesion, values, standards and concern for the public good. Because many people are not given a satisfactory share of the wealth, jobs and resources—and because people are having to work harder, in more insecure circumstances and many are being dumped into 'exclusion'—there is much drug abuse, crime, stress and depression. Public institutions (such as hospitals, universities and public transport) are deprived of sufficient funds. Social attitudes are becoming more selfish and mean. Increasing numbers of people believe the future will be worse than the present. There is little concern for social justice, the public good, the welfare of those least able to cope or the plight of the Third World. 'Losers' are despised as a drain on our taxes. Especially disturbing is the ethos of callous selfishness and greed reinforced by neo-liberalism.

Thus the major global problems are explained

These two basic lines of critical analysis—focusing on sustainability and justice—are crucial for understanding the major problems threatening the planet. For instance, we have an alarming environmental problem essentially because there is far too much production and consumption occurring. We have a problem of Developing World poverty, deprivation and under-development because the rich countries are taking most of the available resources and forcing poor countries to comply with an economic system that devotes their productive capacity to the benefit of advanced industrial societies. We have problems of armed conflict mainly because some are determined to get hold of far more than their fair share of the world's resource wealth. We have increasing inequality because market forces inevitably enable the rich to take more of the available wealth. We have social breakdown and deteriorating quality of life because the top priority is maximising production and consumption, business turnover, individual wealth and GDP. The one factor overwhelmingly responsible for our alarming situation is over-consumption.

Conclusions on the global situation

The foregoing argument has been that the way of life we have in rich countries is grossly unsustainable and unjust—that the basic cause of the alarming global situation is simply over-consumption and that these are inevitable consequences of some of the fundamental structures, values and commitments of our society. Some of the lines of analysis noted above indicate that we should be trying to reduce per capita resource consumption by 90% or more. Nothing like this can be done without huge and radical change to new systems—most obviously to a new economy—but most problematically to a new culture that is not based on competitive, individualistic acquisitiveness.

This general analysis of our situation has been argued for more than 40 years now, but politicians, bureaucrats, teachers, journalists, economists and ordinary people have flatly refused to think about the possibility that the obsession with affluence and growth could be problematic.

The alternative: the simpler way

If the foregoing arguments are basically valid, some of the key requirements for a sustainable society would seem to be clear and indisputable. The basic principles drawn from Trainer (2007b, 2007e) are described below.

Simpler lifestyles

Living more simply does not mean deprivation or hardship. It means focusing on what is sufficient for comfort, hygiene, efficiency, and so on. Most of our basic needs can be met by quite simple and resource-cheap devices and ways, compared with those taken for granted and idolised in consumer society.

Living in ways that minimise resource use should not be seen as an irksome sacrifice that must be made in order to save the planet. These ways can come to be experienced as important sources of life satisfaction. The goal has to be to enable people to find enjoyment in activities such as living frugally, recycling, growing food, 'husbanding' resources, making rather than buying, composting, repairing, giving no-longer-needed things to others, making things last and running a relatively self-sufficient household economy.

Local self-sufficiency

We must develop as much self-sufficiency as we reasonably can at all levels—from the national scale down to that of the household—and especially at the neighbourhood, suburban, town and local regional levels. This means more local production, more use of local resources and more active forms of travel (such as walking and cycling). Neighbourhoods could become 'permaculture jungles'—'edible landscapes' crammed with long-lived, largely self-maintaining productive plants, such as fruit and nut trees. There should be many neighbourhood workshops for all to use. These would include a recycling store, meeting place, surplus exchange, craft rooms, art gallery, shared tools and library. It would be a leisure-rich environment, whereas suburbs at present are leisure deserts. There would be an appropriate blend of 'high tech' and 'high touch' (Naisbitt 2000): The Simpler Way would enable retention of all the high tech and modern ways that made sense; for example, in medicine, in renewable energy, public transport and household appliances.

More communal and cooperative ways

We must share more things. We could have many community tools in the neighbourhood workshop. We would be on various voluntary rosters, committees and working bees to carry out child minding, nursing, basic educating and care of aged and disabled people in the local area. We would also perform most of the functions councils now carry out for us, such as maintaining our own parks and streets. We would therefore need far fewer bureaucrats and professionals, reducing the amount of income we would need to earn to pay for services and pay taxes. Especially important would be the regular voluntary community working bees to build and maintain the commons, edible landscape, energy and water systems.

The new economy

There is no chance of making these changes while we retain the present economic system. The fundamental concern in a satisfactory economy would simply be to apply the available productive capacity to producing what all people need for a good life, with as little bother, resource use, waste and work as possible. Most obviously, there would have to be far less production and consumption occurring, market forces and the profit motive would have to be brought under firm social control and there must be a focus on community development—as distinct from economic growth. The new economy would comprise more small-scale, local economies, so that most of the basic items that populations need are produced closer to where they live, from local resources, by local skill and labour.

Government and politics

These crucial changes must be made in economic, geographical and political structures and systems. They can't be made just by individuals changing their lifestyles.

The political situation would be very different compared with today. There would (have to) be genuine participatory democracy. This would be made possible by the smallness of scale, and it would be vitally necessary. Big centralised governments could not run our small localities. That could only be done by the people who live in them because they are the only ones who would understand the local conditions and could develop policies that will work there.

There would still be some functions for state and national governments—but relatively few—and there will be a role for some international agencies and arrangements.

Our intense dependence on our local ecosystems and social systems would radically transform politics. The focal concern would (have to) be what policies will work best for the town and region, because we as individuals will only live well if our region thrives.

Politics would not be primarily about individuals and groups in zero-sum competition to get what they want from a central state. There would be powerful incentives towards a much more collectivist outlook—to find solutions everyone is content with.

The new values and worldview

The most profound and difficult changes will have to be in values. The present fierce competition for affluent-consumer living standards and endless increases in wealth must be replaced by a strong desire to live simply, cooperatively and self-sufficiently. Only if these alternative values and sources of satisfaction—which contradict much that is at the core of Western culture—become the main factors motivating people can The Simpler Way be achieved.

There is no alternative

If the analysis of our global situation given at the beginning of this chapter is basically correct, then we have no choice but to work for the kind of alternative society sketched here. In rich and poor countries, a sustainable society can only be conceived in terms of simpler lifestyles, mostly in highly self-sufficient and cooperative settlements, zero-growth or steady-state economic systems, mostly small local economies and in terms of new values.

The prospects for transition?

The foregoing arguments constitute such a fundamental challenge to the dominant ideology of consumer society that it is not surprising that they have been almost completely ignored by governments, the corporate world and the general public. The chances of making such a profound transition must be rated as extremely low.

In the last 20 years a 'Global Alternative Society Movement' has developed, in which many people all around the world have begun to build, live in and experiment with new settlements of the kind sketched above. More important, however, will be small groups of ordinary people coming together in their existing neighbourhoods to begin taking the first steps in the eventual transformation of these into thriving local economies under social control. It is important that as consumer society runs into more difficulties, people will be able to see around them examples of the alternative ways. The advent of serious petroleum scarcity will concentrate their minds wonderfully (for more detailed discussion of transition strategy, see Trainer 2007f).

(The themes raised in this chapter are discussed in detail on The Simpler Way website, http://socialwork.arts.unsw.edu.au/tsw.)

References

Augenstein D and Benemann J (2007). *The Cellulosic Ethanol Delusion.* 14 June. <http://www.aiche-norcal.org/.Symposium/Symposium2006/pdfs/EnergySolutions.pdf>

Australian Bureau of Agricultural and Resource Economics (ABARE) (2006). *Technological Development and Economic Growth.* 12 January. ABARE, Canberra.

Baer P and Mastrandrea M (2006). *High Stakes; Designing Emissions Pathways to Reduce the Risk of Dangerous Climate Change.* Institute for Public Policy Research, London.

Bossel U, Eliasson B and Taylor G (2003). *The Future of the Hydrogen Economy: Bright or Bleak?* <http://www.oilcrash.com/articles/h2_eco.htm>

Fulton L (2005). *Biofuels For Transport: An International Perspective.* International Energy Agency, Paris.

International Energy Agency (2006). IEA, Paris. Web site <www.eia.doc>.

Naisbitt J and Aburdene P (2000). *Megatrends 2000.* Morrow, New York.

Trainer FET (2005). Development. the radically alternative view, *Pacific Ecologist* **Summer**, 35–42.

Trainer FET (2007a). *Natural Capitalism Cannot Overcome Resource Limits.* University of New South Wales. <http://ssis.arts.unsw.edu.au/tsw/D50NatCapCannotOvercome.html>

Trainer FET (2007b). *Renewable Energy Cannot Sustain A Consumer Society.* Springer, Dordrecht, Netherlands.

Trainer FET (2007c). *The Economic System; A Radical Critique.* University of New South Wales. <http://ssis.arts.unsw.edu.au/tsw/09c-EC.SY.1.Contents.html>

Trainer FET (2007d). *Our Empire: Its Nature and Maintenance.* University of New South Wales. <http://ssis.arts.unsw.edu.au/tsw/10-Our-Empire.html>

Trainer FET (2007e). *The Alternative; The Simpler Way.* University of New South Wales. <http://ssis.arts.unsw.edu.au/tsw/12c-TheALT.SUS.SOC.long.html>

Trainer FET (2007f). *Thoughts on the Transition.* University of New South Wales. <http://ssis.arts.unsw.edu.au/tsw/15-Transition.html>

Von Weizacker E Lovins A and Lovins LH (1997). *Factor Four: Doubling Wealth, Halving Resource Use.* Allen & Unwin, Sydney.

45

Designing and Innovating a Sustainable Society

Peter Ellyard

In the next three years, humanity will commit itself to reducing emissions to the atmosphere of carbon dioxide, methane and nitrous oxide by more than 50% by the year 2050. If we are not to leave a degraded planet as a legacy to our grandchildren, it has no choice. Although there will be opposition from some nations and vested interests, the tide of global awareness and responsibility has grown to the point that we can be confident this will occur and will be implemented under the UN Framework Convention on Climate Change.

That the world is seeking to reach a goal that is two generations ahead is something to savour and to celebrate. Achieving such a historic reduction in carbon emissions is one essential goal we must achieve by the year 2050. There are other goals we could, and should, achieve by this time as well. But the more fundamental question is: what overarching goal should we seek to accomplish by the year 2050—what concept integrates all these goals into a single vision? For me, this overarching goal is designing, innovating and then building a global sustainable society by the year 2050. The thought of creating a sustainable society in two generations may seem daunting, if not impossible. However, we must keep sight of the heroic changes that have occurred over the past generation and which are now occurring every day. So the prospect that we could complete the rest of the journey to a sustainable society within two generations is more possible than many might think.

The prime reason for asserting this—and the force behind these changes—has been an ongoing shift in values over the past century or so from modernism to postmodernism and then on to what I have called *planetism*. Globalisation (increased interconnectedness in trade and the formation of large multinational communities), tribalisation (the breaking of multicultural nations/empires into small independent cultural entities) and technological interconnectness are collectively shaping this transformation In my book, *Designing 2050: Pathways to Sustainable Prosperity on Spaceship Earth*, I have also called this new planetary awareness the *spaceship culture*, and referred to a person with this culture as a planetist. Planetists give their first allegiance to the planet—to Spaceship Earth—which they share with the rest of humanity. Those who challenge John Howard's not signing the Kyoto Protocol—which he did on the grounds that it was not in the national interest—and who say that our first priority should be to the planet are, in my terminology, planetists rather than nationalists. In contrast to the planetist or cosmonaut culture is the *cowboy culture*—the modernist culture slowly being left behind as the world advances further into the 21st century.

Modernism was the dominant global paradigm of the 19th century and the first two thirds of the 20th century. Being 'modern' was good and being 'old-fashioned' was derided or ridiculed: there was little respect for the old ways of doing and being, and there was an often uncritical embracing of new ways of doing things and of being. Indigenous world views were regarded as inferior and incompatible with western modernist thinking, and modernists set out to 'civilise' other people for their own good. Modernist thinking was dominated by the concept of 'progress' and not long ago virtually every community had a 'progress association' or its equivalent. These are very hard to find today. Moreover, the feelings accompanying the utterance 'you can't stop progress' changed from unbridled enthusiasm to resignation and then to skepticism.

Modernism therefore has now been deconstructed and replaced with a paradigm that, unlike modernism, valued old things as well as new, and encouraged the integration of modern and pre-modern forms and traditions. It also included the embracing of some genuinely new concepts such as sustainability, and was called (logically enough) postmodermism.

In the postmodern era, the merits of many old ways of doing things and of viewing the world are being reaffirmed. Postmodernism fosters the view that the world would be a better place if we paid more respect to pre-modern—including indigenous—wisdom and knowledge. People are doing things in every area that would not have been possible in modernist times. Many farmers are trying to create a new 21st century integration of modern agriculture (with its highly productive, but environmentally threatening practises) with the lower productivity versions based on organic and biodynamic principles, which pre-date the invention of modern agriculture in the 1840s by Justus von Liebig and John Bennet Lawes. Another example that illustrates the switch from modernism to postmodernism is what many people do nowadays if they feel ill. They might consult a conventionally trained doctor; however, they might also consult a Chinese herbalist, an Ayurveda doctor, an acupuncturist, a shiatsu healer, a naturopath or a yoga master. They will appropriate and integrate those trans-cultural approaches they think will be most efficacious to their own personally customised healing solution. This is the essence of postmodernism, which involves sampling the world's cultures and other people's ideas and appropriating for your own needs those practices you find useful.

However, postmodernism, which is essentially a deconstruction of modernism, is a small transitional paradigm that sits between modernism and the next big paradigm, which will be as characteristic of the 21st century as modernism was to the 18th and 20th centuries. Modernism's true successor is planetism. Table 1 indicates the ten ways in which values have changed or are changing as modernism slowly gives ways to planetism.

Most of the terms used in the Table will be familiar, but individualism and communitarianism are concepts central to the concept of sustainability, which need to be briefly discussed. Individualism involves giving priority to the rights of individuals over that of community when these conflict, whereas communitarianism means giving priority to community rights over individual rights in the same situation. Both individualism and communitarianism are present in all cultures to varying degrees, whether these cultures are democratic or autocratic. In Australia, communitarianism is increasingly dominating individualism. Two examples of this are the banning of smoking by individuals in public places to protect community health, and limiting the rights of individual gun owners so that communities can be made safer.

A sustainable society will be one that lives by planetist values—or something very close to them. The degree of adoption of planetist values in any community, and the degree to which governance

Table 1: A comparison of modernism and planetism

From the Cowboy Culture/Modernism (1960)	To the Spaceship Culture/Planetism (2020)
Priority to nation/tribe over planet	Priority to planet over nation/tribe
Individualism	Communitarianism.
Independence	Interdependence
Autocracy	Democracy
Humanity separate from nature	Humanity part of nature
Unsustainable production/consumption /development/ lifestyles	Sustainable production /consumption /development/ lifestyles
Patriarchy	Gender equality
Inter-cultural and inter-religious intolerance/hostility	Inter-cultural and inter-religious tolerance/harmony
Conflict resolution through confrontation/combat	Conflict resolution through cooperation/negotiation
Safekeeping through defence	Safekeeping through security

and business is informed and affected by these values, will probably correlate with the degree of sustainability already achieved.

The values people hold inform and, in the end, define what people value and regard as 'valuable'. They will want more of what they regard as valuable, and so markets will emerge for their provision. Planetist values are the values of the educated middle class—and these values will inform what these people seek in global markets in the 21st century. Globalisation is massively lifting the global membership of the educated middle class. In India and China alone their numbers have now reached half a billion. By understanding the core values which define planetism—the values of the educated middle class—we can predict what products and services will be sought in the next four decades or so. And many of these these are the innovations collectively required to realise and embed a planetist society—a sustainable society—in the 21st century-innovations first demanded by the educated middle class and, over time, by all of us. Essentially, we need two kinds of innovations to reach the destination of a sustainable society by the year 2050. The first group are social innovations: innovations to the way we do things, or changes in our behaviors, to realise a sought end. I call these *ways*. The second group are physical innovations: new things/ tools we use to realise a sought-after end. For example, a *water conservation way* would be a new behaviour such as limiting our time in the shower. A *water conservation ware* would be a new water-saving shower head.

But what is a sustainable society? Readers will have been exposed to many definitions of sustainability and insights into what might constitute a sustainable society. First of all, a sustainable society *could* be more lush, convivial and richer than the unsustainable one we now have. Too many people see a sustainable future as one mostly involving doing without, tightening the belt or narrowing, rather than broadening, future options.

Here is some more new language that enables us to imagine and create such a rich, convivial and lush sustainable society. I want to introduce the concept of *sustainable prosperity*. Many readers will find the pairing of 'sustainable' and 'prosperity' somewhat strange, and even incompatible. John Elkington's introduced the concept of the triple bottom line in his book *Cannibals with Forks*. The triple bottom line invites people to report on, and account for, the performance of an organisation in terms of its economic, social and environment performance. I prefer the notion of a *quadruple* bottom line. This adds culture as a separate entity, instead of implicitly including it under the social category as on Elkington's scale, because it is possible to create an outcome which simultaneously results in a positive social result and a negative cultural result and vice versa. The quadruple bottom line therefore consists of economic, social, environmental and cultural outcomes, and for each of these four domains we can have either prosperity or its opposite—poverty. True prosperity—sustainable prosperity—is achieved when each domain is prosperous. We do not achieve this by creating prosperity in one category while simultaneously creating poverty in another

In the modernist era we did just that: we created economic prosperity while simultaneously creating ecological poverty—yet we called this 'progress'. For example, we logged a forest too rapidly and too severely, beyond its re-growth capability. This does not create sustainable prosperity. On the other hand we can close down a logging operation to protect the biodiversity of an old growth forest. But if we do not provide suitable alternatives for the forestry workers who lose their jobs, we will create social and economic poverty while creating ecological prosperity. This too does not create sustainable prosperity. Alternatively, we might be able to log a forest selectively, thereby increasing the proportion and diversity of sub-climax and disclimax species, so that we increase biodiversity. This would produce sustainable prosperity because it would create economic and ecological prosperity simultaneously. Similarly, we can create economic prosperity by allowing a facility to pollute water—thereby increasing the ecological poverty of the water—and we can subsequently restore that water to high quality again and create more economic prosperity through restoring the ecological prosperity of the water. If we do both together, we can simultaneously create economic prosperity while creating ecological prosperity.

The concept of sustainable prosperity based on the simultaneous realisation of the four forms of prosperity is a framework to help us better articulate and assess 21st-century performance—it is a yardstick for the 21st century. It also provides language which can be used to engage unreconstructed modernists who still and uncritically regard all development as good. When I work with these people, I do not try to convince them to change their ideas. I ask, 'Do you seek to be prosperous in the future?' Naturally, they do. I then ask, 'Do you want to be sustainably prosperous forever? Of course, they love this idea. We then discuss what I mean by sustainable prosperity, and I find I can work constructively with the unconverted defenders of the unsustainable present, as well as the converted initiators of the sustainable future, and even facilitate a constructive dialogue between them.

For decades we have been trying to create concepts that permit us to integrate environment and development and not see them as mutually exclusive entities. This was first established by a conference at Founex, Switzerland in 1971, which was part of the process leading to the 1972 United Nations Conference on the Human Environment. The *Founex Report on Environment and Development* discussed the issue of whether environment and development were mutually exclusive or compatible, antagonistic or synergistic, and elegantly argued the case for compatibility and synergy, saying:

It is evident that, in large measure, the kind of environmental problems that are of importance in developing countries are those that can be overcome by the process of development itself. In advanced countries, it is appropriate to view development as a cause of environmental problems. Badly planned and unregulated development can have a similar result in developing countries as well....

Whilst the concern with the human environment in developing countries can only reinforce the commitment to development, it should serve, however, to provide new dimensions to the development concept itself. In the past, there has been a tendency to equate the development goal with the more narrowly conceived objective of economic growth as measured by the rise in gross national product. It is usually recognised today that high rates of economic growth, necessary and essential as they are, do not by themselves guarantee the easing of urgent social and human problems. Indeed in many countries high growth rates have been accompanied by increasing unemployment, rising disparities in incomes both between groups and between regions, and the deterioration of social and cultural conditions. A new emphasis is thus being placed on the attainment of social and cultural goals as part of the development process. The recognition of environmental issues in developing countries is an aspect of this widening of the development concept. It is part of a more integrated or unified approach to the development objective. (Chapter 1, paragraphs 1.5 and 1.6)

This was the first attempt to overcome the view that conflict between environment and development was an inevitable win/loss situation—a modernist view that still lingers today. Newspapers are still full of stories about conflicts between environmentalists and developers in developed and developing nations alike. But, because the integrative conceptual frameworks have been lacking, the aspiration of the *Founex Declaration* for the integration of economic, ecological and cultural factors has still not occurred. It seems that while John Elkington's concept of the triple bottom line has changed the rhetoric somewhat, the lack of a functioning and acceptable integrative methodology has meant there has been only mediocre progress. I submit that the idea of sustainable prosperity could provide a concept which facilitates a potential win/win framework. It is also an attempt to bury for all time this apparent conflict between environment and development: the modernist idea that if something is to bloom, something else has to die.

In order to create a sustainable society, we must be able to imagine it, model it and understand how it would behave. We must be able to measure our progress through the creation of appropriate indicators and assessment processes. And we must invent what can be collectively called *sustainable society ways and wares* to market to the world's peoples—the tools, products and services that would enable them to make this heroic transformation.

The ways and wares we create should be both problem-centred innovations and mission-directed innovations. Problem-centred innovations are those which set out to reduce collateral damage—to make our societies less unsustainable. Mission-directed innovations are those which set out to avoid collateral damage completely, and so are totally sustainable. The difference between a problem-centred strategy and a mission-directed one is important. For example, treating an illness is a problem-centred approach. It is very different from creating its opposite—wellness—which is a mission-directed approach. To lessen a bad outcome is not the same as creating a good outcome. To improve the efficiency and effectiveness of the fossil fuel engine in order to lessen its impact on global warming is not the same as creating a solar/hydrogen powered alternative that abolishes the impact altogether. To reduce waste is not the same as eliminating it.

Many new innovations are needed to promote ecological prosperity. Two promising lines of approach are biomimicry and eco-mimicry: designing systems, infrastructures and facilities in ways which mimic the behaviour of nature. This can involve mimicking whole ecosystems (eco-mimicry) or the activities of particular organisms or biochemical or physiological pathways (bio-mimicry). Appropriately enough, these are also not new ideas. François Quesnay, physician to King Louis XV, formulated the idea that economic systems were cyclic in nature, and would self-correct and heal themselves just like living organisms and ecosystems if they were left alone to self correct themselves-to *laissez faire*. His discovery of economic cycles was based on mimicking the discovery of the circulation of the blood by Harvey. He established the school of economists called the 'physiocrats', who, in turn, influenced the work of Adam Smith. In the modernist era, the thought that we could learn from nature was sidelined by the anthropocentric hubris of the era, but it has returned in the postmodern era. But, in the post-modern era, people such as biologist and life sciences writer Janine Benyus (1997 pp. 2–3) are pointing out that

> *... doing it natures way has the potential to change the way we grow food, make materials, harness energy, heal ourselves, store information, and conduct business...The biomimics are discovering what works in the natural world, and more important what lasts. After 3.8 billion years of research and development, failures are fossils, and what surrounds us is the secret for survival.*

Imitating natural systems is a good general approach, but we also need a new set of design principles to guide the design of a sustainable society and to provide the design underpinnings for *sustainable society ways and wares*. Below are some possible design principles.

Principle 1: Create zero net collateral damage. Preferably avoid collateral damage

The concept of collateral damage was originally created in the defence and security area. This referred to the unintentional and accidental damage to civic infrastructure and institutions, such as hospitals and schools, which have nothing to do with warfare. The 'smart bomb', in theory, removes the real threat, the enemy and its war machines, and is designed (so far, very imperfectly) to avoid damaging innocent civilians and civil infrastructure. If we are able to accomplish an action with zero net collateral damage, we have achieved a sustainable way of doing it and we can continue doing it forever. Notice that I say zero *net* collateral damage. Some damage cannot be avoided, but the aim should be that if we can't avoid or prevent damage, we should fully ameliorate it afterwards. We can apply the concept of zero collateral damage to other areas.

For example, in the field of medicine, chemotherapy is an unsubtle, unspecific approach to the treatment of cancer because the whole body is poisoned in order to kill cancer cells in one part of it. Zero collateral damage could involve the use of biological systems that intervene in cell biochemistry to turn off or on a critical enzyme causing or healing a disease. Or these biological systems could carry a toxic substance customised to fit into a unique site in the cancer cell. Similarly, in agriculture, the chemical pesticides used to combat insect pests, fungal diseases and weeds can rarely be subtle or selective in their impact. Pesticides are often used to kill specific pest organisms, but because they are not specific for a particular pest and are sprayed indiscriminately on a crop they threaten many non-target organisms, such as beneficial insects, and birds and fish. Moreover, pesticides can be carried by water systems or bio-magnified in food chains so that they can harm organisms quite distant from the spray zone. However, biological systems have the potential to be very subtle in their impacts. More advanced sustainable approaches could use a variety of biological controls, such as using the 4Ps of biological control—parasites, predators, pathogens and pheromones. As well, gene technology can be used to create a similar zero-collateral-damage result. In most cases, the biological control of pests and diseases achieves little or no ecological collateral damage—*if* extensive research is undertaken before these controls are introduced into the environment. Without this, biological control can be as disastrous as any chemical control. The introduction of the cane toad into Australia to control pests of the sugar industry is a notorious example of such a failure. In addition, adherents of organic production might deny this, but gene technology can produce a more sustainable outcome than an organic process. We are still in the T Model Ford stage of this technology.

Principle 2: Live within perpetual solar income

Until now, we have used the product of solar income that arrived on Earth in previous eras—namely fossil fuels. In his film *An Inconvenient Truth*, and in his many public addresses, former US Vice President Al Gore promoted the need for what he called *carbon neutral behaviour* and *zero carbon emissions*. In 2006, these concepts commenced their journey into the mainstream of ideas. Richard Branson announced the Virgin Airlines investment of US$3 billion in renewable technologies to ensure the airline produced no net collateral damage to the planet in terms of global warming. He is being joined in this commitment by many large corporations. In essence, 'carbon neutral' means acting so that there is no net collateral damage to the planet's climate.

There are many ways to produce electricity that will become available in the next few decades, which will help us to live within perpetual solar income. One involves the use of solar/electric conversion systems using banks of concave mirrors that track the movement of the sun and provide a concentrated beam of light to generate electricity through solar/electric cells or heat through thermal/electric conversion. Another is the solar tower, which involves a 10,000 hectare greenhouse that traps hot air. This air rises up a half kilometre tall chimney into colder air, driving electrical turbines as it does so. Such a tower would be sufficient to supply 200,000 homes with electricity. Yet another alternative is OTEC (Ocean Thermal Energy Conversion), which is based on generating power from the temperature differences between the ocean's surface and its lower layers.

Other approaches include two macro forms of marine hydropower. One is a solar powered marine hydro based on using ocean currents and the other is *a lunar powered marine hydro*, which can be based on using either or both tidal and wave power.

Solar marine hydro would work best in so-called western boundary currents. These occur off the east coasts of continents where the currents are narrow, deep, fast and warm. On the other hand, eastern boundary currents off the west coasts of continents are wide, shallow, slow and cold, so this system would not work well there. We can also have a lunar marine hydro equivalent—or tidal power. Such a lunar hydro system could work at the entrance to San Francisco Bay, the Straits of Gibraltar, or any similar place where tides run rapidly through narrow openings. These are only a few examples of ways to live within perpetual solar income

Principle 3: Turn waste into food: realise zero net waste

These words were first used by James McDonough in the Year 2000 Hanover Principles, for the 2000 World Expo, and then by McDonough and Michael Braungart in their book *Cradle to Cradle*. There is no waste in nature: the waste of one species is the food of another. The resource renewal industry facilitates the reduction in use and the re-use and recycling of resources and the management of wastes. This involves mimicking nature and turning all waste into food or resources for further use. I am involved in a new company which is based in Singapore which is able to transform/recycle 100% of the urban waste stream. The organic matter in this waste is converted into compost and then into organic fertiliser which can be customised for particular climatic regimes, soil types and crops. Another example of this is a recycling centre at the Ikea store in Gothenburg, Sweden, where they sell recycled materials such as wood, cardboard, metal and paper that would formerly have gone into landfill.

Principle 4: Avoid overuse with just-enough-in-place-and-time (JEPT)

Just-in-time (JIT) is a principle used in business that involves not pouring finance into the retention of large stockpiles, and purchasing inputs just in time for their immediate use. I have modified JIT for the concept of sustainable production, consumption, lifestyles and development, because excessive levels of inputs in the wrong place at the wrong time are the cause of many environmental problems. I now call it *just-enough-in-place-and-time* (JEPT).

For example, a chemical fertiliser is presently added to the soil in large quantities to increase plant production, but the plants only take up a small proportion of it. Most of it is rendered insoluble and is unavailable to the plant until it is turned into a soluble form by soil micro-organisms. Some of the remaining fertiliser increases the osmotic stress in soil water. This lessens water availability to the plants, kills off earthworms and generally reduces the soil to a mere structure for holding up the plants. Some fertiliser is washed away by rain and causes pollution of the water because these excessive nutrients promote a too nutrient-rich environment (eutrophication). Eutrophication destabilises the water's ecosystem, causing algal blooms and waterweed infestations. These starve the water of oxygen, kill fish and, in tropical oligotrophic (nutrient-poor) marine systems, destroy coral reefs. A more ecologically sustainable approach involves biologically driven, or organic or biodynamic agriculture and horticulture, which is a form of JEPT plant nutrition. Here mineral fertilisers are converted by biological systems in the soil into available nutrients at just the correct rate and in the correct place in the plant's rhizosphere—the zone adjacent to the root hairs where water and nutrients are absorbed. This avoids an excess of soluble nutrients that would be washed out in heavy rainfall to pollute water resources. Even the insoluble fraction of nutrients in the soil can be mobilised by soil microbes which provide access to these nutrients stored in the soil, enabling the crop to gain access to them. This process involves returning to the natural and mineral fertilisers in use before the invention of chemical fertilisers by von Liebig and Lawes, by using advanced technology to lift productivity to the high level equivalent to the productivity of chemical fertilisers. Fertilisers like this already exist, but this possibility is only beginning to be seriously explored.

Principle 5. Protect and nurture biodiversity

The report *The Global Biodiversity Outlook 2* noted that in the last 500 years 844 plants and animals had become extinct because of the rising global population, habit damage and change, over-exploitation, the introduction of alien species and climate change. About 6 million hectares of primary forest are felled each year and about one-third of all mangrove swamps have been lost since the 1980s. Up to 52% of the species of birds studied by the writers of the report faced possible extinction. This is a grim summary indeed. On the positive side, the report noted that 12% of the land surface is now protected. But it is a sad fact in many places where the awareness and the capacity to protect biodiversity is great, such as Europe, the actual biodiversity is poor. And in many countries with a poor capacity to protect biodiversity the biodiversity is rich.

The international trade in endangered species is regulated by CITES—the international Convention in Trade of Endangered Species. This is the regulatory approach, which will work to a degree but will not solve the problem. While human beings want to buy rare parrots and products such as ivory, crocodile skins and rhinoceros horn, people will find a way to beat the system, and species will continue to be endangered. As is the case with drugs such as heroin and cocaine, we will best protect biodiversity via the control of demand, not the choking of supply via law and order. Just as the threats to whales will be solved by convincing Japanese and Norwegian consumers that whale meat should not be consumed, endangered species will stop being threatened when their habitats are sufficiently protected *and* the demand for them, or products related to them, stops. We will never protect our biodiverse world just by creating more national parks or restricting trade in endangered species—important as these initiatives are. We need to protect habitats in places where humans can also live—not just in national parks—by focussing on good habitat design outside national parks and other protected areas. Just as we can, for example, have real estate development built around a golf course or a university, we could build housing or industrial estates around a sanctuary or next to a priceless wetland, where the community accepts a responsibility to become custodians of nature. As with golf courses or marinas, these nature sanctuaries would increase real estate values, not diminish them. In the last decade there has been a significant entry by private sector organisations in biodiversity protection. With the creation of an effective global carbon trading market, biodiversity protection can become a major economic enterprise. This will make afforestation programs more economically viable. Doing economically well by doing ecological good.

We will need many new *biodiversity protection ways and wares* and most of them have yet to be created. One promising innovation is that of a new start-up company with which I have been working. It plans to create intensive forests of a cloned species of tree which would be harvested as a coppice. They would be grown next to, and use the waste water from, urban sewerage plants and will be ready to harvest every six years. By changing the nutrient regime this tree can be customised for either fast growing pulp production or slow growing timber production. This approach will help to make many traditional forms of forestry uneconomic, as this intensive production is undertaken adjacent to its market.

These are just some of the design rules for the ecological prosperity component of sustainable prosperity. In the next two decades it will be necessary to transform these design principles and others yet to be developed into new *green ways and wares.* This is our innovation journey forwards, and it is both a challenge and an opportunity. Those who create ways and wares for ecological, social or cultural prosperity and lessen ecological social and cultural poverty will become the successful providers of 21st century innovation, and will do economically well by doing ecological, social and cultural good. There are similar design rules for the creation of social and cultural prosperity to those I have outlined for ecological prosperity. These could include ways which and wares for mediation, conflict resolution and peace building, for wellness, for learning, for promoting intercultural comfort and gender equality, and for supporting democracy. All these, and many other ways and wares that affirm planetist values, are innovations that will be in demand in 21st century global markets. For example, I recently advised a state electoral office, which is establishing a company to export free and fair elections to the world. At the end of the Second World War there were twelve lonely democracies, one of which was Australia. In other words, Australia got to the future first in establishing a stable democracy. In 2007, there are more than 130 democracies—most of them imperfect ones. But the market in the 21st century for democracy—an essential component of a sustainable society—is now huge. Likewise, Rwanda is interested in establishing an export industry in peace building and the prevention of genocide, and in marketing *peace building ways and wares,* to other parts of Africa, which are seeking to promote intercultural harmony.

At least half of the work to innovate a sustainable society is ahead of us and the only limits are our imagination, our entrepreneurship and our enterprise.

References

Benyus J (1997). *Biomimicry: Innovation Inspired by Nature.* Perennial (Harper Collins), New York.

Elkington J (1999). *Cannibals with Forks: The Triple Bottom Line of 21st Century Business.* Capstone Publishing, London.

Ellyard P (2001). *Ideas for the New Millennium.* 2nd edn. Melbourne University Press, Melbourne.

Ellyard P (2007). *Designing 2050: Pathways to Sustainable Prosperity on Spaceship Earth.* TPNTXT, Melbourne (in press).

Hardin G (1968). The tragedy of the commons. *Science* **162 (3859)**, 1243–1248.

McDonough W (2000). *The Hanover Principles: Design for Sustainability.* 3rd edn. William McDonough Architects, Charlottesville, Virginia.

McDonough W and Braungart M (2002). *Cradle to Cradle.* North Point Press, New York.

Index